Hepatitis C

Antiviral Drug Discovery and Development

Edited by

Seng-Lai Tan

Hoffman-La Roche Inc.
Nutley, NJ
USA

and

Yupeng He

Antiviral Research
Abbott Laboratories
Abbott Park, IL
USA

Caister Academic Press

Caister Academic Press
Norfolk, UK

www.caister.com

British Library Cataloguing-in-Publication Data
A catalogue record for this book is available from the British Library

ISBN: 978-1-904455-78-3

Cover image adapted from Figure 10.3

Printed and bound in Great Britain

Contents

Contributors

Atul Agarwal
Achillion Pharmaceuticals
New Haven, CT
USA

aagarwal@achillion.com

Craig A. Belon
Department of Biochemistry and Molecular Biology
New York Medical College
Valhalla, NY
USA

Craig_Belon@nymc.edu

Brad O. Buckman
InterMune, Inc.
Brisbane, CA
USA

bbuckman@intermune.com

Zhaohui Cai
Southern Research Institute
Frederick, MD
USA

cai@southernresearch.org

Kevin X. Chen
Merck Research Laboratories
Kenilworth, NJ
USA

kevin.chen3@merck.com

Stuart Cockerill
Arrow Therapeutics Ltd
London
UK

stuartcockerill@mac.com

Milind Deshpande
Achillion Pharmaceuticals
New Haven, CT
USA

mdeshpande@achillion.com

Heidi E. Drummer
Centre for Virology
Burnet Institute
Melbourne, Victoria;
Department of Microbiology
Monash University
Clayton, Victoria;
Department of Microbiology and Immunology
University of Melbourne
Melbourne, Victoria
Australia

hdrummer@burnet.edu.au

Menashe Elazar
Department of Medicine
Division of Gastroenterology and Hepatology
Stanford University School of Medicine
Stanford, CA
USA

menashe@stanford.edu

Martijn Fenaux
Department of Biology
Gilead Sciences, Inc.
Foster City, CA
USA

Martijn.Fenaux@gilead.com

David N. Frick
Department of Chemistry and Biochemistry
University of Wisconsin–Milwaukee
Milwaukee, WI
USA

frickd@wwm.edu

Michael Gale, Jr.
Center for the Study of Innate Immunity to HCV Infection
Department of Immunology
School of Medicine
University of Washington
Seattle, WA
USA

mgale@uw.edu

Jeffrey S. Glenn
Department of Medicine
Division of Gastroenterology and Hepatology
Stanford University School of Medicine
Stanford, CA
USA

jeffrey.glenn@stanford.edu

Mingjun Huang
Achillion Pharmaceuticals
New Haven, CT
USA

mhuang@achillion.com

Zhuhui Huang
Southern Research Institute
Frederick, MD
USA

z.huang@southernresearch.org

Klaus Klumpp
Hoffman-La Roche Inc.
Nutley, NJ
USA

klaus.klumpp@roche.com

Karl Kossen
InterMune, Inc.
Brisbane, CA
USA

kklossen@intermune.com

Carla Kuiken
Los Alamos National Laboratory
Los Alamos, NM
USA

kuiken@lanl.gov

Robert E. Lanford
Department of Virology and Immunology
Southwest Foundation for Biomedical Research;
Southwest National Primate Research Center
San Antonio, TX
USA

rlanford@sfbr.org

Stanley M. Lemon
Department of Medicine and the Lineberger Comprehensive Cancer Center
The University of North Carolina at Chapel Hill
Chapel Hill, NC
USA

smlemon@ad.unc.edu

Kai Lin
Novartis Institutes for Biomedical Research, Inc.
Infectious Diseases–Virology
Cambridge, MA
USA

kai.lin@novartis.com

Liangjun Lu
Antiviral Research
Abbott Laboratories
Abbott Park, IL
USA

liangjun.lu@abbott.com

Jane A. McKeating
Institute for Biomedical Research
University of Birmingham
Birmingham
UK

j.a.mckeating@bham.ac.uk

Neil Mathews
Arrow Therapeutics Ltd
London
UK

dr.n.mathews@gmail.com

Hongmei Mo
Department of Biology
Gilead Sciences, Inc.
Foster City, CA
USA

Hongmei.Mo@gilead.com

Akhteruzzaman Molla
Antiviral Research
Abbott Laboratories
Abbott Park, IL
USA

akhter.m.molla@abbott.com

Michael G. Murray
Southern Research Institute
Frederick, MD
USA

murray@southernresearch.org

Pilar Najarro
Arrow Therapeutics Ltd
London
UK

pnajrro@biolty.com

John B. Nicholas
InterMune, Inc.
Brisbane, CA
USA

jnicholas@intermune.com

F. George Njoroge
Merck Research Laboratories
Kenilworth, NJ
USA

george.njoroge@merck.com

Jean-Michel Pawlotsky
National Reference Center for Viral Hepatitis B, C and Delta
Department of Virology and INSERM U955
Hôpital Henri Mondor
Université Paris-Est
Créteil
France

jean-michel.pawlotsky@hmn.aphp.fr

Avinash Phadke
Achillion Pharmaceuticals
New Haven, CT
USA

aphadke@achillion.com

Richard Scheuermann
University of Texas Southwestern Medical Center
Dallas, TX
USA

richard.scheuermann@utsouthernwestern.edu

Scott D. Seiwert
InterMune, Inc.
Brisbane, CA
USA

sseiwert@intermune.com

Mark Smith
Hoffman-La Roche Inc.
Nutley, NJ
USA

mark.smith@roche.com

Kathe Stauber
Achillion Pharmaceuticals
New Haven, CT
USA

kstauber@achillion.com

Deborah R. Taylor
CBER/FDA
Office of Blood Research and Review
Division of Emerging and Transfusion-transmitted Diseases
Laboratory of Chronic Hepatitis and Innate Immunity
Bethseda, MD
USA

Deborah.Taylor@FDA.HHS.gov

Timothy L. Tellinghuisen
The Scripps Research Institute
Scripps Florida
Jupiter, FL
USA

tellint@scripps.edu

Takaji Wakita
Department of Virology II
National Institute of Infectious Diseases
Tokyo
Japan

wakita@nih.go.jp

Christopher Walker
Center for Vaccines and Immunity
The Research Institute at Nationwide Children's Hospital
Columbus, OH
USA

christopher.walker@nationwidechildrens.org

Preface

It is estimated that hepatitis C viruses (HCV) have infected at least 170 million people worldwide, including up to two percent of the population in the USA. Transmitted by blood-borne or sexual routes, chronic HCV infection is a common cause of liver failure and liver cancer. While vaccines are available for hepatitis A and B viruses, this is not the case for HCV, a problem magnified by the lack of an animal disease model that can recapitulate the complex pathogenesis of chronic hepatitis C in humans. In spite of it, progress in our understanding of HCV infection, replication, and molecular biology has been phenomenal, leading to an initial explosion of drug discovery efforts in the search for effective hepatitis C antiviral therapies, a topic covered in our first book.

While traditional antiviral drug discovery approaches have yielded notable successes, anti-HCV drug development has proven to be exceptionally challenging. Not surprisingly, several major pharmaceutical companies have dropped out of the race entirely, a decision that has gone down well with analysts who believe that the hepatitis C markets are already well catered for. But, the reality is the current HCV therapy has limited efficacy and undesirable side effects, and there is a pressing need for new and more effective therapies. Fortunately, as is always the case with drug discovery and development, and thanks to the resilience of those who stayed on the course, we are finally witnessing exciting and promising new progress in the race for new hepatitis C drugs.

This book is devoted to capturing such recent advancements and challenges, as well as new innovations and emerging trends in the field. To that end, we assembled a team of highly regarded experts who collectively have prepared 19 chapters covering various aspects of the current state of the art in anti-HCV drug discovery and development. The first half of the book (Chapters 1–8) provides an updated view of our current understanding of the HCV viral lifecycle, a summary of commonly used assays, animal models, public resources, as well as an overview of the current HCV drug pipeline. The second half of the book (Chapters 9–19) is dedicated to in-depth reviews of established and emerging drug targets for hepatitis C antiviral therapies.

For the emerging small-molecule HCV inhibitors (Chapters 9–16), viral resistance will be a problem. Thus, combination therapy of different drugs with different targets/mechanisms will be necessary to effectively inhibit HCV replication, which in turn necessitates careful clinical trial designs (Chapter 7). It has long been suspected that host genetic factors are key determinants for the control of HCV infection. Host cell factors are gaining ground as drug targets for HCV therapy because they might enable the development of broad-spectrum antiviral drugs with less chance of viral resistance (Chapter 18). Of particular interest are cellular receptors that mediate HCV entry (Chapter 17), factors that facilitate HCV replication and assembly, and intracellular pathways involving lipid biosynthesis and innate immune response (Chapter 19). The challenge will lie in our ability to pharmacologically manipulate such cellular targets for chronic HCV treatment, without being limited by side effects.

The last decade has been unusually exciting and challenging for scientists and clinicians dedicated to novel anti-HCV drug discovery and development, one of the most promising and demanding areas of the entire pharmaceutical/biotechnology industry. Finally, we believe that their dedication and efforts are paying off, with numerous novel anti-HCV drug candidates in various phases of the pipeline. The most advanced drug candidates will reach the market in the coming years, and these emerging novel drugs are bound to revolutionize the practice of hepatitis C therapy and greatly benefit millions of patients. We would like to acknowledge all these heroes who have contributed to anti-HCV drug research, in particular our authors, most of whom are distinguished experts in the field. We feel extremely lucky to have had the opportunity to collaborate with them during the preparation of this timely book that offers a milestone summary of their work.

Seng-Lai Tan, PhD
Yupeng He, PhD

An Overview of the Hepatitis C Virus Life Cycle

1

Timothy L. Tellinghuisen

Abstract

The hepatitis C virus (HCV) is a significant public health problem of international scope. HCV is capable of establishing chronic infections in the majority of those infected, resulting in progressive liver damage and a host of extra-hepatic disorders. Current drug therapies are ineffective in clearing infections from the majority of patients, highlighting the need for better antiviral drugs. Recent developments in HCV genetic systems have led to significant advances in our understanding of the life cycle of this important pathogen, and it is this understanding that will ultimately lead to better antiviral strategies. It is the purpose of this chapter to provide an overview of the life cycle of HCV, with a focus on aspects of the viral life cycle that might be amenable to future drug therapies.

Introduction

Hepatitis C virus (HCV) is a pathogen of global importance, with more than 170 million people chronically infected with this insidious virus, a number representing roughly 3% of the human population (Anonymous, 1997). HCV infections are, by nature, insidious, with some 70% of those infected with the virus developing chronic or persistent infections and those that do reach chronicity fail to spontaneously resolve infection with any appreciable frequency (Alter *et al.*, 1992; Alter and Seeff, 2000). The long-term consequences of decades of virus replication in the hepatocytes of these individuals are chronic hepatitis, progressive cirrhosis, and often hepatocellular carcinoma (Alter and Seeff, 2000). The ability of this simple RNA virus, with no quiescent life cycle stage to establish long-term infections in the host likely results from a complex interplay between the virus and the host. HCV must create a niche for replication in the hepatocyte, evade the innate and adaptive immune system, regulate a number of viral RNA-directed processes, carry out the events of virus replication and production of progeny virions, and balance the rate of virus induced hepatocyte death with liver regeneration to effectively persist, and these events are only beginning to be understood in any significant detail. Deciphering these events is of paramount importance to the development of new therapeutics to treat HCV infections. The current FDA approved 'combination therapy', a mixture of the non-specific antivirals pegylated interferon-α and ribavirin, varies quite dramatically in efficacy depending on the strain, or genotype, of HCV being treated, with clearance rates of about 40% not uncommon for genotypes common to the United States and Europe (Heathcote *et al.*, 2000; Zeuzem *et al.*, 2000). In addition to this moderate efficacy, combination therapy is expensive and is often very poorly tolerated by patients, further highlighting the need for better drugs. Of course, HCV has been an active target for the development of new antiviral compounds since its discovery, with major effort being invested by both academic and industrial laboratories, and this work has yielded the first generation of virus-specific small molecule inhibitors that are currently in clinical trials. Without stealing content from the later chapters of this text, it is clear that many of these molecules are surprisingly effective treatments for HCV in cell culture, animal models, and human trials, and are specifically tailored for the genotypes of HCV that tend to fail to respond to combination therapy. So, from an outside perspective, it would seem that the problem of HCV infection is already, or perhaps about to be, solved. Unfortunately, we are a long way from fulfilling the goal of eliminating the scourge of HCV infection. HCV replication is error prone, with the viral genome in an infected individual, or indeed within a single infected cell, existing as a quasispecies of many closely related RNA molecules, a subset of which may contain mutations that might render virus resistant to an antiviral compound. It is generally believed that a cocktail of several specific antiviral compounds will likely be needed for a widely effective HCV treatment. In addition to this complexity the incredible genotypic diversity of circulating HCV around the globe further necessitates more drug development work, as many small molecules described to date, although effective on their target genotype, lack a pan-genotypic efficacy profile. So, despite the great progress made in regards to HCV drug development in recent years, more work is clearly needed, and it seems this will remain an active and exciting area of research for many years to come. This chapter is focused on providing a brief overview of the molecular aspects of the HCV life cycle, with a focus on the implications of these events to drug design, so that later chapters on specific targets are clear in the context of virus biology. The molecular details of the HCV lifecycle are incredibly complex, and entire books have been written just on this subject alone, so fitting this information into a single chapter necessitates a fairly broad view of events. I apologize at the outset to those authors whose work I could not discuss due to space and scope limitations.

HCV classification and genetic diversity

HCV is classified as a member of the Flaviviridae family of enveloped, single-stranded positive sense RNA viruses, where it occupies and serves as sole member of the genus *Hepacivirus* (Lindenbach, *et al.*, 2007a). The family Flaviviridae is compositionally diverse, containing, in addition to the *Hepacivirus* genus, those of the genera *Flavivirus* and *Pestivirus*. The genus *Flavivirus* contains the classical flaviviruses, such as yellow fever virus and Dengue virus. The genus *Pestivirus* contains a number of viruses of agricultural importance, due to their severe pathogenesis in livestock, most notably cattle and pigs. These viruses include, most notably, classical swine fever virus (CSFV) and bovine viral diarrhoea virus (BVDV). The similarity of BVDV genome organization to that of HCV led to considerable application of BVDV as a surrogate model of HCV replication and infection. In addition, the GB viruses, including GBV-A, B, and C, are designated as members of the *Flaviviridae* family, although not officially classified into a specific genus, and these viruses have been proposed as surrogate models for HCV due to their similar genetic organization and replication strategy to that of HCV, and their replication in small laboratory animals.

It is important to note that although HCV is the sole member of the genus *Hepacivirus*, the virus has considerable genetic diversity among the many known isolates (Simmonds, 2004). These isolates are delineated by an organization system with major groupings designated as genotypes (designated with numbers, including 1, 2, 3, and so on), and variants within genotypes referred to as subtypes (designated with letter modifiers on the parental genotype number, for example, 1a, 1b, 2a, etc.; Kuiken *et al.*, 2006; Simmonds *et al.*, 2005). Within subtypes, individual strains represent the lowest classification designation, and considerable variation exists to the naming of strains, with no coherent plan generally accepted, although they are often given numerical designations. There are currently 6 major genotypes of HCV recognized, many dozens of subtypes, and perhaps hundreds of individual strains. Individual genotypes typically differ from one another by 30–35% at the nucleotide level, and subtypes tend to vary from one another in the range of 20–25% at the nucleotide level (Simmonds *et al.*, 2005). Genotypes and subtypes are loosely distributed geographically, but considerable overlap in distributions exists (Anonymous, 1997; Houghton, 1996; Mondelli and Silini, 1999). Genotypes 1 and 2 are broadly distributed across the globe, and represent the dominant genotypes in North America, Europe, and Japan. Genotype

1 predominates in these areas, specifically with the 1a and 1b subtypes, with genotype 2 and other subtypes of genotype 1 as relatively minor components. Genotype 3 is most common in Southeast Asia and the Indian subcontinent, but also has a broad, global distribution as a minor strain in many areas. Genotype 4 is largely restricted to northern and Central Africa and the Middle East. Genotype 5 is the dominant circulating virus in South Africa. Genotype 6 is found primarily in Asia. Perhaps the most relevant aspect of the different genotypes of HCV, from a clinical perspective, is the vastly different response of individual genotypes of interferon–ribavirin combination therapy. Genotype 1 is the most refractory to treatment (viral clearance rates are typically about 40%), with other genotypes being far more susceptible to therapy (~75–90% clearance; Hnatyszyn, 2005). The mechanism behind this differential susceptibility of the various HCV genotypes to interferon therapy is not currently understood. Genotypes also are likely to play a significant role in another aspect of HCV therapeutics, that of the efficacy of virus-specific small molecule drugs. Due to limitations in the genotypes of HCV amenable to replication in laboratory studies, and the prevalence of genotype 1 infections across the globe, the majority of small molecule inhibitors target genotype 1. The efficacy of these compounds on other genotypes, for the most part, remains poorly explored, but the diversity among HCV genotypes suggests an inhibitor with pan-genotypic antiviral properties is unlikely. Indeed, many compounds developed against a genotype 1b target, have significant potency deficits when assayed even on the closely related genotype 1a virus. Luckily, the high interferon combination therapy response rates of non-genotype 1 HCV partially mitigate some of these concerns.

It is important to consider another aspect of HCV diversity that defies conventional classification schemes, yet has significant impact on future drug design. The HCV RNA-dependent RNA polymerase lacks a proofreading activity, leading to error prone replication of the viral genome. The estimated mutation rate of HCV replication has been estimated at about 10^{-4} misincorporations per nucleotide (Kuiken *et al.*, 2006). HCV replication is reasonably robust, with estimates suggesting an infected individual produces about 10^{12} virions per day (Perelson *et al.*, 2005). The relatively small size of the HCV genome (9.6 kilobases), combined with the error rate of the viral polymerase and the large number of replication events occurring within an infected individual, suggests that an incredibly diverse population of closely related HCV sequences will be generated during infection. This collection of virus mutants has been termed a quasispecies, reflecting the fact that a single fixed sequence of virus does not exist during an infection, but rather a collection of many different sequences (Bukh *et al.*, 1995). The quasispecies nature of HCV infections is particularly relevant when a selective pressure is applied to the population. Perhaps most relevant to this book would be a specific antiviral compound as a selective pressure, but it is important to note other types of selective pressures also exist (host immune recognition, for example). When a selective pressure is applied to an incredibly diverse collection of virus sequences, like those that exist in HCV infections, it is not a matter of the virus evolving resistance to the selective pressure, but rather simply selecting the pre-existing mutant in the quasispecies population that is resistant. This resistant sequence will then quickly become the dominant sequence in the individual, with additional mutations evolving to further enhance the replication of this sequence in the presence of the selective pressure. The end result of this process, in the case of a virus directed small molecule drug, is the rapid development of drug resistance. It is therefore likely that a cocktail of several specific small-molecule HCV inhibitors, in conjunction with non-specific combination therapy, will be likely to provide a sufficient onslaught to overcome the massive genetic diversity of an HCV infection.

Tools for the study of HCV

Before an overview of the HCV life cycle can be presented, it is important to point out that our understanding of HCV biology is still in its infancy. This is, of course, not based on a collective lack of desire to understand the HCV specific details of the life cycle, but rather stems from the significant technical limitations that have plagued

HCV research for the last few decades. It would be remiss to not describe some of these limitations, not only to highlight the monumental effort expended by the HCV community to overcome them, but also because they have led to the development novel tools that have greatly shaped our understanding of HCV biology, and those limitations we have not yet overcome will drive the future directions of HCV research.

The beginning of hepatitis C molecular virology, and our ability to manipulate the virus in model systems, heralds from the 1989 landmark cloning of a large portion of the genome of the infectious agent that caused non-A, non-B post-transfusion associated hepatitis (NANB; Choo *et al.*, 1989). Prior to this amazing feat of molecular biology, research was primarily limited to transmission and limited pathogenesis studies using NANB patient materials in the only available animal model, the chimpanzee, as all attempts at replicating the enigmatic infectious agent in cell culture were unsuccessful (Lanford, 2001a). With a cDNA clone of the NANB infectious agent, now designated hepatitis C virus, there was a great effort to coax replication of HCV RNA produced from this cDNA clone in cell culture, and although this effort provided a few false glimmers of hope, none of this work yielded an authentic, tractable replication system (see Lindenbach and Rice, 2001 for a review). Despite these early failings, the cDNA clone of HCV provided a wealth of information to further our understanding of the virus. Most notable is the primary sequence information contained within the cDNA clone that allowed classification of HCV into the *Flaviviridae* family. This led to an explosion of research on related viruses, most notably BVDV, and later the GB viruses, with a focus of using them as a model system for the understanding of HCV virology (Bukh *et al.*, 1999; Lanford, 2001b; Thiel *et al.*, 1996). These viruses were capable of replication in cell culture or small laboratory animals, and a considerable amount of basic information about the HCV life cycle was generated with these models, at the least by way of putative analogy with the know life cycles of these viruses. The sequence within the HCV cDNA clone also allowed for identification of the various HCV proteins and determination of how these proteins were formed from a viral polyprotein precursor (Choo *et al.*, 1991; Grakoui *et al.*, 1993a,b). Once identified, the individual HCV proteins became targets of intense study, and a considerable amount of functional data, biochemical characterization, enzymology, and even structural biology began to define many aspects of HCV biology. Much of this early work relied on surrogate expression systems that simply generated the HCV protein or proteins of interest in a wide range of laboratory systems in the absence of viral replication, as a replication system remained elusive. It was not until 1997, after the discovery that the initial cloned HCV cDNA was missing a highly conserved 3′-terminal region of the viral genome, that the first functional full-length HCV cDNAs were reported (Kolykhalov *et al.*, 1997). Attempts to demonstrate the replication of RNAs transcribed from these new cDNA clones in a variety of cell culture systems were unsuccessful, but, surprisingly, these RNAs were able to launch an authentic HCV infection when inoculated into chimpanzees by intrahepatic injection (Kolykhalov *et al.*, 1997; Yanagi *et al.*, 1997). This was another critical landmark on the path to understanding the HCV life cycle. The new infectious cDNA clones provided a fixed, validated, functional HCV genome and a model infection system in which replication could be measured, providing the first tool for directed HCV reverse genetic research. Although the chimpanzee model of HCV infection was a powerful tool, and remains in use today for the study of many aspects of HCV biology, it was not the tractable cell culture model the field needed. The ethical and practical issues associated with primate research in chimpanzees severely limited the number and type of experiments that could be conducted.

The next significant breakthrough in tools to study HCV replication came in 1999, with the description of subgenomic replicon technology(Lohmann *et al.*, 1999). In the original description of this system, Lohmann and colleagues generated bicistronic HCV derived RNAs featuring a selectable marker gene (neomycin phosphotransferase) in the first cistron driven by the HCV internal ribosome entry site (IRES), and the components of the viral replication machinery (NS3–NS5B) in the second cistron driven by a heterologous viral IRES element (Lohmann *et al.*, 1999). Both cistrons were then flanked with authentic HCV 5′ and 3′ genomic RNA ends. Delivery of these subgenomic replicon RNAs into the human hepatoma cell line Huh-7 produced low levels of HCV RNA replication, and these rare replication events could be selected using the neomycin phosphotransferase activity encoded in the first cistron. Selection of stable cell clones from these cultures, followed by isolation and sequencing of the HCV replicons contained within, identified a wide range of replicon non-structural protein mutations that could, when re-engineered into the parental replicon, produce very robust RNA replication (Blight *et al.*, 2000; Lohmann *et al.*, 2001). These so called 'adaptive mutations' were capable of increasing subgenomic replicon replication by several orders of magnitude, making the replicon system a potent tool for studying RNA replication. Oddly, introducing these adaptive mutations into the infectious clones of HCV resulted in severe attenuation of infectivity (Bukh *et al.*, 2002), and the nature of how these mutations work in cell culture, and why they do not work in animals remains a mystery (reviewed in Bartenschlager *et al.*, Pietschmann, 2004). Another important development was the selection of highly permissive subclones of Huh-7 cells by selecting individual replicon containing clones and subsequently curing them with interferon alpha, resulting in a cell line with enhanced replicon permissiveness (Blight *et al.*, 2002; Lohmann *et al.*, 2003). Perhaps the most famous of these clones is the Huh-7.5 cell line originally described by Blight and colleagues (Blight *et al.*, 2002). Later characterization of the Huh-7.5 cell line indicated that these cells bear a lesion in the retinoic-acid inducible gene I (RIG-I) viral RNA sensing component of the innate antiviral pathway, explaining their increased capacity for HCV replication (Sumpter *et al.*, 2005). Using adaptive mutations in conjunction with permissive cell lines quickly led to the development of replicons based on the sequence of the H77 strain of genotype 1a found in the original HCV infectious clone (Blight *et al.*, 2003).

Following the rapid development of efficient replicons, these tools were put to a number of important uses to probe the intracellular aspects of the HCV life cycle. Finally, the HCV field had a cell culture system that could be readily manipulated for forward and reverse genetic screens, developed into potent imaging tools, used to provide materials for biochemical characterization, and used as a basis to define host factors involved in the HCV life cycle. Perhaps the most relevant advance provided by subgenomic replicons to the subject matter of this book is the use of replicons as tools for screening for small molecule inhibitors of HCV replication (see Chapter 2). A variety of reporters amenable to high throughput screening assays have been engineered into the first cistron of subgenomic replicons, providing a rapid, convenient, and highly sensitive measure of HCV RNA replication. Indeed, the vast majority of small molecule inhibitors directed against HCV proteins originated from screens using the subgenomic replicon. Additionally, much of what we know about mechanism of action for many of these compounds comes from the evolution and identification of replicon drug resistance mutations. Yet, despite all the advances made possible by replicons, an infectious HCV cell culture system remained elusive. Building monocistronic replicons that essentially recapitulated the structure of the infectious HCV clone, or bicistronic constructs containing the complete HCV polyprotein in the second cistron and a variety of reporters in the first cistron failed to produce infectivity in cell culture for both genotypes 1a- and 1b-based constructs, despite robust levels of RNA replication in these constructs (Blight *et al.*, 2003; Ikeda *et al.*, 2002; Pietschmann *et al.*, 2002). Thus, the HCV community had exceptional tools to study the intracellular aspects of the viral life cycle, but little to probe the important issues of virus particle entry, assembly, and maturation.

Despite the early lack of an infectious cell culture virus system, considerable progress was made in understanding many aspects of the early steps in HCV infection using a variety of surrogate model systems. As the viral E2 glycoprotein was the likely receptor-binding component of virions used in attachment to hepatocytes, much attention was initially paid to the use of various E2 model systems to define HCV receptors (see Chapter 17). Perhaps the simplest, and arguably most successful, of these was the use of carboxy-terminal truncated forms of E2 designed such that they are secreted from cells expressing the protein (Flint *et al.*, 1999a; Michalak *et al.*, 1997; Pileri *et al.*, 1998; Scarselli *et al.*, 2002). Once purified, these secreted E2 proteins served as valuable tools for probing virus cell-interactions and led to the identification of several putative viral receptors involved in infection. Similar concepts for probing virus receptor interactions included the expression of E2 on the surface of cells (Flint *et al.*, 1999c), the production of liposomes containing both the E1 and E2 viral glycoproteins (Cocquerel, 2002), and virus like particles (VLPs) formed by expression of the HCV structural proteins in mammalian or insect cells (Baumert *et al.*, 1998, 1999; Blanchard *et al.*, 2002, 2003a,b; Mizuno *et al.*, 1995). VLPs have also been used to characterize the structure of HCV virions by both biochemical and cryo-electron microscopy image reconstructions, but it is still unclear how similar VLPs are to authentic HCV virions (Yu *et al.*, 2007). Surrogate viral pseudotype systems have also been utilized to identify HCV entry with considerable success. The first notable version of these systems was based on the fusion of the ectodomains of the HCV E1 and E2 glycoproteins to the transmembrane domain of the vesicular stomatitis virus (VSV) G glycoprotein in the context of an infectious pseudotyped VSV particle (Buonocore *et al.*, 2002; Lagging *et al.*, 1998; Matsuura *et al.*, 2001). Perhaps the most significant, and widely used, surrogate pseudotype system is based on the generation of retroviral pseudotypes bearing unmodified HCV E1 and E2 glycoproteins. These pseudotyped particles are generated by transfecting the HCV glycoproteins, the Gag-Pol coding sequence of either murine leukaemia virus (MLV) or human immunodeficiency virus (HIV), and an appropriate retroviral genome reporter construct into 293T cells and simply harvesting the supernatant containing retroviral particles containing the reporter with HCV glycoproteins arrayed on the particle surface (Bartosch *et al.*, 2003; Hsu *et al.*, 2003). When added to susceptible cells, the HCV E1 and E2 glycoproteins mediate entry of the particle, and led to expression of the retrovirally encoded reporter construct as a measure of entry efficiency relative to particles generated without HCV glycoproteins. The pseudotype viruses generated in this manner, regardless of whether HIV or MLV based, are referred to as HCV pseudoparticles (HCVpp). The entry of HCVpps into hepatocyte-like cell lines is dependent on the presence of the HCV glycoproteins and can be blocked by a variety of anti-HCV neutralizing antibodies (Bartosch *et al.*, 2003; Hsu *et al.*, 2003). The entry of HCVpp is thought to represent an accurate model of the entry process of authentic HCV virions, and these particles have been very successful in allowing the identification of claudin-1 and occludin as novel HCV entry receptors (Evans *et al.*, 2007; Ploss *et al.*, 2009). Although very useful for understanding some aspects of receptor interaction, antibody neutralization, and early virion attachment and entry of HCV, these systems are far from monitoring actual HCV infections, and make things like understanding HCV nucleocapsid disassembly impossible.

Although significant progress was made on understanding HCV entry using surrogate model systems, and the replicon system allowed analysis of RNA replication events, the inability to generate authentic infectious HCV in cell culture remained a significant block to obtaining a complete picture of the HCV life cycle. As mentioned previously, full-length replicons based on the H77 genotype 1a infectious clone, despite being nearly identical to the original HCV infectious clone, fail to generate infectious virus particles in cell culture (Blight *et al.*, 2003). Several potential ideas were proposed to explain this apparent disconnect between cell culture and chimpanzee infections. The first of these ideas was that the cell culture environment was missing something required for the production of infectious virus, be it a host factor, metabolic process, cellular architecture, or organ culture environment was not known. Unfortunately, no such missing factor or process could be identified. Another prevalent idea was that the adaptive mutations required to attain measurable RNA replication in cell culture might disrupt processes required for virus production. The observation that viral genomes containing adaptive mutations were attenuated in chimpanzees, and efficient replication in cell culture was inversely proportional to productive infection in chimpanzees, provided some basis for the hypothesis that adaptive mutations were responsible for the loss of virus production (Bukh *et al.*, 2002). The quest was on to understand adaptive mutations, both in how they increase replication, and how they might affect virion production. It was known from the earliest days of the identification of adaptive that despite their location in any number of the viral non-structural proteins, the majority seemed to affected the phosphorylation of the enigmatic viral NS5A protein (see Chapter 14). NS5A exists in two distinct phosphorylation states, designated hypophosphorylated and hyperphosphorylated, and adaptive mutations tend to favour the production of hypophosphorylated NS5A and reduce the presence of the hyperphosphorylated form of the protein (Tanji *et al.*, 1995b). Mutations that eliminate putative sites of hyperphosphorylation tended to increase viral RNA replication (Appel *et al.*, 2005; Blight *et al.*, 2000; Lohmann *et al.*, 2001), and chemical inhibition of one of the proposed NS5A directed kinases responsible for NS5A hyperphosphorylation, casein kinase I, rendered replicons capable of efficient RNA replication in the absence of any adaptive mutations (Neddermann *et al.*, 2004; Quintavalle *et al.*, 2006, 2007). The mechanism of how phosphorylation of NS5A controls RNA replication is not known, but the general trend that emerged from these and other experiments was that NS5A hypophosphorylation was essential for RNA replication, and that hyperphosphorylation served as a negative regulator of the process, and this interplay between phospho-forms was somehow tied to adaptive mutations (reviewed in Huang *et al.*, 2007b). One model of the role of phosphorylation in RNA replication efficiency involves the differential interaction of NS5A with cellular factors involved in replication in a manner dependent on the phosphorylation state of NS5A. This differential, phosphorylation-dependent interaction has been described for NS5A and the host tSNARE protein hVAP-A. h-VAPA interacts with NS5A and the viral RNA-dependent RNA polymerase and may control the assembly and stability of the viral RNA replication machinery, but the interaction with NS5A is lost in the context of NS5A hyperphosphorylation (Evans, Rice, and Goff, 2004). This loss of interaction with hVAP-A might serve to destabilize the viral RNA replication complex sufficiently that NS5A could exit the complex to participate in virus assembly events. Adaptive mutations that prevent NS5A hyperphosphorylation would therefore serve to stabilize the replication complex by maintaining the interaction with factors like hVAP-A, locking the complex in a processive RNA replication mode, potentially at the expense of other life cycle events such as virion assembly. This model fits well with the aforementioned observations that full-length genomes based on infectious clones that bear adaptive mutations do not yield infectious particles in cell culture despite efficient replication (Blight *et al.*, 2003; Ikeda *et al.*, 2002; Pietschmann *et al.*, 2002), and are severely impaired or non-infectious in chimpanzees (Bukh *et al.*, 2002). Another important prediction of this model is that genomes capable of efficient replication in cell culture without adaptive mutations might be capable of producing infectious virus particles in this environment. Unfortunately, at the time these models were being widely discussed in the HCV community, no such genomes had yet been identified.

The hunt was on for HCV isolates that could replicate efficiently in cell culture in the absence of adaptive mutations. A genotype 2a isolate from a rare Japanese case of fulminant hepatitis (designated JFH-1) was identified that, amazingly, fit this requirement of efficient cell culture replication without adaptive mutations when present as a subgenomic replicon (Date *et al.*, 2004; Kato *et al.*, 2003; Kato *et al.*, 2005). Generating full-length viral genomes from this JFH-1 isolate, and delivering these RNAs into Huh-7 cells, resulted in the release of HCV infectivity into the cell culture supernatant that could be passaged to naïve cultures of Huh-7 cells resulting in a productive infection of the naïve cells (Wakita *et al.*, 2005). This infectious HCV was designated HCVcc (for cell culture-derived HCV) (see Chapter 5). The initial titres of HCVcc produced were low, but subsequent use of Huh-7.5 and similar adapted cell lines in place of Huh-7 led to increases in titre (Lindenbach *et al.*, 2005; Zhong *et al.*, 2005). Furthermore, genetic manipulation of the JFH-1 genome replacing the structural region and the p7 and NS2 portion of the non-structural region with corresponding regions from the infectious genotype 2a clone J6 increased virus production (Lindenbach *et al.*, 2005). Further optimization altering the fusion junction between J6 and JFH-1 genome fragments to a junction between the first and second transmembrane segment of the viral NS2 protein (Pietschmann *et al.*, 2006). Passage of these chimeric viruses in cell culture has identified a number of mutations that increase the release of infectivity, but the mechanistic details of how these mutations function in this regard remains unknown. Using the optimized junction sites defined for genotype 2a, chimeric viruses containing structural regions from other HCV genotypes can be created, and these produce infectious particles, but less efficiently than the parental 2a constructs. HCVcc virus was infectious in both the chimpanzee model and in models based on immunodeficient mice bearing chimeric human livers (Wakita *et al.*, 2005; Lindenbach *et al.*, 2006). Infectivity isolated from these animal infections could be passed back into naïve Huh-7 cell cultures where they initiated robust infections. For the first time in nearly two decades, HCV researchers had an infectious virus that could replicate in cell culture as well as animal models, thereby allowing experiments to understand the complete life cycle of the virus. It remains a mystery as to how JFH-1 isolates are able to produce infectious virus whereas other isolates cannot. Attempts at identifying other fulminant hepatitis isolates that produce infectivity in cell culture have been unsuccessful, suggesting that the acute clinical stage of the original JFH-1 isolate has little to do with the ability to generate infectivity in cell culture. The original models of HCV infectivity predicted that the lack of adaptive mutations in JFH-1 was responsible for the production of infectivity, but the identification of a genotype 1a isolate designated H77S that can produce low levels of cell culture infectivity, despite encoding five cell culture adaptive mutations, stands in conflict to the original model (Yi and Lemon, 2004; Yi *et al.*, 2006). One could argue that the low levels of infectivity might be increased by eliminating the H77S adaptive mutations, but the requirement of these mutations for efficient RNA replication, and the requirement for RNA replication for virus production, makes this an untestable hypothesis at the present time. Regardless of the correctness of the original ideas about adaptive mutations and their impact on assembly, the idea of their importance led to the identification of the HCVcc system. Since the development of HCVcc a number of DNA launched infectious culture systems have been described, including those for genotypes 1a, 1b, and 2a (Cai *et al.*, 2005; Kato *et al.*, 2007). In these systems, cDNA encoding the HCV genome, under control of cellular transcription, are delivered to cells in plasmid form or integrated into the host chromosome. One possibility as to why these systems are capable of producing infectious virus might be that the cellular transcription creating free genome, couple with HCV RNA replication, provides an excess of non-replication complex constrained viral RNA and shifts life cycle events towards particle assembly. Regardless of how these systems work, it is increasingly clear that subtle alterations in the balance of RNA levels and replication machinery stability can have dramatic effects on virus production in a manner we have yet to decipher. Having an infectious culture system for HCV makes addressing these issues, as well as a full characterization of the HCV life cycle possible.

General overview of the HCV life cycle

The best way to discuss the HCV life cycle is to provide a general overview of the process here, then focus in on specific steps of the process in subsequent sections. This overview is also presented graphically in Fig. 1.1, with individual steps in the process designated by letters. The HCV life cycle begins with circulating HCV infectivity in the serum, with virions complexed with host lipoprotein particles (A). These particles attach to cells via interaction with primary attachment receptor molecules, and follow a series of receptor oligomerization and transfer steps, resulting in the arrival of virions at the cellular tight junction complex. Following a clathrin mediated endocytosis of the receptor–virus complex, endosomal acidification activates a fusion activity in the virion resulting in the viral membrane fusing with the endosomal vesicle. This results in the release of the viral nucleocapsid,

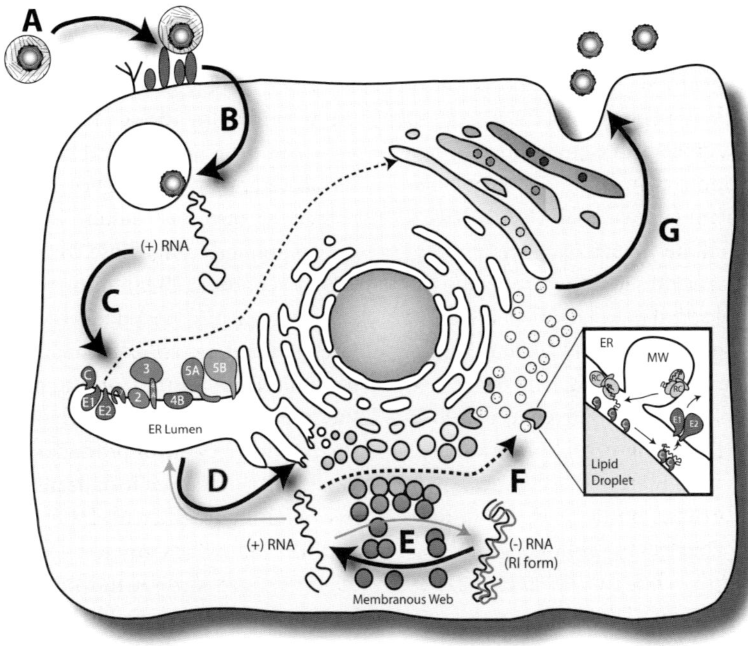

Figure 1.1 Graphical representation of the HCV life cycle. The diagram represents an idealized hepatocyte cell and the broad steps of the HCV lifecycle are indicated. These include (A) free virions complexed with lipoproteins, (B) receptor interactions and endocytosis, (C) translation of viral RNA and protein processing, (D) creation of the site of viral RNA replication from ER membranes, (E) RNA replication, (F, and inset box) early assembly events involving RNA release from replicase complex (RC) in membranous web compartments (MW) to core protein(labelled C) on lipid droplets and subsequent budding into the ER via interaction with the E1 and E2 glycoproteins and (G), the maturation and release of virions through the host secretory pathway.

or naked viral RNA, into the cytoplasm of the newly infected cell (B). The positive sense viral genome serves directly as messenger RNA, and is infectious alone, suggesting no viral proteins are necessary to initiate infection. This viral RNA serves directly as messenger RNA, and, via the activity of an internal ribosome entry site (IRES) at the 5′ end of the genome, can drive translation of the HCV polyprotein. HCV encodes a single large protein of roughly 3000 amino acids that is extensively processed to produce the 10 mature HCV proteins, all of which are membrane associated (C). At this point the viral proteins move to appropriate compartments within the cell to carry out various aspects of the virus life cycle, with the glycoproteins moving to the ER, the nucleocapsid core moving to lipid droplets (LDs), and the viral non-structural proteins inducing a membranous expansion of the ER to create the site of RNA replication, a structure designated the membranous web (MW) (D). RNA replication, as with all positive sense RNA viruses, occurs associated with altered cellular membranes. Replication uses a subset of the viral non-structural proteins to catalyse the conversion of the incoming positive sense RNA genome into a minus-strand replicative intermediate which then serves as template for the production of numerous new positive-strand genomic RNAs (E). Some of these RNAs re-enter the translation process, some serve as template for creating more negative strands for replication, and others serve as substrates for virion assembly. What regulates the fate of viral RNA is not known. The next stage of the viral life cycle is the production of progeny virus particles. This process begins with the release of viral RNA from replication into the assembly process. HCV assembly is incredibly complex, and involved the recruitment of cellular lipid droplets to viral replicase containing membrane invaginations of the membranous web. The HCV non-structural proteins play an important, but undefined, role in the release of RNA from replication events, but the end result of this process is the transfer of viral genomic RNA from the membranous web to the nucleocapsid core protein that coats cellular LDs (F, and inset box). Once viral RNA is associated with core, viral nucleocapsids form and bud into the ER via association of the core with the matured viral glycoprotein spikes that reside in this organelle. Newly budded virions then transit the host secretory pathway, associates with host lipoproteins, and are secreted from the cell to infect new hosts (G). The process appears deceptively simple, yet each step of the life cycle is a carefully orchestrated event involving a multitude of viral and host factors. The following sections will describe the life cycle stages illustrated in Fig. 1.1 in more detail.

Virions

Perhaps the best place to begin a detailed discussion of the HCV life cycle is with the extracellular infectious virus particle. The difficulties in the characterization of the HCV virion rival those described for generating model systems for the study of HCV. Perhaps the most significant problem is that very few convincing electron micrographs of infectious virus particles from actual patient samples have been generated due to a variety of technical limitations. From the limited electron microscopic analysis available, and a variety of filtration experiments, HCV virions appear as roughly spherical particles of 30–80 nm in size (Bradley *et al.*, 1985; He *et al.*, 1987; Yuasa *et al.*, 1991). Electron microscopic analysis of HCVcc virions is in good agreement with these previous findings, with images showing 50 nm spherical particles (Wakita *et al.*, 2005). The HCVcc particles contain an electron dense core, presumably representing the nucleocapsid, surrounded by an outer lipid bilayer membrane protein shell that lacks obvious surface features. A low-resolution cryo-electron microscopy image reconstruction of VLPs has been generated (Yu *et al.*, 2007), and these particles appear smooth and mostly featureless, much like authentic virion reconstructions seen for several flaviviruses (Ferlenghi *et al.*, 2001; Kuhn *et al.*, 2002; Mukhopadhyay *et al.*, 2003), but it is unclear how closely these VLPs mimic the actual HCV virion. Collectively, the evidence to date suggests that HCV virions are more similar to those of the flaviviruses, with envelope proteins lying flat on the surface of the virion rather than arranged in prominent spikes projecting from the surface of the particle, such as exemplified by the alphaviruses (Cheng *et al.*, 1995). Compositionally, little is known about the HCV virion. The HCV virion contains an internal nucleocapsid, presumably composed of the viral genomic RNA surrounded by multiple copies of the basic viral core protein. This nucleocapsid in surrounded by a host derived lipid bilayer that is likely embedded heterodimeric oligomers of the two virally encoded envelope glycoproteins, E1 and E2, and by analogy to the flaviviruses, these oligomers are likely arranged in an icosahedral lattice. It is not known if other viral proteins are present as minor components of the virion, but the recent demonstration that several viral non-structural proteins can dramatically affect the release of virus infectivity makes this a real possibility. It is also unclear if host proteins are incorporated into virions as essential components of the infectious particle. A detailed proteomic analysis of highly purified virions is clearly warranted, but the diverse and heterogeneous nature of HCV particles complicates purification.

HCV from cell culture, chimpanzee infections, or human samples exists as a heterogeneous mixture of particles of a broad range of buoyant densities. This mixture likely contains free virions, free nucleocapsids, virions complexed with immunoglobulins, virions complexed with low-density lipoproteins (LDL), and virions complexed with very-low-density lipoproteins (vLDL). Early analysis of HCV infectivity fractionated on sucrose gradients suggested infectious particles had an unusually low buoyant density of 1.10 g/ml (Bradley *et al.*, 1991; Miyamoto *et al.*, 1992). By comparison, the classical flaviviruses have a buoyant density range of 1.19–1.20 g/ml, and the pestiviruses in the 1.12–1.16 g/ml range (Miyamoto *et al.*, 1992). Stripping the membranes off of HCV particles with chloroform or detergents releases what is believed to be the nucleocapsid (buoyant density of 1.17–1.25 g/ml; Miyamoto *et al.*, 1992; Hijikata *et al.*, 1993b;

Kanto *et al.*, 1994). Monitoring both HCV RNA and infectivity in density gradients from HCVcc infections provided an interesting result, in that the peak of viral RNA had a buoyant density of 1.15 g/ml, whereas the peak of infectivity was in a lighter 1.10 g/ml fraction (Cai *et al.*, 2005; Lindenbach *et al.*, 2005; Wakita *et al.*, 2005; Zhong *et al.*, 2005). Higher buoyant density fractions were heterogeneous and of reduced infectivity, despite containing viral RNA. This is in good agreement with clinical observations of HCV samples from humans. Interestingly, passing HCVcc virus through chimpanzees or humanized chimeric mice increased the percentage of viral RNA in the 1.10 g/ml fraction and increased the overall specific infectivity of the virus (Lindenbach *et al.*, 2006). Collectively, these data suggest that the 1.15 g/ml fraction might be the free HCV virion, and the lighter 1.10 g/ml fraction is the virion associated with host lipoproteins. Alternatively, these fractions might represent virions associated with different types of lipoproteins or even complexed with different amounts of lipoproteins. Whatever the case, the association of virus with lipoproteins is important for infectivity and is more readily accomplished in natural infections than in cell culture. Attempts at manipulating lipoprotein levels with the large number of available pharmaceuticals designed for this purpose have met with mixed results when used for HCV therapy (reviewed in Argo *et al.*, 2008). Clearly, a more accurate picture of what constitutes an infectious virion, how lipoproteins function in infectivity, and if this can be better explored therapeutically are of great importance.

Receptors and entry

The HCV life cycle really begins with the virion engaging receptor molecules on the surface of the target cell to facilitate entry of the virus. The primary target cell infected by HCV is the hepatocyte, although infections of dendritic cells, B cells, and a number of other cell types have been suggested (reviewed in Lindenbach *et al.*, 2007a). These non-hepatocyte infections may be important to the evasion of the immune system by HCV, but none of these putative reservoirs of non-hepatic infection appear to be very robust in terms of virus replication, making detailed interpretation of this concept difficult. HCV infection produces significant circulating viraemia in most patients, yet, aside from the noted putative exceptions, appears to target hepatocytes, suggesting a specific targeting of virions to the hepatocytes, a restriction of replication to these cells, or both. Whatever the case, the factors required for productive HCV infection are sufficiently narrow in terms of genetic diversity to restrict infection to humans and chimpanzees. Indeed, even supplying all of the known receptor molecules for HCV on the surface of murine cell lines only allows viral entry, but not efficient replication (Ploss *et al.*, 2009). It is therefore likely that the narrow host range, and hepatocyte specificity of HCV infection is the result of both receptor availability and the presence of unknown intracellular replication factors or replication permissive host cell environments. The following section of this chapter will discuss the candidate receptors required for HCV binding and entry. This field has seen remarkable progress in the last few decades, with a fairly complete picture of the molecules required for HCV entry emerging.

The search for the cellular receptor for HCV binding and entry has a long history, beginning with demonstration of the binding of a soluble form of E2 to cells and potential receptor molecules, and eventually leading to the use of artificial HCV like particles, including artificial membrane vesicles containing the HCV glycoproteins, VLPs, and HCVpp, as well as authentic HCVcc infections (reviewed in Flint *et al.*, 2001; Zeisel *et al.*, 2009). Current experimental data have implicated low density lipoprotein receptor (LDLR), dendritic cell-specific intracellular adhesion molecule 3 grabbing non-integrin (DC-SIGN), liver/lymph node specific intracellular adhesion molecule 3 grabbing integrin (L-SIGN), the glycosaminoglycan (GAG) heparan sulfate, scavenger receptor class B type I (SRBI), tetraspanin CD81, claudin-1, and occludin as potential HCV receptor molecules (Flint *et al.*, 1999a; Scarselli *et al.*, 2002; Barth, 2003; Bartosch *et al.*, 2003; Gardner *et al.*, 2003; Pohlmann *et al.*, 2003; Evans *et al.*, 2007; Ploss *et al.*, 2009). Many of these interactions can be blocked by virus glycoprotein E2 neutralizing antibodies, suggesting the specific nature of the observed interactions. An emerging body of evidence suggests that a number of these molecules are required in a 'receptor complex' for productive binding and entry of HCV, and that no one molecule alone represents the HCV receptor.

The HCV 'receptor complex' likely functions in a semi-sequential manner; with concentration, primary attachment, receptor recognition, and receptor endocytosis occurring in stepwise fashion to promote virion entry into the cell. The first of these steps is the initial interaction of the virion with the target hepatocyte. For this step to occur, the virion must reach the host liver and exit from circulation to engage hepatocytes. Hepatocytes are isolated from the blood by a layer of cells referred to as liver sinusoidal epithelial cells. The DC-SIGN and L-SIGN molecules are calcium-dependent lectins that are highly expressed on these cells and possess a strong affinity for the HCV virion glycoproteins (Gardner *et al.*, 2003; Lozach *et al.*, 2003; Pohlmann *et al.*, 2003). Virion binding to DC-SIGN and L-SIGN on the liver sinusoidal epithelial cells may serve to concentrate viral particles in the liver at a site where the circulation of blood is closest to the hepatocytes, making infection of the hepatocytes possible (Cormier *et al.*, 2004a; Gardner *et al.*, 2003; Lai *et al.*, 2006; Lozach *et al.*, 2003). The precise details of how virus concentrated on liver sinusoidal epithelial cells reaches hepatocytes is not entirely clear, but a model similar to the presentation of human immunodeficiency virus on dendritic cells by DC-SIGN has been proposed (Lai *et al.*, 2006). This model is logical, as DC-SIGN and L-SIGN are unable to directly support entry of virions, but bind the particles with a high affinity.

Once concentrated in the liver sinusoids, and in close proximity to hepatocytes permissive for HCV infection, the virion makes initial interactions with the target cell via interaction with a variety of primary hepatocyte attachment receptors. These include the GAG heparan sulfate, and/or the LDLR. Heparan sulfate comprises a diverse family of linear polysaccharides that functions as a receptor for a large number of viruses (reviewed in Chen *et al.*, 2008). Early research with heparan sulfate, and the related molecule heparin, identified a high affinity interaction of these GAGs with purified forms of the HCV E2 glycoprotein (Barth, 2003; Barth *et al.*, 2006), and to a lesser extent the E1 glycoprotein (Barth *et al.*, 2006). Pretreatment of HCVpp or HCVcc particles with heparin, or cells with heparinase, inhibited HCV attachment and reduces, but does not eliminate, infectivity (Barth *et al.*, 2006; Basu *et al.*, 2007; Koutsoudakis *et al.*, 2006; Morikawa *et al.*, 2007). It is clear heparan sulfate plays some role in HCV binding and entry, but the presence of this molecule on a large number of cell types, and the hepatotrophic nature of HCV, combined with the moderate impact of the manipulation of heparan sulfate on HCV infections, suggests GAGs function as accessory attachment factors that stabilize virion/cell interactions rather than drive them. In a similar fashion, the interaction of virions with the LDLR, likely via the virion associated lipoprotein particles rather than direct contact with a viral protein, might serve as a means of primary attachment to hepatocytes (Monazahian *et al.*, 1999). Hepatocytes are rich in LDLR, and HCV infectivity is often associated with LDL and vLDL particles, making a potential interaction of virus associated LDL/vLDL and LDLR an attractive route of virus attachment. Pre-incubation of cultured cells and primary hepatocytes with LDL can prevent HCV from patient isolates from binding to the cell surface (Monazahian *et al.*, 1999), and in one report, prevent the infection of primary hepatocytes by HCV in a manner similar to antibodies against the LDLR (Molina *et al.*, 2007). Unfortunately, these experiments have not yet been confirmed in the more robust entry and infection system of the HCVcc virus and the precise mechanism of how LDLR might promote infection remains nebulous. It certainly is possible that both of LDLR and GAGs play a role in the initial attachment of HCV to hepatocytes, as well as other molecules that have yet to be identified.

Once initial attachment of virions to the hepatocyte surface has occurred, a range of post-binding receptor molecules mediate the entry of HCV. The tetraspanin CD81 was the first of these molecules to be identified, via an observed *in vitro* interaction between CD81 and soluble form of the HCV E2 glycoprotein (Drummer *et al.*, 2002; Drummer *et al.*, 2005; Flint *et al.*, 1999a; Higginbottom *et al.*, 2000; Pileri *et al.*, 1998). CD81 is a tetraspanin, and as this name suggests, it contains four transmembrane segments, yielding two extracellular loops (designated as long extracellular loop (LEL), and short extracellular loop), a short intracellular loop, and cytoplasmic extensions at the amino and carboxyl termini. CD81 is a small (25 kDa), non-glycosylated protein that is widely expressed on most, if not all, human cells. The LEL contains two conserved disulfide bonds that generate the fold indicative of members of the tetraspanin family, and this LEL and the disulfide bonds within the LEL are essential for the interaction with the HCV E2 glycoprotein (Petracca *et al.*, 2000). Soluble forms of the LEL and antibodies against CD81, or the LEL, can block HCVpp and HCVcc infections in cell culture as well as *in vivo* HCV infections in chimpanzees, clearly indicating CD81 is essential for the HCV lifecycle (Bartosch, Dubuisson, and Cosset, 2003; Lindenbach *et al.*, 2005; Wakita *et al.*, 2005; Zhong *et al.*, 2005; Bertaux and Dragic, 2006). Silencing of CD81 expression using siRNA mediated approaches have also shown this protein to be essential for HCV infections (Zhang *et al.*, 2004; Koutsoudakis *et al.*, 2006). Expression of CD81 on some hepatoma cell lines that lack endogenous CD81 expression renders these cells permissive to HCV infection, but not all CD81-negative cell lines can be rescued in this manner, indicating CD81 is not the sole HCV receptor (Bartosch *et al.*, 2003; Zhang *et al.*, 2004; Lindenbach *et al.*, 2005). The interaction of HCV with CD81 appears to be specific, as other members of the tetraspanin family show negligible interaction with soluble E2 glycoproteins or HCV virions (Flint *et al.*, 1999b). CD81 appears to function as a post-binding receptor, as CD81 directed antibodies are capable of blocking infection when added after initial virion binding has occurred (Cormier *et al.*, 2004b; Flint *et al.*, 2006; Koutsoudakis *et al.*, 2006). A recent study has suggested that HCV engagement of CD81 can induce Rho GTPases that result in the organization of microtubules for the lateral movement of CD81 to deliver virions to downstream receptors, and further suggests activation of cellular signalling cascades in the Raf pathway that may be involved in later infection events (Brazzoli *et al.*, 2008).

Another molecule proposed as an HCV entry receptor is the human scavenger receptor class B type I (SRBI; Scarselli *et al.*, 2002). SRBI is expressed on a wide variety of human tissues, but it is enriched in liver and steroidogenic tissues (Acton *et al.*, 1996; Rodrigueza *et al.*, 1999). SRBI is an 82 kDa glycoprotein featuring amino and carboxyl-terminal cytoplasmic domains connected by a large extracellular domain that contains a number of binding sites for a variety of lipoprotein molecules. SRBI liganded with lipoproteins serves as a source for the selective uptake of cholesterol esters from lipoprotein particles, and does so in a manner that does not destroy the proteinaceous components of the particle (Acton *et al.*, 1996; Rodrigueza *et al.*, 1999; Silver *et al.*, 2001). The well-documented association of HCV infectivity with lipoproteins makes this an attractive candidate for an HCV receptor, much like LDLR. Unfortunately, the story that has developed for SRBI is far more complex than the simple interaction of HCV virions with SRBI via the affinity of their associated LDLs with the SRBI receptor. Although adding LDL to cell culture facilitates HCVcc and HCVpp entry, it appears to do so with no apparent requirement for an interaction with the virion (Bartosch *et al.*, 2005; Kapadia *et al.*, 2007; Voisset *et al.*, 2005). Indeed, the interaction between SRBI and HCV appears to be mediated by direct interaction of the viral E2 glycoprotein with SRBI, rather than an indirect interaction via LDLs (Scarselli *et al.*, 2002). Neutralizing antibodies, an genetic manipulation of the HCV E2 glycoprotein can reduce binding to SRBI and impair HCVpp and HCVcc infectivity, and in the case of antibodies, do so in a dose-dependent manner. Potent SRBI directed antibodies can block infection in cell culture and infection in chimpanzees (Catanese *et al.*, 2007; Grove *et al.*, 2007). Further supporting a direct binding model, the expression level of SRBI can modulate HCVcc infectivity and can be manipulated artificially to promote entry (Grove *et al.*, 2007). SRBI is considered as a post-binding receptor for HCV, as SRBI directed antibodies can neutralize HCV infection when added as much as an hour post initial virus binding to the cell (Cormier *et al.*, 2004b). SRBI and the related molecule SRBII may play a role in the cell–cell transmission of HCV (Grove *et al.*, 2007). Despite all the evidence indicating SRBI is an HCV entry receptor that functions via a direct interaction with the viral E2 glycoprotein, the issue of how LDL promotes HCV infection still remains unanswered. Curiously, the ligand effects of SRBI and HCV entry appear quite complex, with LDL promoting infection (Bartosch *et al.*, 2005; Meunier *et al.*, 2005; Voisset *et al.*, 2005), but vLDL and oxidized LDL inhibiting the process (von Hahn *et al.*, 2006). The conformational 'activation' of the SRBI protein by interaction with the exogenous LDL particles has been proposed as a potential mechanism of action to explain these results, as has the addition of cholesterol to the plasma membrane near the site of the HCV receptor complex via the cholesterol transfer activity of SRBI (Kapadia *et al.*, 2007). Interestingly, depletion of cholesterol from cell membranes significantly reduces the expression of CD81, but not SRBI, and reduces HCV infection. This suggests that CD81 requires a cholesterol rich membrane, and suggests, with other data, that SRBI and CD81 may form a complex in which SRBI provides the necessary membrane alterations for CD81 function. The formation of such a receptor complex between HCV E2, CD81, and SRBI has been demonstrated, and removal of any one protein abrogated formation of any complex between the remaining proteins (Heo *et al.*, 2006). Perhaps the interaction of virus with SRBI promotes this process of CD81 activation and receptor oligomerization, resulting in the generation of an entry permissive receptor complex. Whatever the precise mechanism by which SRBI and LDL promote HCV entry, the data to date clearly implicate as an HCV entry receptor.

Claudin-1 is a member of the claudin family of transmembrane protein components of the cellular tight junction. These proteins are highly expressed in epithelial tissues as well as the liver, where they are important in maintaining cell polarity. The general topology of this family is conserved in nature, with four transmembrane helices producing two extracellular and one intracellular loop, with amino and carboxy-terminal cytoplasmic extensions (Van Itallie and Anderson, 2006). Using a cyclic expression cloning strategy, claudin-1 was identified as a gene that conferred susceptibility to HCVpp and HCVcc entry on non-permissive cell lines and allowed for HCVcc infection on non-permissive hepatocyte cell lines lacking endogenous claudin-1 expression (Evans *et al.*, 2007). The HCV entry-dependent activity of claudin-1 can be attributed to the first large extracellular loop of the protein (Evans *et al.*, 2007). The claudin-1-dependent HCV entry process has been shown to be dependent on the HCV glycoproteins, CD81, and endosomal acidification (Evans *et al.*, 2007). Down regulating the expression of claudin-1 by shRNA or siRNA-based approaches led to an inhibition of infection by HCVcc and HCVpp (Liu *et al.*, 2009). Antibody inhibition studies suggest that claudin-1 functions at a step after virion binding to the target cell has occurred, and suggests claudin-1 functions after the virus interacts with CD81 (Koutsoudakis *et al.*, 2006). Cellular fusion assays suggest claudin-1 is essential for HCV envelope mediated membrane fusion, but the mechanism of these events is poorly understood (Evans *et al.*, 2007). Imaging studies suggest that the localization of claudin-1 to tight junction complexes correlates with permissiveness to HCV infection (Liu *et al.*, 2009; Yang *et al.*, 2008). Additional Claudin family members, claudin-6 and clauding-9, have been shown to function in HCV entry as well, but in the case of these molecules their function in entry appears dependent on the cell type utilized (Meertens *et al.*, 2008; Zheng *et al.*, 2007). The precise details of how virus transits from receptor molecules like CD81 and SRBI to the tight junction to interact with Claudin are largely unknown.

After the identification of claudin-1, it was known that at least one additional HCV entry receptor must exist, as expression of claudin-1 on human cell lines that express CD81 and SRBI but lack claudin-1 failed to allow productive HCV entry (Evans *et al.*, 2007). Recent studies have implicated occludin (OCLN) as the missing entry receptor (Benedicto *et al.*, 2009; Ploss *et al.*, 2009). OCLN is a 60-kDa protein that shares topological similarity with claudin-1, having four transmembrane segments, two large extracellular loops and cytoplasmically oriented termini (reviewed in Feldman *et al.*, 2005). OCLN, like claudin-1, is a molecule involved in the formation and/or maintenance of cellular tight junctions. Silencing OCLN using siRNA or shRNA blocks

HCVpp and HCVcc entry and infection of Huh-7.5 cells (Ploss *et al.*, 2009). Similar to claudin-1, overexpression of OCLN on HCV permissive cells does not seem to increase the rate or magnitude of HCV entry, suggesting other factors are limiting in the entry process. The expression of OCLN on a number of cell lines resistant to HCV infection, including some murine cell lines, allows for HCV entry into these cells (Ploss *et al.*, 2009). Immunoprecipitation and colocalization studies suggest OCLN forms complexes with the viral E2 glycoprotein, which is consistent with the role of E2 as the receptor-binding component of the HCV virion. Collectively, these data suggest OCLN is an essential HCV entry receptor, and again suggest the importance of cellular tight junctions to the HCV lifecycle. The levels of claudin-1 and OCLN appear to be down regulated by infection, suggesting HCV infection actively manipulates the level of these proteins (Liu *et al.*, 2009). As is the case with claudin-1, the intricacies of how and why HCV manipulates OCLN and tight junctions remain to be determined.

Once HCV virions have attached to hepatocytes and transited through the complex series of receptor interactions the virions are taken into the cell via a clathrin-dependent endocytic process (Blanchard *et al.*, 2006; Codran *et al.*, 2006; Meertens *et al.*, 2006). These early virion containing endosomes are then acidified, resulting in the activation of the viral fusion protein activity. Viral entry is clearly pH dependent, with an optimum fusion activity at pH 5.5 (Kobayashi *et al.*, 2006). Inhibition of the endosome acidification process can block the productive entry of both HCVpp and HCVcc, presumably trapping virion within vesicles by preventing activation of the viral fusion activity (Bartosch *et al.*, 2003; Hsu *et al.*, 2003; Tscherne *et al.*, 2006). Surprisingly, low pH treatment of free HCVcc virions does not alter their infectivity, suggesting that receptor interactions may alter the virus such that it becomes pH sensitive (Koutsoudakis *et al.*, 2006; Meertens, Bertaux, and Dragic, 2006; Tscherne *et al.*, 2006). This is not the case for HCVpp, which are inactivated by low-pH treatment, highlighting a key difference between HCVcc and HCVpp in entry (Hsu *et al.*, 2003; Op De Beeck *et al.*, 2004). The entry process is dependent on the host microtubule network, both for receptor virus complex lateral movement on the cell surface as particles transition through the various steps of the receptor complex interaction to reach tight junctions, and for the later fusion process itself (Roohvand *et al.*, 2009).

The location of the HCV membrane fusion activity is not known. Both sequences in E1 and E2 have been proposed as viral fusion peptides based on sequence analysis, but neither protein has been demonstrated to be fusogenic. Recent disulfide mapping and antibody probing analysis has allowed the modelling of the HCV E2 glycoprotein on the general scaffold of a *Flavivirus* type II fusion protein, and these data, and earlier modelling studies, suggest that the fusion peptide resides within E2 protein (Krey *et al.*, 2010; Yagnik *et al.*, 2000). Considerable data exists on the fusion activity of the classical flaviviruses, and it is likely that HCV uses a similar mechanism. Flaviviruses contain a type II fusion peptide that is buried in the mature virion (reviewed in Harrison, 2008). Upon acidification of the endosome, the *Flavivirus* envelope protein undergoes a complex pH-dependent domain rearrangement and protein trimerization, resulting in the exposure of the function fusion peptide. It is unclear if this mechanism occurs in HCV entry, but the structural similarities between HCV E2 and the *Flavivirus* E proteins makes this a likely mechanism for HCV entry. HCV infection exhibits a pronounced superinfection exclusion, and this phenomenon appears to occur at a step post entry, as exclusion requires viral RNA translation (Schaller *et al.*, 2007; Tscherne *et al.*, 2007). The end result of the entry and fusion process is the deposition of the viral nucleocapsid, or perhaps naked viral RNA, into the cytoplasm of the newly infected cell.

HCV genome organization

The HCV genome is a 9.6 kb positive-sense, monopartite RNA that contains a single, large open reading frame (ORF) flanked by highly structured, and highly conserved, 5′ and 3′ non-coding regions (NCRs; Choo *et al.*, 1991). The RNA genome lacks a 5′ cap structure and does not encode a polyadenosine tail, and translation, replication, and possibly packaging events are mediated by a number of *cis*-acting replication elements (CREs) found in the NCRs and elsewhere in the viral genome (Kolykhalov *et al.*, 1996; Tanaka *et al.*, 1995; Tanaka *et al.*, 1996). The mechanisms that regulate these various genomic RNA directed processes are unknown, but they likely involve a combination of viral and host factors, CRE elements in the RNA, and interactions between these molecules. A schematic representation of the HCV genome is shown as Fig. 1.2.

The HCV 5′ NCR is a highly structured, well-conserved, 341-nucleotide element that contains both multiple CREs involved in RNA replication as well as the viral IRES element required for genome translation. The IRES will be discussed in later sections, and here we will focus on the CREs in this region of the genome. The 5′NCR is organized into four domains (numbered I through IV), and these domains are further separated into individual stem–loop structures that are designated with lower case letters (e.g. IIIb is domain III, stem–loop b). The first of these domains is not required for the IRES activity in the 5′ NCR, but is an essential CRE for genome replication (Friebe *et al.*, 2001; Kim *et al.*, 2002; Rijnbrand *et al.*, 2005). Extensive mutagenesis has mapped the minimal 5′ CRE required for genome replication to domains I and II of the 5′ NCR, although

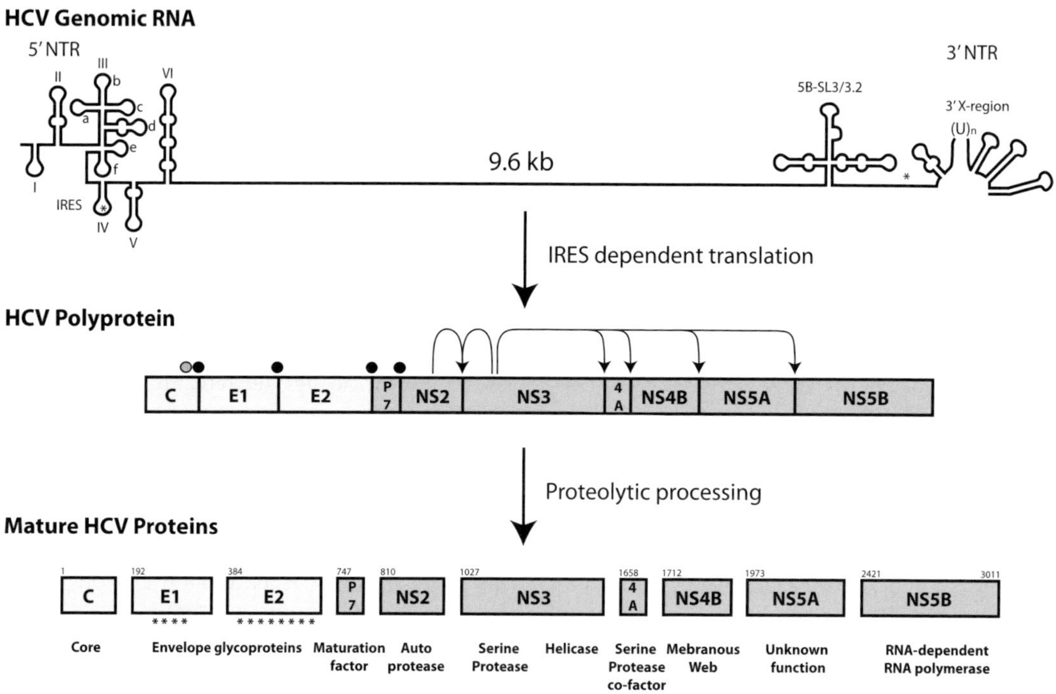

Figure 1.2 Schematic of the HCV genomic RNA, Polyprotein processing scheme, and mature HCV proteins. Top panel, HCV genomic RNA schematic showing the locations of the 5′ and 3′ NCR regions and other CREs. The 5′ NCR domains are labelled in Roman numerals, and the IRES element within this region is designated. The asterisk in the loop of the stem labelled IV corresponds to the AUG initiation codon of the HCV open reading frame. The 3′ NCR shows the location of the 3′ X tail and the poly (U/UC) tract. The asterisk in this region represents the stop codon of the viral open reading frame. The CRE located within the NS5B coding region is also shown (5B-SL3/3.2). Middle panel, the HCV polyprotein with structural proteins shaded in light grey, and non-structural proteins in dark grey. Dots represent cleavage by host signal peptidase (black dots), and signal peptide peptidase (grey dot). Arrows represent cleavage events mediated by NS2 (double arrow) and NS3 (all other arrows). Bottom panel, mature HCV proteins designating amino acid sequence numbers that delineate the proteins. Asterisks indicate putative glycosylation sites. The putative functions of each HCV protein is listed below the box labelled with the protein identity.

efficient replication essentially requires the entire 5′ NCR region (Friebe *et al.*, 2001; Kim *et al.*, 2002). Experiments suggest the 5′ NCR region of the genome is essential for new positive-strand RNA synthesis (Binder *et al.*, 2007). The loop region between domains I and II contains two conserved sequences that bind the liver-specific cellular micro RNA miR-122, and this interaction enhances HCV genome replication (Jopling *et al.*, 2005). Other reports suggest miR-122 may affect translation and virus production as well (Jangra *et al.*, 2010). Antisense gene silencing strategies targeting miR-122 in chimpanzees have fairly potent anti-HCV effects, suggesting that targeting this micro-RNA may have some therapeutic potential (Lanford *et al.*, 2010). The demonstration that HCV utilizes a micro RNA for replication, and the requirement of the cellular micro-RNA processing machinery for replication (Randall *et al.*, 2007), is a particularly unusual aspect of the virus life cycle, and these events makes for an attractive antiviral target.

The 3′NCR is located immediately downstream of the genomic RNA's open reading frame. This region of the genome consists of a short variable region, a polyuridinde/polypyrimidine (U/UC) tract with an average length of 80 nucleotides, and a highly conserved 98 nucleotide region designated the 3′ X-tail which contains three stem–loop structures (Blight and Rice, 1997; Kolykhalov *et al.*, 1996; Tanaka *et al.*, 1995, 1996). The variable region seems to influence the efficiency of viral RNA replication, whereas the poly(U/UC) tract and 3′ X-tail are essential for this process both *in vitro* and *in vivo* (Yanagi *et al.*, 1999; Friebe and Bartenschlager, 2002; Yi and Lemon, 2003a,b). Mutagenesis studies indicate the poly(U/UC) tract has a minimum size of about 25 nucleotides and requires uninterrupted blocks of polyuridine (Friebe and Bartenschlager, 2002; Yi and Lemon, 2003a). Intergenotypic chimeras suggest the 3′ X-tail region is involved in minus-strand synthesis (Binder *et al.*, 2007).

In addition to the CREs in the NCRs, a number of CREs have been identified within the open reading frame of the HCV genome. Perhaps the most studied of these elements lies within the coding sequence of the NS5B RNA-dependent RNA polymerase, a region designated 5B-SL3 (Friebe *et al.*, 2005; Lee *et al.*, 2004; You *et al.*, 2004). 5B-SL3 is a large cruciform structure, and one stem–loop, designated 5B-SL3.2, is an essential CRE (You *et al.*, 2004). 5B-SL3.2 appears to make a long-distance 'kissing' interaction with a loop in the 3′ X-tail region to generate a pseudoknot structure that appears to be essential for RNA replication (Friebe *et al.*, 2005). Another important CRE exists in the core protein coding sequence. This element has been designated stem–loop VI. This region overlaps a putative alternate reading frame within the core gene, and mutagenesis data suggest that stem–loop VI, and not the alternate reading frame, is important for the virus life cycle (McMullan *et al.*, 2007). Interestingly, this stem–loop is not required for replication in subgenomic replicons, but alteration of the loop in the context of full-length viral genomes attenuates virus replication *in vitro* and *in vivo* (McMullan *et al.*, 2007).

Translation and protein processing

The HCV genome contains a single, large ORF encoding a polyprotein of approximately 3000 amino acids. The polyprotein is organized with structural proteins comprising the amino-terminal third of the polyprotein and non-structural proteins making up the remaining 2/3 of the polyprotein. The proteins are organized in the polyprotein the order NH2-C, E1, E2, p7, NS2, NS3, NS4A, NS4B, NS5A, and NS5B-COOH (Choo *et al.*, 1991). A schematic of the polyprotein can be found in Fig. 1.2. The translation of the HCV ORF is dependent on the viral IRES element located in the 5′ NCR (reviewed in Fraser and Doudna, 2007). The IRES activity of the 5′ NCR requires domains II, III and IV of the 5′ NCR and 24–40 nucleotides of the core protein coding sequence (Bartenschlager *et al.*, 2004). High-resolution structures of a number of the essential portion of the IRES activity have been determined (Collier *et al.*, 2002; Kieft *et al.*, 2002; Klinck *et al.*, 2000; Lukavsky *et al.*, 2003; Lukavsky *et al.*, 2000; Rijnbrand *et al.*, 2004). The HCV IRES is unusual in that it alone can recruit the direct binding of vacant 40S ribosomal subunits and form a binary complex through an extensive series of IRES–ribosome interaction

(Kieft *et al.*, 20001, 2002; Lukavsky *et al.*, 2000; Lytle *et al.*, 2002; Otto *et al.*, 2002; Otto and Puglisi, 2004; Pisarev *et al.*, 2005). Domain III of the IRES appears to direct the 40S subunit binding such that the AUG start codon is positioned in the ribosomal P site. A cryo-electron microscopy image reconstruction of the HCV IRES complexed with the 40S ribosome has been determined, and this structure indicates the IRES induces large conformational changes to the 40S subunit upon binding (Spahn *et al.*, 2001). Once this binary complex has been formed, the eukaryotic initiation factor 3 (EIF3), and the ternary complex consisting of eukaryotic initiation factor 2 (EIF2), methionine $tRNA_i$, and GTP, are recruited to the complex, and a 48S-like complex is formed at the AUG start codon (Otto and Puglisi, 2004). A cryo-electron microscopy image reconstruction of the IRES complexed with the 40S ribosomal subunit and EIF3 suggests that EIF3 uses the HCV IRES to position the viral RNA near the 40S ribosomal subunit, thereby promoting the assembly of the initiation complex (Siridechadilok *et al.*, 2005). The 60S ribosomal subunit then associates with the 48S-like complex in a GTP-dependent manner to form the complete mature, translation competent 80S complex (visualized in (Boehringer *et al.*, 2005). A number of additional cellular factors appear to participate in HCV IRES function and modulators of translation. These include the La autoantigen, nucleolin, proteome subunit PSMA7, heterogeneous ribonucleoprotein L, polycytidine binding protein, and pyrimidine tract binding protein (Anwar *et al.*, 2000; Fukushi *et al.*, 2001; Hwang *et al.*, 2009; Izumi *et al.*, 2001). Once initiation has occurred, the 3′ end of the viral RNA appears to promote efficient translation of the ORF, possibly in a manner similar to ribosome recycling mediated by the polyadenosine tail of cellular mRNAs (Bradrick *et al.*, 2006; Ito *et al.*, 1998; McCaffrey *et al.*, 2002; Song *et al.*, 2006). Inhibition of the IRES activity by small molecules has been described, and this is an attractive antiviral strategy (Parsons *et al.*, 2009).

The end result of HCV genomic RNA translation is the production of a large polyprotein precursor of approximately 3000 amino acids. This polyprotein is co- and post-translationally processed by a combination of viral and host cell protease activities to generate the ten individual mature HCV proteins. The amino-terminal structural protein-encoding region of the polyprotein is cleaved in association with cellular membranes via the action of signal peptidases. Signal peptidases cleave the C/E1, E1/E2, E2/p7, and p7/NS2 junctions. The E2/p7 bond cleavage is often incomplete, but the significance of this observation to the virus lifecycle is not clear (Lin *et al.*, 1994a). An additional cleavage at the carboxy terminus of the core protein by signal peptide peptidase occurs within the interior endoplasmic reticulum membrane and results in the removal of the remaining E1 signal sequence from the core (McLauchlan *et al.*, 2002; Okamoto *et al.*, 2004; Santolini *et al.*, 1994; Yasui *et al.*, 1998). The non-structural proteins are processed by two distinct virally encoded protease activities. The first of these activities resides in the NS2 protein, and functions as an auto-catalytic papain-like cysteine protease, resulting in the cleavage of the NS2/NS3 junction (Grakoui *et al.*, 1993a). Once the NS2/NS3 junction is processed, a second protease activity in the amino-terminal 180 amino acids of the NS3 protein is liberated. The serine protease activity encoded by NS3 then cleaves all of the remaining junctions in the viral polyprotein, including NS3/NS4A, NS4A/NS4B, NS4B/NS5A, and NS5A/NS5B (Bartenschlager *et al.*, 1993, 1994, 1995; Failla *et al.*, 1995; Lin *et al.*, 1994b). The cleavage of the NS3/NS4A site occurs in an intramolecular manner, and all other cleavages are intermolecular in nature. The protease activity of NS3 is greatly enhances by the presence of the NS4A protein, which serves as a co-factor for the protease, is a part of the chymotrypsin like serine protease fold of this part of NS3, and tethers NS3 to cellular membranes (Bartenschlager *et al.*, 1994; Failla *et al.*, 1994; Lin *et al.*, 1995; Tanji *et al.*, 1995a). Once processing has occurred, the 10 individual HCV proteins undergo complex intracellular trafficking to participate in various aspects of the virus life cycle. A diagram of the topology of the mature HCV proteins relative to the ER/MW membrane can be seen in Fig. 1.3.

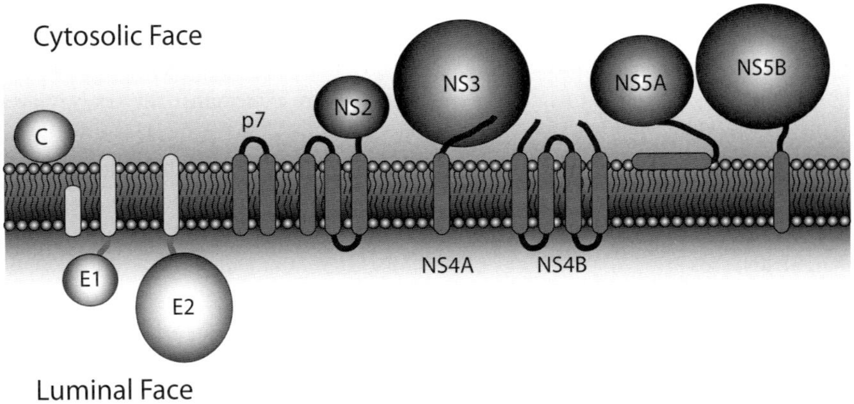

Figure 1.3 Topology of the mature HCV proteins. The topology of the HCV proteins are shown relative to the ER/MW membrane. The cytoplasmic and lumen face of the membrane are indicated. Individual proteins are labelled and collared as in Fig. 1.2.

Structure and function of the HCV proteins

Core

The HCV core protein is the most amino-terminal product of the HCV polyprotein, and is believed to be the major structural protein component of the viral nucleocapsid. Core is produced as a 23 kDa precursor protein of 191 amino acids via a carboxyl-terminal cleavage by signal peptidase in a co-translational manner at the C/E1 junction (Hijikata *et al.*, 1991). This librates the amino-terminus of E1, and leaves a portion of a hydrophobic signal sequence attached to the carboxyl-terminus of the immature core, tethering this precursor to the ER membrane. An additional cleavage event by signal peptide peptidase removes this signal sequence, generating the mature 21 kDa core protein (McLauchlan *et al.*, 2002; Okamoto *et al.*, 2004; Santolini *et al.*, 1994; Yasui *et al.*, 1998). The precise cleavage site has not been determined, and the exact size of the core protein is not known. Once the mature protein is generated, core exhibits a complex intracellular localization, and has been observed associated with ER, in the MW, on mitochondria, in the nucleus, in the cytosol, and coating the surface of LDs, although differences in distribution are observed by different laboratories (Barba *et al.*, 1997; Harada *et al.*, 1991; Hope *et al.*, 2002; Moradpour *et al.*, 1996; Moriya *et al.*, 1998; Okuda *et al.*, 2002; Selby *et al.*, 1993; Suzuki *et al.*, 1995, 2005; Yasui *et al.*, 1998). Recent data suggest that core on the surface of LDs represent the site of nucleocapsid assembly, and core alters the localization, composition, and formation of these structures (reviewed in McLauchlan, 2009). The functions of core in the other locations it has been observed are largely unknown. The core protein is organized into three domains; the amino-terminal hydrophilic/basic domain designated D1, the carboxy-terminal hydrophobic domain designated D2, and the tail domain (reviewed in McLauchlan, 2009). The D1 domain contains a large number of basic amino acids and is responsible for the RNA binding activity of the core protein (Klein *et al.*, 2005; Santolini, Migliaccio, and La Monica, 1994). This region is also important for homotypic interactions leading to the oligomerization of the core protein (Matsumoto *et al.*, 1996; Nolandt *et al.*, 1997). The hydrophobic D2 domain contains two amphipathic helices (HI and HII) separated by a hydrophobic loop region. Hydrophobic residues in HI, HII and the loop region are essential for the interaction of core with lipid droplets (Boulant *et al.*, 2006). The mechanism of core assembly into nucleocapsids is poorly understood, although a number of *in vitro* nucleocapsid assembly assays have begun to decipher this process (Acosta-Rivero *et al.*, 2001;

Kunkel *et al.*, 2001; Klein *et al.*, 2005; Majeau *et al.*, 2004). In addition to the role of core in assembly, this protein appears to impact a number of processes in the host cell, including the alteration of signalling pathways, transcriptional regulation, modulation of apoptosis, and carcinogenesis. The interactions and function of core in these diverse cellular processes have been reviewed elsewhere (McLauchlan, 2000).

E1 and E2 glycoproteins

The E1 (~30 kDa) and E2 (~70 kDa) glycoproteins represent the major protein components of the viral envelope. By comparison with other flaviviruses, they likely form a densely packed icosahedral array on the surface of the virion, with little exposure of the underlying host-derived lipid bilayer (Ferlenghi *et al.*, 2001; Kuhn *et al.*, 2002; Mukhopadhyay *et al.*, 2003). E1 and E2 are integral membrane proteins, and the cytoplasmic extensions of these proteins likely make contacts with the viral nucleocapsid inside the particle membrane. The cytoplasmic tail of E1 appears the likely candidate for this interaction, although this area is not well explored (Lo *et al.*, 1996; Ma *et al.*, 2002). E1 and E2 are processed from the polyprotein precursor by signal peptidase and oriented such that the bulk of both proteins, what will become their ectodomains on virions, are oriented into the lumen of the ER. Both proteins are type I membrane proteins, possessing carboxy-terminal transmembrane anchors that exist as two short hydrophobic helices with an intervening polar region separating the helices. The fact that these proteins are adjacent to one another in the polyprotein, yet both have single transmembrane sequences and exhibit the same type I membrane orientation is unusual, and the insertion of these proteins into membranes is rather unique. The two helices and polar loop regions of the membrane anchors likely adopt a hairpin like shape within the membrane during translocation to maintain proper topological threading of the polyprotein, and these hairpins are resolved following cleavage of E1 and E2 and become conventional transmembrane anchor elements (Cocquerel, 2002). The transmembrane regions play an important role in the later oligomerization of E1 and E2 and contain ER retention signals that are important for proper maturation of these proteins (Cocquerel *et al.*, 1998, 1999 Op De Beeck *et al.*, 2000). The maturation of E1 and E2 are extremely complex (reviewed in Lavie *et al.*, 2007). The process involves glycosylation, disulfide bond formation, association with ER chaperones, and extensive interactions between the E1 and E2 transmembrane anchors and ectodomains (Op De Beeck *et al.*, 2000; Voisset and Dubuisson, 2004; Yi *et al.*, 1997). The end result is the formation of a non-covalent, heterodimer of E1 and E2, although considerable amounts of aggregated, disulfide-linked 'dead end' products are formed. Analysis of the glycosylation state of E1 and E2 suggest these proteins contain up to 5 and 11 glycans, respectively, and these tend to be high mannose for E1 and complex for E2, suggesting late maturation of the complex occurs, in part, the Golgi (Op De Beeck *et al.*, 2004). Little is known structurally about the glycoproteins, but models suggest that E2 is similar to *Flavivirus* E proteins (Krey *et al.*, 2010). Functionally, E2 is involved in contacting the large number of cellular receptors involved in HCV entry that were described in previous sections. Modelling suggests E2 might contain the fusion peptide activity required for virion entry (Krey *et al.*, 2010; Yagnik *et al.*, 2000), although this function has also been attributed to E1 (Flint and McKeating, 1999; Garry and Dash, 2003; Rey *et al.*, 1995). The function of E1 is less clear, but it might be involved in interactions with core during assembly (Lo *et al.*, 1996; Ma *et al.*, 2002). The advent of HCVpp and HCVcc systems have allowed very detailed examination of E1 and E2 over the last several years, and more detailed roles for these proteins in HCV biology are sure to come. Blocking the proper maturation of these proteins, inhibiting the yet to be characterized fusion activity, or blocking receptor interactions are potential avenues of inhibiting HCV, but all of these targets are poorly understood (see Chapter 17).

p7

p7 is a small, 63 amino acid, protein that consists of two transmembrane helices and a short cytoplasmic loop connecting these elements (Saint *et al.*, 2009). The p7 protein is a member of the viroporin family of small, virally encoded proteins that oligomerize to form a pore or channel that alters membrane permeability to ions or other small molecules (Gonzalez and Carrasco, 2003). p7 oligomerizes to form hexa- or heptameric pore-like complexes with a cation-selective ion channel activity (Clarke *et al.*, 2006; Griffin *et al.*, 2003). The cryo-electron microscopy image reconstruction of p7 shows a hexameric molecule with a flower-like shape with the petals of the flower facing into the ER lumen (Luik *et al.*, 2009). Although the pore of the channel is not completely resolved in this structure, docking of models of p7 into the electron density of the reconstruction shows a clear central channel lined by the amino-terminal membrane-spanning helix (Luik *et al.*, 2009). This model has been confirmed by mutagenesis of the putative channel lining residues (Chew *et al.*, 2009). Although p7 is not required for HCV RNA replication in cell culture (Lohmann *et al.*, 2003), it is essential for infection in chimpanzees and production of infectious virus in the HCVcc system (Jones *et al.*, 2007; Pietschmann *et al.*, 2006; Sakai *et al.*, 2003; Steinmann *et al.*, 2007a). A number of different inhibitor molecules have been shown to block the ion channel activity of p7 in artificial membrane-based assays, and these compounds block the production of infectious virus particles (Griffin *et al.*, 2003, 2008; Pavlovic *et al.*, 2003; Premkumar *et al.*, 2004; Steinmann *et al.*, 2007b). It is generally accepted that p7 plays a role late in infectious virus production, although the precise function of this protein is unknown (Pietschmann *et al.*, 2006; Jones *et al.*, 2007; Steinmann *et al.*, 2007a). One report suggests a role of p7 in virus entry (Griffin *et al.*, 2008). The ability of several small molecule inhibitors of p7 to block virus production has made this channel an attractive, and novel, antiviral target.

NS2

NS2 is a 23kDa transmembrane protein that represents the viral autoprotease activity. The transmembrane segments of NS2 are clustered in the amino-terminal domain of the protein, but the exact topology of this region is not clear. An active debate exists as to whether there are three or four membrane spanning helices in this region, but topologically three makes more sense, as this would position the NS2/NS3 junction and NS3 on the cytoplasmic face of the ER membrane (Lorenz *et al.*, 2006; Santolini *et al.*, 1995; Yamaga, 2002). A detailed and definitive mapping of this portion of the protein is clearly needed. The carboxy-terminal domain of NS2 contains the viral autoprotease activity. The protease activity of NS2 requires the first 180 amino acids of the NS3 protein, although it is unclear why (Grakoui *et al.*, 1993a; Hijikata *et al.*, 1993a; Reed *et al.*, 1995). Site directed mutagenesis experiments indicate a cysteine protease activity for NS2, with the catalytic triad residues His-952, Glu-972, and Cys-993 essential for cleavage activity (Grakoui *et al.*, 1993a; Hijikata *et al.*, 1993a; Lorenz *et al.*, 2006). The structure of NS2 is a novel protein fold, yet it retains the active site geometry shared by all papain-like cysteine proteases (Lorenz *et al.*, 2006). The structure is that of a dimeric protease featuring extensive contacts among the dimers and an unusual composite active site, with two residues of the catalytic triad contributed by one monomer, and the third active site residue from the other monomer (Lorenz *et al.*, 2006). The dimeric nature of the active site has been confirmed by complementation experiments. NS2 is not required for RNA replication in the context of subgenomic replicons that initiate with NS3, but it is essential for replication in full-length viral genomes (reviewed in Bartenschlager, *et al.*, 2004). In addition to its role in processing the NS2/3 junction, NS2 has emerged as an important determinant for infectious virus production (Lindenbach *et al.*, 2005; Pietschmann *et al.*, 2006). Alteration of the site of joining chimeric viral genomes within the NS2 protein in the genotype 2a HCVcc system can have dramatic effects on the output of infectious virus (Pietschmann *et al.*, 2006), and similar recombinants have been seen in infected patient samples (Kalinina *et al.*, 2002). The role of NS2 in particle assembly is not well understood, but a number of sophisticated complementation experiments have confirmed this protein as an essential assembly factor (Jirasko *et al.*, 2008; Jones *et al.*, 2007; Pietschmann *et al.*, 2006). NS2 would seem to be an ideal target for antiviral compound development, but to date no molecules targeting this protein have been described. Perhaps the fact that the carboxyl-terminal residue of NS2 remains in the protease active site after cleavage, thereby potentially blocking access to inhibitors, might complicate specific targeting of the protease active site, but other potential targets in NS2 exists, including its extensive dimer interface and role in particle morphogenesis (Lorenz *et al.*, 2006). As will be seen for other viral non-structural proteins in subsequent sections, NS2 also can form a complex with cyclophilin proteins, suggesting these host proteins play an important role in the life cycle of HCV (Ciesek *et al.*, 2009).

NS3–NS4A

NS3 and NS4A function as an intimate complex in the HCV life cycle, so they will be discussed as a single entity. NS3–NS4A is a multifunctional complex enzyme, with NS3 containing the functional components of the complex, and NS4A functioning as a co-factor for these functions as well as a membrane anchor for the NS3 protein (Failla *et al.*, 1994; Tanji *et al.*, 1995a; Wölk *et al.*, 2000; Pang *et al.*, 2002). NS3 is a 69-kDa hydrophobic protein that contains two distinct functionalities; the amino-terminal region of the protein functions as a serine protease, while the carboxyl-terminal region contains an RNA helicase activity. Both of these activities are modulated by the presence of the NS4A co-factor, and are essential for RNA replication (Bartenschlager, 1994; Failla, 1994; Lin, 1995; Tanji, 1995a; Pang *et al.*, 2002; Beran *et al.*, 2007). Like several other non-structural proteins, NS3 also plays an undefined role in virus particle formation (Ma *et al.*, 2008). Numerous crystal structures exist of the protease domain, the helicase domain, and complete NS3 proteins complexed with NS4A, and NS3 is probably the best characterized protein both structurally and enzymatically (Cho *et al.*, 1998; Kim *et al.*, 1996, 1998; Love *et al.*, 1996; Yan *et al.*, 1998; Yao *et al.*, 1997, 1999). The serine protease domain is the major viral polyprotein processing enzyme, cleaving the NS3/NS4A junction *in cis*, and all the remaining downstream junctions in the polyprotein *in trans* (Bartenschlager *et al.*, 1993, 1994, 1995; Failla *et al.*, 1995; Lin *et al.*, 1994b). The central portion of the cytoplasmic domain of the NS4A co-factor form an integral part of the NS3 protease fold, contributing a β-strand to the chymotrypsin like fold of the protease domain (reviewed in Penin *et al.*, 2004b). This interaction orders the amino-terminus of NS3, stabilizing the protein and stimulating protease activity. As mentioned, NS3 has a chymotrypsin like fold consisting of two 'squashed' β-barrel domains with a central cleft containing the active site catalytic triad (His-1083, Asp-1107, and Ser 1165). The cleft/substrate pocket for NS3 is unusually shallow and hydrophobic, and this has confounded targeting of NS3 by conventional protease inhibitors (reviewed in De Francesco, 2003). Large substrate-like peptidomimetic compounds that bind, in part, outside of the active site to obtain specific targeting have been much more successful than conventional inhibitors, and these are discussed in detail in other chapter of this text. The protease activity of NS3 requires zinc for maximal activity, and the protease fold contains a zinc-binding site that likely stabilizes the fold of this domain of NS3 (Love *et al.*, 1996; Stempniak *et al.*, 1997). In addition to cleaving the viral polyprotein, NS3 has emerged as an important protein for the ability of HCV to escape the innate immune system. NS3 cleaves the cellular signalling molecules IPS-1 and TRIF, resulting in a 'short-circuiting' of signal transduction pathways from the RIG-I and TLR3 pathogen sensors to the transcription factors IRF-3 and NFκβ (Li *et al.*, 2005; Meylan *et al.*, 2005). In light of this role of NS3 in antagonizing the innate immune system, compounds targeting NS3 could both inhibit processing and lead to increased immune recognition and response to HCV infection, making for very potent virus inhibition (see Chapters 9–11).

The carboxy-terminal domain of NS3 constitutes a helicase domain classified as a member of the superfamily 2 DExH/D-box helicases (Cho *et al.*, 1998; Kim *et al.*, 1998; Yao *et al.*, 1997, 1999). This activity is capable of unwinding double stranded RNA, or highly structured single-stranded RNA molecules in a 3′ to 5′ direction in an ATP-dependent manner. The unwinding of RNA by NS3 has been studied extensively, and in appears the helicase unwinds relatively long 11–18 base pair RNA segments, and functions in a stepwise manner with short pauses that occur at regular intervals along the substrate (Dumont *et al.*, 2006; Levin *et al.*, 2005; Levin, Wang, and Patel, 2004; Serebrov and Pyle, 2004). Although the helicase domain alone is enzymatically active, full activity requires the NS3 protease domain, which functions, in part, to properly position the RNA substrate (Beran *et al.*, 2007; Pang *et al.*, 2002). Evidence of cross-talk regulation of protease and helicase activity suggests additional, yet to be defined, functional interactions exist between these domains of NS3 (Gu *et al.*, 2005; Howe *et al.*, 1999; Kuang *et al.*, 2004; Pang *et al.*, 2002). The NS3 helicase activity appears to function as a higher order oligomer (Levin *et al.*, 2004; Locatelli *et al.*, 2002; Mackintosh *et al.*, 2006; Serebrov and Pyle, 2004), and recent structural evidence confirms an oligomeric arrangement helicase associated with a single large nucleic acid substrate (Mackintosh *et al.*, 2006). The natural substrate for NS3 in the HCV infected cell is not known, but it has been hypothesized that the helicase unwinds double stranded replicative intermediates to expose CREs on the RNA, or unpairs the extensive structures on the NCRs of the genome to allow access of the polymerase to these regions. As the helicase is an essential enzymatic activity for genome replication, one might expect this to be a prime target for antiviral compound development, but progress in this area has been slow and compounds targeting NS3 are few relative to those targeting other aspects of HCV biology (see Chapter 12) (Kwong *et al.*, 2005).

As mentioned, NS4A constitutes an important co-factor for the enzymatic activities of NS3 and contributes a portion of the fold of the NS3 protease. This small, 54 amino acid, hydrophobic integral membrane protein also serves as the membrane tether, via its amino-terminal transmembrane helix, that holds NS3 in close proximity to cellular membranes (Tanji *et al.*, 1995a; Wölk *et al.*, 2000). NS4A also plays additional roles in the HCV life cycle that are only beginning to be defined. Mutations in NS4A can clearly affect replication and seem to impact the phosphorylation of the NS5A protein, particularly those in the acidic carboxyl-terminal region of NS4A (Lindenbach *et al.*, 2007b). The significance of these observations is not yet clear, but they highlight that this small co-factor is an important component of the larger replicase machine.

NS4B

NS4B is a small (27 kDa protein), hydrophobic integral membrane protein that contains at least four central transmembrane spanning helices, but the precise topology of NS4B is complex and not fully understood (reviewed in Gouttenoire *et al.*, 2010). Early studies suggested the fairly large amino and carboxyl termini of NS4B are oriented towards the cytoplasm, but subsequent work suggested the amino-terminal region of NS4B is translocated into the ER lumen (Elazar *et al.*, 2004; Lundin *et al.*, 2003, 2006). The expression of other HCV non-structural proteins, in particular NS5A, may influence the orientation of this region of NS4B, favouring a cytoplasmic-facing amino terminus (Elazar *et al.*, 2004; Lundin *et al.*, 2003, 2006). The precise meaning of the apparent dynamics of this region of NS4B is not known, but this portion of the protein is known to play a role in the assembly of the replication complex (Lundin *et al.*, 2006). The NS4B is palmitoylated on two cysteine residues near the carboxyl terminus of the protein and this post-translational modification appears to be important for the NS4B oligomerization (Yu *et al.*, 2006). NS4B is also reported to contain a nucleotide binding motif in a short cytoplasmically oriented loop between two of the transmembrane segments, and a GTPase and ATPase activity has been reported for NS4B purified from a bacterial and insect cell expression system (Einav *et al.*, 2004). The fact that a cell culture adaptive mutation has been identified that destroys the nucleotide binding motif, and the general lack of conservation of the residues comprising this loop among all HCV isolates, raises some questions about the importance of this activity to the HCV life cycle (Bartenschlager *et al.*, 2004). It is, of course, possible that this activity is only found in some isolates, and is a negative regulator of replication (so that mutations destroying the activity are adaptive in nature), but this remains an open area of investigation. Perhaps the best-characterized role for NS4B in the HCV life cycle is the activity of this protein in the alteration of cellular membranes to produce the site of replication, a structure designated the membranous web (Egger *et al.*, 2002; Gosert *et al.*, 2003). Interactions between NS4B and all of the replicase proteins have been described in a variety of systems, and this protein has been proposed to serve as a scaffold upon which the replicase assembles (Dimitrova *et al.*, 2003). NS4B is clearly an essential component of the replicase, and numerous mutations that destroy RNA replication have been generated in NS4B (reviewed in Gouttenoire *et al.*, 2010). Additionally, a number of cell culture adaptive mutants have been mapped to NS4B (reviewed in Bartenschlager *et al.*, 2004), and allelic variation within NS4B correlates with RNA replication efficiency in replicons (Blight, 2007). In addition to its role in RNA replication, NS4B has recently been shown to influence the production of infectious virus particles (Jones *et al.*, 2009). NS4B has also gained considerable interest in recent years, as a small molecule inhibitor that is directed against NS4B can alter the localization of this protein and inhibit RNA replication (Einav *et al.*, 2008). The precise mechanism of how this compound functions is only beginning to be elucidated, but the fact that potent inhibitors can be developed targeting NS4B provides a new potential therapeutic target (see Chapter 13).

NS5A

NS5A is a large hydrophilic phosphoprotein that exists in two distinct phospho-forms, a hypophosphorylated form of 56 kDa, and a hyperphosphorylated form of 58 kDa (Tanji *et al.*, 1995b). NS5A has no defined role in the virus life cycle, but it is absolutely required for RNA replication and particle assembly. NS5A is peripherally anchored to membranes by an amino-terminal amphipathic helix that is partially embedded in outer ER/MW membrane leaflet (Brass *et al.*, 2002, 2007; Elazar *et al.*, 2003; Penin *et al.*, 2004a; Sapay *et al.*, 2006). Deletion or mutation of this helix is lethal for replication and results in a diffuse cellular localization of NS5A (Elazar *et al.*, 2003; Penin *et al.*, 2004a). Following this membrane anchor, the bulk of the NS5A protein is organized into three domains, separated by two repetitive low complexity sequence stretches that are likely surface exposed loops (Tellinghuisen *et al.*, 2004). The amino-terminal domain contains a conserved zinc-binding site that appears to be a structural metal ion required for NS5A's role in RNA replication (Tellinghuisen *et al.*, 2004). This region also contains residues essential for the interaction with lipid droplets (Miyanari *et al.*, 2007). The first low complexity sequence block, separating domain I from domain II, is a serine rich region of the protein that represents a hot spot for cell culture adaptive mutations, but how these function is not clear (Blight *et al.*, 2000). Domain II contains a number of essential residues for RNA replication, but the function of this domain is obscure (Tellinghuisen *et al.*, 2008a). The LCS sequence connecting domains II and III contains a poly-proline sequence that interacts with a variety of cellular SH3 containing proteins (reviewed in Macdonald and Harris, 2004). These interactions appear to play a role in replication and virion assembly (Hughes *et al.*, 2009a). Domain III is more variable among HCV isolates and is poorly understood. Deletions of domain III and point mutants of serine phosphoacceptor sites near the carboxyl-terminus of NS5A have dramatic particle assembly defects with little replication deficit, implicating NS5A in assembly, although this may be somewhat isolate and context specific (Appel *et al.*, 2008; Tellinghuisen *et al.*, 2008a,b; Hughes *et al.*, 2009b). The structure of the amino-terminal domain I of NS5A is dimeric in nature, and represents a novel class of protein fold (Tellinghuisen *et al.*, 2005). The dimer is oriented such that a large groove exists between the two monomers, and the overall shape of the dimer resembles a claw. The groove is lined with basic amino acids, and has been proposed as an RNA binding interaction motif. NS5A has RNA binding activity, but this activity has not yet been linked to RNA replication or virion assembly (Huang *et al.*, 2005). An alternative dimeric arrangement of domain I has been determined, and this interface lacks the large groove, or any obvious RNA interaction surface (Love *et al.*, 2009). As is the case with the original structure, the validity of this conformation of NS5A to virus biology has yet to be determined. Attempts at determination of the complete structure of NS5A, or those of domains II and III, have been unsuccessful, as these regions appear to be natively unfolded (Hanoulle *et al.*, 2009b, 2010; Liang *et al.*, 2006, 2007).

NS5A is phosphorylated on multiple serine residues in both the hypophosphorylated and hyper-phosphorylated forms (Tanji *et al.*, 1995b). The identification of which of the many potential phosphoacceptor sites in NS5A are relevant to these forms is not known. Nonetheless, evidence has accumulated suggesting that hypophosphorylation primarily targets serine residues in domains II and III, whereas hyperphosphorylation sites cluster in and around domain I (reviewed in Huang *et al.*, 2007b). Cell culture adaptive mutations that increase RNA replication tend to decrease hyperphosphorylation, suggesting NS5A phosphorylation may regulate replication (Blight *et al.*, 2000). Perhaps the most compelling evidence supporting this idea comes from the use of compounds selected to decrease the hyperphosphorylation of NS5A (Neddermann *et al.*, 2004). These drugs increase RNA replication such that they eliminate the need for cell culture adaptive mutations in HCV replicons for RNA replication. These compounds are directed against casein kinase 1α (CKIα), and genetic silencing of CKIα produces results similar to the inhibitors (Quintavalle *et al.*, 2006). These data provide strong evidence that CKIα is a regulator of RNA replication. A host of other kinases target NS5A (Huang *et al.*, 2007b). A number of biochemical experiments have implicated casein kinase II (CKII) as an NS5A directed kinase (Huang *et al.*, 2004; Kim *et al.*, 1999; Reed *et al.*, 1997). *In vitro* phosphorylation of NS5A by CKII suggests at least five CKII sites exist in NS5A (Huang *et al.*, 2004), and these are located in domain III (Kim *et al.*, 1999). CKII is likely involved in hypophosphorylation of NS5A (Quintavalle *et al.*, 2007). The mutation of a number of potential CKII acceptors does not alter RNA replication (Appel, Pietschmann, and Bartenschlager, 2005), but recent genetic and biochemical evidence has suggested CKII phosphorylation regulates virion production (Tellinghuisen *et al.*, 2008b). Clearly, NS5A phosphorylation has emerged as a critical event in several HCV life cycle events, placing NS5A in a potential role as a central regulator of the HCV life cycle.

In addition to roles in RNA replication and virion production, NS5A also interacts with a large number of host proteins, presumably to modify the cell to create an environment more permissive for virus infection. The list of these interactions is extensive, and many have not been fully explored. This topic has been reviewed in the literature, and the reader is directed to a review in this area for a more complete picture (Macdonald and Harris, 2004). Perhaps the most well known of these interactions is the interaction of NS5A with the double-stranded RNA activated protein kinase (PKR; Gale, Jr, *et al.*, 1998). Presumably, the interaction of NS5A eliminates the antiviral activity of PKR by sequestering this protein from regulating translation in response to viral replication. NS5A also clearly interacts with apolipoproteins, and these interactions may account for the role of NS5A in replication and particle assembly, as these proteins are required for both of these events (Benga *et al.*, 2010; Jiang and Luo, 2009; Shi *et al.*, 2002). The interaction of NS5A with hVAP proteins, a property shared with the NS5B protein, might have a role in organization of the viral replicase (Evans *et al.*, 2004; Gao *et al.*, 2004; Tu *et al.*, 1999). Another interaction common to NS5A and NS5B is the association with the cyclophilin peptidyl-prolyl isomerases. These interactions are essential for RNA replication, but seem to be somewhat genotype dependent, limiting their broad therapeutic potential (Fernandes *et al.*, 2007; Hanoulle *et al.*, 2009a; Ishii *et al.*, 2006; Watashi *et al.*, 2003). Nonetheless, compounds targeting cyclophilins can prevent the interaction with HCV viral proteins, and inhibit replication, suggest these may have some value in the future.

After lagging behind the other large viral replicase components for several years, NS5A has finally emerged as antiviral target for small molecule design. A recent publication describes compounds that inhibit replication with extreme potency, highlighting the validity of NS5A as a target (Gao *et al.*, 2010; Lemm *et al.*, 2010). The patent literature demonstrates several different classes of compounds that target NS5A, but the most potent of these seem to bind at the NS5A dimer interface, based on current models of NS5A structure and analysis of drug resistance mutants (Fridell *et al.*, 2010). One could imagine targeting NS5A, a protein with roles in RNA replication, assembly, and modulation of the host cell, might be a very potent, multifactorial therapy (see Chapter 14).

NS5B

NS5B is a large, 68 kDa protein that constitutes the RNA-dependent RNA polymerase activity of the HCV replicase. Predictably, this protein is essential for replication *in vitro* and *in vivo*, with mutations of the hallmark GDD signature motif of polymerases producing lethal phenotypes (Blight *et al.*, 2000; Kolykhalov *et al.*, 2000; Lohmann *et al.*, 1999). NS5B has been extensively characterized biochemically (for review, see De Francesco *et al.*, 1996; Hagedorn *et al.*, 1999; Lohmann *et al.*, 2000) and structurally (for review, see De Francesco, 2003; Hagedorn *et al.*, 1999). The structure of NS5B is similar to that of other single chain polymerases, with some notable differences. The overall shape of the molecule is analogous to that of a right hand, with clear palm, thumb and fingers domains readily apparent; a fold common to this class of proteins. NS5B is unusual in that it has a fully encircled active site due to extensive contacts made between the fingers and thumb domains, resulting in a conformation capable of directly binding nucleotides without additional conformational changes (Ago *et al.*, 1999; Bressanelli *et al.*, 1999; Lesburg *et al.*, 1999). This has been proposed as the conformation of NS5B during initiation events. An open conformation, with movement of the thumb domain that disrupts contact with the finger domain, also exists in the structural record, and this is believed to represent the post-initiation processive elongation form of the polymerase (Biswal *et al.*, 2005). Another unusual feature of the NS5B structure is the presence of a β-hairpin loop structure near the active site, and this feature has been proposed to function in properly positioning the 3′ end of RNA templates relative to the active site to favour *de novo* initiation (Hong, 2001; Kim *et al.*, 2005; O'Farrell *et al.*, 2003; Ranjith-Kumar *et al.*, 2003). NS5B also contains an unusual carboxyl-terminal loop that encircles the thumb domain and is positioned close to the active site (Leveque *et al.*, 2003). This loop decreases RNA binding *in vitro* and might play a role similar to the β-hairpin loop in proper template selection and favouring *de novo* initiation over 'copy back' or 'primer extension' replicase activities that have been described *in vitro* (Leveque *et al.*, 2003; Ranjith-Kumar *et al.*, 2003). Another unusual feature of NS5B, and one particularly relevant to antiviral drug design, is the presence of a GTP binding allosteric regulator site in the thumb domain (Bressanelli *et al.*, 2002). A number of small molecule non-nucleoside inhibitors of NS5B have been described that target this site, and these compounds may function by preventing conformational movements of the thumb domain that lead from a closed active site initiation form of the enzyme to the more open elongation form. More conventional active site directed nucleoside-based inhibitors of the polymerase active site have also been extensively described, and a number of these molecules appear to have significant therapeutic potential. Both allosteric and active site targeting compounds are discussed in detail elsewhere in this text, and the reader is directed to Chapters 15 and 16 and references therein for detailed discussion of these fascinating antiviral compounds. NS5B is also unusual in that it is a member of the emerging family of tail-anchored proteins (Moradpour *et al.*, 2004). The carboxyl-terminal region of NS5B forms a 21 amino acid helix that is inserted into host membranes post-translationally to tether the polymerase to the membrane (Moradpour *et al.*, 2004; Schmidt-Mende *et al.*, 2001). This helix is not required for polymerase activity *in vitro*, but is essential for RNA replication in cells (Ferrari *et al.*, 1999; Lohmann *et al.*, 1997, 1998; Yamashita *et al.*, 1998). Much like the case for the NS3 helicase activity, biochemical and structural evidence suggests that NS5B forms large oligomers that may have an important role in replication, but the precise meaning of these complexes has yet to be elucidated (Qin *et al.*, 2002; Wang *et al.*, 2002a). The activity of NS5B is clearly modified by homotypic interactions in these complexes, as well as by interactions with other viral proteins in the replicase. NS3–NS4A positively stimulates the activity of the polymerase (Piccininni, 2002), whereas NS4B and NS5A down regulate NS5B (Piccininni, 2002; Shirota *et al.*, 2002). As no clear picture exists of the complexes formed in an authentic replicase, it is unclear how these interactions modulate the polymerase, or what the significance of these observations are to *bona fide* RNA replication within a cell, but protein–protein interaction inhibitors that alter these contacts might have therapeutic benefit. NS5B, like NS5A, interacts with both hVAP proteins and cyclophilins, and these interactions are essential for RNA replication (Hamamoto *et al.*, 2005; Tu *et al.*, 1999; Watashi *et al.*, 2005). Although not fully understood, it is intriguing to think of these host factors as modulating the interaction between the NS5A and NS5B proteins in the assembly and regulation of the RNA replicase.

ARFP/F proteins

The alternative reading frame protein (ARFP), also designated the frameshift (F) protein, was identified as a product produced from a predicted alternate reading frame in the core coding sequence (Baril and Brakier-Gingras, 2005; Branch *et al.*, 2005; Choi *et al.*, 2003; Varaklioti *et al.*, 2002; Vassilaki and Mavromara, 2003; Walewski *et al.*, 2001; Xu *et al.*, 2001, 2003). In the case of genotype 1a, a −2/+1 frame shift at codon 11 of the core protein results in the production of a number of products of variable size, reaching up to 160 amino acids. This protein product is highly variable in size and sequence for different HCV genotypes, and in some cases involves multiple frame shifting events and alternative initiation sites (reviewed in Branch *et al.*, 2005). In some instances the frameshift phenomena are described solely in bacterial expression systems, making application of these findings to authentic HCV infection difficult. Although serologic and T-cell responses against ARFP/F have been detected in HCV patients (Troesch *et al.*, 2005), suggesting these proteins are made in infections, the lack of evolutionary constraints among different HCV isolates in ARFP/F sequence argues against a specific defined function (Cristina *et al.*, 2005). This alternate reading frame that generates the ARFP/F proteins contains a number of conserved CRE elements, most notably domain VI of the 5′ NCR (Branch *et al.*, 2005; McMullan *et al.*, 2007; Tuplin *et al.*, 2002, 2004; Walewski *et al.*, 2001). Analysis of this region in chimpanzee infections suggests that the domain VI stem–loop is important for HCV replication, whereas the maintenance of the alternate reading frame is not (McMullan *et al.*, 2007). What function these proteins may play in infection remain to be determined, but they are not required for virus replication *in vitro* or *in vivo*.

Formation of the site of RNA replication

All positive-strand RNA viruses studied to date induce significant membrane alterations within the infected cell as sites of viral genome replication, and HCV is no exception. HCV replication induces a dense net of perinuclear vesicles and tubules referred to as the membranous web (MW; Egger *et al.*, 2002). This MW structure is strikingly similar to the sponge-like inclusions observed in early studies of HCV infected chimpanzee liver using electron microscopy (Pfeifer, 1980). The MW has been observed in cells expressing the HCV polyprotein in the absence of replication, suggesting that the formation of the MW does not require RNA replication and is generated through the activity of the HCV viral proteins (Egger *et al.*, 2002). Expression of the HCV NS4B protein alone can induce the formation of structures identical to the MW observed in cells containing replicons, suggesting NS4B is the major, and perhaps only, viral determinant for MW formation. The carboxyl-terminal region of NS4A appears to be critical for creating the MW (Aligo *et al.*, 2009). In cells containing HCV replicons, the MW contains the HCV non-structural proteins, viral genomic RNA, and is the site of nascent RNA synthesis (Gosert, 2003; Moradpour *et al.*, 2004). Partially purified membrane fractions believed to represent the MW have been generated, and the RNA-dependent RNA polymerase activity within these complexes is largely resistant to nuclease and protease digestion (Aizaki *et al.*, 2004; Miyanari *et al.*, 2003; Quinkert *et al.*, 2005; Yang *et al.*, 2004). By analogy to other positive-strand RNA virus, it has been proposed that the observed protection of the replicase by the MW is due to replication occurring within partially enclosed membrane invaginations that might represent the spherical vesicle components of the MW. Current models suggest the viral replication machinery lines the inner surface of these vesicles and is only partially accessible to the cytoplasm via a restrictive neck or pore like shape that contacts the tubule like portions of the MW and/or the ER. A schematic representation of a MW invagination can be seen in the inset image of Fig. 1.1. Low-resolution confocal microscopy and image reconstruction analysis has visualized the MW, ER, viral RNA, and LDs in infected cells, and the complex network of vesicles and tubules in the MW are largely contiguous, and are in close proximity to, or intimate contact with, both ER and LDs (Targett-Adams *et al.*, 2008). The cellular origin of the MW membranes is not entirely clear, but most studies suggest the MW is derived from modified rough ER membranes, although the MW appears compositionally somewhat different from this parent organelle (Egger *et al.*, 2002; Gosert, 2003). The Golgi apparatus has also been proposed as a source of the MW, based on the resistance of MW membranes to non-ionic detergents, but many Golgi markers are not seen in the MW (Aizaki *et al.*, 2004). Regardless of the parent organelle used for MW formation, it is clear that HCV infection induces massive rearrangement of membranes to create the MW, and this activity likely requires the manipulation of a number of proteins involved in lipid metabolism, lipid transport, and vesicle trafficking. HCV induces a number of genes involved in lipid metabolism such as ATP citrate lyase and acetyl-CoA synthase (Kapadia and Chisari, 2005; Su *et al.*, 2002). Replication is stimulated by saturated and monounsaturated fatty acids and inhibited by polyunsaturated fatty acids (Kapadia and Chisari, 2005). Inhibitors of lipid synthesis are potent inhibitors of HCV replication. Additionally, NS4B appears to regulate the activity of sterol regulator element binding proteins (SREBPs), regulators of lipid metabolism, and this might alter the production of lipids to favour the formation of the membranous web structure (Park *et al.*, 2009). Collectively, these data suggest that lipids that favour membrane fluidity are important for the function or formation of the HCV replication complex. Cholesterol is likely important to replicase membrane fluidity as well, but the requirement for the geranyl-geranylation of the FBL2 protein, a protein that plays an important but largely uncharacterized role in replication, has complicated the analysis of the requirements for cholesterol in MW function (Kapadia and Chisari, 2005; Ye *et al.*, 2003). Recent high-content genetic screens from several laboratories have identified phosphotidylinositol-4-kinase (PI4K) as an essential host factors for HCV replication (Berger *et al.*, 2009; Borawski *et al.*, 2009; Trotard *et al.*, 2009; Vaillancourt *et al.*, 2009). Data suggest that the the MW is enriched in phopshotidylinositol-4,5-bisphosphate, the product of PI4K enzymatic activity, and infection of cells by HCV dramatically alters the localization of phosphatidylinositols and PI4K within the cell. The precise meaning of these observations is not yet clear, but one interpretation is that phosphatidylinositols are required for the formation of the MW structure and/or are important lipid components of this structure. Recent data also implicate Rab5 and Rab7 are potential vesicle transporters important for the formation of the MW (Manna *et al.*, 2010; Stone *et al.*, 2007). Targeting the host factors involved in the formation of the MW is particularly attractive as a therapeutic avenue for HCV, as compounds that inhibit MW formation should be potent inhibitors of virus replication and avoid the generation of viral resistance mutants (see Chapter 18). The challenge in this area is to inhibit MW formation while not inducing host cell toxicity by disrupting normal cellular physiology.

RNA replication

The mechanisms and regulation of HCV RNA replication are enigmatic, with only a scant few clues of what controls the transitions of an HCV genomic RNA through the various stages of the virus life cycle. The complexity of intracellular events associated with HCV infection is staggering, with RNA involved in active RNA replication on positive and negative strand templates, serving as message for translation of viral proteins, and serving as a substrate for progeny virus assembly. The commitment of viral RNAs to these processes in the cell must be regulated to avoid functional conflicts. Clearly, a viral RNA cannot serve as a substrate for translation with ribosomes moving down the genome in the 5′ to 3′ direction, and a template for new minus-strand synthesis, with the viral replicase reading the RNA in the 3′ to 5′ direction. In some instances, such as translation versus replication, events can be controlled by compartmentalization of replication within the MW vesicles. Similar conflicts exist at the RNA level as well, but within the replicase itself within the MW. What controls the switch between generating minus-strand replicative form genomes versus new progeny positive-strand RNAs is unknown. In this instance, compartmentalization theories provide little help, as both plus and minus-strand synthesis likely occur in the same MW vesicle. Nonetheless, it is clear that all of these processes are regulated in some manner, as vastly different amounts of positive-strand genome, minus-strand RNA, and HCV proteins are generated within the infected cell.

Quantification of the viral RNA and proteins within cells bearing subgenomic replicons indicates that each cell contains about 1000 positive-strand RNAs, 100 negative strands, and about 1,000,000 copies of each viral protein (Quinkert *et al.*, 2005). Thus, translation of the viral RNA is far more frequent than RNA replication events. An excess of viral structural proteins makes sense, as at least 180 copies of each structural protein and a single copy of viral positive-strand RNA is likely required for each infectious virion (assuming a T=3 particle). All of the HCV proteins should be generated in equimolar amounts, as they are generated from a single polyprotein precursor. The function of excess non-structural proteins is less clear than that of the structural proteins, but increasing evidence from other positive-strand virus systems suggests replication occurs in enclosed micro-environments consisting of convoluted vesicles (perhaps like the HCV MW) often induced by and completely lined with various viral non-structural proteins (Mackintosh *et al.*, 2006). Some evidence exists suggesting the NS5B polymerase may function in large, sheet like arrays, and these make for an attractive lining of the MW vesicle. Another possible function for the excess of non-structural proteins may lie in the increasing body of evidence for the extra-replicative activities of some of the HCV non-structural proteins, including the manipulation of innate immunity and cell signalling pathways (see Chapter 19). Recent structural and biochemical analyses have demonstrated that many of the non-structural proteins exist as functional oligomers, again requiring more than one copy of each protein per genome replication event (Lorenz *et al.*, 2006; Mackintosh *et al.*, 2006; Tackett *et al.*, 2005; Tellinghuisen *et al.*, 2005; Wang *et al.*, 2002b; Yu *et al.*, 2006). When all of these factors are considered the excess of viral proteins generated, both structural and non-structural, over the number of viral RNAs does not seem so large.

Even if the differences in translation versus replication events can be reconciled, it is clear that some regulation must occur to favour translation, yet still allow for unhindered replication of RNA. A number of possible mechanisms have emerged to describe this orchestration. The best developed of these involve the use of cellular factors to promote one activity, such as translation, while simultaneously inhibiting another, replication. The polypyrimidine tract binding protein (PTB) seems to fit these criteria, as PTB binds to both the 5′ NCR-core coding region of the HCV RNA, where it modulates translation from the viral IRES, and to the 3′ NCR where it may suppress RNA replication (Anwar *et al.*, 2000; Ito and Lai, 1999; Tischendorf *et al.*, 2004; Tsuchihara *et al.*, 1997). A number of host proteins bind to one or both HCV NCRs, including autoantigen La and PCBP-2, proteins implicated in regulation of translation and replication of other RNA viruses, but their role in HCV regulation is not clear (Fukushi *et al.*, 2001; Pudi *et al.*, 2003; Spangberg and Schwartz, 1999; Spangberg *et al.*, 2001). miR122 also may function as an important regulatory element in the HCV lifecycle. As mentioned previously, miR122 binds to the 5′ NCR of the HCV genome and this interaction is required for RNA replication (Jopling *et al.*, 2005). It is intriguing to imagine that the association of miR122 with the HCV genome leads to translational silencing, much like what is observed for cellular messages regulated by this micro RNA. This would, presumably, favour RNA replication over translation. Further to this model, the HuR protein depresses the silencing of cellular messages by miR122 (Bhattacharyya *et al.*, 2006), and this protein is an essential hot factor for HCV replication that binds to both positive and negative strands of the HCV genome (Korf *et al.*, 2005; Randall *et al.*, 2007; Spangberg *et al.*, 2000). Perhaps the regulation of translation is more simple, involving a simple feedback inhibition, and indeed, the HCV core protein appears to have the ability to regulate translation efficiency in manner dependent with its level of expression (Boni *et al.*, 2005; Zhang *et al.*, 2002). Despite these interesting observations, the switch that regulates the move of RNA from translation to replication remains to be demonstrated. Once translation has occurred, the viral RNA must move to the site of replication, the membranous web, and perhaps this recruitment of RNA to the web represents another mechanism of regulation, the physical sequestration of RNA in the replicase microenvironment (Gosert *et al.*, 2003).

One the viral RNA has served as substrate for the production of sufficient viral proteins to generate a functional viral replicase, the RNA is recruited out of translation to the site of RNA replication, the MW (Gosert *et al.*, 2003). As discussed previously, the MW is likely a highly specialized compartment optimized specifically for viral RNA replication. Once RNA has reached the web, it enters the replication cycle. The minimal viral protein components required for productive RNA replication include NS3, NS4A, NS4B, NS5A, and NS5B, but the organization of these proteins within a complex replicase structure is unknown. A large number of host factor have been implicated in the replication process, and these have been discussed, in part, in previous sections describing the individual HCV proteins where possible. Once bound by the replicase, HCV genomic RNA serves as template for the production of a complementary genome-length negative strand RNA. Presumably this is directed by signals within the 3′ NCR that assemble the replicase in a position near the end of the viral genomic RNA. Based on *in vitro* polymerase experiments with NS5B and experiments using permeabilized cells, cell lysates, and purified membrane fractions, it is likely that the HCV polymerase uses a primer-free *de novo* initiation mechanism for RNA synthesis (Aizaki *et al.*, 2004; Ali *et al.*, 2002; Hardy *et al.*, 2003; Lai *et al.*, 2003; Miyanari *et al.*, 2003; Quinkert *et al.*, 2005; Yang *et al.*, 2004). Once the minus-strand RNA is synthesized, it is found associated with the positive-strand RNA, either in partially double stranded replicative intermediates or fully double stranded replicate forms (Ali *et al.*, 2002). The MW structure may play a role in preventing the host cell recognition of these double stranded RNA intermediates and activation of innate immune responses. The negative strand RNA then serves as a template to direct the synthesis of multiple positive-strand genomic RNAs. This nascent positive-strand genomic RNA then serves as template for additional rounds of minus-strand synthesis, substrate for translation, and substrate for encapsidation into virions. The process of genome replication is clearly regulated, as synthesis of positive and negative strand RNAs is asymmetric, with approximately 10 positive strands generated for each minus strand (Aizaki *et al.*, 2004; Lanford *et al.*, 1995; Lohmann *et al.*, 1999; Miyanari *et al.*, 2003; Ranjith-Kumar *et al.*, 2004). The regulation of replication events that leads to this asymmetry is unknown, but it could potentially be at the level of replicase recognition of CREs on the various RNAs, the composition of the viral replicase, or the differential processivity of the replicase on the different templates. Experiments using chimeric replicons have suggested that contacts required for minus-strand synthesis are made between the 3′ X-tail and the NS5B polymerase, whereas assembly of the replicase on the 5′ end of the RNA (i.e. the 3′ end of the minus-strand RNA) requires NS3, NS5A, and NS5B (Binder *et al.*, 2007). These data provide the first evidence of differential requirements of replicase proteins for replication at the 5′ and 3′ ends of the genome, and as such, may be important events in the regulation of strand synthesis asymmetry. Regardless of the mechanism by which replication is controlled, the end result of this process is the formation of an excess of positive-strand RNA, some of which is fed back into translation to generate more viral proteins, and some of which serves as genomic RNA for the assembly of new virions.

Particle assembly

Little is know about the assembly of HCV virions, largely due to the inability to study assembly related events in a cell culture environment until quite recently. The advent of HCVcc systems has revolutionized the study of HCV assembly, and this is an exciting developing area in HCV research. Overall, the process appears quite different than what was imagined even few years ago, and seems to be somewhat unique among RNA viruses. Perhaps most exciting are the apparent requirement of numerous non-structural proteins in the assembly process as well as the requirement of LDs for assembly. This is very much an emerging field, with significant gaps in our understanding of the assembly process, but the following section of this chapter attempts to present the current model of virion formation.

The assembly process likely begins as soon as translation and processing has occurred, with the release of the mature form of the core protein from the viral polyprotein (McLauchlan *et al.*, 2002). The core protein, in contrast to the other viral proteins, localizes to cellular LDs (in addition to ER and the MW), and can be found on LDs early in HCV infections (reviewed in McLauchlan, 2009). The association of core with LDs is dependent on the central, relatively hydrophobic, domain 2 of the protein (Boulant *et al.*, 2006; Hope *et al.*, 2002). Mutations that disrupt the association of core with LDs prevent the release of infectious virus particles (Boulant *et al.*, 2007). A number of papers suggest core functions as an oligomer of some kind, and dimerization/oligomerization of core requires the basic, amino-terminal domain 1 of the core, suggesting RNA interaction might drive oligomerization (Klein *et al.*, 2005; Lu *et al.*, 2001; McLauchlan, 2000; Santolini *et al.*, 1994). It is unclear if core on LDs exists in a monomeric or oligomeric form prior to the assembly of virions. It is interesting to note that the core LD interaction is not passive, and core active promotes the production of LDs and affects their localization as well. This may have clinical significance in the steatosis observed in some HCV patients (for review, see Lonardo *et al.*, 2004). LDs coated with core protein tend to distribute and aggregate near the periphery of the nucleus, near the perinuclear ER membrane, the MW, and the microtubule-organizing centre (Boulant *et al.*, 2008). This LD localization requires the microtubule network and can be blocked by disrupting microtubules or inhibiting dynein activity (Boulant *et al.*, 2008). Core also appears to alter the proteomic content of lipid droplets, but the significance of this to the assembly process is not clear (Boulant *et al.*, 2008). The intimate association of LDs with the MW and ER membranes has been visualized in several microscopic studies of infected cells (Dubuisson, 1999; Miyanari *et al.*, 2007; Roingeard *et al.*, 2008; Targett-Adams *et al.*, 2008). With core coated LDs in close proximity to the MW, the next step in assembly must be the release of viral RNA from replicative events to allow packaging. Little is known about what controls the release of RNA from replication, but the phosphorylation of NS5A by casein kinase 2 (CK2) on a serine residue in domain III has been proposed as a possible mechanism (Tellinghuisen *et al.*, 2008b). Chemical or genetic manipulation of CK2 can inhibit the release of infectious virus, and phosphomimetic mutations of the putative serine phosphorylation acceptor site decrease replication without altering virus production. A number of additional deletions and serine mutations in domain III of NS5A affect virus production, generally supporting the concept that NS5A functions as a phosphorylation-dependent switch between replication and assembly (Appel *et al.*, 2008; Masaki *et al.*, 2008; Tellinghuisen *et al.*, 2008b; Hughes *et al.*, 2009b). NS5A localizes to LDs in infected cells, and mutations in domain I of NS5A block this localization and inhibit virus production (Miyanari *et al.*, 2007). Surprisingly, mutations in domain III of NS5A disrupt virus production still localize to LDs, but fail to co-localize with core protein on the LDs (Appel *et al.*, 2008; Masaki *et al.*, 2008). One model of how this might work might be that NS5A is released from replication events via phosphorylation of residues in domain III, and exits the replicase with the viral RNA and delivers the RNA to the core coated LDs. Many questions remain as to how NS5A functions in assembly, and what controls the switch for replication to assembly. Indeed, a general theme has emerged in the past few years that several replicase proteins can affect virion production, including NS3 and NS4B (Jones *et al.*, 2009; Ma *et al.*, 2008). It is possible that these proteins also function in assembly by destabilizing the replicase, thereby promoting the release of viral RNA for assembly, but as is the case with NS5A, these events are poorly understood. Whatever the signal for the switch between replication and assembly may be, the next step in the assembly process is likely the movement of the replicase complex, or a subset of replicase proteins, and the viral RNA from the MW membrane invaginations to the surface of the MW membrane that is proximal to the LD. The transfer of RNA from the replicase to the core presumably triggers the process of nucleocapsid assembly. No packaging signal for HCV has yet been defined, but interactions of the core with the 5′ NCR might be important for this process (Hwang *et al.*, 1995; Tanaka *et al.*, 2000). HCV nucleocapsid assembly is poorly understood. Nucleocapsid assembly system based on recombinant protein and core over-expression have been developed, but it is unclear how similar *in vitro* assembled nucleocapsids are to the authentic particles (Acosta-Rivero *et al.*, 2001; Kunkel *et al.*, 2001; Majeau *et al.*, 2004; Klein *et al.*, 2005). These, and other *in vitro* studies suggest the amino-terminal half of the core protein is essential for particle formation and RNA interaction, and the hydrophobic domain and more carboxyl-terminal residues may regulate the assembly process (Yan *et al.*, 1998). These findings have recently begun to be incorporated into experiments involving HCVcc systems, and a number of residues that are essential for virus production in the core have been identified, but a detailed understanding of nucleocapsids remains elusive.

The HCV glycoproteins also follow a different route after processing than most of the HCV non-structural proteins, with these proteins undergoing a complex maturation process in the ER (reviewed in Dubuisson, 1999). The end result of this process is a the production of glycosylated, non-covalent heterodimers of E1 and E2 with the ectodomains facing into the ER lumen, and small cytoplasmic tails oriented outward from the ER surface, presumably to make contacts with the core protein during budding events. With LDs positioned close to both MW and the ER, it is likely that core on LDs accepts RNA from the replicase, assembles nucleocapsids, and then buds into the ER through glycoprotein core interactions (Dubuisson, 1999; Miyanari *et al.*, 2007; Roingeard *et al.*, 2008; Targett-Adams *et al.*, 2008). Studies suggest an interaction between the carboxyl-terminal domain of core and the cytoplasmic tail of E1 are important for particle budding, but this has yet to be evaluated fully in the HCVcc system (Lo *et al.*, 1996; Ma *et al.*, 2002). Nonetheless, the general concept makes good topological sense for the current models of assembly. Whatever the mechanism, nucleocapsids then bud into the ER, acquiring a complete outer shell of host membrane and glycoproteins (Roingeard *et al.*, 2008). Late assembly events and/or virion maturation events are largely unknown, but it is clear that the viroporin p7 and the amino-terminal portion of the NS2 autoprotease are essential for the release of infectious particles, and mutations in these proteins block late assembly or maturation events (Jirasko *et al.*, 2008; Jones *et al.*, 2007; Pietschmann *et al.*, 2006; Sakai *et al.*, 2003; Steinmann *et al.*, 2007a). Unfortunately, the precise role of these proteins in assembly remains to be determined. Assembled particles then follow the host secretory machinery through the ER and Golgi apparatus where they associate with LDL and/or vLDL particles before being released as infectious virions (Andre *et al.*, 2002, 2005). This is in good agreement with the observation that intracellular virus exists in a high-density immature form, and then associates with lipoproteins during or soon after secretion from the cell (Gastaminza *et al.*, 2006). Studies also clearly show the requirements of apolipoproteins and microsomal transfer protein for production of infectivity, suggesting this alteration of virus density occurs during secretion (Gastaminza *et al.*, 2008; Huang *et al.*, 2007a).

Perspectives

The last decade and a half has seen remarkable progress in understanding the life cycle of HCV, and several of this effort has begun providing specific antiviral compounds. First to emerge from laboratory into clinical trials have been the conventional compounds targeting the enzymatic activities of the virus, most notably the viral serine protease and RNA-dependent RNA polymerase. A number of these 'first generation' compounds are likely to reach the clinic in the next few years. The error prone nature of HCV replication, and the incredible sequence diversity this creates, suggests that a cocktail of specific inhibitors will be required for the long-term control of HCV infection in humans. To achieve this control, an ever-expanding panel of new antivirals will be needed, and our recent HCV discoveries will fuel the development of new compounds. Surprising new targets and compounds have already begun to emerge, such as those targeting NS5A and NS4B, and the development of these compounds into therapeutics is of great importance. It is critical that we, collectively, do not lose focus on this ever-moving viral target and continue to develop new and better antivirals.

References

Acosta-Rivero, N., Aguilar, J.C., Musacchio, A., Falcon, V., Vina, A., de la Rosa, C., and Morales, J. (2001). Characterization of the HCV core virus-like particles produced in the methylotrophic yeast *Pichia pastoris*. Biochem. Biophys. Res. Commun. *287*, 122–125.

Acton, S., Rigotti, A., Landschulz, K.T., Xu, S., Hobbs, H., and Krieger, M. (1996). Identification of scavenger receptor SR-BI as a high density lipoprotein receptor. Science *271*, 518–520.

Ago, H., Adachi, T., Yoshida, A., Yamamoto, M., Habuka, N., Yatsunami, K., and Miyano, M. (1999). Crystal structure of the RNA-dependent RNA polymerase of hepatitis C virus. Structure Fold. Des. *7*, 1417–1426.

Aizaki, H., Lee, K.J., Sung, V.M., Ishiko, H., and Lai, M.M. (2004). Characterization of the hepatitis C virus RNA replication complex associated with lipid rafts. Virology *324*, 450–461.

Ali, N., Mori, K., and Siddiqui, A. (2002). Cell-free replication of the hepatitis C virus subgenomic replicon. J. Virol. *76*, 12001–12007.

Aligo, J., Jia, S., Manna, D., and Konan, K.V. (2009). Formation and function of hepatitis C virus replication complexes require residues in the carboxy-terminal domain of NS4B protein. Virology *393*, 68–83.

Alter, J., and Seeff, L.B. (2000). Recovery, persistence and sequelae in hepatitis C virus infection: a perspective on the long-term outcome. Semin. Liver Dis. *20*, 17–25.

Alter, M.J., Margolis, H.S., Krawczynski, K., Judson, F.N., Mares, A., Alexander, W.J., Hu, P.Y., Miller, J.K., Gerber, M.A., Sampliner, R.E., Meeks, L., and Beach, M.J. (1992). The natural history of community-acquired hepatitis C in the United States. N. Engl. J. Med. *32*, 1899–1905.

Andre, P., Komurian-Pradel, F., Deforges, S., Perret, M., Berland, J.L., Sodoyer, M., Pol, S., Brechot, C., Paranhos-Baccala, G., and Lotteau, V. (2002). Characterization of low- and very-low-density hepatitis C virus RNA-containing particles. J. Virol. *76*, 6919–6928.

Andre, P., Perlemuter, G., Budkowska, A., Brechot, C., and Lotteau, V. (2005). Hepatitis C virus particles and lipoprotein metabolism. Semin. Liver Dis. *25*, 93–104.

Anonymous (1997). World Health Organization – Hepatitis C: global prevalence. Wkly Epidemiol. Rec. *72*, 341–344.

Anwar, A., Ali, N., Tanveer, R., and Siddiqui, A. (2000). Demonstration of functional requirement of polypyrimidine tract-binding protein by SELEX RNA during hepatitis C virus internal ribosome entry site-mediated translation initiation. J. Biol. Chem. *275*, 34231–34235.

Appel, N., Pietschmann, T., and Bartenschlager, R. (2005). Mutational analysis of hepatitis C virus nonstructural protein 5A: potential role of differential phosphorylation in RNA replication and identification of a genetically flexible domain. J. Virol. *79*, 3187–3194.

Appel, N., Zayas, M., Miller, S., Krijnse-Locker, J., Schaller, T., Friebe, P., Kallis, S., Engel, U., and Bartenschlager, R. (2008). Essential role of domain III of nonstructural protein 5A for hepatitis C virus infectious particle assembly. PLoS Pathog. *4*, e1000035.

Argo, C.K., Loria, P., Caldwell, H., and Lonardo, A. (2008). Statins in liver disease: a molehill, an iceberg, or neither? Hepatology *48*, 662–669.

Barba, G., Harper, F., Harada, T., Kohara, M., Goulinet, S., Matsuura, Y., Eder, G., Schaff, Z., Chapman, M.J., Miyamura, T., and Bréchot, C. (1997). Hepatitis C virus core protein shows a cytoplasmic localization and associates to cellular lipid storage droplets. Proc. Natl. Acad. Sci. U.S.A. *94*, 1200–1205.

Baril, M., and Brakier-Gingras, L. (2005). Translation of the F protein of hepatitis C virus is initiated at a non-AUG codon in a +1 reading frame relative to the polyprotein. Nucleic Acids Res. 33, 1474–1486.

Bartenschlager, R., Ahlborn-Laake, L., Mous, J., and Jacobsen, H. (1993). Nonstructural protein 3 of the hepatitis C virus encodes a serine-type proteinase required for cleavage at the NS3/4 and NS4/5 junctions. J. Virol. *67*, 3835–3844.

Bartenschlager, R., Ahlborn-Laake, L., Mous, J., and Jacobsen, H. (1994). Kinetic and structural analyses of hepatitis C virus polyprotein processing. J. Virol. *68*, 5045–5055.

Bartenschlager, R., Ahlborn-Laake, L., Yasargil, K., Mous, J., and Jacobsen, H. (1995). Substrate determinants for cleavage in cis and in trans by the hepatitis C virus NS3 proteinase. J. Virol. *69*, 198–205.

Bartenschlager, R., Frese, M., and Pietschmann, T. (2004). Novel insights into hepatitis C virus replication and persistence. Adv. Virus Res. *63*, 71–180.

Barth, H., Schafer, C., Adah, M.I., Zhang, F., Linhardt, R.J., Toyoda, H., Kinoshita-Toyoda A., Toida, T., Van Kuppevelt T.H., Depla, E., *et al.* (2003). Cellular binding of hepatitis C virus envelope glycoprotein E2 requires cell surface heparan sulfate. J. Biol. Chem. *278*, 41003–41012.

Barth, H., Schnober, E.K., Zhang, F., Linhardt, R.J., Depla, E., Boson, B., Cosset, F.L., Patel, A.H., Blum, E., and Baumert, T.F. (2006). Viral and cellular determinants of the hepatitis C virus envelope-heparan sulfate interaction. J. Virol. *80*, 10579–10590.

Bartosch, B., Dubuisson, J., and Cosset, F.L. (2003). Infectious hepatitis C virus pseudo-particles containing functional E1–E2 envelope protein complexes. J. Exp. Med. *197*, 633–642.

Bartosch, B., Verney, G., Dreux, M., Donot, P., Morice, Y., Penin, F., Pawlotsky, J.M., Lavillette, D., and Cosset, F.L. (2005). An interplay between hypervariable region 1 of the hepatitis C virus E2 glycoprotein, the scavenger receptor BI, and high-density lipoprotein promotes both enhancement of infection and protection against neutralizing antibodies. J. Virol. *79*, 8217–8229.

Basu, A., Kanda, T., Beyene, A., Saito, K., Meyer, K., and Ray, R. (2007). Sulfated homologues of heparin inhibit hepatitis C virus entry into mammalian cells. J. Virol. *81*, 3933–3941.

Baumert, T.F., Ito, S., Wong, T., and Liang, T.J. (1998). Hepatitis C virus structural proteins assemble into viruslike particles in insect cells. J. Virol. *72*, 3827–3836.

Baumert, T.F., Vergalla, J., Satoi, J., Thomson, M., Lechmann, M., Herion, D., Greenberg, H.B., Ito, S., and Liang, T.J. (1999). Hepatitis C virus-like particles synthesized in insect cells as a potential vaccine candidate. Gastroenterology *117*, 1397–1407.

Benedicto, I., Molina-Jimenez, F., Bartosch, B., Cosset, F.L., Lavillette, D., Prieto, J., Moreno-Otero, R., Valenzuela-Fernandez, A., Aldabe, R., Lopez-Cabrera, M., and Majano, P.L. (2009). The tight junction-associated protein occludin is required for a postbinding step in hepatitis C virus entry and infection. J. Virol. *83*, 8012–8020.

Benga, W.J., Krieger, S.E., Dimitrova, M., Zeisel, M.B., Parnot, M., Lupberger, J., Hildt, E., Luo, G., McLauchlan, J., Baumert, F., and Schuster, C. (2010). Apolipoprotein E interacts with hepatitis C virus nonstructural protein 5A and determines assembly of infectious particles. Hepatology *51*, 43–53.

Beran, R.K., Serebrov, V., and Pyle, A.M. (2007). The serine protease domain of hepatitis C viral NS3 activates RNA helicase activity by promoting the binding of RNA substrate. J. Biol. Chem. *282*, 34913–34920.

Berger, K.L., Cooper, J.D., Heaton, N.S., Yoon, R., Oakland, T.E., Jordan, T.X., Mateu, G., Grakoui, A., and Randall, G. (2009). Roles for endocytic trafficking and phosphatidylinositol 4-kinase III alpha in hepatitis C virus replication. Proc. Natl. Acad. Sci. U.S.A. *106*, 7577–7582.

Bertaux, C., and Dragic, T. (2006). Different domains of CD81 mediate distinct stages of hepatitis C virus pseudoparticle entry. J. Virol. *80*, 4940–4948.

Bhattacharyya, S.N., Habermacher, R., Martine, U., Closs, I., and Filipowicz, W. (2006). Stress-induced reversal of microRNA repression and mRNA P-body localization in human cells. Cold Spring. Harb. Symp. Quant. Biol. *71*, 513–521.

Binder, M., Quinkert, D., Bochkarova, O., Klein, R., Kezmic, N., Bartenschlager, R., and Lohmann, V. (2007). Identification of determinants involved in initiation of hepatitis C virus RNA synthesis by using intergenotypic replicase chimeras. J. Virol. *81*, 5270–5283.

Biswal, B.K., Cherney, M.M., Wang, M., Chan, L., Yannopoulos, C.G., Bilimoria, D., Nicolas, O., Bedard, J., and James, M.N. (2005). Crystal structures of the RNA-dependent RNA polymerase genotype 2a of hepatitis C virus reveal two conformations and suggest mechanisms of inhibition by non-nucleoside inhibitors. J. Biol. Chem. *280*, 18202–18210.

Blanchard, E., Brand, D., Trassard, S., Goudeau, A., and Roingeard, P. (2002). Hepatitis C virus-like particle morphogenesis. J. Virol. *76*, 4073–4079.

Blanchard, E., Brand, D., and Roingeard, P. (2003a). Endogenous virus and hepatitis C virus-like particle budding in BHK-21 cells. J. Virol. 77, 3888–3889; author reply 3889.

Blanchard, E., Hourioux, C., Brand, D., Ait-Goughoulte, M., Moreau, A., Trassard, S., Sizaret, P.Y., Dubois, F., and Roingeard, P. (2003b). Hepatitis C virus-like particle budding: role of the core protein and importance of its Asp111. J. Virol. *77*, 10131–10138.

Blanchard, E., Belouzard, S., Goueslain, L., Wakita, T., Dubuisson, J., Wychowski, C., and Rouille, Y. (2006). Hepatitis C virus entry depends on clathrin-mediated endocytosis. J. Virol. *80*, 6964–6972.

Blight, K.J. (2007). Allelic variation in the hepatitis C virus NS4B protein dramatically influences RNA replication. J. Virol. *81*, 5724–5736.

Blight, K.J., and Rice, C.M. (1997). Secondary structure determination of the conserved 98-base sequence at the 3′ terminus of hepatitis C virus genome RNA. J. Virol. *71*, 7345–7352.

Blight, K.J., Kolykhalov, A., and Rice, C.M. (2000). Efficient initiation of HCV RNA replication in cell culture. Science *290*, 1972–1974.

Blight, K.J., McKeating, A., and Rice, C.M. (2002). Highly permissive cell lines for hepatitis C virus genomic and subgenomic RNA replication. J. Virol. *76*, 13001–13014.

Blight, K.J., McKeating, J.A., Marcotrigiano, J., and Rice, C.M. (2003). Efficient RNA replication of hepatitis C virus genotype 1a in cell culture. J. Virol. *77*, 3181–3190.

Boehringer, D., Thermann, R., Ostareck-Lederer, A., Lewis, D., and Stark, H. (2005). Structure of the hepatitis C virus IRES bound to the human 80S ribosome: remodeling of the HCV IRES. Structure *13*, 1695–1706.

Boni, S., Lavergne, J.P., Boulant, S., and Cahour, A. (2005). Hepatitis C virus core protein acts as a transmodulating factor on internal translation initiation of the viral RNA. J. Biol. Chem. *280*, 17737–17748.

Borawski, J., Troke, P., Puyang, X., Gibaja, V., Zhao, S., Mickanin, C., Leighton-Davies, J., Wilson, C.J., Myer, V., Cornellataracido, I., Baryza, J., Tallarico, J., Joberty, G., Bantscheff, M., Schirle, M., Bouwmeester, T., Mathy, J.E., Lin, K., Compton, T., Labow, M., Wiedmann, B., and Gaither, L.A. (2009). Class III phosphatidylinositol 4-kinase alpha and beta are novel host factor regulators of hepatitis C virus replication. J. Virol. *83*, 10058–10074.

Boulant, S., Montserret, R., Hope, R.G., Ratinier, M., Targett-Adams, P., Lavergne, J.P., Penin, F., and McLauchlan, J. (2006). Structural determinants that target the hepatitis C virus core protein to lipid droplets. J. Biol. Chem. *281*, 22236–22247.

Boulant, S., Targett-Adams, P., and McLauchlan, J. (2007). Disrupting the association of hepatitis C virus core protein with lipid droplets correlates with a loss in production of infectious virus. J. Gen. Virol. *88*, 2204–2213.

Boulant, S., Douglas, M.W., Moody, L., Budkowska, A., Targett-Adams, P., and McLauchlan, J. (2008). Hepatitis C virus core protein induces lipid droplet redistribution in a microtubule- and dynein-dependent manner. Traffic *9*, 1268–1282.

Bradley, D., McCaustland, K., Krawczynski, K., Spelbring, J., Humphrey, C., and Cook, E.H. (1991). Hepatitis C virus: Buoyant density of the factor VIII-derived isolate in sucrose. J. Med. Virol. *34*, 206–208.

Bradley, D.W., McCaustland, K.A., Cook, E.H., Schable, C.A., Ebert, W., and Maynard, J.E. (1985). Posttransfusion non-A, non-B hepatitis in Chimpanzees: Physicochemical evidence that the tubule-forming agent is a small, enveloped virus. Gastroenterology *88*, 773–779.

Bradrick, S.S., Walters, W., and Gromeier, M. (2006). The hepatitis C virus 3′-untranslated region or a poly(A) tract promote efficient translation subsequent to the initiation phase. Nucleic Acids Res. *34*, 1293–1303.

Branch, A.D., Stump, D.D., Gutierrez, J.A., Eng, F., and Walewski, J.L. (2005). The hepatitis C virus alternate reading frame (ARF) and its family of novel products: the alternate reading frame protein/F-protein, the double-frameshift protein, and others. Semin. Liver Dis. *25*, 105–117.

Brass, V., Bieck, E., Montserret, R., Wolk, B., Hellings, J.A., Blum, H.E., Penin, F., and Moradpour, D. (2002). An Amino-terminal Amphipathic alpha -Helix Mediates Membrane Association of the Hepatitis C Virus Nonstructural Protein 5A. J. Biol. Chem. *277*, 8130–8139.

Brass, V., Pal, Z., Sapay, N., Deleage, G., Blum, H.E., Penin, F., and Moradpour, D. (2007). Conserved determinants for membrane association of nonstructural protein 5A from hepatitis C virus and related viruses. J. Virol. *81*, 2745–2757.

Brazzoli, M., Bianchi, A., Filippini, S., Weiner, A., Zhu, Q., Pizza, M., and Crotta, S. (2008). CD81 is a central regulator of cellular events required for hepatitis C virus infection of human hepatocytes. J. Virol. *82*, 8316–8329.

Bressanelli, S., Tomei, L., Roussel, A., Incitti, I., Vitale, R.L., Mathieu, M., De Francesco, R., and Rey, F.A. (1999). Crystal structure of the RNA-dependent RNA polymerase of hepatitis C virus. Proc. Natl. Acad. Sci. U.S.A. *96*, 13034–13039.

Bressanelli, S., Tomei, L., Rey, F.A., De Francesco, R. (2002). Structural analysis of the hepatitis C virus RNA polymerase in complex with ribonucleotides. J. Virol. *76*, 3482–3492.

Bukh, J., Miller, H., and Purcell, R.H. (1995). Genetic heterogeneity of hepatitis C virus: Quasispecies and genotypes. Sem. Liver Dis. *15*, 41–63.

Bukh, J., Apgar, L., and Yanagi, M. (1999). Toward a surrogate model for hepatitis C virus: An infectious molecular clone of the GB virus-B hepatitis agent. Virology *262*, 470–478.

Bukh, J., Pietschmann, T., Lohmann, V., Krieger, N., Faulk, K., Engle, R.E., Govindarajan, S., Shapiro, M., St. Claire, M., and Bartenschlager, R. (2002). Mutations that permit efficient replication of hepatitis C virus RNA in Huh-7 cells prevent productive replication in chimpanzees. Proc. Natl. Acad. Sci. U.S.A. *99*, 14416–14421.

Buonocore, L., Blight, K.J., Rice, M., and Rose, J.K. (2002). Characterization of vesicular stomatitis virus recombinants that express and incorporate high levels of hepatitis C virus glycoproteins. J. Virol. *76*, 6865–6872.

Cai, Z., Zhang, C., Chang, K.S., Jiang, J., Ahn, B.C., Wakita, T., Liang, J., and Luo, G. (2005). Robust production of infectious hepatitis C virus (HCV) from stably HCV cDNA-transfected human hepatoma cells. J. Virol. *79*, 13963–13973.

Catanese, M.T., Graziani, R., von Hahn, T., Moreau, M., Huby, T., Paonessa, G., Santini, C., Luzzago, A., Rice, C.M., Cortese, R., Vitelli, A., and Nicosia, A. (2007). High-avidity monoclonal antibodies against the human scavenger class B type I receptor efficiently block hepatitis C virus infection in the presence of high-density lipoprotein. J. Virol. *81*, 8063–8071.

Chen, Y., Gotte, M., Liu, J., and Park, P.W. (2008). Microbial subversion of heparan sulfate proteoglycans. Mol. Cells *26*, 415–426.

Cheng, R.H., Kuhn, R.J., Olson, N.H., Rossmann, M.G., Choi, H.K., Smith, J., and Baker, T.S. (1995). Nucleocapsid and glycoprotein organization in an enveloped virus. Cell *80*, 621–630.

Chew, C.F., Vijayan, R., Chang, J., Zitzmann, N., and Biggin, P.C. (2009). Determination of pore-lining residues in the hepatitis C virus p7 protein. Biophys. J. *96*, L10–12.

Cho, H.S., Ha, N.C., Kang, L.W., Chung, K.M., Back, S.H., Jang, K., and Oh, B.H. (1998). Crystal structure of RNA helicase from genotype 1b hepatitis C virus. A feasible mechanism of unwinding duplex RNA. J. Biol. Chem. *273*, 15045–15052.

Choi, J., Xu, Z., and Ou, J.H. (2003). Triple decoding of hepatitis C virus RNA by programmed translational frameshifting. Mol. Cell Biol. *23*, 1489–1497.

Choo, Q.-L., Kuo, G., Weiner, A.J., Overby, L.R., Bradley, W., and Houghton, M. (1989). Isolation of a cDNA clone derived from a blood-borne non-A, non-B viral hepatitis genome. Science *244*, 359–362.

Choo, Q.-L., Richman, K.H., Han, J.H., Berger, K., Lee, C., Dong, C., Gallegos, C., Coit, D., Medina-Selby, A., Barr, P.J., Weiner, A.J., Bradley, D.W., Kuo, G., and Houghton, M. (1991). Genetic organization and diversity of the hepatitis C virus. Proc. Natl. Acad. Sci. U.S.A. *88*, 2451–2455.

Ciesek, S., Steinmann, E., Wedemeyer, H., Manns, M.P., Neyts, J., Tautz, N., Madan, V., Bartenschlager, R., von Hahn, T., and Pietschmann, T. (2009). Cyclosporine A inhibits hepatitis C virus nonstructural protein 2 through cyclophilin A. Hepatology *50*, 1638–1645.

Clarke, D., Griffin, S., Beales, L., Gelais, C.S., Burgess, S., Harris, M., and Rowlands, D. (2006). Evidence for the formation of a heptameric ion channel complex by the hepatitis C virus p7 protein in vitro. J. Biol. Chem. *281*, 37057–37068.

Cocquerel, L., Meunier, J.-C., Pillez, A., Wychowski, C., and Dubuisson, J. (1998). A retention signal necessary and sufficient for endoplasmic reticulum localization maps to the transmembrane domain of hepatitis C virus glycoprotein E2. J. Virol. *72*, 2183–2191.

Cocquerel, L., Duvet, S., Meunier, J.-C., Pillez, A., Cacan, R., Wychowski, C., and Dubuisson, J. (1999). The transmembrane domain of hepatitis C virus glycoprotein E1 is a signal for static retention in the endoplasmic reticulum. J. Virol. *73*, 2641–2649.

Cocquerel, L., Op de Beek, A., Lambot, M., Roussel, J., Delgrange, D., Pillez, A., Wychowski, C., Penin, F., and Dubuisson, J. (2002). Topological changes in the transmembrane domains of hepatitis C virus envelope glycoproteins. EMBO J. *21*, 2893–2902.

Codran, A., Royer, C., Jaeck, D., Bastien-Valle, M., Baumert, T.F., Kieny, M.P., Pereira, A., and Martin, J.P. (2006). Entry of hepatitis C virus pseudotypes into primary human hepatocytes by clathrin-dependent endocytosis. J. Gen. Virol. *87*, 2583–2593.

Collier, A.J., Gallego, J., Klinck, R., Cole, P.T., Harris, S.J., Harrison, G.P., Aboul-Ela, F., Varani, G., and Walker, S. (2002). A conserved RNA structure within the HCV IRES eIF3-binding site. Nat. Struct. Biol. *9*, 375–380.

Cormier, E.G., Durso, R.J., Tsamis, F., Boussemart, L., Manix, C., Olson, W.C., Gardner, P., and Dragic, T. (2004a). L-SIGN (CD209L) and DC-SIGN (CD209) mediate transinfection of liver cells by hepatitis C virus. Proc. Natl. Acad. Sci. U.S.A. *101*, 14067–14072.

Cormier, E.G., Tsamis, F., Kajumo, F., Durso, R.J., Gardner, P., and Dragic, T. (2004b). CD81 is an entry coreceptor for hepatitis C virus. Proc. Natl. Acad. Sci. U.S.A. *101*, 7270–7274.

Cristina, J., Lopez, F., Moratorio, G., Lopez, L., Vasquez, S., Garcia-Aguirre, L., and Chunga, A. (2005). Hepatitis C virus F protein sequence reveals a lack of functional constraints and a variable pattern of amino acid substitution. J. Gen. Virol. *86*, 115–120.

Date, T., Kato, T., Miyamoto, M., Zhao, Z., Yasui, K., Mizokami, M., and Wakita, T. (2004). Genotype 2a hepatitis C virus subgenomic replicon can replicate in HepG2 and IMY-N9 cells. J. Biol. Chem. *279*, 22371–22376.

De Francesco, R., Behrens, S.E., Tomei, L., Altamura, S., and Jiricny, J. (1996). RNA-dependent RNA polymerase of hepatitis C virus. Methods Enzymol. *275*, 58–67.

De Francesco R., T.L., Altamura, S., Summa, V., and Migliaccio, G. (2003). Approaching a new era for hepatitis C virus therapy: inhibitors of the NS3–4A serine protease and the NS5B RNA-dependent RNA polymerase. Antiviral Res. *58*, 1–16.

Dimitrova, M., Imbert, I., Kieny, P., and Schuster, C. (2003). Protein–protein interactions between hepatitis C virus nonstructural proteins. J. Virol. *77*, 5401–5414.

Drummer, H.E., Wilson, A., and Poumbourios, P. (2002). Identification of the hepatitis C virus E2 glycoprotein binding site on the large extracellular loop of CD81. J. Virol. *76*, 11143–11147.

Drummer, H.E., Wilson, A., and Poumbourios, P. (2005). Determinants of CD81 dimerization and interaction with hepatitis C virus glycoprotein E2. Biochem. Biophys. Res. Commun. *328*, 251–257.

Dubuisson, J. (1999). Folding, assembly, and subcellular localization of the hepatisis C virus glycoproteins. Curr. Top. Microbiol. Immunol. *242*, 135–148.

Dumont, S., Cheng, W., Serebrov, V., Beran, R.K., Tinoco, I., Jr., Pyle, M., and Bustamante, C. (2006). RNA translocation and unwinding mechanism of HCV NS3 helicase and its coordination by ATP. Nature *439*, 105–108.

Egger, D., Wolk, B., Gosert, R., Bianchi, L., Blum, H.E., Moradpour, D., and Bienz, K. (2002). Expression of hepatitis C virus proteins induces distinct membrane alterations including a candidate viral replication complex. J. Virol. *76*, 5974–5984.

Einav, S., Elazar, M., Danieli, T., and Glenn, J.S. (2004). A nucleotide binding motif in hepatitis C virus (HCV) NS4B mediates HCV RNA replication. J. Virol. *78*, 11288–11295.

Einav, S., Gerber, D., Bryson, P.D., Sklan, E.H., Elazar, M., Maerkl, S.J., Glenn, S., and Quake, S.R. (2008). Discovery of a hepatitis C target and its pharmacological inhibitors by microfluidic affinity analysis. Nat. Biotechnol. *26*, 1019–1027.

Elazar, M., Cheong, K.H., Liu, P., Greenberg, H.B., Rice, M., and Glenn, J.S. (2003). Amphipathic helix-dependent localization of NS5A mediates hepatitis C virus RNA replication. J. Virol. *77*, 6055–6061.

Elazar, M., Liu, P., Rice, M., and Glenn, J.S. (2004). An N-terminal amphipathic helix in hepatitis C virus (HCV) NS4B mediates membrane association, correct localization of replication complex proteins, and HCV RNA replication. J. Virol. *78*, 11393–400.

Evans, M.J., Rice, M., and Goff, S.P. (2004). Phosphorylation of hepatitis C virus nonstructural protein 5A modulates its protein interactions and viral RNA replication. Proc. Natl. Acad. Sci. U.S.A. *101*, 13038–13043.

Evans, M.J., von Hahn, T., Tscherne, D.M., Syder, A.J., Panis, M., Wolk, B., Hatziioannou, T., McKeating, J.A., Bieniasz, D., and Rice, C.M. (2007). Claudin-1 is a hepatitis C virus co-receptor required for a late step in entry. Nature. *446*, 801–805.

Failla, C., Tomei, L., De Francesco, R. (1994). Both NS3 and NS4A are required for proteolytic processing of hepatitis C virus nonstructural proteins. J. Virol. *68*, 3753–3760.

Failla, C., Tomei, L., and DeFrancesco, R. (1995). An amino-terminal domain of the hepatitis C virus NS3 protease is essential for interaction with NS4A. J. Virol. *69*, 1769–1777.

Feldman, G.J., Mullin, M., and Ryan, M.P. (2005). Occludin: structure, function and regulation. Adv. Drug Deliv. Rev. *57*, 883–917.

Ferlenghi, I., Clarke, M., Ruttan, T., Allison, S.L., Schalich, J., Heinz, F.X., Harrison, S.C., Rey, A., and Fuller, S.D. (2001). Molecular organization of a recombinant subviral particle from tick-borne encephalitis virus. Mol. Cell *7*, 593–602.

Fernandes, F., Poole, D.S., Hoover, S., Middleton, R., Andrei, A.C., Gerstner, J., and Striker, R. (2007). Sensitivity of hepatitis C virus to cyclosporine A depends on nonstructural proteins NS5A and NS5B. Hepatology *46*, 1026–1033.

Ferrari, E., Wright-Minogue, J., Fang, J.W., Baroudy, B.M., Lau, Y., and Hong, Z. (1999). Characterization of soluble hepatitis C virus RNA-dependent RNA polymerase expressed in Escherichia coli. J. Virol. *73*, 1649–1654.

Flint, M., and McKeating, J.A. (1999). The C-terminal region of the hepatitis C virus E1 glycoprotein confers localization within the endoplasmic reticulum. J. Gen. Virol. *80*, 1943–1947.

Flint, M., Maidens, C., Loomis-Price, L.D., Shotton, C., Dubuisson, J., Monk, P., Higginbottom, A., Levy, S., and McKeating, J.A. (1999a). Characterization of hepatitis C virus E2 glycoprotein interaction with a putative cellular receptor, CD81. J. Virol. *73*, 6235–6244.

Flint, M., Maidens, C., Loomis-Price, L.D., Shotton, C., Dubuisson, J., Monk, P., Higginbottom, A., Levy, S., and McKeating, J.A. (1999b). Characterization of hepatitis C virus E2 glycoprotein interaction with a putative cellular receptor, CD81. J. Virol. *73*, 6235–6244.

Flint, M., Thomas, J.M., Maidens, C.M., Shotton, C., Levy, S., Barclay, S., and McKeating, J.A. (1999c). Functional analysis of cell surface-expressed hepatitis C virus E2 glycoprotein. J. Virol. *73*, 6782–6790.

Flint, M., Quinn, R., and Levy, S. (2001). In search of hepatitis C virus receptor(s). Clin. Liver Dis. *5*, 873–893.

Flint, M., von Hahn, T., Zhang, J., Farquhar, M., Jones, C.T., Balfe, P., Rice, M., and McKeating, J.A. (2006). Diverse CD81 proteins support hepatitis C virus infection. J. Virol. *80*, 11331–11342.

Fraser, S., and Doudna, J.A. (2007). Structural and mechanistic insights into hepatitis C viral translation initiation. Nat. Rev. Microbiol. *5*, 29–38.

Fridell, R.A., Qiu, D., Wang, C., Valera, L., and Gao, M. (2010). Resistance analysis of the HCV NS5A inhibitor, BMS-790052, in the in vitro replicon system. Antimicrob. Agents Chemother. *54*, 3641–3650.

Friebe, P., and Bartenschlager, R. (2002). Genetic analysis of sequences in the 3′ nontranslated region of hepatitis C virus that are important for RNA replication. J. Virol. *76*, 5326–5338.

Friebe, P., Lohmann, V., Krieger, N., and Bartenschlager, R. (2001). Sequences in the 5′ nontranslated region of hepatitis C virus required for RNA replication. J. Virol. *75*, 12047–12057.

Friebe, P., Boudet, J., Simorre, P., and Bartenschlager, R. (2005). Kissing-loop interaction in the 3′ end of the hepatitis C virus genome essential for RNA replication. J. Virol. *79*, 380–392.

Fukushi, S., Okada, M., Kageyama, T., Hoshino, F.B., Nagai, K., and Katayama, K. (2001). Interaction of poly(rC)-binding protein 2 with the 5′-terminal stem–loop of the hepatitis C-virus genome. Virus Res. *73*, 67–79.

Gale Jr., M., Blakely, C.M., Kwieciszewski, B., Tan, S.L., Dossett, M., Tang, N.M., Korth, M.J., Polyak, S.J., Gretch, R., and Katze, M.G. (1998). Control of PKR protein kinase by hepatitis C virus nonstructural 5A protein: molecular mechanisms of kinase regulation. Mol. Cell Biol. *18*, 5208–5218.

Gao, L., Aizaki, H., He, W., and Lai, M.M. (2004). Interactions between viral nonstructural proteins and host protein hVAP-33 mediate the formation of hepatitis C virus RNA replication complex on lipid raft. J. Virol. *78*, 3480–3488.

Gao, M., Nettles, R.E., Belema, M., Snyder, L.B., Nguyen, V.N., Fridell, R.A., Serrano-Wu, M.H., Langley, D.R., Sun, J.H., O'Boyle, D.R., 2nd, Lemm, J.A., Wang, C., Knipe, J.O., Chien, C., Colonno, R.J., Grasela, D.M., Meanwell, A., and Hamann, L.G. (2010). Chemical genetics strategy identifies an HCV NS5A inhibitor with a potent clinical effect. Nature *465*, 96–100.

Gardner, J.P., Durso, R.J., Arrigale, R.R., Donovan, G.P., Maddon, P.J., Dragic, T., and Olson, W.C. (2003). L-SIGN (CD 209L) is a liver-specific capture receptor for hepatitis C virus. Proc. Natl. Acad. Sci. U.S.A. *100*, 4498–4503.

Garry, F., and Dash, S. (2003). Proteomics computational analyses suggest that hepatitis C virus E1 and pestivirus E2 envelope glycoproteins are truncated class II fusion proteins. Virology *307*, 255–265.

Gastaminza, P., Kapadia, B., and Chisari, F.V. (2006). Differential biophysical properties of infectious intracellular and secreted hepatitis C virus particles. J. Virol. *80*, 11074–11081.

Gastaminza, P., Cheng, G., Wieland, S., Zhong, J., Liao, W., and Chisari, F.V. (2008). Cellular determinants of hepatitis C virus assembly, maturation, degradation and secretion. J. Virol. *82*, 2120–2129.

Gonzalez, E., and Carrasco, L. (2003). Viroporins. FEBS Lett. *552*, 28–34.

Gosert, R., Egger, D., Lohmann, V., Bartenschlager, R., Blum, H.E., Bienz, K., and Moradpour, D. (2003). Identification of the hepatitis C virus RNA replication complex in Huh-7 cells harboring subgenomic replicons. J. Virol. *77*, 5487–5492.

Gouttenoire, J., Penin, F., and Moradpour, D. (2010). Hepatitis C virus nonstructural protein 4B: a journey into unexplored territory. Rev. Med. Virol. *20*, 117–129.

Grakoui, A., McCourt, D.W., Wychowski, C., Feinstone, M., and Rice, C.M. (1993a). A second hepatitis C virus-encoded proteinase. Proc. Natl. Acad. Sci. U.S.A. *90*, 10583–10587.

Grakoui, A., Wychowski, C., Lin, C., Feinstone, M., and Rice, C.M. (1993b). Expression and identification of hepatitis C virus polyprotein cleavage products. J. Virol. *67*, 1385–1395.

Griffin, S., Stgelais, C., Owsianka, A.M., Patel, A.H., Rowlands, D., and Harris, M. (2008). Genotype-dependent sensitivity of hepatitis C virus to inhibitors of the p7 ion channel. Hepatology *48*, 1779–1790.

Griffin, S.D., Beales, L.P., Clarke, D.S., Worsfold, O., Evans, S.D., Jaeger, J., Harris, P., and Rowlands, D.J. (2003). The p7 protein of hepatitis C virus forms an ion channel that is blocked by the antiviral drug, Amantadine. FEBS Lett. *535*, 34–38.

Grove, J., Huby, T., Stamataki, Z., Vanwolleghem, T., Meuleman, P., Farquhar, M., Schwarz, A., Moreau, M., Owen, J.S., Leroux-Roels, G., Balfe, P., and McKeating, J.A. (2007). Scavenger receptor BI and BII expression levels modulate hepatitis C virus infectivity. J. Virol. *81*, 3162–3169.

Gu, B., Pruss, C.M., Gates, T., and Khandekar, S.S. (2005). The RNA-unwinding activity of hepatitis C virus non-structural protein 3 (NS3) is positively modulated by its protease domain. Protein Pept. Lett. *12*, 315–321.

Hagedorn, C.H., van Beers, E.H., De Staercke, C. (1999). Hepatitis C virus RNA-dependent RNA polymerase (NS5B polymerase). In Hepatitis C virus, Hagedorn, C, Rice, CM, eds (Springer-Verlag, Berlin), vol. 242, pp. 225–260.

Hamamoto, I., Nishimura, Y., Okamoto, T., Aizaki, H., Liu, M., Mori, Y., Abe, T., Suzuki, T., Lai, M.M., Miyamura, T., Moriishi, K., and Matsuura, Y. (2005). Human VAP-B is involved in hepatitis C virus replication through interaction with NS5A and NS5B. J. Virol. *79*, 13473–13482.

Hanoulle, X., Badillo, A., Wieruszeski, J.M., Verdegem, D., Landrieu, I., Bartenschlager, R., Penin, F., and Lippens, G. (2009a). Hepatitis C virus NS5A protein is a substrate for the peptidyl-prolyl cis/trans isomerase activity of cyclophilins A and B. J. Biol. Chem. *284*, 13589–13601.

Hanoulle, X., Verdegem, D., Badillo, A., Wieruszeski, J.M., Penin, F., and Lippens, G. (2009b). Domain 3 of non-structural protein 5A from hepatitis C virus is natively unfolded. Biochem. Biophys. Res. Commun. *381*, 634–638.

Hanoulle, X., Badillo, A., Verdegem, D., Penin, F., and Lippens, G. (2010). The domain 2 of the HCV NS5A protein is intrinsically unstructured. Protein Pept. Lett. *17*, 1012–1018.

Harada, S., Watanabe, Y., Takeuchi, K., Suzuki, T., Katayama, T., Takebe, Y., Saito, I., and Miyamura, T. (1991). Expression of processed core protein of hepatitis C virus in mammalian cells. J. Virol. *65*, 3015–3021.

Hardy, R.W., Marcotrigiano, J., Blight, K.J., Majors, E., and Rice, C.M. (2003). Hepatitis C virus RNA synthesis in a cell-free system isolated from replicon-containing hepatoma cells. J. Virol. *77*, 2029–2037.

Harrison, S.C. (2008). Viral membrane fusion. Nat. Struct. Mol. Biol. *15*, 690–698.

He, L.F., Alling, D., Popkin, T., Shapiro, M., Alter, J., and Purcell, R.H. (1987). Determining the size of non-A, non-B hepatitis by filtration. J. Infect. Dis. *156*, 636–640.

Heathcote, E.J., Shiffman, M.L., Cooksley, W.G., Dusheiko, G.M., Lee, S.S., Balart, L., Reindollar, R., Reddy, R.K., Wright, T.L., Lin, A., Hoffman, J., De Pamphilis, J. (2000). Peginterferon alfa-2a in patients with chronic hepatitis C and cirrhosis. N. Engl. J. Med. *343*, 1673–1680.

Heo, T.H., Lee, S.M., Bartosch, B., Cosset, L., and Kang, C.Y. (2006). Hepatitis C virus E2 links soluble human CD81 and SR-B1 protein. Virus Res. *121*, 58–64.

Higginbottom, A., Quinn, E.R., Kuo, C.C., Flint, M., Wilson, L.H., Bianchi, E., Nicosia, A., Monk, P.N., McKeating, A., and Levy, S. (2000). Identification of amino acid residues in CD81 critical for interaction with hepatitis C virus envelope glycoprotein E2. J. Virol. *74*, 3642–3649.

Hijikata, M., Kato, N., Ootsuyama, Y., Nakagawa, M., and Shimotohno, K. (1991). Gene mapping of the putative structural region of the hepatitis C virus genome by *in vitro* processing analysis. Proc. Natl. Acad. Sci. U.S.A. *88*, 5547–5551.

Hijikata, M., Mizushima, H., Akagi, T., Mori, S., Kakiuchi, N., Kato, N., Tanaka, T., Kimura, K., and Shimotohno, K. (1993a). Two distinct proteinase activities required for the processing of a putative nonstructural precursor protein of hepatitis C virus. J. Virol. *67*, 4665–4675.

Hijikata, M., Shimizu, Y.K., Kato, H., Iwamoto, A., Shih, J.W., Alter, H.J., Purcell, H., and Yoshikura, H. (1993b). Equilibrium centrifugation studies of hepatitis C virus: evidence for circulating immune complexes. J. Virol. *67*, 1953–1958.

Hnatyszyn, H.J. (2005). Chronic hepatitis C and genotyping: the clinical significance of determining HCV genotypes. Antivir. Ther. *10*, 1–11.

Hong Z., C.C.E., Walker, M.P., Castro, C., Yao, N., Lau, J.Y., and Zhong, W. (2001). A novel mechanism to ensure terminal initiation by hepatitis C virus NS5B polymerase. Virology *285*, 6–11.

Hope, R.G., Murphy, J., and McLauchlan, J. (2002). The domains required to direct core proteins of hepatitis C virus and GB virus-B to lipid droplets share common features with plant oleosin proteins. J. Biol. Chem. *277*, 4261–4270.

Houghton, M. (1996). Hepatitis C viruses. In Fields Virology, 3rd edn, Fields BN, Knipe DM, Howley PM, eds (Raven Press, Philadelphia, PA), vol. 1, pp. 1035–1058.

Howe, A.Y., Chase, R., Taremi, S.S., Risano, C., Beyer, B., Malcolm, B., and Lau, J.Y. (1999). A novel recombinant single-chain hepatitis C virus NS3-NS4A protein with improved helicase activity. Protein Sci. *8*, 1332–1341.

Hsu, M., Zhang, J., Flint, M., logvinoff, C., Cheng-Mayer, C., Rice, M., and McKeating, J.A. (2003). Hepatitis C virus glycoproteins mediate pH-dependent cell entry of pseudotyped retroviral particles. Proc. Natl. Acad. Sci. U.S.A. *100*, 7271–7276.

Huang, L., Sineva, E.V., Hargittai, M.R., Sharma, S.D., Suthar, M., Raney, D., and Cameron, C.E. (2004).

Purification and characterization of hepatitis C virus non-structural protein 5A expressed in *Escherichia coli*. Protein Expr. Purif. *37*, 144–153.

Huang, L., Hwang, J., Sharma, S.D., Hargittai, M.R., Chen, Y., Arnold, J.J., Raney, D., and Cameron, C.E. (2005). Hepatitis C virus nonstructural protein 5A (NS5A) is an RNA-binding protein. J. Biol. Chem. *280*, 36417–36428.

Huang, H., Sun, F., Owen, D.M., Li, W., Chen, Y., Gale, M., Jr. and Ye, J. (2007a). Hepatitis C virus production by human hepatocytes dependent on assembly and secretion of very low-density lipoproteins. Proc. Natl. Acad. Sci. U.S.A. *104*, 5848–5853.

Huang, Y., Staschke, K., De Francesco, R., and Tan, S.L. (2007b). Phosphorylation of hepatitis C virus NS5A nonstructural protein: A new paradigm for phosphorylation-dependent viral RNA replication? Virology. *364*, 1–9.

Hughes, M., Gretton, S., Shelton, H., Brown, D.D., McCormick, C.J., Angus, A.G., Patel, A.H., Griffin, S., and Harris, M. (2009a). A conserved proline between domains II and III of hepatitis C virus NS5A influences both RNA replication and virus assembly. J. Virol. *83*, 10788–10796.

Hughes, M., Griffin, S., and Harris, M. (2009b). Domain III of NS5A contributes to both RNA replication and assembly of hepatitis C virus particles. J. Gen. Virol. *90*, 1329–1334.

Hwang, B., Lim, J.H., Hahm, B., Jang, K., and Lee, S.W. (2009). hnRNP L is required for the translation mediated by HCV IRES. Biochem. Biophys. Res. Commun. *378*, 584–588.

Hwang, S.B., Lo, S.-Y., Ou, J.-H., and Lai, M.M.C. (1995). Detection of cellular proteins and viral core protein interacting with the 5′ untranslated region of hepatitis C virus RNA. J. Biomed. Sci. *2*, 227–236.

Ikeda, M., Yi, M., Li, K., and Lemon, S.M. (2002). Selectable subgenomic and genome-length dicistronic RNAs derived from an infectious molecular clone of the HCV-N strain of hepatitis C virus replicate efficiently in cultured Huh7 cells. J. Virol. *76*, 2997–3006.

Ishii, N., Watashi, K., Hishiki, T., Goto, K., Inoue, D., Hijikata, M., Wakita, T., Kato, N., and Shimotohno, K. (2006). Diverse effects of cyclosporine on hepatitis C virus strain replication. J. Virol. *80*, 4510–4520.

Ito, T., Tahara, M., and Lai, M.M.C. (1998). The 3′-untranslated region of hepatitis C virus RNA enhances translation from an internal ribosomal entry site. J. Virol. 72, 8789–8796.

Ito, T., and Lai, M.M. (1999). An internal polypyrimidine-tract-binding protein-binding site in the hepatitis C virus RNA attenuates translation, which is relieved by the 3′-untranslated sequence. Virology *254*, 288–296.

Izumi, R.E., Valdez, B., Banerjee, R., Srivastava, M., and Dasgupta, A. (2001). Nucleolin stimulates viral internal ribosome entry site-mediated translation. Virus Res. *76*, 17–29.

Jangra, R.K., Yi, M., and Lemon, S.M. (2010). Regulation of hepatitis C virus translation and infectious virus production by the microRNA miR-122. J. Virol. *84*, 6615–6625.

Jiang, J., and Luo, G. (2009). Apolipoprotein E but not B is required for the formation of infectious hepatitis C virus particles. J. Virol. *83*, 12680–12691.

Jirasko, V., Montserret, R., Appel, N., Janvier, A., Eustachi, L., Brohm, C., Steinmann, E., Pietschmann, T., Penin, F., and Bartenschlager, R. (2008). Structural and functional characterization of nonstructural protein 2 for its role in hepatitis C virus assembly. J. Biol. Chem. *283*, 28546–28562.

Jones, C.T., Murray, C.L., Eastman, D.K., Tassello, J., and Rice, C.M. (2007). Hepatitis C virus p7 and NS2 proteins are essential for production of infectious virus. J. Virol. *81*, 8374–8383.

Jones, D.M., Patel, A.H., Targett-Adams, P., and McLauchlan, J. (2009). The hepatitis C virus NS4B protein can trans-complement viral RNA replication and modulates production of infectious virus. J. Virol. *83*, 2163–2177.

Jopling, C.L., Yi, M., Lancaster, A.M., Lemon, M., and Sarnow, P. (2005). Modulation of hepatitis C virus RNA abundance by a liver-specific MicroRNA. Science *309*, 1577–1581.

Kalinina, O., Norder, H., Mukomolov, S., and Magnius, L.O. (2002). A natural intergenotypic recombinant of hepatitis C virus identified in St. Petersburg. J. Virol. *76*, 4034–4043.

Kanto, T., Hayashi, N., Takehara, T., Hagiwara, H., Mita, E., Naito, M., Kasahara, A., Fusamoto, H., and Kamada, T. (1994). Buoyant density of hepatitis C virus recovered from infected hosts: two different features in sucrose equilibrium density-gradient centrifugation related to degree of liver inflammation. Hepatology *19*, 296–302.

Kapadia, B., and Chisari, F.V. (2005). Hepatitis C virus RNA replication is regulated by host geranylgeranylation and fatty acids. Proc. Natl. Acad. Sci. U.S.A. *102*, 2561–2566.

Kapadia, S.B., Barth, H., Baumert, T., McKeating, A., and Chisari, F.V. (2007). Initiation of hepatitis C virus infection is dependent on cholesterol and cooperativity between CD81 and scavenger receptor B type I. J. Virol. *81*, 374–383.

Kato, T., Date, T., Miyamoto, M., Furusaka, A., Tokushige, K., Mizokami, M., and Wakita, T. (2003). Efficient replication of the genotype 2a hepatitis C virus subgenomic replicon. Gastroenterology *125*, 1808–1817.

Kato, T., Date, T., Miyamoto, M., Zhao, Z., Mizokami, M., and Wakita, T. (2005). Nonhepatic cell lines HeLa and 293 support efficient replication of the hepatitis C virus genotype 2a subgenomic replicon. J. Virol. *79*, 592–596.

Kato, T., Matsumura, T., Heller, T., Saito, S., Sapp, R.K., Murthy, K., Wakita, T., and Liang, T.J. (2007). Production of infectious hepatitis C virus of various genotypes in cell culture. J. Virol. *81*, 4405–4411.

Kieft, J.S., Zhou, K., Jubin, R., and Doudna, J.A. (2001). Mechansim of ribosome recruitment by hepatitis C IRES RNA. RNA 7, 194–206.

Kieft, J.S., Zhou, K., Grech, A., Jubin, R., and Doudna, J.A. (2002). Crystal structure of an RNA tertiary domain essential to HCV IRES-mediated translation initiation. Nat. Struct. Biol. *9*, 370–374.

Kim, J., Lee, D., and Choe, J. (1999). Hepatitis C virus NS5A protein is phosphorylated by casein kinase II. Biochem. Biophys. Res. Commun. *257*, 777–781.

Kim, J.L., Morgenstern, K.A., Lin, C., Fox, T., Dwyer, M.D., Landro, J.A., Chambers, S.P., Markland, W., Lepre, C.A., O'Malley, E.T., *et al.* (1996). Crystal structure of the hepatitis C virus NS3 protease domain complexed with a synthetic NS4A cofactor peptide. Cell *87*, 343–355.

Kim, J.L., Morgenstern, K.A., Griffith, J.P., Dwyer, M.D., Thomson, J.A., Murcko, M.A., Lin, C., and Caron, P.R. (1998). Hepatitis C virus NS3 RNA helicase domain with a bound oligonucleotide: the crystal structure provides insights into the mode of unwinding. Structure *6*, 89–100.

Kim, Y.C., Russell, W.K., Ranjith-Kumar, C.T., Thomson, M., Russell, H., and Kao, C.C. (2005). Functional analysis of RNA binding by the hepatitis C virus RNA-dependent RNA polymerase. J. Biol. Chem. *280*, 38011–38019.

Kim, Y.K., Kim, C.S., Lee, H., and Jang, S.K. (2002). Domains I and II in the 5′ nontranslated region of the HCV genome are required for RNA replication. Biochem. Biophys. Res. Commun. *290*, 105–112.

Klein, K.C., Dellos, R., and Lingappa, J.R. (2005). Identification of residues in the hepatitis C virus core protein that are critical for capsid assembly in a cell-free system. J. Virol. *79*, 6814–6826.

Klinck, R., Westhof, E., Walker, S., Afshar, M., Collier, A., and Aboul-Ela, F. (2000). A potential RNA drug target in the hepatitis C virus internal ribosomal entry site. RNA *6*, 1423–1431.

Kobayashi, M., Bennett, M.C., Bercot, T., and Singh, I.R. (2006). Functional analysis of hepatitis C virus envelope proteins, using a cell–cell fusion assay. J. Virol. *80*, 1817–1825.

Kolykhalov, A.A., Feinstone, M., and Rice, C.M. (1996). Identification of a highly conserved sequence element at the 3′ terminus of hepatitis C virus genome RNA. J. Virol. *70*, 3363–3371.

Kolykhalov, A.A., Agapov, E.V., Blight, K.J., Mihalik, K., Feinstone, M., and Rice, C.M. (1997). Transmission of hepatitis C by intrahepatic inoculation with transcribed RNA. Science *277*, 570–574.

Kolykhalov, A.A., Mihalik, K., Feinstone, M., and Rice, C.M. (2000). Hepatitis C virus-encoded enzymatic activities and conserved RNA elements in the 3′ nontranslated region are essential for virus replication in vivo. J. Virol. *74*, 2046–2051.

Korf, M., Jarczak, D., Beger, C., Manns, P., and Kruger, M. (2005). Inhibition of hepatitis C virus translation and subgenomic replication by siRNAs directed against highly conserved HCV sequence and cellular HCV cofactors. J. Hepatol. *43*, 225–234.

Koutsoudakis, G., Kaul, A., Steinmann, E., Kallis, S., Lohmann, V., Pietschmann, T., and Bartenschlager, R. (2006). Characterization of the early steps of hepatitis C virus infection by using luciferase reporter viruses. J. Virol. *80*, 5308–5320.

Krey, T., d'Alayer, J., Kikuti, C.M., Saulnier, A., Damier-Piolle, L., Petitpas, I., Johansson, D.X., Tawar, R.G., Baron, B., Robert, B., England, P., Persson, M.A., Martin, A., and Rey, F.A. (2010). The disulfide bonds in glycoprotein E2 of hepatitis C virus reveal the tertiary organization of the molecule. PLoS Pathog. *6*, e1000762.

Kuang, W.F., Lin, Y.C., Jean, F., Huang, Y.W., Tai, C.L., Chen, D.S., Chen, J., and Hwang, L.H. (2004). Hepatitis C virus NS3 RNA helicase activity is modulated by the two domains of NS3 and NS4A. Biochem. Biophys. Res. Commun. *317*, 211–217.

Kuhn, R.J., Zhang, W., Rossmann, M.G., Pletnev, S.V., Corver, J., Lenches, E., Jones, C.T., Mukhopadhyay, S., Chipman, P.R., Strauss, E.G., Baker, S., and Strauss, J.H. (2002). Structure of Dengue virus: implications for flavivirus organization, maturation, and fusion. Cell *108*, 717–725.

Kuiken, C., Combet, C., Bukh, J., Shin, I.T., Deleage, G., Mizokami, M., Richardson, R., Sablon, E., Yusim, K., Pawlotsky, M., and Simmonds, P. (2006). A comprehensive system for consistent numbering of HCV sequences, proteins and epitopes. Hepatology *44*, 1355–1361.

Kunkel, M., Lorinczi, M., Rijnbrand, R., Lemon, M., and Watowich, S.J. (2001). Self-assembly of nucleocapsid-like particles from recombinant hepatitis C virus core protein. J. Virol. *75*, 2119–2129.

Kwong, A.D., Rao, G., and Jeang, K.T. (2005). Viral and cellular RNA helicases as antiviral targets. Nat. Rev. Drug Discov. *4*, 845–853.

Lagging, L.M., Meyer, K., Owens, J., and Ray, R. (1998). Functional role of hepatitis C virus chimeric glycoproteins in the infectivity of pseudotyped virus. J. Virol. *72*, 3539–3546.

Lai, V.C., Dempsey, S., Lau, J.Y., Hong, Z., and Zhong, W. (2003). In vitro RNA replication directed by replicase complexes isolated from the subgenomic replicon cells of hepatitis C virus. J. Virol. *77*, 2029–2037.

Lai, W.K., Sun, P.J., Zhang, J., Jennings, A., Lalor, P.F., Hubscher, S., McKeating, A., and Adams, D.H. (2006). Expression of DC-SIGN and DC-SIGNR on human sinusoidal endothelium: a role for capturing hepatitis C virus particles. Am. J. Pathol. *169*, 200–208.

Lanford, R.E., Chavez, D., Chisari, V., and Sureau, C. (1995). Lack of detection of negative-strand hepatitis C virus RNA in peripheral blood mononuclear cells and other extrahepatic tissues by the highly strand-specific rTth reverse transcriptase PCR. J. Virol. *69*, 8079–8083.

Lanford, R.E., Bigger, C., Bassett, S., and Klimpel, G. (2001a). The chimpanzee model of hepatitis C virus infections. ILAR J. *42*, 117–126.

Lanford R.E., Chavez, D., Guerra, B., Lau, J.Y., Hong, Z., Brasky, K.M., and Beames, B. (2001b). Ribavirin induces error-prone replication of GB virus B in primary tamarin hepatocytes. J. Virol. 75, 8074–8081.

Lanford, R.E., Hildebrandt-Eriksen, E.S., Petri, A., Persson, R., Lindow, M., Munk, M.E., Kauppinen, S., and Orum, H. (2010). Therapeutic silencing of microRNA-122 in primates with chronic hepatitis C virus infection. Science *327*, 198–201.

Lavie, M., Goffard, A., and Dubuisson, J. (2007). Assembly of a functional HCV glycoprotein heterodimer. Curr. Issues Mol. Biol. *9*, 71–86.

Lee, H., Shin, H., Wimmer, E., and Paul, A.V. (2004). cis-acting RNA signals in the NS5B C-terminal coding sequence of the hepatitis C virus genome. J. Virol. *78*, 10865–10877.

Lemm, J.A., O'Boyle, D., 2nd, Liu, M., Nower, P.T., Colonno, R., Deshpande, M.S., Snyder, L.B., Martin, S.W., St Laurent, D.R., Serrano-Wu, M.H., Romine, J.L., Meanwell, A., and Gao, M. (2010). Identification of hepatitis C virus NS5A inhibitors. J. Virol. *84*, 482–491.

Lesburg, C.A., Cable, M.B., Ferrari, E., Hong, Z., Mannarino, F., and Weber, P.C. (1999). Crystal structure of the RNA-dependent RNA polymerase from hepatitis C virus reveals a fully encircled active site. Nat. Struct. Biol. *6*, 937–943.

Leveque, V.J., Johnson, R.B., Parsons, S., Ren, J., Xie, C., Zhang, F., and Wang, Q.M. (2003). Identification of a C-terminal regulatory motif in hepatitis C virus RNA-dependent RNA polymerase: structural and biochemical analysis. J. Virol. *77*, 9020–9028.

Levin, M.K., Wang, H., and Patel, S.S. (2004). The functional interaction of the hepatitis C virus helicase molecules is responsible for unwinding processivity. J. Biol. Chem. *279*, 26005–26012.

Levin, M.K., Gurjar, M., and Patel, S.S. (2005). A Brownian motor mechanism of translocation and strand separation by hepatitis C virus helicase. Nat. Struct. Mol. Biol. *12*, 429–435.

Li, K., Foy, E., Ferreon, J.C., Nakamura, M., Ferreon, A.C., Ikeda, M., Ray, S.C., Gale, M., Jr. and Lemon, S.M. (2005). Immune evasion by hepatitis C virus NS3/4A protease-mediated cleavage of the Toll-like receptor 3 adaptor protein TRIF. Proc. Natl. Acad. Sci. U.S.A. *102*, 2992–2997.

Liang, Y., Kang, B., and Yoon, H.S. (2006). Molecular and structural characterization of the domain 2 of hepatitis C virus non-structural protein 5A. Mol. Cells *22*, 13–20.

Liang, Y., Ye, H., Kang, B., and Yoon, H.S. (2007). Domain 2 of nonstructural protein 5A (NS5A) of hepatitis C virus is natively unfolded. Biochemistry *46*, 11550–1158.

Lin, C., Lindenbach, B.D., Prágai, B., McCourt, W., and Rice, C.M. (1994a). Processing of the hepatitis C virus E2-NS2 region: identification of p7 and two distinct E2-specific products with different C termini. J. Virol. *68*, 5063–5073.

Lin, C., Prágai, B., Grakoui, A., Xu, J., and Rice, C.M. (1994b). Hepatitis C virus NS3 serine proteinase: *trans*-cleavage requirements and processing kinetics. J. Virol. *68*, 8147–8157.

Lin, C., Thomson, A., and Rice, C.M. (1995). A central region in the hepatitis C virus NS4A protein allows formation of an active NS3–NS4A serine proteinase complex in vivo and in vitro. J. Virol. *69*, 4373–4380.

Lindenbach, D., and Rice, C.M. (2001). *Flaviviridae*: The viruses and their replication. In Fields Virology, 5th edn, Knipe DM, Howley PM, eds (Lippincott-Raven Publishers, Philadelphia, PA), vol. 1, pp. 991–1041.

Lindenbach, B.D., Evans, M.J., Syder, A.J., Wolk, B., Tellinghuisen, T.L., Liu, C.C., Maruyama, T., Hynes, R.O., Burton, D.R., McKeating, A., and Rice, C.M. (2005). Complete replication of hepatitis C virus in cell culture. Science *309*, 623–626.

Lindenbach, B.D., Meuleman, P., Ploss, A., Vanwolleghem, T., Syder, A.J., McKeating, J.A., Lanford, R.E., Feinstone, S.M., Major, M.E., Leroux-Roels, G., and Rice, C.M. (2006). Cell culture-grown hepatitis C virus is infectious in vivo and can be recultured in vitro. Proc. Natl. Acad. Sci. U.S.A. *103*, 3805–3809.

Lindenbach, B.D., Thiel, J., and Rice, C.M. (2007a). Flaviviridae: The viruses and their replication. In Fields Virology, 5th edn, Knipe DM, Howley PM, eds (Lippincott-Raven Publishers, Philadelphia, PA) vol. 1, pp. 1101–1152.

Lindenbach, B.D., Pragai, B.M., Montserret, R., Beran, R.K., Pyle, A.M., Penin, F., and Rice, C.M. (2007b). The C terminus of hepatitis C virus NS4A encodes an electrostatic switch that regulates NS5A hyperphosphorylation and viral replication. J. Virol. *81*, 8905–8918.

Liu, S., Yang, W., Shen, L., Turner, J.R., Coyne, B., and Wang, T. (2009). Tight junction proteins claudin-1 and occludin control hepatitis C virus entry and are downregulated during infection to prevent superinfection. J. Virol. *83*, 2011–2014.

Lo, S.-Y., Selby, M.J., Ou, J.-H. (1996). Interaction between hepatitis C virus core protein and E1 envelope protein. J. Virol. *70*, 5177–5182.

Locatelli, G.A., Spadari, S., and Maga, G. (2002). Hepatitis C virus NS3 ATPase/helicase: an ATP switch regulates the cooperativity among the different substrate binding sites. Biochemistry *41*, 10332–10342.

Lohmann, V., Körner, F., Herian, U., and Bartenschlager, R. (1997). Biochemical properties of hepatitis C virus NS5B RNA-dependent RNA polymerase and identification of amino acid sequence motifs essential for enzymatic activity. J. Virol. *71*, 8416–8428.

Lohmann, V., Roos, A., Korner, F., Koch, O., and Bartenschlager, R. (1998). Biochemical and kinetic analyses of NS5B RNA-dependent RNA polymerase of the hepatitis C virus. Virology *249*, 108–118.

Lohmann, V., Korner, F., Koch, J.O., Herian, U., Theilmann, L., and Bartenschlager, R. (1999). Replication of subgenomic hepatitis C virus RNAs in a hepatoma cell line. Science *285*, 110–113.

Lohmann, V., Roos, A., Korner, F., Koch, O., and Bartenschlager, R. (2000). Biochemical and structural analysis of the NS5B RNA-dependent RNA polymerase of the hepatitis C virus. J. Viral Hepat. *7*, 167–174.

Lohmann, V., Korner, F., Dobierzewska, A., and Bartenschlager, R. (2001). Mutations in hepatitis C virus RNAs conferring cell culture adaptation. J. Virol. *75*, 1437–1449.

Lohmann, V., Hoffmann, S., Herian, U., Penin, F., and Bartenschlager, R. (2003). Viral and cellular determinants of hepatitis C virus RNA replication in cell culture. J. Virol. *77*, 3007–3019.

Lonardo, A., Adinolfi, L.E., Loria, P., Carulli, N., Ruggiero, G., and Day, C.P. (2004). Steatosis and hepatitis C virus: mechanisms and significance for hepatic and extrahepatic disease. Gastroenterology *126*, 586–597.

Lorenz, I.C., Marcotrigiano, J., Dentzer, G., and Rice, C.M. (2006). Structure of the catalytic domain of the hepatitis C virus NS2–3 protease. Nature *442*, 831–835.

Love, R.A., Parge, H.E., Wickersham, J.A., Hostomsky, Z., Habuka, N., Moomaw, E.W., Adachi, T., and Hostomska, Z. (1996). The crystal structure of hepatitis C virus NS3 proteinase reveals a trypsin-like fold and a structural zinc binding site. Cell *87*, 331–342.

Love, R.A., Brodsky, O., Hickey, M.J., Wells, A., and Cronin, C.N. (2009). Crystal structure of a novel dimeric form of NS5A domain I protein from hepatitis C virus. J. Virol. *83*, 4395–4403.

Lozach, P.Y., Lortat-Jacob, H., De Lacroix De Lavalette, A., Staropoli, I., Foung, S., Amara, A., Houles, C., Fieschi, F., Schwartz, Virelizier, J.L., Arenzana-Seisdedos, F., and Altmeyer, R. (2003). DC-SIGN and L-SIGN are high-affinity binding receptors for hepatitis C Virus glycoprotein E2. J. Biol. Chem. *278*, 20358–20366.

Lu, W., Strohecker, A., Ou Jh, J.H. (2001). Post-translational modification of the hepatitis C virus core protein by tissue transglutaminase. J. Biol. Chem. *276*, 47993–47999.

Luik, P., Chew, C., Aittoniemi, J., Chang, J., Wentworth, P., Jr., Dwek, R.A., Biggin, P.C., Venien-Bryan, C., and Zitzmann, N. (2009). The 3-dimensional structure of a hepatitis C virus p7 ion channel by electron microscopy. Proc. Natl. Acad. Sci. U.S.A. *106*, 12712–12716.

Lukavsky, P.J., Otto, G.A., Lancaster, A.M., Sarnow, P., and Puglisi, J.D. (2000). Structures of two RNA domains essential for hepatitis C virus internal ribosome entry site function. Nat. Struct. Biol. *7*, 1105–1110.

Lukavsky, P.J., Kim, I., Otto, A., and Puglisi, J.D. (2003). Structure of HCV IRES domain II determined by NMR. Nat. Struct. Biol. *10*, 1033–1038.

Lundin, M., Monne, M., Widell, A., Von Heijne, G., and Persson, M.A. (2003). Topology of the membrane-associated hepatitis C virus protein NS4B. J. Virol. *77*, 5428–5438.

Lundin, M., Lindstrom, H., Gronwall, C., and Persson, M.A. (2006). Dual topology of the processed hepatitis C virus protein NS4B is influenced by the NS5A protein. J. Gen. Virol. *87*, 3263–3272.

Lytle, J.R., Wu, L., and Robertson, H.D. (2002). Domains on the hepatitis C virus internal ribosome entry site for 40s subunit binding. RNA *8*, 1045–1055.

Ma, H.C., Ke, C.H., Hsieh, Y., and Lo, S.Y. (2002). The first hydrophobic domain of the hepatitis C virus E1 protein is important for interaction with the capsid protein. J. Gen. Virol. *83*, 3085–3092.

Ma, Y., Yates, J., Liang, Y., Lemon, M., and Yi, M. (2008). NS3 helicase domains involved in infectious intracellular hepatitis C virus particle assembly. J. Virol. *82*, 7624–7639.

McCaffrey, A.P., Ohashi, K., Meuse, L., Shen, S., Lancaster, A.M., Lukavsky, P.J., Sarnow, P., and Kay, M.A. (2002). Determinants of hepatitis C translational initiation in vitro, in cultured cells and mice. Mol. Ther. *5*, 676–684.

Macdonald, A., and Harris, M. (2004). Hepatitis C virus NS5A: tales of a promiscuous protein. J. Gen. Virol. *85*, 2485–2502.

Mackintosh, S.G., Lu, J.Z., Jordan, J.B., Harrison, M.K., Sikora, B., Sharma, S.D., Cameron, C.E., Raney, D., and Sakon, J. (2006). Structural and biological identification of residues on the surface of NS3 helicase required for optimal replication of the hepatitis C virus. J. Biol. Chem. *281*, 3528–3535.

McLauchlan, J. (2000). Properties of the hepatitis C virus core protein: a structural protein that modulates cellular processes. J. Viral Hepat. *7*, 2–14.

McLauchlan, J., Lemberg, M.K., Hope, G., and Martoglio, B. (2002). Intramembrane proteolysis promotes trafficking of hepatitis C virus core protein to lipid droplets. EMBO J. *21*, 3980–3988.

McLauchlan, J. (2009). Hepatitis C virus: viral proteins on the move. Biochem. Soc. Trans. *37*, 986–990.

McMullan, L.K., Grakoui, A., Evans, M.J., Mihalik, K., Puig, M., Branch, A.D., Feinstone, M., and Rice, C.M. (2007). Evidence for a functional RNA element in the hepatitis C virus core gene. Proc. Natl. Acad. Sci. U.S.A. *104*, 2879–2884.

Majeau, N., Gagne, V., Boivin, A., Bolduc, M., Majeau, J.A., Ouellet, D., and Leclerc, D. (2004). The N-terminal half of the core protein of hepatitis C virus is sufficient for nucleocapsid formation. J. Gen. Virol. *85*, 971–981.

Manna, D., Aligo, J., Xu, C., Park, W.S., Koc, H., Do Heo, W., and Konan, K.V. (2010). Endocytic Rab proteins are required for hepatitis C virus replication complex formation. Virology *398*, 21–37.

Masaki, T., Suzuki, R., Murakami, K., Aizaki, H., Ishii, K., Murayama, A., Date, T., Matsuura, Y., Miyamura, T., Wakita, T., and Suzuki, T. (2008). Interaction of hepatitis C virus nonstructural protein 5A with core protein is critical for the production of infectious virus particles. J. Virol. *82*, 7964–7976.

Matsumoto, M., Hwang, S.B., Jeng, K.-S., Zhu, N., and Lai, M.M.C. (1996). Homotypic interaction and multimerization of hepatitis C virus core protein. Virology *218*, 43–51.

Matsuura, Y., Tani, H., Suzuki, K., Kimura-Someya, T., Suzuki, R., Aizaki, H., Ishii, K., Moriishi, K., Robison, C.S., Whitt, A., and Miyamura, T. (2001). Characterization of pseudotype VSV possessing HCV envelope proteins. Virology *286*, 263–275.

Meertens, L., Bertaux, C., and Dragic, T. (2006). Hepatitis C virus entry requires a critical postinternalization step and delivery to early endosomes via clathrin-coated vesicles. J. Virol. *80*, 11571–11578.

Meertens, L., Bertaux, C., Cukierman, L., Cormier, E., Lavillette, D., Cosset, L., and Dragic, T. (2008). The tight junction proteins claudin-1, -6, and -9 are entry cofactors for hepatitis C virus. J. Virol. *82*, 3555–3560.

Meunier, J.C., Engle, R.E., Faulk, K., Zhao, M., Bartosch, B., Alter, H., Emerson, S.U., Cosset, F.L., Purcell, H., and Bukh, J. (2005). Evidence for cross-genotype neutralization of hepatitis C virus pseudo-particles and enhancement of infectivity by apolipoprotein C1. Proc. Natl. Acad. Sci. U.S.A. *102*, 4560–4565.

Meylan, E., Curran, J., Hofmann, K., Moradpour, D., Binder, M., Bartenschlager, R., and Tschopp, J. (2005). Cardif is an adaptor protein in the RIG-I antiviral pathway and is targeted by hepatitis C virus. Nature *437*, 1167–1172.

Michalak, J.P., Wychowski, C., Choukhi, A., Meunier, J.C., Ung, S., Rice, M., and Dubuisson, J. (1997). Characterization of truncated forms of the hepatitis C virus glycoproteins. J. Gen. Virol. *78*, 2299–2306.

Miyamoto, H., Okamoto, H., Sato, K., Tanaka, T., and Mishiro, S. (1992). Extraordinarily low density of hepatitis C virus estimated by sucrose density centrifugation and polymerase chain reaction. J. Gen. Virol. *73*, 715–718.

Miyanari, Y., Hijikata, M., Yamaji, M., Hosaka, M., Takahashi, H., and Shimotohno, K. (2003). Hepatitis C virus non-structural proteins in the probable membranous compartment function in viral genome replication. J. Biol. Chem. *278*, 50301–50308.

Miyanari, Y., Atsuzawa, K., Usuda, N., Watashi, K., Hishiki, T., Zayas, M., Bartenschlager, R., Wakita, T., Hijikata, M., and Shimotohno, K. (2007). The lipid droplet is an important organelle for hepatitis C virus production. Nat. Cell Biol. *9*, 1089–1097.

Mizuno, M., Yamada, G., Tanaka, T., Shimotohno, K., Takatani, M., and Tsuji, T. (1995). Virion-like structures in Hela G cells transfected with the full-length sequence of the hepatitis C virus genome. Gastroenterology *109*, 1933–1940.

Molina, S., Castet, V., Fournier-Wirth, C., Pichard-Garcia, L., Avner, R., Harats, D., Roitelman, J., Barbaras, R., Graber, P., Ghersa, P., Smolarsky, M., Funaro, A., Malavasi, F., Larrey, D., Coste, J., Fabre, J.M., Sa-Cunha, A., and Maurel, P. (2007). The low-density lipoprotein receptor plays a role in the infection of primary human hepatocytes by hepatitis C virus. J. Hepatol. *46*, 411–419.

Monazahian, M., Bohme, I., Bonk, S., Koch, A., Scholz, C., Grethe, S., and Thomssen, R. (1999). Low density lipoprotein receptor as a candidate receptor for hepatitis C virus. J. Med. Virol. *57*, 223–229.

Mondelli, U., and Silini, E. (1999). Clinical significance of hepatitis C genotypes. J. Hepatol. *31*, 65–70.

Moradpour, D., Englert, C., Wakita, T., and Wands, J.R. (1996). Characterization of cell lines allowing tightly regulated expression of hepatitis C virus core protein. Virology *222*, 51–63.

Moradpour, D., Brass, V., Bieck, E., Friebe, P., Gosert, R., Blum, H.E., Bartenschlager, R., Penin, F., and Lohmann, V. (2004). Membrane association of the RNA-dependent RNA polymerase is essential for hepatitis C virus RNA replication. J. Virol. *78*, 13278–13284.

Morikawa, K., Zhao, Z., Date, T., Miyamoto, M., Murayama, A., Akazawa, D., Tanabe, J., Sone, S., and Wakita, T. (2007). The roles of CD81 and glycosaminoglycans in the adsorption and uptake of infectious HCV particles. J. Med. Virol. *79*, 714–723.

Moriya, K., Fujie, H., Shintani, Y., Yotsuyanagi, H., Tsutsumi, T., Ishibashi, K., Matsuura, Y., Kimura, S., Miyamura, T., and Koike, K. (1998). The core protein of hepatitis C virus induces hepatocellular carcinoma in transgenic mice. Nat. Med. *4*, 1065–1067.

Mukhopadhyay, S., Kim, B.S., Chipman, P.R., Rossmann, G., and Kuhn, R.J. (2003). Structure of West Nile virus. Science *302*, 248.

Neddermann, P., Quintavalle, M., Di Pietro, C., Clementi, A., Cerretani, M., Altamura, S., Bartholomew, L., De Francesco, R. (2004). Reduction of hepatitis C virus NS5A hyperphosphorylation by selective inhibition of cellular kinases activates viral RNA replication in cell culture. J. Virol. *78*, 13306–13314.

Nolandt, O., Kern, V., Muller, H., Pfaff, E., Theilmann, L., Welker, R., and Krausslich, H.G. (1997). Analysis of hepatitis C virus core protein interaction domains. J. Gen. Virol. *78*, 1331–1340.

O'Farrell, D., Trowbridge, R., Rowlands, D., and Jager, J. (2003). Substrate complexes of hepatitis C virus RNA polymerase (HC-J4): structural evidence for nucleotide import and de-novo initiation. J. Mol. Biol. *326*, 1025–1035.

Okamoto, K., Moriishi, K., Miyamura, T., and Matsuura, Y. (2004). Intramembrane proteolysis and endoplasmic reticulum retention of hepatitis C virus core protein. J. Virol. *78*, 6370–6380.

Okuda, M., Li, K., Beard, M.R., Showalter, L.A., Scholle, F., Lemon, M., and Weinman, S.A. (2002). Mitochondrial injury, oxidative stress, and antioxidant gene expression are induced by hepatitis C virus core protein. Gastroenterology *122*, 366–375.

Op De Beeck, A., Montserret, R., Duvet, S., Cocquerel, L., Cacan, R., Barberot, B., Le Maire, M., Penin, F., and Dubuisson, J. (2000). The transmembrane domains of hepatitis C virus envelope glycoproteins E1 and E2 play a major role in heterodimerization. J. Biol. Chem. *275*, 31428–31437.

Op De Beeck, A., Voisset, C., Bartosch, B., Ciczora, Y., Cocquerel, L., Keck, Z., Foung, S., Cosset, L., and Dubuisson, J. (2004). Characterization of functional hepatitis C virus envelope glycoproteins. J. Virol. *78*, 2994–3002.

Otto, G.A., Lukavsky, P.J., Lancaster, A.M., Sarnow, P., and Puglisi, J.D. (2002). Ribosomal proteins mediate the hepatitis C virus IRES-HeLa 40S interaction. RNA *8*, 913–923.

Otto, G.A., and Puglisi, J.D. (2004). The pathway of HCV IRES-mediated translation initiation. Cell *119*, 369–380.

Pang, P.S., Jankowsky, E., Planet, J., and Pyle, A.M. (2002). The hepatitis C viral NS3 protein is a processive DNA helicase with cofactor enhanced RNA unwinding. EMBO J. *21*, 1168–1176.

Park, C.Y., Jun, H.J., Wakita, T., Cheong, H., and Hwang, S.B. (2009). Hepatitis C virus nonstructural 4B protein modulates sterol regulatory element-binding protein signaling via the AKT pathway. J. Biol. Chem. *284*, 9237–9246.

Parsons, J., Castaldi, M.P., Dutta, S., Dibrov, S.M., Wyles, L., and Hermann, T. (2009). Conformational inhibition of the hepatitis C virus internal ribosome entry site RNA. Nat. Chem. Biol *5*, 823–825.

Pavlovic, D., Neville, D.C., Argaud, O., Blumberg, B., Dwek, R.A., Fischer, B., and Zitzmann, N. (2003). The hepatitis C virus p7 protein forms an ion channel that is inhibited by long-alkyl-chain iminosugar derivatives. Proc. Natl. Acad. Sci. U.S.A. *100*, 6104–6108.

Penin, F., Brass, V., Appel, N., Ramboarina, S., Montserret, R., Ficheux, D., Blum, H.E., Bartenschlager, R., and Moradpour, D. (2004a). Structure and function of the membrane anchor domain of hepatitis C virus nonstructural protein 5A. J. Biol. Chem. *279*, 40835–43.

Penin, F., Dubuisson, J., Rey, F.A., Moradpour, D., and Pawlotsky, J.M. (2004b). Structural biology of hepatitis C virus. Hepatology *39*, 5–19.

Perelson, A.S., Herrmann, E., Micol, F., and Zeuzem, S. (2005). New kinetic models for the hepatitis C virus. Hepatology *42*, 749–754.

Petracca, R., Falugi, F., Galli, G., Norais, N., Rosa, D., Campagnoli, S., Burgio, V., Di Stasio, E., Giardina, B., Houghton, M., Abrignani, S., and Grandi, G. (2000). Structure–function analysis of hepatitis C virus envelope-CD81 binding. J. Virol. *74*, 4824–4830.

Pfeifer U., T.R., . Legler K., Bottcher, U., Gerlich, W., Weinmann, E., and Klinge, O. (1980). Experimental non-A, non-B hepatitis: four types of cytoplasmic alteration in hepatocytes of infected chimpanzees. Virchows Arch. B Cell Pathol. Incl. Mol. Pathol. *33*, 233–243.

Piccininni S., V.A., Nardelli, M., Dave, B., Raney, K.D., McCarthy J.E. (2002). Modulation of the hepatitis C virus RNA-dependent RNA polymerase activity by the non-structural (NS) 3 helicase and the NS4B membrane protein. J. Biol. Chem. *277*, 45670–45679.

Pietschmann, T., Lohmann, V., Kaul, A., Krieger, N., Rinck, G., Rutter, G., Strand, D., and Bartenschlager, R. (2002). Persistent and transient replication of full-length hepatitis C virus genomes in cell culture. J. Virol. *76*, 4008–4021.

Pietschmann, T., Kaul, A., Koutsoudakis, G., Shavinskaya, A., Kallis, S., Steinmann, E., Abid, K., Negro, F., Dreux, M., Cosset, L., and Bartenschlager, R. (2006). Construction and characterization of infectious intragenotypic and intergenotypic hepatitis C virus chimeras. Proc. Natl. Acad. Sci. U.S.A. *103*, 7408–7413.

Pileri, P., Uematsu, Y., Compagnoli, S., Galli, G., Falugi, F., Petracca, R., Weiner, A.J., Houghton, M., Rosa, D., Grandi, G., and Abrignani, S. (1998). Binding of hepatitis C virus to CD81. Science *282*, 938–941.

Pisarev, A.V., Shirokikh, E., and Hellen, C.U. (2005). Translation initiation by factor-independent binding of eukaryotic ribosomes to internal ribosomal entry sites. C. R. Biol. *328*, 589–605.

Ploss, A., Evans, M.J., Gaysinskaya, V.A., Panis, M., You, H., de Jong, P., and Rice, C.M. (2009). Human occludin is a hepatitis C virus entry factor required for infection of mouse cells. Nature *457*, 882–886.

Pohlmann, S., Zhang, J., Baribaud, F., Chen, Z., Leslie, G.J., Lin, G., Granelli-Piperno, A., Doms, R.W., Rice, M., and McKeating, J.A. (2003). Hepatitis C Virus Glycoproteins Interact with DC-SIGN and DC-SIGNR. J. Virol. *77*, 4070–4080.

Premkumar, A., Wilson, L., Ewart, D., and Gage, P.W. (2004). Cation-selective ion channels formed by p7 of hepatitis C virus are blocked by hexamethylene amiloride. FEBS Lett. *557*, 99–103.

Pudi, R., Abhiman, S., Srinivasan, N., and Das, S. (2003). Hepatitis C virus internal ribosome entry site-mediated translation is stimulated by specific interaction of independent regions of human La autoantigen. J. Biol. Chem. *278*, 12231–12240.

Qin, W., Luo, H., Nomura, T., Hayashi, N., Yamashita, T., and Murakami, S. (2002). Oligomeric interaction of hepatitis C virus NS5B is critical for catalytic activity of RNA-dependent RNA polymerase. J. Biol. Chem. *277*, 2132–2137.

Quinkert, D., Bartenschlager, R., and Lohmann, V. (2005). Quantitative analysis of the hepatitis C virus replication complex. J. Virol. *79*, 13594–135605.

Quintavalle, M., Sambucini, S., Di Pietro, C., De Francesco, R., and Neddermann, P. (2006). The alpha isoform of protein kinase CKI is responsible for hepatitis C virus NS5A hyperphosphorylation. J. Virol. *80*, 11305–11312.

Quintavalle, M., Sambucini, S., Summa, V., Orsatti, L., Talamo, F., De Francesco, R., and Neddermann, P. (2007). Hepatitis C virus NS5A is a direct substrate of casein kinase I-alpha, a cellular kinase identified by inhibitor affinity chromatography using specific NS5A hyperphosphorylation inhibitors. J. Biol. Chem. *282*, 5536–5544.

Randall, G., Panis, M., Cooper, J.D., Tellinghuisen, T.L., Sukhodolets, K.E., Pfeffer, S., Landthaler, M., Landgraf, P., Kan, S., Lindenbach, B.D., *et al.* (2007). Cellular cofactors affecting hepatitis C virus infection and replication. Proc. Natl. Acad. Sci. U.S.A. *104*, 12884–12889.

Ranjith-Kumar, C.T., Gutshall, L., Sarisky, T., and Kao, C.C. (2003). Multiple interactions within the hepatitis C virus RNA polymerase repress primer-dependent RNA synthesis. J. Mol. Biol. *330*, 675–685.

Ranjith-Kumar, C.T., Sarisky, R.T., Gutshall, L., Thomson, M., and Kao, C.C. (2004). De novo initiation pocket mutations have multiple effects on hepatitis C virus RNA-dependent RNA polymerase activities. J. Virol. *78*, 12207–12217.

Reed, K.E., Grakoui, A., and Rice, C.M. (1995). The hepatitis C virus NS2–3 autoproteinase: cleavage site mutagenesis and requirements for bimolecular cleavage. J. Virol. *69*, 4127–4136.

Reed, K.E., Xu, J., and Rice, C.M. (1997). Phosphorylation of the hepatitis C virus NS5A protein *in vitro* and *in vivo*: properties of the NS5A-associated kinase. J. Virol. *71*, 7187–7197.

Rey, F.A., Heinz, F.X., Mandl, C., Kunz, C., and Harrison, S.C. (1995). The envelope glycoprotein from tick-borne encephalitis virus at 2 A resolution. Nature *375*, 291–298.

Rijnbrand, R., Thiviyanathan, V., Kaluarachchi, K., Lemon, M., and Gorenstein, D.G. (2004). Mutational and structural analysis of stem-loop IIIC of the hepatitis C virus and GB virus B internal ribosome entry sites. J. Mol. Biol. *343*, 805–817.

Rijnbrand, R., Yang, Y., Beales, L., Bodola, F., Goettge, K., Cohen, L., Lanford, R.E., Lemon, M., and Martin, A. (2005). A chimeric GB virus B with 5′ nontranslated RNA sequence from hepatitis C virus causes hepatitis in tamarins. Hepatology *41*, 986–994.

Rodrigueza, W.V., Thuahnai, S.T., Temel, R.E., Lund-Katz, S., Phillips, C., and Williams, D.L. (1999). Mechanism of scavenger receptor class B type I-mediated selective uptake of cholesteryl esters from high density lipoprotein to adrenal cells. J. Biol. Chem. *274*, 20344–20350.

Roingeard, P., Hourioux, C., Blanchard, E., and Prensier, G. (2008). Hepatitis C virus budding at lipid droplet-associated ER membrane visualized by 3D electron microscopy. Histochem. Cell Biol. *130*, 561–566.

Roohvand, F., Maillard, P., Lavergne, J.P., Boulant, S., Walic, M., Andreo, U., Goueslain, L., Helle, F., Mallet, A., McLauchlan, J., and Budkowska, A. (2009). Initiation of hepatitis C virus infection requires the dynamic microtubule network: role of the viral nucleocapsid protein. J. Biol. Chem. *284*, 13778–13791.

Saint, N., Montserret, R., Chipot, C., and Penin, F. (2009). Structural and functional analysis of the HCV p7 protein. Methods Mol. Biol. *510*, 125–143.

Sakai, A., Claire, M.S., Faulk, K., Govindarajan, S., Emerson, S.U., Purcell, H., and Bukh, J. (2003). The p7 polypeptide of hepatitis C virus is critical for infectivity and contains functionally important genotype-specific sequences. Proc. Natl. Acad. Sci. U.S.A. *100*, 11646–11651.

Santolini, E., Migliaccio, G., La Monica, N. (1994). Biosynthesis and biochemical properties of the hepatitis C virus core protein. J. Virol. *68*, 3631–3641.

Santolini, E., Pacini, L., Fipaldini, C., Migliaccio, G., and Monica, N. (1995). The NS2 protein of hepatitis C virus is a transmembrane polypeptide. J. Virol. *69*, 7461–7471.

Sapay, N., Montserret, R., Chipot, C., Brass, V., Moradpour, D., Deleage, G., and Penin, F. (2006). NMR structure and molecular dynamics of the in-plane membrane anchor of nonstructural protein 5A from bovine viral diarrhea virus. Biochemistry *45*, 2221–2233.

Scarselli, E., Ansuini, H., Cerino, R., Roccasecca, R.M., Acali, S., Filocamo, G., Traboni, C., Nicosia, A., Cortese, R., and Vitelli, A. (2002). The human scavenger receptor class B type I is a novel candidate receptor for the hepatitis C virus. EMBO J. *21*, 5017–5025.

Schaller, T., Appel, N., Koutsoudakis, G., Kallis, S., Lohmann, V., Pietschmann, T., and Bartenschlager, R. (2007). Analysis of hepatitis C virus superinfection exclusion by using novel fluorochrome gene-tagged viral genomes. J. Virol. *81*, 4591–4603.

Schmidt-Mende, J., Bieck, E., Hugle, T., Penin, F., Rice, C.M., Blum, E., and Moradpour, D. (2001). Determinants for membrane association of the hepatitis C virus RNA-dependent RNA polymerase. J. Biol. Chem. *276*, 44052–44063.

Selby, M.J., Choo, Q.-L., Berger, K., Kuo, G., Glazer, E., Eckart, M., Lee, C., Chien, D., Kuo, C., and Houghton, M. (1993). Expression, identification and subcellular localization of the proteins encoded by the hepatitis C viral genome. J. Gen. Virol. *74*, 1103–1113.

Serebrov, V., and Pyle, A.M. (2004). Periodic cycles of RNA unwinding and pausing by hepatitis C virus NS3 helicase. Nature *430*, 476–480.

Shi, S.T., Polyak, S.J., Tu, H., Taylor, D.R., Gretch, R., and Lai, M.M. (2002). Hepatitis C virus NS5A colocalizes with the core protein on lipid droplets and interacts with apolipoproteins. Virology *292*, 198–210.

Shirota, Y., Luo, H., Qin, W., Kaneko, S., Yamashita, T., Kobayashi, K., and Murakami, S. (2002). Hepatitis C virus (HCV) NS5A binds RNA-dependent RNA polymerase (RdRP) NS5B and modulates RNA-dependent RNA polymerase activity. J. Biol. Chem. *277*, 11149–11155.

Silver, D.L., Wang, N., Xiao, X., and Tall, A.R. (2001). High density lipoprotein (HDL) particle uptake mediated by scavenger receptor class B type 1 results in selective sorting of HDL cholesterol from protein and polarized cholesterol secretion. J. Biol. Chem. *276*, 25287–93.

Simmonds, P. (2004). Genetic diversity and evolution of hepatitis C virus – 15 years on. J. Gen. Virol. *85*, 3173–3188.

Simmonds, P., Bukh, J., Combet, C., Deleage, G., Enomoto, N., Feinstone, S., Halfon, P., Inchauspe, G., Kuiken, C., Maertens, G., Mizokami, M., Murphy, D.G., Okamoto, H., Pawlotsky, J.M., Penin, F., Sablon, E., Shin, I.T., Stuyver, L.J., Thiel, H.J., Viazov, S., Weiner, J., and Widell, A. (2005). Consensus proposals for a unified system of nomenclature of hepatitis C virus genotypes. Hepatology *42*, 962–973.

Siridechadilok, B., Fraser, C.S., Hall, R.J., Doudna, A., and Nogales, E. (2005). Structural roles for human translation factor eIF3 in initiation of protein synthesis. Science *310*, 1513–1515.

Song, Y., Friebe, P., Tzima, E., Junemann, C., Bartenschlager, R., and Niepmann, M. (2006). The hepatitis C virus RNA 3′-untranslated region strongly enhances translation directed by the internal ribosome entry site. J. Virol. *80*, 11579–11588.

Spahn, C.M.T., Kieft, J.S., Grassucci, R.A., Penczek, P.A., Zhou, K., Doudna, A., and Frank, J. (2001). Hepatitis C virus IRES RNA-induced changes in the conformation of the 40S ribosomal subunit. Science *291*, 1959–1962.

Spangberg, K., and Schwartz, S. (1999). Poly(C)-binding protein interacts with the hepatitis C virus 5′ untranslated region. J. Gen. Virol. *80*, 1371–1376.

Spangberg, K., Wiklund, L., and Schwartz, S. (2000). HuR, a protein implicated in oncogene and growth factor mRNA decay, binds to the 3′ ends of hepatitis C virus RNA of both polarities. Virology *274*, 378–390.

Spangberg, K., Wiklund, L., and Schwartz, S. (2001). Binding of the La autoantigen to the hepatitis C virus 3′ untranslated region protects the RNA from rapid degradation in vitro. J. Gen. Virol. *82*, 113–120.

Steinmann, E., Penin, F., Kallis, S., Patel, A.H., Bartenschlager, R., and Pietschmann, T. (2007a). Hepatitis C Virus p7 Protein Is Crucial for Assembly and Release of Infectious Virions. PLoS Pathog. *3*, e103.

Steinmann, E., Whitfield, T., Kallis, S., Dwek, R.A., Zitzmann, N., Pietschmann, T., and Bartenschlager, R. (2007b). Antiviral effects of amantadine and iminosugar derivatives against hepatitis C virus. Hepatology *46*, 330–338.

Stempniak, M., Hostomska, Z., Nodes, R., and Hostomsky, Z. (1997). The NS3 proteinase domain of hepatitis C virus is a zinc-containing enzyme. J. Virol. *71*, 2881–2886.

Stone, M., Jia, S., Do Heo, W., Meyer, T., and Konan, K.V. (2007). Participation of Rab5, an early endosome protein, in hepatitis C virus RNA replication machinery. J. Virol. *81*, 4551–4563.

Su, A.I., Pezacki, J.P., Wodicka, L., Brideau, A.D., Supekova, L., Thimme, R., Wieland, S., Bukh, J., Purcell, R.H., Schultz, G., and Chisari, F.V. (2002). Genomic analysis of the host response to hepatitis C virus infection. Proc. Natl. Acad. Sci. U.S.A. *99*, 15669–15674.

Sumpter, R., Jr., Loo, Y.M., Foy, E., Li, K., Yoneyama, M., Fujita, T., Lemon, S.M., Gale, M., Jr. (2005). Regulating intracellular antiviral defense and permissiveness to hepatitis C virus RNA replication through a cellular RNA helicase, RIG-I. J. Virol. *79*, 2689–2699.

Suzuki, R., Matsuura, Y., Susuki, T., Ando, A., Chiba, J., Harada, S., Saito, I., and Miyamura, T. (1995). Nuclear localization of the truncated hepatitis C virus core protein with its hydrophobic C terminus deleted. J. Gen. Virol. *76*, 53–61.

Suzuki, R., Sakamoto, S., Tsutsumi, T., Rikimaru, A., Tanaka, K., Shimoike, T., Moriishi, K., Iwasaki, T., Mizumoto, K., Matsuura, Y., Miyamura, T., and Suzuki, T. (2005). Molecular determinants for subcellular localization of hepatitis C virus core protein. J. Virol. *79*, 1271–1281.

Tackett, A.J., Chen, Y., Cameron, E., and Raney, K.D. (2005). Multiple full-length NS3 molecules are required for optimal unwinding of oligonucleotide DNA in vitro. J. Biol. Chem. *280*, 10797–10806.

Tanaka, T., Kato, N., Cho, M.-J., and Shimotohno, K. (1995). A novel sequence found at the 3′ terminus of hepatitis C virus genome. Biochem. Biophys. Res. Commum. *215*, 744–749.

Tanaka, T., Kato, N., Cho, M.-J., Sugiyama, K., and Shimotohno, K. (1996). Structure of the 3′ terminus of the hepatitis C virus genome. J. Virol. *70*, 3307–3312.

Tanaka, Y., Shimoike, T., Ishii, K., Suzuki, R., Suzuki, T., Ushijima, H., Matsuura, Y., and Miyamura, T. (2000). Selective binding of hepatitis C virus core protein to synthetic oligonucleotides corresponding to the 5′ untranslated region of the viral genome. Virology *270*, 229–236.

Tanji, Y., Hijikata, M., Satoh, S., Kaneko, T., and Shimotohno, K. (1995a). Hepatitis C virus-encoded nonstructural protein NS4A has versatile functions in viral protein processing. J. Virol. *69*, 1575–1581.

Tanji, Y., Kaneko, T., Satoh, S., and Shimotohno, K. (1995b). Phosphorylation of hepatitis C virus-encoded nonstructural protein NS5A. J. Virol. *69*, 3980–3986.

Targett-Adams, P., Boulant, S., and McLauchlan, J. (2008). Visualization of double-stranded RNA in cells supporting hepatitis C virus RNA replication. J. Virol. *82*, 2182–2195.

Tellinghuisen, T.L., Marcotrigiano, J., Gorbalenya, E., and Rice, C.M. (2004). The NS5A protein of hepatitis C virus is a zinc metalloprotein. J. Biol. Chem. *279*, 48576–48587.

Tellinghuisen, T.L., Marcotrigiano, J., and Rice, C.M. (2005). Structure of the zinc-binding domain of an essential component of the hepatitis C virus replicase. Nature *435*, 374–379.

Tellinghuisen, T.L., Foss, K.L., Treadaway, C., and Rice, C.M. (2008a). Identification of residues required for RNA replication in domains II and III of the hepatitis C virus NS5A protein. J. Virol. *82*, 1073–1083.

Tellinghuisen, T.L., Foss, L., and Treadaway, J. (2008b). Regulation of hepatitis C virion production via phosphorylation of the NS5A protein. PLoS Pathog. *4*, e1000032.

Thiel, H.-J., Plagemann, W., and Moennig, V. (1996). Pestiviruses. In Fields Virology, 3rd edn, Fields BN, Knipe DM, Howley PM, eds (Raven Press, New York), vol. 1, pp. 1059–1073.

Tischendorf, J.J., Beger, C., Korf, M., Manns, P., and Kruger, M. (2004). Polypyrimidine tract-binding protein (PTB) inhibits Hepatitis C virus internal ribosome entry site (HCV IRES)-mediated translation, but does not affect HCV replication. Arch. Virol. *149*, 1955–1970.

Troesch, M., Jalbert, E., Canobio, S., Boulassel, M.R., Routy, J.P., Bernard, N.F., Bruneau, J., Lapointe, N., Boucher, M., and Soudeyns, H. (2005). Characterization of humoral and cell-mediated immune responses directed against hepatitis C virus F protein in subjects co-infected with hepatitis C virus and HIV-1. Aids *19*, 775–784.

Trotard, M., Lepere-Douard, C., Regeard, M., Piquet-Pellorce, C., Lavillette, D., Cosset, F.L., Gripon, P., Le Seyec, J. (2009). Kinases required in hepatitis C virus entry and replication highlighted by small interference RNA screening. FASEB J, *23*, 3780–3789.

Tscherne, D.M., Jones, C.T., Evans, M.J., Lindenbach, B.D., McKeating, A., and Rice, C.M. (2006). Time- and temperature-dependent activation of hepatitis C virus for low-pH-triggered entry. J. Virol. *80*, 1734–1741.

Tscherne, D.M., Evans, M.J., von Hahn, T., Jones, C.T., Stamataki, Z., McKeating, J.A., Lindenbach, D., and Rice, C.M. (2007). Superinfection exclusion in cells infected with hepatitis C virus. J. Virol. *81*, 3693–3703.

Tsuchihara, K., Tanaka, T., Hijikata, M., Kuge, S., Toyoda, H., Nomoto, A., Yamamoto, N., and Shimotohno, K. (1997). Specific interaction of polypyrimidine tract-binding protein with the extreme 3′-terminal structure of the hepatitis C virus genome, the 3′X. J. Virol. *71*, 6720–6726.

Tu, H., Gao, L., Shi, S.T., Taylor, D.R., Yang, T., Mircheff, A.K., Wen, Y., Gorbalenya, A.E., Hwang, B., and Lai, M.M.C. (1999). Hepatitis C virus RNA polymerase and NS5A complex with a SNARE-like protein. Virology *263*, 30–41.

Tuplin, A., Wood, J., Evans, D.J., Patel, H., and Simmonds, P. (2002). Thermodynamic and phylogenetic prediction of RNA secondary structures in the coding region of hepatitis C virus. RNA *8*, 824–841.

Tuplin, A., Evans, J., and Simmonds, P. (2004). Detailed mapping of RNA secondary structures in core and NS5B-encoding region sequences of hepatitis C virus by RNase cleavage and novel bioinformatic prediction methods. J. Gen. Virol. *85*, 3037–3047.

Vaillancourt, F.H., Pilote, L., Cartier, M., Lippens, J., Liuzzi, M., Bethell, R.C., Cordingley, G., and Kukolj, G. (2009). Identification of a lipid kinase as a host factor involved in hepatitis C virus RNA replication. Virology *387*, 5–10.

Van Itallie, M., and Anderson, J.M. (2006). Claudins and epithelial paracellular transport. Annu. Rev. Physiol. *68*, 403–429.

Varaklioti, A., Vassilaki, N., Georgopoulou, U., and Mavromara, P. (2002). Alternate translation occurs within the core coding region of the hepatitis C viral genome. J. Biol. Chem. *277*, 17713–17721.

Vassilaki, N., and Mavromara, P. (2003). Two alternative translation mechanisms are responsible for the expression of the HCV ARFP/F/core+1 coding open reading frame. J. Biol. Chem. *278*, 40503–40513.

Voisset, C., and Dubuisson, J. (2004). Functional hepatitis C virus envelope glycoproteins. Biol. Cell *96*, 413–420.

Voisset, C., Callens, N., Blanchard, E., Op De Beeck, A., Dubuisson, J., and Vu-Dac, N. (2005). High density lipoproteins facilitate hepatitis C virus entry through the scavenger receptor class B type I. J. Biol. Chem. *280*, 7793–7799.

von Hahn, T., Lindenbach, B.D., Boullier, A., Quehenberger, O., Paulson, M., Rice, M., and McKeating, J.A. (2006). Oxidized low-density lipoprotein inhibits hepatitis C virus cell entry in human hepatoma cells. Hepatology *43*, 932–942.

Wakita, T., Pietschmann, T., Kato, T., Date, T., Miyamoto, M., Zhao, Z., Murthy, K., Habermann, A., Krausslich, H.G., Mizokami, M., Bartenschlager, R., and Liang, T.J. (2005). Production of infectious hepatitis C virus in tissue culture from a cloned viral genome. Nat. Med. *11*, 791–796.

Walewski, J.L., Keller, T.R., Stump, D., and Branch, A.D. (2001). Evidence for a new hepatitis C virus antigen encoded in an overlapping reading frame. RNA *7*, 710–721.

Wang, Q.M., Hockman, M.A., Staschke, K., Johnson, R.B., Case, K.A., Lu, J., Parsons, S., Zhang, F., Rathnachalam, R., Kirkegaard, K., and Colacino, J.M. (2002a). Oligomerization and cooperative RNA synthesis activity of hepatitis C virus RNA-dependent RNA polymerase. J. Virol. *76*, 3865–3872.

Wang, Q.M., Hockman, M.A., Staschke, K., Johnson, R.B., Case, K.A., Lu, J., Parsons, S., Zhang, F., Rathnachalam, R., Kirkegaard, K., and Colacino, J.M. (2002b). Oligomerization and cooperative RNA synthesis activity of hepatitis C virus RNA-dependent RNA polymerase. J. Virol. *76*, 3865–3872.

Watashi, K., Hijikata, M., Hosaka, M., Yamaji, M., and Shimotohno, K. (2003). Cyclosporin A suppresses replication of hepatitis C virus genome in cultured hepatocytes. Hepatology *38*, 1282–1288.

Watashi, K., Ishii, N., Hijikata, M., Inoue, D., Murata, T., Miyanari, Y., and Shimotohno, K. (2005). Cyclophilin B is a functional regulator of hepatitis C virus RNA polymerase. Mol. Cell *19*, 111–122.

Wölk, B., Sansonno, D., Kräusslich, H.G., Dammacco, F., Rice, C.M., Blum, E., and Moradpour, D. (2000). Subcellular localization, stability, and trans-cleavage competence of the hepatitis C virus NS3–NS4A complex expressed in tetracycline-regulated cell lines. J. Virol. *74*, 2293–2304.

Xu, Z., Choi, J., Yen, T.S., Lu, W., Strohecker, A., Govindarajan, S., Chien, D., Selby, J., and Ou, J. (2001). Synthesis of a novel hepatitis C virus protein by ribosomal frameshift. EMBO J. *20*, 3840–3848.

Xu, Z., Choi, J., Lu, W., and Ou, J.H. (2003). Hepatitis C virus f protein is a short-lived protein associated with the endoplasmic reticulum. J. Virol. *77*, 1578–1583.

Yagnik, A.T., Lahm, A., Meola, A., Roccasecca, R.M., Ercole, B.B., Nicosia, A., and Tramontano, A. (2000). A model for the hepatitis C virus envelope glycoprotein E2. Proteins *40*, 355–366.

Yamaga A.K., O.J.H. (2002). Membrane topology of the hepatitis C virus NS2 protein. J. Biol. Chem. *277*, 33228–33234.

Yamashita, T., Kaneko, S., Shirota, Y., Qin, W., Nomura, T., Kobayashi, K., and Murakami, S. (1998). RNA-dependent RNA polymerase activity of the soluble recombinant hepatitis C virus NS5B protein truncated at the C-terminal region. J. Biol. Chem. *273*, 15479–15486.

Yan, B.S., Tam, H., and Syu, W.J. (1998). Self-association of the C-terminal domain of the hepatitis-C virus core protein. Eur. J. Biochem. *258*, 100–106.

Yan, Y., Li, Y., Munshi, S., Sardana, V., Cole, J.L., Sardana, M., Steinkuehler, C., Tomei, L., De Francesco, R., Kuo, C., and Chen, Z. (1998). Complex of NS3 protease and NS4A peptide of BK strain hepatitis C virus: a 2.2 A resolution structure in a hexagonal crystal form. Protein Sci. *7*, 837–847.

Yanagi, M., Purcell, R.H., Emerson, U., and Bukh, J. (1997). Transcripts from a single full-length cDNA clone of hepatitis C virus are infectious when directly transfected into the liver of a chimpanzee. Proc. Natl. Acad. Sci. U.S.A. *94*, 8738–8743.

Yanagi, M., St Claire, M., Emerson, S.U., Purcell, H., and Bukh, J. (1999). In vivo analysis of the 3′ untranslated region of the hepatitis C virus after in vitro mutagenesis of an infectious cDNA clone. Proc. Natl. Acad. Sci. U.S.A. *96*, 2291–2295.

Yang, G., Pevear, D.C., Collett, M.S., Chunduru, S., Young, D.C., Benetatos, C., and Jordan, R. (2004). Newly synthesized hepatitis C virus replicon RNA is protected from nuclease activity by a protease-sensitive factor(s). J. Virol. *78*, 10202–10205.

Yang, W., Qiu, C., Biswas, N., Jin, J., Watkins, S.C., Montelaro, R.C., Coyne, B., and Wang, T. (2008). Correlation of the tight junction-like distribution of Claudin-1 to the cellular tropism of hepatitis C virus. J. Biol. Chem. *283*, 8643–8653.

Yao, N., Hesson, T., Cable, M., Hong, Z., Kwong, A.D., Le, V., and Weber, P.C. (1997). Structure of the hepatitis C virus RNA helicase domain. Nat. Struct. Biol. *4*, 463–467.

Yao, N., Reichert, P., Taremi, S.S., Prosise, W., and Weber, P.C. (1999). Molecular views of viral polyprotein processing revealed by the crystal structure of the hepatitis C virus bifunctional protease-helicase. Structure Fold. Des. *7*, 1353–1363.

Yasui, K., Wakita, T., Tsukiyama-Kohara, K., Funahashi, S.I., Ichikawa, M., Kajita, T., Moradpour, D., Wands, R., and Kohara, M. (1998). The native form and maturation process of hepatitis C virus core protein. J. Virol. *72*, 6048–6055.

Ye, J., Wang, C., Sumpter, R., Jr., Brown, M.S., Goldstein, J.L., Gale, M., Jr. (2003). Disruption of hepatitis C virus RNA replication through inhibition of host protein geranylgeranylation. Proc. Natl. Acad. Sci. U.S.A. *100*, 15865–15870.

Yi, M. and Lemon, S.M. (2003a). 3′ nontranslated RNA signals required for replication of hepatitis C virus RNA. J. Virol. *77*, 3557–3568.

Yi, M., and Lemon, S.M. (2003b). Structure–function analysis of the 3′ stem-loop of hepatitis C virus genomic RNA and its role in viral RNA replication. RNA *9*, 331–345.

Yi, M., and Lemon, S.M. (2004). Adaptive mutations producing efficient replication of genotype 1a hepatitis C virus RNA in normal Huh7 cells. J. Virol. *78*, 7904–7915.

Yi, M., Nakamoto, Y., Kaneko, S., Yamashita, T., and Murakami, S. (1997). Delineation of regions important for heteromeric association of hepatitis C virus E1 and E2. Virology *231*, 119–129.

Yi, M., Villanueva, R.A., Thomas, D.L., Wakita, T., and Lemon, S.M. (2006). Production of infectious genotype 1a hepatitis C virus (Hutchinson strain) in cultured human hepatoma cells. Proc. Natl. Acad. Sci. U.S.A. *103*, 2310–2315.

You, S., Stump, D.D., Branch, D., and Rice, C.M. (2004). A cis-acting replication element in the sequence encoding the NS5B RNA-dependent RNA polymerase is required for hepatitis C virus RNA replication. J. Virol. *78*, 1352–1366.

Yu, G.Y., Lee, K.J., Gao, L., and Lai, M.M. (2006). Palmitoylation and polymerization of hepatitis C virus NS4B protein. J. Virol. *80*, 6013–6023.

Yu, X., Qiao, M., Atanasov, I., Hu, Z., Kato, T., Liang, J., and Zhou, Z.H. (2007). Cryo-electron microscopy and three-dimensional reconstructions of hepatitis C virus particles. Virology *367*, 126–134.

Yuasa, T., Ishikawa, G., Manabe, S., Sekiguchi, S., Takeuchi, K., and Miyamura, T. (1991). The particle size of hepatitis C virus estimated by filtration through microporous regenerated cellulose fibre. J. Gen. Virol. *72*, 2021–2024.

Zeisel, M.B., Barth, H., Schuster, C., and Baumert, T.F. (2009). Hepatitis C virus entry: molecular mechanisms and targets for antiviral therapy. Front. Biosci. *14*, 3274–3285.

Zeuzem, S., Feinman, S.V., Rasenack, J., Heathcote, E.J., Lai, M.Y., Gane, E., O'Grady, J., Reichen, J., Diago, M., Lin, A., Hoffman, J., and Brunda, M.J. (2000). Peginterferon alfa-2a in patients with chronic hepatitis C. N. Engl. J. Med. *343*, 1666–1672.

Zhang, J., Yamada, O., Yoshida, H., Iwai, T., and Araki, H. (2002). Autogenous translational inhibition of core protein: implication for switch from translation to RNA replication in hepatitis C virus. Virology *293*, 141–150.

Zhang, J., Randall, G., Higginbottom, A., Monk, P., Rice, M., and McKeating, J.A. (2004). CD81 is required for hepatitis C virus glycoprotein-mediated viral infection. J. Virol. *78*, 1448–1455.

Zheng, A., Yuan, F., Li, Y., Zhu, F., Hou, P., Li, J., Song, X., Ding, M., and Deng, H. (2007). Claudin-6 and claudin-9 function as additional coreceptors for hepatitis C virus. J. Virol. *81*, 12465–12471.

Zhong, J., Gastaminza, P., Cheng, G., Kapadia, S., Kato, T., Burton, D.R., Wieland, S.F., Uprichard, S.L., Wakita, T., and Chisari, F.V. (2005). Robust hepatitis C virus infection in vitro. Proc. Natl. Acad. Sci. U.S.A. *102*, 9294–9299.

Hepatitis C Virus-related Assays

2

Zhuhui Huang, Zhaohui Cai and Michael G. Murray

Abstract

The use of modern molecular virology techniques was successfuly applied to the identification of the etiologic agent of non-A, non-B hepatitis, i.e. hepatitis C virus (HCV), and played an integral part in the initial discovery and development of potential anti-HCV therapies. The development of biochemical assays for HCV-encoded protease, polymerase, and helicase activities, combined with the elucidation of the enzymes' three dimensional structures, has facilitated and guided the chemistry of rational approaches to antiviral drug development and allowed for the initiation of antiviral drug screening campaigns. In addition, cell-based assays developed to study co-translational proteolytic processing of the viral polyprotein and virus–host interactions have been utilized in the development of potential therapeutics for HCV intervention. An important milestone was achieved with the development of HCV molecular clones that were fully infectious in cell culture. These clones produced viruses *in vitro* that were shown to be infectious in chimpanzees and in immunodeficient mice models which carry chimeric human livers. Viruses derived from these HCV-infected animals were shown to be infectious in cell culture assays, fulfilling Koch's postulate and allowing classical virological techniques to be applied to the study of HCV antiviral targets. This led to further development of HCV-related assays employed in drug discovery and development. In this chapter, biochemical, cell- and virus-based assays, as well as surrogate viral and animal model systems useful to the anti-HCV drug discovery process, are highlighted.

Background

The importance of choosing the appropriate assay for discovery research was clearly illustrated in the identification of the etiologic agent of non-A, non-B hepatitis (NANBH). The discovery of HCV was achieved through the use of modern molecular virology techniques, as opposed to long-standing and well-proven classical virology techniques. HCV was discovered in 1989 through the use of an immunoscreening approach applied to a recombinant cDNA library (Choo *et al.* 1989). The library was derived from RNA extracted from infectious chimpanzee plasma; the antibodies used for the immunoscreening were from a patient diagnosed with NANBH. Proof that the cDNA clone isolated using this screening technique encoded the etiologic agent of NANBH was provided by the demonstration that the clone was not derived from the host chimpanzee genome. Thus, molecular cloning combined with a blind immunoscreening assay was successful in identifying the HCV genome. Based on the isolated molecular clones, the virus was determined to be a positive-strand RNA virus containing a genome of about 9.6 kb with significant sequence similar to members of the Flaviviridae family of viruses (Choo *et al.* 1991). The molecular cloning of the HCV genome allowed for the production of recombinant proteins that led to sensitive blood screening and diagnostic assays for HCV infection. Diagnostic kits were quickly produced to screen blood donors in a successful campaign that further reduced transfusion-associated hepatitis C from approximately 4.1% to nearly zero percent (Alter and Houghton 2000).

Despite these advances and the rapid molecular characterization of the virus that followed, many attempts to passage the disease in animal model systems or to propagate the virus in cell culture systems have failed, which in turn has hampered progress in anti-HCV therapy development. At this time, HCV infection represents a major area of unmet medical need. It is estimated that over 3% of the world population (4–5 million people in the United States alone) are infected with HCV. Acute infection with HCV is usually asymptomatic, but the majority of the infected individuals develop a persistent infection that results in permanent scarring of the liver (cirrhosis) and eventually hepatocellular carcinoma. Currently in the United States, HCV infection is the leading cause of liver transplantation; however, liver transplantation does not cure the infection. Even after successful surgery, recurrent infection of the transplanted liver occurs in almost all instances after transplantation, frequently leading to damage of the new liver and occasionally a more severe condition for the patient. Patients have limited treatment options; pegylated type 1 interferon-alpha 2a or -alpha 2b, either alone or in combination with the guanosine analogue, ribavirin, is the current standard of care. While improvements were made recently, the interferon-based therapy contains essentially the same components as it did when originally introduced in the late 1990s. The therapeutic regime remains expensive, poorly tolerated, and partially effective. Furthermore, resistance to this treatment and to treatment with antiviral drugs directed at the viral protease occurs quickly and is frequently observed in patients infected with the common viral genotypes. While there are experimental therapeutics under clinical investigation, including a protease inhibitor in phase III clinical studies, there is no vaccine to prevent HCV infection.

Development of HCV-related assays began with the molecular characterization of the virus and the selection of virally encoded targets for inhibitors. The genomic RNA of HCV consists of a long open reading frame (ORF) flanked by non-translated regions at both the 5′ and 3′ ends of the genome (Bukh *et al.* 1992; Kolykhalov *et al.* 1996). The ORF encodes a large viral polyprotein, which is co- or post-translationally processed by cellular and virally encoded proteases to yield at least 10 different structural (C, E1, E2, and possibly p7) and non-structural (NS) proteins (NS2, NS3, NS4A, NS4B, NS5A and NS5B; Reed *et al.* 2000; Luo 2004). The NS3 to NS5B proteins are sufficient for HCV RNA replication (Lohmann *et al.* 1999; Blight *et al.* 2000), which occurs in a membrane bound replication complex. This replication complex consists of viral RNA and non-structural proteins as well as cellular proteins (El-Hage *et al.* 2003; Moradpour *et al.* 2003; Evans *et al.* 2004; Wang *et al.* 2005). By the early 1990's, HCV presented a target-rich environment for assay development. The putative HCV protease was rapidly cloned, expressed and purified for development of assays incorporating purified reagents, including synthetic peptide substrates. Subsequently, truncated versions of the HCV RNA dependent RNA polymerase (RdRp) were cloned and expressed as soluble forms for development of functional assays and for three dimensional structure determinations.

Biochemical assays for the viral enzymes' activity were initially developed to explore protein function and their role in virus replication. Molecular cloning of the viral enzymes, followed by protein expression, purification, and crystallization, led to a vast amount of information on the structure of these proteins and eventually to the elucidation of the three dimensional structure. With the three dimensional structure information available, biochemical assays were subsequently used to study the functionality of the enzymes and to guide the design of targeted inhibitors for HCV intervention. These innovative, structure-based approaches were applied successfully for the development of inhibitors targeting HCV NS3/4A protease (see Chapters 9–11) and NS5B RNA RdRp (see Chapters 15 and 16), several of which are in clinical evaluation.

The application of molecular techniques led to successful biochemical assay development, used for antiviral discovery and the study of HCV structure and replication. However there was a lack of an *in vitro* virus infection system considered necessary to evaluate antiviral compounds in a cellular environment. In 1999, a major breakthrough occurred when Lohmann *et al.* established the HCV replicon system. In this system, a sub-genomic piece of the HCV RNA is capable of autonomous replication in a liver cell line *in vitro*. With this breakthrough, it became possible to measure HCV RNA replication in a cellular environment. Use of the replicon system allowed characterization of the HCV replication complex and the identification of host cell factors involved in virus replication. The HCV replicon was further developed into more robust assay systems including reporter gene readouts that were amenable to high throughput screenings. Today, the HCV replicon system is an important tool for the evaluation of HCV inhibitors and antiviral resistance studies.

An important finding occurred in 2003 when Kato and colleagues isolated an HCV clone from a Japanese patient with fulminant hepatitis C, which they used to establish an *in vitro* HCV infection based system. This clone, designated JFH-1 for Japanese fulminant hepatitis 1, was found to replicate in liver cell lines without cell culture adaptive mutations. Subsequently, it was reported that when the JFH-1 genome was transfected into Huh-7 cells, infectious virus was produced. This virus, designated HCVcc for cell-culture derived HCV, allowed for the establishment of a powerful tool to study the entire HCV life-cycle in the laboratory. The efficiency of the system was improved by creating hybrid clones with a different virus clone designated J6, and the virus produced *in vitro* was found to be infectious in chimpanzees. Furthermore, the development of Alb-uPA/SCID mice transplanted with human liver cells permissive to HCV infection provided a small animal model system for unravelling virus–host interactions.

The molecular cloning of the HCV genome, the elucidation of the three-dimensional structures of the HCV proteins, and the development of enzymatic assays, cell-based assays and a small animal model of HCV replication have allowed for the discovery and evaluation of several HCV inhibitors which are now in clinical trials. While not meant to be comprehensive, this chapter provides a summary of biochemical and cell-based assays, and animal models that are used in the discovery of anti-HCV drugs.

HCV-related assays

A vast array of assays that have contributed to the current understanding of HCV biology and HCV drug discovery, including conventional *in vitro* biochemical assays and cell-based assays, are reviewed in the sections that follow. When choosing an assay for drug screening, several aspects need to be considered. Investigators must consider whether there is a requirement for the assay to include host factors, membranes, or the unknown factors present in the cellular environment in order to balance the relevance of the natural environment versus the purified protein and buffer systems that provide a controlled reaction in a biochemical system. In general, the biochemical assay examines a single biochemical or enzymatic reaction. Cell-based assays are preferred for targets expressed on the cell membranes, such as the HCV replication complex, or for screening against several targets in the replication pathway. However, the complexity of cell-based assays can be a disadvantage in that the mechanism of action of the identified inhibitor may be unknown at first, and subsequent biochemical assays are needed to further characterize the inhibitor. Each approach has its advantages and disadvantages. Well-defined biochemical assays serve as excellent and powerful tools to study the formation of protein complexes and specific interactions between individual components of these complexes. These assays allow for the direct measurement of kinetic parameters of a given complex and evaluation of compounds that inhibit the complex formation or enzyme activity. Biochemical assays are often the best choice to characterize HCV receptor binding or to evaluate HCV protease, helicase and polymerase activities. Moreover, because of reagent production and stability considerations, biochemical assays are generally more amenable to high-throughput screening campaigns for discovery of potential lead compounds and for guidance in the choice of constructs for crystallographic studies.

While biochemical methods offer several advantages the question remains as to whether the outcome of the *in vitro* assay can be translated to a cellular setting. As such, there is increasing use of cell-based assays to elucidate virus–host interactions at multiple levels that cannot be predicted using biochemical assays. Moreover, in the cell-based format, anti-HCV compounds must be able to penetrate the cell membrane. Perhaps the greatest advantage of using a cell-based screen is the capability of generating cytotoxicity data in parallel with efficacy data, which is helpful to medicinal chemistry and to the elucidation of more complete structure activity relationships of compounds in development. Thus, cell-based assays are very powerful at revealing new information about HCV replication and the interaction of the virus with cellular factors. While certain target classes are better suited to a certain assay format, biochemical and cell-based assays are often used to complement each other.

Biochemical assays

The HCV NS3 and NS5B proteins are two viral enzymes that have been well characterized and are amenable to evaluation using biochemical assays. HCV NS5B is an RNA-dependent RNA polymerase (RdRp). HCV NS3 is a multifunctional protein with three different enzymatic activities. The N-terminal one-third of NS3 serves as a serine protease that associates with NS4A to proteolytically process the viral polyprotein at the junctions between NS3 and NS4A, NS4A and NS4B, NS4B and NS5A, and NS5A and NS5B (De Francesco *et al.* 2000). The C-terminal two-thirds of the protein possess helicase/nucleoside triphosphatase (NTPase) activities (Kim *et al.* 1995). The enzymatic activities of both NS3 and NS5B are essential for HCV replication, making both of these proteins ideal targets for antiviral drug discovery. With the availability of protein structure and function information, biochemical assays for NS3 and NS5B were developed to support the evaluation and screening of small molecules as HCV inhibitors. As a result of these efforts, several inhibitors targeting the NS3 serine protease and NS5B RdRp are currently in clinical trials.

The expression of recombinant protein with functional enzymatic activity is a prerequisite for the development of *in vitro* biochemical assays. Several protein expression systems are available for this purpose, including cell-based systems using *E. coli*, yeast, insect and mammalian cells, as well as cell-free systems for *in vitro* transcription and translation. While many systems are available, *E. coli* is the most widely used host for the production of recombinant HCV proteins. Using an *E. coli* system, HCV NS3 or NS5B are cloned into an expression vector with an isopropyl-β-D-thiogalactopyranoside (IPTG) promoter. *E. coli* transformed with the NS3 or NS5B expression vectors are then cultured in normal conditions and protein expression is induced by treatment with IPTG. The recombinant proteins produced are then purified using fast protein liquid chromatography (FPLC) and commercially available columns (e.g. HiTrap chelating column; Amersham). Using these methods, HCV NS3 and NS5B proteins can be expressed and purified to greater than 95% purity.

Prior to producing recombinant proteins for use in biochemical assays, it is important to consider the final form of the proteins that are desired for use in the assays. For example, assays evaluating NS3 protease activity can use either full-length NS3 protein or a truncated form of the protein made up of the protease domain (aa 1–181) only. In contrast, it has been reported that the protease domain of NS3 increases the helicase activity of the protein (Zhang *et al.* 2005). Therefore, when evaluating NS3 helicase activity in a biochemical assay, the use of full-length NS3 protein instead of a truncated form of the protein (i.e. the helicase domain only) will result in better enzyme activity. On the other hand, the expression of truncated NS5B protein lacking the C-terminal 21 amino acids leads to increased protein solubility without affecting enzyme activity is another example of the importance of the desired final form of the protein.

NS3 protease assay

Biochemical assays for NS3 protease activity are based on the NS3-dependent proteolytic cleavage of peptide substrates derived from the NS4A/NS4B, NS4B/NS5A or NS5A/NS5B cleavage sites in the HCV polyprotein. NS4A, a cofactor for NS3 proteolytic activity, is generally included in the protease assay for optimal performance. Notably, it has been reported that the central region of NS4A (residues 21–34; GSVVIVGRIILSGR) is sufficient for its cofactor activity (Lin *et al.* 1995; Shimidzu *et al.* 1996).

Several approaches are used to develop NS3 protease assays. An early approach was based on the use of HPLC to analyze processed peptide substrates following cleavage by NS3. However, this method was not suitable for high-throughput screening (HTS) and is no longer used. One alternative to the HPLC-based assay is an ELISA assay. Using this approach, an N-terminally digoxigenin-labelled peptide substrate is immobilized on the ELISA plate. The amount of digoxigenin label remaining following cleavage by NS3 is subsequently analyzed using an immunological reaction with anti-digoxigenin–alkaline phosphatase conjugated antibodies, allowing for colorimetric detection and quantification of the digoxigenin label using standard ELISA reagents (Takeshita *et al.* 1997). Although HTS can be performed using this approach, the assay is time-consuming, thereby limiting its utility.

Fluorogenic (Taliani *et al.* 1996) and chromogenic (Zhang *et al.* 1999) assays are the most widely used assay formats for NS3 protease activity. Both assays can be performed in 96-well, 384-well, or higher density microplates. Fluorogenic assays are based on the principle of fluorescence resonance energy transfer (FRET). The peptide substrate contains a fluorescent dye as well as a quencher that absorbs the fluorescent signal when in proximity of the dye. Cleavage by NS3 releases the quencher from the substrate, resulting in recovery of the fluorescent signal, which is analyzed using a fluorometric plate reader. Similarly, chromogenic assays use a peptide substrate derived from the P site of the NS5A–NS5B junction sequence [Ac-DTEDVVX(L-norvaline), where X = A or P] that has the C-terminal carboxyl groups esterified with one of four different chromophoric alcohols, allowing for substrate detection and quantification using a spectrophotometric microtitre plate reader (Zhang *et al.* 1999). The SensoLyte®520 HCV protease Assay Kit (AnaSpec Inc., San Jose, CA) is one example of a fluorogenic NS3 protease assay that is commercially available.

As an alternative to traditional biochemical protease assays, a cell-based molecular target assay for HCV NS3 was developed (Pacini *et al.* 2004; Ludmerer *et al.* 2008). This assay does not require purification of recombinant NS3 protein, so it is suitable to compare NS3 protease activity from different samples, such as HCV clinical samples and NS3 site-directed mutants. In this system, NS3/4A is cloned into an expression vector. The NS3/4A cDNA clone is then co-transfected into cells with a reporter construct that encodes a modified form of secreted alkaline phosphatase (SEAP). The SEAP reporter is secreted into the culture medium only upon proteolytic release from an ER-localizing transmembrane domain that is linked to the SEAP through an intervening NS3/4A cleavage site. As SEAP secretion depends on the NS3 protease, quantification of the alkaline phosphatase activity in the culture medium provides an indirect measure of NS3 proteolytic activity.

NS3 helicase assay

The NS3 helicase unwinds dsRNA, dsDNA and RNA–DNA hybrids that have a 3′ single-stranded sequence of more than three nucleotides. This helicase activity is mediated through the disruption of the hydrogen bonds that keep the two strands of nucleic acid together, and is associated with the hydrolysis of nucleoside triphosphates (NTPs) by NS3. The helicase activity of NS3 is necessary for viral replication. Therefore, the NS3 helicase domain is an important target for the discovery and development of anti-HCV agents. Different approaches are used to assess HCV helicase activity. The HCV helicase is able to unwind partially duplex RNA-RNA, RNA-DNA and DNA-DNA without exhibiting substrate specificity *in vitro*. A dsDNA substrate is usually chosen for use in biochemical helicase assays because it is less likely to degrade than dsRNA or RNA-DNA duplexes, and the HCV NS3 enzyme has a higher affinity for dsDNA. The HCV helicase requires a 3′-single-stranded overhang to unwind. The partially double-stranded substrate is designed to be significantly larger than the unwound product–capture oligonucleotide duplex so that it can be easily separated by gel electrophoresis. A typical helicase assay reaction includes NS3 helicase, a partial double stranded nucleic acid substrate, a capture oligonucleotide, nucleoside triphosphate and a divalent cation, such as Mg^{2+} or Mn^{2+}. The shorter strand of the double stranded substrate is 5′-end labelled with [^{32}P]. When the NS3 helicase separates the two strands of the substrate, the [^{32}P] 5′-end labelled product is captured by the capture oligonucleotide in order to prevent it from reforming the original substrate. After incubation, the products of the reaction are separated by electrophoresis through a non-denaturing polyacrylamide gel, subjected to autoradiography and quantified by phosphor imaging (Zhang *et al.* 2005). However, this assay is not suitable for testing large numbers of compounds because of the low-throughput nature of gel-based systems. Moreover, the substrate needs to be prepared frequently because of the short half-life of [^{32}P]. More time is needed to prepare the gel before loading and to dry the gel after running. As such, scintillation proximity (SPA; Kyono *et al.* 1998), ELISA-based (Hsu *et al.* 1998; Artsaenko *et al.* 2003), Flashplate®-based (Hicham *et al.* 2000) and FRET-based assays (Boguszewska-Chachulska *et al.* 2004) have been developed for higher throughput screening. All of these assays are designed to avoid the need for separation of single-stranded reaction products from the duplex substrate by electrophoresis, but there are advantages and disadvantages to the use of each of these assays. For example, the assays that use radiolabels for detection of product are sensitive; they require the tedious preparation of substrates and require handling and disposal of radioactivity. ELISA-based assays are non-radioactive, can be readily adapted to higher throughput, and are easy to perform, but they are time-consuming. FRET-based assays can be completed very rapidly, but can lead to the identification of false-positive results owing to autofluorescence of test compounds.

SPA is a homogeneous technique for radioactive assays that eliminates the need for separating input substrates from reaction products prior to quantifying the product. SPA formats are adaptable to higher throughput and have been used to assay the HCV helicase and polymerase activities. In SPA, scintillant is incorporated into solid microscopic beads. If a radioactive molecule is bound to the bead, it is brought into proximity close enough to stimulate the scintillant contained within the bead to emit light. The radioactivity from unbound molecules is too distant from the bead and is therefore not detected because the energy released dissipates before reaching the bead. SPA-based NS3 helicase assays (Kyono *et al.* 1998) are high-throughput and use an approach similar to the one used for gel-based assays. In this case, the 3′-end of the single stranded product released from the substrate is ^{3}H-labeled. Once unwound and released as a result of NS3 helicase activity, it is annealed to a capture oligonucleotide that is biotin-labelled. This newly formed complex is then captured by streptavidin coated SPA beads. The proximity of the ^{3}H-labeled complex to the beads leads to stimulation of the scintillant that is incorporated into the beads. The resulting signal is then quantified using a scintillation detector. There is no need to separate unwound reaction products from the double stranded substrate by washing and the substrate half-life of [^{3}H]-labelled substrate used in the SPA-based approach is much longer than that of the [^{32}P]-labelled substrate used in the gel-based assay format. Therefore, larger batches of substrate can be prepared and used over time, which helps to reduce the variation caused by the use of different batches of substrate. The helicase [^{3}H] SPA assay system from Amersham is one example of a commercially available kit that can be used for characterizing HCV NS3 helicase inhibitors.

An approach similar to the SPA-based assay is used for the FlashPlate®-based assay. FlashPlates® are white polystyrene microplates designed for high-volume, homogeneous radiometric assays, and are based on the principle of scintillation proximity. The interior of each well is permanently coated with a thin layer of a polystyrene-based scintillant, which provides a platform for non-separation assays using a variety of isotopes without the addition of liquid scintillation cocktail. FlashPlates® are available in both 96- and 384-well formats (Perkin Elmer). In the FlashPlate®-based assay (Hicham *et al.* 2000), the plate is pre-coated with neutravidin. The partial dsDNA substrate has a 5′-biotin labelled template strand and a 5′-radiolabeled release strand. The substrate is immobilized on the plate surface via the 5′-biotin molecule on the template strand. Unwinding of the dsDNA substrate by NS3 helicase releases the radiolabeled strand, resulting in a reduction in the scintillation signal.

For the ELISA-based assay (Hsu *et al.* 1998; Artsaenko *et al.* 2003), a partially double stranded substrate with a 5′-biotin-labelled template strand and a 5′-digoxigenin (DIG)-labelled release strand is added to streptavidin-coated microtitre plates. After the enzymatic reaction, substrate that is not unwound retains the DIG-labelled release strand, which is detected by anti-DIG antibodies coupled with horseradish peroxidase or alkaline phosphatase using standard ELISA techniques. Because no radioisotope is used in the ELISA-based assay, it eliminates the regulatory, tracking, disposal, and potential contamination issues associated with radioisotopes. The substrate is very stable and can be stored and used over a long period of time.

The FRET-based helicase assay (Boguszewska-Chachulskaa *et al.* 2004) uses an approach similar to the one previously described for the fluorogenic NS3 protease assays. In this case, the double stranded helicase substrate has a cyanine dye (Cy3) incorporated on the 5′-end of the template strand and a quencher (Black Hole Quencher-2; BHQ-2) incorporated on the 3′-end of the release strand. Fluorescence emitted by the Cy3 dye is quenched as a result of FRET or contact-mediated quenching between the Cy3 donor and the BHQ-2 quencher when the molecules are only separated by 10–100Å. NS3 helicase activity separates the two strands and increases the distance between the Cy3 and BHQ-2, leading to the arrest of FRET and a detectable increase in fluorescence emission.

NS5B RdRp assay

The HCV NS5B RNA-dependent, RNA polymerase (RdRp) catalyzes the polymerization of ribonucleotides during viral RNA replication. It utilizes RNA as a template, and the polymerase activity is initiated either in a primer-dependent or independent manner (i.e. *de novo* initiation). Assessment of the enzymatic activity of NS5B is typically determined by measuring the incorporation of radiolabeled ribonucleotide triphosphates into RNA. For biochemical assays, the RNA template can be poly(rA) primed with oligo(rU) or *in vitro* transcribed viral RNA that is complementary to the HCV 5′UTR region. Gel-based RdRp assays usually use RNA templates derived from the 3′-end of either the plus- or minus-strand of the viral genome, while filter-based screening assays commonly use synthetic homopolymeric templates/primers such as poly(rA)/oligo(rU_{12}) (McKercher *et al.* 2004). However, viral RNA-based templates also are used for filter-based screening assays, particularly when the incorporation of alternative rNTPs other than radiolabeled UTP is of interest. For gel-based assays, reaction products are analyzed by polyacrylamide gel electrophoresis and quantified with a phosphor imager. Although this method allows for the direct visualization of product purity and integrity, and for the analysis of the nature and pattern of the products synthesized by the enzyme, it is not suitable for the screening of large numbers of compounds. Therefore, several alternative assay methods have been developed for high-throughput screening. For example, in a SPA format, the oligo(rU) primer is biotinylated, which allows for the binding of the radiolabeled product to streptavidin-coated SPA beads once the enzymatic reaction has been completed. To enhance the sensitivity of the assay, the beads can be easily washed to remove unincorporated [^{3}H] UTP (McKercher *et al.* 2004). Alternatively, a filter-based assay can be used. For this method, upon completion of the enzymatic reaction the sample is quenched by the addition of ice-cold 5% (v/v) trichloroacetic acid (TCA) containing 0.5 mM pyrophosphate. The quenched reaction mixtures are then incubated at –20 °C for one hour in order to precipitate the denatured polymeric RNA products, which are subsequently transferred to glass fibre (GF)-B filters and washed to remove unincorporated [^{3}H] UTP (Kaushik-Basu *et al.* 2008). For both approaches, the incorporated radioactivity is quantified using a microplate scintillation counter.

Cell-based assays

HCV replicon

The first HCV cDNA clone was reported in 1989 (Choo *et al.* 1989). Our basic understanding of the HCV life cycle was slow to develop owing to a lack of a cell culture system capable of supporting the growth of the virus *in vitro*. The establishment of the HCV genotype 1b (Con1 strain) subgenomic replicon system, which allows for efficient HCV RNA replication in the human hepatoma cell line Huh7 (Lohmann *et al.* 1999), provided for the first time a powerful tool that could be used to study the RNA replication of HCV in a cell-based system. This bicistronic replicon consists of a selectable marker, the neomycin phosphotransferase gene, as the upstream cistron, and the HCV non-structural genes NS2 or NS3 through NS5B as the, downstream cistron. Translation of the selectable marker is controlled by the HCV internal ribosome entry site (IRES), while translation of the HCV polyprotein is regulated by the IRES of encephalomyocarditis virus (EMCV). Upon transfection of the replicon into Huh7 cells and selection using G418, colonies that maintain the replicon through efficient HCV RNA replication are obtained.

Adaptive mutations and the use of highly permissive cell lines are two factors that were found to be determinants of HCV RNA replication efficiency in the replicon system (Bartenschlager *et al.* 2003). Characterization of G418-selected RNA replicons identified several cell culture adaptive mutations that emerge in the replicons during the process of G418 selection. Most of these mutations are located in the NS5A, NS3 and NS4B coding regions. Usually, more than one mutation is found within a single colony. For example, a replicon containing a combination of the single S2197P mutation in NS5A and the double E1202G and T1280I mutations in NS3 replicates more efficiently than the wild-type (wt) replicon originally derived from the Con1 strain of HCV. Similarly, the more recently developed H77 genotype 1a replicon has mutations in the NS3 helicase domain (P1496L) and in NS5A (S2204I) that lead to a synergistic increase in RNA replication (Bartenschlager *et al.* 2003). In addition, cellular factors are now known to contribute to the efficiency of RNA replication. Most notably, Huh7.5 cells, a sub-line of Huh7, were isolated and found to be highly permissive for HCV RNA replication as a result of a mutation that inactivates the retinoic acid inducible gene I (RIG-I), a protein that plays a role in the intrinsic cellular antiviral response pathway (Blight *et al.* 2002; Sumpter *et al.* 2005). Furthermore, it was reported that replication of subgenomic and full-length H77 genotype 1a replicons require the use of the highly permissive Huh7.5 cell line in addition to cell culture adaptive mutations (Blight *et al.* 2003). Different passages of the same Huh7 cell culture have shown as high as 130-fold differences in their ability to support replicon replication (Lohmann *et al.* 2003). Although the cellular mechanism(s) leading to the differences in replication capacity between the passages is not known, at a minimum, these results indicate that host cell factors are key components that affect HCV RNA replication.

Interestingly, the genotype 2a JFH1 subgenomic replicon replicates efficiently in Huh7 cells without the requirement for cell culture adaptive mutations (Kato *et al.* 2001, 2003). The replication of this replicon is about 200- to 400-fold higher than that exhibited by wt genotype 1b replicons or even more recent improved versions of the genotype 1b replicon containing the poliovirus IRES (Targett-Adams *et al.* 2005). In addition, the genotype 2a JFH1 replicon can replicate in other cell lines including HepG2 and IMY-N9 (Date *et al.* 2004), HuH6 (Windisch *et al.* 2005), 293 (Ali *et al.* 2004; Kato *et al.* 2005), and HeLa (Kato *et al.* 2005). Replication is also found in mouse embryonic fibroblasts (MEF; Chang *et al.* 2006), albeit with lower efficiency. Based on the variety of cell lines in which the JFH1 replicon replicates, studying the differences in cellular factors and HCV RNA replication in these cell lines compared to Huh7 cells and primary hepatocytes should lead to a better understanding of the cellular factors involved in HCV RNA replication. Overall, the generation of a variety of different replicon assay systems, including replicons with reporters or markers for replication such as luciferase (Lohmann *et al.* 2003), beta-lactamase (Murray *et al.* 2003), secreted alkaline phosphatase (Bourne *et al.* 2005), and green fluorescent protein (Nelson *et al.* 2006; Dansako *et al.* 2008), and different HCV genotypes 1a (Blight *et al.* 2000; 2002), 1b (Lohmann *et al.* 1999) and 2a (Kato *et al.* 2003) represents a major advancement in HCV research and antiviral drug discovery and development.

Surrogate model systems

Because the replicon system does not represent the whole virus life cycle, and does not produce infectious virions it cannot be used to study viral attachment and entry into cells, the viral uncoating, or virion maturation and release from the cells. As a result, replicon systems cannot be used for the identification and characterization of antiviral agents targeting these early and late stages in the virus life cycle. Thus, surrogate models of HCV replication are required to study such events.

Bovine viral diarrhoea virus (BVDV) is a member of the *Pestivirus* genus of the Flaviviridae family and has a genetic structure similar to that of HCV. BVDV causes chronic infection, although the mechanisms of establishment and maintenance of viral persistence in the infected host are different from HCV. Owing to the similarities in the replication mechanisms and comparable genome structure and function of gene products with HCV, as well as the ability of BVDV replication to generate infectious virus *in vitro*, BVDV is used as a surrogate model for antiviral drug discovery (Buckwold *et al.* 2003). Another virus used as a surrogate model in the study of HCV is yellow fever virus (YFV). However, YFV is less commonly used because of its genomic structure. YFV is more distantly related to HCV than BVDV, and YFV employs a cap-dependent mechanism to drive the translation of its polyprotein instead of the IRES-mediated mechanism used by HCV.

GBV-B is another suitable surrogate model to study HCV replication and to evaluate antiviral agents. GBV-B is a hepatotrophic virus that can undergo a complete life cycle in primary hepatocytes obtained from non-human primates. Moreover, phylogenetic analysis indicates GBV-B is the virus most closely related to HCV since it has the highest level of genomic sequence identity to HCV compared to other viruses. Like HCV, GBV-B undergoes IRES-mediated initiation of polyprotein translation, and it has a similar replication cycle and genetic organization to HCV. In addition, both GBV-B and HCV use homologous NS3 proteins for polyprotein processing and homologous NS5B proteins for RNA replication (Beames *et al.* 2001; Buckwold *et al.* 2003; Akari *et al.* 2009). Furthermore, the GBV-B NS3 protease functionally cleaves the HCV polyprotein at the junctions of NS4A/NS4B, NS4B/NS5A and NS5A/NS5B, while the HCV NS3 protease is also able to cleave the GBV-B polyprotein (Beames *et al.* 2001). These findings suggest that protease inhibitors active against GBV-B NS3 may have activity against HCV NS3. Therefore, GBV-B is a valuable surrogate model for HCV and can be used to screen compounds targeting all stages in the virus life cycle. Establishment of chimeric GBV-B/HCV viruses allows for a more direct approach to screening anti-HCV inhibitors. However, when considering the use of GBV-B as a surrogate model for HCV, the researcher should also be aware of the differences between the viruses. GBV-B is infectious in tamarins, but not in chimpanzees, and it only causes an acute infection in the animals. In contrast, HCV is infectious in chimpanzees, but not in tamarins, and it causes both acute and chronic infections (Beames *et al.* 2001; Akari *et al.* 2009). These observations suggest there may be significant differences in the replication cycles of the two viruses, as well as in the immune systems of the animals, which might affect the potential use of GBV-B as a surrogate model of HCV (Akari *et al.* 2009).

HCV pseudotyped particle (HCVpp) system

The establishment of retroviral pseudotyped particles harbouring HCV envelope glycoproteins (HCVpp) was an important breakthrough in the study of HCV entry. HCVpp are produced by cotransfection of 293 cells with an HCV E1E2 expression vector and an envelope-deficient human immunodeficiency virus (HIV) or murine leukemia virus (MLV) genome engineered to encode a reporter gene (e.g. luciferase). These pseudotypes are infectious to human liver cells; their entry into cells is pH-dependent; their infectivity can be neutralized by anti-HCV E2 and anti-CD81 antibodies, indicating the entry of HCVpp resembles the cell entry properties of authentic HCV virions (Hsu *et al.* 2003; Zhong *et al.* 2006).

The HCVpp system can be used to study the mechanism of virus entry and identify cellular receptors mediating virus attachment and fusion (Bartosch *et al.* 2003a; Hsu *et al.* 2003; Codran *et al.* 2006; Keck *et al.* 2008). In addition, it can be used to evaluate neutralizing antibodies from HCV infected patients and chimpanzees (Bartosch *et al.* 2003b; Logvinoff *et al.* 2004) and to develop antiviral drugs targeting HCV entry. For example, this system was used to demonstrate that HCV entry is blocked by a recombinant soluble form of CD81, by a monoclonal antibody specific for CD81, and by knockdown of CD81 expression through siRNA. In addition, the ectopic expression of CD81 confers permissiveness to infection with HCVpp in normally non-permissive HepG2 cells that lack expression of CD81. These data further confirmed that CD81 is a component of the receptor complex used by HCV for attachment and entry into cells (Hsu *et al.* 2003; Bartosch *et al.* 2006). Similarly, using the HCVpp system in conjunction with the HCV infectious virus system, Evans *et al.* (2007) demonstrated that claudin-1 is an HCV co-receptor for virus entry.

Another advantage of the HCVpp system is that pseudotypes can be made using envelope glycoproteins from diverse HCV isolates. As a result, it can be used to study virus tropism and compare differences in the process of virus entry between the various HCV genotypes (McKeating *et al.* 2004; Owsianka *et al.* 2005; Cukierman *et al.* 2009). However, the HCVpp system has limitations. Unlike authentic HCV, which has HCV RNA replication and virion assembly dependent on cholesterol metabolism in the infected host cells (Ye *et al.* 2003; Kapadia *et al.* 2005; Chang *et al.* 2007; Huang *et al.* 2007; Ye, 2007; Jiang *et al.* 2009), HCVpp are not produced in human liver cells, and therefore they are not associated with lipoproteins. In addition, HCVpp are replication-defective and do not produce progeny virus from infected Huh7 cells, thus re-infection is not possible. Recently, a replication-competent recombinant vesicular stomatitis virus (VSV) pseudotyped with unmodified E1 and E2 glycoproteins from HCV genotypes 1a and 1b was established. This newer system addresses the limitation of the previous HCVpp systems that are replication-defective, and provides an alternative approach to study HCV entry and find novel inhibitors targeting the viral entry process (Tani *et al.* 2007).

HCV infectious virus (HCVcc) system

While the HCV subgenomic replicon system has contributed significantly to the study of HCV RNA replication and screening of potential antiviral compounds, it is not sufficient for the study of the complete life cycle of HCV. Multiple potential targets for therapeutic intervention, including the early steps of virus attachment and entry, and the late steps of virion assembly and release, cannot be evaluated in the replicon system. Similarly, the HCVpp system can only be used to study virus attachment and entry, while surrogate models are only relevant as close approximations of HCV replication, which may have many differences compared to authentic HCV replication. To address these limitations, several groups (Cai *et al.* 2005; Lindenbach *et al.* 2005; Wakita *et al.* 2005; Zhong *et al.* 2005) have developed an infectious, replication competent cell culture system (HCVcc) of the HCV genotype 2a strain JFH1 that can reliably produce infectious virus *in vitro* without the requirement for adaptive mutations. The development of the HCVcc system represents a breakthrough that makes the study of all aspects of the HCV life cycle possible in a cell culture setting, which in turn ultimately expands the number of potential antiviral drug targets that can be evaluated *in vitro*.

The JFH1 strain of HCV was originally isolated from a Japanese patient with fulminant hepatitis. The JFH1 genome was cloned and upon transfection into Huh7 cells was found to replicate efficiently and produced infectious virus.

Infectivity of the JFH1 clone was neutralized by human CD81 antibody, as well as by immunoglobulins from individuals chronically infected with HCV (Cai *et al.* 2005; Wakita *et al.* 2005; Zhong *et al.* 2005), demonstrating that virus produced by the cell culture system is authentic. Virus produced *in vitro* in this system was infectious in chimpanzees and in uPA-SCID mice transplanted with human liver grafts (Wakita *et al.* 2005; Lindenbach *et al.* 2006). Parental JFH1 produced moderate levels of infectious virus from cultured Huh7 cells, however, further development of the system resulted in an intragenotypic chimeric clone, FL-J6/JFH1, that showed much higher infectivity. This chimeric clone contained the structural genes core through p7 and the non-structural gene NS2 from HCV J6, another genotype 2a strain, combined with the non-structural genes NS3 through NS5B from JFH1 (Lindenbach *et al.* 2005). Additional advancements in the HCVcc system led to even higher virus titres (more than 100-fold) when alternative chimeric clones were created by altering the fusion junction between JFH1 and J6 from NS2/NS3 to a site located directly after the first transmembrane domain of NS2 (Koutsoudakis *et al.* 2006; Pietschmann *et al.* 2006) or to the junctions between E2/p7 or p7/NS2 (Mateu *et al.* 2008). Subsequently, another important achievement was made by Yi *et al.* (2006) who demonstrated that the genotype 1a strain H77-S full-length replicon containing five adaptive mutations located within the NS3, NS4A and NS5A genes produced infectious virus, although with approximately 400-fold lower virus infectivity than JFH1.

Even though HCVcc systems for the production of infectious virus were developed for some strains, full-length, replication competent HCVcc clones for the most predominant strains found in the clinic, genotypes 1a and 1b, are either still not available (genotype 1b) or have very low infectivity (genotype 1a). As a result, the generation of additional chimeric viruses provides an alternative way to study genotype 1a and 1b viruses. Intergenotypic chimeras created by fusing genotype 1a, 1b, or 3a structural genes with the JFH1 non-structural genes via a site directly after the first transmembrane domain of NS2 have been established. With the exception of the GT3a/JFH1 chimera, these chimeric clones efficiently produce infectious virus, albeit with lower virus yields than the intragenotypic 2a chimera J6CF/JFH1 (Pietschmann *et al.* 2006). These results suggest that viral structural genes may control virus assembly and release. In addition, other observations indicate that virus infectivity is determined by other factors. For example, continuous passage of transfected cells carrying replicating JFH1 RNA results in an increase in virus titre (Zhong *et al.* 2005). Cell culture adaptive mutations that accelerate virus replication kinetics occur in the virus (e.g. G451R mutation in E2) under these conditions of persistent replication. Similarly, host cell factors can affect the permissibility of the cells to HCV infection and replication (Zhong *et al.* 2006); Huh7 cells show delayed virus production kinetics compared to the highly permissive Huh7.5 subclone. Furthermore, Huh7.5.1 cells, which are derived from Huh7.5 cells, produce ~50-fold higher titres of virus than naïve Huh-7 cells (Bartenschlager *et al.* 2005).

An interesting observation related to the HCVcc system is that cell culture adaptive mutations that enhance HCV replicon RNA replication may actually inhibit virus production (Bartenschlager *et al.*, 2003). In support of this observation, intrahepatic inoculation of chimpanzees with the parental HCV Con-1 genomic RNA caused a persistent infection, but similar inoculations using genomic RNA containing two cell culture adaptive mutations associated with enhanced HCV replicon RNA replication (Bukh *et al.* 2002) did not lead to persistent infection. Also, HCV JFH1 replicates efficiently without adaptive mutations, and it can produce infectious virus in cells, as well as in chimpanzees (Wakita *et al.* 2005). However, this observation is not supported by results obtained with the full-length genotype 1a H77 replicon (H77-S), which has five adaptive mutations and still produces infectious virus, although with low titre (Yi *et al.* 2006). Whether this low titre is caused by the adaptive mutations or whether the mutations are required for virus infectivity deserves further investigation.

The infectious HCVcc system provides a platform for the study of the early and late steps of the HCV life cycle. This valuable tool enables the identification of viral and cellular factors required for HCV binding, entry, assembly, and release, which are all potential targets for antiviral drug discovery. For example, a single-cycle replication assay in CD81 deficient cells that dissects the mechanism of virus entry and release has been described (Russell *et al.* 2008). The role of a scavenger receptor B type I (SR-BI) as an HCV entry receptor was confirmed using this single-cycle system. In addition, it was demonstrated that both CD81 and SR-BI function cooperatively to initiate HCV infection (Kapadia *et al.* 2007). Recent studies also have provided evidence that another tight junction protein, occludin (OCLN), interacts with HCV E2 to facilitate viral entry, suggesting OCLN is a key factor for HCV entry (Liu, S. *et al.* 2009). Advancements in understanding of late events in the virus life cycle also were made using the HCVcc system. It is now known that HCV virion assembly occurs in the endoplasmic reticulum (ER), while maturation occurs in the post-ER compartments (Gastaminza *et al.* 2008), and both of these processes depend on microsomal transfer protein (MTP) and apolipoprotein (Apo) E and/or ApoB (Chang *et al.* 2007; Huang *et al.* 2007; Jiang *et al.* 2009). Furthermore, ApoC1 also plays an important role during viral morphogenesis, an effect that is mediated by intracellular association of its C-terminal region with surface components of HCV virions (Meunier *et al.* 2008).

The availability of infectious HCVcc systems engineered to express convenient and assayable reporter proteins (e.g. luciferase) makes the rapid and quantitative analysis of HCV replication possible. Although the HCV RNA replication of these HCVcc reporter constructs is attenuated and the virus titres produced by these constructs are lower than the titres produced by parental virus without the reporter (Wakita *et al.* 2005; Koutsoudakis *et al.* 2006), the sensitivity and low cost of detection of such reporter systems makes them viable options for use in evaluating the activity of HCV inhibitors. In fact, a monocistronic chimeric genotype 1b/2a HCVcc luciferase reporter system has been established and could be used for high-throughput screening, and is especially useful for evaluating compounds targeting the genotype 1 structural proteins that play roles in virus attachment, entry, and assembly (Zhang, Y. *et al.* 2008). Similarly, JFH1-based infectious systems that express selectable markers instead of reporter proteins are used to continually produce infectious virus *in vitro* (Kato *et al.* 2007; Date *et al.* 2007; Dansako *et al.* 2008), which allows the use of these systems for the study of HCV tropism and the identification of host factors important for virus infection and replication.

HCV animal model systems

The natural species tropism of HCV is limited to humans. The chimpanzee is the only experimental animal model of HCV infection that is capable of reproducing most aspects of the human infection after HCV inoculation (Kolykhalov *et al.* 1997; Yanagi *et al.* 1997; Bassett *et al.* 1998; Major *et al.* 2004; Wakita *et al.* 2005; Lindenbach *et al.* 2006). The chimpanzee model was instrumental in defining the clinical course of HCV infection because samples could be collected from the infected animal from the time of first exposure to the virus, a situation that is difficult to reproduce clinically with infected patients. The model continues to be essential to the elucidation of the mechanisms of viral clearance and immune response to infection, and will continue to be useful to the development of vaccines. Unfortunately, the model is limited by restricted availability of animals, ethical concerns, and high cost. Owing to these issues, a small-animal model for HCV infection is urgently needed to support the development of antiviral therapies and to help elucidate the mechanisms underlying viral pathogenesis. However, after years of searching for a susceptible animal species, none has been identified. Therefore, researchers have increasingly been working toward developing a small animal model for HCV infection through artificial means (Ploss *et al.* 2009). Two general approaches have been envisioned. One involves adapting the virus to rodent cells *in vitro* and then using the cell-culture adapted virus to infect the animals. Alternatively, rodent cells could be humanized to accept infection with HCV. For example, Mercer and colleagues reported the transplantation of normal human hepatocytes into immunodeficient mice that carry a human plasminogen activator transgene (Alb-*uPA*), thereby generating mice with hybrid mouse/human livers. These mice developed prolonged HCV infections with high viral titres after inoculation with infected human serum (Mercer *et al.* 2001) or with the chimeric HCVcc strain FL-J6/JFH (Lindenbach *et al.*, 2006). Importantly, virus recovered from these animals was highly infectious in naïve mice (Mercer *et al.* 2001) and cell culture (Lindenbach *et al.*, 2006). The biophysical characteristics of the viral particles produced in the chimeric mouse closely resembled those of particles isolated from infected humans and chimpanzees, but differed from viral particles produced in cell culture (Lindenbach *et al.*, 2006). This model was used to demonstrate the antiviral activity of the HCV protease inhibitor BILN-2061, and an HCV polymerase inhibitor. The viral load in genotype 1b infected mice exhibited a 2.5 $\log_{10}$ reduction after a 4-day administration of BILN-2061; however, toxicity was also observed. These results were similar to results obtained in human clinical trials (Meuleman *et al.* 2005; 2008). Using a similar approach, Knetman and colleagues developed a SCID/bg mouse harbouring HCV-permissive tumours generated by a tumorigenic population of the human hepatocarcinoma-derived Huh7 cell line. These mice also exhibited active viral replication following infection. The virus replication was inhibited by interferon-α and by the protease inhibitor BILN-2061, with results similar to those seen in human clinical trials (Kneteman *et al.* 2006; Guévin *et al.* 2009). Thus, these models are critical for the evaluation of *in vivo* efficacy of antiviral compounds and are useful in providing a level of confidence in moving compounds toward clinical development. Unfortunately, because of the technical complexity and logistics involved in producing large numbers of mice and the immune-deficient nature of these chimeric mice, they are not widely used. Moreover, these animals are not suitable for evaluation of immunotherapeutic agents or vaccines owing to the lack of a functioning immune system. Other alternative approaches are therefore being developed with the goals of overcoming the HCV species barrier and creating a robust animal model that can be easily produced.

Conclusion

As a result of the establishment of various innovative biochemical, cell-based and virus-based assays, significant progress has been made in understanding HCV replication and the development of antiviral therapies. These assays are complimentary to each other, and together they have further enhanced our understanding of the HCV life cycle, virus–host interactions, and disease progression, which has greatly facilitated development of antiviral therapeutics. The recently established infectious HCVcc system represents a milestone in HCV research that will provide further insight into the process of HCV cell entry, uncoating, genome packaging, virion assembly, maturation and release from cells. However, many questions still remain. For example, while genotype 1 virus represents the most predominant strains in a clinical setting, the levels of infectious virus observed in the HCVcc system are surprisingly low. Although the development of the chimeric infectious genotype 1b/JFH1 virus is an important step towards understanding this question, it is genetically engineered and is not found in nature; therefore, it is unlikely to represent HCV replication in an infected individual. By the same token, it remains to be seen how the various adaptive mutations identified in the genotype 1b subgenomic replicon cell culture system will translate to virus replication *in vivo*. Finally, studies of small animal models of HCV replication and pathogenesis have been limited to immunodeficient mice. As a result, these are not the appropriate models for studying the interactions between HCV and the host immune system. However, it is probable that the use of systematic approaches will eventually lead to breakthroughs in discovering the genetic barrier to HCV infection, which will eventually allow for the development of a robust *in vivo* model.

Acknowledgements

The authors thank Roger Ptak, Andrew Penman and Connie Penman for critical review of the manuscript.

References

Akari, H., Iwasaki, Y., Yoshida, T., and Lijima, S. (2009). Non-human primate surrogate model of hepatitis C virus infection. Microbiol. Immunol. *53*, 53–57.

Ali, S., Pellerin, C., Lamarre, D., and Kukolj, G. (2004). Hepatitis C virus subgenomic replicons in the human embryonic kidney 293 cell line. J. Virol. *78*, 491–501.

Alter, H.J. and Houghton, M. (2000). Hepatitis C Virus and eliminating post-transfusion hepatitis. Nat. Med. *6*, 1082–1086.

Artsaenko, O., Tessmann, K., Sack, M., Häussinger, D., and Heintges, T. (2003). Abrogation of hepatitis C virus NS3 helicase enzymatic activity by recombinant human antibodies. J. Gen. Virol. *84*, 2323–2332.

Bartenschlager, R. and Pietschmann, T. (2005). Efficient hepatitis C virus cell culture system: what a difference the host cell makes. Proc. Natl. Acad. Sci. U.S.A. *102*, 9739–9740.

Bartenschlager, R., Kaul, A., and Sparacio, S. (2003). Replication of the hepatitis C virus in cell culture. Antiviral Res. *60*, 91–102.

Bartosch, B. and Cosset, F.L. (2006). Cell entry of hepatitis C virus. Virology *348*, 1–12.

Bartosch, B., Dubuisson, J., and Cosset, F.L. (2003a). Infectious hepatitis C virus pseudo-particles containing functional E1-E2 envelope protein complex. J. Exp. Med. *197*, 633–642.

Bartosch, B., Bukh, J., Meunier, J.C., Granier, C., Engle, R.E., Blackwelder, W.C., Emerson, S.U., Cosset, F.L., and Purcell, R.H. (2003b). *In vitro* assay for neutralizing antibody to hepatitis C virus: evidence for broadly concerved neutralization epitopes. Proc. Natl. Acad. Sci. U.S.A. *100*, 14199–14204.

Bassett, S.E., Rasky, K.M., and Lanford, R.E. (1998). Analysis of hepatitis C virus-inoculated chimpanzees reveals unexpected clinical profiles. J. Virol. *72*, 2589–2599.

Beames, B., Chavez, D., Lanford, R.E., (2001). GBvirus B as a model for hepatitis C virus. ILAR J. *42*, 152–160.

Blight, K.J., Kolykhalov, A.A., and Rice, C.M. (2000). Efficient initiation of HCV RNA replication in cell culture. Science. *290*, 1972–1974.

Blight, K.J., McJeating, J.A., and Rice, C.M. (2002). Highly permissive cell lines for subgenomic and genomic hepatitis C virus RNA replication. *76*, 13001–13014.

Blight, K.J., McKeating, J.A., Marcotrigiano, J., and Rice, C.M. (2003). Efficient replication of hepatitis C virus genotype 1a RNAs in cell culture. J. Virol. *77*, 3181–3190.

Boguszewska-Chachulskaa, A. M., Krawczyka, M., Stankiewicza, A., Gozdeka, A., Haennia, A., and Strokovskayaa, L. (2004). Direct fluorometric measurement of hepatitis C virus helicase activity. FEBS Lett. *567*, 253–258.

Bourne, N., Pyles, R.B., Yi, M., Veselenak, R.L., Davis, M.M., and Lemon, S.M. (2005). Screening for hepatitis C virus antiviral activity with a cell-based secreted alkaline phosphatase replicon system. Antivir. Res. *67*, 76–82.

Buckwold, V.E., Beer, B.E., and Donis, R.O. (2003). Bovine viral diarrhea virus as a surrogate model of hepatitis C virus for the evaluation of antiviral agents. Antiviral. Res. *60*, 1–15.

Bukh, J., Purcell, R.H., and Miller, R.H. (1992). Sequence analysis of the 5′ noncoding region of hepatitis C virus. Proc. Natl. Acad. Sci. U.S.A. *89*, 4942–4946.

Bukh, J., Pietschmann, T., Lohmann, V., Krieger, N., Faulk, K., Engle, R.E., Govindarajan, S., Shapiro, M., St Claire, M., and Bartenschlager, R. (2002). Mutations that permit efficient replication of hepatitis C virus RNA in Huh-7 cells prevent productive replication in chimpanzees. Proc. Natl. Acad. Sci. U.S.A. *99*, 14416–14421.

Cai, Z., Zhang, C., Chang, K.S., Jiang, J., Ahn, B.C., Wakita, T., Liang, T. J., and Luo, G. (2005). Robust production of infectious hepatitis C virus (HCV) from stably HCV cDNA-transfected human hepatoma cells. J. Virol. *79*, 13963–13973.

Chang, K.S., Cai, Z., Zhang, C., Sen, G.C., Williams, B.R., and Luo, G. (2006). Replication of hepatitis C virus (HCV) RNA in mouse embryonic fibroblast: protein kinase R (PKR) – dependent and PKR-independent mechanisms for controlling HCV RNA replication and mediating interferon activities. J. Virol. *80*, 7364–7374.

Chang, K.S., Jiang, J., Cai, Z., and Luo, G. (2007). Human apolipoprotein E is required for infectivity and production of hepatitis C virus in cell culture. J. Virol. *81*, 13783–13793.

Choo, Q.L., Kuo, G., Weiner, A.J., Overby, L.R., Bradley, D.W., and Houghton, M. (1989). Isolation of a cDNA clone derived from a blood-borne non-A, non-B viral hepatitis genome. Science *244*, 359–362.

Choo, Q.L., Richman, K.H., Han, J.H., Berger, K., Lee, C., Dong, C., Gallegos, C., Coit, D., Medina-Selby, A., Barr, P.J., *et al.* (1991). Genetic organization and diversity of the hepatitis C virus. Proc. Natl. Sci. U.S.A. *88*, 2451–2455.

Codran, A., Royer, C., Jaeck, D., Bastien-Valle, M., Baumert, T., F., Kieny, M.P., Pereira, C.A., and Martin, J.P. (2006). Entry of hepatitis C virus pseudotypes into primary human hepatocytes by clathrin-dependent endocytosis. J. Gen. Virol. *87*, 2583–2593.

Cukierman, L., Meertens, L., Bertaux, C., Kajumo, F., and Dragic, T. (2009). Residues in a highly conserved claudin-1 motif are required for hepatitis C virus entry and mediate the formation of cell–cell contacts. J. Virol. *83*, 5477–5484.

Dansako, H., Ikeda, M., Abe, K., Mori, K., Takemoto, K., Ariumi, Y., and Kato, N. (2008). A new living cell-based assay system for monitoring genome-length hepatitis C virus RNA replication. Virus Res. *137*, 72–79.

Date, T., Kato, T., Miyamoto, M. Zhao, Z., Yasui, K., Mizokami, M., and Wakita, T. (2004). Genotype 2a hepatitis C virus subgenomic replicon can replicate in HepG2 and IMY-N9 cells. J. Biol. Chem. *279*, 22371–22376.

Date, T., Miyamoto, M., Kato, T., Morikawa, K., Murayama, A., Akazawa, D., Tanabe, J., Sone, S., Mizokami, M., and Wakita, T. (2007). An infectious and selectable full-length replicon system with hepatitis C virus JFH1 strain. Hepatology Res. *37*, 433–443.

De Francesco, R. and Steinkuhler, C. (2000). Structure and function of the hepatitis C virus NS3–4A serine proteinase. Curr. Top. Microbiol. Immunol. *242*, 149–169.

El-Hage, N. and Luo, G. (2003). Replication of hepatitis C virus RNA occurs in a membrane-bound replication complex containing nonstructural viral proteins and RNA. J. Gen. Virol. *84*, 2761–2769.

Evans, M.J., Rice, C.M., and Goff, S.P. (2004). Phosphorylation of hepatitis C virus nonstructural protein 5A modulates its protein interactions and viral RNA replication. Proc. Natl. Acad. Sci. U.S.A. *101*, 13038–13043.

Evans, M.J., von Hahn, T., Tscherne, D.M., Syder, A.J., Panis, M., Wölk, B., Hatziioannou, T., McKeating, J.A., Bieniasz, P.D., and Rice, C.M. (2007). Claudin-1 is a hepatitis C virus co-receptor required for a late step in virus entry. Nature, 446, 801–805.

Gastaminza, P., Cheng, G., Wieland, S., Zhong, J., Liao, W., and Chisari, F.V. (2008). Cellular determinants of hepatitis C virus assembly, maturation, degradation, and secretion. J. Virol. *82*, 2120–2129.

Guévin, C., Lamarre, A., and Labonté, P. (2009). Novel HCV replication mouse model using human hepatocellular carcinoma xenografts. Antivir. Res. *84*, 14–32.

Hicham Alaoui-Ismaili, M., Gervais, C., Brunette, S., Gouin, G., Hamel, M., Rando, R.F., and Bedard, J. (2000). A novel high throughput screening assay for HCV NS3 helicase activity. Antivir. Res. *46*, 181–193.

Hsu, M., Zhang, J., Flint, M., Logvinoff, C., Cheng-Mayer, C., and Rice, C.M. (2003). Hepatitis C virus glycoproteins mediate pH-dependent cell entry of pseudotyped retroviral particle. Proc. Natl. Acad. Sci. U.S.A. *100*, 7271–7276.

Huang, H., Sun, F., Owen, D.M., Li, W., Chen, Y., Gale, M.Jr., and Ye, J. (2007). Hepatitis C virus production by human hepatocytes dependent on assembly and secretion of very low-density lipoproteins. Proc. Natl. Acad. Sci. U.S.A. *104*, 5848–5853.

Jiang, J. and Luo, G. (2009). Apolipoprotein E but not B is required for the formation of infectious hepatitis C virus particles. J. Virol. *83*, 12680–12691.

Kapadia, S.B. and Chisari, F.V. (2005). Hepatitis C virus RNA replication is regulated by host protein geranylgeranylation and fatty acids. Proc. Natl. Acad. Sci. U.S.A. *102*, 2561–2566.

Kapadia, S.B., Barth, H., Baumert, T., McKeating, J.A., and Chisari, F.V. (2007). Initiation of hepatitis C virus infection is dependent on cholesterol and cooperativity between CD81 and scavenger receptor B type I. J. Virol. *81*, 374–383.

Kato, T., Furusaka, A., Miyamoto, M., Date, T., Yasui, K., Hiramoto, J., Nagayama, K., Tanaka, T., and Wakita, T. (2001). Sequence analysis of hepatitis C virus isolated from a fulminant hepatitis patient. J. Med. Virol. *64*, 334–339.

Kato, T., Date, T., Miyamoto, M., Furusaka, A., Tokushige, K., Mizokami, M., and Wakita, T. (2003). Efficient replication of the genotype 2a hepatitis C virus subgenomic replicon. Gastroenterology *125*, 1808–1817.

Kato, T., Date, T., Miyamoto, M., Zhao, Z., Mizokami, M., and Wakita, T. (2005). Nonhepatic cell line HeLa and 293 support efficient replication of the hepatitis C virus genotype 2a subgenomic replicon. J. Virol. *79*, 592–596.

Kato, T., Matsumura, T., Heller, T., Saito, S., Sapp, R.K., Murthy, K., Wakita, T., and Liang, T.J. (2007). Production of infectious hepatitis C virus of various genotypes in cell culture. J. Virol. *81*, 4405–4411.

Kaushik-Basu, N., Bopda-waffo, A., Talele, T.T., Basu, A., Costa, P.R.R., Da Silva, A.J.M., Sarafianos, S.G., and Noel, F. (2008). Identification and characterization of coumestans as novel HCV NS5B polymerase inhibitors. Nucleic Acids Res. *36*, 1482–1496.

Keck, Z.Y., Li, T.K., Xia, J., Gal-Tanamy, M., Olson, O., Li, S.H., Patel, A.H., Ball, J.K., Lemon, S.M., and Foung, S.K. (2008). Definition of a conserved immunodominant domain on hepatitis C virus E2 glycoprotein by neutralizing human monoclonal antibodies. J. Virol. *82*, 6061–6066.

Kim, D.W., Gwack, Y., Han, J.H., and Choe, J. (1995). C-terminal domain of the hepatitis C virus NS3 protein contains an RNA helicase activity. Biochem. Biophys. Res. Commun. *215*, 160–166.

Kneteman, N.M., Weiner, A.J., O'Connell, J., Connell, M., Gao, T., Aukerman, L., Kovelsky, R., Ni, Z.J., Zhu, Q., Hashash, A., *et al.* (2006). Anti-HCV therapies in chimeric scid-Alb/uPA mice parallel outcomes in human clinical application. Hepatology, 43, 1346–1353.

Kolykhalov, A.A., Feinstone, S.M., and Rice, C.M. (1996). Identification of a highly conserved sequence element at the 3′ terminus of hepatitis C virus genome RNA. J. Virol. *70*, 3363–3371.

Kolykhalov, A. A., Agapov, E.V., Blight, K.J., Mihalik, K., Feinstone, S.M., and Rice, C.M. (1997). Transmission of hepatitis C by intrahepatic inoculation with transcribed RNA. Science *277*, 570–574.

Koutsoudakis, G., Kaul, A., Steinmann, E., Kallis, S., Lohmann, V., Pietschmann, T., and Bartenschlager, R. (2006). Characterization of the early steps of hepatitis C virus infection by using luciferase reporter viruses. J. Virol. *80*, 5308–5320.

Kyono, K., Miyashiro, M., and Taguchil, I. (1998). Detection of hepatitis C virus helicase activity using the scintillation proximity assay system. Anal. Biochem. *257*, 120–126.

Lin, C., Thomson, J.A., and Rice, C.M. (1995). A central region in the hepatitis C virus NS4A protein allows formation of an active NS3-NS4A serine proteinase complex *in vivo* and *in vitro*. J. Virol. *69*, 4373–4380.

Lindenbach, B.D., Evans, M.J., Syder, A.J., Wölk, B., Tellinghuisen, T.L., Liu, C.C., Maruyama, T., Hynes,

R.O., Burton, D.R., Mckeating, J.A., *et al.* (2005). Complete replication of hepatitis C virus in cell culture. Science *309*, 623–626.

Lindenbach, B.D., Meuleman, P., Ploss, A., Vanwolleghem, T., Syder, A.J., McKeating, J.A., Lanford, R.E., Feinstone, S.M., Major, M.E., Leroux-Roels, G., *et al.* (2006). Cell culture-grown hepatitis C virus is infectious *in vivo* and can be recultured *in vitro*. Proc. Natl. Acad. Sci. U.S.A. *103*, 3805–3809.

Liu, S., Yang, W., Shen, L., Turner, J.R., Coyne, C.B., and Wang, T. (2009). Tight junction proteins claudin-1 and occludin control hepatitis C virus entry and are downregulated during infection to prevent superinfection. J. Virol. *83*, 2011–2014.

Logvinoff, C., Major, M.E., Oldach, D., Heyward, S., Talal, A., Balfe, P., Feinstone, S.M., Alter, H., Rice, C.M., and McKeating, J.A. (2004). Neutralizing antibody response during acute and chronic hepatitis C virus infection. Proc. Natl. Acad. Sci. U.S.A. *101*, 10149–10154.

Lohmann, V., Korner, F., Koch, J.O., Herian, U., Theilmann, L., and Bartenschlager, R. (1999). Replication of subgenomic hepatitis C virus RNAs in a hepatoma cell line. Science *285*, 110–113.

Lohmann, V., Hoffmann, S., Herian, U., Penin, F., and Bartenschlager, R. (2003). Viral and cellular determinants of hepatitis C virus RNA replication in cell culture. J. Virol. 77, 3007–3019.

Ludmerer, S.W., Graham, D.J., Patel, M., Gilbert, K., Stahlhut, M., and Olsen, D.B. (2008). A transient cell-based phenotype assay for hepatitis C NS3/4A protease: application to potency determinations of a novel macrocyclic inhibitor against diverse protease sequences isolated from plasma infected with HCV. J. Virol. Methods *151*, 301–307.

Luo, G. (2004). Molecular virology of hepatitis C virus. In Hepatitis Prevention and Treatment, Colacino JM and Heinz, BA, eds (Birkhauser, Basel, Switzerland), pp. 67–85.

McKeating, J.A., Zhang, L.Q., Logvinoff, C., Flint, M., Zhang, J., Yu, J., Butera, D., Ho, D.D., Dustin, L.B., and Rice, C.M. (2004). Diverse hepatitis C virus glycoproteins mediate viral infection in a CD81-dependent manner. J. Virol. *78*, 8496–8505.

McKercher, G., Beaulieu, P.L., Lamarre, D., LaPlante, S., Lefebvre, S., Pellerin, C., Thauvette, L., and Kukoij, G. (2004). Specific inhibitors of HCV polymerase identified using an NS5B with lower affinity for template/primer substrate. Nucleic Acids Res. *32*, 422–431.

Major, M.E., Dahari, H., Mihalik, K., Puig, M., Rice, C.M., Neumann, A.U., and Feinstone, S.M. (2004). Hepatitis C virus kinetics and host responses associated with disease and outcome of infection in chimpanzees. Hepatology *39*, 1709–1720.

Mateu, G., Donis, R.O., Wakita, T., Bukh, J., and Grakoui, A. (2008). Intragenotypic JFH1 based recombinant hepatitis C virus produces high levels of infectious particles but causes increased cell death. Virology. *376*, 397–407.

Mercer, D.F., Schiller, D.E., Elliott, J.F., Douglas, D.N., Hao, C., Rinfret, A., Addison, W.R., Fischer, K.P., Churchill, T.A., Lakey, J.R., *et al.* (2001). Hepatitis C virus replication in mice with chimeric human livers. Nat. Med. *7*, 927–933.

Meuleman, P. and Leroux-Roels, G. (2008). The human liver-uPA-SCID mouse: a model for the evaluation of antiviral compounds against HBV and HCV. Antiviral Res. *80*, 231–238.

Meuleman, P., Libbrecht, L., De Vos, R., de Hemptinne, B., Gevaert, K., Vandekerckhove, J., Roskams, T., and Leroux-Roels, G. (2005). Morphological and biochemical characterization of a human liver in a uPA-SCID mouse chimera. Hepatology *41*, 847–856.

Meunier, J., Russell, R.S., Engle, R.E., Faulk, K.N., Purcell, R.H., and Emerson, S.U. (2008). Apolipoprotein C1 association with hepatitis C virus. J. Virol. *82*, 9647–9656.

Moradpour, D., Gosert, R., Egger, D., Penin, F., Blum, H.E., and Bienz, K. (2003). Membrane association of hepatitis C virus nonstructural proteins and identification of the membrane alteration that harbors the viral replication complex. Antivir. Res. *60*, 103–109.

Murray, E.M., Grobler, J.A., Markel, E.J., Pagnoni, M.F., Paonessa, G., Simon, A.J., and Flores, O.A. (2003). Persistent replication of hepatitis C virus replicons expressing the beta-lactamase reporter in subpopulations of highly permissive Huh7 cells. J. Virol. *77*, 2928–2935.

Nelson, H.B. and Tang, H. (2006). Effects of cell growth on hepatitis C virus (HCV) replication and a mechanism of cell confluence-based inhibition of HCV RNA and protein expression. J. Virol. *80*, 1181–1190.

Owsianka, A., Tarr, A.W., Juttla, V.S., Lavillette, D., Bartosch, B., Cosset, F.L., Ball, J.K., and Patel, A.H. (2005). Monoclonal antibody AP33 defines a broadly neutralizing epitope on the hepatitis C virus E2 envelope glycoprotein. J. Virol. 79, 11095–11104.

Pacini, L., Bartholomew, L., Vitelli, A., and Migliaccio, G. (2004). Reporter substrates for assessing the activity of hepatitis C virus NS3–4A serine protease in living cells. Anal. Biochem. *331*, 46–59.

Pietschmann, T., Kaul, A., Koutsoudakis, G., Shavinskaya, A., Kallis, S., Steinmann, E., Abid, K., Negro, F., Dreux, M., Cosset, F.L., *et al.* (2006). Construction and characterization of infectious intragenotypic and intergenotypic hepatitis C virus chimeras. Proc. Natl. Acad. Sci. U.S.A. *103*, 7408–7413.

Ploss, A. and Rice, C.M. (2009). Towards a small animal model for hepatitis C. EMBO Rep. *10*, 1220–1227.

Reed, K.E. and Rice, C.M. (2000). Overview of hepatitis C virus genome structure, polyprotein processing, and protein properties. Curr. Top. Microbiol. Immunol. *242*, 55–84.

Russel, R.S., Meunier, J.C., Takikawa, S., Faulk, K., Engle, R.E., Bukh, J., Purcell, R.H., and Emerson, S.U. (2008). Advantages of single-cycle production assay to study cell culture-adaptive mutations of hepatitis C virus. Proc. Natl. Acad. Sci. U.S.A. *105*, 4370–4375.

Shimidzu, Y., Yamanji, K., Masuho, Y., Yokota, T., Inoue, H., Sudo, K., Satoh, S., and Shimotohno, K. (1996). Identification of the sequence of NS4A required for enhanced cleavage of the NS5A/5B site by hepatitis C virus NS3 protease. J. Virol. *70,* 127–132.

Sumpter, R. Jr., Loo, Y.M., Foy, E., Li, K., Yoneyama, M., Fujita, T., Lemon, S.M., and Gale, M. Jr. (2005). Regulating intracellular antiviral defense and permissiveness to hepatitis C virus RNA replication through a cellular RNA helicase, RIG-I. J. Virol. *79,* 2689–2699.

Takeshita, N., Kakiuchi, N., Kanazawa, T., Konno, Y., Nishizawa, M., Tani, T., and Shmotohno, K. (1997). An enzyme-linked immunosorbent assay for detecting protolytic activity of hepatitis C virus proteinase. Anal. Biochem. *247,* 242–246.

Taliani, M., Bianchi, E., Narjes, F., Fossatelli, M., Rubani, A., Steinkuhler, C., De Francesco, R., and Pessi, A. (1996). Continuous assay of hepatitis C virus protease based on resonance energy transfer depsipeptide substrates. Anal. Biochem. *240,* 60–67.

Tani, H., Komoda, Y., Matsuo, E., Suzuki, K., Hamamoto, I., Yamashita, T., Moriishi, K., Fujiyama, K., Kanto, T., Hayashi, N., *et al.* (2007). Replication-competent recombinant vesicular stomatitis virus encoding hepatitis C virus envelope proteins. J. Virol. *81,* 8601–8612.

Targett-Adams, P. and McLauchlan, J. (2005). Development and characterization of a transient-replication assay for the genotype 2a hepatitis C virus subgenomic replicon. J. Gen. Virol. *86,* 3075–3080.

Wakita, T., Pietschmann, T., Kato, T., Date, T., Miyamoto, M., Zhao, Z., Murthy, K., Habermann, A., Krausslich, H.G., Mizokami, M., *et al.* (2005). Production of infectious hepatitis C virus in tissue culture from a cloned viral genome. Nat. Med. *11,* 791–796.

Wang, C., Gale, M., Jr., Keller, B.C., Huang, H., Brown, M.S., Goldstein, J.L., and Ye, J. (2005). Identification of FBL2 as a geranylgeranylated cellular protein required for hepatitis C virus RNA replication. Mol. Cell. *18,* 425–434.

Windisch, M.P., Frese, M., Kaul, A., Trippler, M., Lohmann, V., and Bartenschlager, R. (2005). Dissecting interferon-induced inhibition of hepatitis C virus replication by using a novel host cell line. J. Virol. *79,* 13778–13793.

Yanagi, M., Purcell, R.H., Emerson, S.U., and Bukh, J. (1997). Transcrips from a single full-length cDNA clone of hepatitis C virus are infectious when directly transfected into the liver of a chimpanzee. Porc. Natl. Acad. Sci. U.S.A. *94,* 8738–8743.

Ye, J. (2007). Reliance of host cholesterol metabolic pathways for the life cycle of hepatitis C virus. PLoS Pathog. *3,* 1017–1022.

Ye, J., Wang, C., Sumpter, R.Jr., Brown, M.S., Goldstein, J.L., and Gale, M. Jr. (2003). Disruption of hepatitis C virus RNA replication through inhibition of host protein geranylgeranylation. Proc. Natl. Acad. Sci. U.S.A. *100,* 15865–15870.

Yi, M., Villanueva, R.A., Thomas, D.L., Wakita, T., and Lemon, S.M. (2006). Production of infectious genotype 1a hepatitis C virus (Hutchinson strain) in cultured human hepatoma cells. Proc. Natl. Acad. Sci. U.S.A. *103,* 2310–2315.

Zhang, C., Cai, Z., Kim, Y., Kumar, R., Yuan, F., Shi, P., Kao, C., and Luo, G. (2005). Stimulation of Hepatitis C Virus (HCV) Nonstructural protein 3 (NS3) helicase activity by the NS3 protease domain and by HCV RNA-dependent RNA polymerase. J. Virol. 79, 8687–8697.

Zhang, R., Beyer, B.M., Durkin, J., Ingram, R., Njoroge, F.G., Windsor, W.T., and Malcolm, B.A. (1999). A Continuous spectrophotometric assay for the hepatitis C virus serine protease. Anal. Biochem. *270,* 268–275.

Zhang, Y., Weady, P., Duggal, R., and Hao, W. (2008). Novel chimeric genotype 1b/2a hepatitis C virus suitable for high-through screening. Antimocrob. Agents Chemother. *52,* 666–674.

Zhong, J., Gastaminza, P., Cheng, G., Kapadia, S., Kato, T., Burton, D.R., Wieland, S.F., Uprichard, S.L., Wakita, T., and Chisari, F.V. (2005). Robust hepatitis C virus infection *in vitro*. Proc. Natl. Acad. Sci. U.S.A. *102,* 9294–9299.

Zhong, J., Gastaminza, P., Chung, J., Stamataki, Z., Isogawa, M., Cheng, G., McKeating, J.A., and Chisari, F.V. (2006). Persistent hepatitis C virus infection *in vitro*: coevolution of virus and host. J. Virol. *80,* 11082–11093.

Hepatitis C Virus: Public Resources

3

Deborah R. Taylor

Abstract

The tremendous progress in the field of HCV research has led to an explosion of resources, ranging from research tools and scientific reagents to information sources, including internet tools, meetings and conferences, and regulation policies. Access to such valuable HCV resources is important to facilitate HCV pathogenesis research as well as drug discovery and development. This chapter summarizes the current major resources that are publicly available for scientists, physicians, health professionals, educators, and patients.

Information

National Library of Medicine

A comprehensive HCV information source that enables scientists, physicians, educators, and other health professionals gain rapid access to current information on research, diagnostics, treatment, and prevention is available at the National Library of Medicine (NLM). NLM, the world's largest medical library, has information services available online and on site (NLM at 9000 Rockville Pike, Bethesda, MD 20892; website: www.nlm.nih.gov). Research publications are available through MEDLINE/PubMed, a bibliographic database managed by the NLM and covering medicine and the preclinical sciences. MEDLINE/PubMed provides access to life science journal citations. LocatorPlus is an online catalogue of books, audiovisuals, and meeting abstracts. NLM also provides access to a wide variety of selected and reviewed links to additional resources; including MedlinePlus and MedlinePlus en Español, and GenBank for DNA and amino acid sequences.

Health resources and services administration

As part of the Health and Human Services Department, the Health Resources and Services Administration (HRSA) comprise six bureaus and 13 offices. HRSA provides leadership and financial support with 90% of its funding used for grants to health care providers in every state and US territory. HRSA focuses on uninsured, underserved, and special-needs populations and they train health professionals and improve systems of care in rural communities. HRSA oversees organ, bone marrow and cord blood donation. It supports programmes that prepare against bioterrorism, compensate individuals harmed by vaccination, and maintains databases that protect against health care malpractice and health care waste, fraud and abuse. HRSA provides programme resources and services needed to improve access to culturally competent, quality health care. The HRSA website (Table 3.1) contains information about clinical research on HCV.

National Center for Research Resources

The National Center for Research Resources (NCRR; www.ncrr.nih.gov) supports research resources that create, develop, and supply a variety of high-quality animal models, biological

Table 3.1 Government websites

Centers for Disease Control and Prevention (CDC)	http://www.cdc.gov
Department of Health and Human Services	http://www.dhhs.gov
Health Resources and Services Administration (HRSA)	http://www.hrsa.gov
Food and Drug Administration (FDA)	http://www.fda.gov
National Institutes of Health	http://www.nih.gov
World Health Organization (WHO)	http://www.who.int/home-page/

materials, and similar resources. The purpose of the resources directory is to guide investigators, who have projects supported by the National Institutes of Health and other peer-reviewed funding sources. NCRR provides clinical and translational researchers with the training and tools they need to transform basic discoveries into improved human health. NCRR offers funding, job and training opportunities, provides information on management of core facilities and the availability of animal models. Research training opportunities at the NIH can be found at their website (http://www.training.nih.gov/).

The University of Alabama at Birmingham has launched a new website, funded by the Center for Disease Control and Prevention (CDC; Table 3.1). The website is directed towards health-care and social service providers with the goal of increasing knowledge and awareness of chronic hepatitis. Online webinars are available (Table 3.2), including one entitled 'Viral Hepatitis 101' and another given by Dr John Ward, director of CDC's Division of Viral Hepatitis. He discusses the Institute of Medicine report recommendations made in 2009. The site can be found at www.KnowHepatitis.org.

Immunology reagents

Tetramers

The NIH Tetramer Facility was established in 1999 and provides custom synthesis and distribution of soluble MHC-peptide tetramer reagents that can be used to identify antigen-specific T cells. These tetramers are produced at Emory University, where the NIH Tetramer Facility site is located. Reagents are provided free of charge to investigators except for shipping costs. The facility offers tetramers with fluorophores, including Pacific Blue, fluorescein, Alexa488, Alexa647 and Alexa680. While most of the tetramers are class I tetramers, both human and mouse CD1d monomers and tetramers and select human and mouse class II MHC tetramers are available. (The NIH Tetramer Core Facility at Emory University is located at 954 Gatewood Road, Atlanta, GA 30329; http://tetramer.yerkes.emory.edu/).

Antibodies

The Antibody Resource Page is a guide designed by scientists for scientists to find companies that sell custom monoclonal, polyclonal, and non-custom/stock antibodies (http://www.antibodyresource.com/). Bionity.com also provides a database of over 170,000 antibodies from 20 different manufacturers and vendors. It has a convenient search engine with direct links to suppliers at the website: http://www.Bionity.com/antibodies/.

Serology and standards

Both serology and nucleic acid standard panels are available from the National Institute for Biological Standards and Control (NIBSC), South Mimms, UK. NIBSC is a centre of the Health Protection Agency (HPA) in the UK. The Institute has a close relationship with the World Health Organization (WHO) and is a WHO International Laboratory for Standards (http://www.nibsc.ac.uk). WHO biological standards are used to ensure the quality of biological products worldwide. These International Standards are used by laboratories and manufacturers to calibrate their own working standards. A collaborative study was performed to evaluate HCV genotype panels against the international

Table 3.2 Internet resources

Education and teaching
http://www.virology.net/garryfavwebindex.html#General
http://www.nlm.nih.gov/medlineplus/hepatitisc.html
http://www3.niaid.nih.gov/topics/hepatitis/hepatitisC/
Pathogenesis
http://hcv.lanl.gov/content/sequence/HCV/ToolsOutline.html
Virology books
http://www.virology.net/Bookpage/Virobooks.html
Slides and photographs
http://www.hepfi.org/learning/library/index.htm (Hepatitis Foundation International)
http://www.nlm.nih.gov/medlineplus/tutorials/hepatitisc/htm/_no_50_no_0.htm
Health-care worker and patient information resources
http://www.cdc.gov/hepatitis/index.htm
http://www.hhs.state.ne.us/epi/epihep_c.htm (Nebraska Dept Health)
http://www.who.int/mediacentre/factsheets/fs164/en/index.html (WHO)
http://www.rocheexchange.com/hepatology (treatment, Roche)
http://www.hepatitis-central.com/
http://digestive.niddk.nih.gov/ddiseases/pubs/hepc_ez/
http://www3.niaid.nih.gov/
http://clinicaltrials.gov/search/open/condition=%22Hepatitis+C%22
http://www.fda.gov/ForConsumers/ByAudience/ForPatientAdvocates/ucm153441.htm
http://www.KnowHepatitis.org
Foundations and societies
http://www.hepfi.org/learning/index.htm
http://www.liverfoundation.org/
http://www.hepcassoc.org/
http://publish.aasld.org/Pages/Default.aspx

standard for genotype 1 (Saldanha *et al.*, 2003). Standard panels are provided for a small charge.

Serology panels are also available by contacting the Office for Blood Research and Review, Center for Biologics Evaluation and Research (CBER), Food and Drug Administration (see Table 3.1). While these panels have been available in the past for research and validation studies, due to limited resources, CBER makes them available for lot release only.

Proteins and peptides

Biodefense and Emerging Infections Research Resources Repository (BEI Resources) provides a peptide array for HCV. The 71-peptide array spans the NS5A protein from the H77 strain of genotype 1a, peptides from the J4 strain are also available. Lyophilized peptides (at 1 mg/vial) are 13 to 19-mers with an 11 amino acid overlap. Users are limited to ordering one time. Registration with BEI is required and more information is available at their website: http://www.beiresources.org/Catalog/tabid/248/Default.aspx.

Epitope database

The Immune Epitope Data Base (IEDB) contains searchable data related to B and T cell epitopes for humans, non-human primates, and small mammals. Peptidic epitope data, including epitope analysis tools relating to all infectious diseases is current. Information by HCV isolate/genotype is also available at this website: http://www.immuneepitope.org/.

Therapeutic reagents

Interferons of many types are available free of charge from BEI Resources. The cost of shipping and registration is required and only one vial of each type is provided without the approval of NIAID (http://www.beiresources.org/tabid/716/stabid/716/PathogenLinkID/13/Default.aspx).

Cells and viruses

While many cell lines and human viruses are available through American Type Culture Collection (ATCC), only expression vectors containing viral cDNAs are available. A few hepatocyte cell lines are available as well at http://www.atcc.org.

The NIH AIDS reagent programme provides expression vectors and replicons containing HCV sequences and can be contacted through the website https://www.aidsreagent.org/index.cfm.

Some cell lines are available through BEI Resources for the cost of shipping (http://www.beiresources.org/tabid/716/stabid/716/PathogenLinkID/13/Default.aspx).

Genetics

Nucleic acid standards

Nucleic acid standards are available from NIBSC (see above) or from the Food and Drug Administration, Center for Biologics Evaluation and Research (CBER). CBER HCV RNA Panel consists of 10 members with HCV genotype 1b RNA levels ranging from 0 to 100,000 copies/ml. The panel members and their concentrations in copies/ml are: #1. 1000, #2. 0, #3. 100,000, #4. 10,000, #5. 0, #6. 500, #7. 100, #8. 50, #9. 10 and #10. 5 copies/ml. Panel member #1 was assigned as 250 IU/ml (Saldanha *et al.*, 2000). Each vial contains a 0.75 ml aliquot of the corresponding panel member. (Saldanha *et al.*, 2000). HCV RNA panels are offered for use in lot release or assay standardization and are available by contacting NIBSC (http://www.nibsc.ac.uk) or the Office of Blood Research and Review, CBER, FDA.

Microarray

The Pathogen Functional Genomics Resource Center (PFGRC) provides researchers with free-of-charge functional genomic research resources and training to aid functional genomics research on a variety of pathogens for which genomic sequence information is currently available. At this time, HCV is not listed as an available array.

Animals and animal models

Mice

Knockout Mouse Project (KOMP) repository

The KOMP Repository was established in 2007 to archive and distribute vectors and ES cell clones. These reagents are produced by KOMP teams at Children's Hospital Oakland Research Institute, The Sanger Institute, University of California at Davis and Regeneron, Inc. The KOMP Repository ensures the preservation and availability of products for use by the scientific community. KOMP archives, maintains, and distributes all ES cell clones, live mouse lines, frozen embryos and sperm, and all KOMP vectors for a reasonable cost. In addition to obtaining KOMP products, investigators have full access to resources and services, including microinjection of ES cells into live mice and conversion of a cryopreserved archive into live mice. The KOMP repository provides an online searchable catalogue and ordering system for all KOMP products, services, and related information. Further, the KOMP repository can be reached by telephone (1–888-KOMP-MICE), website (www.KOMP.org) and e-mail (service@komp.org) for information and assistance in finding, ordering, and using KOMP products.

Mutant Mouse Regional Resource Center at the University of Missouri/Harlan
This centre distributes inbred, hybrid, and mutant mice to investigators. The MH-MMRRC selects and import mouse strains and stocks important to the biomedical research community in a pathogen-free state, cryopreserves gametes and embryos, performs genotyping, phenotyping, and infectious disease monitoring and distributes mice to investigators as live mice, cryopreserved germplasm, or tissues. The MMRC is located at the University of Missouri, Comparative Medicine Center and Research Animal Diagnostic and Investigative Laboratory in Columbia, MO 65211, USA, and at their website: www.mmrrc.org.

Chimpanzee

The NIH Chimpanzee Management Program (ChiMP) supports long-term, cost-effective housing and maintenance at NCRR-supported facilities for chimpanzees. NCRR provides programmatic oversight of the facilities and ensures they comply with the Animal Welfare Act, and policies concerning laboratory animal care and use. While NCRR acknowledges the continuing importance of chimpanzees to biomedical research, after careful review of existing chimpanzee resources, NCRR determined that it does not have the financial resources to support the breeding of chimpanzees. However, NCRR supports existing chimpanzee facilities, including the federal sanctuary for chimpanzees that are no longer needed in biomedical research. The chimpanzee sanctuary, Chimp Haven, is a private, non-profit organization that provides lifetime care for federally owned or supported chimpanzees that are no longer needed for biomedical research. The Department of Health and Human Services published standards of care for chimpanzees held in the sanctuary system. The standards apply to only those facilities receiving funds as a part of the federally funded chimpanzee sanctuary system (DHHS, 2008).

Vaccine development and safety

There has been difficulty in developing a hepatitis C vaccine that induces adequate or appropriate immunity from the infection. The genetic variability of the virus, lack of a convenient animal model and unknown factors that relate to natural immunity have posed barriers to vaccine development. New opportunities for vaccine development now exist with the recent development of cell culture methods to grow this virus, the use of pseudoviruses to measure the blocking of infection and promising small animal models may change the outlook of vaccine development. PATH's Vaccine Resource Library (VRL) provides information and guidance on vaccines and global health related to the lack of effective vaccination programmes at (www.path.org/vaccineresources/about.php). In addition, proposals for lowering the risk of unsafe injections and other issues such as waste disposal and improper administration of vaccines are important problems related to injection safety that are addressed by this organization. According to the World Health Organization (WHO), 16 billion injections are given each year in developing countries, and the risk of unsafe injections is a serious concern. A report issued by WHO in 2000 is entitled *Technologies for Vaccine Delivery in the 21st Century* (Lloyd, 2000). This document is available online at: www.who.int/vaccines-documents/.

Regulations

United States guidelines

Federal Register
The Federal Register is a daily publication for rules, proposed rules, and notices of federal agencies and organizations, as well as executive orders and other presidential documents. It is updated daily and is published Monday through Friday, except Federal holidays. Federal Register volumes from 1994 to the present are available at http://www.gpoaccess.gov/fr/index.html.

US Code of Federal Regulations (CFR)
The CFR is the book of laws and permanent rules published in the Federal Register by agencies of the Federal Government. It is divided into 50 titles that represent broad areas subject to Federal regulation. Each volume of the CFR is updated once each calendar year. The online CFR is a joint

project authorized by the National Archives and Records Administration's Office of the Federal Register (OFR), and the Government Printing Office (GPO) and can be accessed online at the website: http://www.gpoaccess.gov/cfr/index.html. Regulations covering investigational new drugs (INDs) and licensing of new biologicals and drugs are covered under 21 CFR.

The FedWorld website is a gateway to government information. This site is managed by the National Technical Information Service (NTIS), an agency of the US Department of Commerce. This is an online locator service for information disseminated by the Federal Government on all topics, including information related to hepatitis C virus. The website provides a search engine for government reports that can be obtained online at http://www.fedworld.gov/.

Clinical trials and institutional review boards

FDA's Bioresearch Monitoring Program works with FDA's Office of Regulatory Affairs to oversee institutional review boards (IRBs) with respect to clinical trials. The monitoring programme provides information on regulations and draft guidance, FDA basics, laws enforced by FDA, approvals of FDA-regulated products and FDA clinical trial requirements. The make-up of an IRB for the FDA's requirements is described in 21 CFR 56.107. Online information is available at http://www.fda.gov/ScienceResearch/SpecialTopics/RunningClinicalTrials/default.htm.

Canada

Canada requires that before marketing, all drug submissions are reviewed and must meet scientific requirements under the Food and Drug Regulations. Guidance documents are available for the purpose of assisting in the interpretation of policies and regulations.

Health Canada provides a list of international regulatory agencies and contains links to their websites (http://www.hc-sc.gc.ca/dhp-mps/compli-conform/int/agenreg/index-eng.php)

A PDF version of 'Access to Therapeutic Products: The Regulatory Process in Canada' can be found at http://www.hc-sc.gc.ca/ahc-asc/pubs/hpfb-dgpsa/access-therapeutic_acces-therapeutique-eng.ph.

European Union Guidelines

Paul-Ehrlich-Institute

The Paul Ehrlich Institute (PEI) was first established as the Institute for Sera Research and Serum Testing in 1896. The institute sets criteria for verifying efficacy, quality and safety of therapeutic products and reviews applications for marketing approvals. PEI is an agency of the Federal Republic of Germany and is mandated with regulation of medicines and biologics before being brought to market in Germany. PEI research and control activities promote the quality, efficacy and safety of biological products for human use. The institute has a website that provides detailed information in English at http://www.pei.de/nn_162428/EN/infos-en/pu-en/pu-node-en.html?__nnn=true. The website contains information on scientific advice, marketing authorization and batch control.

The CE mark

Manufacturers place the CE mark on products to demonstrate that the product meets all the appropriate provisions of the relevant legislation implementing certain European directives and gives companies easier access into the European market. The European Commission's 'Blue Guide' (Guide to the Implementation of Directives Based on the New Approach and Global Approach) lists products where the CE marking is required. It is available on the Commission's website at http://ec.europa.eu/enterprise/policies/single-market-goods/documents/blue-guide/. More information on European regulations can be found in English at: http://europa.eu/index_en.htm.

Meetings and conferences

Support groups

Information about traditional and alternative medicines and patient support groups can be found on websites, including healthyhepper.com. This website provides links to information about alternative medicines, dosages, therapeutic goals and living with HCV. http://www.healthyhepper.com/events.htm

Liver research

The 45th Annual Meeting of the European Association for the Study of the Liver (EASL, 2010) was held in Vienna, Austria in 2010. The International Liver Congress hosts nearly 8000 scientists and clinicians from around the globe. The goal of this meeting is to share the latest research on pathogenesis and treatment of HCV among leading experts in the field.

The Annual Meeting of the American Association for the Study of Liver Diseases (AASLD). Also known as the Liver Meeting is a large conference in the science and practice of hepatology. Presenters address topics, including the latest findings on new drugs, novel treatments, and the results from clinical studies.

HCV research

International Meeting on Hepatitis C Virus and Related Viruses has been held annually since 1993. The scope of this meeting is: to cover all aspects of HCV and related viruses, including basic and translational research.

Antiviral research

International Society for Antiviral Research holds an Annual International Conference on Antiviral Research (ICAR). The scope of this meeting is to provide state-of-the-art reports for scientists, governmental authorities and healthcare workers. The society also hopes to accelerate progress in the development of antivirals, including small molecules and biologics. The congress serves as an interface between academics, governmental agencies, charities, and health and industry professionals.

HEP DART assembles clinicians, researchers and basic scientists together to advance knowledge of the ongoing drug development processes in the treatment of hepatitis B and hepatitis C and other viruses affecting the liver, including co-infections with HIV. HEP DART provides the scientific community with an increased understanding of the current and future challenges in therapeutics for hepatitis infection.

Summary

There seems to be an abundance of public resources for AIDS and HIV research and development, but HCV resources have been slow in coming. Is this because the research community has been unwilling to share reagents and materials or has the funding from public sources been smaller? The NIH funding for HIV/AIDS research in the US for 2009 was $319 million while the funding for HCV research was $12 million (data obtained from NIH research portfolio online reporting tools; http://report.nih.gov/rcdc/categories/default.aspx). With 0.1% of the US population currently infected with HIV and 1.7% of the US population infected with HCV demonstrates an ongoing disproportionate allocation in funds that is common in the public and private sectors. Under funding of HCV may result in a public health crisis and economic burden for the US and local governments in the future. The future of public health would benefit by a boost in cooperation, productive collaboration and an increase in public resources provided by federal, state and local governments and universities.

Acknowledgements

'The findings and conclusions in this chapter have not been formally disseminated by the Food and Drug Administration and should not be construed to represent any Agency determination or policy.'

References

Code of Federal Regulations (CFR), published annually by the Office of the Federal Register. IRB guidelines with FDA's requirements. In 21 CFR 56.107 (Office of the Federal Register, Washington, DC).

D.H.H.S. (2008). Federal Register Notice: Standards of Care for Chimpanzees Held in the Federally Supported Chimpanzee System. 42 CFR; 42 U.S.C. 287a–3a, pp. 106–551.

Lloyd, J. (2000). Technologies for vaccine delivery in the 21st Century, World Health Organization, available online at: www.who.int/vaccines-documents/

Saldanha J., Heath, A., Lelie, N., Pisani, G., Nübling, M., and Yu, M. (2000). Calibration of HCV working reagents for NAT assays against the HCV international standard. The Collaborative Study Group. Vox Sang. 78, 217–224.

Saldanha J., Cleland, A., Jarvis, L., Chudy, D., Gärtner, T., Wienand, S., Gerlich, W., Gierman, T., Gross, P., Zerlauth, G., *et al.* (2003). Collaborative study to calibrate hepatitis C virus genotypes 2–6 against the HCV International Standard, 96/790 (genotype 1). Vox Sang. 84:20–27.

HCV Databases

4

Carla Kuiken and Richard Scheuermann

Abstract

Sequence databases provide a resource for scientific researchers worldwide that is especially important for highly variably pathogens such as HCV. There are currently a number of HCV sequence databases, two of which will be discussed in detail in this review. The Los Alamos HCV sequence database (http://hcv.lanl.gov) has been widely used since 2004, and still is a commonly used resource. A new addition is the Virus Pathogen Resource (ViPR, http://www.viprbrc.org/), which contains sequences from viral pathogens from many different families, and does not focus on HCV. We compare mission, features, and future plans of both databases and attempt to provide some guidance to help users decide which resource is more useful for specific data analysis problems.

Background

HCV is the cause of a worldwide pandemic with an estimated 170 million people infected (Armstrong *et al.*, 2000; Lauer and Walker, 2001). Acute HCV infection leads to chronic infection in an estimated 70–80% of individuals (Alter and Mast, 1994; Conry-Cantilena *et al.*, 1996), which if untreated can result in liver cirrhosis, liver failure and hepatocellular carcinoma (Lauer and Walker, 2001). While the current treatment of pegylated interferon and the nucleoside analogue ribavirin (Davis *et al.*, 1998; Zeuzem *et al.*, 2000; Manns *et al.*, 2001; Fried *et al.*, 2002) has an improved success rate, the therapy is expensive and requires sophisticated medical management, which is out of reach for the majority of infected people. Newly developed specific antivirals offer promise of improved treatment results, but there is little doubt that they will also rapidly lead to drug resistance (Thompson and McHutchison, 2009). Development of a prophylactic vaccine to prevent the continued spread of HCV infection remains paramount.

In contrast to the situation in HIV, there are currently several sequence databases for hepatitis C. Two are maintained outside the US, one in Japan and one in France. These were previously described (Kuiken *et al.*, 2006; Combet *et al.*, 2007; Shin *et al.*, 2008). Two US-based HCV databases are described in this chapter, the new Virus Pathogen Database and Analysis Resource (ViPR), which also contains many other viral pathogen sequences, and the Los Alamos HCV database, established in 2004.

The role of sequence databases in research

Sequence databases are vital resources for genetic research, especially in highly variable organisms. In order to determine the evolutionary relationships between strains, an up-to-date and well-maintained sequence repository is essential. New strains and recombinants are necessarily classified in relation to existing strains, and defined by their distance (and its genomic distribution) to previously sequenced isolates. In the case of HIV, absence of a good repository was one of the factors leading to the delay in recognizing its vast genetic variability (Srinivasan *et al.*, 1989) and its propensity to recombination (Sabino *et al.*, 1994), both hallmarks of the virus.

For HCV, the vast genetic variation is already leading to nomenclature and classification issues. Genotype 7 has recently attained 'confirmed' status, but precious few data are available. For genotype 2, on the other hand, the number of recognized subtypes has increased so quickly that it was decided to only assign subtype numbers to strains that caused clearly visible epidemics, and not to those that were found only occasionally (Simmonds *et al.*, 2005). For a concise overview of currently recognized genotypes and subtypes, see http://hcv.lanl.gov/content/sequence/HCV/classification/genotable.html.

Having a good sequence repository is also important for a wide variety of treatment-related issues. For example, for keeping track of developing drug resistance mutations, including their interactions and fitness implications, a sequence database is indispensable. The Los Alamos HCV database is used by Roche Diagnostics, both in their FDA applications to validate their HCV detection and typing tools for all currently known genetic variants, and as a sentinel system to detect new variants.

Development of a vaccine against HCV is a high priority, and one important consideration is the ability of a vaccine to generate protective responses against a wide variety (and hopefully all) variants of the viruses. One of the major challenges for developing an effective HCV vaccine is its considerable sequence diversity (Houghton and Abrignani, 2005). Sequence databases clearly have an important role in the study of HCV variability. A new vaccine reagent design method was created by the Los Alamos HIV database and analysis team (Fischer *et al.*, 2007), and recently applied to HCV, with very promising results (Yusim *et al.*, 2010).

Another HIV field that is currently receiving much attention is the study of the interaction of (HLA-restricted) human cellular immune response and viral evolution (Bhattacharya *et al.*, 2007; Carlson and Brumme, 2008). The HLA-dependent selection of certain escape amino acids (and therefore possibly of certain strains) in certain populations has by now been well documented for HIV, but in HCV, too, this effect has been documented (e.g. Gaudieri *et al.*, 2006; Neumann-Haefelin *et al.*, 2008; Timm *et al.*, 2007). Analysis of sequence variation in relation to host HLA characteristics and diversity quickly becomes very complex, and requires sophisticated statistical techniques. In HIV, inclusion of patient HLA data in publications has become routine. For HCV it still happens only rarely, which may hamper progress in this area.

Below we will highlight some features of the ViPR database and the Los Alamos database. Rather than an exhaustive enumeration of the features, the capabilities of each database and associated website will be highlighted based on sample 'use cases'.

ViPR

History

In 2009, NIAID renewed its Bioinformatics Resource Center's programme to both consolidate major organism categories into single resources, and to focus on extending the traditional sequence-based databases to capture more primary experiment data about pathogen phenotype characteristics, host–pathogen interactions and other data related to the development of improved diagnostics therapeutics and vaccines. The Virus Pathogen Database and Analysis Resource (ViPR; www.viprbrc.org; Fig. 4.1A) was established to realize these objectives for human pathogenic viruses, including hepatitis C. Initially ViPR has focused on the capture and analysis of sequence data and the sequence features that can be defined therein (e.g. predicted immune epitopes, enzyme active sites, protein domains and motifs, etc.). Classical analysis tools for sequence similarity determination (BLAST), multiple sequence alignment (MUSCLE), and phylogenetic tree construction (PhyML) are supported. For the purposes of this chapter, we focus on those features of ViPR that are relatively unique among HCV-related resources. Since ViPR was first released in December of 2009, both current and future features will be summarized.

Workbench

One of the unique features of ViPR is the availability of a workbench function that allow users to save sequence sets, search criteria and analysis results in their own private workspaces.

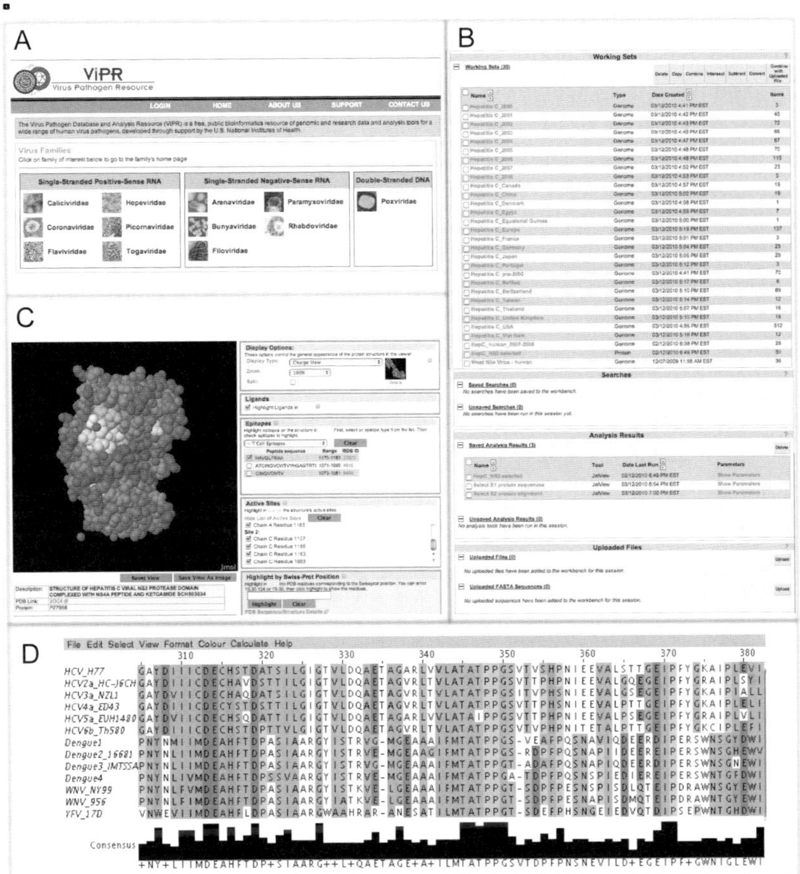

Figure 4.1 Virus Pathogen Database and Analysis Resource (ViPR). (A) ViPR provide a comprehensive resource of research data about a wide variety of different human pathogenic viruses, including HCV and related members of the Flaviviridae family. (B) An example of a private workbench containing saved working sets of sequence records and analysis results is shown. HCV complete genome sequence sets for viruses isolated in recent years or in defined country locations can be maintained. Working sets can be combined to generate unions (Combine), intersections (Intersect) and differences (Subtract) sets (e.g. Hepatitis C_Europe). (C) An example of data integration using the ViPR 3D protein structure viewer is shown. The viewer allows users to customize the display to highlight different protein features, including the location of enzyme active sites (orange), bound ligands (green) and immune epitopes (yellow). (D) Access to data about all flaviviruses allows for the comparative analysis within a virus family. The alignment of NS3 protein sequences of NCBI Reference Sequences for HCV, Dengue, West Nile virus and Yellow Fever virus is shown together with the level of sequence conservation (bar graph) and the resulting consensus sequence. A colour version of this figure is located in the plate section at the back of the book.

For example, datasets of whole HCV genome sequences can be constructed based on year of isolation or country of isolation for focused analysis (Fig. 4.1B). Datasets can also be combined to generate datasets that reflect higher order grouping (e.g. European datasets – Hepatitis C_Europe). Sequence sets and analysis results persist on the ViPR site and allow users to go back and modify and re-analyse sequences and other data using the analysis tools supported by ViPR. The workbench feature thus obviates the need to download results for various search and analysis tasks, and allows users to take these data directly into further analysis.

Data integration

One of the major goals of the ViPR resource is to serve as a one-stop shop for data related to these human pathogenic viruses, which provides an opportunity to support the integration of related data types. One example is the implementation of the JMol 3D protein structure visualization tool that allows users to colour code amino acid residues in protein structures based on the location of known sequence features and other sequence characteristics (e.g. sequence conservation). In this ViPR component, the location of immune epitopes can be highlighted in the context of virus protein 3D structures, and the spatial relation with other important protein features (e.g. enzyme active sites) can be visualized (Fig. 4.1C). Information about co-location between immune epitopes and critical functional features can be helpful for targeted vaccine development.

Comparative genomics

Since ViPR also contains sequence data for other *Flaviviridae*, the workbench can also be used for comparative analysis between HCV and related viral species. Sequences of interest can be identified using the ViPR search interface based on a number of different characteristics, for example those sequences that are derived from NCBI Reference Sequences. The selected sequences can be saved to a workbench working set for further analysis, including multiple sequence alignment using MUSCLE and phylogenetic tree construction using PhyML. Multiple sequence alignments can be visualized using ViPR JalView to identify regions of sequence conservation across viral species (Fig. 4.1D).

Experiment data

ViPR is now expanding to serve as a repository of experiment data related to the molecular characterization of these viruses. Through its Driving Biological Project programme and other outreach activities, ViPR is actively capturing valuable experimental data about the phenotypic characteristics of individual viruses, especially as they relate to host range, transmission, virulence and pathogenesis. The availability of experimental phenotype data linked with sequence records will facilitate genotype–phenotype association studies that identify important functional sequence features of the viral genome.

As ViPR continues to grow, emphasis on host–pathogen interaction will drive future development. Experimental data examining the response to viral infection at the host mRNA level using gene expression microarrays, host gene requirements from RNAi screens, and clinical data from human surveillance projects will be added. The ultimate goal will be to provide a warehouse of experiment data related to these human pathogenic viruses for further use by the virology research community.

The LANL HCV sequence database

The Los Alamos HCV database has been set up following the model of the HIV database, and it focuses on the same type of analysis issues – ones that are often specific to highly variable pathogens with compact genomes. Among such problems are alignment, translation to amino acid (especially in genomes that contain reverse-oriented or dually coding regions), residue or codon use bias, and generally analysis types that require very detailed study of the sequences in question. The HIV and HCV databases have developed some 20 unique tools for this type of sequence analysis, and we will illustrate some of them by means of a demonstration analysis.

The dataset consists of 34 E1/E2 sequences from one patient, sampled over four time points. Two of these samples were taken before the patient seroconverted for HIV, two were taken after the seroconversion. The purpose of the study was to investigate if the HIV infection (with its associated effects on the immune system) had a measurable influence on the within-patient evolution of HCV.

The first step is to align the 34 sequences. This can be done by hand, but the HCV database has an HCV-specific alignment programme that uses a curated HMM model to align the sequences: HCValign. The sequences can be downloaded *in toto*, or split into separate genes. For now, we will use the original sequences. The alignment is also codon-aligned, which means that it can be immediately translated. Fig. 4.2 shows the result page of this operation. It shows that our alignment overlaps the genes Core, E1, E2, and NS2. We can download all those regions separately,

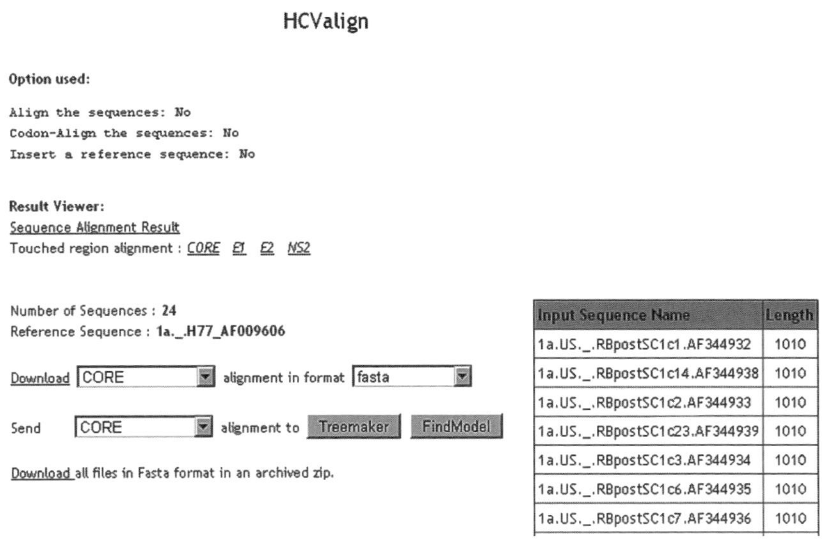

Figure 4.2 Output from HCValign tool.

or download the entire codon alignment, which we will opt for now. The tool also allows users to immediately process HCValign's output to TreeMaker (to make a simple phylogenetic tree) or Findmodel (to determine the best evolutionary model for this alignment), but it is often better to first look at the resulting alignment to make sure it doesn't contain any obvious problems, which can greatly distort the analysis results.

After manually inspecting the alignment for gross errors using an alignment editor (e.g. BioEdit, available from http://www.mbio.ncsu.edu/BioEdit/bioedit.html), we proceed to use it to find other sequences stored in the database that may also be useful for our study.

The HCV search interface (shown in Fig. 4.3) as it opens provides access to several database fields, but by clicking on the orange + signs the interface can be 'unfolded' and it will show many more search fields, ordered by subject. Clicking on the field labels will open the Help file, which explains which searches and values are possible.

The search interface (Fig. 4.3) contains a window where users can paste or upload their alignment. This will then be re-aligned against a model sequence, from which the region is calculated that the user's sequences cover. Other sequences can then be found that cover the same region, and fulfil other criteria that can be specified in the search interface. We will search for other sequences from samples where the patient's HIV status was known. A set of >200 sequences that meet these criteria is found, and they can be downloaded and used for further analysis. Note that the search results by default exclude 'problematic' sequences (an example is shown in Fig. 4.4). They can be excluded for many reasons, including being synthetic sequences (e.g. artificial recombinants), patent sequences, containing large numbers of N's, and being too short (<50 nucleotides).

These problematic sequences can be added to the search results by clicking the 'Include problematic' button. Also of note in this output:

- The yellow icon shows where in the genome this sequence is located.
- For each sequence, links are provided to
  - an immediate Blast search against the HCV database;
  - the associated patient data;
  - the GenBank record, which in turn provides a link to all sequences from the same publication.
- At the top, the menu bar provides access to all available datasets and tools.

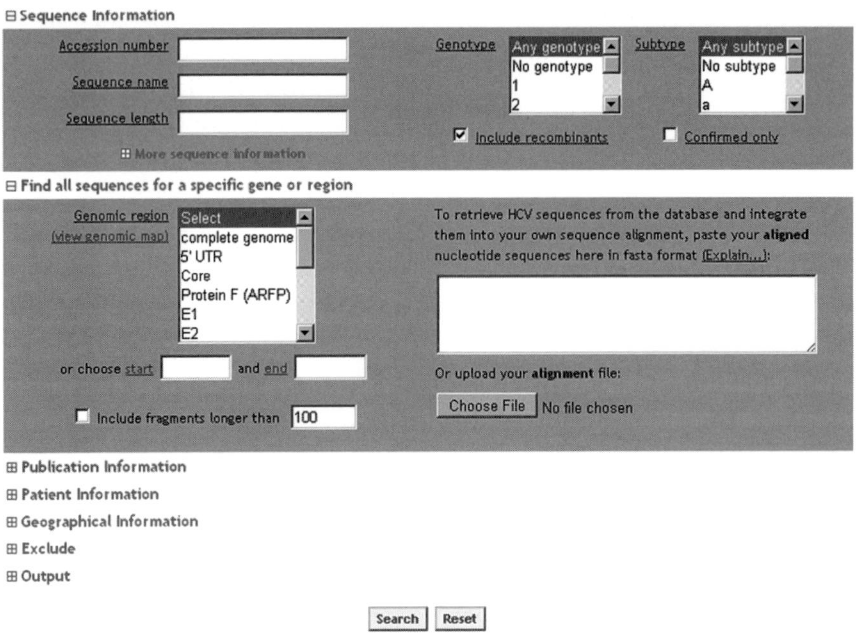

Figure 4.3 HCV database search interface, default view. A colour version of this figure is located in the plate section at the back of the book.

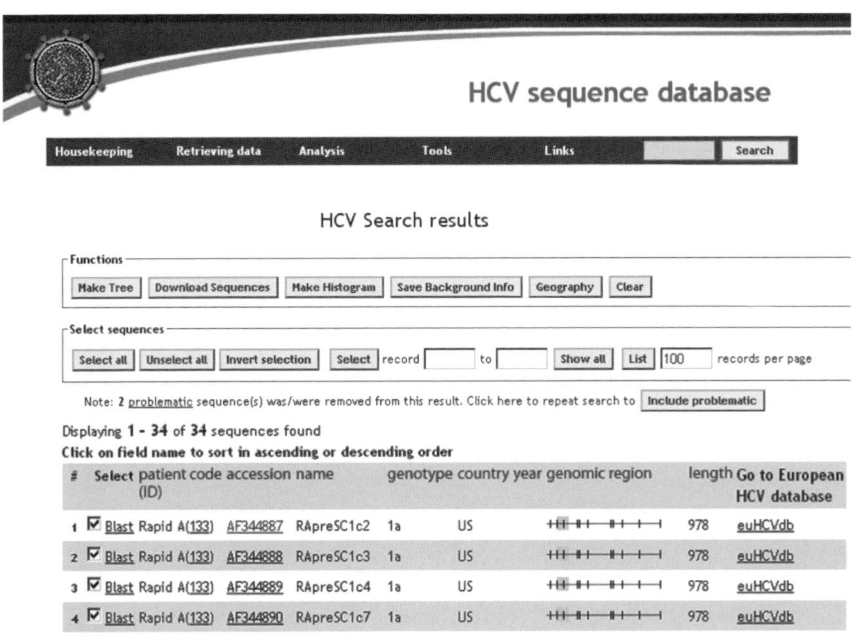

Figure 4.4 LANL HCV database search output. A colour version of this figure is located in the plate section at the back of the book.

- The function buttons allow download of the sequences (aligned on the fly) and further analysis options.
- Results can be sorted by any of the columns by clicking on the column headers.

For now, we will first make a phylogenetic tree (Fig. 4.5) using the original set of 34 sequences. The interface can automatically include a 'genotype reference set' that can be used to determine which clades are available in the dataset. An alternative way to do this is to use the 'SynchAligns' tool, which can synchronize alignments for the same region by aligning both to the reference sequence H77; one option is to align you alignment to the HCV reference alignment rather than a second user-provided alignment. We include all genotype 1 reference sequences and a genotype 2 sequence as an out-group.

The tree shows that all sequences are genotype 1a except for one, which appears to be 1b. This sequence will be removed, since it clearly is not immediately related to the other sequences and will not provide any information about in-patent evolution. The other noticeable feature is the 1a sequence that clusters far outside the other

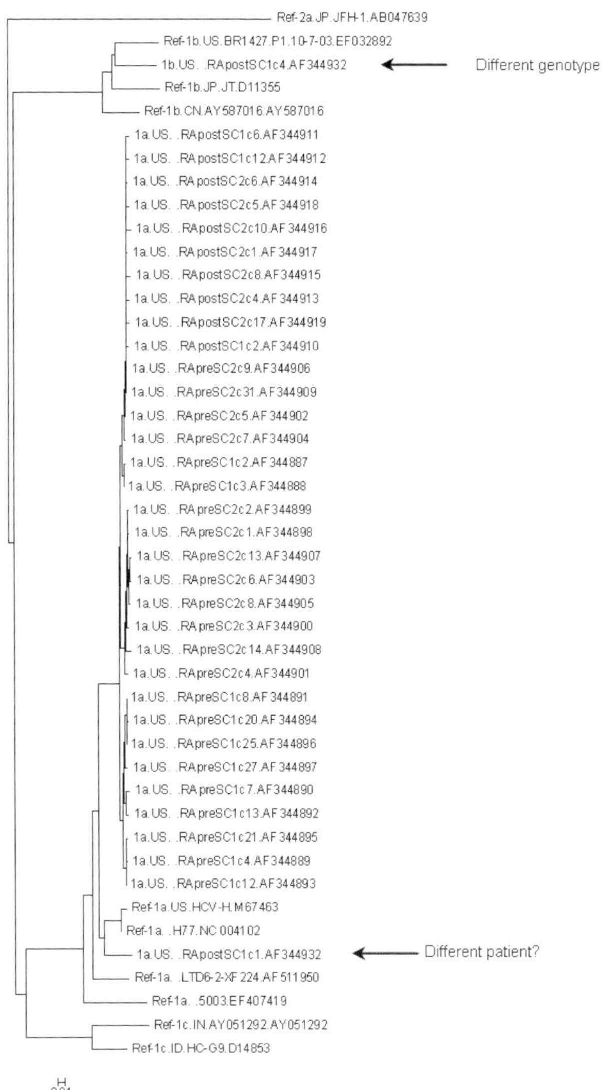

Figure 4.5 Initial phylogenetic tree of the test dataset.

patient sequences. It is likely that this sequence is not from this patient, but results from a labelling error. This sequence too will be removed. The easiest way to do this is to select 'download alignment in tree order', since the sequences that are the outliers in the tree will be separate, and can easily be removed.

When the sequence quality control step has been completed and the uncovered issues have been addressed, we can further analyse the cleaned alignment. The next step could be to make an improved tree that uses a better phylogenetic model, rather than the 'quick and dirty' tree that is made by the search interface extension. The first step could be to run the FindModel tool(Fig. 4.6), to calculate which evolutionary model best fits the data. The tool can create a start tree using three different tree-building methods. Using different methods should not significantly affect the model that is identified as the best one. For this dataset, the model selected is the most complex one, called general time-reversible (GTR) with an estimated gamma parameter.

This model can now be used to generate a maximum likelihood tree via the PhyML interface (not shown). The next step is to investigate more closely the mutations that can be seen in this sequence set. The Highlighter tool allows a visual assessment of the mutation pattern, which can be plotted relative to a 'master sequence'. In the

FindModel Results

Initial tree constructed using Weighbor

MODELS CONSIDERED:

Model name	AIC	LnL
JC: Jukes-Cantor (model 1)	16733.206244	-8366.603122
JC: Jukes-Cantor plus Gamma (model 3)	16203.129808	-8100.564904
F81: Felsenstein 1981 (model 5)	16688.910728	-8341.455364
F81: Felsenstein 1981 plus Gamma (model 7)	16129.619562	-8060.809781
K80: Kimura 2-parameter (model 9)	16154.530594	-8076.265297
K80: Kimura 2-parameter plus Gamma (model 11)	15514.6553	-7755.327650
HKY: Hasegawa-Kishino-Yano (model 13)	16083.32436	-8037.662180
HKY: Hasegawa-Kishino-Yano plus Gamma (model 15)	15404.524192	-7697.262096
TrN: Tamura-Nei (model 21)	16083.255278	-8036.627639
TrN: Tamura-Nei plus Gamma (model 23)	15405.534532	-7696.767266
GTR: General Time Reversible (model 53)	16057.575514	-8020.787757
GTR: General Time Reversible plus Gamma (model 55)	15393.817872	-7687.908936

Click one of the model names in the table to display the parameter matrix for that model.

AIC-SELECTED MODEL: GTR: General Time Reversible plus Gamma (model 55)
LnL = -7687.908936
AIC = 15393.817872
Parameter details

The matrix on the right illustrates the different models considered by Findmodel.
Rate Parameters with the same color are assumed to have the same value.
Base Frequencies are the same when they are shown as f_N, and different otherwise.

General Time Reversible + γ

	T	C	A	G
T	f_T	a	b	c
C	a	f_C	d	e
A	b	d	f_A	f
G	c	e	f	f_G

Figure 4.6 FindModel output showing the tested models, their scores, and a dynamic matrix showing the features of each selected model. A colour version of this figure is located in the plate section at the back of the book.

case of longitudinal data, the easiest and often the most effective graphs are obtained by using (one of) the earliest sequences as the master; all mutations will be plotted relative to this sequence. This is the option we use here. Another sensible possibility is to use a consensus of all the sequences of the first time point as the master. The resulting graph is shown in Fig. 4.7. As expected, the number of mutations relative to the first sequence increases over time, indicating a 'drift' away from the earliest available sequences.

Another feature of Highlighter is to plot the mutations divided into silent and non-silent mutations. This distinction can not always be made exactly (see Nei and Gojobori, 1986, for an explanation), but the tool distinguishes mutations that are and are not certain to cause synonymous or non-synonymous changes. In the current situation the number of ambiguous mutations is small, and the resulting graph (Fig. 4.7B) shows an interesting pattern.

The distribution of the two types of mutations over the sequenced stretch is completely non-uniform, with mutations 5′ of codon 200 being almost entirely silent, and those beyond codon 200 almost entirely non-silent. If we look at the location of the sequence in the HCV genomes (the tool Sequence Locator can be used for this; see Fig. 4.8) we see that the region where silent changes predominate corresponds to E1, while the non-silent changes are located in E2.

We can further study the pattern of synonymous and non-synonymous mutations using SNAP (Silent/Nonsilent Analysis Program), which analyses the mutations in a codon-alignment by their location in the sequences. The graph in Fig. 4.9 shows the cumulative silent and non-silent changes on our alignment, as calculated by SNAP (it uses the same algorithm as the Highlighter tool). The difference between the two genes can be seen very clearly in this graph: initially the red line showing synonymous changes climbs quickly, but slightly after codon 200 the non-synonymous changes accumulate while the synonymous changes slow down somewhat. SNAP will also calculate statistics for the

Figure 4.7 Highlighter output; left: standard settings; right: silent (green) and non-silent (red) changes. A colour version of this figure is located in the plate section at the back of the book.

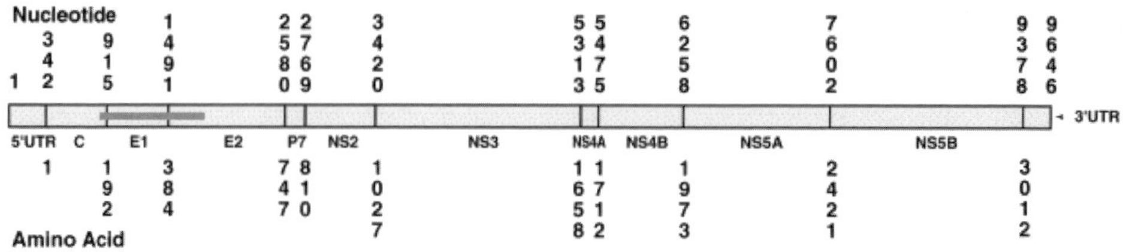

Figure 4.8 Sequence Locator output showing the location of the query sequence in the alignment. A colour version of this figure is located in the plate section at the back of the book.

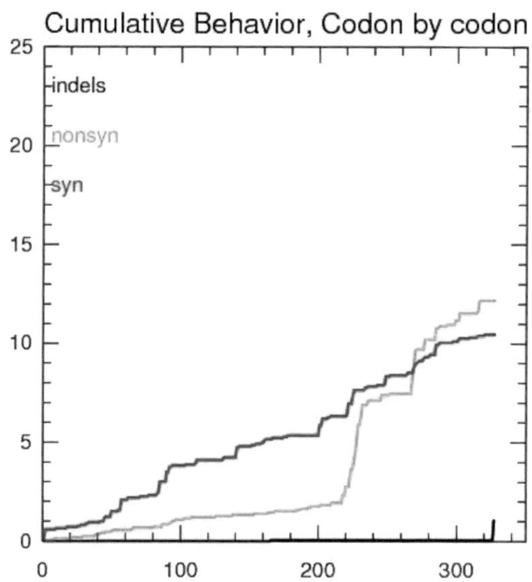

Figure 4.9 Graphical output of SNAP, the Silent/Nonsilent Analysis Program. A colour version of this figure is located in the plate section at the back of the book.

synonymous and non-synonymous changes, and generate phylogenetic trees for each mutation category separately.

It would be helpful if we could determine if the visible difference between the silent and non-silent mutations in genes E1 and E2 is statistically significant. SNAP provides the numbers for this test, but does not actually perform it. First, we use the HCV-align tool to divide our alignment into an E1 and an E2 portion (both are already codon-aligned) and save both separately. Next we run SNAP on each of the alignments. Checking the 'statistics' box on the SNAP input page will generate a summary of the silent and non-silent mutation means, standard deviations, and variances, both with Jukes–Cantor correction for multiple hits (ds and dn) and without (ps and pn; Table 4.1).

By comparing the means and the standard deviations, we see that the distance between the means for the silent mutations (both corrected and uncorrected) is less than two standard deviations, but this is not true for the non-silent changes (two standard deviations correspond to a *P*-value of approximately 0.05). Thus, the difference between the non-silent mutations in the two genes is significant.

To assess if the total number of mutations is evenly spread over the alignment, we can study the total mutation output of Highlighter (see Fig. 4.7), but a more sophisticated method is offered by the Entropy tool, which will plot the entropy

Table 4.1 SNAP statistics for the number of silent and non-silent changes in the E1 and E2 genes

	E1				E2			
	Silent		Non-silent		Silent		Non-silent	
	Uncorrected (ps)	Corrected (ds)	Uncorrected (pn)	Corrected (dn)	Uncorrected (ps)	Corrected (ds)	Uncorrected (pn)	Corrected (dn)
Average	0.039166	0.040481	0.003915	0.003932	0.050998	0.053389	0.037851	0.039104
Variance	0.000077	0.000088	0.000001	0.000001	0.000141	0.000171	0.000061	0.000072
Standard deviation	0.008768	0.009394	0.000708	0.000747	0.011878	0.013067	0.007788	0.008462

(a measure of variability) of the alignment for each position (Fig. 4.10). Both analyses show that there is no indication of a different overall distribution of the mutations; the only difference is in their nature.

It would be interesting to repeat this analysis on other sequence sets containing longitudinal intrapatient sets for the same region. For this, we can select the same region (E1/E2) in the search interface, and also set 'number of sequences from patient' to >10. This does not ensure that the samples are longitudinal, but limits the results to sets that contain more than 10 sequences per patient, which gives a reasonable chance to find longitudinal sets. Any of the sets that are retrieved this way can be used for a background analysis. Similar analysis of randomly selected set containing sequences from two visits for each of 15 patients (slow and rapid progressors) shows that the pattern of uneven distribution seen in our initial patient is not typical.

A final analysis for our intrapatient set is to study the location of the sequenced region in the 3D structure of the E1 and E2 proteins, to see if the structure of these proteins can shed some light on the distribution of the silent and non-silent mutations. Unfortunately, as can be seen in the menu of the Protein Feature Accent tool, no complete structures are currently available for either of the proteins. The only structure that is available is a for 29-aa partial transmembrane region of the E1 protein in solution (Fig. 4.11). For demonstration purposes, we used this fragment (PDB ID 2KNU) as a basis to map the entropy (i.e. sequence variability) scores of the alignment, a function that is built into the Protein Feature Accent tool. The colours follow the rainbow, from blue (highly conserved) to red (highly variable).

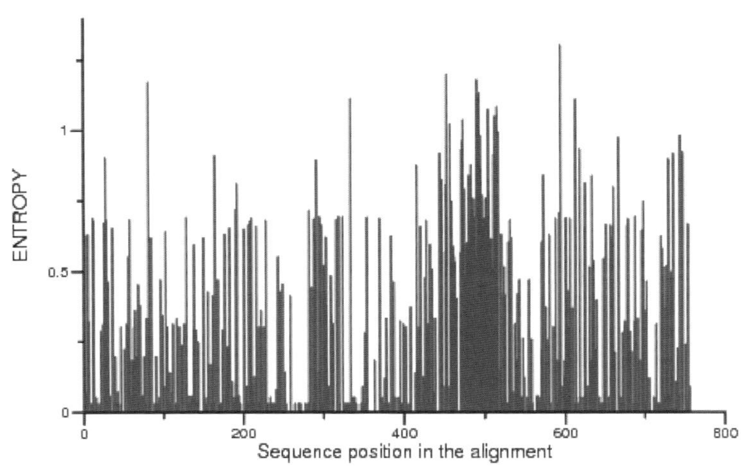

Figure 4.10 Entropy of the sequences in the sample dataset.

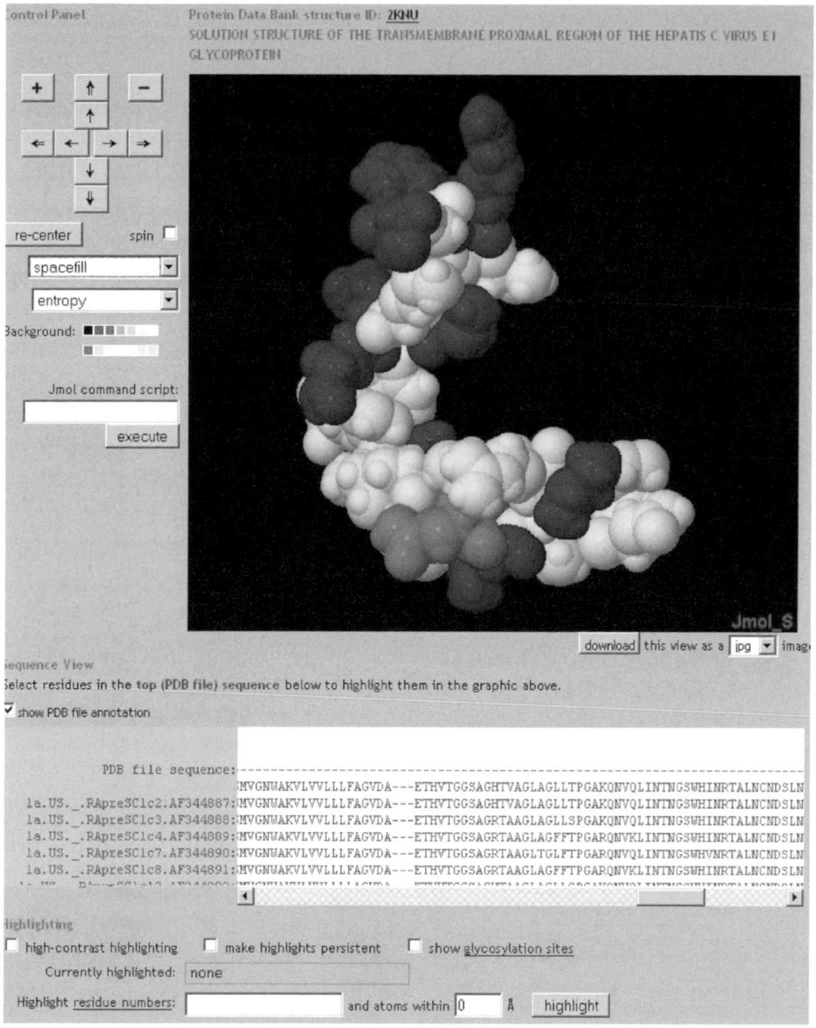

Figure 4.11 'Protein Feature Accent', the protein visualization and annotation tool, showing the entropy source alignment and the calculated entropy, mapped onto an E2 fragment. A colour version of this figure is located in the plate section at the back of the book.

Summary

Since the discovery of the hepatitis C virus in the early 1990s, some progress has been made in preventing and treating the infection, but there is little doubt that the main battle has not yet been won. One of the many issues that impede progress is the astonishing genetic variability and adaptability of the virus. Sequence databases play a very important role in the study of the variability, be it in epidemiology, vaccine and drug design, resistance tracking, and host immune interactions. Other than for HIV, there are currently several sequence databases for HCV, of which the two based in the USA are discussed at length in this chapter. They differ in many areas, including the types of metadata they offer and analysis tools that are available. However, both are based on the same sequence data collection, namely those deposited in GenBank by scientists worldwide. Thus, whether users choose to use one or the other, or both, they will have access to the same datasets, and have the ability to analyse those data using tools offered by both services.

Acknowledgements

The Los Alamos HCV database owes its existence and almost all its tools to the efforts of the LANL HIV database staff, in particular its PI, Dr Bette Korber.

References

Alter, J., and Mast, E.E. (1994). The epidemiology of viral hepatitis in the United States. Gastroenterol. Clin. North Am. *23*, 437–455.

Armstrong, G.L., Alter, M.J., McQuillan, M., and Margolis, H.S. (2000). The past incidence of hepatitis C virus infection: implications for the future burden of chronic liver disease in the United States. Hepatology *31*, 777–782.

Bhattacharya, T., Daniels, M., Heckerman, D., Foley, B., Frahm, N., Kadie, C., Carlson, J., Yusim, K., McMahon, B., Gaschen, B., *et al.* (2007). Founder effects in the assessment of HIV polymorphisms and HLA allele associations. Science *315*, 1583–1586.

Carlson, M., and Brumme, Z.L. (2008). HIV evolution in response to HLA-restricted CTL selection pressures: a population-based perspective. Microbes Infect. *10*, 455–461.

Combet, C., Garnier, N., Charavay, C., Grando, D., Crisan, D., Lopez, J., Dehne-Garcia, A., Geourjon, C., Bettler, E., Hulo, C., *et al.* (2007). euHCVdb: the European hepatitis C virus database. Nucleic Acids Res. *35*, D363–366.

Conry-Cantilena, C., VanRaden, M., Gibble, J., Melpolder, J., Shakil, A.O., Viladomiu, L., Cheung, L., DiBisceglie, A., Hoofnagle, J., Shih, J.W., *et al.* (1996). Routes of infection, viremia, and liver disease in blood donors found to have hepatitis C virus infection. N. Engl. J. Med. *334*, 1691–1696.

Davis, G.L., Esteban-Mur, R., Rustgi, V., Hoefs, J., Gordon, S.C., Trepo, C., Shiffman, M.L., Zeuzem, S., Craxi, A., Ling, H., and Albrecht, J. (1998). Interferon alfa-2b alone or in combination with ribavirin for the treatment of relapse of chronic hepatitis C. International Hepatitis Interventional Therapy Group. N. Engl. J. Med. *339*, 1493–1499.

Fischer, W., Perkins, S., Theiler, J., Bhattacharya, T., Yusim, K., Funkhouser, R., Kuiken, C., Haynes, B., Letvin, N.L., Walker, B.D., *et al.* (2007). Polyvalent vaccines for optimal coverage of potential T-cell epitopes in global HIV-1 variants. Nat. Med. *13*, 100–106.

Fried, M.W., Shiffman, M.L., Reddy, K.R., Smith, C., Marinos, G., Goncales, F.L., Jr., Haussinger, D., Diago, M., Carosi, G., Dhumeaux, D., *et al.* (2002). Peginterferon alfa-2a plus ribavirin for chronic hepatitis C virus infection. N. Engl. J. Med. *347*, 975–982.

Gaudieri, S., Rauch, A., Park, L.P., Freitas, E., Herrmann, S., Jeffrey, G., Cheng, W., Pfafferott, K., Naidoo, K., Chapman, R., Battegay, M., Weber, R., Telenti, A., Furrer, H., James, I., Lucas, M., and Mallal, S.A. (2006). Evidence of viral adaptation to HLA class I-restricted immune pressure in chronic hepatitis C virus infection. J. Virol. *80*, 11094–11104.

Houghton, M., and Abrignani, S. (2005). Prospects for a vaccine against the hepatitis C virus. Nature *436*, 961–966.

Kuiken, C., Mizokami, M., Deleage, G., Yusim, K., Penin, F., Shin, I.T., Charavay, C., Tao, N., Crisan, D., Grando, D., *et al.* (2006). Hepatitis C databases, principles and utility to researchers. Hepatology *43*, 1157–1165.

Lauer, M., and Walker, B.D. (2001). Hepatitis C virus infection. N. Engl. J. Med. *345*, 41–52.

Manns, M.P., McHutchison, J.G., Gordon, S.C., Rustgi, V.K., Shiffman, M., Reindollar, R., Goodman, Z.D., Koury, K., Ling, M., and Albrecht, J.K. (2001). Peginterferon alfa-2b plus ribavirin compared with interferon alfa-2b plus ribavirin for initial treatment of chronic hepatitis C: a randomised trial. Lancet *358*, 958–965.

Nei, M., and Gojobori, T. (1986). Simple methods for estimating the numbers of synonymous and nonsynonymous nucleotide substitutions. Mol. Biol. Evol. *3*, 418–426.

Neumann-Haefelin, C., Frick, D.N., Wang, J.J., Pybus, O.G., Salloum, S., Narula, G.S., Eckart, A., Biezynski, A., Eiermann, T., Klenerman, P., *et al.* (2008). Analysis of the evolutionary forces in an immunodominant CD8 epitope in hepatitis C virus at a population level. J. Virol. *82*, 3438–3451.

Sabino, E.C., Shpaer, E.G., Morgado, M.G., Korber, B.T., Diaz, R.S., Bongertz, V., Cavalcante, S., Galvao-Castro, B., Mullins, I., and Mayer, A. (1994). Identification of human immunodeficiency virus type 1 envelope genes recombinant between subtypes B and F in two epidemiologically linked individuals from Brazil. J. Virol. *68*, 6340–6346.

Shin, I.T., Tanaka, Y., Tateno, Y., and Mizokami, M. (2008). Development and public release of a comprehensive hepatitis virus database. Hepatol. Res. *38*, 234–243.

Simmonds, P., Bukh, J., Combet, C., Deleage, G., Enomoto, N., Feinstone, S., Halfon, P., Inchauspe, G., Kuiken, C., Maertens, G., Mizokami, M., Murphy, D.G., Okamoto, H., Pawlotsky, J.M., Penin, F., Sablon, E., Shin, I.T., Stuyver, L.J., Thiel, H.J., Viazov, S., Weiner, J., and Widell, A. (2005). Consensus proposals for a unified system of nomenclature of hepatitis C virus genotypes. Hepatology *42*, 962–973.

Srinivasan, A., York, D., Butler, D., Jr., Jannoun-Nasr, R., Getchell, J., McCormick, J., Ou, C.Y., Myers, G., Smith, T., Chen, E., *et al.* (1989). Molecular characterization of HIV-1 isolated from a serum collected in 1976: nucleotide sequence comparison to recent isolates and generation of hybrid HIV. *AIDS* Res. Hum. Retroviruses 5, 121–129.

Thompson, J. and McHutchison, J.G. (2009). Antiviral resistance and specifically targeted therapy for HCV (STAT-C). J. Viral Hepat. *16*, 377–387.

Timm, J., Li, B., Daniels, M.G., Bhattacharya, T., Reyor, L.L., Allgaier, R., Kuntzen, T., Fischer, W., Nolan, B.E., Duncan, J., Schulze zur Wiesch, J., Kim, A.Y., Frahm, N., Brander, C., Chung, R.T., Lauer, G.M., Korber, T., and Allen, T.M. (2007). Human leukocyte antigen-associated sequence polymorphisms in hepatitis C virus reveal reproducible immune responses and constraints on viral evolution. Hepatology *46*, 339–349.

Yusim, K., Fischer, W., Yoon, H., Thurmond, J., Fenimore, P.W., Lauer, G., Korber, B., and Kuiken, C. (2010). Genotype 1 and global hepatitis C T-cell vaccines designed to optimize coverage of genetic diversity. J. Gen. Virol. *91*, 1194–1206.

Zeuzem, S., Feinman, S.V., Rasenack, J., Heathcote, E.J., Lai, M.Y., Gane, E., O'Grady, J., Reichen, J., Diago, M., Lin, A., *et al.* (2000). Peginterferon alfa-2a in patients with chronic hepatitis C. N. Engl. J. Med. *343*, 1666–1672.

HCV Infection Systems

5

Takaji Wakita

Abstract

Hepatitis C virus (HCV) infection causes chronic liver disease and is a health problem worldwide. Recent rapid progress in HCV research has been largely dependent on the development of an HCV culture system and infectious small animal model. The most important part of the HCV infection system is the discovery of the JFH-1 clone isolated from a fulminant hepatitis C patient. The JFH-1strain fits into the cluster of genotype 2a, and its viral genome sequence is monoclonal. The JFH-1 strain replicates efficiently in cultured cell lines without acquiring adaptive mutations, and it secretes infectious viral particles into culture medium. This infectious HCV system provides for the first time a powerful tool to study the viral life cycle, construct antiviral strategies and develop effective vaccines.

Introduction

The hepatitis C virus (HCV), one of the plus-strand RNA viruses, is a principal agent in post-transfusion and sporadic acute hepatitis (Choo et *et al.*, 1989; Kuo *et al.*, 1989; Table 5.1). Infection with HCV leads to chronic liver diseases, including cirrhosis and hepatocellular carcinoma (Kiyosawa *et al.*, 1990). Although HCV belongs to the *Flaviviridae* family and has a genome structure similar to the other flaviviruses (Fig. 5.1), efficient virus propagation in cultured cells had proved to be difficult (Bartenschlager and Lohmann, 2000). Many attempts have been made to develop a system for HCV infection and replication in cell culture, and some advances have been reported (Bertolini *et al.*, 1993; Ito *et al.*, 1996; Mizutani *et al.*, 1996; Iacovacci *et al.*, 1997; Fournier *et al.*, 1998; Rumin *et al.*, 1999; Ito *et al.*, 2001; Zhao *et al.*, 2002; Zhu *et al.*, 2003). However, the viral replication efficiencies reported in these studies were modest, requiring detection by a reverse transcription polymerase chain reaction (RT-PCR). An important development was the creation of a subgenomic HCV RNA replicon system (Lohmann *et al.*, 1999; Table 5.1) that could be used to assess the HCV replication in cultured cells. Although this represents a powerful tool in the study of HCV replication mechanisms and the search

Table 5.1 History of HCV research

1989	Discovery of HCV cDNA
1993	Analysis of virus genome structure
1996	Crystal structure of NS3 protease
1997	Infectious HCV cDNA clone
1998	Introduction of interferon and ribavirin therapy
1999	Subgenomic replicon of Con1 strain
2003	Pseudotype virus system
2005	Virus culture using JFH-1 clone

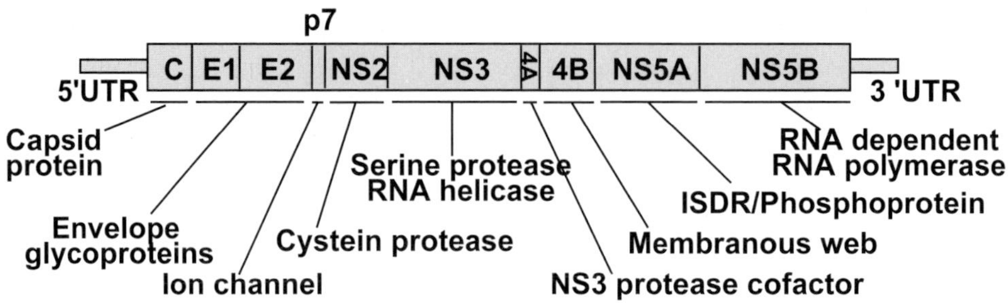

Figure 5.1 Genome structure of hepatitis C virus. At least 10 viral proteins are processed from large polypeptides encoded by the ORF. Each viral protein has a single or multiple functions.

for potential antiviral agents, infectious viral particles were not produced from replicon cells even though the structural genes were included in the replicon constructs (Ikeda *et al.*, 2002; Pietschmann *et al.*, 2002). An analysis of HCV replication capacities revealed that wild-type viral genomes had low replication capacities (Pietschmann *et al.*, 2002). Adaptive mutations substantially enhanced the replication of subgenomic and full-length HCV RNA; however, the introduction of these adaptive mutations into the full-length genome resulted in a loss of *in vivo* infectivity (Pietschmann *et al.*, 2002; Pietschmann T *et al.*, 2009). On the other hand, HCV strain JFH-1 was cloned from a patient with fulminant hepatitis, and its sequence deviated from chronic hepatitis isolates (Kato *et al.*, 2001). A subgenomic replicon containing JFH-1 cDNA replicated with high colony-formation efficiency in Huh7 cells and other cell lines without adaptive mutations (Kato *et al.*, 2003; Date *et al.*, 2004). Thus, the wild-type JFH-1 strain appeared to have greater replication capacities in cultured cells compared with other HCV strains. Therefore, it was considered to be important to test the JFH-1 strain for full-length viral RNA replication and infectious viral particle formation (Table 5.1). When we transfected Huh7 cells with *in vitro*-transcribed full-length JFH-1 HCV RNA, the JFH-1 RNA efficiently replicated, and the cells produced viral particles capable of infecting cultured cells and a chimpanzee (Wakita *et al.*, 2005). A robust HCV culture system was established with the JFH-1 clone and permissive cells (Lindenbach *et al.*, 2005; Zhong *et al.*, 2005).

Isolation of strain JFH-1 from a fulminant hepatitis patient

The JFH-1 (Jikei or Japanese Fulminant Hepatitis patient) strain was isolated from a 32-year-old male patient with acute liver failure (Kato *et al.*, 2001). At the time of hospital admission, the patient's serum aspartate aminotransferase (AST) and alanine aminotransferase (ALT) concentrations were 9160 IU/l and 6970 IU/l, respectively, and the minimum prothrombin time was 16%. The patient developed stage II encephalopathy five days after admission, and he was then diagnosed with fulminant hepatitis. HCV RNA was detected by RT-PCR performed with sera obtained during the acute phase. A second-generation enzyme linked immunosorbent assay (Ortho Diagnostics, Tokyo, Japan) did not detect anti-HCV antibody at the time of admission. All viral markers indicating exposure to other hepatitis viruses were negative. The patient's liver function and clinical condition recovered with conservative treatment, and anti-HCV antibody was detected 6 weeks after admission. These findings suggest that the fulminant hepatitis was in fact due to HCV infection. To analyse the infectious strain of HCV, 12 sets of nested RT-PCR, as well as 5′ RACE and 3′ RACE RT-PCR, were performed to cover the entire HCV genome. Five clones of each PCR fragment were sequenced, and the consensus sequence was determined. Based on the sequence analysis, the JFH-1 strain belongs to genotype 2a, and its sequence deviates slightly from other genotype 2a clones isolated from patients with chronic hepatitis (Kato *et al.*, 2001).

Replicon experiments

Subgenomic replicon and full-length constructs were assembled using the cloned PCR fragments (Kato *et al.*, 2001; Uprichard *et al.*, 2006; Date *et al.*, 2007). The colony formation efficiency of the JFH-1 replicon was much greater than that of the Con1 replicon with adaptive mutations. Furthermore, the transient transfection of replicon RNA into Huh7 cells resulted in autonomous RNA replication, as determined by Northern blot analysis (Kato *et al.*, 2003; 2005a). Importantly, adaptive mutations were not necessary for efficient JFH-1 replicon replication in Huh7 cells. In addition, the JFH-1 replicon produced colonies in several other cell lines, including HepG2, IMY-N9, HeLa, and 293 human cells, as well as mouse NIH3T3 fibroblast cells and mouse AML12, MMHD3, and MMH1-1 hepatocytes (Date *et al.*, 2004; Kato *et al.*, 2005b, Uprichard *et al.*, 2006). The JFH-1 subgenomic replicon was compared with the genotype 1b HCV replicon, Con1 NK5.1, with adaptive mutations (Miyamoto *et al.*, 2006). Cell growth rates and HCV RNA replication of isolated replicon cells were evaluated. The saturation densities of JFH-1 replicon cells were lower than those of Con1 NK5.1 replicon cells and Huh7 cells, and the mean doubling time of JFH-1 was longer than that for Huh7 cells. HCV RNA replication levels in isolated clones were similar between JFH-1 and Con1 NK5.1. RNA replication levels decreased in inverse proportion to cell densities with both JFH-1 and Con1 NK5.1 replicon cells. HCV RNA replication was more resistant to interferon in JFH-1 replicon cells than in Con1 NK5.1 replicon cells. In the clinical setting, genotype 1 infections are more resistant to interferon therapy than genotype 2 infections. These discrepancies may result from strain-specific characteristics, such as the viral source, rather than genotypic characteristics. Interestingly, the sensitivities to Cyclosporine (CsA) treatment are also different between genotype 1 and 2 replicon cells (Ishii *et al.*, 2006). CsA has a suppressive effect on HCV replication (Watashi *et al.*, 2003). The replication of genotype 1b replicons was strongly suppressed by treatment with CsA. In contrast, JFH-1 replication was less sensitive to CsA. Replication of JFH-1 did not require the cellular replication cofactor, cyclophilin B (CyPB). CyPB stimulated the RNA-binding activity of NS5B in the genotype 1b replicon but not in the genotype 2a JFH-1 strain. The mechanisms of diversity that govern virus–cell interactions and the sensitivity of these strains to antiviral agents are still not clear.

HCV infection system and permissive cell lines

JFH-1 demonstrated a markedly enhanced replication efficiency compared with other HCV clones. To take advantage of this characteristic of the JFH-1 clone, we examined full-length JFH-1 with regard to replication in Huh7 cells transfected with synthesized RNA (Fig. 5.2; Wakita *et al.*, 2005). Transfected full-length RNA replicated efficiently in Huh7 cells, and surprisingly, infectious virus particles were secreted into the culture medium. Culture supernatant from the transfected cells was cleared by centrifugation and filtration and then used to inoculate naïve Huh7 cells. Following inoculation, several infected cells were identified by immunostaining with HCV-specific antibodies. The infection efficiency in standard Huh7 cells was less than 0.5%. However, the infection was specific because anti-CD81 antibody and anti-E2 antibody inhibited infection (Wakita *et al.*, 2005; Zhong *et al.*, 2005). Furthermore, the virus particles secreted into the culture supernatant were infectious in chimpanzees (Wakita *et al.*, 2005; Kato *et al.*, 2008). These results strongly suggest that the secreted virus particles were authentic HCV. This infectious HCV system has already provided a variety of applications in the HCV research field.

Although the infection efficiency using original Huh7 cells maintained in our laboratory was quite low, efficient viral infection was achieved using cured cell lines such as Huh7.5, Huh7.5.1, and Lunet/CD81 cells (Blight *et al.*, 2002; Zhong *et al.*, 2005; Koutsoudakis *et al.*, 2007). These cell lines were produced by interferon treatment of subgenomic replicons. Huh7.5 cells have a T55I mutation in the first domain of the RIG-I gene, resulting in defective intracellular interferon signalling against HCV RNA replication (Sumpter *et al.*, 2005). Transfection of the JFH-1 genome or inoculation of infectious JFH-1 virus into Huh7.5 and Huh7.5.1 cells thus produces robust

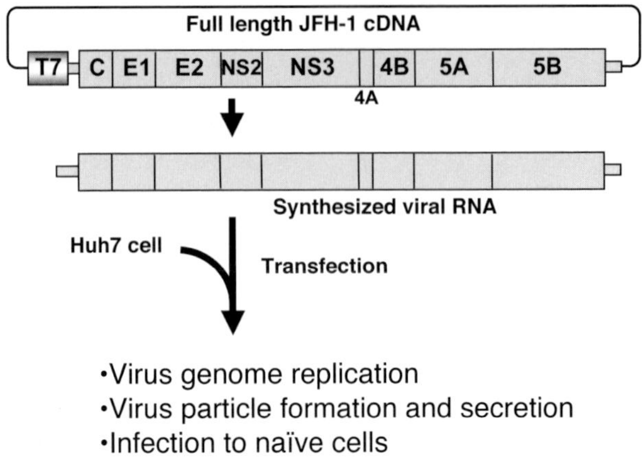

Figure 5.2 Infectious HCV production with JFH-1 cDNA. Viral RNA is synthesized from a full-length JFH-1 cDNA and transfected into Huh7 cells. Infectious HCV particles are produced in the transfected cells and secreted into culture medium.

replication and HCV virus infection (Lindenbach *et al.*, 2005; Zhong *et al.*, 2005). Therefore, we tested the permissiveness of JFH-1 replication in these cell lines by transient transfection of a subgenomic JFH-1 replicon, and we found that JFH-1 RNA replication was not enhanced in these cell lines, compared with the original Huh7 cells (T. Wakta, unpublished data). However, there was a stable increase in the cell-surface expression of CD81 by these cell lines, compared with the original Huh7 cells. Of note, Koutsoudakis *et al.* reported that the level of CD81 cell-surface expression is a key determinant of HCV infection (Koutsoudakis *et al.*, 2007). In fact, we have observed that the Huh7 cells used in the initial infection assay were a mixture of cell clones with varying levels of CD81 expression and infectivity (Akazawa *et al.*, 2007). Interestingly, we have isolated several Huh7 subclones without cell-surface CD81 expression. Among these subclones, Huh7–25 supported highly efficient subgenomic replicon replication. Thus, we transfected a CD81 expression vector into these cell clones to produce stable cells with a high level of ectopic CD81 expression. These cells supported a greater degree of infectivity of the JFH-1 virus than the original Huh7 cells, as well as the cured cell lines (Akazawa *et al.*, 2007). As mentioned above, Huh7.5 cells have a defective mutation in the RIG-I gene that prevents interferon induction through the double-strand RNA signalling pathway. Huh7–25 cells are also defective in the interferon induction pathway; however, the defect is at the level of TRIM25, which is a ubiquitin ligase of RIG-I (Arnaud *et al.*, 2010). Thus, to achieve a high level of infectivity, it is important to check the interferon induction upon double-strand RNA stimulation and cell-surface expression of CD81. Liang's group recently reported that CD81 expression levels are also important for HCV replication, which suggests that CD81 plays more roles in the HCV life cycle (Zhang *et al.*, 2010). Huh7 cells were first isolated more than 20 years ago (Nakabayashi *et al.*, 1982), and were distributed worldwide. The phenotype of Huh7 cells, including permissiveness for JFH-1 virus infectivity, may vary among the subclones maintained in different laboratories.

Infectious HCV clones other than JFH-1 and the mechanism of efficient JFH-1 replication

It has been difficult to propagate other non-JFH-1 HCV strains in cultured cells, despite the development of chimeric viruses that contain structural proteins from other strains and genotypes. H77S, a genotype 1a strain that contains 5 adaptive mutations, produces infectious virus after synthesized RNA transfection into Huh7 cells, albeit with limited efficiency (Yi *et al.*,

2006). Continuous virus passages were difficult, probably due to low infectivity or unknown reasons. However, it is clear that JFH-1 is not a single HCV strain that can be propagated in cultured cells. Further study is necessary to develop other genotypic HCV strains that are infectious in cell culture. Determining the mechanism of JFH-1 replication in cultured cells should provide critical information for the establishment of other HCV culture systems. The JFH-1 strain belongs to genotype 2a, and the J6 HCV strain is the prototype for genotype 2a. J6CF was established as an infectious HCV clone for chimpanzees (Yanagi *et al.*, 1999), however, neither the full-length nor the subgenomic replicon of J6CF replicates in Huh7 cells (Murayama *et al.*, 2007). Therefore, we compared JFH-1 and J6CF, which share similar sequences over the entire HCV genome. A chimeric replicon analysis revealed that the NS3 helicase and NS5B to 3′X regions are important for efficient replication and virus particle formation of the JFH-1 strain (Murayama *et al.*, 2007). There are 29 amino acid differences between JFH-1 and J6CF NS5B, and the different *in vitro* NS5B activity levels are dependent on 6 amino acid mutations at the C terminus (Murayama *et al.*, 2010). Furthermore, a single point mutation at amino acid 561 is important for both the NS5B polymerase activity and the RNA structure important for replicon replication. The coding sequence for 561 amino acids of NS5B overlaps with the loop structure of 5BSL3.2, which is important for the kissing interaction with 3′X SL2. Thus, JFH-1 type NS3 helicase, high polymerase activity, enhanced kissing-loop interactions and optimal viral RNA structure of the 3′UTR are required for JFH-1 replication in cultured cells. Among the mutations, phenylalanine at amino acid 561 is the most important. Interestingly, insertions of this mutation into genotype1b N or 1a H77S strain subgenomic replicons resulted in enhanced RNA replication compared with wild type. Further study is necessary to achieve efficient replication levels with genotype 1 strains. These studies are important to obtain robust cell culture-adapted HCV strains.

Wild-type JFH-1 is the only HCV strain without adaptive mutations that is infectious to chimpanzees and cultured cells. Upon transfection or infection of Huh7 cells, several mutations may arise in its genome to increase its infection efficiency (Zhong *et al.*, 2006; Delgrange *et al.*, 2007; Russell *et al.*, 2008; Bungyoku *et al.*, 2009; Kang *et al.*, 2009; Takahashi *et al.*, 2010). However, most of the point mutations usually do not increase the replication efficiency but do increase the virus particle formation. A possible explanation for this observation is that point mutations that increase the already high replication efficiency of JFH-1 in cultured cells may be lethal by the following reasons. The higher level of replication may be more easily sensed by an intracellular double-stranded RNA sensor such as RIG-I, or it may induce a stress signal that leads to cell cycle arrest or apoptosis. Another explanation is that a greater accumulation of the viral genome in the infected cell may select the viral mutations that induce viral release. Thus, it would be unlikely to obtain a mutated JFH-1 strain with a much higher replication capacity with the current infection system. To induce or select such mutations, it may be necessary to use more permissive cells in which the interferon induction pathways are completely destroyed.

In vivo infectivity and pathogenesis

Recombinant JFH-1 virus was produced from cells transfected with viral RNA synthesized from cloned viral cDNA, and the recombinant virus was infectious for Huh7 cells. The virus particles that were produced were inoculated into a chimpanzee to determine if they were identical to the viruses in the circulation of patients. Although the recombinant JFH-1 virus was infectious for the chimpanzee, the initial experiment resulted in transient viraemia, no antibody seroconversions and no pathological changes in the liver (Wakita *et al.*, 2005). These results may be due to the inoculation of the minimal infectious dose of JFH-1 virus or the monoclonality of the inoculated virus, which was produced from cloned cDNA. The virus genome sequence recovered from the inoculated chimpanzee was identical to the inoculated virus. To further understand the *in vivo* characteristics of the JFH-1 virus, we inoculated either an original JFH-1 patient serum containing ~10^7 copies of HCV RNA or a similar amount of recombinant JFH-1 virus produced from the transfected cells (Kato *et al.*, 2008). Although this chimpanzee developed extended viraemia for 8 weeks after inoculation of the recombinant JFH-1 virus, it did not exhibit liver pathology or seroconversion. Interestingly, the JFH-1 patient serum-inoculated chimpanzee developed prolonged but intermittent viraemia up to 34 weeks after inoculation. The full-length virus genome was sequenced from sera isolated 2 and 7 weeks after HCVcc inoculation, and 2, 19 and 23 weeks after patient serum inoculation. No mutations were found in week 2 serum from the HCVcc-inoculated chimpanzee; however, several mutations accumulated in the viral genome, especially in the patient serum-infected chimpanzee, and the number of mutations increased with time. The mutations were found mainly in the NS2 and NS5A regions, and a single Q-to-R mutation in NS2 at amino acid 838 was commonly found among the viruses from both the HCVcc-inoculated and patient-serum inoculated chimpanzees. A functional analysis of this Q838R mutation revealed that it increases viral particle formation efficiency but not virus release. Mutations found in NS5A may also affect viral RNA replication and/or virus particle formation and release. Further analysis is necessary to determine the effects of the accumulated mutations. Alterations in viral replication and viral particle formation/release may be important for viral persistence mechanisms. A human liver–transplanted uPA/SCID mouse model was established for HCV infection. This mouse is also permissive to JFH-1 virus infection, and infected mice exhibit persistently higher levels of viraemia (Lindenbach *et al.*, 2006; Morikawa *et al.*, unpublished data). These results clearly indicate that the host cellular immunity is quite important for the control of HCV infection. Furthermore, the IL28B haplotype was recently reported to be important for the efficacy of interferon and ribavirin therapy (Ge *et al.*, 2009; Suppiah *et al.*, 2009; Tanaka *et al.*, 2009). Cellular immune responses are mainly regulated by host gene expression. Further studies are necessary in this field.

The cytotoxic effect of viral infection is also an important aspect of the host–virus interaction. HCV has been considered to be a non-cytopathic virus, and liver injury in the chronic HCV carrier has been considered to be a result of the host immune responses against infected hepatocytes. However, immune-compromised patients infected with HCV frequently develop more rapid fibrotic changes in their liver (Firpi *et al.*, 2009). Thus, it is reasonable to consider the possibility that the cytotoxicity may be dependent on the levels of viral replication or other viral factors in the infected liver. In fact, cultures of HCV-infected cells sometimes exhibit massive cell death, which suggests the presence of HCV-induced cytopathic effects. Sekine-Osajima developed an HCV plaque assay to determine the cytopathic effect of HCV infection, and they isolated a JFH-1 variant with high cytotoxicity (Sekine-Osajima *et al.*, 2008). The plaque-purified virus had nine amino acid substitutions, five of which were clustered in the C-terminal of the NS5B region. Furthermore, the ER stress markers, GRP78 and phosphorylated eIF2-alpha, were overexpressed in the infected cells, and the cells in the plaques were strongly positive for an apoptosis marker, annexin V. These results suggest that the cytopathic effect of HCV infection involves ER stress-induced apoptotic cell death. Further research should analyse the importance of cytotoxic effects in acute and persistent HCV infection.

HCVpp, HCVcc and HCV receptor molecules

It has been difficult to analyse the HCV entry step due to the lack of experimental HCV infection systems. CD81, SR-BI and DC-SIGN have been considered as candidate molecules for HCV entry because they bind to recombinant E2 proteins. To overcome the lack of a conventional cell culture system for the propagation of infectious HCV particles, researchers have generated pseudotype viruses expressing the HCV envelope gps (Bartosch *et al.*, 2003; Hsu *et al.*, 2003). This method is based on the retrovirus packaging system of either murine leukaemia virus (MLV) or human immunodeficiency virus (HIV). The HCV E1/E2 protein expression vector was co-transfected into HEK293T cells with a gag-pol plasmid without a packaging signal and a transfer vector encoding a reporter gene with a packaging signal. Secreted pseudotype viruses were infectious in a CD81-dependent manner. This HCV pseudotype system enabled a detailed analysis of the early steps of HCV infection and neutralizing antibodies. There may be some differences in the infection process of HCVpp and HCVcc. Infectious HCVcc is assembled at the surface of lipid droplets, and the association with VLDL and apolipoproteins is important for its infectivity (Huang *et al.*, 2007; Gastaminza *et al.*, 2008). The association between HCVpp, VLDL and apolipoproteins is not clear, and the assembly site of HCVpp should be confirmed. Furthermore, HCVpp may be used for other applications such as tissue-specific gene delivery and vaccine platforms.

Both HCVpp and HCVcc systems provide strong tools to study cellular receptor molecules for HCV infection. Interestingly, CD81 may play an important role after virus binds to the cell surface (Koutsoudakis *et al.*, 2006, 2007; Morikawa *et al.*, 2007). Furthermore, SR-BI and SR-BII are thought to play a role in early infection, but the role of SR-BI/II receptors in HCV infection remains unclear (Grove *et al.*, 2007; Kapadia *et al.*, 2007). Heparan sulfate and heparinase reduce cell-surface HCV binding, and anti-E2 and -E1 antibodies also block HCV infection in cultured cells. These results suggest that HCV first binds to the cell surface via a heparan sulfate proteoglycan at low affinity, after which it may be transferred to high-affinity receptors, such as CD81, which may facilitate virus internalization (Koutsoudakis *et al.*, 2006, 2007; Morikawa *et al.*, 2007). However, CD81 may also be involved in initial virus attachment to the cell surface. Blanchard *et al.* reported that HCV enters target cells by clathrin-mediated endocytosis, followed by fusion within an acidic endosomal compartment (Blanchard *et al.*, 2006). Evans *et al.* reported that claudin-1 (CLDN1), a tight junction component, is a co-receptor for HCV infection (Evans *et al.*, 2007). 293T cells are both CD81- and SR-B1-positive but not permissive for HCV infection. 293T cells became permissive for HCV infection when CLDN1 was ectopically expressed (Evans *et al.*, 2007). Recently, an additional cellular factor, occludin (OCLN), was reported to be another receptor molecule for HCV infection (Ploss *et al.*, 2009). Importantly, the expression of OCLN enabled non-human cells to be permissive for HCV infection. HCVpp can infect NIH3T3 and CHO cells with ectopic expression of CD81, SR-BI, CLDN1 and OCLN. Furthermore, the mouse versions of SR-BI and CLDN1 have the same function as their human counterparts for HCV infection. This result suggests that a double transgenic mouse with human CD81 and OCLN may become a potential infectious animal model of HCV infection. Further studies are necessary to understand how these molecules are cooperatively involved in the process of virus entry. These observations may also explain the tissue tropism of HCV.

Viral life cycle

HCV has a positive-strand RNA genome encoding RNA-dependent RNA polymerase, which transcribes the negative-strand RNA from the positive-strand template and then the positive strand from the negative strand, respectively, and the amplified viral RNA pool in the infected cells. HCV RNA replication occurs at intracellular membrane structures. The viral replication alters the membrane structure, which appears as a membranous web in the replicon cells by electron microscopic analysis (Gosert *et al.*, 2003). This membranous web was induced by only NS4B protein expression (Egger *et al.*, 2002) and was also observed in JFH-1 virus-infected cells along with increased lipid droplets (Rouillé *et al.*, 2006). We found that transiently expressed core protein accumulated on the surface of lipid droplets in human osteosarcoma cells (Darius Virology, 1996). The intracellular membrane structure usually consists of a double leaf structure. However, lipid droplets are surrounded by a single leaf membrane. HCV core proteins are processed by signal peptidase at its C terminus and further cleaved by signal peptide peptidase. Thus, the HCV core protein has only a short membrane-anchoring C terminus, and it can easily move onto the surface of lipid droplets (McLauchlan *et al.*, 2002). In the JFH-1 virus-infected Huh7 cells, lipid droplets were surrounded by core proteins and further surrounded by NS proteins (Miyanari *et al.*, 2007). This association at the surface of lipid droplets, especially between core proteins and NS5A, was quite important for HCV infectious virus particle formation. NS5A is a phosphorylated protein that has hyper and basal phosphorylated forms. Serine residue clusters at the C terminus of NS5A are important for its basal phosphorylated form. Although GFP gene insertion around this serine cluster did not affect viral RNA replication, it did decrease viral particle efficiency. Masaki *et al.* found that mutation or deletion of this serine cluster reduced the phosphorylation level of the basal phosphorylated form and thus abolished the interaction between core and NS5A proteins (Masaki *et al.*, 2008). NS5A is also an RNA-binding protein, and it may function as a cargo protein for the viral genomic RNA. The basal phosphorylated form of NS5A binds to core and then transfers the bound genomic RNA to core to form viral nucleocapsid, although the details of this mechanism remain to be elucidated.

Primary hepatocyte culture and 3D culture system

Immortalized human hepatocytes (IHH) that express HCV core protein were also used to develop a permissive cell line for HCV infection. A similar degree of virus particle production was observed upon transfection of RNA from genotype 1a (H77) and genotype 2a (JFH1) into IHH, suggesting that IHH may support HCV genome replication and virus assembly (Kanda *et al.*, 2006). These results suggest that a number of host factors are involved in the virus–host interaction, thus determining the permissiveness of a host cell for HCV infection. In fact, up-regulation of several genes activated by peroxisome proliferator-activated receptor alpha (PPARalpha) signalling is important for efficient HCV replication in 3D cultures of an immortalized primary human hepatocyte (HuS-E/2) cell line (Aly *et al.*, 2009a,b). This cell culture system supports blood-borne HCV infection and proliferation. Recently, micropatterned cocultures (MPCCs) of primary human hepatocytes and supportive stroma in a multiwell format were described (Ploss *et al.*, 2010). MPCCs form polarized cell layers expressing all known HCV entry factors and sustain viral replication for several weeks. MPCCs have potential as a high-throughput platform for the simultaneous assessment of *in vitro* efficacy and toxicity profiles of anti-HCV therapeutics.

Because the liver is a highly differentiated and structured organ, cultured primary hepatocytes easily lose their liver-specific enzymatic activity such as CYP3A4. A radial flow bioreactor (RFB) was used for three-dimensional (3D) perfusion culture of hepatocellular carcinoma cells supported by the up-regulation of CYP34A expression (Iwahori *et al.*, 2003). Interestingly, full-length HCV RNA-transfected FLC4 cells in an RFB supported the production and release of infectious HCV, although the virus titre in the secreted culture medium was not high (Aizaki *et al.*, 2003). Infectious HCV production was only observed in the 3D cell culture system but not in monolayer cell cultures (Murakami *et al.*, 2006). 3D cell culture was achieved by RFB, the thermoreversible gelatin polymer (3D/TGP) systems and the hollow-fibre culture system (Aizaki *et al.*, 2003; Murakami *et al.*, 2006; Aly *et al.*, 2009b). Thus, it is necessary to elucidate the minimal necessary gene expression to support efficient production of HCV in hepatocytes by the analysis of 3D hepatocyte culture.

Engineering of HCV infection systems: reporter virus, cellular reporter system and trans-complementation system

Recombinant HCV constructs were produced by inserting a reporter gene into the HCV genome. At the 5′ end of the genome, the luciferase gene was inserted like a replicon construct with EMCV IRES gene insertion for expression of HCV ORF (Wakita *et al.*, 2005). This reporter virus enabled virus infection to be monitored by measuring the luciferase activity in the cell lysate of infected cells. Insertion of the reporter gene at the C terminus of NS5A region allowed the HCV replicon to replicate. Infectious reporter virus was also constructed with GFP or luciferase gene insertions (Kim *et al.*, 2007; Masaki *et al.*, 2008). This system is particularly useful for monitoring HCV replication in individual infected live cells. Recently, a cell-based fluorescent reporter system was developed to allow the sensitive detection of individual HCV-infected cells using fusion protein of the reporter and IPS-1 gene connected by the NS3 protease processing sequence (Jones *et al.*, 2010).

Viral RNA replication occurs at the membrane structure in the infected cells, and virus particles are formed at the surface of the lipid droplet. Both regions are very closely located. Viral particle formation is not an independent process from viral genome replication. Interestingly, we can separate the viral genome into two parts. By the subgenomic replicon, which lacks core, E1, E2 and NS2, virus particles are not produced in the replicon cells. Upon co-transfection with core-to-NS2 region expression vector, the subgenomic replicon cells secreted infectious virus particles (Ishii *et al.*, 2008; Steinmann *et al.*, 2008; Adair *et al.*, 2009). However, co-transfection with core-to-E2 and p7 did not recover the infectious virus production from subgenomic replicon cells. Other non-structural viral genes are also rescued *in trans*. The defective JFH-1 genome in NS2, 4B or 5A was complemented by the co-expression of each viral protein *in trans* (Appel *et al.*, 2008; Jirasko *et al.*, 2008; Jones *et al.*, 2009). These *trans*-complementation systems are important tools for a detailed analysis of viral gene functions.

Neutralizing antibody and vaccine development

Since the discovery of HCV, the question of whether HCV infection can be prevented by prophylaxis vaccination has remained. This question has been very difficult to address, because the experimental HCV infection models are limited to chimpanzees. Unlike HBV, HCV infection in adults usually leads to persistent infection. HCV also causes persistent infection in some chimpanzees, although it can be cleared from the circulation after acute infection in other chimpanzees. Interestingly, repeated HCV infections are possible by identical or different HCV inoculations (Farci *et al.*, 1992). Therefore, it is believed that HCV infection does not induce protective immunity. In another chimpanzee experiment, repeated inoculations of synthesized infectious or defective HCV RNA into the liver induced protection from identical HCV challenge (Weiner *et al.*, 2001).

Using an *in vitro* HCV infection system, researchers found that HCV carriers developed neutralizing antibody in their circulation. The reason for the co-existence of infectious HCV and neutralizing antibodies is unclear. However, most established neutralizing antibodies prevent the interaction between HCV E2 protein and CD81, indicating the importance of this interaction in the early HCV infection process, although it includes multiple complex steps. It is still not clear if HCV infection can be prevented with neutralizing antibody induced by vaccination or the passive transfer of immunoglobulin. However, a great deal of research has been focused on the development of an HCV vaccine (Houghton and Abrignani, 2005). Among a cohort of humans who experienced acute HCV infection caused by contaminated blood products, a virus-cleared patient developed a high titre of neutralizing antibody within 6 months after infection, while the persistently infected group of patients developed a lower titre of neutralizing antibody. This result may indicate that neutralizing antibody is important to prevent persistent infection.

The sera of rodents that were experimentally vaccinated with recombinant E1 and E2 protein contained neutralizing antibodies against HCVcc infection in cultured cells (Stamataki *et al.*, 2007). Polyclonal and monoclonal neutralizing antibody protected human liver–transplanted mice from HCV challenge (Law *et al.*, 2008; Vanwolleghem *et al.*, 2008). We recently found that mice immunized with highly purified HCV particles developed antibodies against HCV E2 proteins (Takahashi *et al.*, 2010). Chiron researchers developed a prototype prophylaxis vaccine using recombinant E1 and E2 proteins expressed in CHO cells (Houghton *et al.*, 2005; Stamataki *et al.*, 2007). Vaccinated chimpanzees had a reduced ratio of persistent infection compared with control chimpanzees. It will be very interesting to determine if vaccination with a viral component, such as envelope proteins, or inactivated virus particles can prevent the development of persistent HCV infection. A prophylactic HCV vaccine and immunoglobulin would be beneficial for high-risk groups including healthcare workers, doctors, individuals with certain medical conditions and patient family members. We have developed an HCV particle production and purification system using the JFH-1 clone

(Takahashi *et al.*, 2010; Morikawa *et al.*, unpublished data). It will be important to determine if inactivated purified HCV particles can induce protective immunity. To develop efficient vaccines and immunotherapies, it is important to determine how HCV evades the host immune surveillance and how protective immunity against HCV infection develops.

Antiviral development

Efficient therapies must be developed to eliminate HCV infections from chronic carriers. Interferon and ribavirin have been used in clinics with limited efficacy. NS3 protease inhibitors, NS5B polymerase inhibitors and other drugs are undergoing clinical trials but have not yet been approved for therapeutic use (reviewed in Francesco and Migliaccio, 2005). The HCV infection system that uses the JFH-1 clone may provide a good method for screening new antiviral agents. Furthermore, stable JFH-1 cDNA–transfected cell lines capable of producing infectious virus may be suitable for screening antiviral agents (Cai *et al.*, 2005; Kato *et al.*, 2007). Recently, 2 groups performed antiviral screenings with HCVcc-infected cells (Chockalingam *et al.*, 2010; Gastaminza *et al.*, 2010). Both groups identified compounds that block multiple steps in the HCV life cycle, such as entry, replication and release. These applications are very important to develop many more classes of HCV antivirals. Future combination therapies with multiple classes of antivirals are anticipated to combat HCV infection in a shorter amount of time with less side-effects and a reduced cost. It remains important to further understand the HCV life cycle, because each step needs to be targeted in order to control HCV infection and replication. HCV-infected cell systems enable us to characterize the subcellular localization of HCV structural proteins in the context of the infectious cycle. Interestingly, Rouillé *et al.* have reported the co-localization of core and NS3 proteins in infected cells (Rouillé *et al.*, 2006). This finding may suggest that the interaction of structural and non-structural proteins is important for infectious virus formation. On the other hand, Shirakura *et al.* have identified ubiquitin ligase E6AP as an HCV core-binding protein (Shirakura *et al.*, 2007). E6AP has been observed to bind to the core protein and promote its degradation. The exogenous expression of E6AP decreases intracellular core protein levels, as well as supernatant HCV infectivity titres, in HCV JFH1-infected Huh-7 cells. Furthermore, the knockdown of endogenous E6AP by RNA interference increases intracellular core protein levels and supernatant HCV infectivity titres in HCV JFH1-infected cells. These studies suggest several novel targets for the control of HCV infection.

Conclusions

A robust HCV infection system was first reported in 2005, and the JFH-1 strain is still the only HCV strain that can be propagated in Huh7 cells. This system has been useful for multiple purposes in the HCV research field. The development of antivirals and vaccines is an important goal of research that utilizes HCV infection systems. Furthermore, an HCV infection system is critical for the determination of the viral life cycle. An HCV infection system with genotype 1 HCV has not been sufficiently robust. Future studies are needed to establish a robust infectious genotype 1 HCV clone.

Acknowledgements

This work was supported by a grant-in-aid for Scientific Research from the Japan Society for the Promotion of Science, from the Ministry of Health, Labor and Welfare of Japan, and from the Ministry of Education, Culture, Sports, Science and Technology and by the Research on Health Sciences Focusing on Drug Innovation from the Japan Health Sciences Foundation.

References

Adair, R., Patel, A.H., Corless, L., Griffin, S., Rowlands, D.J., and McCormick, C.J. (2009). Expression of hepatitis C virus (HCV) structural proteins in trans facilities encapsidation and transmission of HCV subgenomic RNA. J. Gen. Virol. *90*, 833–842.

Aizaki H., Nagamori, S., Matsuda, M., Kawakami, H., Hashimoto, O., Ishiko, H., Kawada, M., Matsuura, T., Hasumura, S., Matsuura, Y., *et al.* (2003). Production and release of infectious hepatitis C virus from human liver cell cultures in the three-dimensional radial-flow bioreactor. Virology. *314*, 16–25.

Akazawa D., Date, T., Morikawa, K., Murayama, A., Miyamoto, M., Kaga, M., Barth, H., Baumert, T.F., Dubuisson, J., and Wakita, T. (2007). CD81 expression is important for heterogeneous HCV permissiveness of Huh7 cell clones. J. Virol. *81*, 5036–5045.

Aly H.H., Qi, Y., Atsuzawa, K., Usuda, N., Takada, Y., Mizokami, M., Shimotohno, K., and Hijikata, M. (2009a). Strain-dependent viral dynamics and virus-cell interactions in a novel in vitro system supporting the life cycle of blood-borne hepatitis C virus. Hepatology *50*, 689–696.

Aly H.H., Shimotohno, K., and Hijikata, M. (2009b). 3D cultured immortalized human hepatocytes useful to develop drugs for blood-borne HCV. Biochem Biophys. Res. Commun. *379*, 330–334.

Appel N., Zayas, M., Miller, S., Krijnse-Locker J., Schaller, T., Friebe, P., Kallis, S., Engel, U., and Bartenschlager, R. (2008). Essential role of domain III of nonstructural protein 5A for hepatitis C virus infectious particle assembly. PLoS Pathog. *4*, e1000035.

Arnaud N., Dabo, S., Maillard, P., Budkowska, A., Kalliampakou, K.I., Mavromara, P., Garcin, D., Hugon, J., Gatignol, A., Akazawa, D., *et al.* (2010). Hepatitis C virus controls interferon production through PKR activation. PLoS One *5*, e10575.

Bartenschlager R. and Lohmann, V. (2000). Replication of hepatitis C virus. J. Gen. Virol. *81*, 1631–1648.

Bartosch B., Dubuisson, J., and Cosset, F.L. (2003). Infectious hepatitis C virus pseudo-particles containing functional E1-E2 envelope protein complexes. J. Exp. Med. *197*, 633–642.

Bertolini L., Iacovacci, S., Ponzetto, A., Gorini, G., Battaglia, M., and Carloni, G. (1993). The human bone-marrow-derived B-cell line CE, susceptible to hepatitis C virus infection. Res. Virol. *144*, 281–285.

Blanchard E., Belouzard, S., Goueslain, L., Wakita, T., Dubuisson, J., Wychowski, C., and Rouillé, Y. (2006). Hepatitis C virus entry depends on clathrin-mediated endocytosis. J Virol. *80*, 6964–6972.

Blight K.J., McKeating J.A., and Rice, C.M. (2002). Highly permissive cell lines for subgenomic and genomic hepatitis C virus RNA replication. J Virol. *76*, 13001–13014.

Bungyoku Y., Shoji, I., Makine, T., Adachi, T., Hayashida, K., Nagano-Fujii M., Ide, Y.H., Deng, L., and Hotta, H. (2009). Efficient production of infectious hepatitis C virus with adaptive mutations in cultured hepatoma cells. J. Gen. Virol. *90*, 1681–1691.

Cai Z., Zhang, C., Chang, K.-Y., Jiang, J., Ahn, B.-C., Wakita, T., Liang, J.T., and Luo, G. (2005). Robust production of infectious hepatitis C virus (HCV) from stably HCV cDNA-transfected human hepatoma cells. J. Virol. *79*, 13963–13973.

Chockalingam K., Simeon, R.L., Rice, C.M., and Chen, Z. (2010). A cell protection screen reveals potent inhibitors of multiple stages of the hepatitis C virus life cycle. Proc. Natl. Acad. Sci. U.S.A. *107*, 3764–3769.

Choo Q.L., Kuo, G., Weiner, A.J., Overby, L.R., Bradley, D.W., and Houghton, M. (1989). Isolation of a cDNA clone derived from a blood-borne non-A, non-B viral hepatitis genome. Science *244*, 359–362.

Date T., Kato, T., Miyamoto, M., Zhao, Z., Yasui, K., Mizokami, M., and Wakita, T. (2004). Genotype 2a Hepatitis C Virus Subgenomic Replicon Can Replicate in HepG2 and IMY-N9 cells. J. Biol. Chem. *279*, 22371–22376.

Date T., Miyamoto, M., Kato, T., Morikawa, K., Murayama, A., Akazawa, D., Tanabe, J., Sone, S., Mizokami, M., and Wakita, T. (2007). An infectious and selectable full-length replicon system with hepatitis C virus JFH-1 strain. Hepatol. Res. *37*, 433–443.

Delgrange D., Pillez, A., Castelain, S., Cocquerel, L., Rouillé, Y., Dubuisson, J., Wakita, T., Duverlie, G., and Wychowski, C. (2007). Robust production of infectious viral particles in Huh-7 cells by introducing mutations in hepatitis C virus structural proteins. J. Gen. Virol. *88*, 2495–2503.

Egger D., Wölk, B., Gosert, R., Bianchi, L., Blum, H.E., Moradpour, D., and Bienz, K. (2002). Expression of hepatitis C virus proteins induces distinct membrane alterations including a candidate viral replication complex. J. Virol. *76*, 5974–5984.

Evans M.J., von Hahn T., Tscherne, D.M., Syder, A.J., Panis, M., Wolk, B., Hatziioannou, T., McKeating J.A., Bieniasz, P.D., and Rice, C.M. (2007). Claudin-1 is a hepatitis C virus co-receptor required for a late step in entry. Nature. *446*, 801–805.

Farci P., Alter, H.J., Govindarajan, S., Wong, D.C., Engle, R., Lesniewski, R.R., Mushahwar, I.K., Desai, S.M., Miller, R.H., Ogata, N., *et al.* (1992). Lack of protective immunity against reinfection with hepatitis C virus. Science. *258*, 135–140.

Firpi R.J., Clark, V., Soldevila-Pico C., Morelli, G., Cabrera, R., Levy, C., Machicao, V.I., Chaoru, C., and Nelson, D.R. (2009). The natural history of hepatitis C cirrhosis after liver transplantation. Liver Transpl. *15*, 1063–1071.

Francesco R.D. and Migliaccio, G. (2005). Challenges and successes in developing new therapies for hepatitis C. Nature. *436*, 953–960.

Fournier C., Sureau, C., Coste, J., Ducos, J., Pageaux, G., Larrey, D., Domergue, J., and Maurel, P. (1998). *In vitro* infection of adult normal human hepatocytes in primary culture by hepatitis C virus. J. Gen. Virol. *79*, 2367–2374.

Gastaminza P., Cheng, G., Wieland, S., Zhong, J., Liao, W., and Chisari, F.V. (2008). Cellular determinants of hepatitis C virus assembly, maturation, degradation, and secretion. J. Virol. *82*, 2120–2129.

Gastaminza P., Whitten-Bauer C., and Chisari, F.V. (2010). Unbiased probing of the entire hepatitis C virus life cycle identifies clinical compounds that target multiple aspects of the infection. Proc. Natl. Acad. Sci. U.S.A. *107*, 291–296.

Ge D., Fellay, J., Thompson, A.J., Simon, J.S., Shianna, K.V., Urban, T.J., Heinzen, E.L., Qiu, P., Bertelsen, A.H., Muir, A.J., *et al.* (2009). Genetic variation in IL28B predicts hepatitis C treatment-induced viral clearance. Nature. *461*, 399–401.

Gosert R., Egger, D., Lohmann, V., Bartenschlager, R., Blum, H.E., Bienz, K., and Moradpour, D. (2003). Identification of the hepatitis C virus RNA replication complex in Huh-7 cells harboring subgenomic replicons. J. Virol. *77*, 5487–5492.

Grove J., Huby, T., Stamataki, Z., Vanwolleghem, T., Meuleman, P., Farquhar, M., Schwarz, A., Moreau, M., Owen, J.S., Leroux-Roels G., *et al.* (2007). Scavenger receptor BI and BII expression levels modulate hepatitis C virus infectivity. J. Virol. *81*, 3162–3169.

Houghton M. and Abrignani, S. (2005). Prospects for a vaccine against the hepatitis C virus. Nature. *436*, 961–966.

Hsu M., Zhang, J., Flint, M., logvinoff C., Cheng-Mayer C., Rice, C.M., and McKeating J.A. (2003). Hepatitis C virus glycoproteins mediate pH-dependent cell entry of pseudotyped retroviral particles. Proc. Natl. Acad. Sci. U.S.A. *100*, 7271–7276.

Huang H., Sun, F., Owen, D.M., Li, W., Chen, Y., Gale M Jr, Ye, J. (2007). Hepatitis C virus production by human hepatocytes dependent on assembly and secretion of very low-density lipoproteins. Proc. Natl. Acad. Sci. U.S.A. *104*, 5848–5853.

Iacovacci S., Manzin, A., Barca, S., Sargiacomo, M., Serafino, A., Valli, M.B., Macioce, G., Hassan, H.J., Ponzetto, A., Clementi, M., *et al.* (1997). Molecular characterization and dynamics of hepatitis C virus replication in human fetal hepatocytes infected *in vitro*. Hepatology *26*, 1328–1337.

Ikeda M., Yi, M., Li, K., and Lemon, S.M. (2002). Selectable subgenomic and genome-length dicistronic RNAs derived from an infectious molecular clone of the HCV-N strain of hepatitis C virus replicate efficiently in cultured Huh7 cells. J. Virol. *76*, 2997–3006.

Ishii K., Murakami, K., Hmwe, S.S., Zhang, B., Li, J., Shirakura, M., Morikawa, K., Suzuki, R., Miyamura, T., Wakita, T., and Suzuki, T. (2008). Trans-encapsidation of hepatitis C virus subgenomic replicon RNA with viral structure proteins. Biochem. Biophys. Res. Commun. *371*, 446–450.

Ishii N., Watashi, K., Hishiki, T., Goto, K., Inoue, D., Hijikata, M., Wakita, T., Kato, N., and Shimotohno, K. (2006). Diverse effects of cyclosporine on hepatitis C virus strain replication. J. Virol. *80*, 4510–4520.

Ito T., Mukaigawa, J., Zuo, J., Hirabayashi, Y., Mitamura, K., and Yasui, K. (1996). Cultivation of hepatitis C virus in primary hepatocyte culture from patients with chronic hepatitis C results in release of high titre infectious virus. J. Gen. Virol. *77*, 1043–1054.

Ito T., Yasui, K., Mukaigawa, J., Katsume, A., Kohara, M., and Mitamura, K. (2001). Acquisition of susceptibility to hepatitis C virus replication in HepG2 cells by fusion with primary human hepatocytes: establishment of a quantitative assay for hepatitis C virus infectivity in a cell culture system. Hepatology. *34*, 566–572.

Iwahori T., Matsuura, T., Maehashi, H., Sugo, K., Saito, M., Hosokawa, M., Chiba, K., Masaki, T., Aizaki, H., Ohkawa, K., and Suzuki, T. (2003). CYP3A4 inducible model for in vitro analysis of human drug metabolism using a bioartificial liver. Hepatology *37*, 665–673.

Jirasko V., Montserret, R., Appel, N., Janvier, A., Eustachi, L., Brohm, C., Steinmann, E., Pietschmann, T., Penin, F., and Bartenschlager, R. (2008). Structural and functional characterization of nonstructural protein 2 for its role in hepatitis C virus assembly. J. Biol. Chem. *283*, 28546–28562.

Jones C.T., Catanese, M.T., Law, L.M., Khetani, S.R., Syder, A.J., Ploss, A., Oh, T.S., Schoggins, J.W., MacDonald M.R., Bhatia, S.N., and Rice, C.M. (2010). Real-time imaging of hepatitis C virus infection using a fluorescent cell-based reporter system. Nat. Biotechnol. *28*, 167–171.

Jones D.M., Patel, A.H., Targett-Adams P., McLauchlan J. (2009). The hepatitis C virus NS4B protein can trans-complement viral RNA replication and modulates production of infectious virus. J. Virol. *83*, 2163–2177.

Kanda T., Basu, A., Steele, R., Wakita, T., Ryerse, J.S., Ray, R., and Ray, R.B. (2006). Generation of infectious hepatitis C virus in immortalized human hepatocytes. J. Virol. *80*, 4633–4639.

Kang J.I., Kim, J.P., Wakita, T., and Ahn, B.Y. (2009). Cell culture-adaptive mutations in the NS5B gene of hepatitis C virus with delayed replication and reduced cytotoxicity. Virus Res. *144*, 107–116.

Kapadia S.B., Barth, H., Baumert, T., McKeating J.A., and Chisari, F.V. (2007). Initiation of hepatitis C virus infection is dependent on cholesterol and cooperativity between CD81 and scavenger receptor B type I. J. Virol. *81*, 374–383.

Kato T., Furusaka, A., Miyamoto, M., Date, T., Yasui, K., Hiramoto, J., Nagayama, K., Tanaka, T., and Wakita, T. (2001). Sequence analysis of hepatitis C virus isolated from a fulminant hepatitis patient. J. Med. Virol. *64*, 334–339.

Kato T., Date, T., Miyamoto, M., Furusaka, A., Tokushige, K., Mizokami, M., and Wakita, T. (2003). Efficient replication of the genotype 2a hepatitis C virus subgenomic replicon. Gastroenterology *125*, 1808–1817.

Kato T., Date, T., Miyamoto, M., Sugiyama, M., Tanaka, Y., Orito, E., Ohno, T., Sugihara, K., Hasegawa, I., Fujiwara, K., *et al.* (2005a). Detection of anti-hepatitis C virus effects of interferon and ribavirin by a sensitive replicon system. J. Clin. Microbiol. *43*, 5679–5684.

Kato T., Date, T., Miyamoto, M., Zhao, Z., Mizokami, M., and Wakita, T. (2005b). Non-hepatic cell lines HeLa and 293 cells support efficient replication of hepatitis C virus genotype 2a subgenomic replicon. J. Virol. *79*, 592–596.

Kato T., Matsumura, T., Heller, T., Saito, S., Sapp, R.K., Murthy, K., Wakita, T., and Liang, T.J. (2007). Production of infectious hepatitis C virus of various genotypes in cell cultures. J. Virol. *81*, 4405–4411.

Kato T., Choi, Y., Elmowalid, G., Sapp, R.K., Barth, H., Furusaka, A., Mishiro, S., Wakita, T., Krawczynski, K., and Liang, T.J. (2008). Hepatitis C virus JFH-1 strain infection in chimpanzees is associated with low pathogenicity and emergence of an adaptive mutation. Hepatology *48*, 732–740.

Kim C.S., Jung, J.H., Wakita, T., Yoon, S.K., and Jang, S.K. (2007). Monitoring the antiviral effect of alpha interferon on individual cells. J. Virol. *81*, 8814–8820.

Kiyosawa K., Sodeyama, T., Tanaka, E., Gibo, Y., Yoshizawa, K., Nakano, Y., Furuta, S., Akahane, Y., Nishioka, K., Purcell, R.H., *et al.* (1990). Interrelationship of blood transfusion, non-A, non-B hepatitis and hepatocellular carcinoma: analysis by detection of antibody to hepatitis C virus. Hepatology *12*(4 Pt 1), 671–675.

Koutsoudakis G., Kaul, A., Steinmann, E., Kallis, S., Lohmann, V., Pietschmann, T., and Bartenschlager, R. (2006). Characterization of the early steps of hepatitis C virus infection by using luciferase reporter viruses. J. Virol. *80*, 5308–5320.

Koutsoudakis G., Herrmann, E., Kallis, S., Bartenschlager, R., and Pietschmann, T. (2007). The level of CD81 cell surface expression is a key determinant for productive entry of hepatitis C virus into host cells. J. Virol. *81*, 588–598.

Kuo G., Choo, Q.L., Alter, H.J., Gitnick, G.L., Redeker, A.G., Purcell, R.H., Miyamura, T., Dienstag, J.L., Alter, M.J., Stevens, C.E., *et al.* (1989). An assay for circulating antibodies to a major etiologic virus of human non-A, non-B hepatitis. Science *244*, 362–364.

Law M., Maruyama, T., Lewis, J., Giang, E., Tarr, A.W., Stamataki, Z., Gastaminza, P., Chisari, F.V., Jones, I.M., Fox, R.I., *et al.* (2008). Broadly neutralizing antibodies protect against hepatitis C virus quasispecies challenge. Nat. Med. *14*, 25–27.

Lindenbach B.D., Evans, M.J., Syder, A.J., Wolk, B., Tellinghuisen, T.L., Liu, C.C., Maruyama, T., Hynes, R.O., Burton, D.R., McKeating J.A., and Rice, C.M. (2005). Complete replication of hepatitis C virus in cell culture. Science *309*, 623–626.

Lindenbach B.D., Meuleman, P., Ploss, A., Vanwolleghem, T., Syder, A.J., McKeating J.A., Lanford, R.E., Feinstone, S.M., Major, M.E., Leroux-Roels G., and Rice, C.M. (2006). Cell culture-grown hepatitis C virus is infectious in vivo and can be recultured in vitro. Proc. Natl. Acad. Sci. U.S.A. *103*, 3805–3809.

Lohmann V., Körner, F., Koch, J., Herian, U., Theilmann, L., and Bartenschlager, R. (1999). Replication of subgenomic hepatitis C virus RNAs in a hepatoma cell line. Science. *285*, 110–113.

McLauchlan J., Lemberg, M.K., Hope, G., and Martoglio, B. (2002). Intramembrane proteolysis promotes trafficking of hepatitis C virus core protein to lipid droplets. EMBO J. *21*, 3980–3988.

Masaki T., Suzuki, R., Murakami, K., Aizaki, H., Ishii, K., Murayama, A., Date, T., Matsuura, Y., Miyamura, T., Wakita, T., and Suzuki, T. (2008). Interaction of hepatitis C virus nonstructural protein 5A with core protein is critical for the production of infectious virus particles. J. Virol. *82*, 7964–7976.

Miyamoto M., Kato, T., Date, T., Mizokami, M., and Wakita, T. (2006). Comparison between subgenomic replicons of hepatitis C virus genotypes 2a (JFH-1) and 1b (Con1 NK5.1). Intervirology *49*, 37–43.

Miyanari Y., Atsuzawa, K., Usuda, N., Watashi, K., Hishiki, T., Zayas, M., Bartenschlager, R., Wakita, T., Hijikata, M., and Shimotohno, K. (2007). The lipid droplet is an important organelle for hepatitis C virus production. Nat. Cell Biol. *9*, 1089–1097.

Mizutani T., Kato, N., Saito, S., Ikeda, M., Sugiyama, K., and Shimotohno, K. (1996). Characterization of hepatitis C virus replication in cloned cells obtained from a human T-cell leukemia virus type 1-infected cell line, MT-2. J. Virol. *70*, 7219–7223.

Morikawa K., Zhao, Z., Date, T., Miyamoto, M., Murayama, A., Akazawa, D., Tanabe, J., Sone, S., and Wakita, T. (2007). The Roles of CD81 and glycosaminoglycans in the adsorption and uptake of infectious HCV particles. J. Med. Virol. *79*, 714–723.

Murakami K., Ishii, K., Ishihara, Y., Yoshizaki, S., Tanaka, K., Gotoh, Y., Aizaki, H., Kohara, M., Yoshioka, H., Mori, Y., *et al.* (2006). Production of infectious hepatitis C virus particles in three-dimensional cultures of the cell line carrying the genome-length dicistronic viral RNA of genotype 1b. Virology *351*, 381–392.

Murayama A., Date, T., Morikawa, K., Akazawa, D., Miyamoto, M., Kaga, M., Ishii, K., Suzuki, T., Kato, T., Mizokami, M., and Wakita, T. (2007). The NS3 helicase and NS5B-to-3′ X regions are important for efficient hepatitis C virus strain JFH-1 replication in Huh7 cells. J. Virol. *81*, 8030–8040.

Murayama A., Weng, L., Date, T., Akazawa, D., Gao, Y., Chen, W., Suzuki, T., Kato, T., Mizokami, M., Wakita, T., and Toyoda, T. (2010). Specific RNA structures and mutations implicated for HCV RNA replication and virus particle formation in cultured cells. PLoS Pathog. *6*, e1000885.

Nakabayashi H., Taketa, K., Miyano, K., Yamane, T., and Sato, J. (1982). Growth of human hepatoma cells lines with differentiated functions in chemically defined medium. Cancer Res. *42*, 3858–3863.

Pietschmann T., Lohmann, V., Kaul, A., Krieger, N., Rinck, G., Rutter, G., Strand, D., and Bartenschlager, R. (2002). Persistent and transient replication of full-length hepatitis C virus genomes in cell culture. J. Virol. *76*, 4008–4021.

Pietschmann T., Zayas, M., Meuleman, P., Long, G., Appel, N., Koutsoudakis, G., Kallis, S., Leroux-Roels G., Lohmann, V., and Bartenschlager, R. (2009). Production of infectious genotype 1b virus particles in cell culture and impairment by replication enhancing mutations. PLoS Pathog. *5*, e1000475.

Ploss A., Evans, M.J., Gaysinskaya, V.A., Panis, M., You, H., de Jong Y.P., and Rice, C.M. (2009). Human occludin is a hepatitis C virus entry factor required for infection of mouse cells. Nature.*457*, 882–886.

Ploss A., Khetani, S.R., Jones, C.T., Syder, A.J., Trehan, K., Gaysinskaya, V.A., Mu, K., Ritola, K., Rice, C.M., and Bhatia, S.N. (2010). Persistent hepatitis C virus infection in microscale primary human hepatocyte cultures. Proc. Natl. Acad. Sci. U.S.A. *107*, 3141–3145.

Rouillé Y., Helle, F., Delgrange, D., Roingeard, P., Voisset, C., Blanchard, E., Belouzard, S., McKeating J., Patel, A.H., Maertens, G., *et al.* (2006). Subcellular localization of hepatitis C virus structural proteins in a cell culture system that efficiently replicates the virus. J. Virol. *80*, 2832–2841.

Rumin S., Berthillon, P., Tanaka, E., Kiyosawa, K., Trabaud, M.A., Bizollon, T., Gouillat, C., Gripon, P., Guguen-Guillouzo C., Inchauspe, G., and Trepo, C. (1990). Dynamic analysis of hepatitis C virus replication and quasispecies selection in long-term cultures of adult human hepatocytes infected *in vitro*. J. Gen. Virol. *80*, 3007–3018.

Russell R.S., Meunier, J.C., Takikawa, S., Faulk, K., Engle, R.E., Bukh, J., Purcell, R.H., and Emerson, S.U. (2008). Advantages of a single-cycle production assay to study cell culture-adaptive mutations of hepatitis C virus. Proc. Natl. Acad. Sci. U.S.A. *105*, 4370–4375.

Sekine-Osajima Y., Sakamoto, N., Mishima, K., Nakagawa, M., Itsui, Y., Tasaka, M., Nishimura-Sakurai Y., Chen, C.H., Kanai, T., Tsuchiya, K., *et al.* (2008). Development of plaque assays for hepatitis C virus-JFH1 strain and isolation of mutants with enhanced cytopathogenicity and replication capacity. Virology *371*, 71–85.

Shirakura M., Murakami, K., Ichimura, T., Suzuki, R., Shimoji, T., Fukuda, K., Abe, K., Sato, Fukasawa, M., Yamakawa, Y., Nishijima, M., Moriishi, K., Matsuura, Y., Wakita, T., Suzuki, T., *et al.* (2007). The E6AP ubiquitin ligase mediates ubiquitylation and degradation of hepatitis C virus core protein. J. Virol. *81*, 1174–1185.

Stamataki Z., Coates, S., Evans, M.J., Wininger, M., Crawford, K., Dong, C., Fong, Y.L., Chien, D., Abrignani, S., Balfe, P., *et al.* (2007). Hepatitis C virus envelope glycoprotein immunization of rodents elicits cross-reactive neutralizing antibodies. Vaccine *25*, 7773–7784.

Steinmann, E., Brohm, C., Kallis, S., Bartenschlager, R., and Pietschmann, T. (2008). Efficient trans-encapsidation of hepatitis C virus RNAs into infectious virus-like particles. J. Virol. *82*, 7034–7046.

Sumpter R Jr, Loo, Y.M., Foy, E., Li, K., Yoneyama, M., Fujita, T., Lemon, S.M., and Gale M Jr. (2005). Regulating intracellular antiviral defense and permissiveness to hepatitis C virus RNA replication through a cellular RNA helicase, RIG-I. J. Virol. *79*, 2689–2699.

Suppiah V., Moldovan, M., Ahlenstiel, G., Berg, T., Weltman, M., Abate, M.L., Bassendine, M., Spengler, U., Dore, G.J., Powell, E., *et al.* (2009). IL28B is associated with response to chronic hepatitis C interferon-alpha and ribavirin therapy. Nat. Genet. *41*, 1100–1104.

Takahashi H., Akazawa, D., Kato, T., Date, T., Shirakura, M., Nakamura, N., Mochizuki, H., Tanaka-Kaneko K., Sata, T., Tanaka, Y., *et al.* (2010). Biological properties of purified recombinant HCV particles with an epitope-tagged envelope. Biochem. Biophys. Res. Commun. 395, 565–571.

Tanaka Y., Nishida, N., Sugiyama, M., Kurosaki, M., Matsuura, K., Sakamoto, N., Nakagawa, M., Korenaga, M., Hino, K., Hige, S., *et al.* (2009). Genome-wide association of IL28B with response to pegylated interferon-alpha and ribavirin therapy for chronic hepatitis C. Nat. Genet. *41*, 1105–1109.

Uprichard S.L., Chung, J., Chisari, F.V., and Wakita, T. (2006). Replication of a hepatitis C virus replicon clone in mouse cells. Virol. J. 3, 89.

Vanwolleghem T., Bukh, J., Meuleman, P., Desombere, I., Meunier, J.C., Alter, H., Purcell, R.H., Leroux-Roels G. (2008). Polyclonal immunoglobulins from a chronic hepatitis C virus patient protect human liver-chimeric mice from infection with a homologous hepatitis C virus strain. Hepatology *47*, 1846–1855.

Wakita T., Pietschmann, T., Kato, T., Date, T., Miyamoto, M., Zhao, Z., Murthy, K., Habermann, A., Krausslich, H.G., Mizokami, M., Bartenschlager, R., and Liang, T.J. (2005). Production of infectious hepatitis C virus in tissue culture from a cloned viral genome. Nat Med. *11*, 791–796.

Watashi K., Hijikata, M., Hosaka, M., Yamaji, M., and Shimotohno, K. (2003). Cyclosporin A suppresses replication of hepatitis C virus genome in cultured hepatocytes. Hepatology *38*, 1282–1288.

Weiner A.J., Paliard, X., Selby, M.J., Medina-Selby A., Coit, D., Nguyen, S., Kansopon, J., Arian, C.L., Ng, P., Tucker, J., Lee, C.T., *et al.* (2001). Intrahepatic genetic inoculation of hepatitis C virus RNA confers cross-protective immunity. J. Virol. *75*, 7142–7148.

Yanagi M., Purcell, R.H., Emerson, S.U., and Bukh, J. (1999). Hepatitis C virus: an infectious molecular clone of a second major genotype (2a) and lack of viability of intertypic 1a and 2a chimeras. Virology. *262*, 250–263.

Yi M., Villanueva, R.A., Thomas, D.L., Wakita, T., and Lemon, S.M. (2006). Production of infectious genotype 1a hepatitis C virus (Hutchinson strain) in cultured human hepatoma cells. Proc. Natl. Acad. Sci. U.S.A. *103*, 2310–2315.

Zhang Y.Y., Zhang, B.H., Ishii, K., and Liang, T.J. (2010). Novel function of CD81 in controlling hepatitis C virus replication. J. Virol. *84*, 3396–3407.

Zhao X., Tang, Z.Y., Klumpp, B., Wolff-Vorbeck G., Barth, H., Levy, S., von Weizsacker F., Blum, H.E., and Baumert, T.F. (2002). Primary hepatocytes of *Tupaia belangeri* as a potential model for hepatitis C virus infection. J. Clin. Invest. *109*, 221–232.

Zhong J., Gastaminza, P., Cheng, G., Kapadia, S., Kato, T., Burton, D.R., Wieland, S.F., Uprichard, S.L., Wakita, T., and Chisari, F.V. (2005). Robust hepatitis C virus infection in vitro. Proc. Natl. Acad. Sci. U.S.A. *102*, 9294–9299.

Zhong J., Gastaminza, P., Chung, J., Stamataki, Z., Isogawa, M., Cheng, G., McKeating J.A., and Chisari, F.V. (2006). Persistent hepatitis C virus infection in vitro: coevolution of virus and host. J. Virol. *80*, 11082–11093.

Zhu Q., Guo, J.T., and Seeger, C. (2003). Replication of hepatitis C virus subgenomes in nonhepatic epithelial and mouse hepatoma cells. J. Virol. 77, 9204–9210.

The Chimpanzee Model of Hepatitis C Infections and Small Animal Surrogates

6

Robert E. Lanford, Stanley M. Lemon and Christopher Walker

Abstract

The chimpanzee model of HCV infection has been instrumental in many of the key advances in HCV research and therapy. The demonstration of an infectious agent for non-A, non-B hepatitis, the propensity for persistent infections, and the physical properties of the virus were all determined in the chimpanzee prior to the isolation of HCV. The cloning of HCV was dependent on high titre chimpanzee plasma, and the verification of infectious clones could be accomplished only in this animal model. Immunological analyses in HCV infected chimpanzees have been essential in defining the immune correlates of viral clearance and the failed immune response in persistent infections. The chimpanzee is the only animal suitable for vaccine development, and the knowledge gained from chimpanzee studies has brought us closer to that realization. Analysis of hepatic gene expression in the chimpanzee has revealed signature changes in the innate immune response to HCV and the basis for the lack of response to IFN in null responders. Chimpanzees are often the last stage of preclinical development for antivirals and new therapeutics progressing to human trials. Although the chimpanzee is indispensable for many studies, the marmoset model of GBV-B infection and immunodeficient mice with mouse/human chimeric livers are important surrogate models. The identification of HCV receptors and many host factors essential for replication suggest that the future may hold a mouse fully susceptible to HCV infection.

Introduction

HCV: the disease

The prevalence of hepatitis C virus (HCV) infection is 3% worldwide with an estimated 170 million people persistently infected. Although the overall prevalence of HCV chronicity is 2% in the US, the National Health and Nutrition Exam and Survey (NHANES III; Alter *et al.*, 1999) indicates that certain groups have an increased risk. The adult population (~40–60 years of age) has a prevalence of approximately 4%. The virus has a high propensity for causing persistent infections. Historically, it has been estimated that chronic infections develop in up to 85% of exposed individuals. However, the rate of persistent infection is context dependent and influenced by host genetics. Chimpanzees have a persistent infection rate of approximately 33% (Bassett *et al.*, 1998), and data from some human cohorts suggest that healthy adults with a single exposure may have persistent rates close to 50% (see adaptive immunity below). The disease is typically asymptomatic or mild for several decades, however up to 20% of infected individuals eventually develop cirrhosis, and these individuals are at risk for progression to hepatocellular carcinoma at an estimated rate of 5% per year (Alter and Seeff, 2000; Thomas and Seeff, 2005). End-stage liver disease due to HCV infection is the leading cause of liver transplantation in the US, and HCV-associated liver cancer is now the most rapidly increasing cause of death due to cancer in the US. HCV liver disease is also a leading cause of morbidity and mortality in individuals co-infected with HIV, whose HIV infection is under control with antiviral therapy, but who will experience more rapid progression of liver disease.

Blood products were one of the primary means of transmission prior to the introduction of tests for antibodies to HCV. Although blood products are no longer a significant mode of transmission in developed countries, many individuals with chronic infections today were infected by this route. IVDU is the primary means of transmission today and accounts for most new infections. Sexual transmission appears to be uncommon. Little to no transmission occurs between partners in a monogamous relationship. In general men that have sex with men are not at increased risk, but outbreaks within this population have identified the potential of transmission with certain risk behaviours. Household transmission occurs at a low frequency and is difficult to estimate. The spouse of an infected individual only has a slightly higher risk than other household members, and this increased risk could be explained by other factors. Household transmission probably reflects a lifetime likelihood of being exposed to contaminated blood possibly from cuts and abrasions. Maternal transmission occurs at a rate of 4–5% and is increased in mothers co-infected with HIV. Community acquired HCV is a term used for individuals with no known risk factor. Many of these individuals may have used i.v. drugs at some time in their past but are not willing to divulge this information. However, other methods of transmission occur, including hospital and clinic outbreaks due to improper use of i.v. procedures. No evidence of chimpanzee to chimpanzee transmission has been observed even in instances where infected and uninfected animals have been group housed.

HCV: the virus

HCV is classified within the genus *Hepacivirus* in the family *Flaviviridae*. There are six major genotypes, and extensive genetic heterogeneity within the genus. The genome is single-stranded RNA of positive sense with a single open reading frame (reviewed in Lindenbach and Rice, 2005; Tellinghuisen *et al.*, 2007; Fig. 6.1). The 5′ non-coding region contains an internal ribosome entry site (IRES) that directs the expression of a lengthy polyprotein from a large single open reading frame in the genomic RNA. The amino terminus of the viral polyprotein contains the structural proteins, including the capsid (core) and 2 envelope proteins, E1 and E2. The amino terminus of E2 has a 30 amino acid hypervariable

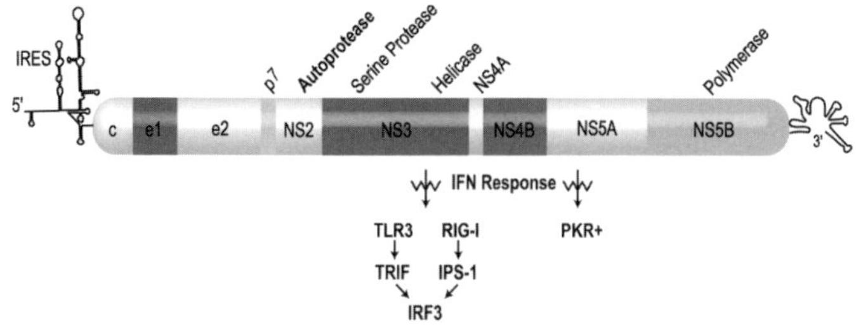

Figure 6.1 HCV genome structure. The single-stranded RNA structure of the HCV genome is illustrated. The single open reading frame and the individual proteins processed from the polyprotein are indicated by a cylinder. The 5′ NCR (non-coding region) containing the IRES (internal ribosome entry site) precede the coding sequences which are followed by a 3′ NCR including a highly conserved 98 nucleotide terminus. The names of the individual proteins are indicated with the structural proteins (core, e1 and e2) located at the amino terminus, and the non-structural proteins (NS2-NS5B) situated at the carboxy-terminus. The functions for some of the proteins are show above the cylinder. Two of the proteins involved in subverting the innate immune response are highlighted below with the cellular pathways inhibited by these proteins. NS3 protease cleaves TRIF and IPS-1, thus blocking IRF3 activation, while NS5A has been shown to interact with and inhibit PKR.

region that is highly divergent among different strains. Antibody responses to this region may result in the selection of neutralizing antibody escape variants that may contribute to persistence of the virus. However, this sequence element is remarkably stable in chronically infected chimpanzees (Bassett *et al.*, 1999; Ray *et al.*, 2000) and can be deleted without eliminating infectivity in the chimpanzee (Forns *et al.*, 2000), and thus is not essential for viability or viral persistence. These structural proteins are processed from the polyprotein by host signal peptidases.

Adjacent to E2 in the polyprotein is a small porin-like molecule, p7, the function of which is essential for production of infectious virus. This is followed by NS2, which plays an essential role in viral assembly through interactions with other viral proteins, and together with the N-terminal sequence of NS3 contains an autoprotease activity that processes the polyprotein at the NS2/NS3 junction. NS3 contains a serine protease in its N-terminal domain that is responsible for polyprotein cleavages between the non-structural proteins and an RNA helicase in its C-terminal domain. The NS3 protease is a major target for antiviral drug development, and its role in subverting the host innate immune response is discussed below. NS4A is a cofactor for the NS3 serine protease. NS4B is a hydrophobic protein involved in membrane association of the replication complex. NS5A is a unique, multi-domain zinc-finger protein that is highly phosphorylated and likely forms a dimer. It has RNA binding activity, is involved in the adaptation of viral RNAs to high replication in cultured cells, and plays an essential role in assembly of infectious particles. It appears to possess several distinct functions in the viral life cycle, none of which is well understood. NS5A binds a number of host proteins, including hVAP which seems to be critical in the adaptation to replication in cell culture, and may also contribute to interferon resistance through interactions with protein kinase R (PKR). NS5B is the viral RNA-dependent RNA polymerase. Most of the non-structural proteins (NS3-NS5B) are thought to participate in the viral replicase complex, while several of these proteins (p7, NS2, NS3 and NS5A) play essential roles in viral assembly as well (Jones *et al.*, 2007, 2009; Appel *et al.*, 2008; Masaki *et al.*, 2008; Dentzer *et al.*, 2009; Hughes *et al.*, 2009; Yi *et al.*, 2009;). It is likely that RNA replication and viral assembly are highly coordinated processes. Following the open reading frame (ORF) is a long 3′ non-coding region that has a variable region, a long polyU/polypyrimidine stretch, and a highly conserved 98 nucleotide terminus.

Chimpanzees as an experimental model

The genetic, physiological, biochemical, and immunological similarities of chimpanzees to humans make them unique models to study several infectious diseases; to test the effects of drugs and novel therapies specifically designed for use in humans; and to test the safety and efficacy of human monoclonal antibodies (VandeBerg and Zola, 2005). Chimpanzees (*Pan troglodytes*) have greater than 98% DNA sequence identity with humans and are unique among experimental animal models in being susceptible to infection with certain human pathogens. Analysis of the chimpanzee genome was published in 2005 (Nature, September 2005, volume 46, multiple articles). The genetic differences between human and chimpanzee genes include about 35 million single-nucleotide changes, five million insertion/deletion events, and various chromosomal rearrangements. Gene duplications account for most of the sequence differences between humans and chimps. Single base pair substitutions account for about half as much genetic change as does gene duplication. Typical human and chimpanzee homologues of proteins differ on average by only two amino acids. About 30% of all human proteins are identical in sequence to the corresponding chimpanzee protein. About 600 genes have been identified that may have undergone strong positive selection in the human and chimpanzee lineages; many of these genes are involved in immune system defence against microbial disease. By comparing human and chimpanzee genes to the genes of other mammals, it has been found that genes coding for transcription factors, such as forkhead-box P2 (FOXP2), have often evolved faster in the human in comparison to the chimpanzee; relatively small changes in these genes may account for some of the morphological differences between humans and chimps. A set of 348 transcription factor genes on average have a 50 per cent increase in amino acid changes relative to other genes when comparing the human and chimpanzee lineages.

The use of chimpanzees in infectious disease research can be traced to the 1960s (Table 6.1). Gibbs *et al.* (1968) demonstrated that Creutzfeldt–Jakob disease could be experimentally transmitted to chimpanzees. In the early 1970s, chimpanzees were found to be susceptible to human hepatitis viral infections (Maynard *et al.*, 1972; Barker *et al.*, 1975; Deinhardt, 1976) and were initially used to demonstrate the safety and efficacy of immunoglobulin preparations against hepatitis A virus (HAV). Subsequently, the safety and efficacy of inactivated and recombinant hepatitis B virus (HBV) vaccine preparations were demonstrated in the chimpanzee model (Table 6.2). The initial vaccine was produced from the plasma of infected individuals. There were obvious concerns over the safety of this vaccine. The current vaccine is manufactured from yeast. It was among the first recombinant DNA products used in man and again the safety of such a product was a major concern. Without safety and efficacy studies in the chimpanzees, these vaccines may have never overcome regulatory and public concerns. The HBV vaccine is now being used to prevent transmission of HBV worldwide, and is one of the most successful strategies to combat cancer every deployed. HBV-associated hepatocellular carcinoma is the fourth leading cause of death due to cancer worldwide.

The emergence of acquired immunodeficiency syndrome (AIDS) in the early 1980s and the demonstration that chimpanzees were susceptible to infection further emphasized their critical role in biomedical research. The first human to animal transmission of infection occurred when blood and plasma from AIDS patients were injected into chimpanzees (Alter *et al.*, 1984).

Table 6.1 Chimpanzees in research

1960	NIH funded chimpanzee breeding; end of importation and implementation of CITES
1986	Increased breeding for HIV vaccine development
1997	NIH imposed moratorium on breeding
2005	Chimp Haven for retired NIH chimpanzees
2007	Fewer than 1000 chimpanzees in captivity nationally; rapidly ageing population
Future	Endangered existence in wild; population reduced from 2 million to 150,000
Threats	Habitat encroachment, poaching, Ebola disease

CITES: Convention on International Trade in Endangered Species; HIV: human immunodeficiency virus; NIH: National Institutes of Health.

Table 6.2 Highlights of chimpanzee research

Development of HBV vaccine used worldwide
Development of HAV vaccine
Testing of HIV vaccines that advanced to human trials
Characterization of NANBH infections
Cloning of HCV from chimpanzee serum
Validation of infectious clones of HCV
Vaccine trials for HCV
Last stage of preclinical testing for HCV antivirals
Safety testing of humanized monoclonal antibodies

HAV: hepatitis A virus; HVB, hepatitis B virus; HCV: hepatitis C virus; NANBH: non-A, non-B hepatitis.

Additional studies established the vital role of the chimpanzee model in understanding the dynamics of human immunodeficiency virus (HIV) infection (Murthy *et al.*, 1996), and for evaluating the safety and efficacy of passive and active vaccine strategies and gene therapy for AIDS. Safety and efficacy of several strategies to protect against HIV were first tested in chimpanzees and went on to phase I and II trials, including active/passive vaccination (Murthy *et al.*, 1996), recombinant gp120 subunit vaccines (Berman *et al.*, 1996; el-Amad *et al.*, 2007), and human anti-HIV immunoglobulin (HIVIG) preparations. Chimpanzees are susceptible to infection with HIV as well as SIVcpz; however, no depletion of $CD4^+$ T cells occurs and there is no progression to AIDS. Today, most preclinical trials for AIDS vaccines focus on the SHIV infected rhesus model. Experimental infection with HIV results in relatively robust viraemia that wanes over 6–12 months when most infected animals become aviremic. However, cross-sectional analysis of a large number of chimpanzees infected with HIV more than a decade previously has revealed that many animals have viral loads of up to 10^5 GE/ml (Lanford, unpublished data). It is still not certain whether long-term infections in chimpanzees with HIV will result in adverse outcomes in some animals later in life.

Chimpanzees will continue to be the model of choice for many research programmes where susceptibility to human pathogens is critical and for several novel immune-based strategies that are being developed for clinical application in organ transplantation, cancer, and autoimmune diseases. Many of these strategies are based on monoclonal antibodies for blocking the activation of specific cell types, inducing the ablation of a given cell population, or neutralization of cytokines and chemokines. Because of the close genetic relationship between humans and chimpanzees, antibodies developed against human antigens normally cross-react with the corresponding antigens in chimpanzees. Similarly, chimpanzees respond to most recombinant human cytokines/chemokines in a manner identical to that observed in humans. The failure to evaluate safety in chimpanzees of a monoclonal anti-CD28 led to the near fatal activation of cytokine pathways (cytokine storm) in healthy human volunteers with resulting organ failure (Suntharalingam *et al.*, 2006). Chimpanzees continue to be the only animal model susceptible to HCV infection (reviewed in Lanford *et al.*, 2001a; Lanford, 2006; Walker, 1997; Prince and Brotman, 1994). Similar to humans, both viral clearance and persistent viraemia in HCV-infected chimpanzees have been observed. There are a number of advantages to studying HCV infections in chimpanzees, including the opportunity to induce infections with well-characterized HCV inocula, the ability to obtain liver tissue samples during the course of infection, and most importantly, the availability of specimens from the time of infection for the analysis of events early in infection.

The NANBH era

The development of assays capable of discerning HAV and HBV infections in the clinic lead to the observation that many infections were due to uncharacterized agents to be designated simply as non-A, non-B hepatitis (NANBH; Feinstone *et al.*, 1975). A major advance in defining the agent was the demonstration of infectivity in the chimpanzee, thus establishing an animal model (Alter *et al.*, 1978; Hollinger *et al.*, 1978; Table 6.3). Infections were defined by several difficult to characterize changes including a rise in serum levels of the liver enzyme ALT, histological changes consistent with inflammatory disease in the liver, and perhaps the most definitive change, the appearance of membranous tubules in the cytoplasm of hepatocytes (Shimizu *et al.*, 1979). The difficulty of using membranous tubules as the primary diagnostic tool is highlighted by the requirement for both liver biopsy and electron microscopy for detection. Nonetheless, the availability of an animal model led to a characterization of the transmission of the agent from various blood products, pathogenesis of the acute and persistent infection, and the physical properties of the viral particle, all in the absence of an assay for detection of the causative agent. The loss of infectivity following treatment with organic solvents suggested that the virus possessed a lipid envelope (Bradley *et al.*, 1983; Feinstone *et al.*, 1983), while filtration studies determined

Table 6.3 Non-A, non-B hepatitis to HCV

1975	NANBH transmission recognized in patients
1978	NANBH agent transmitted to chimpanzees
1989	HCV cloned from chimpanzee serum
1990	HCV antibody assay developed to screen donor blood
1997	Validation of infectious clones in chimpanzees
1999	Development of replicon model in Huh7 cells
2005	*In vitro* replication of HCV-JFH1 in Huh7.5
2006	Transmission from culture to chimp and back to culture

HCV: hepatitis C virus; NANBH: non-A, non-B hepatitis.

the size to be 50–80 nM (Bradley, 1985; He *et al.*, 1987). The demonstration of a lipid envelope spurred development of methods to inactivate the agent in blood products. Coagulation products used for the treatment of haemophilia and immunoglobulins were routinely contaminated with NANBH, since large pools of plasma were used to produce these products, and it was estimated that 10% of donor blood was contaminated. The early methods of UV light in the presence of beta-propiolactone (Prince *et al.*, 1985), would evolve to the solvent detergent methods still used today, all of which relied on safety testing in chimpanzees (Prince *et al.*, 1984, 1987; Horowitz *et al.*, 1993). Long-term follow-up of individuals in the clinic as well as chimpanzees revealed that NANBH was capable of persistent infections (Bradley *et al.*, 1981) and progression of liver disease. These studies were not without problems, repeated claims of antigen and antibody detection that could not be confirmed, reports of multiple agents being involved, and belief that at least one of the agents was a retrovirus, all caused considerable confusion in the field at different times.

A clone at last

Despite the progress in defining the characteristics of the virus inducing non-A, non-B hepatitis by *in vivo* infectivity studies in chimpanzees, the isolation of the virus, detection of an antibody to the virus or cloning of a fragment of the genome eluded many scientific groups around the world for over ten years. The eventual cloning would come from a breakthrough at Chiron laboratories by a team led by Michael Houghton. Using lambda gt11 phage expression libraries produced from high titre chimpanzee serum and NANBH patient serum with elevated ALT as the antibody, a single clone of approximately 150 nucleotides (clone 5-1-1) was detected (Choo *et al.*, 1989). The dependence of this feat on the exceptional skill and perseverance on the part of the scientist involved is without doubt, but the essential role of the chimpanzee must also be appreciated. The cloning was accomplished using chimpanzee serum that had been show to have a particularly high infectious titre by Dan Bradley at the CDC. The fact that a single clone was detected out of millions, suggests that the clone would have been missed if the viral titre of the starting material had not been of such high level. In fact, production of additional libraries from the same source failed to produce a second clone of HCV (Houghton, 2009). Today, detection of serum with higher HCV RNA levels is not unusual, but when one does not know which serum to use in quest for an unknown virus, the validated chimpanzee serum was an invaluable resource. The use of the 5-1-1 clone permitted selection of overlapping clones until the entire open reading frame of the polyprotein had been captured.

The initial clone was rapidly validated by expression of the antigen and detection of antibodies from multiple cohorts of NANBH patient (Alter *et al.*, 1989). The most influential proof came from cracking the elusive panel compiled by Harvey Alter (Alter and Houghton, 2000). The claims and hopes of a number of investigators had ended with the inability to correctly decipher the Alter Panel, a blinded panel comprised serum from patients with hepatitis of different diagnosis. Chiron developed the first generation antibody test to screen the blood supply using only the 5-1-1 antigen, but this assay was soon replaced by a more sensitive multiple antigen ELISA, thus brining to a close the era of transfusion associated NANBH.

Enter molecular biology and infectious clones

The isolation of the first clone containing a fragment of HCV led to the rapid development of full-length clones of the open reading frame and studies on polyprotein processing and viral functions (reviewed in Lindenbach and Rice, 2005). However, the development of an infectious cDNA clone of HCV proved to be a particularly arduous task for a number of reasons. Initial attempts failed due to the lack of the 3′-terminal sequence. The cloning of the 3′ terminus of HCV was hampered due to the low levels of starting RNA for amplification, and the lack of a polyA tail from which to prime cDNA synthesis (Tanaka *et al.*, 1995, 1996; Kolykhalov *et al.*, 1996; Blight and Rice, 1997). However, even after the entire sequence was available, the validation of infectious clones was not straightforward. In the absence of a tissue culture system, the only assay for infectivity was the direct intrahepatic inoculation of a chimpanzee with synthetic RNA derived from the clones. Kolykhalov *et al.* (1997) inoculated 34 full-length clones into chimpanzees without identification of an infectious clone. Molecular research was clearly hampered by the unknown validity of reagents in the lab. Ultimately, the first infectious clones were produced by making a consensus sequence based on multiple full-length sequences from the same starting material (Kolykhalov *et al.*, 1997; Yanagi *et al.*, 1997). Subsequently, several infectious clones were constructed and validated in the chimpanzee by correcting non-consensus residues identified by alignment with unrelated HCV sequences of the same genotype present in GenBank (Hong *et al.*, 1999; Beard *et al.*, 1999; Lanford *et al.*, 2000; Yanagi *et al.*, 1998, 1999).

The development of infectious cDNA clones of the virus provided a number of new opportunities in HCV research including reverse genetics for analysis of sequence motifs essential for replication; however, a tissue culture system would not be forthcoming simply by the identification of these clones. The testing of mutant clones continued to be dependent on intrahepatic inoculation of chimpanzees with synthetic RNA. The essential nature of the conserved 98 nucleotide stem–loop structure at the 3′ NCR was demonstrated in the chimpanzee in this manner (Yanagi *et al.*, 1999; Kolykhalov *et al.*, 2000). Even after the development of the replicon model, the essential nature of p7 (Sakai *et al.*, 2003) and the non-essential nature of the E2 HVR1 (Forns *et al.*, 2000) were demonstrated in the chimpanzee, since these elements were not required in the replicon.

The replicon model (Lohmann *et al.*, 1999) was the first system truly useful for reverse genetics, but was restricted to the analysis of functions required for RNA replication. Perhaps, the greatest importance of the replicon model was in drug discovery and analysis of resistance mutants. The vast majority of drugs entering human clinical trials were either developed or perfected using this model. The limitations of the replicon model were evident from studies in chimpanzees demonstrating that three adaptive mutations which provided high levels of RNA replication *in vitro* rendered the Con1 infectious clone non-infectious *in vivo*. Even a single NS5A mutation severely attenuated *in vivo* replication and rapidly reverted to the WT sequence (Bukh *et al.*, 2002). Although the genotype 2a JFH1 strain was isolated from a patient with fulminant hepatitis and led to the full replication cycle in HuH7.5 cells without a requirement for adaptation (Wakita *et al.*, 2005), the *in vitro*-derived virus and related derivatives were severely attenuated *in vivo* in chimpanzees with low level replication and very early clearance (Lindenbach *et al.*, 2006; Kato *et al.*, 2008). Adaptation of the H77c strain for replication in Huh7 cells was accomplished using a series of mutations; this modified sequence (H77S) produces infectious virus when transfected as RNA into permissive cells, but with infectious virus yields well below JFH1 (Yi *et al.*, 2006). The limitations that prevent this sequence from fully evolving for more efficient replication in cell culture are unknown, but cell culture-derived H77S virus was infectious in a chimpanzee, in which it replicated to levels similar to JFH1 after a period of re-adaptation to the chimpanzee. In contrast to the rapid clearance of JFH1 and related chimeric viruses from chimpanzees, H77-S induced persistent infection (Yi, Bourne, Lanford and Lemon, unpublished data). Thus, despite the advances in molecular biology and *in vitro* cultivation methods, no virus produced *in vitro* initiates a typical infection *in vivo* and natural isolates cannot be grown in the current culture systems. The chimpanzee will remain an essential method of validation until this can be accomplished.

Innate immune response

The host response to viral RNA and subversion of the response by HCV proteins

A short introduction to the recognition of viruses by the innate immune system and the interaction of HCV viral proteins with this system is warranted to completely appreciate the studies conducted in chimpanzees on the innate response to HCV infection. A number of host proteins sense the presence of viral RNA and stimulate the production of type I IFN (IFNα and IFNβ) through complex signalling pathways. In infected hepatocytes, viral RNAs containing different 'pathogen-associated molecular patterns' (PAMPs) are recognized in the endosomal compartment by Toll-like receptor 3 (TLR3) and in the cytosol by DExD-box helicase receptors (RIG-I and MDA5). These pathogen recognition receptors act as sensors, and signal the activation of the transcription factors interferon regulatory factor 3 (IRF-3) and NF-κB through specific adapter proteins and kinases of the IKK complex. Recent studies have revealed the complexity of the RNA structures recognized by these sensor proteins (Rehwinkel and Reis e Sousa, 2010). For example, RIG-I recognizes RNA with a 5′ triphosphate and double-stranded RNA structure, or in the case of HCV, a poly-U stretch in the 3′ non-coding RNA (Saito *et al.*, 2008). In contrast, MDA5 recognizes lengthy segments of dsRNA, including lengthier forms of the synthetic agonist poly-I:C. TLR3 recognizes dsRNA segments of 40–50 bps. In addition, PKR binds dsRNA and becomes activated by phosphorylation; it contributes to the activation of NF-κB. IRF-3 and NF-κB act in a coordinate fashion with other transcription factors to promote the transcription of IFNβ. IRF-3 also directly induces the transcription of a subset of 'interferon stimulated genes' (ISGs), while NF-κB promotes the expression of proinflammatory cytokines. Thus, in hepatocytes and other cells of epithelial origin, there are two major pathways by which RNA viruses are sensed (the endosomal TLR3 pathway and the cytosolic RIG-I-like helicase pathway), both of which lead to common signalling pathways downstream of IRF-3 and NF-κB that induce expression of IFNs and other cytokines. Other cell types have different sensing pathways, and it is important to recognize that plasmacytoid dendritic cells (pDCs) are the major producers of IFNα in many viral infections. The recognition of viral RNAs by pDCs occurs through a very different signalling pathway involving TLR7.

HCV infection, both in chimpanzees and humans, evokes a strong IFN response within the liver (Su *et al.*, 2002; Bigger *et al.*, 2001, 2004; Sarasin-Filipowicz *et al.*, 2008), as described in greater detail below. Despite this, *in vitro* studies show that HCV has evolved a number of mechanisms for subverting these redundant innate immune responses. The NS3/4A protease blocks both TLR3 and RIG-I pathways and prevents the activation of IRF-3 (Foy *et al.*, 2003, 2005; Li *et al.*, 2005a) and NF-κB. The TLR3 pathway is blocked by cleavage of the adaptor protein TRIF (Li *et al.*, 2005a; Wang *et al.*, 2009) while the RIG-I pathway is blocked by NS3/4A cleavage of the adapter protein MAVS (aka IPS-1, CARDIF, or VISA; Li *et al.*, 2005b; Meylan *et al.*, 2005). Studies of IFNα induction in primary hepatocytes demonstrated expression of both TLR3 and RIG-I by hepatocytes and their induction by IFNα (Lanford *et al.*, 2006). This positive feedback of IFNα on TLR3 and RIG-I expression in the liver may contribute to the continued high ISG expression levels observed in many patients during HCV infection, especially by sensitizing uninfected or newly infected cells prior to suppression of the signalling pathways by NS3/4A. However, as mentioned above, in other viral systems pDCs produce most of the type I IFN that is expressed systemically, via sensing of viral

RNAs by TLR7. It is likely that pDCs are induced to express type I IFN by HCV infection, but there are few data available specific to this question. There is no evidence that HCV has evolved a specific mechanism to block TLR7 signalling within pDCs, and pDCs remain a very likely suspect for the origin of IFN responses in the HCV-infected liver.

NS5A may contribute to the subversion of the innate response as well. NS5A is a phosphoprotein that contains a sequence known as the 'interferon sensitivity-determining region' (ISDR). Sequence diversity in this region has been shown to correlate with the outcome of IFNα therapy (Enomoto *et al.*, 1996; Miyamura, 1996). This region of NS5A interacts with and inhibits PKR (Gale *et al.*, 1998), but NS5A also exhibits PKR-independent mechanisms of blocking the IFN response (Podevin *et al.*, 2001; Polyak *et al.*, 2001; Tan and Katze, 2001; Girard *et al.*, 2002). Recently studies have suggested, rather counter-intuitively, that HCV also suppresses ISG protein translation via PKR activation and general suppression of host protein translation (Garaigorta and Chisari, 2009). Thus, NS5A may play multiple roles in blocking the response to exogenous IFNα, while NS3/4a blocks signalling pathways inducing the production of type I IFN by infected hepatocytes.

The secretion of INFβ and IFNα result in their binding to the type 1 IFN receptor, a heterodimer of IFNAR1 and IFNAR2, and induction of JAK1 and TYK2 tyrosine autophosphorylation. This leads to the phosphorylation of STAT1 and STAT2 which heterodimerize and interact with IRF9 to form ISGF3, which is the primary transcription factor inducing genes with an ISRE. Hundreds of ISGs are stimulated by this pathway, while others genes are repressed. Altered expression of over 1000 ISGs in peripheral blood mononuclear cells (PBMCs) and/or liver were detected by microarray analysis following IFNα dosing of uninfected chimpanzees (Lanford *et al.*, 2006). Nonetheless, there is considerable evidence that JAK-STAT signalling may be directly compromised in HCV-infected cells by one or more specific mechanisms (Lemon, 2010). IRF-3 and IRF-7 play roles in maintaining the production of IFN. IFN production is controlled through both positive and negative feedback regulation at multiple levels within these pathways.

Hepatic innate immune response during acute and chronic HCV infection

Surprisingly, the first reliable marker of infection with NANBH, the cytoplasmic membranous tubules detected by electron microscopy (Shimizu *et al.*, 1979), is induced by the innate response to the virus. This was determined following isolation and cloning of the host protein responsible for these changes (Shimizu and Purcell, 1989; Shimizu *et al.*, 1990). More than two decades after the description of this first HCV-induced ISG in chimpanzee liver, microarray analysis would reveal dramatic up-regulation of hundreds of genes during HCV infection. The analysis of an HCV-infected chimpanzee liver for changes in gene expression by high-density oligonucleotide microarray analysis represented the first use of this new technology to examine the progressive changes in the transcriptome during an infectious disease in an animal model (Bigger *et al.*, 2001; Su *et al.*, 2002). Microarrays had been used to monitor changes in cellular gene expression associated with infection of tissue culture cells with human immunodeficiency virus (HIV), human papillomavirus (HPV) and human cytomegalovirus (HCMV; Geiss *et al.*, 2000; Chang and Laimins, 2000; Zhu *et al.*, 1998). These studies revealed that a relatively small number of genes (e.g. 258 for HCMV) changed in response to viral infection, indicating that the technique provided a useful and manageable volume data, but sceptics still believed it would be difficult to obtain reproducible data *in vivo*.

One of the advantages of the chimpanzee model of HCV infection is the uniformity of the animals. They are genetically more closely related than members of various racial human populations, the inoculum is identical between subjects and well defined, and it is administered in a defined manner and at a defined time. None of these latter conditions hold for infected humans. Moreover, infected animals can be maintained under highly standardized environmental conditions, in contrast to the diversity of factors human cohorts are exposed to that confound clinical studies (obesity, insulin resistance, alcohol consumption, smoking, iron intake, numerous prescription and non-prescription medications including natural remedies, to mention just a few factors impacting the liver). Even the time of day for the biopsy can have tremendous impact on liver gene expression, since it is closely involved in the circadian rhythm. For these reasons, analysis of very small populations of chimpanzees under well-controlled conditions can often provide insight that is difficult to obtain from larger, more diverse human cohorts. In addition, while protocols involving multiple liver biopsies are possible in human subjects, these are extraordinarily difficult to recruit patients to and the number of biopsies possible is typically limited to two. Most important, it is simply not possible to acquire pre-infection baseline samples of liver tissue, or serial samples of liver during acute HCV infection in humans, since the time of infection is rarely known in humans.

Studies of hepatic gene expression during acute HCV infection revealed a rapid innate immune response occurring days after infection involving the induction of hundreds of ISG transcripts, some of which are elevated by 100-fold or more (Bigger *et al.*, 2001; Su *et al.*, 2002). The ISG response parallels the rise and decline of viraemia during acute-resolving infection, leaving no doubt that the changes are in response to viral RNA. Viraemia reaches a plateau early in acute infection, despite the fact that the T cell response is delayed by 8–12 weeks. The percentage of infected cells in the liver has been extrapolated from the level of viral RNA, suggesting that less than 10% of hepatocytes are infected during the plateau phase (Bigger *et al.*, 2004; Lanford *et al.*, 2006, 2007). These estimates are in agreement with calculations based on the production viral particles per day (10^{12} or approximately one per hepatocyte; Neumann *et al.*, 1998), and recent advances in detection of infected cells using multiphoton microscopy (Liang *et al.*, 2009). One possible conclusion from these studies is that host responses limit the spread of the virus in the liver. While adaptive immunity is a possibility (see below), the innate immune response and induction of antiviral ISGs are also likely candidates. A second wave of changes in gene expression accompanies T cell responses. This first becomes apparent 8 or more weeks after infection (see below) and correlates with a decline in viraemia. This phase of infection involves changes in many genes associated with the adaptive immune response, including IFNγ, CXCL9, lymphocyte markers, proinflammatory cytokines and chemokines, and HLA class II genes.

The analysis of hepatic gene expression during chronic HCV infection in chimpanzees has revealed a pattern of ISG expression that is remarkably similar to that observed during acute HCV infection. However, in contrast to the rise and fall of ISG expression in parallel with viral RNA in acute infection, the magnitude of induction of various ISGs during chronic infection was similar in animals with a 1000-fold variation in viraemia, suggesting that the response is maximally induced at relatively low levels of viral RNA (Bigger *et al.*, 2004). In contrast, studies in HCV-infected humans revealed highly divergent hepatic ISG responses, generally classified as high or low ISG response. Individuals with high pretreatment ISG responses have weaker antiviral responses to therapy, typically pegylated IFN (Peg-IFN) and ribavirin (Chen *et al.*, 2005; Feld *et al.*, 2007; Sarasin-Filipowicz *et al.*, 2008). The data suggest that antiviral ISG effector functions are disrupted, or that ISGs are not being expressed specifically within infected cells, in individuals who have high-level pretreatment ISG expression and poor therapeutic responses to Peg-IFN. Similar observations were made by measuring the serum levels of IP-10 (CXCL10); high serum levels of IP-10 negatively correlate with the outcome of IFNα therapy (Butera *et al.*, 2005; Diago *et al.*, 2006; Lagging *et al.*, 2006; Romero *et al.*, 2006). Serum levels of IP-10 can be considered a surrogate marker for hepatic ISG levels during HCV infection, since most IP-10 produced is secreted by the liver.

Chimpanzees chronically infected with HCV typically have no measurable decline in viraemia in response to IFNα therapy (Lanford *et al.*, 2000, 2007; Huang *et al.*, 2007). Gene expression studies were undertaken to elucidate the mechanism responsible for the lack of antiviral response. Following a single dose of IFNα, increases in ISG transcript abundance were observed only in PBMCs from chronically infected animals, and not in the liver, while robust ISG responses were observed in both PBMCs and liver of uninfected animals (Lanford *et al.*, 2007; Huang *et al.*, 2007). This is consistent with the concept that the ISG response is maximally induced in the liver of chronically infected chimpanzees, and thus cannot be further induced by exogenous IFNα. Thus, chimpanzees represent the extreme in the non-responsive, high ISG phenotype and are a model for the human 'null IFN response' phenotype. Studies in humans with paired liver biopsies have since confirmed that ISG expression is induced from a low baseline in rapid responders to IFN, while high levels of ISG expression prior to treatment define patients who are unresponsive to exogenous IFN (Sarasin-Filipowicz *et al.*, 2008). The factors contributing to the relationship between hepatic ISG levels and response to therapy may precede the development of a chronic infection, and may be reflective of genetic polymorphisms in innate immune response genes, such as IFN-λ3 (aka IL28B; Ge *et al.*, 2009; Suppiah *et al.*, 2009; Tanaka *et al.*, 2009). However, it is the individuals who lack a pretreatment ISG response to the virus (low ISG phenotype) who respond favourably to IFNα therapy for reasons that are poorly understood.

Adaptive immune response

The correlates of a successful adaptive immune response to HCV have been reviewed in depth elsewhere and will only be briefly discussed here (Bowen and Walker, 2005b; Rehermann and Nascimbeni, 2005; Rehermann, 2009). The adaptive response of chimpanzees to HCV infection appears to be similar to humans in most aspects, with both acute-resolving and persistent infections being observed. The rate of viral clearance may be higher in chimpanzees, but this could be a reflection of the cohorts examined. Analysis of several hundred chimpanzees at different facilities for viral RNA and antibody response to HCV suggested that only 33% of the animals with a documented infection (antibody positive) progressed to persistent infections (Bassett *et al.*, 1998; Lanford, 2006). Although many of these animals may have been experimentally manipulated; vaccine studies did not appear to have influenced the outcome of the analysis. This persistent rate is considerably lower than what is observed in most human cohorts. Some of the difference may be due to the nature of the exposure. Intravenous drug users are multiply exposed and transfusion recipients are exposed to high levels of virus at a time when they are medically compromised. The dose of virus appears to play an important role in clearance in humans. Analysis of polymorphisms in HLA-C and NK cell KIR receptors that influence viral clearance following low dose exposure, suggest that the dose of virus received during transfusion overwhelms the innate immune response (Khakoo *et al.*, 2004) and leads to persistent infections. Data from several types of human cohorts at least partially support a higher rate of viral clearance in humans (Koziel *et al.*, 1997; Barrett *et al.*, 1999; Scognamiglio *et al.*, 1999; Vogt *et al.*, 1999; Wiese *et al.*, 2000; Larghi *et al.*, 2002) and a few of these studies suggest that the clearance rate in young, healthy individuals may approach 50% (Barrett *et al.*, 1999; Kenny-Walsh and Irish, 1999; Wiese *et al.*, 2000; Larghi *et al.*, 2002).

The adaptive immune response to HCV including the antibody and T cell response is highly delayed for reasons that are not understood and normally does not become detectable until at least 8 weeks after infection, coincident with the appearance of liver disease or a rise in serum ALT levels. The delay in adaptive immune response is not likely due to a failure of the innate response to orchestrate the adaptive response, since a robust ISG response is detected in the liver within days. T cell immune responses are very similar in humans and chimpanzees. In both species acute resolution of infection is correlated with a sustained, robust virus-specific $CD4^+$ and $CD8^+$ T cell response (Ishii and Koziel, 2008; Thimme *et al.*, 2008). In chimpanzees that have cleared HCV and are rechallenged, the magnitude and duration of viraemia are dramatically reduced or undetectable (Bassett *et al.*, 2001; Major *et al.*, 2002; Nascimbeni *et al.*, 2003; Lanford *et al.*, 2004; Prince *et al.*, 2005; Bukh *et al.*, 2008;). Humans with prior exposure and clearance have some protection from chronic infection following subsequent exposures as well (Mehta *et al.*, 2002; Osburn *et al.*, 2009). These studies alone suggest that the adaptive immune response plays an essential role in viral clearance; however, a strong innate response is detected in the liver during rechallenge including the chemokines that recruit non-specific macrophages, NK cells and dendritic cells that are involved in the secretion of proinflammatory cytokines. Thus, it is not possible to entirely discount the role of the innate response in the elimination of infected cells or clearance of viral RNA by non-cytolytic mechanisms. The most compelling evidence of the essential nature of the T cell response comes from antibody depletion of T cell subsets in the setting of rechallenge studies in the chimpanzee. Antibody mediated depletion of CD4 T cells prior to rechallenge in an animal that has previously cleared virus results in persistent infection, thus there is a continuous requirement for CD4 help of the CD8 CTLs (Grakoui *et al.*, 2003). Depletion of CD8 T cells in the same setting prolongs the viremic phase until HCV-specific CD8 T cells return to the liver (Shoukry *et al.*, 2003).

Although persistence was originally thought to be associated with a weak $CD8^+$ T cell response, characterized by low frequencies of effector cells that targeted few epitopes (Diepolder *et al.*, 1995; Cooper *et al.*, 1999; Lechner *et al.*, 2000; Thimme *et al.*, 2001), there is an emerging consensus that priming and expansion of the $CD8^+$ T cells is not necessarily defective in all persistent HCV infections. Indeed, the strength of acute phase $CD8^+$ T cell activity is highly variable in animals with a persistent outcome and probably governs the pattern of virus replication early in the course of infection. For instance in some chimpanzees, T cells target multiple epitopes and are present at frequencies similar to those measured in individuals who ultimately clear the infection. Under these circumstances virus replication is often transiently controlled for several weeks to months before persistence is established (Thimme *et al.*, 2002; Gottwein *et al.*, 2010). In others, T cells target few epitopes, are present at low frequency, and provide limited if any control of HCV replication during acute infection (Thimme *et al.*, 2002; Gottwein *et al.*, 2010). The basis for T cell failure also appears to be similar in chimpanzees and man. Early collapse of $CD4^+$ T cell help is perhaps the most reliable predictor of persistence in both species, although mechanisms underlying this defect have not been identified. Even during persistent HCV infection multiple MHC class I epitopes are often targeted (Neumann-Haefelin *et al.*, 2007; Rehermann, 2009). $CD8^+$ T cells are primed and survive for decades in the persistently infected liver of humans and chimpanzees. The failure of this response may involve mutational escape of targeted epitopes and functional anergy induced by signalling through co-inhibitory receptors.

Mutational escape

Infection of chimpanzees with a genetically defined HCV inoculum provided the first statistical evidence that $CD8^+$ T cells exert Darwinian selection pressure against some (but not all) class I epitopes (Erickson *et al.*, 2001). This mechanism is also operational in humans where a persistent outcome of infection has been correlated with emergence of HCV escape variants (Bowen and Walker, 2005a). Immune escape of class I epitopes is common (Bowen and Walker, 2005a), while escape of class II epitopes in uncommon (Fuller *et al.*, 2009; Fleming *et al.*, 2010).

Studies in chimpanzees (Fernandez *et al.*, 2004) and humans (Cox *et al.*, 2005a; Kuntzen *et al.*, 2007) indicate that the rate of mutation is highest in the first few weeks or months of persistent infection, consistent with the view that $CD8^+$ T cells can exert selective pressure for a limited period of time before loss of function (Urbani *et al.*, 2005). There is strong evidence for both divergent evolution caused by $CD8^+$ T cell selective pressure and for reversion of those changes that impair fitness to a more fit ancestral HCV sequence (Timm *et al.*, 2004; Ray *et al.*, 2005; Tester *et al.*, 2005; Salloum *et al.*, 2008). Mutations occurring outside of T cell epitopes that are not within the envelope region may represent conversion to a more fit consensus sequence (Cox *et al.*, 2005b). Mutations that appeared in HCV genomes at the earliest time points after infection of the chimpanzee were least fit for replication in the cell culture model, suggesting that the initial escape mutations may have a high fitness cost and are replaced by changes that are more stable *in vivo* (Uebelhoer *et al.*, 2008). Impaired HCV replication due to mutational escape from CD8$^+$ T cell responses has also been observed in humans (Salloum *et al.*, 2008; Dazert *et al.*, 2009), although in many instances these inhibitory amino acid substitutions are probably counterbalanced by adaptive compensating mutations in genomic sequences that flank the epitope.

Inhibitory receptor signalling

Some HCV epitopes remain intact during chronic infection even though cognate CD8$^+$ T cells persist in the liver, sometimes at high frequency. This suggests that a second mechanism to silence CD8$^+$ T cells is operational. Signalling through inhibitory receptors has emerged as the leading possibility. PD-1 expression is perhaps the best studied to date. HCV-specific CD8$^+$ T cells in the blood and liver of persistently infected humans display high levels of PD-1 (Urbani *et al.*, 2006; Golden-Mason *et al.*, 2007; Penna *et al.*, 2007; Bowen *et al.*, 2008; Urbani *et al.*, 2008), and the intrahepatic populations appear prone to apoptosis (Radziewicz *et al.*, 2007). Importantly, antibody-mediated blockade of PD-1 signalling in cell culture models restores proliferation if not function to these anergized CD8$^+$ T cells (Golden-Mason *et al.*, 2007; Nakamoto *et al.*, 2008, 2009; Urbani *et al.*, 2008). The relationship between PD-1 expression on CD8$^+$ T cells and infection outcome is not yet fully understood. High levels of this inhibitory receptor are expressed on CD8$^+$ T cells during the acute phase of HCV infection regardless of outcome (Kasprowicz *et al.*, 2008). Establishing an association between PD-1 expression and the course of infection is complicated because this receptor is naturally upregulated on activated CD8$^+$ T cells. One recent study indicates that co-expression of the co-stimulatory molecule CD86 (B7.2) can override the inhibitory effect of PD-1 on HCV-specific CD8$^+$ T cells during acute HCV infection (Radziewicz *et al.*, 2010). Sustained T cell expression of PD-1 in the absence of CD86 during the chronic phase of infection marks exhausted populations that no longer respond to cytokines like IL-2 (Radziewicz *et al.*, 2010).

PD-1 positive CD8$^+$ T cells also lack expression of CD127, the IL-7α receptor that is a marker of functional memory populations. Studies in the chimpanzee model indicated that PD-1 expression is absent or tempered on CD127-positive memory T cell populations that emerge after the infection clears (Bowen *et al.*, 2008). This pattern is clearly distinct from persistent infections where PD-1 levels are elevated and CD127 required for T cell self-renewal is low or absent. Together these observations suggest that constant antigen stimulation is required to maintain PD-1 positive CD8$^+$ T cells in chronic HCV infection. This possibility is consistent with increased PD-1 and low CD127 on CD8$^+$ T cells targeting intact versus mutated class I epitopes in acute and chronic HCV infections (Rutebemberwa *et al.*, 2008; Kasprowicz *et al.*, 2010).

Studies in the LCMV model of persistent infection predict that multiple inhibitory receptors co-regulate CD8$^+$ T cell anergy (Blackburn *et al.*, 2009). The inhibitory NK receptor KLRG1 was detected on the surface of anergized HCV-specific CD8$^+$ T cells (Bengsch *et al.*, 2007). Blockade of CTLA-4 and TIM3 signalling have also been shown to reverse CD8$^+$ T cell defects in chronic hepatitis C (Golden-Mason *et al.*, 2009; Nakamoto *et al.*, 2009). Other negative receptors like LAG3 (lymphocyte activation gene 3) could also potentially play a role in regulating CD8$^+$ T cell activity. CD8$^+$ T cells from chronically infected individuals are usually fully anergic, lacking the ability to proliferate or produce antiviral cytokines like IFNγ.

Regulatory CD8$^+$ T cell activity

HCV-specific CD8$^+$ T cells targeting intact and mutated epitopes can survive in the persistently infected liver without any apparent mechanism for self-renewal. It is likely that they require constant antigen stimulation for maintenance. It is not entirely clear that these cells are simply anergic. HCV-specific CD8$^+$ T cells that produce IL-10 (Accapezzato *et al.*, 2004; Abel *et al.*, 2006) and/or TGF-β (Alatrakchi *et al.*, 2007) have been observed in the blood and liver of persistently infected subjects. Some HCV-specific CD8$^+$ T cells also express CD161 that marks liver-homing populations (Northfield *et al.*, 2008). These cells, which produce IL-17 and IL-22, belong to the Tc17 subset (Billerbeck *et al.*, 2010). In HCV infection, an increased frequency of intrahepatic CD161-positive Tc17 populations is associated with less severe liver disease (Billerbeck *et al.*, 2010). Together these observations suggesting that the cytokine environment is complex at the site of virus replication and that HCV-specific CD8+ T cells can exert anti-inflammatory effects to slow progressive liver disease.

CD4+ T cells and the outcome of HCV infection

CD4+ T cells are particularly important to the outcome of infection. As noted above antibody-mediated depletion of the helper subset in chimpanzees converts a rapidly resolving HCV infection into one that persists (Grakoui *et al.*, 2003). There is also a striking temporal relationship between control of acute viraemia and expansion of a functional, multi-specific CD4+ T cell response in blood (Semmo and Klenerman, 2007). Moreover, if CD4+ T cell responses are not sustained past the point of apparent control of virus replication, a chronic infection is predictable (Semmo and Klenerman, 2007). Indeed, transient control of HCV replication is lost if CD4+ T cells lose the ability to proliferate or produce antiviral (pro-inflammatory) cytokines like IFN-γ or IL-2 (Gerlach *et al.*, 1999; Ulsenheimer *et al.*, 2003, 2006).

Explanations for spontaneous CD4+ T cell failure in some infections remain poorly developed. Ineffective antigen presentation (Szabo and Dolganiuc, 2008) or direct inhibitory effects of viral proteins on helper cells (Cummings *et al.*, 2009) are possibilities. Remarkably HCV-specific CD4+ T cells can be expanded from the persistently infected liver, but their function in that organ is poorly understood. It is conceivable that they are simply anergized, but the possibility that they exert regulatory function can't be excluded. Indeed, immunohistochemical staining has revealed that a high proportion of CD4+ T cells in the liver express the transcriptional regulator FoxP3 associated with regulatory activity (Ward *et al.*, 2007). There is some evidence that they are HCV-specific. A small percentage of circulating CD4+ T cells visualized with HLA class II tetramers express FoxP3 during the acute (Heeg *et al.*, 2009) and chronic (Ebinuma *et al.*, 2008) phases of infection. Production of anti-inflammatory mediators like TGF-β and IL-10 by these circulating antigen-specific CD4+ T cells was documented in the chronically infected subjects (Ebinuma *et al.*, 2008). One recent study demonstrated that intrahepatic CD4+ T regulatory cells targeting HCV antigens can dampen effector T cell activity (Franceschini *et al.*, 2009). Interestingly the regulatory activity was governed by the co-inhibitory molecule programmed cell death 1 (PD-1; Franceschini *et al.*, 2009). Whether these regulatory cells develop early in infection when they could influence outcome is not clear. In one recent study, HLA class II tetramers were used to visualize low frequencies of FoxP3-positive CD4+ T cells in acute phase blood samples but they were not correlated with clearance or persistence of the infection (Heeg *et al.*, 2009). Given the strong homing of HCV-specific T cells to liver, it is possible that access to that compartment would provide more definitive insight into their influence on effector T cell responses and infection outcome.

Vaccines: the challenge

Recent advances in antiviral therapy suggest that combination therapies will be available to cure most infected individuals in the future. However, these cocktails will be expensive, delivery will be difficult to implement in some settings, and resistant populations will likely be significant and difficult to treat. A protective vaccine remains an elusive but essential goal for future research. Chimpanzees remain the only animal susceptible to HCV infection with an intact immune system for the evaluation of vaccine efficacy. After two decades of research on vaccines and the correlates of protective immunity, the definition of an acceptable candidate vaccine may be changing. The goal of sterilizing immunity from broadly reactive neutralizing antibodies may be too lofty for a disease where neutralizing antibody does not appear to play a major role in resolution of acute infection or protection from re-exposure. Conventional vaccines from inactivated virus are not possible due to low level production of virus *in vitro*, and an attenuated vaccine is unlikely considering the risk of persistent infection and disease progression. With a more realistic goal of preventing disease through reduced viraemia during acute phase and lack of progression to chronic infection, recent T-cell-based vaccines have provided significant progress. The use of chimpanzees for vaccine research presents significant obstacles, since the limited size of vaccine groups complicates interpretation due to variation in the rate of persistent infection among control groups. Although promising candidates are emerging, it is unknown how well they will perform in real world settings with the potential for high dose exposure, frequent low dose exposure and extensive diversity. If anything can be learned from HIV vaccine trials, it is that the real world will provide an immense challenge for a virus with great diversity that commonly induces persistent infection. Clearly, vaccine research with conceptually new approaches warrants the continued use of the chimpanzee model.

Early observations in the chimpanzee provided critical evidence for the protective nature of neutralizing antibody (Farci *et al.*, 1994). Virus exposed *in vitro* to homologous antibody under ideal conditions failed to induce infection in chimpanzees. However, recent studies with HCV pseudoparticles (HCVpp) have revealed that broadly reactive neutralizing antibody normally appears after prolonged periods of persistent infection (von Hahn *et al.*, 2007; Liu *et al.*, 2010) and seems to provide little benefit to the patient. Nonetheless, broadly reactive neutralizing antibodies should remain a goal of future vaccines in concert with potent T cell activity. Recent studies have demonstrated protection of chimpanzees challenged after infusion with a humanized monoclonal antibody (Blair *et al.*, 2009). Such antibodies have an obvious role in protection in the transplant setting, but also provide hope for the role of antibody in vaccines.

Chiron conducted the first vaccine studies, soon after the cloning of HCV. The initial vaccine was based on E1 and E2 proteins purified from HeLa cells and provided protection with apparent lack of viraemia for 5 of 7 animals challenged with 10 CID_{50} (chimpanzee infectious dose 50%) of homologous virus. Two vaccinated chimpanzees with viraemia cleared the virus, while all four controls progressed to persistent infection (Choo *et al.*, 1994). The same vaccine produced in yeast provided similar results with combined studies showing protection from viraemia in 5 animals, protection from persistence in an additional five and persistent infection in only two animals (Houghton and Abrignani, 2005). Challenge with a closely related strain of the same subgenotype 1a, protected 8 of 9 animals from chronic infection, although all animals developed viraemia. In contrast, 63% of the controls progressed to chronic infection. Thus, this vaccine was highly efficacious in preventing chronic infection to low dose homologous challenge. Good correlation was observed in the neutralization of binding assay for antibodies that block binding of E2 to cell receptors (Pileri *et al.*, 1998). This research would eventually lead to the identification of CD81 as the HCV receptor involved in the NOB assay.

Expression of core-E2 in insect cells using baculovirus results in the production of virus-like particles (VLPs) that have been used as a vaccine that mimics the HCV particle. Challenge of four immunized chimpanzees resulted in reduced viraemia and early viral clearance, although virus was detectable at late times by more sensitive TMA assays (Elmowalid *et al.*, 2007). The protection provided by this vaccine appeared to be due to T cell responses rather than neutralizing antibody. Use of a vaccinia vector encoding HCV core, E1, E2, p7, NS2, and NS3 provided equivocal results in chimpanzees using a homologous challenge. All 4 vaccinated animals experienced high level viraemia, although it was 1.3 $\log_{10}$ lower than the two controls. None of the vaccinated animals became chronically infected, while both controls developed chronic infection (Youn *et al.*, 2008). A vaccine based on DNA prime and vaccinia boost (modified vaccinia Ankara; MVA) encoding core-E2 and NS3 resulted in chronic infection in 3 of 4 chimpanzees challenged with low dose, homologous virus; however, the level of viraemia was dramatically reduced in both the acute and chronic phases (Rollier *et al.*, 2007).

Perhaps the most successful vaccine to date involved multiple immunizations with Adenovirus and DNA expressing NS3–NS5B. This is the first vaccine to provide good protection to a significantly divergent, heterologous challenge; genotypes 1b and 1a for vaccination and challenge, respectively. Although all animals experienced viraemia, as would be expected with a T cell vaccine, four of five vaccinated animals were protected from chronic infection, in comparison to three of five unvaccinated animals. In addition, the vaccinated animals that cleared infection had very low levels and much shorter duration of viraemia (Folgori *et al.*, 2006). Okairos has entered human clinical trials with a related vaccine using two serotypes of adenovirus for prime and boost, Ad6 not commonly seen in humans and chimpanzee Ad3 (www.clinicaltrials.gov) to avoid suppression of the vaccine by pre-existing neutralizing antibodies to adenovirus.

The use of HCV vaccines as a therapeutic treatment has not been explored as extensively in chimpanzees, since these vaccines will likely be used in combination with antiviral therapy. As discussed above, chimpanzees do not respond to IFNα and may not be suitable for testing of vaccines in combination with IFNα-based therapies. Therapeutic vaccines may become a critical area of research if IFNα is to be eliminated from future antiviral cocktails. The potential role played by IFNα in enhancing the immune response to eliminate the few remaining infected cells, a potential reservoir of resistant virus, during combination antiviral therapy may only be appreciated once it is removed from the cocktail. Numerous candidate vaccines are being developed for therapeutic purposes and most if not all will advance to human trials with no efficacy data from chimpanzees. An E1 protein-based vaccine developed by Innogentics was the first to enter human trials and achieved good immune responses, but failed to reduce viral RNA levels, although some decrease in ALT and histological activity correlated with anti-E1 titres (Nevens *et al.*, 2003). GlobeImmune is currently conducting trials with yeast expressing an HCV NS3 core fusion protein in patients on pegIFN–ribavirin therapy with preliminary data suggesting some increase in early virological response (Habersetzer *et al.*, 2009). Use of the IC41 peptide-based vaccine during pegIFN-ribavirin therapy did not increase the rate of viral clearance (Wedemeyer *et al.*, 2009). A DNA-based vaccine encoding the structural proteins administered with core protein, CIGB-230, failed to reduce viraemia but increased neutralizing antibody responses and may have had an effect on reducing hepatitis (Alvarez-Lajonchere *et al.*, 2009). Transgene is currently progressing a vaccine (TG4040) based on vaccinia MVA encoding the HCV NS3/4/5B into humans trials (www.clinicaltrials.gov). The use of immunomodulators in combination with vaccines and antivirals could represent a direction for future therapeutic vaccines. Whether such heroic efforts will be required to remove IFNα from the current antiviral cocktail will not be known until such trials are conducted in humans or chimpanzees.

Antivirals: the last preclinical step

Efficacy and safety testing of new therapeutics in chimpanzees often represent the last preclinical step prior to initiation of human clinical trials. Testing of new HCV therapies is currently one of the primary uses of chimpanzees in research. Although not all new therapies are tested in chimpanzees prior to human trials, it should be emphasized that there have been many failures in clinic trials, and trials for a number of drugs were halted by the FDA due to serious adverse events and even fatal outcomes. This often raises the two-sided ethical debate of whether new therapies should be tested in the chimpanzee model prior to human trials to provide greater safety and efficacy for the patients enrolling in trials. The ethics of use of chimpanzees to protect humans is not within the scope of this review, so the discussion will focus on the advantages of the use of chimpanzees in HCV antiviral research. Small molecule antivirals targeting viral functions can proceed to human trials without efficacy testing in chimpanzees; however, the knowledge gained in preclinical testing with chimpanzees is often invaluable in the design and success of human trials. An inability to deliver the molecule to the liver at sufficient level and duration will result in failure, even if the molecule is highly potent in the replicon model and appears to have good pharmacokinetics in smaller non-human primates. The use of chimpanzees becomes more critical for novel therapies that modulate the immune response in an unpredictable manner or that target host factors essential for viral replication or persistence.

Although the chimpanzee has been used extensively in antiviral testing, a paucity of literature is available to cite in this review, since most of the research is conducted under strict confidentiality agreements and involves intellectual property highly valued by pharmaceutical companies. However, much of the data are presented at scientific meetings specializing in the treatment of liver disease (AASLD and EASL) and significant findings can be gleamed from these abstracts. Currently, there are four major classes of antiviral compounds being developed for HCV under the general term of STAT-C drugs (specifically targeted antiviral therapy for HCV) including: protease inhibitors (PI), non-nucleoside polymerase inhibitors (NNI), nucleoside analogues, and NS5A inhibitors. The future will likely see more antiviral functions being targeted, as well as host factors essential for replication, such as cyclophilins and microRNAs. This review will not attempt to cover the numerous compounds under development, but rather focus on those that used the chimpanzee model and for which there are publicly available data. A number of excellent reviews on HCV antivirals have been published recently (Kwong *et al.*, 2008; Shimakami *et al.*, 2009; Thompson *et al.*, 2009; Sarrazin and Zeuzem, 2010).

Typically, efficacy testing of new antiviral molecules in chimpanzees is conducted in two phases: pharmacokinetic studies in uninfected animals and efficacy testing in infected animals. These studies are preceded by extensive *in vitro* potency testing with viral proteins, the replicon model and more recently the JFH1 tissue culture model. Thus, prior to chimpanzee studies, a great deal of knowledge has been acquired on the potency of the molecule, synergistic or additive interactions with other antivirals, and the type of resistance mutations to anticipate *in vivo*. In addition, extensive safety and toxicology data are normally available in species such as rat and macaques. Thus, chimpanzee trials are typically very short and focused on the kinetics of decline in viraemia and rebound following emergence of resistance. The decline in viraemia is partially reflected by the potency, but more often involves delivery of the molecule to the liver which is not predictable based on plasma pharmacokinetic alone or studies in small non-human primates. To ensure proper exposure to the liver, pharmacokinetic studies are often conducted in uninfected animals first, to reserve the limited number of chronically infected animals for the essential test of efficacy. Pharmacokinetics in uninfected animals can determine the appropriate dose and frequency of dosing, as well as, the serum to liver ratio during dosing which may be invaluable during human trials, since the sampling of liver will not likely be permitted. Thus, serum concentration in human trials will be used to predict liver concentration based on serum to liver ratios determined in the chimpanzee. A similar understanding can be empirically determined by extrapolation of serum pharmacokinetics with viral decline in patients during a dose escalation study with the understanding that inadequate dosing in the low dose phase may result in viral resistance prior to reaching doses with maximum suppression of replication.

The greatest limitation in the use of chimpanzees, beyond the limited availability and expense, is the method of delivery of the molecule. Frequently, the antiviral is delivered by oral dosing in a flavoured beverage such as Tang or Kool-Aid. The determination of acceptable taste to the chimpanzee is a determining factor on how to dose during efficacy and is often optimized during pharmacokinetic studies in uninfected animals. In some cases, the drug cannot be delivered in a flavoured beverage, and alternatives are pursued. Mixture of micro-encapsulated drug with applesauce has been successful in many instances. Chimpanzees can be trained to take a tablet or pill if sufficient time is available to conduct the training and if the molecule can be formulated in a pill to provide adequate pharmacokinetics. In some instances, the drug is delivered by oral gavage in sedated animals, but this is limited to very short studies by the Institutional Animal Care and Use Committee (IACUC). However, if all parameters are determined in uninfected animals prior to efficacy, very short trials (2–3 days) can provide the kinetics of decline in viraemia and will accurately predict how well the molecule will perform in human trials.

One of the early key publications on HCV antiviral studies in chimpanzees involved monotherapy with a benzothiadiazine non-nucleoside NS5B polymerase inhibitor (Chen *et al.*, 2007a). A-837093 demonstrated sub-nanomolar potency against genotype 1b enzyme and an IC_{50} of less than 2 nM for genotype 1a enzyme. EC_{50} potency in the replicon varied from 3 to 11 nM. Although the bioavailability was >60% in rats and dogs with a half-life of 7 hours, in small monkeys the half-life was only 1.9 hours with a 15% bioavailability. Extrapolation of the dose used in monkeys to that used in chimpanzees suggests an improved Cmax and bioavailability in the chimpanzee. Dosing was performed twice per day for 14 days, but the maximum viral decline occurred within 2 days with 2.5 $\log_{10}$ drop in viral load for the genotype 1b animal at which time the 67% of the isolates contained at least one resistant mutation, and by day 11 only 16% of isolates had WT sequence. The rapid emergence of resistant viruses during viral decline suggests that these mutations not only pre-exist in the quasispecies, but exist as replication complexes in hepatocytes and are the primary source of viral production early after antiviral treatment when the decline of WT from the circulation correlates with the half-life of viral clearance. The rapid rebound; however, suggests that new replication complexes containing resistance mutations are emerging, perhaps by viral spread. Alternatively, the decline in viral RNA decreases the stimulus for the innate immune response and thus releases some pressure on viral production. The resistance mutations to A-837093 included Y448H, G554D, and D559G in the genotype 1a-infected chimpanzee and C316Y and G554D in the genotype 1b-infected chimpanzee. It should be noted that the NS5B polymerase has five distinct allosteric binding sites for NNI compounds and that resistance is not shared by compounds binding to different sites. Proof of concept studies of this nature highlighted the need to develop multiple drug cocktails to avoid resistance, similar to those used for HIV.

Viral production has been estimated at 10^{12} viruses per day and the error rate of the polymerase is estimated at 10–^{5} suggesting that single mutations providing resistance to drugs are produced 10 million times per day, and that most double mutations providing resistance to two drugs are produced at high levels with some triple mutant viruses being produced, as well. Recent studies examined treatment-naïve patients for pre-existing resistance mutations to 27 antiviral drugs currently in development that target either NS3 or NS5B (protease inhibitors, nucleosides and non-nucleoside inhibitors). Remarkably, 45% of genotype 1b patients have at least one detectable mutation at a site associated with resistance and 2.7% had mutations in both NS3 and NS5B (Gaudieri *et al.*, 2009). With the use of deep sequencing technology or possibly a SNP chip, the resistance pattern of patients could theoretically be determined prior to therapy. Although multiple mutations will impose serious fitness costs on the virus, given time, second site compensatory mutations would likely restore at least part of the lost replication capacity.

Nucleoside analogues have provided a greater barrier to resistance than the NNI and PI drugs. The only resistance mutations are within the binding site for the polymerase which is highly conserved and mutations have very high fitness costs. Carroll *et al.* examined MK-0608 (2 C-methyl-7-deaza-adenosine) in chimpanzees (Carroll *et al.*, 2009). Intravenous dosing at 2 mg/kg once per day for 7 days provided approximately 6 logs of decline in viral load in both animals. At end of dosing and the beginning of viral rebound, the S282T mutation known to confer resistance was transiently detected in one animal. Oral dosing of the same compound at 1 mg/kg once per day for 37 days provided equally impressive results. One animal with a low prestudy viral load became undetectable within 2 days with standard assays (<20 IU/ml) but was still detectable by TMA (10 IU/ml) until day 10. Virus remained undetectable for the first two weeks after dosing was terminated. Variable sequence including S282T was detected on day 49 at viral rebound, but all sequences were WT 1 week later. The other animal started therapy with a higher viral load and gradually decreased through the entire 37-day dosing period to achieve a decline of 4.6 $\log_{10}$. No resistance mutations were noted in this animal.

These studies illustrate the high barrier to resistance observed for nucleoside analogues. Longer trials will be required to determine if second site mutations can restore partial fitness to resistance mutations in nucleoside analogues.

A number of other noteworthy trials have been conducted in chimpanzees and the data presented at international meetings. Merck has preformed chimpanzee trials with both nucleoside inhibitors and protease inhibitors (Olsen *et al.*, 2007), showing potent antiviral activity with emergence of resistant mutations S282T for the nucleoside analogue and R155K for the PI. The NNI ANA598 was orally dosed at 30 mg/kg q.d. for 4 days in two genotype 1b chimpanzees and reduced viraemia by 2.1–2.6 $\log_{10}$ within 24–48 hours (Steffy *et al.*, 2008). The liver targeted nucleoside analogue IDX184 (2′ methyl guanosine monophosphate prodrug) was orally dosed q.d. at 10 mg/kg and achieved a 1.4–3.8 $\log_{10}$ decline in viraemia on days 3 and 4.

One of the more exciting studies (submitted for publication; David Olsen, personal communication) involved a combination of the Merck protease inhibitor MK-7009 and a nucleoside analogue MK-0608. In this study, three chimpanzees were treated for 37 days with a combination of the two drugs, followed by additional dosing of the PI for a total of 3 months. All three animals dropped to undetectable levels of viral RNA quickly after dosing initiated. The chimpanzee with the highest viral load had breakthrough viraemia after the 37 days of combination therapy during the PI monotherapy. Another animal rebounded after cessation of dosing. However, the animal with the lowest baseline viral load remained undetectable 6 months after dosing stopped. This is an important proof of concept study demonstrating that a cure can be obtained in the absence of IFNα.

Trials in man with two antivirals in combination for 14 days, nucleoside analogue and PI ($R7128^{+}R7227$), followed by pegIFNα and ribavirin standard of care, are currently underway. Preliminary data, indicate that 63% of patients had undetectable viraemia on day 14 (Gane *et al.*, 2009).

The use of chimpanzees in safety and efficacy studies becomes more critical when the inhibitor targets host factors essential for viral replication, involve novel therapies not used in man, or modulate the immune system. A recent success in this class of studies involved the inhibition of miR122. miR-122 is a highly abundant, liver specific microRNA. HCV has two binding sites for miR-122 in the 5′ non-coding region of the RNA genome and both are essential for maintaining RNA abundance (Jopling *et al.*, 2005, 2008). The drug SPC3649, developed by the Danish company Santaris Pharma, is an antisense oligonucleotide that targets miR122. This drug is the first example of a DNA-based therapy that is highly efficacious when administered systemically, and the success of the drug relies on the use of the Santaris technology called Locked Nucleic Acid, which provides enhanced stability and higher binding affinity for the oligonucleotide. SPC3649 was administered once per week for 12 weeks and induced a 2.6 log drop in viral RNA levels that persisted for months after the last dose of drug (Lanford *et al.*, 2010). The stability of the SPC3649–miR122 complex in the liver was demonstrated by both Northern hybridization and quantitative RT-PCR. The drug provided an extremely high genetic barrier to resistance. No adaptive or antiviral resistant mutations were detected in the 5′ NCR even with the use of deep sequencing technology to probe the virus for minor variations. An additional benefit of the drug was the reduction of total serum cholesterol by 45% due to inhibition of the known function of miR122 in cholesterol metabolism. In addition to the proof of concept for HCV, success with Locked Nucleic Acids antisense suggests that this DNA-based therapy can be used to target critical genes involved in other diseases.

Hepatocellular carcinoma in chimpanzees

Three decades have passed since the first experimental transmission of non-A, non-B hepatitis to chimpanzees (Alter *et al.*, 1978; Hollinger *et al.*, 1978). The long-term consequences of persistent HCV infection in this animal model have been poorly studied. Long-term follow-up of chimpanzees persistently infected with HCV (n = 46) has been ongoing at the Southwest National Primate Research Center for the last decade. With increasing follow-up and duration of infections, it has become clear that chimpanzees experience progressive liver disease. Despite the lack of contributing cofactors such as alcohol consumption, obesity, and high iron diet, an increasing percentage of the animals exhibit significant liver disease, including elevated ALT, GGT, bilirubin, moderate histological changes, and mild to moderate fibrosis. During this follow-up, two animals developed hepatocellular carcinoma (HCC), one each with HCV and HBV persistent infections (Lanford *et al.*, 2008). The HCV infected animal developed several distinct tumours with the largest being 14 cm in diameter. Although some mild portal fibrosis was noted, cirrhosis was not evident. The HBV-infected animal had metastatic HCC with spread to most organs. Advanced cirrhosis was present in this animal, yet a biopsy from two years earlier exhibited only mild to moderate fibrosis, suggesting rapid progression of disease after decades of mild infection. Two chimpanzees with HCC were also reported at the LEMSIP facility, but only one of the tumours was subsequently confirmed as HCC (Muchmore *et al.*, 1988; Tabor *et al.*, 1994). Confounding the suspected viral association with the tumours discussed above, HCC has been reported in two captive chimpanzees without known or detected viral exposure (Porter *et al.*, 2004; Lanford, Dick, Lammey and Lee, unpublished data). Considering that fewer than 100 chimpanzees are persistently infected with HCV at the various facilities housing infected animals and most have been infected for less than two decades, it is surprising that this number of HCC cases have been detected. With the exception of the one HBV HCC, all of the tumours arose in the absence of cirrhosis. Although cirrhosis is often considered a prerequisite for development of HCV associated HCC in man, recent studies have revealed an increasing number of human cases of HCC in the absence of cirrhosis (Yeh *et al.*, 2010). This raises obvious questions of whether HCV possess direct oncogenic activity. Interestingly, the first transgenic mice expressing HCV proteins developed spontaneous HCC, one with the entire ORF and one with core protein (Moriya *et al.*, 1998; Lerat *et al.*, 2002).

Small animal surrogates

GBV-B in new world monkeys

History of the GB agents

In the 1960s, Deinhardt and co-workers injected tamarins, *Saguinus* species, with sera from patients with hepatitis in an attempt to pass the human hepatitis viruses to primates. These studies were conducted before the development of diagnostic assays for HAV, HBV, and HCV. Among the sera injected into tamarins was the serum from a 34 year old surgeon (initials GB) who presented with jaundice that persisted for 4 weeks (reviewed in Beames *et al.*, 2001). Serum from this patient caused hepatitis in tamarins based on increases in the serum levels of liver enzymes. The agent was maintained through serial passage in tamarins but eluded attempts at isolation until a new approach in RT/PCR was used to clone nucleic acids from three viruses: two from tamarins and a third from humans.

In 1995, in search for non-ABC hepatitis viruses, investigators at Abbott Laboratories identified two viruses in the serum of a tamarin infected with an aliquot of the GB agent using representational difference analysis (Schlauder *et al.*, 1995; Simons *et al.*, 1995). The viral agents, termed GB virus A (GBV-A) and GB virus B (GBV-B) showed similarity to the *Flaviviridae*. Subsequent studies cloned a related virus from human serum, and it was designated GBV-C. It was later recognized that hepatitis G virus and GBV-C were the same virus. Since GBV-A and GBV-C are not hepatotropic viruses and cause no known disease, they will not be further discussed here.

Sequence analysis of the GBV-B genome revealed a single ORF of 2864 amino acids (Muerhoff *et al.*, 1995) with 28% amino acid similarity with HCV genotype 1 over the whole length of its ORF (Ohba *et al.*, 1996). Although clearly distinct from HCV, phylogenetically GBV-B is the virus most closely related to HCV (Muerhoff *et al.*, 1995; Ohba *et al.*, 1996). Additional similarities between GBV-B and HCV can be found in the 5′ and 3′ ends of the genomes. GBV-B possesses an IRES that resembles HCV (Grace *et al.*, 1999). While the GBV-B 3′ NTR is structurally similar to the 3′ end of HCV, it has a much shorter poly-U tract and a region of three times greater length following the poly-U tract than in HCV (Bukh *et al.*, 1999).

At the time of the original GB studies, it was assumed that the hepatitis agent originated from the serum of the surgeon; however, in retrospect, it seems that the agent was already present in the tamarin and that GBV-B represents a tamarin virus which if isolated today would certainly be designated tamarin HCV. GBV-B has not been recovered from humans, and GBV-B has a very narrow host range for tamarins and other closely related New World monkeys (Bukh *et al.*, 2001; Lanford *et al.*, 2003a). The fact that it has not been recovered a second time from tamarins may be due to the rapid resolution of the acute infection in tamarins and the limited number of wild caught tamarins that have been examined immediately upon introduction into captivity.

Rationale for GBV-B as a surrogate for HCV

The major advantage of GBV-B is the small animal model. Although tamarins (*Saguinus* sp.) appear to be the natural host, the virus has been adapted to the closely related marmoset (*Callithrix jacchus*; see below). Marmosets are readily available and adults weigh 300g, or approximately 200 times less than a chimpanzee. In addition, the level of viraemia in infected animals often approaches 10^{10} GE/ml, more than 1000-fold what is observed in most HCV infected chimpanzees and humans. The nature of this remarkable difference could be replication efficiency, but it may involve the efficiency in suppressing the innate response. Finally, GBV-B is an acute resolving infection, with only a few examples of persistent infections. Although infection normally resolves in 10–20 weeks, at least four long-term persistent infections (1–2 years) have been documented (Martin *et al.*, 2004; Nam *et al.*, 2004; Takikawa *et al.*, 2006, 2010). Insight into the differences between GBV-B and HCV in the development of persistent infections may provide critical insight into vaccine strategies.

Unlike HCV, the GBV-B can be cultured in primary hepatocytes which more accurately reflect the *in vivo* environment (Beames *et al.*, 2000). In addition, there are attributes of GBV-B distinct from HCV biology such that the comparison and contrast of the two systems should yield valuable insights. The virus can move from animal to culture and back to animals without requiring adaptive mutations. The primary hepatocyte culture system for GBV-B (Beames *et al.*, 2000) has an intact innate immune response (Lanford *et al.*, 2003b; Chavez *et al.*, 2009) providing a more authentic environment to evaluate the ability of viral proteins to subvert the innate response. Although modulation of the innate response by HCV proteins has been extensively studied, the GBV-B model provides an alternate perspective that may yet reveal critical viral host interactions being missed by using the current HCV culture system.

The functional similarity of GBV-B to HCV was first highlighted when HCV and GBV-B proteases were shown to process each others polyproteins (Scarselli *et al.*, 1997; Butkiewicz *et al.*, 2000). The high level of homology between HCV and GBV-B led to the anticipation that antiviral compounds developed for HCV would be active against GBV-B. Although this has not been fully explored, HCV protease inhibitors were shown to inhibit GBV-B replication in infected marmosets (Bright *et al.*, 2004) and infected primary marmoset hepatocytes (Lanford *et al.*, 2003a). Perhaps of greater current interest is the finding that similar to HCV, the GBV-B protease cleaves MAVS, the mitochondrial protein in the RIGI/MDA5 pathway that is involved in activation of IRF3 and induction of IFNβ synthesis (Chen *et al.*, 2007b), revealing a common strategy for subverting the host innate immune response. Marmoset and tamarin IFNα and IFNγ have potent antiviral activity against GBV-B *in vitro* (Chavez *et al.*, 2009). The antiviral nature of ribavirin was explored using the GBV-B model as well (Lanford *et al.*, 2001b). Virions harvested from ribavirin-treated cultures exhibited a dramatically reduced specific infectivity. These data suggested that incorporation of ribavirin triphosphate induces error prone replication with concomitant reduction in infectivity. The induction of error prone replication with ribavirin has been demonstrated in the HCV replicon model (Lanford *et al.*, 2003b), but there is no conclusive that error prone replication is involved in the antiviral effect of ribavirin for HCV *in vivo*.

Infectious cDNA clones of GBV-B have been produced that induced hepatitis upon intrahepatic inoculation of tamarins and marmosets with *in vitro* transcribed RNA (Bukh *et al.*, 1999; Sbardellati *et al.*, 2001; Martin *et al.*, 2003). This has permitted reverse genetic studies, but they have been limited to *in vivo* infectivity studies, due the current inability to initiate infection *in vitro* with synthetic RNA. These studies and their relevance to HCV will be briefly summarized here. Chimeric viruses replacing a portion of the GBV-B IRES with domain III of the HCV IRES (Rijnbrand *et al.*, 2005) were viable *in vivo*. Interestingly, the virus failed to replicate at detectable levels for 12 weeks and then emerged with WT replication capacity. A number of sequence changes were present, but subsequent studies indicated that 2 nucleotide changes in the 3′ NCR region were compensatory mutations that allowed the chimeric to replicate efficiently (Warter *et al.*, 2009). GBV-B encodes a p13 protein that undergoes processing to a p7 protein that closely resembles HCV p7. A GBV-B construct engineered to encoded a p7 replicated efficiently *in vivo* (Takikawa *et al.*, 2006), and replacement of the GBV-B p13 sequence with HCV p7 also yielded an infectious virus (Griffin *et al.*, 2008). A chimeric virus encoding the HCV e2 hypervariable region was infectious *in vivo* as well (Haqshenas *et al.*, 2007). The processing of core protein signal peptide by signal protein protease has been evaluated with GBV-B mutant genomes, demonstrating the sequence requirements for cleavage and essential nature of the cleavage for infectivity (Targett-Adams *et al.*, 2006). The essential nature of the 3′ NCR was determined by deletion studies, but surprisingly deletion of the poly-U stretch resulted in chronic infection persisting for 90 weeks (Nam *et al.*, 2004).

GBV-B adaptation to the marmoset

Although the natural host for GBV-B appears to be the tamarin, several New World primates are susceptible to infection (Bukh *et al.*, 2001; Lanford *et al.*, 2003a). Marmosets are the only alternative that has been adequately explored. Two phenotypes are observed in marmosets infected with GBV-B, susceptible and partially resistant (Bright *et al.*, 2004; Weatherford *et al.*, 2009a). The different phenotypes are likely due to a genetic polymorphism in the marmoset population that affects a critical viral–host interaction, possibly involving suppression of the innate response. Serial passage of WT GBV-B in resistant animals results in sequence changes, adaptation, and a uniformly susceptible phenotype. The fifth passage in marmosets contained 21 nucleotide changes and 6 amino acid changes. A clone engineered to contain the minimal adaptive changes may provide insight into the genetic polymorphism regulating susceptibility in marmosets; however, the fifth passage marmoset adapted virus permits the use of this species without a complete understanding of the steps required for adaptation. Although WT GBV-B could be adapted for efficient replication in marmosets, this was not true for the chimeric virus containing HCV domain III of the IRES (Weatherford *et al.*, 2009b). Although this virus replicated well in tamarins following the acquisition of compensatory changes in the 3′ NCR, it was not able to adapt for replication in the marmosets having the resistant phenotype. Serial passage in marmosets resulted in a loss of fitness, possibly due to reduction in the quasispecies during a bottleneck event. Whether the marmoset will pose additional barriers to the analysis of various mutants of GBV-B that are not observed in the tamarin remains to be determined.

Mice with chimeric mouse–human livers and the potential for an HCV mouse

The ultimate animal model for HCV would be a mouse susceptible to infection and fully permissive for replication. Despite the rapid progress in identification of the four receptors required for entry and the numerous host factors influencing replication, this goal at times appears distant and possibly unobtainable. Two mouse models with chimeric mouse/human livers have been developed that can be infected with HCV. The first chimeric liver mouse model utilized a transgenic mouse with urokinase plasminogen activator (uPA) over expressed via the albumin promoter, leading to severe hepatotoxicity at birth that can be compensated by transplantation of non-transgenic hepatocytes, or when in an immunodeficient background ($RAG^{-/-}$) with hepatocytes from many species including human (reviewed in Meuleman and Leroux-Roels, 2008). Transplanted hepatocytes gradually replace the dying liver with the development of appropriate vascularization and liver architecture. The first successful transplantation of human hepatocytes in this model provided approximately 15% replacement of the mouse liver with human hepatocytes (Dandri *et al.*, 2001), while use of homozygous Alb-uPA mice eventually provide near complete engraftment by the transplanted hepatocytes (Meuleman *et al.*, 2005). The mice were susceptible to infection with both HBV (Dandri *et al.*, 2001; Meuleman *et al.*, 2006) and HCV (Mercer *et al.*, 2001; Meuleman *et al.*, 2005). These mice developed very high levels of viraemia perhaps due to the lack of an immune system to control viraemia and have been very useful in studies with IFNα (Kneteman *et al.*, 2006) and STAT-C antivirals (Meuleman and Leroux-Roels, 2008). Tissue culture produced virus was capable of robust replication in this model as well (Lindenbach *et al.*, 2006). There are a number of limitations in using this model. The immunodeficient background requires that the animals be maintained in a sterile environment. The microsurgical technique required for engraftment is difficult, since homozygous mice require transplantation within days after birth, and this transgenic mouse is prone to severe bleeding. Another challenge is having a reliable source of human hepatocytes such that cells are available within days of the birth of litters.

An alternative chimeric liver mouse model has been developed that allows transplantation of adult mice. Mice with a deletion in fumarylacetoacetate hydrolase (FAH) accumulate toxic concentrations of tyrosine metabolites within hepatocytes (Grompe *et al.*, 1998). Mice can be raised to adulthood by placing NTBC (2-(2-nitro-4-trifluoro-methylbenzoyl)-1,3-cyclohexanedione) in the drinking water which blocks the enzyme hydroxyphenylpyruvate dioxygenase upstream of FAH, and thus blocks accumulation of hepatotoxic tyrosine metabolites. In an immunodeficient background, these mice can be transplanted with human hepatocytes (Azuma *et al.*, 2007; Bissig *et al.*, 2007), and recent improvements in the model have allowed robust infection with HBV and HCV (Bissig *et al.*, 2010). The ability to use adult mice avoids the issues of timing the availability of hepatocytes to the birth of litters, and the microsurgical techniques required with newborn mice. In addition, the FAH model avoids the bleeding problems associated with uPA mice. This mouse model may rapidly become the preferred model for human/mouse chimeric liver studies with HCV.

The potential to develop a mouse susceptible to infection and permissive for replication of HCV may require many steps. The issues surrounding entry appear to be within grasp with the discovery of the four receptors involved in HCV infection: CD81, SCARB1, CLDN1, OCLN; of which CD81 and OCLN must be of human origin (Ploss *et al.*, 2009; Ploss and Rice, 2009). The list of host factors involved in replication grows daily and has recently been reviewed in the context of making an HCV mouse (Ploss and Rice, 2009). The critical question is how many of the factors are truly essential and how many can be of mouse origin? The number that are truly essential and must be of human origin may be limited, since HCV replicons can be adapted for replication in mouse hepatocytes (Zhu *et al.*, 2003; Uprichard *et al.*, 2006). However, reconstruction of genomes with mutations acquired in mouse cells, failed to yield a mouse adapted clone. RNA viruses are extremely adept at adaptation to a new environment due to the extensive sequence variation within the quasispecies, thus even low level replication may lead to progressive adaptation to the new host. On the other hand, HCV has not been successfully transferred to an Old World primate outside of the great apes. As discussed above, transfer of a chimeric GBV-B/HCV virus between two very closely related species (tamarins and marmosets) lead to a loss of fitness after a few serial passages (Weatherford *et al.*, 2009b). Clearly, we will not understand how close we are until we have arrived.

References

Abel, M., Sene, D., Pol, S., Bourliere, M., Poynard, T., Charlotte, F., Cacoub, P., and Caillat-Zucman, S.

(2006). Intrahepatic virus-specific IL-10-producing CD8 T cells prevent liver damage during chronic hepatitis C virus infection. Hepatology *44*, 1607–1616.

Accapezzato, D., Francavilla, V., Paroli, M., Casciaro, M., Chircu, L.V., Cividini, A., Abrignani, S., Mondelli, M.U., and Barnaba, V. (2004). Hepatic expansion of a virus-specific regulatory CD8(+) T cell population in chronic hepatitis C virus infection. J. Clin. Invest. *113*, 963–972.

Alatrakchi, N., Graham, C.S., van der Vliet, H.J.J., Sherman, K.E., Exley, M.A., and Koziel, M.J. (2007). Hepatitis C virus (HCV)-specific $CD8^+$ cells produce transforming growth factor beta that can suppress HCV-specific T-cell responses. J. Virol. *81*, 5882–5892.

Alter, H.J. and Houghton, M. (2000). Hepatitis C virus and eliminating post-transfusion hepatitis. Nat. Med. *6*, 1082–1086.

Alter, H.J., and Seeff, L.B. (2000). Recovery, persistance, and sequelae in hepatitis C virus infection: a perspective on long-term outcome. Semin. Liver Dis. *20*, 17–35.

Alter, H.J., Purcell, R.H., Holland, P.V., and Popper, H. (1978). Transmissible agent in non-A, non-B hepatitis. Lancet *1*, 459–463.

Alter H.J., Eichberg, J.W., Masur, H., Saxinger, W.C., Gallo, R.C., Macher, A.M., Lane, H.C., and Fauci, A.S. (1984). Transmission of HTLV-III infection from human plasma to chimpanzees: an animal model for AIDS. Science *226*, 549–552.

Alter, H.J., Purcell, R.H., Shih, J.W., Melpolder, J.C., Houghton, M., Choo, Q.-L., and Kuo, G. (1989). Detection of antibody to hepatitis C virus in prospectively followed transfusion recipients with acute and chronic non-A, non-B hepatitis. N. Engl. J. Med. *321*, 1494–1500.

Alter, M.J., Kruszon-Moran, D., Nainan, O.V., McQuillan, G.M., Gao, F.X., Moyer, L.A., Kaslow, R.A., and Margolis, H.S. (1999). The prevalence of hepatitis C virus infection in the United States, 1988 through 1994. N. Engl. J. Med. *341*, 556–562.

Alvarez-Lajonchere, L., Shoukry, N.H., Gra, B., Amador-Canizares, Y., Helle, F., Bedard, N., Guerra, I., Drouin, C., Dubuisson, J., Gonzalez-Horta, E.E., Martinez, G., Marante, J., Cinza, Z., Castellanos, M., and Duenas-Carrera, S. (2009). Immunogenicity of CIGB-*230*, a therapeutic DNA vaccine preparation, in HCV-chronically infected individuals in a Phase I clinical trial. J. Viral Hepat. *16*, 156–167.

Appel, N., Zayas, M., Miller, S., Krijnse-Locker, J., Schaller, T., Friebe, P., Kallis, S., Engel, U., and Bartenschlager, R. (2008). Essential role of domain III of nonstructural protein 5A for hepatitis C virus infectious particle assembly. PLoS Pathog. *4*, e1000035.

Azuma, H., Paulk, N., Ranade, A., Dorrell, C., Al-Dhalimy, M., Ellis, E., Strom, S., Kay, M.A., Finegold, M., and Grompe, M. (2007). Robust expansion of human hepatocytes in Fah–/–/Rag2–/–/Il2rg–/– mice. Nat. Biotechnol. *25*, 903–910.

Barker, L.F., Maynard, J.E., Purcell, R.H., Hoofnagle, J.H., Berquist, K.R., and London, W.T. (1975). Viral hepatitis, type B, in experimental animals. Am. J. Med. Sci. *270*, 189–195.

Barrett, S., Ryan, E., and Crowe, J. (1999). Association of the HLA-DRB1*01 allele with spontaneous viral clearance in an Irish cohort infected with hepatitis C virus via contaminated anti-D immunoglobulin. J. Hepatol. *30*, 979–983.

Bassett, S.E., Brasky, K.M., and Lanford, R.E. (1998). Analysis of hepatitis C virus-inoculated chimpanzees reveals unexpected clinical profiles. J. Virol. *72*, 2589–2599.

Bassett, S.E., Thomas, D.L., Brasky, K., and Lanford, R.E. (1999). Viral persistence, antibody to E1 and E2, and hypervariable region 1 sequence stability in hepatitis C virus-inoculated chimpanzees. J. Virol. *73*, 1118–1126.

Bassett, S.E., Guerra, B., Brasky, K., Miskovsky, E., Houghton, M., Klimpel, G.R., and Lanford, R.E. (2001). Protective immune response to hepatitis C virus in chimpanzees rechallenged following clearance of primary infection. Hepatology *33*, 1479–1487.

Beames, B., Chavez, D., Guerra, B., Notvall, L., Brasky, K.M., and Lanford, R.E. (2000). Development of a primary tamarin hepatocyte culture system for GB virus-B: a surrogate model for hepatitis C virus. J. Virol. *74*, 11764–11772.

Beames, B., Chavez, D., and Lanford, R.E. (2001). GB virus B as a model for hepatitis C virus. ILAR J. *42*, 152–160.

Bengsch, B., Spangenberg, H.C., Kersting, N., Neumann-Haefelin, C., Panther, E., von Weizsacker, F., Blum, H.E., Pircher, H., and Thimme, R. (2007). Analysis of CD127 and KLRG1 expression on hepatitis C virus-specific $CD8^+$ T cells reveals the existence of different memory T-cell subsets in the peripheral blood and liver. J. Virol. *81*, 945–953.

Berman, J.W., Murthy, K.K., Wrin, T., Vennari, J.C., Cobb, E.K., Eastman, D.J., Champe, M., Nakamura, G.R., Davison, D., Powell, M.F., Bussiere, J., Francis, D.P., Mathews, T., Gregory, T.J., and Obijeski, J.F. (1996). Protection of MN-rgp120-immunized chimpanzees from heterologous infection with a primary isolate of HIV type 1. J. Infect. Dis. *173*, 52–59.

Bigger, C.B., Brasky, K.M., and Lanford, R.E. (2001). DNA microarray analysis of chimpanzee liver during acute resolving hepatitis C virus infection. J. Virol. *75*, 7059–7066.

Bigger, C.B., Guerra, B., Brasky, K.M., Hubbard, G., Beard, M.R., Luxon, B.A., Lemon, S.M., and Lanford, R.E. (2004). Intrahepatic gene expression during chronic hepatitis C virus infection in chimpanzees. J. Virol. *78*, 13779–13792.

Billerbeck, E., Kang, Y.H., Walker, L., Lockstone, H., Grafmueller, S., Fleming, V., Flint, J., Willberg, C.B., Bengsch, B., Seigel, B., *et al.* (2010). Analysis of CD161 expression on human $CD8^+$ T cells defines a distinct functional subset with tissue-homing properties. Proc. Natl. Acad. Sci. U.S.A. *107*, 3006–3011.

Bissig, K.D., Le, T.T., Woods, N.B., and Verma, I.M. (2007). Repopulation of adult and neonatal mice with human hepatocytes: a chimeric animal model. Proc. Natl. Acad. Sci. U.S.A. *104*, 20507–20511.

Bissig, K.D., Wieland, S.F., Tran, P., Isogawa, M., Le, T.T., Chisari, F.V., and Verma, I.M. (2010). Human liver chimeric mice provide a model for hepatitis B and C virus infection and treatment. J. Clin. Invest *120*, 924–930.

Blackburn, S.D., Shin, H., Haining, W.N., Zou, T., Workman, C.J., Polley, A., Betts, M.R., Freeman, G.J., Vignali, D.A., and Wherry, E.J. (2009). Coregulation of CD8+ T cell exhaustion by multiple inhibitory receptors during chronic viral infection. Nat. Immunol. *10*, 29–37.

Blair, B.M., Broering, T.J., Babcock, G.J., Szabo, G., Finberg, R.W., Cheslock, P.S., Knauber, M., Leav, B.A., Lanford, R., Purcell, R.H., *et al.* (2009). A novel human monoclonal antibody directed against the e2 glycoprotein of hepatitis c virus (hcv) prevents infection in chimpanzees. J. Hepatol. *50*, S381.

Bowen, D.G. and Walker, C.M. (2005a). Mutational escape from CD8+ T cell immunity: HCV evolution, from chimpanzees to man. J. Exp. Med. *201*, 1709–1714.

Bowen, D.G. and Walker, C.M. (2005b). Adaptive immune responses in acute and chronic hepatitis C virus infection. Nature *436*, 946–952.

Bowen, D.G., Shoukry, N.H., Grakoui, A., Fuller, M.J., Cawthon, A.G., Dong, C., Hasselschwert, D.L., Brasky, K.M., Freeman, G.J., Seth, N.P., *et al.* (2008). Variable patterns of programmed death-1 expression on fully functional memory T cells after spontaneous resolution of hepatitis C virus infection. J. Virol. *82*, 5109–5114.

Bradley, D.W., Maynard, J.E., Popper, H., Cook, E.H., Ebert, J.W., McCaustland, K.A., Schable, C.A., and Fields, H.A. (1983). Posttransfusion nonA nonB hepatitis: physicochemical properties of two distinct agents. J. Infect. Dis. *148*, 254–265.

Bradley, D.W. (1985). The agents of non-A, non-B viral hepatitis. J. Virol. Methods *10*, 307–319.

Bright, H., Carroll, A.R., Watts, P.A., and Fenton, R.J. (2004). Development of a GB virus B marmoset model and its validation with a novel series of hepatitis C virus NS3 protease inhibitors. J. Virol. *78*, 2062–2071.

Bukh, J., Apgar, C.L., Govindarajan, S., and Purcell, R.H. (2001). Host range studies of GB virus-B hepatitis agent, the closest relative of hepatitis C virus, in New World monkeys and chimpanzees. J. Med. Virol. *65*, 694–697.

Bukh, J., Apgar, C.L., and Yanagi, M. (1999). Toward a surrogate model for hepatitis C virus: an infectious molecular clone of the GB virus-B hepatitis agent. Virology *262*, 470–478.

Bukh, J., Pietschmann, T., Lohmann, V., Krieger, N., Faulk, K., Engle, R.E., Govindarajan, S., Shapiro, M., St Claire, M., and Bartenschlager, R. (2002). Mutations that permit efficient replication of hepatitis C virus RNA in Huh-7 cells prevent productive replication in chimpanzees. Proc. Natl. Acad. Sci. U.S.A. *99*, 14416–14421.

Bukh, J., Thimme, R., Meunier, J.C., Faulk, K., Spangenberg, H.C., Chang, K.M., Satterfield, W., Chisari, F.V., and Purcell, R.H. (2008). Previously infected chimpanzees are not consistently protected against reinfection or persistent infection after reexposure to the identical hepatitis C virus strain. J. Virol. *82*, 8183–8195.

Butera, D., Marukian, S., Iwamaye, A.E., Hembrador, E., Chambers, T.J., Di Bisceglie, A.M., Charles, E.D., Talal, A.H., Jacobson, I.M., Rice, C.M., and Dustin, L.B. (2005). Plasma chemokine levels correlate with the outcome of antiviral therapy in patients with hepatitis C. Blood *106*, 1175–1182.

Butkiewicz, N., Yao, N., Zhong, W., Wright-Minogue, J., Ingravallo, P., Zhang, R., Durkin, J., Strandring, D.N., Baroudy, B.M., Sangar, D.V. *et al.* (2000). Virus-specific cofactor requirement and chimeric hepatitis C virus/GB virus B nonstructural protein 3. J. Virol. *74*, 4291–4301.

Carroll, S.S., Ludmerer, S., Handt, L., Koeplinger, K., Zhang, N.R., Graham, D., Davies, M.E., MacCoss, M., Hazuda, D., and Olsen, D.B. (2009). Robust antiviral efficacy upon administration of a nucleoside analog to hepatitis C virus-infected chimpanzees. Antimicrob. Agents Chemother. *53*, 926–934.

Chavez, D., Guerra, B., and Lanford, R.E. (2009). Antiviral activity and host gene induction by tamarin and marmoset interferon-alpha and interferon-gamma in the GBV-B primary hepatocyte culture model. Virology *390*, 186–196.

Chen, L.M., Borozan, I., Feld, J., Sun, J., Tannis, L.L., Coltescu, C., Heathcote, J., Edwards, A.M., and McGilvray, I.D. (2005). Hepatic gene expression discriminates responders and nonresponders in treatment of chronic hepatitis C viral infection. Gastroenterology *128*, 1437–1444.

Chen, C.M., He, Y., Lu, L., Lim, H.B., Tripathi, R.L., Middleton, T., Hernandez, L.E., Beno, D.W.A., Long, M.A., Kati, W.M. *et al.* (2007a). Activity of a potent hepatitis C virus polymerase inhibitor in the chimpanzee model. Antimicrob. Agents Chemother. *51*, 4290–4296.

Chen, Z., Benureau, Y., Rijnbrand, R., Yi, J., Wang, T., Warter, L., Lanford, R.E., Weinman, S.A., Lemon, S.M., Martin, A., and Li, K. (2007b). GB virus B disrupts RIG-I signaling by NS3/4A-mediated cleavage of the adaptor protein MAVS. J. Virol. *81*, 964–976.

Choo, Q.-L., Kuo, G., Weiner, A.J., Overby, L.J., Bradley, D.W., and Houghton, M. (1989). Isolation of a cDNA clone derived from blood-borne non-A, non-B viral hepatitis genome. Science *244*, 359–362.

Choo, Q.-L., Kuo, G., Ralston, R., Weiner, A., Chien, D., Van Nest, G., Han, J., Berger, K., Thudium, K., Kuo, C., *et al.* (1994). Vaccination of chimpanzees against infection by the hepatitis C virus. Proc. Natl. Acad. Sci. U.S.A. *91*, 1294–1298.

Cooper, S., Erickson, A.L., Adams, E.J., Kansopon, J., Weiner, A.J., Chien, D.Y., Houghton, M., Parham, P., and Walker, C.M. (1999). Analysis of a successful immune response against hepatitis C virus. Immunity. *10*, 439–449.

Cox, A.L., Mosbruger, T., Lauer, G.M., Pardoll, D., Thomas, D.L., and Ray, S.C. (2005a). Comprehensive analyses of CD8+ T cell responses during longitudinal study of acute human hepatitis C. Hepatology *42*, 104–112.

Cox, A.L., Mosbruger, T., Mao, Q., Liu, Z., Wang, X.H., Yang, H.C., Sidney, J., Sette, A., Pardoll, D., Thomas, D.L., and Ray, S.C. (2005b). Cellular immune selection with hepatitis C virus persistence in humans. J. Exp. Med. *201*, 1741–1752.

Cummings, K.L., Rosen, H.R., and Hahn, Y.S. (2009). Frequency of gC1qR+CD4+ T cells increases during acute hepatitis C virus infection and remains elevated in patients with chronic infection. Clin. Immunol. *132*, 401–411.

Dandri, M., Burda, M.R., Török, E., Pollok, J.M., Iwanska, A., Sommer, G., Rogiers, X., Rogler, C.E., Gupta, S., Will, H., *et al.* (2001). Repopulation of mouse liver with human hepatocytes and *in vivo* infection with hepatitis B virus. Hepatology *33*, 981–988.

Dazert, E., Neumann-Haefelin, C., Bressanelli, S., Fitzmaurice, K., Kort, J., Timm, J., McKiernan, S., Kelleher, D., Gruener, N., Tavis, J.E. *et al.* (2009). Loss of viral fitness and cross-recognition by CD8+ T cells limit HCV escape from a protective HLA-B27-restricted human immune response. J. Clin. Invest. *119*, 376–386.

Deinhardt, F. (1976). Hepatitis in primates. Adv. Virus Res. *20*, 113–157.

Dentzer, T.G., Lorenz, I.C., Evans, M.J., and Rice, C.M. (2009). Determinants of the hepatitis C virus nonstructural protein 2 protease domain required for production of infectious virus. J. Virol. *83*, 12702–12713.

Diago, M., Castellano, G., Garcia-Samaniego, J., Perez, C., Fernandez, I., Romero, M., Iacono, O.L., and Garcia-Monzon, C. (2006). Association of pretreatment serum interferon gamma inducible protein 10 levels with sustained virological response to peginterferon plus ribavirin therapy in genotype 1 infected patients with chronic hepatitis C. Gut 55, 374–379.

Diepolder, H.M., Zachoval, R., Hoffmann, R.M., Wierenga, E.A., Santantonio, T., and Jung, M.-C. (1995). Possible mechanism involving T-lymphocyte response to non-structural protein 3 in viral clearance in acute hepatitis C virus infection. Lancet *346*, 1006–1007.

Ebinuma, H., Nakamoto, N., Li, Y., Price, D.A., Gostick, E., Levine, B.L., Tobias, J., Kwok, W.W., and Chang, K.M. (2008). Identification and in vitro expansion of functional antigen-specific CD25+ FoxP3+ regulatory T cells in hepatitis C virus infection. J. Virol. *82*, 5043–5053.

el-Amad, Z., Murthy, K.K., Higgins, K., Cobb, E.K., Haigwood, N.L., Levy, A., and Steimer, K.S. (2007). Resistance of chimpanzees immunized with recombinant gp120SF2 to challenge by HIV-SF2. AIDS *9*, 1313–1322.

Elmowalid, G.A., Qiao, M., Jeong, S.H., Borg, B.B., Baumert, T.F., Sapp, R.K., Hu, Z., Murthy, K., and Liang, T.J. (2007). Immunization with hepatitis C virus-like particles results in control of hepatitis C virus infection in chimpanzees. Proc. Natl. Acad. Sci. U.S.A. *104*, 8427–8432.

Enomoto, N., Sakuma, I., Asahina, Y., Kurosaki, M., Murakami, T., Yamamoto, C., and Izumi, N. (1996). Interferon sensitivity determining sequence of the hepatitis C virus genome. Hepatology *24*, 460–461.

Erickson, A.L., Kimura, Y., Igarashi, S., Eichelberger, J., Houghton, M., Sidney, J., McKinney, D., Sette, A., Hughes, A.L., and Walker, C.M. (2001). The outcome of hepatitis C virus infection is predicted by escape mutations in epitopes targeted by cytotoxic T lymphocytes. Immunity *15*, 883–895.

Farci, P., Alter, H.J., Wong, D.C., Miller, R.H., Govindarajan, S., Engle, R., Shapiro, M., and Purcell, R.H. (1994). Prevention of hepatitis C virus infection in chimpanzees after antibody-mediated in vitro neutralization. Proc. Natl. Acad. Sci. U.S.A. *91*, 7792–7796.

Feinstone, S.M., Kapikian, A.Z., Purcell, R.H., Alter, H.J., and Holland, P.V. (1975). Transfusion-associated hepatitis not due to viral hepatitis type A or B. N. Engl. J. Med. *292*, 767–770.

Feinstone, S.M., Mihalik, K.B., Kamimura, T., Alter, H.J., London, W.T., and Purcell, R.H. (1983). Inactivation of hepatitis B virus and non-A, non-B hepatitis by chloroform. Infect. Immun. *41*, 816–821.

Feld, J.J., Nanda, S., Huang, Y., Chen, W., Cam, M., Pusek, S.N., Schweigler, L.M., Theodore, D., Zacks, S.L., Liang, T.J., and Fried, M.W. (2007). Hepatic gene expression during treatment with peginterferon and ribavirin: Identifying molecular pathways for treatment response. Hepatology *46*, 1548–1563.

Fernandez, J., Taylor, D., Morhardt, D.R., Mihalik, K., Puig, M., Rice, C.M., Feinstone, S.M., and Major, M.E. (2004). Long-term persistence of infection in chimpanzees inoculated with an infectious hepatitis C virus clone is associated with a decrease in the viral amino acid substitution rate and low levels of heterogeneity. J. Virol. *78*, 9782–9789.

Fleming, V.M., Harcourt, G., Barnes, E., and Klenerman, P. (2010). Virologic footprint of CD4+ T cell responses during chronic HCV infection. J. Gen. Virol. *91*, 1396–1406.

Folgori, A., Capone, S., Ruggeri, L., Meola, A., Sporeno, E., Ercole, B.B., Pezzanera, M., Tafi, R., Arcuri, M., Fattori, E., Lahm, A., Luzzago, A., Vitelli, A., Colloca, S., Cortese, R., and Nicosia, A. (2006). A T-cell HCV vaccine eliciting effective immunity against heterologous virus challenge in chimpanzees. Nat. Med. *12*, 190–197.

Forns, X., Thimme, R., Govindarajan, S., Emerson, S.U., Purcell, R.H., Chisari, F.V., and Bukh, J. (2000). Hepatitis C virus lacking the hypervariable region 1 of the second envelope protein is infectious and causes acute resolving or persistent infection in chimpanzees. Proc. Natl. Acad. Sci. U.S.A. *97*, 13318–13323.

Foy, E., Li, K., Wang, C.F., Sumpter, R., Ikeda, M., Lemon, S.M., and Gale, M. (2003). Regulation of interferon regulatory factor-3 by the hepatitis C virus serine protease. Science *300*, 1145–1148.

Foy, E., Li, K., Sumpter, R., Loo, Y.M., Johnson, C.L., Wang, C.F., Fish, P.M., Yoneyama, M., Fujita, T., Lemon, S.M., and Gale, M. (2005). Control of antiviral defenses through hepatitis C virus disruption of retinoic acid-inducible gene-I signaling. Proc. Natl. Acad. Sci. U.S.A. *102*, 2986–2991.

Franceschini, D., Paroli, M., Francavilla, V., Videtta, M., Morrone, S., Labbadia, G., Cerino, A., Mondelli, M.U., and Barnaba, V. (2009). PD-L1 negatively regulates $CD4^{+}CD25+Foxp3+$ Tregs by limiting STAT-5 phosphorylation in patients chronically infected with HCV. J. Clin. Invest. *119*, 551–564.

Fuller, M.J., Shoukry, N.H., Gushima, T., Bowen, D.G., Callendret, B., Campbell, K.J., Hasselschwert, D.L., Hughes, A.L., and Walker, C.M. (2009). Selection-driven immune escape is not a significant factor in the failure of CD4 T cell responses in persistent hepatitis C virus infection. Hepatology *51*, 378–387.

Gale, M., Blakely, C.M., Kwieciszewski, B., Tan, S.L., Dossett, M., Tang, N.M., Korth, M.J., Polyak, S.J., Gretch, D.R., and Katze, M.G. (1998). Control of PKR protein kinase by hepatitis C virus nonstructural 5A protein: Molecular mechanisms of kinase regulation. Mol. Cell. Biol. *18*, 5208–5218.

Gane, E.J., Roberts, S.K., Stedman, C., Angus, P.W., Ritchie, B., Elston, R., Ipe, D., Baher, L., Morcos, P., Najera, I., *et al.* (2009). First-in man demonstration of potent antiviral activity with a nucleoside ploymerase (R7128) and protease (R7227/ITMN-191) inhibitor or combination in HCV: safety, pharmacokinetics, and virological results from INFORM-1. J. Hepatol. *50*, S380.

Garaigorta, U., and Chisari, F.V. (2009). Hepatitis C virus blocks interferon effector function by inducing protein kinase R phosphorylation. Cell Host. Microbe *6*, 513–522.

Gaudieri, S., Rauch, A., Pfafferott, K., Barnes, E., Cheng, W., McCaughan, G., Shackel, N., Jeffrey, G.P., Mollison, L., Baker, R. *et al.* (2009). Hepatitis C virus drug resistance and immune-driven adaptations: relevance to new antiviral therapy. Hepatology *49*, 1069–1082.

Ge, D., Fellay, J., Thompson, A.J., Simon, J.S., Shianna, K.V., Urban, T.J., Heinzen, E.L., Qiu, P., Bertelsen, A.H., Muir, A.J., Sulkowski, M., McHutchison, J.G., and Goldstein, D.B. (2009). Genetic variation in IL28B predicts hepatitis C treatment-induced viral clearance. Nature *461*, 399–401.

Gerlach, J.T., Diepolder, H.M., Jung, M.C., Gruener, N.H., Schraut, W.W., Zachoval, R., Hoffmann, R., Schirren, C.A., Santantonio, T., and Pape, G.R. (1999). Recurrence of hepatitis C virus after loss of virus-specific CD4(+) T-cell response in acute hepatitis C. Gastroenterology *117*, 933–941.

Girard, S., Shalhoub, P., Lescure, P., Sabile, A., Misek, D.E., Hanash, S., Brechot, C., and Beretta, L. (2002). An altered cellular response to interferon and up-regulation of interleukin-8 induced by the hepatitis C viral protein NS5A uncovered by microarray analysis. Virology *295*, 272–283.

Golden-Mason, L., Palmer, B., Klarquist, J., Mengshol, J.A., Castelblanco, N., and Rosen, H.R. (2007). Upregulation of PD-1 expression on circulating and intrahepatic hepatitis c virus-specific $CD8^{+}$ T cells associated with reversible immune dysfunction. J. Virol. *81*, 9249–9258.

Golden-Mason, L., Palmer, B.E., Kassam, N., Townshend-Bulson, L., Livingston, S., McMahon, B., Castelblanco, N., Kuchroo, V., Gretch, D.R., and Rosen, H.R. (2009). Negative immune regulator TIM-3 is over-expressed in hepatitis C infection and its blockade rescues dysfunctional $CD4^{+}$ and $CD8^{+}$ T cells. J. Virol. *83*, 9122–9130.

Gottwein, J.M., Scheel, T.K., Callendret, B., Li, Y.P., Eccleston, H.B., Engle, R.E., Govindarajan, S., Satterfield, W., Purcell, R.H., Walker, C.M., and Bukh, J. (2010). Novel infectious cDNA clones of hepatitis C virus genotype 3a (strain S52) and 4a (strain ED43): genetic analyses and in vivo pathogenesis studies. J. Virol. *84*, 5277–5293.

Grace, K., Gartland, M., Karayiannis, P., McGarvey, M.J., and Clarke, B. (1999). The 5′ untranslated region of GB virus B shows functional similarity to the internal ribosome entry site of hepatitis C virus. J. Gen. Virol. *80*, 2337–2341.

Grakoui, A., Shoukry, N.H., Woollard, D.J., Han, J.-H., Hanson, H.L., Ghrayeb, J., Murthy, K.K., Rice, C.M., and Walker, C.M. (2003). HCV persistence and immune evasion in the absence of memory T cell help. Science *302*, 659–662.

Griffin, S., Trowbridge, R., Thommes, P., Parry, N., Rowlands, D., Harris, M., and Bright, H. (2008). Chimeric GB virus B genomes containing hepatitis C virus p7 are infectious in vivo. J. Hepatol. *49*, 908–915.

Grompe, M., Overturf, K., Al-Dhalimy, M., and Finegold, M. (1998). Therapeutic trials in the murine model of hereditary tyrosinaemia type I: a progress report. J. Inherit. Metab. Dis. *21*, 518–531.

Habersetzer, F., Baumert, T.F., and Stoll-Keller, F. (2009). GI-5005, a yeast vector vaccine expressing an NS3-core fusion protein for chronic HCV infection. Curr. Opin. Mol. Ther. *11*, 456–462.

Haqshenas, G., Dong, X., Netter, H., Torresi, J., and Gowans, E.J. (2007). A chimeric GB virus B encoding the hepatitis C virus hypervariable region 1 is infectious in vivo. J. Gen. Virol. *88*, 895–902.

He, L.-F., Alling, D., Popkin, T., Shapiro, M., Alter, H.J., and Purcell, R.H. (1987). Determining the size of non-A, non-B hepatitis virus by filtration. J. Infect. Dis. *156*, 636–640.

Heeg, M.H., Ulsenheimer, A., Gruner, N.H., Zachoval, R., Jung, M.C., Gerlach, J.T., Raziorrouh, B., Schraut, W., Horster, S., Kauke, T., *et al.* (2009). FOXP3 expression in hepatitis C virus-specific $CD4^{+}$ T cells during acute hepatitis C. Gastroenterology *137*, 1280–1288.

Hollinger, F.B., Gitnick, G.L., Aach, R.D., Szmuness, W., Mosley, J.W., and Stevens, C.E. (1978). Non-A, non-B hepatitis transmission in chimpanzees: a project of the transfusion-transmitted viruses study group. Intervirology *10*, 60–68.

Horowitz, B., Prince, A.M., Horowitz, M.S., and Watklevicz, C. (1993). Viral safety of solvent-detergent treated blood products. Dev. Biol. Stand. *81*, 147–161.

Houghton, M., and Abrignani, S. (2005). Prospects for a vaccine against the hepatitis C virus. Nature *436*, 961–966.

Huang, Y., Feld, J.J., Sapp, R.K., Nanda, S., Lin, J.H., Blatt, L.M., Fried, M.W., Murthy, K., and Liang, T.J. (2007). Defective hepatic response to interferon and activation of suppressor of cytokine signaling 3 in chronic hepatitis C. Gastroenterology *132*, 733–744.

Hughes, M., Griffin, S., and Harris, M. (2009). Domain III of NS5A contributes to both RNA replication and assembly of hepatitis C virus particles. J. Gen. Virol. *90*, 1329–1334.

Ishii, S., and Koziel, M.J. (2008). Immune responses during acute and chronic infection with hepatitis C virus. Clin. Immunol. *128*, 133–147.

Jones, C.T., Murray, C.L., Eastman, D.K., Tassello, J., and Rice, C.M. (2007). Hepatitis C virus p7 and NS2 proteins are essential for production of infectious virus. J. Virol. *81*, 8374–8383.

Jones, D.M., Patel, A.H., Targett-Adams, P., and McLauchlan, J. (2009). The hepatitis C virus NS4B protein can trans-complement viral RNA replication and modulates production of infectious virus. J. Virol. *83*, 2163–2177.

Jopling, C.L., Yi, M.K., Lancaster, A.M., Lemon, S.M., and Sarnow, P. (2005). Modulation of hepatitis C virus RNA abundance by a liver-specific microRNA. Science *309*, 1577–1581.

Jopling, C.L., Schutz, S., and Sarnow, P. (2008). Position-dependent function for a tandem microRNA miR-122-binding site located in the hepatitis C virus RNA genome. Cell Host. Microbe *4*, 77–85.

Kasprowicz, V., Schulze Zur, W.J., Kuntzen, T., Nolan, B.E., Longworth, S., Berical, A., Blum, J., McMahon, C., Reyor, L.L., Elias, N., *et al.* (2008). High level of PD-1 expression on hepatitis C virus (HCV)-specific $CD8^+$ and $CD4^+$ T cells during acute HCV infection, irrespective of clinical outcome. J. Virol. *82*, 3154–3160.

Kasprowicz, V., Kang, Y.H., Lucas, M., Schulze Zur, W.J., Kuntzen, T., Fleming, V., Nolan, B.E., Longworth, S., Berical, A., Bengsch, B., *et al.* (2010). Hepatitis C virus (HCV) sequence variation induces an HCV-specific T-cell phenotype analogous to spontaneous resolution. J. Virol. *84*, 1656–1663.

Kato, T., Choi, Y., Elmowalid, G., Sapp, R.K., Barth, H., Furusaka, A., Mishiro, S., Wakita, T., Krawczynski, K., and Liang, T.J. (2008). Hepatitis C virus JFH-1 strain infection in chimpanzees is associated with low pathogenicity and emergence of an adaptive mutation. Hepatology *48*, 732–740.

Kenny-Walsh, E., and Irish, H.R.G. (1999). Clinical outcomes after hepatitis C infection from contaminated anti-D immune globulin. N. Engl. J. Med. *340*, 1228–1233.

Khakoo, S.I., Thio, C.L., Martin, M.P., Brooks, C.R., Gao, X.J., Astemborski, J., Cheng, J., Goedert, J.J., Vlahov, D., Hilgartner, M., *et al.* (2004). HLA and NK cell inhibitory receptor genes in resolving hepatitis C virus infection. Science *305*, 872–874.

Kneteman, N.M., Weiner, A.J., O'Connell J., Collett, M., Gao, T., Aukerman, L., Kovelsky, R., Ni, Z., Hashash, A., Kline, J., *et al.* (2006). Anti-HCV therapies in chimeric *scid*-Alb/uPA mice parallel outcomes in human clinical application. Hepatology *43*, 1346–1353.

Koziel, M.J., Wong, D.K., Dudley, D., Houghton, M., and Walker, B.D. (1997). Hepatitis C virus-specific cytolytic T lymphocyte and T helper cell responses in seronegative persons. J. Infect. Dis. *176*, 859–866.

Kuntzen, T., Timm, J., Berical, A., Lewis-Ximenez, L.L., Jones, A., Nolan, B., Schulze zur Wiesch, J., Li, B., Schneidewind, A., Kim, A.Y., *et al.* (2007). Viral sequence evolution in acute hepatitis C virus infection. J. Virol. *81*, 11658–11668.

Kwong, A.D., McNair, L., Jacobson, I., and George, S. (2008). Recent progress in the development of selected hepatitis C virus NS3.4A protease and NS5B polymerase inhibitors. Curr. Opin. Pharmacol. *8*, 522–531.

Lagging, M., Romero, A., Westin, J., Norkrans, G., Dhillon, A., Pawlotsky, J., Zeuzem, S., Von Wagner, M., Negro, F., Schalm, S.W., *et al.*, for the Ditto-H.C.V Study Group (2006). IP-10 predicts viral response and therapeutic outcome in difficult-to-treat patients with HCV genotype 1 infection. Hepatology *44*, 1617–1625.

Lanford, R.E. (2006). The Chimpanzee Model of HCV. In Framing The Knowledge of Viral Hepatitis, Schinazi R.F. and Schiff, E.R., eds (IHL Press, College Park, GA, USA), pp. 197–218.

Lanford, R.E., Guerra, B., Demers, G.W., Chang, S., Sugarman, B.J., Maneval, D.C., Hong, Z., and Lau, J.Y.N. (2000). Gene Ther. for chronic hepatitis C based on adenoviral-mediated transfer of interferon-2b gene study based on the chimpanzee model. Hepatology *32*, 1141.

Lanford, R.E., Bigger, C., Bassett, S., and Klimpel, G.R. (2001a). The chimpanzee model of hepatitis c virus infections. ILAR J. *42*, 117–126.

Lanford, R.E., Chavez, D., Guerra, B., Lau, J.Y.N., Hong, Z., Brasky, K.M., and Beames, B. (2001b). Ribavirin induces error-prone replication of GB virus B in primary tamarin hepatocytes. J. Virol. *75*, 8074–8081.

Lanford, R.E., Chavez, D., Notvall, L., and Brasky, K.M. (2003a). Comparison of tamarins and marmosets as hosts for GBV-B infections and the effect of immunosuppression on duration of viremia. Virology *311*, 72–80.

Lanford, R.E., Guerra, B., Lee, H., Averett, D.R., Pfeiffer, B., Chavez, D., Notvall, L., and Bigger, C. (2003b). Antiviral effect and virus–host interactions in response to alpha interferon, gamma interferon, poly(I)-poly(C), tumor necrosis factor alpha, and ribavirin in hepatitis C virus subgenomic replicons. J. Virol. *77*, 1092–110.

Lanford, R.E., Guerra, B., Chavez, D., Bigger, C.B., Brasky, K.M., Wang, X.H., Ray, C., and Thomas, D.L. (2004).

Cross-genotype immunity to hepatitis C virus. J. Virol. *78*, 1575–1581.

Lanford, R.E., Guerra, B., Lee, H., Chavez, D., Brasky, K., and Bigger, C.B. (2006). Genomic response to interferon-alpha in chimpanzees: implications of rapid downregulation for hepatitis C kinetics. Hepatology *43*, 961–972.

Lanford, R.E., Guerra, B., Bigger, C.B., Lee, H., Chavez, D., and Brasky, K.M. (2007). Lack of response to exogenous interferon-alpha in the liver of HCV chronically infected chimpanzees. Hepatology *46*, 999–1008.

Lanford, R.E., Chavez, D., Guerra, B., Hubbard, G., and Brasky, K. (2008). Hepatocellular carcinoma as a long term outcome of HCV infection in chimpanzees. Abstracts of the 15th International Symposium on Hepatitis C Virus, 2008.

Lanford, R.E., Hildebrandt-Eriksen, E.S., Petri, A., Persson, R., Lindow, M., Munk, M.E., Kauppinen, S., and Orum, H. (2010). Therapeutic silencing of microRNA-122 in primates with chronic hepatitis C virus infection. Science *327*, 198–201.

Larghi, A., Zuin, M., Crosignani, A., Ribero, M.L., Pipia, C., Battezzati, P.M., Binelli, G., Donato, F., Zanetti, A.R., Podda, M., and Tagger, A. (2002). Outcome of an outbreak of acute hepatitis C among healthy volunteers participating in pharmacokinetics studies. Hepatology *36*, 993–1000.

Lechner, F., Wong, D.K., Dunbar, P.R., Chapman, R., Chung, R.T., Dohrenwend, P., Robbins, G., Phillips, R., Klenerman, P., and Walker, B.D. (2000). Analysis of successful immune responses in persons infected with hepatitis C virus. J. Exp. Med. *191*, 1499–1512.

Lemon, S.M. (2010). Induction and evasion of innate antiviral responses by hepatitis C virus 8783. J. Biol. Chem. *285*, 22741–22747.

Lerat, H., Honda, M., Beard, M.R., Loesch, K., Sun, J., Yang, Y., Okuda, M., Gosert, R., Xiao, S.Y., Weinman, A., and Lemon, S.M. (2002). Steatosis and liver cancer in transgenic mice expressing the structural and nonstructural proteins of hepatitis C virus. Gastroenterology *122*, 352–365.

Li, K., Foy, E., Ferreon, J.C., Nakamura, M., Ferreon, A.C.M., Ikeda, M., Ray, S.C., Gale, M., Jr. and Lemon, S.M. (2005a). Immune evasion by hepatitis C virus NS3/4A protease-mediated cleavage of the Toll-like receptor 3 adaptor protein TRIF. Proc. Natl. Acad. Sci. U.S.A. *102*, 2992–2997.

Li, X.D., Sun, L., Seth, R.B., Pineda, G., and Chen, Z.J. (2005b). From the cover: hepatitis C virus protease NS3/4A cleaves mitochondrial antiviral signaling protein off the mitochondria to evade innate immunity. Proc. Natl. Acad. Sci. U.S.A. *102*, 17717–17722.

Liang, Y., Shilagard, T., Xiao, S.Y., Snyder, N., Lau, D., Cicalese, L., Weiss, H., Vargas, G., and Lemon, S.M. (2009). Visualizing hepatitis C virus infections in human liver by two-photon microscopy. Gastroenterology *137*, 1448–1458.

Lindenbach, B.D. and Rice, C.M. (2005). Unravelling hepatitis C virus replication from genome to function. Nature *436*, 933–938.

Lindenbach, B.D., Meuleman, P., Ploss, A., Vanwolleghem, T., Syder, A.J., McKeating, J.A., Lanford, R.E., Feinstone, S.M., Major, M.E., Leroux-Roels, G., and Rice, C.M. (2006). Cell culture-grown hepatitis C virus is infectious in vivo and can be recultured in vitro. Proc. Natl. Acad. Sci. U.S.A. *103*, 3805–3809.

Liu, L., Fisher, B.E., Dowd, K.A., Astemborski, J., Cox, A.L., and Ray, S.C. (2010). Acceleration of hepatitis C virus envelope evolution in humans is consistent with progressive humoral immune selection during the transition from acute to chronic infection. J. Virol. *84*, 5067–5077.

Major, M.E., Mihalik, K., Puig, M., Rehermann, B., Nascimbeni, M., Rice, C.M., and Feinstone, S.M. (2002). Previously infected and recovered chimpanzees exhibit rapid responses that control hepatitis C virus replication upon rechallenge. J. Virol. *76*, 6586–6595.

Martin, A., Bodola, F., Sangar, D.V., Goettge, K., Popov, V., Rijnbrand, R., Lanford, R.E., and Lemon, S.M. (2003). Chronic hepatitis associated with GB virus B persistence in a tamarin after intrahepatic inoculation of synthetic viral RNA. Proc. Natl. Acad. Sci. U.S.A. *100*, 9962–9967.

Martin, A., Rijnbrand, R., Yang, Y., Cohen, L., Goettge, K., Bodola, F., Lanford, R.E., and Lemon, S.M. (2004). Adaptation of a chimeric GBV-B RNA containing HCV IRES sequence to efficient replication in tamarins. In Viral Hepatitis and Liver Disease, Jilbert, A.R., Grgacic, E.V.L., Vickery, K., Burrell, C.J., and Cossart, Y.E., eds (Australian Centre for Hepatitis Virology, Melbourne), pp. 361–363.

Masaki, T., Suzuki, R., Murakami, K., Aizaki, H., Ishii, K., Murayama, A., Date, T., Matsuura, Y., Miyamura, T., Wakita, T., and Suzuki, T. (2008). Interaction of hepatitis C virus nonstructural protein 5A with core protein is critical for the production of infectious virus particles. J. Virol. *82*, 7964–7976.

Maynard, J.E., Berquist, K.R., Krushak, H., and Purcell, R.H. (1972). Experimental infection of chimpanzees with the virus of hepatitis B. Nature *237*, 514–515.

Mehta, S.H., Cox, A., Hoover, D.R., Wang, X.H., Mao, Q., Ray, S., Strathdee, S.A., Vlahov, D., and Thomas, D.L. (2002). Protection against persistence of hepatitis C. Lancet *359*, 1478–1483.

Mercer, D.F., Schiller, D.E., Elliott, J.F., Douglas, D.N., Hao, C., Rinfret, A., Addison, W.R., Fischer, K.P., Churchill, T.A., Lakey, J.R., *et al.* (2001). Hepatitis C virus replication in mice with chimeric human livers. Nat. Med. *7*, 927–933.

Meuleman, P. and Leroux-Roels, G. (2008). The human liver-uPA-SCID mouse: a model for the evaluation of antiviral compounds against HBV and HCV. Antiviral Res. *80*, 231–238.

Meuleman, P. and Libbrecht, L., De, V.R., de, H.B., Gevaert, K., Vandekerckhove, J., Roskams, T., and Leroux-Roels, G. (2005). Morphological and biochemical characterization of a human liver in a uPA-SCID mouse chimera. Hepatology *41*, 847–856.

Meuleman, P., Libbrecht, L., Wieland, S., De Vos, R., Habib, N., Kramvis, A., Roskams, T., and

Leroux-Roels, G. (2006). immune suppression uncovers endogenous cytopathic effects of the hepatitis B virus. J. Virol. *80*, 2797–2807.

Meylan, E., Curran, J., Hofmann, K., Moradpour, D., Binder, M., Bartenschlager, R., and Tschopp, J. (2005). Cardif is an adaptor protein in the RIG-I antiviral pathway and is targeted by hepatitis C virus. Nature *437*, 1167–1172.

Miyamura, T. (1996). Interferon sensitivity determining sequence of the hepatitis C virus genome. Hepatology *24*, 460–461.

Moriya, K., Fujie, H., Shintani, Y., Yotsuyanagi, H., Tsutsumi, T., Ishibashi, K., Matsuura, Y., Kimura, S., Miyamura, T., and Koike, K. (1998). The core protein of hepatitis C virus induces hepatocellular carcinoma in transgenic mice. Nat. Med. *4*, 1065–1067.

Muchmore, E., Popper, H., Peterson, D.A., Miller, M.F., and Lieberman, H.M. (1988). Non-A, non-B hepatitis-related hepatocellular carcinoma in a chimpanzee. J. Med. Primatol. *17*, 235–246.

Muerhoff, A.S., Leary, T.P., Simons, J.N., Pilot-Matias, T.J., Dawson, G.J., Erker, J.C., Chalmers, M.L., Schlauder, G.G., Desai, S.M., and Mushahwar, I.K. (1995). Genomic organization of GB viruses A and B: Two new members of the Flaviviridae associated with GB agent hepatitis. J. Virol. *69*, 5621–5630.

Murthy, K.K., Cobb, E.K., Rouse, S.R., Lunceford, S.M., Johnson, D.E., and Galvan, A.R. (1996). Correlates of protective immunity against HIV-1 infection in immunized chimpanzees. Immunol. Lett. *51*, 121–124.

Nakamoto, N., Kaplan, D.E., Coleclough, J., Li, Y., Valiga, M.E., Kaminski, M., Shaked, A., Olthoff, K., Gostick, E., Price, D.A., *et al.* (2008). Functional restoration of HCV-specific CD8 T cells by PD-1 blockade is defined by PD-1 expression and compartmentalization. Gastroenterology 134, 1927–37, 1937.

Nakamoto, N., Cho, H., Shaked, A., Olthoff, K., Valiga, M.E., Kaminski, M., Gostick, E., Price, D.A., Freeman, G.J., Wherry, E.J., and Chang, K.M. (2009). Synergistic reversal of intrahepatic HCV-specific CD8 T cell exhaustion by combined PD-1/CTLA-4 blockade. PLoS Pathog. *5*, e1000313.

Nam, J.H., Faulk, K., Engle, R.E., Govindarajan, S., St Claire, M., and Bukh, J. (2004). In vivo analysis of the 3 ' untranslated region of GB virus B after in vitro mutagenesis of an infectious cDNA clone: Persistent infection in a transfected tamarin. J. Virol. *78*, 9389–9399.

Nascimbeni, M., Mizukoshi, E., Bosmann, M., Major, M.E., Mihalik, K., Rice, C.M., Feinstone, S.M., and Rehermann, B. (2003). Kinetics of CD4+ and CD8+ memory T-cell responses during hepatitis C virus rechallenge of previously recovered chimpanzees. J. Virol. *77*, 4781–4793.

Neumann, A.U., Lam, N.P., Dahari, H., Gretch, D.R., Wiley, T.E., Layden, T.J., and Perelson, A.S. (1998). Hepatitis C viral dynamics in vivo and the antiviral efficacy of interferon-α therapy. Science *282*, 103–107.

Neumann-Haefelin, C., Spangenberg, H.C., Blum, H.E., and Thimme, R. (2007). Host and viral factors contributing to CD8+ T cell failure in hepatitis C virus infection. World J. Gastroenterol. *13*, 4839–4847.

Nevens, F., Roskams, T., Van Vlierberghe, H., Horsmans, Y., Sprengers, D., Elewaut, A., Desmet, V., Leroux-Roels, G., Quinaux, E., Depla, E., *et al.* (2003). A pilot study of therapeutic vaccination with envelope protein E1 in 35 patients with chronic hepatitis C. Hepatology *38*, 1289–1296.

Northfield, J.W., Kasprowicz, V., Lucas, M., Kersting, N., Bengsch, B., Kim, A., Phillips, R.E., Walker, B.D., Thimme, R., Lauer, G., *et al.* (2008). CD161 expression on hepatitis C virus-specific CD8+ T cells suggests a distinct pathway of T cell differentiation. Hepatology *47*, 396–406.

Ohba, K., Mizokami, M., Lau, J.Y.N., Orito, E., Ikeo, K., and Gojobori, T. (1996). Evolutionary relationship of hepatitis C, pesti-, flavi-, plantviruses, and newly discovered GB hepatitis agents. FEBS Lett. *378*, 232–234.

Olsen, D.B., Carroll, S.S., Handt, L., Ludmerer, S., Graham, D., Fandozzi, C., DeLuca, J., Liverton, N., Vacca, J., and Hazuda, D. (2007). HCV antiviral activity and resistance analysis in chronically infected chimpanzees treated with NS3/4a protease and NS5B polymerase inhibitors. J. Hepatol. *46*, S298.

Osburn, W.O., Fisher, B.E., Dowd, K.A., Urban, G., Liu, L., Ray, S.C., Thomas, D.L., and Cox, A.L. (2009). Spontaneous control of primary hepatitis C virus infection and immunity against persistent reinfection. Gastroenterology *138*, 315–324.

Penna, A., Pilli, M., Zerbini, A., Orlandini, A., Mezzadri, S., Sacchelli, L., Missale, G., and Ferrari, C. (2007). Dysfunction and functional restoration of HCV-specific CD8 responses in chronic hepatitis C virus infection. Hepatology *45*, 588–601.

Pileri, P., Uematsu, Y., Campagnoli, S., Galli, G., Falugi, F., Petracca, R., Weiner, A.J., Houghton, M., Rosa, D., Grandi, G., and Abrignani, S. (1998). Binding of hepatitis C virus to CD81. Science *282*, 938–941.

Ploss, A., and Rice, C.M. (2009). Towards a small animal model for hepatitis C. EMBO Rep. *10*, 1220–1227.

Ploss, A., Evans, M.J., Gaysinskaya, V.A., Panis, M., You, H., de Jong, Y.P., and Rice, C.M. (2009). Human occludin is a hepatitis C virus entry factor required for infection of mouse cells. Nature *457*, 882–886.

Podevin, P., Sabile, A., Gajardo, R., Delhem, N., Abadie, A., Lozach, P.Y., Beretta, L., and Brechot, C. (2001). Expression of hepatitis C virus NS5A natural mutants in a hepatocytic cell line inhibits the antiviral effect of interferon in a PKR- independent manner. Hepatology *33*, 1503–1511.

Polyak, S.J., Khabar, K.S.A., Rezeiq, M., and Gretch, D.R. (2001). Elevated levels of interleukin-8 in serum are associated with hepatitis C virus infection and resistance to interferon therapy. J. Virol. *75*, 6209–6211.

Porter, B.F., Denise Goens, S., Brasky, K.M., and Hubbard, G.B. (2004). A case report of hepatocellular carcinoma and focal nodular hyperplasia with a myelolipoma in two chimpanzees and a review of spontaneous hepatobiliary tumors in non-human primates. J. Med. Primatol. 33, 38–47.

Prince, A.M. and Brotman, B. (1994). The biology of hepatitis C virus infection. Curr. Stud. Hematol. Blood Transf. *61*, 195–207.

Prince, A.M., Horowitz, B., Brotman, B., Huima, T., Richardson, L., van den Ende, M.C. (1984). Inactivation of hepatitis B and Hutchinson strain non-A, non-B hepatitis viruses by exposure to Tween 80 and ether. Vox Sang. *46*, 36–43.

Prince, A.M., Horowitz, B., Horowitz, M.S., and Zang, E. (1987). The development of virus-free labile blood derivatives – a review. Eur. J. Epidemiology 3, 103.

Prince, A.M., Stepan, W., Dichtelmuller, H., Brotman, B., and Huima, T. (1985). Inactivation of the Hutchinson strain of Non-A, Non-B hepatitis virus by combined use of beta-propiolactone and ultraviolet irradiation. J. Med. Virol. *16*, 119–125.

Prince, A.M., Brotman, B., Lee, D.H., Pfahler, W., Tricoche, N., Andrus, L., and Shata, M.T. (2005). Protection against Chronic Hepatitis C Virus Infection after Rechallenge with Homologous, but Not Heterologous, Genotypes in a Chimpanzee Model. J. Infect. Dis *192*, 1701–1709.

Radziewicz, H., Ibegbu, C.C., Fernandez, M.L., Workowski, K.A., Obideen, K., Wehbi, M., Hanson, H.L., Steinberg, J.P., Masopust, D., Wherry, E.J., *et al.* (2007). Liver-Infiltrating lymphocytes in chronic human hepatitis C virus infection display an exhausted phenotype with high levels of PD-1 and Low levels of CD127 expression. J. Virol. *81*, 2545–2553.

Radziewicz, H., Ibegbu, C.C., Hon, H., Bedard, N., Bruneau, J., Workowski, K.A., Knechtle, S.J., Kirk, A.D., Larsen, C.P., Shoukry, N.H., and Grakoui, A. (2010). Transient CD86 expression on hepatitis C virus-specific CD8^{+} T cells in acute infection is linked to sufficient IL-2 signaling. J. Immunol. *184*, 2410–2422.

Ray, S.C., Fanning, L., Wang, X.H., Netski, D.M., Walsh-Kenny, E., and Thomas, D.L. (2005). Divergent and convergent evolution after a common-source outbreak of hepatitis C virus. J. Exp. Med. *201*, 1753–1759.

Ray, S.C., Mao, Q., Lanford, R.E., Bassett, S., Laeyendecker, O., Wang, Y.M., and Thomas, D.L. (2000). Hypervariable region 1 sequence stability during hepatitis C virus replication in chimpanzees. J. Virol. *74*, 3058–3066.

Rehermann, B. (2009). Hepatitis C virus versus innate and adaptive immune responses: a tale of coevolution and coexistence. J. Clin. Invest. *119*, 1745–1754.

Rehermann, B., and Nascimbeni, M. (2005). Immunology of hepatitis B virus and hepatitis C virus infection. Nat. Rev. Immunol. *5*, 215–229.

Rehwinkel, J., and Reis e Sousa (2010). RIGorous detection: exposing virus through RNA sensing. Science *327*, 284–286.

Rijnbrand, R., Yang, Y., Beales, L., Bodola, F., Goettge, K., Cohen, L., Lanford, R.E., Lemon, S.M., and Martin, A. (2005). A chimeric GB virus B with 5′ nontranslated RNA sequence from hepatitis C virus causes hepatitis in tamarins. Hepatology *41*, 986–994.

Rollier, C.S., Paranhos-Baccala, G., Verschoor, E.J., Verstrepen, B.E., Drexhage, J.A.R., Fagrouch, Z., Berland, J.L., Komurian-Pradel, F., Duverger, B., Himoudi, N., *et al.* (2007). Vaccine-induced early control of hepatitis C virus infection in chimpanzees fails to impact on hepatic PD-1 and chronicity. Hepatology *45*, 602–613.

Romero, A.I., Lagging, M., Westin, J., Dhillon, A.P., Dustin, L.B., Pawlotsky, J.M., Neumann, A.U., Ferrari, C., Missale, G., Haagmans, B.L., *et al.* (2006). Interferon (IFN)-γ-inducible protein-10: association with histological results, viral kinetics, and outcome during treatment with pegylated IFN-α2a and ribavirin for chronic hepatitis C virus infection. J. Infect. Dis *194*, 895–903.

Rutebemberwa, A., Ray, S.C., Astemborski, J., Levine, J., Liu, L., Dowd, K.A., Clute, S., Wang, C., Korman, A., Sette, A., *et al.* (2008). High-programmed death-1 levels on hepatitis C virus-specific T cells during acute infection are associated with viral persistence and require preservation of cognate antigen during chronic infection. J. Immunol. *181*, 8215–8225.

Saito, T., Owen, D.M., Jiang, F., Marcotrigiano, J., Gale, M., Jr. (2008). Innate immunity induced by composition-dependent RIG-I recognition of hepatitis C virus RNA. Nature *454*, 523–527.

Sakai, A., St Claire, M.S., Faulk, K., Govindarajan, S., Emerson, S.U., Purcell, R.H., and Bukh, J. (2003). The p7 polypeptide of hepatitis C virus is critical for infectivity and contains functionally important genotype-specific sequences. Proc. Natl. Acad. Sci. U.S.A. *100*, 11646–11651.

Salloum, S., Oniangue-Ndza, C., Neumann-Haefelin, C., Hudson, L., Giugliano, S., aus dem Siepen., M., Nattermann, J., Spengler, U., Lauer, G.M., Wiese, M., *et al.*. (2008). Escape from HLA-B*08-restricted CD8 T cells by hepatitis C virus is associated with fitness costs. J. Virol. *82*, 11803–11812.

Sarasin-Filipowicz, M., Oakeley, E.J., Duong, F.H., Christen, V., Terracciano, L., Filipowicz, W., and Heim, M.H. (2008). Interferon signaling and treatment outcome in chronic hepatitis C. Proc. Natl. Acad. Sci. U.S.A. *105*, 7034–7039.

Sarrazin, C., and Zeuzem, S. (2010). Resistance to direct antiviral agents in patients with hepatitis C virus infection. Gastroenterology *138*, 447–462.

Sbardellati, A., Scarselli, E., Verschoor, E., DeTomassi, A., Lazzaro, D., and Traboni, C. (2001). Generation of infectious and transmissible virions from a GB virus B full-length consensus clone in tamarins. J. Gen. Virol *82*, 2437–2448.

Scarselli, E., Urbani, A., Sbardellati, A., Tomei, L., De Francesco, R., and Traboni, C. (1997). GB virus B and hepatitis C virus NS3 serine proteases share substrate specificity. J. Virol. *71*, 4985–4989.

Schlauder, G.G., Dawson, G.J., Simons, J.N., Pilot-Matias, T.J., Gutierrez, R.A., Heynan, C.A., Kniggo, M.F., Kurpiewski, G.S., Buijk, S.L., Leary, T.P., *et al.* (1995). Molecular and serologic analysis in the transmission of the GB hepatitis agents. J. Med. Virol. *46*, 81–90.

Scognamiglio, P., Accapezzato, D., Casciaro, M.A., Cacciani, A., Artini, M., Bruno, G., Chircu, M.L., Sidney, J., Southwood, S., Abrignani, S., Sette, A., and Barnaba, V. (1999). Presence of effector CD8^{+} T cells in hepatitis C virus-exposed healthy seronegative donors. J. Immunol. *162*, 6681–6689.

Semmo, N., and Klenerman, P. (2007). CD4+ T cell responses in hepatitis C virus infection. World J. Gastroenterol. *13*, 4831–4838.

Shimakami, T., Lanford, R.E., and Lemon, S.M. (2009). Hepatitis C: recent successes and continuing challenges in the development of improved treatment modalities. Curr. Opin. Pharmacol. *9*, 537–544.

Shimizu, Y.K., and Purcell, R.H. (1989). Cytoplasmic antigen in hepatocytes of chimpanzees infected with non-A, non-B hepatitis virus or hepatitis delta virus: relationship to interferon. Hepatology *10*, 764–768.

Shimizu, Y.K., Feinstone, S.M., Purcell, R.H., Alter, H.J., and London, W.T. (1979). Non-A, non-B hepatitis: ultrastructural evidence for two agents in experimentally infected chimpanzees. Science *205*, 197–200.

Shimizu, Y.K., Weiner, A.J., Rosenblatt, J., Wong, D.C., Shapiro, M., Popkin, T., Houghton, M., Alter, H.J., and Purcell, R.H. (1990). Early events in hepatitis C virus infection of chimpanzees. Proc. Natl. Acad. Sci. U.S.A. *87*, 6441–6444.

Shoukry, N.H., Grakoui, A., Houghton, M., Chien, D.Y., Ghrayeb, J., Reimann, K.A., and Walker, C.M. (2003). Memory CD8+ T cells are required for protection from persistent hepatitis C virus infection. J. Exp. Med. *197*, 1645–1655.

Simons, J.N., Pilot-Matias, T.J., Leary, T.P., Dawson, G.J., Desai, S.M., Schlauder, G.G., Muerhoff, A.S., Erker, J.C., Buijk, S.L., Chalmers, M.L., *et al.* (1995). Identification of two flavivirus-like genomes in the GB hepatitis agent. Proc. Natl. Acad. Sci. U.S.A. *92*, 3401–3405.

Steffy, K., Kirkovsky, L., Lanford, R.E., Showalter, R.E., Sergeeva, M.V., Zhao, J., Averett, D.R., and Appleman, J.R. (2008). Antiviral efficacy of the HCV RNA polymerase inhibitor ANA598 in the chimpanzee model of HCV infection. Hepatology *48*, 1163A.

Su, A.I., Pezacki, J.P., Wodicka, L., Brideau, A.D., Supekova, L., Thimme, R., Wieland, S., Bukh, J., Purcell, R.H., Schultz, P.G., and Chisari, F.V. (2002). Genomic analysis of the host response to hepatitis C virus infection. Proc. Natl. Acad. Sci. U.S.A. *99*, 15669–15674.

Suntharalingam, G., Perry, M.R., Ward, S., Brett, S.J., Castello-Cortes, A., Brunner, M.D., and Panoskaltsis, N. (2006). Cytokine storm in a phase 1 trial of the anti-CD28 monoclonal antibody TGN1412. N. Engl. J. Med. *355*, 1018–1028.

Suppiah, V., Moldovan, M., Ahlenstiel, G., Berg, T., Weltman, M., Abate, M.L., Bassendine, M., Spengler, U., Dore, G.J., Powell, E., *et al.* (2009). IL28B is associated with response to chronic hepatitis C interferon-alpha and ribavirin therapy. Nat. Genet. *41*, 1100–1104.

Szabo, G., and Dolganiuc, A. (2008). Hepatitis C and innate immunity: recent advances. Clin. Liver Dis. 12, *675–92*, x.

Tabor, E., Hsia, C.C., and Muchmore, E. (1994). Histochemical and immunohistochemical similarities between hepatic tumors in two chimpanzees and man. J. Med. Primatol. *23*, 271–279.

Takikawa, S., Engle, R.E., Emerson, S.U., Purcell, R.H., St. Claire, M., and Bukh, J. (2006). Functional analyses of GB virus B p13 protein: development of a recombinant GB virus B hepatitis virus with a p7 protein. Proc. Natl. Acad. Sci. U.S.A. *103*, 3345–3350.

Takikawa, S., Engle, R.E., Faulk, K.N., Emerson, S.U., Purcell, R.H., and Bukh, J. (2010). Molecular evolution of GB virus B hepatitis virus during acute resolving and persistent infections in experimentally infected tamarins. J. Gen. Virol. *91*, 727–733.

Tan, S.L. and Katze, M.G. (2001). How hepatitis C virus counteracts the interferon response: the jury is still out on NS5A. Virology *284*, 1–12.

Tanaka, Y., Nishida, N., Sugiyama, M., Kurosaki, M., Matsuura, K., Sakamoto, N., Nakagawa, M., Korenaga, M., Hino, K., Hige, S., *et al.* (2009). Genome-wide association of IL28B with response to pegylated interferon-alpha and ribavirin therapy for chronic hepatitis C. Nat. Genet. *41*, 1105–1109.

Targett-Adams, P., Schaller, T., Hope, G., Lanford, R.E., Lemon, S.M., Martin, A., and McLauchlan, J. (2006). Signal peptide peptidase cleavage of GB virus B core protein is required for productive infection in vivo. J. Biol. Chem. *281*, 29221–29227.

Tellinghuisen, T.L., Evans, M.J., von Hahn, T., You, S., and Rice, C.M. (2007). Studying hepatitis C virus: making the best of a bad virus. J. Virol. *81*, 8853–8867.

Tester, I., Smyk-Pearson, S., Wang, P., Wertheimer, A., Yao, E., Lewinsohn, D.M., Tavis, J.E., and Rosen, H.R. (2005). Immune evasion versus recovery after acute hepatitis C virus infection from a shared source. J. Exp. Med. *201*, 1725–1731.

Thimme, R., Oldach, D., Chang, K.M., Steiger, C., Ray, S.C., and Chisari, F.V. (2001). Determinants of viral clearance and persistence during acute hepatitis C virus infection. J. Exp. Med. *194*, 1395–1406.

Thimme, R., Bukh, J., Spangenberg, H.C., Wieland, S., Pemberton, J., Steiger, C., Govindarajan, S., Purcell, R.H., and Chisari, F.V. (2002). Viral and immunological determinants of hepatitis C virus clearance, persistence, and disease. Proc. Natl. Acad. Sci. U.S.A. *99*, 15661–15668.

Thimme, R., Neumann-Haefelin, C., Boettler, T., and Blum, H.E. (2008). Adaptive immune responses to hepatitis C virus: from viral immunobiology to a vaccine. Biol. Chem. *389*, 457–467.

Thomas, D., and Seeff, L.B. (2005). Natural history of hepatitis C. Clin. Liver Dis. *9*, 383–398.

Thompson, A., Patel, K., Tillman, H., and McHutchison, J.G. (2009). Directly acting antivirals for the treatment of patients with hepatitis C infection: a clinical development update addressing key future challenges. J. Hepatol. *50*, 184–194.

Timm, J., Lauer, G.M., Kavanagh, D.G., Sheridan, I., Kim, A.Y., Lucas, M., Pillay, T., Ouchi, K., Reyor, L.L., zur Wiesch, J.S., *et al.* (2004). CD8 epitope escape and reversion in acute HCV infection. J. Exp. Med. *200*, 1593–1604.

Uebelhoer, L., Han, J.H., Callendret, B., Mateu, G., Shoukry, N.H., Hanson, H.L., Rice, C.M., Walker, C.M., and Grakoui, A. (2008). Stable cytotoxic T cell escape mutation in hepatitis C virus is linked to maintenance of viral fitness. PLoS. Pathog. *4*, e1000143.

Ulsenheimer, A., Gerlach, J.T., Gruener, N.H., Jung, M.C., Schirren, C.A., Schraut, W., Zachoval, R., Pape, G.R., and Diepolder, H.M. (2003). Detection of functionally altered hepatitis C virus-specific CD4 T cells in acute and chronic hepatitis C. Hepatology *37*, 1189–1198.

Ulsenheimer, A., Lucas, M., Seth, N.P., Tilman, G.J., Gruener, N.H., Loughry, A., Pape, G.R., Wucherpfennig, K.W., Diepolder, H.M., and Klenerman, P. (2006). Transient immunological control during acute hepatitis C virus infection: ex vivo analysis of helper T-cell responses. J. Viral Hepat. *13*, 708–714.

Uprichard, S., Chung, J., Chisari, F., and Wakita, T. (2006). Replication of a hepatitis C virus replicon clone in mouse cells. Virol. J. *3*, 89.

Urbani, S., Amadei, B., Tola, D., Pedrazzi, G., Sacchelli, L., Cavallo, M.C., Orlandini, A., Missale, G., and Ferrari, C. (2008). Restoration of HCV-specific T cell functions by PD-1/PD-L1 blockade in HCV infection: effect of viremia levels and antiviral treatment. J. Hepatol. *48*, 548–558.

Urbani, S., Amadei, B., Cariani, E., Fisicaro, P., Orlandini, A., Missale, G., and Ferrari, C. (2005). The impairment of CD8 responses limits the selection of escape mutations in acute hepatitis C virus infection. J. Immunol. *175*, 7519–7529.

Urbani, S., Amadei, B., Tola, D., Massari, M., Schivazzappa, S., Missale, G., and Ferrari, C. (2006). PD-1 expression in acute hepatitis C is associated with HCV-specific CD8 exhaustion. J. Virol. *80*, 11398–11403

VandeBerg, J.L., and Zola, S.M. (2005). A unique biomedical resource at risk 8785. Nature *437*, 30–32.

Vogt, M., Lang, T., Frosner, G., Klingler, C., Sendl, A.F., Zeller, A., Wiebecke, B., Langer, B., Meisner, H., and Hess, J. (1999). Prevalence and clinical outcome of hepatitis C infection in children who underwent cardiac surgery before the implementation of blood-donor screening. N. Engl. J. Med. *341*, 866–870.

von Hahn, T., Yoon, J.C., Alter, H., Rice, C.M., Rehermann, B., Balfe, P., and McKeating, J.A. (2007). Hepatitis C virus continuously escapes from neutralizing antibody and t-cell responses during chronic infection in vivo. Gastroenterology *132*, 667–678.

Wakita, T., Pietschmann, T., Kato, T., Date, T., Miyamoto, M., Zhao, Z., Murthy, K., Habermann, A., Krausslich, H.G., Mizokami, M., *et al.* (2005). Production of infectious hepatitis C virus in tissue culture from a cloned viral genome. Nat. Med. *11*, 791–796.

Walker, C.M. (1997). Comparative features of hepatitis C virus infection in humans and chimpanzees. Springer Semin. Immunopathol. *19*, 85–98.

Wang, N., Liang, Y., Devaraj, S., Wang, J., Lemon, S.M., and Li, K. (2009). Toll-like receptor 3 mediates establishment of an antiviral state against hepatitis C virus in hepatoma cells 8784. J. Virol. *83*, 9824–9834.

Ward, S.M., Fox, B.C., Brown, P.J., Worthington, J., Fox, S.B., Chapman, R.W., Fleming, K.A., Banham, A.H., and Klenerman, P. (2007). Quantification and localisation of FOXP3+ T lymphocytes and relation to hepatic inflammation during chronic HCV infection. J. Hepatol. *47*, 316–324.

Warter, L., Cohen, L., Benureau, Y., Chavez, D., Yang, Y., Bodola, F., Lemon, S.M., Traboni, C., Lanford, R.E., and Martin, A. (2009). A cooperative interaction between nontranslated RNA sequences and NS5A protein promotes in vivo fitness of a chimeric hepatitis C/GB virus B. PLoS One. *4*, e4419.

Weatherford, T., Chavez, D., Brasky, K.M., and Lanford, R.E. (2009a). The marmoset model of GB virus B infections: adaptation to host phenotypic variation. J. Virol. *83*, 5806–5814.

Weatherford, T., Chavez, D., Brasky, K.M., Lemon, S.M., Martin, A., and Lanford, R.E. (2009b). Lack of adaptation of chimeric GB virus B/hepatitis C virus in the marmoset model: possible effects of bottleneck. J. Virol. *83*, 8062–8075.

Wedemeyer, H., Schuller, E., Schlaphoff, V., Stauber, R.E., Wiegand, J., Schiefke, I., Firbas, C., Jilma, B., Thursz, M., Zeuzem, S., *et al.* (2009). Therapeutic vaccine IC41 as late add-on to standard treatment in patients with chronic hepatitis C. Vaccine *27*, 5142–5151.

Wiese, M., Berr, F., Lafrenz, M., Porst, H., and Oesen, U. (2000). Low frequency of cirrhosis in a hepatitis C (genotype 1b) single-source outbreak in Germany: a 20-year multicenter study. Hepatology *32*, 91–96.

Yanagi, M., St, C.M., Shapiro, M., Emerson, S.U., Purcell, R.H., and Bukh, J. (1998). Transcripts of a chimeric cDNA clone of hepatitis C virus genotype 1b are infectious in vivo. Virology *244*, 161–172.

Yanagi, M., Purcell, R.H., Emerson, S.U., and Bukh, J. (1999). Hepatitis C virus: An infectious molecular clone of a second major genotype (2a) and lack of viability of intertypic 1a and 2a chimeras. Virology *262*, 250–263.

Yeh, M.M., Daniel, H.D., and Torbenson, M. (2010). Hepatitis C-associated hepatocellular carcinomas in non-cirrhotic livers. Mod. Pathol. *23*, 276–283.

Yi, M., Villanueva, R.A., Thomas, D.L., Wakita, T., and Lemon, S.M. (2006). Production of infectious genotype 1a hepatitis C virus (Hutchinson strain) in cultured human hepatoma cells. Proc. Natl. Acad. Sci. U.S.A. *103*, 2310–2315.

Yi, M.K., Ma, Y.H., Yates, J., and Lemon, S.M. (2009). trans-Complementation of an NS2 Defect in a Late Step in Hepatitis C Virus (HCV) Particle Assembly and Maturation. PLoS Pathog. *5*, e1000403.

Youn, J.W., Hu, Y.W., Tricoche, N., Pfahler, W., Shata, M.T., Dreux, M., Cosset, F.L., Folgori, A., Lee, D.H., Brotman, B., and Prince, A.M. (2008). Evidence for protection against chronic hepatitis C virus infection in chimpanzees by immunization with replicating recombinant vaccinia virus. J. Virol. *82*, 10896–10905.

Zhu, Q., Guo, J.T., and Seeger, C. (2003). Replication of hepatitis C virus subgenomes in nonhepatic epithelial and mouse hepatoma cells. J. Virol. *77*, 9204–9210.

Hepatitis C Virus Drug Pipeline Overview and Clinical Trial Design

7

Jean-Michel Pawlotsky

Abstract

In the last decade, insights into the virology of hepatitis C virus (HCV) have unravelled several targets for potential novel therapeutics that, unlike interferon (IFN)-α and ribavirin, are specifically targeted to HCV. Many such direct-acting antiviral (DAA) drugs are at the preclinical developmental stage and several are in clinical development. Initial clinical trials using some of these inhibitors, either alone or in combination with pegylated IFN-α and ribavirin, have yielded encouraging results.

Direct-acting antivirals in development

Several DAAs have reached clinical development (Table 7.1). The most advanced ones include inhibitors of the HCV non-structural (NS) 3/4A protease; two have reached phase III development in combination with pegylated IFN-α and ribavirin: telaprevir and boceprevir.

NS3/4A protease inhibitors

Highly selective, potent peptidomimetic inhibitors of the HCV NS3/4A protease have been derived from the viral substrate of the enzyme using structure-based drug design techniques (see Chapters 9–11). A number of such molecules are in clinical development (Table 7.1) and they can be categorized into two groups: macrocyclic and linear protease inhibitors. The most advanced NS3/4A protease inhibitors, currently in phase III trials with pegylated IFN-α and ribavirin, are telaprevir (VX-950, Vertex Pharmaceuticals) and boceprevir (SCH503034, Schering-Plough), both of which are linear inhibitors.

Telaprevir

Telaprevir has shown antiviral activity against HCV *in vitro*. In a phase Ib clinical trial, HCV genotype-1-infected patients receiving telaprevir monotherapy for 14 days all had at least a 2 log decrease in HCV RNA levels. Viral breakthroughs occurred in the patients with lower exposure to the drug, due to selection of telaprevir-resistant HCV variants (Sarrazin *et al.*, 2007a). The following telaprevir resistance substitutions have been reported, by order of increasing resistance: V36A/M, T54A, R155K/T, V36A/M + R155K/T, A156V/T, and V36A/M + A156V/T (Kieffer *et al.*, 2007; Sarrazin *et al.*, 2007a). Differences in resistance profiles between HCV subtypes 1a and 1b have been reported. For instance, variants with changes at amino acid position R155 are often selected by telaprevir in patients infected with subtype 1a, whereas such variants are rarely selected in patients infected with HCV subtype 1b. HCV resistance to telaprevir is significantly less frequent when the drug is administered in combination with either pegylated IFN-α or both pegylated IFN-α and ribavirin (Forestier *et al.*, 2007; Hezode *et al.*, 2009; McHutchison *et al.*, 2009).

The results of two phase II clinical trials were recently reported. The PROVE 1 trial tested the combination of telaprevir with pegylated IFN-α and ribavirin in treatment-naïve American patients infected with HCV genotype 1 versus standard-of-care (McHutchison *et al.*, 2009). A

Table 7.1 *In vitro* antiviral efficacy of DAAs that have reached clinical development as of June 2010 and for which clinical trial data have been published or presented at international medical or scientific conferences

Drug	Phase	Dose	Duration (days)	Median/mean log HCV RNA level reduction
NS3/4A protease inhibitors				
Telaprevir (Vertex and Tibotec)	III	750 mg q.8h	14	–4.4
Boceprevir (Merck)	III	400 mg t.i.d[a]	7	–1.6
TMC435 (Tibotec)	II	200 mg q.d.	7	–4.1
Danoprevir (Genentech/Roche)	II	200 mg q.8h	14	–3.8
Vaniprevir (Merck)	II	700 mg b.i.d.	8	–4.7
BI201335 (Boehringer-Ingelheim)	II	240 mg q.d.	14	–4.0
Narlaprevir (Merck)	II	400 mg b.i.d.	7	–4.2
BMS-650032 (Bristol-Myers Squibb)	Ib	300 mg b.i.d.	3	–3.3
ACH-1625 (Achillion)	Ib	600 mg b.i.d.	4.5	–3.9
GS-9256 (Gilead)	Ib	200 mg b.i.d.	3	–2.8
Nucleoside/nucleotide analogue inhibitors of HCV RNA-dependent RNA polymerase				
RG7128 (Genentech/Roche)	II	1500 mg b.i.d.	14	–2.7
IDX184	II	100 mg q.d.	3	–0.7
Non-nucleoside inhibitors of HCV RNA-dependent RNA polymerase				
GS-9190 (Gilead)	II	40 mg b.i.d.	8	–1.4
Filibuvir (Pfizer)	II	300 mg b.i.d.	8	–2.1
ANA598 (Anadys)	II	800 mg b.i.d.	3	–2.9
BI207127 (Boehringer-Ingelheim)	II	800 mg q.h	3	–3.1
ABT-333	II	600 mg b.i.d.	2	–1.5
VX-222	Ib	750 mg b.i.d.	3	–3.2
NS5A inhibitors				
BMS790052	II	10 mg q.d.	1	–3.2
AZD7295	Ib	233 mg q.d.	5	–2.1 (only 1b)
Cyclophilin inhibitors				
Alisporivir (Novartis)	II	1200 mg b.i.d.	14	–3.6
SCY-465	1b	900 mg q.d.	15	–2.2

[a]A higher dose of 800 mg t.i.d. of boceprevir is used in phase III clinical trials.

DAA: direct-acting antiviral; HCV: hepatitis C virus.

The log HCV RNA reductions are median or mean reductions achieved with different doses after different durations of administration for the different drugs.

sustained virological response (SVR), defined as an undetectable HCV RNA 24 weeks after the end of therapy, was observed in 41% of the patients receiving the combination of pegylated IFN-α and ribavirin for 48 weeks. In contrast, SVR was achieved in 61% of those receiving a triple combination of telaprevir, 750 mg every 8 hours, pegylated IFN-α and ribavirin for 12 weeks followed by 12 weeks of pegylated IFN-α and ribavirin, and in 67% of those receiving the triple combination for 12 weeks, followed by 36 weeks of pegylated IFN-α and ribavirin ($P = 0.02$ and $P = 0.001$, respectively; McHutchison *et al.*, 2009).

In the PROVE 2 trial, performed in treatment-naïve European patients infected with HCV genotype 1, the SVR rates were 46% in the control arm, 60% in the patients receiving the triple combination for 12 weeks, and 69% in the patients receiving the triple combination for 12 weeks followed by 12 weeks of pegylated IFN-α and ribavirin ($P = 0.12$ and $P = 0.001$, respectively; Hezode *et al.*, 2009). The SVR rate reached only 36% of the patients receiving the combination of pegylated IFN-α and telaprevir for 12 weeks, essentially due to a high rate of relapse. In almost all patients receiving telaprevir who experienced a virological breakthrough on treatment or a relapse after treatment cessation, the dominant viral population at the time of relapse comprised telaprevir-resistant variants (Hezode *et al.*, 2009). In both trials, dermatological side-effects, including rash and pruritus, were frequent (more than 50%). Other side-effects included gastrointestinal symptoms and moderate anaemia (Hezode *et al.*, 2009; McHutchison *et al.*, 2009).

The results of another trial, the C208 study, were recently presented. Treatment-naïve patients infected with HCV genotype 1 were randomized into four groups. Two groups received pegylated IFN-α2a, 180 μg/week, and ribavirin, 1000–1200 mg/day according to body weight, with telaprevir at either 750 mg every 8 hours or 1250 mg every 12 hours. The two other groups received pegylated IFN-α2b, 1.5 μg/kg/week, and ribavirin, 800–1200 mg/day according to body weight, with telaprevir at either 750 mg every 8 hours, or 1250 mg every 12 hours. The SVR rates ranged from 81% to 85% and no significant difference was observed between the groups (Forns *et al.*, 2010). Larger studies in more difficult-to-treat populations will be needed to confirm these results.

In the PROVE 3 trial, performed in American and European patients infected with HCV genotype 1 who did not achieve an SVR after a prior treatment with pegylated IFN-α and ribavirin, the SVR rates were 14% only in the control arm (retreatment with a standard dose of pegylated IFN-α and ribavirin), 51% in the patients receiving the triple combination for 12 weeks followed by 12 weeks of pegylated IFN-α and ribavirin, and 53% in the patients receiving the triple combination for 24 weeks followed by 24 weeks of pegylated IFN-α and ribavirin ($P < 0.001$ and $P < 0.001$ vs. control, respectively; McHutchison *et al.*, 2010). Interestingly, the patients who relapsed after the first course of treatment achieved an SVR in 69% and 76% of cases, respectively, whereas those who did not respond to therapy (less than a 2 log drop at week 12 of treatment) achieved an SVR in 39% and 38% of cases, respectively. The SVR rate reached only 24% of patients receiving the combination of pegylated IFN-α and telaprevir without ribavirin for 24 weeks (McHutchison *et al.*, 2010). The side-effect profiles were not different from those reported in treatment-naïve patients (Hezode *et al.*, 2009; McHutchison *et al.*, 2009).

The results of rollover study C107 were recently presented. Patients included in one of the control arms of the PROVE1, PROVE2 or PROVE3 studies who failed to eliminate HCV on pegylated IFN-α and ribavirin were retreated with a triple combination of pegylated IFN-α, ribavirin and telaprevir, 750 mg every 8 hours, for 12 weeks, followed by 12 weeks of pegylated IFN-α and ribavirin. Overall, 59% of them achieved an SVR. The likelihood of an SVR was strongly related to the outcome of prior standard-of-care therapy. Indeed, the SVR rates were 97%, 75%, 55% and 37% in prior relapsing patients, in patients with a virological breakthrough, in partial responders and in null responders, respectively (Berg *et al.*, 2010). Breakthroughs on treatment were observed essentially in prior null responders, defined as patients who achieved a less than 1 log drop at week 4 or a less than 2 log drop at week 12 on pegylated IFN-α and ribavirin. Telaprevir-resistant variants were dominant at the time of breakthrough in these patients.

Boceprevir

Boceprevir is a potent inhibitor of HCV replication *in vitro* (Malcolm *et al.*, 2006; Venkatraman *et al.*, 2006). One week of boceprevir monotherapy in patients who had previously failed to respond to pegylated IFN-α alone or in combination with ribavirin was well tolerated and induced an average HCV RNA level reduction of 1.0 and 1.6 log IU/ml at doses of 200 mg and 400 mg three times per day, respectively (Sarrazin *et al.*, 2007b). A higher dose of boceprevir (800 mg, three times per day) was evaluated in phase II and is currently used in phase III trials. Few data are available on boceprevir resistance *in vivo*. When treated with the triple combination of pegylated IFN-α, ribavirin and a relatively low dose of boceprevir-resistant variants, boceprevir associated with virological breakthrough was frequently observed in prior non-responders to pegylated IFN-α and ribavirin. The reported substitutions so far include those found in telaprevir resistance, with the exception of V170A (Curry *et al.*, 2008).

The results of a phase II clinical trial (SPRINT-1) have been recently reported (Kwo *et al.*, 2009a). SVR rates in patients receiving the triple combination of boceprevir, 800 mg three times per day, pegylated IFN-α and ribavirin for 28 weeks were not significantly different in the presence or absence of a 'lead-in' phase with pegylated IFN-α and ribavirin during the first 4 weeks of therapy (54% and 56%, respectively). In the patients treated for 48 weeks, SVR rates were 67% and 75% (not significantly different) without or with a 4-week 'lead-in' phase (Kwo *et al.*, 2009a). A lower dose of ribavirin was associated with a lower SVR rate in combination with pegylated IFN-α and boceprevir. This study did not show a beneficial effect of the 'lead-in' phase in subsequent SVR. However, the 'lead-in' phase provided interesting information; the SVR rates were strongly related to the ability of pegylated IFN-α and ribavirin to substantially reduce HCV RNA levels without boceprevir. Whatever the treatment duration, the highest SVR rates were achieved when pegylated IFN-α and ribavirin reduced HCV RNA levels by 4 log or more during the 4 weeks 'lead-in' phase, whereas the highest failure rates were observed in patients with a less than 1.5 log HCV RNA level reduction during the 'lead-in' phase (Kwo *et al.*, 2009b). The main side-effect associated with boceprevir administration was anaemia, and no skin events were reported (Kwo *et al.*, 2009a).

Other NS3/4A protease inhibitors

Other NS3/4A protease inhibitors in phase Ib or II development (Table 7.1) in combination with pegylated IFN-α and ribavirin and/or with other DAAs include TMC435 (Tibotec), danoprevir (ITMN-191/RG7227, Roche), vaniprevir (MK-7009, Merck), BI201335 (Boehringer-Ingelheim), narlaprevir (Merck), BMS650032 (Bristol-Myers Squibb), ACH-1625 (Achillion) and GS-9256 (Gilead). These drugs are structurally close to each other and share extensive cross-resistance. When administered alone, they induce a 3.5–4.5 log HCV RNA level reduction on average in patients infected with HCV genotype 1 within a few days of administration (Seiwert *et al.*, 2008; Liverton *et al.*, 2010; Reesink *et al.*, 2010a,b). Vaniprevir and TMC435 appear to be well tolerated, whereas BI201335 occasionally induces hyperbilirubinaemia and photosensitivity, and danoprevir has been associated with ALT elevations. Danoprevir and narlaprevir can be boosted by ritonavir in order to use lower doses and reduce side-effects (Gane *et al.*, 2010).

Other inhibitors of HCV replication

HCV replication inhibitors that are currently in early clinical evaluation target HCV viral proteins include the HCV RNA-dependent RNA polymerase (RdRp) and the NS5A protein, as well as the host protein cyclophilin. HCV RdRp enzyme inhibitors include nucleoside (see Chapter 15) and non-nucleoside analogue inhibitors (see Chapter 16).

Nucleoside/nucleotide RdRp inhibitors

RG7128 (Genentech/Roche) is a 2′-methyl nucleoside analogue capable of reducing HCV RNA level by 2.7 log mean (at a dose of 1500 mg, three times per day). When the same dose of RG7128 was administered in combination with pegylated IFN-α and ribavirin, a 5.1 log mean reduction in HCV RNA level was observed; 85% of the patients had undetectable HCV RNA at week 4. Pegylated IFN-α and ribavirin alone (without RG7128) induced an average HCV RNA level reduction of 2.5 log. RG7128 appeared to be safe and well tolerated over the 4 weeks (Lalezari *et al.*, 2008). A single S282T substitution within the active site of the RdRp has been reported to confer a 3- to 6-fold loss of sensitivity *in vitro* to PSI-6130, the active derivative of RG7128. However, HCV variants bearing this substitution replicate poorly *in vivo*, explaining why this variant did not emerge on RG7128 monotherapy at various doses for a few weeks. Longer administration of RG7128 in combination with pegylated IFN-α and ribavirin is currently under study.

Recently, increasing doses of both RG7128 and danoprevir (ITMN-191/RG7227) were administrated to treatment-naïve patients infected by HCV genotype 1 (INFORM-1 trial). The 2-week combination therapy was well tolerated and induced at least an additive antiviral effect that was sustained over the (short) treatment period, without any selection of resistant variants (Gane *et al.*, 2009). This study provides proof of concept for IFN-α-ribavirin sparing regimens. However, a longer treatment regimen of the RG7128–danoprevir combination therapy is needed to assess SVR rates, long-term safety and viral resistance.

Non-nucleoside RdRp inhibitors

Non-nucleoside inhibitors target allosteric sites of the RdRp. Five such sites, named A to E, have been identified so far within the RdRp structure. Various antiviral efficacies and resistance profiles have been described *in vitro* in the replicon system for the different families of molecules targeting these sites. Several molecules have reached clinical development (Table 7.1), including GS-9190 (Gilead), filibuvir (Pfizer), ANA598 (Anadys), BI207127 (Boehringer-Ingelheim), VX-222 (Vertex), and ABT-333 (Abbott). Some of them have been used in combination with pegylated IFN-α and ribavirin. Treatment failures have been associated with the selection of HCV variants resistant to the non-nucleoside inhibitor. For instance, filibuvir targets the thumb 2 domain of the RdRp. In combination with pegylated IFN-α and ribavirin, virological breakthroughs were characterized by the selection of variants bearing a substitution at position M423, which was shown to interfere with filibuvir binding to its target site. Cross-resistance has been described between drugs targeting different sites. The amino acid substitutions probably induce conformational changes that impact the binding affinity of the other class of drugs.

NS5A inhibitors

NS5A is a non-structural HCV protein, the role of which in HCV life cycle is unclear (see Chapters 1 and 4). It appears to be an important component of the replication complex and may regulate the enzymatic function of RdRp. However, the three-dimensional structure of the entire protein has not been determined, and no functional assay is available. In spite of this, NS5A inhibitors have been identified based on their inhibitory effect in replicon-based assays. BMS-790052 (Bristol-Myers Squibb) is a potent inhibitor of HCV replication with picomolar activity *in vitro* and broad genotype coverage (Gao *et al.*, 2010). BMS-790052 has been shown *in vitro* to have a low genetic barrier to resistance, selecting for single mutant viruses with high-level and low-level resistance for subtypes 1a and 1b, respectively (Gao *et al.*, 2010). However, the intrinsic antiviral potency of the drug makes it likely that *in vivo* exposure will be above the *in vitro* EC50 of the resistant viruses. Indeed, a single administration of this drug at the dose of 10 mg is able to reduce HCV RNA levels by more than 3 log and maintain low-level replication over a week (Gao *et al.*, 2010). BMS-790052 is currently tested in combination with pegylated IFN-α and ribavirin and with a protease inhibitor in an IFN-α and ribavirin-free regimen. Since NS5A has emerged as a promising new target for HCV DAAs, other NS5A inhibitors have begun to enter the preclinical developmental stage.

Cyclophilin inhibitors

Alisporivir (DEBIO-025, Novartis) is a cyclophilin inhibitor which potently inhibits HCV replication (Paeshuyse *et al.*, 2006). In a phase Ib trial, alisporivir was administered to treatment-naïve patients co-infected with human immunodeficiency virus at the dose of 1200 mg twice daily for 15 days (Flisiak *et al.*, 2006). All treated patients experienced an HCV RNA level decrease of more than 2 log IU/ml upon administration, with a maximal HCV RNA level reduction of 3.6 log IU/ml, independent of the HCV genotype (Flisiak *et al.*, 2006). Hyperbilirubinaemia was, however, frequent and led to treatment withdrawal in several cases. In a recent phase II trial, alisporivir at daily doses of 600 mg and 1000 mg demonstrated an additive antiviral effect in combination with pegylated IFN-α. However, hyperbilirubinaemia was also observed at higher doses (Flisiak *et al.*, 2009). Current development of alisporivir in combination with pegylated IFN-α and/or ribavirin explores the use of lower doses of 600–800 mg for twice daily. Resistance to alisporivir in the replicon assay appears to be mostly associated with amino acid substitutions in the NS2 and NS5A coding regions. However, the levels of resistance conferred by these substitutions *in vitro* are extremely low and it is unclear whether mutants that bear these substitutions may actually be selected *in vivo* and replicate at high levels. No selection of resistant viruses *in vivo* has been reported so far.

Another cyclophilin inhibitor, SCY-635 (Scynexis), has been shown to reduce HCV RNA levels by more than 2 log over 15 days of administration at 900 mg/day. The safety and resistance profiles of SCY-635 have not been reported.

Other potential DAA targets

HCV entry inhibitors

Inhibition of HCV entry can be achieved by using specific polyclonal or monoclonal antibodies that neutralize infectious particles and prevent their attachment to the receptor molecules. Polyclonal hepatitis C immune globulins have shown no efficacy on HCV replication in HCV-infected liver transplant recipients (Davis *et al.*, 2005). Anti-HCV monoclonal antibodies have shown moderate and transient antiviral properties *in vivo* (Schiano *et al.*, 2006; Galun *et al.*, 2007). Studies evaluating more frequent daily dosing at higher doses and/or combinations with other DAAs of different mode of actions will need to be undertaken.

The development of specific small molecule inhibitors of HCV entry has been hampered by our rudimentary understanding of HCV entry mechanisms. Fusion could also become an interesting target for novel therapies once its mechanisms are better understood (see Chapter 17).

Inhibitors of HCV translation

Antisense DNA or RNA oligonucleotides, the sequences of which are complementary to the target mRNA sequence, can prevent translation of viral proteins. Several oligonucleotides targeting the 5′ UTR, the most conserved region of HCV genome, have been reported to inhibit HCV gene expression in cell culture systems. Their clinical development was, however, halted due to lack of antiviral efficacy and concerns about their tolerance (Soler *et al.*, 2004; McHutchison *et al.*, 2006) A chemically modified ribozyme has been shown to cleave HCV RNA within the IRES, but early phase clinical trials were halted because of animal toxicity. RNA interference (RNAi) technology targeting the 5′ UTR has been shown to inhibit HCV IRES-mediated translation in various cell culture models. However, because of their size and chemical composition, RNAi are currently not orally bioavailable and require parenteral administration.

The three-dimensional functional IRES complexed with cellular and viral proteins and the ribosomal subunits offer a promising target for small molecule inhibitors. Future improvement of our understanding of the mechanisms of HCV polyprotein translation will boost the development of small molecule inhibitors targeting this function.

Inhibitors of HCV assembly and release

Iminosugars have been suggested to interfere with the assembly of different viruses via the ability to competitively inhibit endoplasmic reticulum-resident alpha-glucosidases, alter envelope proteins glycosylation. Celgosivir (Migenix), an iminosugar derivative, has shown a modest antiviral effect on HCV in monotherapy.

Phase III trial design and future standard of care

Telaprevir phase III trials

Three phase III trials are ongoing with telaprevir in patients infected with HCV genotype 1, including the ADVANCE and ILLUMINATE trials in treatment-naïve patients and the REALIZE trial in non-responders to a prior course of pegylated IFN-α and ribavirin. Their designs are shown

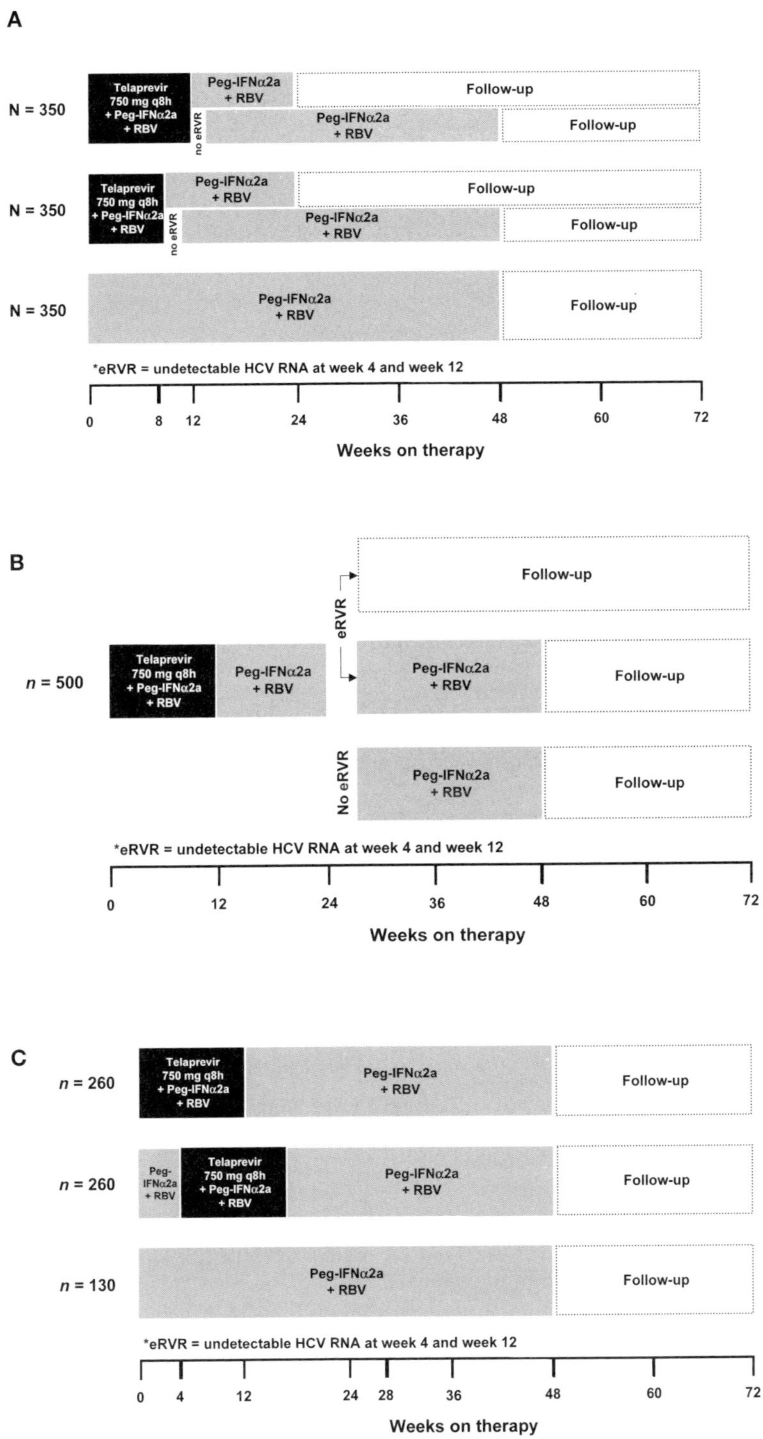

Figure 7.1 Telaprevir phase III trial design, including ADVANCE trial (A), ILLUMINATE trial (B), and REALIZE trial (C). eRVR, extended rapid virological response; Peg-IFNα2a, pegylated interferon-α2a; RBV, ribavirin.

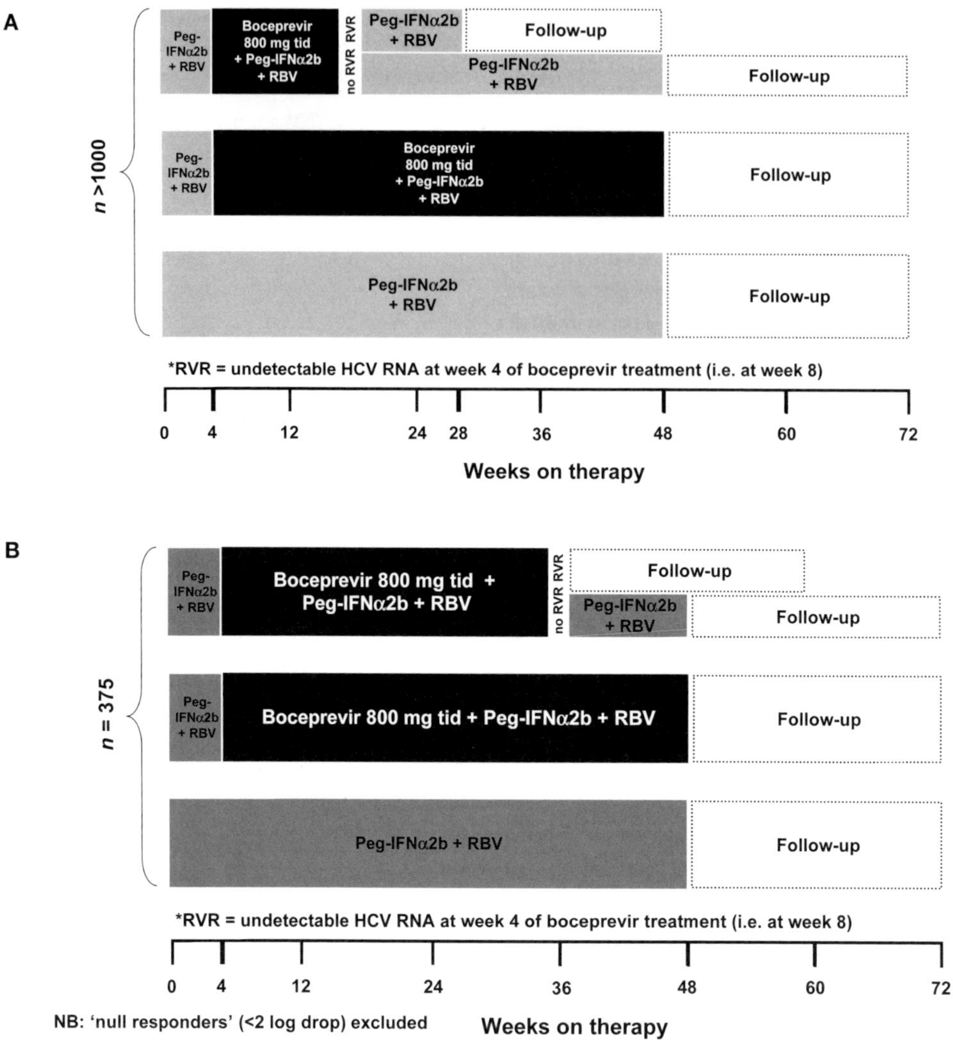

Figure 7.2 Boceprevir phase III trial design, including SPRINT-2 trial (A) and RESPOND-2 trial (B).

in Fig. 7.1A–C. Preliminary results from the ADVANCE trial have been reported in a recent press release (Vertex press release, 25 May 2010). The SVR rate in the control arm was 44% after 48 weeks of pegylated IFN-α and ribavirin. In the patients receiving the triple combination of pegylated IFN-α, ribavirin and telaprevir for 12 weeks, followed by 12 or 36 weeks of pegylated IFN-α and ribavirin according to the achievement of an extended rapid virological response (defined by an undetectable HCV RNA at weeks 4 and 12), the SVR rate was 75%. The SVR rate was 69% in the patients who received only 8 weeks of the triple combination. Further results will be presented at the end of 2010 and in 2011.

Boceprevir phase III trials

Two phase III trials are ongoing with boceprevir in patients infected with HCV genotype 1. The SPRINT-2 and RESPOND-2 trials included treatment-naïve and non-responder patients, respectively. Their designs are shown in Fig. 7.2A and B. Results are anticipated in late 2010/early 2011.

Future standard of care and expectations for future developments

The future standard-of-care for patients with HCV genotype 1 infection is expected to be the combination of pegylated IFN-α, ribavirin and either telaprevir or boceprevir. This combination should be approved in 2011 and become available to patients soon after. Response-guided therapy is likely to be used in treatment-naïve patients, who will need longer than 6 months of therapy if they fail to achieve a rapid virological response (undetectable HCV RNA at week 4). Side-effect management will be crucial as telaprevir has been associated with frequent dermatological symptoms, and boceprevir induces anaemia. Other drugs are currently assessed in clinical trials in combination with pegylated IFN-α and ribavirin. Better tolerated drugs may replace telaprevir and boceprevir within a few years after they have been evaluated in phase III trials. Other approaches for clinical trials include quadruple therapies (pegylated IFN-α, ribavirin, and two DAAs without cross-resistance) in order to increase the genetic barrier to resistance of the DAAs, as well as achieve IFN-α-independent regimens, with or without ribavirin. Exploratory studies are urgently needed in order to test this option.

Conclusion

In conclusion, within the next few months, the current standard-of-care treatment with pegylated IFN-α and ribavirin will be replaced by a triple combination of pegylated IFN-α and ribavirin with a potent specific HCV inhibitor, either telaprevir or boceprevir. Given the clinical data obtained thus far and the number of potential new therapeutic approaches on the horizon, it is likely that oral therapy based on a combination of specific HCV DAAs, akin to the HIV 'cocktail' inhibitors, will replace the current standard of care in the not-too-distant future, at least for a substantial number of patients.

References

Berg, T., McHutchison, J.G., Adda, N., Poordad, F., Shiffman, M.L., Ferenci, P., Heathcote, J., Pawlotsky, J.M., Zeuzem, S., Reesink, H.W., *et al.* (2010). SVR with telaprevir, peginterferon alfa-2a and ribavirin in HCV patients with well-characterized prior null-response, partial response, viral breakthrough or relapse after PR. J. Hepatol. *52*, S2.

Curry ,S., Qiu, P., and Tong, X. (2008). Analysis of HCV resistance mutations during combination therapy with protease inhibitor boceprevir and PEG-IFN alpha-2b using TaqMan mismatch amplification mutation assay. J. Virol. Methods *153*, 156–162.

Davis, G.L., Nelson, D.R., Terrault, N., Pruett, T.L., Schiano, T.D., Fletcher, C.V., Sapan, C.V., Riser, L.N., Li, Y., Whitley, R.J., and Gnann, J.W., Jr. (2005). A randomized, open-label study to evaluate the safety and pharmacokinetics of human hepatitis C immune globulin (Civacir) in liver transplant recipients. Liver Transpl. *11*, 941–949.

Flisiak, R., Orban, A., Kierkus, J., Stanczak, J., Cielnak, I., Stanczak, G.P., Wiercinska-Drapalo, A., Siwak, E., Higersberger, J., Aeschlimann, C., *et al.* (2006). The cyclophilin inhibitor DEBIO-025 has a potent dual anti-HIV and anti-HCV activity in treatment-naïve HIV/HCV co-infected subjects. Hepatology *44*, 609A.

Flisiak, R., Feinman, S.V., Jablkowski, M., Horban, A., Kryczka, W., Pawlowska, M., Heathcote, J.E., Mazzella, G., Vandelli, C., Nicolas-Metral, V. *et al.* (2009). The cyclophilin inhibitor Debio-025 combined with PEG IFN-alpha2a significantly reduces viral load in treatment-naive hepatitis C patients. Hepatology *49*, 1460–1468.

Forns, X., Marcellin, P., Ferenci, P., Goser, T., Nevens, F., Carosi, G., Drenth, J.P., Serfaty, L., De Backer, K., van Heeswijk, R. *et al.* (2010). On-treatment response-guided therapy with telaprevir q8h or q12h combined with poeginterferon alfa-2a or peginterferon alfa-2b and ribavirin in treatment-naive genotype 1 hepatitis C (study C208). J. Hepatol. *52*, S26.

Forestier, N., Reesink, H.W., Weegink, C.J., McNair, L., Kieffer, T.L., Chu, H.M., Purdy, S., Jansen, P.L., and Zeuzem, S. (2007). Antiviral activity of telaprevir (VX-950) and peginterferon alfa-2a in patients with hepatitis C. Hepatology *46*, 640–648.

Galun, E., Terrault, N.A., Eren, R., Zauberman, A., Nussbaum, O., Terkieltaub, D., Zohar, M., Buchnik, R., Ackerman, Z., Safadi, R.,. *et al.* (2007). Clinical evaluation (Phase I) of a human monoclonal antibody against hepatitis C virus: safety and antiviral activity. J. Hepatol. *46*, 37–44.

Gane, E.J., Roberts, S.K., Stedman, C.A., Angus, P.W., Ritchie, B., Elston, R., Ipe, D., Morcos, P.N.,. Najera, I., Chu, T. *et al.* (2009). Combination therapy with a nucleoside polymerase (R7128) and protease (R7227/ITMN-191) inhibitor in HCV: safety, pharmacokinetics, and virologic results from INFORM-1. Hepatology *50*, 394A.

Gane E.J., Rouzier, R., Stedman, C.A., Wiercinska-Drapalo A., Horban, A., Chang, L., Zhang, Y., Sampeur, P., Najera, I., Shulman, N., Smith, P.F., and Tran, J. (2010). Ritonavir boosting of low dose RG7227/ITMN-*191*, HCV RNA protease inhibitor, results in robust reduction in HCV RNA at lower exposures than provided by unboosted regimens. J. Hepatol. *52*, S16.

Gao, M., Nettles, R.E., Belema, M., Snyder, L.B., Nguyen, V.N., Fridell,R.A., Serrano-Wu ,M.H., Langley, D.R.,

Sun, J.H.,. O'Boyle, D.R., 2nd *et al.* (2010). Chemical genetics strategy identifies an HCV NS5A inhibitor with a potent clinical effect. Nature *465*, 96–100.

Hezode, C., Forestier, N., Dusheiko, G., Ferenci, P., Pol, S., Goeser, T., Bronowicki, J.P., Bourliere, M., Gharakhanian, S., Bengtsson, L., *et al.* (2009). Telaprevir and peginterferon with or without ribavirin for chronic HCV infection. N. Engl. J. Med. *360*, 1839–1850.

Kieffer T.L., Sarrazin, C., Miller, J.S., Welker, M.W., Forestier, N., Reesink, H.W., Kwong, A.D., and Zeuzem, S. (2007). Telaprevir and pegylated interferon-alpha-2a inhibit wild-type and resistant genotype 1 hepatitis C virus replication in patients. Hepatology *46*, 631–639.

Kwo, P., Lawitz, E., McCone, J., Schiff, E., Vierling, J., Pound, D., Davis, M., Galati, J., Gordon, S.C. *et al.* (2009a). HCV SPRINT-1 final results: SVR 24 from a phase 2 study of boceprevir plus PegIntron (peginterferon alfa-2b)/ribavirin in treatment-naive subjects with genotype 1 chronic hepatitis C. J. Hepatol. *50*, S4.

Kwo, P., Lawitz, E., McCone, J., Schiff, E., Vierling, J., Pound, D., Davis, M., Galati, J., Gordon, S.C., Ravendhran, N. *et al.* (2009b). High sustained virologic response (SVR) in genotype 1 (G1) null responders to peg-interferon alfa-2b (P) plus ribavirin (R) when treated with boceprevir (BOC) combination therapy. Hepatology *50*, 331A.

Lalezari, J., Gane, E., Rodriguez-Torres, M., De Jesus, E., Nelson, D., Everson, G., Jacobson, I., Reddy, R., Hill, G.Z., Beard, A. *et al.* (2008). Potent antiviral activity of the HCV nucleoside polymerase inhibitor R7128 with peg-IFN and ribavirin: interim results of R7128, 500 mg BID for 28 days. J. Hepatol. *48*, S29.

Liverton, N.J., Carroll, S.S., Dimuzio, J., Fandozzi, C., Graham, D.J., Hazuda, D., Holloway,M.K., ,.W., McCauley J.A., McIntyre, C.J. *et al.* (2010). MK-7*009*, a potent and selective inhibitor of hepatitis C virus NS3/4A protease. Antimicrob. Agents Chemother. *54*, 305–311.

McHutchison, J.G., Patel, K., Pockros, P., Nyberg, L., Pianko, S., Yu, R.Z., Dorr, F.A., and Kwoh, T.J. (2006). A phase I trial of an antisense inhibitor of hepatitis C virus (ISIS 14803), administered to chronic hepatitis C patients. J. Hepatol. *44*, 88–96.

McHutchison, J.G., Everson, G.T., Gordon, S.C., Jacobson, I.M., Sulkowski, M., Kauffman, R., McNair, L., Alam, J., and Muir, A.J. (2009). Telaprevir with peginterferon and ribavirin for chronic HCV genotype 1 infection. N. Engl. J. Med. *360*, 1827–1838.

McHutchison, J.G., Manns, M.P., Muir, A.J., Terrault, N.A., Jacobson, I.M., Afdhal, N.H., Heathcote, E.J., Zeuzem, S., Reesink, H.W., Garg, J., *et al.* (2010). Telaprevir for previously treated chronic HCV infection. N. Engl. J. Med. *362*, 1292–1303.

Malcolm, B.A., Liu, R., Lahser, F., Agrawal, S., Belanger, B., Butkiewicz, N., Chase, R., Gheyas, F., Hart, A., *et al.* (2006). SCH 503*034*, a mechanism-based inhibitor of hepatitis C virus NS3 protease, suppresses polyprotein maturation and enhances the antiviral activity of alpha interferon in replicon cells. Antimicrob. Agents Chemother. *50*, 1013–1020.

Paeshuyse, J., Kaul, A., De Clercq, E., Rosenwirth, B., Dumont, J.M., Scalfaro, P., Bartenschlager, R., and Neyts, J. (2006). The non-immunosuppressive cyclosporin DEBIO-025 is a potent inhibitor of hepatitis C virus replication in vitro. Hepatology *43*, 761–770.

Reesink, H.W., Fanning, G.C., Farha, K.A., Weegink, C., Van Vliet, A., Van't Klooster, G., Lenz, O., Aharchi, F., Marien, K., Van Remoortere, P., *et al.* (2010a). Rapid HCV-RNA decline with once daily TMC435: a phase I study in healthy volunteers and hepatitis C patients. Gastroenterology *138*, 913–921.

Reesink, H.W., Fanning, G.C., Farha, K.A., Weegink, C., Van Vliet, A., Van't Klooster, G., Lenz, O., Aharchi, F., Marien, K., Van Remoortere, P., *et al.* (2010b). Rapid HCV-RNA decline with once daily TMC435: a phase I study in healthy volunteers and hepatitis C patients. Gastroenterology *138*, 913–921.

Sarrazin, C., Kieffer, T.L., Bartels, D., Hanzelka, B., Muh, U., Welker, M., Wincheringer, D., Zhou, Y., Chu, H.M., Lin, C., *et al.* (2007a). Dynamic HCV genotypic and phenotypic changes in patients treated with the protease inhibitor telaprevir (VX-950). Gastroenterology *132*, 1767–1777.

Sarrazin, C., Rouzier, R., Wagner, F., Forestier, N., Larrey, D., Gupta, S., Hussain, M., Shah, A., Cutler, D., Zhang, J. and Zeuzem, S. (2007b). SCH 503*034*, a novel hepatitis C virus protease inhibitor, plus pegylated interferon α–2b for genotype 1 non-responders. Gastroenterology *132*, 1270–1278.

Seiwert, S.D., Andrews, S.W., Jiang, Y., Serebryany, V., Tan, H., Kossen, K., Rajagopalan, P.T., Misialek, S., Stevens, S.K., Stoycheva, A. *et al.* (2008). Preclinical characteristics of the hepatitis C virus NS3/4A protease inhibitor ITMN-191 (R7227). Antimicrob. Agents Chemother. *52*, 4432–4441.

Schiano, T.D., Charlton, M., Younossi, Z., Galun, E., Pruett, T., Tur-Kaspa, R., Eren, R., Dagan, S., Graham, N., Williams, P.V., and Andrews, J. (2006). Monoclonal antibody HCV-AbXTL68 in patients undergoing liver transplantation for HCV: results of a phase 2 randomized study. Liver Transpl. *12*, 1381–1389.

Soler, M., McHutchison, J.G., Kwoh, T.J., Dorr, F.A., and Pawlotsky, J.M. (2004). Virological effects of ISIS 14*803*, an antisense oligonucleotide inhibitor of hepatitis C virus (HCV) internal ribosome entry site (IRES), on HCV IRES in chronic hepatitis C patients and examination of the potential role of primary and secondary HCV resistance in the outcome of treatment. Antivir. Ther. *9*, 953–968.

Venkatraman, S., Bogen, S.L., Arasappan, A., Bennett, F., Chen, K., Jao, E., Liu, Y.T., Lovey, R., Hendrata, S., Huang ,Y., *et al.* (2006). Discovery of (1R, 5S)-N-[3-amino-1-(cyclobutylmethyl)–2, 3-dioxopropyl]- 3-[2(S)-[[[(1, 1-dimethylethyl)amino] carbonyl]amino]–3, 3-dimethyl-1-oxobuty l]- 6, 6-dimethyl-3-azabicyclo[3.1.0]hexan-2(S)-carboxamide (SCH 503034), a selective, potent, orally bioavailable hepatitis C virus NS3 protease inhibitor: a potential therapeutic agent for the treatment of hepatitis C infection. J. Med. Chem. *49*, 6074–6086.

Clinical Virology Support for HCV Drug Development

8

Yupeng He, Liangjun Lu and Akhteruzzaman Molla

Abstract

Clinical virology studies are integral to the antiviral drug development and approval process by providing critical information and guidance to allow optimal trial designs, timely treatment decisions, and effective therapy monitoring. A main focus and high priority of clinical virology research is to monitor and characterize the development of viral drug resistance in drug-treated patients. Numerous viral titre, subtype, genotypic (sequencing) and phenotypic assays, both certified and investigational, are currently available for HCV clinical virology usage. Genotypic and phenotypic analyses of baseline, on- and post-treatment clinical samples are critical to detection and monitoring of drug resistance. Characterization of drug resistance *in vitro* helps to guide design of clinical studies and help predict or interpret *in vivo* drug resistance. In this chapter we discuss standard guidelines for HCV clinical virology, followed by a review of the different assays and methods required for a clinical virology study (Fig. 8.1). In the last part, we discuss *in vitro* HCV antiviral drug resistance study methods and results, and their implications for clinical studies.

Part I: Policies and guidelines related to HCV clinical virology research

In this section we review guidelines related to HCV clinical virology studies from two major regulatory agencies, the FDA (Food and Drug Administration) and the EMA (European Medicines Agency), as well as recommendations from the HCV DRAG (drug resistance advisory group), an *ad hoc* opinion leader group comprised industrial, clinical, and academic experts. Central to all the guideline and recommendation documents is the analysis of drug resistance during clinical trials. Due to the dynamic landscape of HCV drug development, all current guidelines and recommendations from these organizations are subject to revisions and updates as new information becomes available.

FDA guidelines

FDA has two official guidance documents specifically related to clinical virology studies. One is the '*Guidance for Industry Antiviral Product Development – Conducting and Submitting* Virology *Studies to the Agency*' (hereby referred to as the 'FDA Guidance'; FDA, 2006a), and the other is '*Guidance for Submitting HCV Resistance Data*' as an Attachment to the first Guidance (hereby referred to as the 'Attachment to Guidance'; FDA, 2006b). Both documents were published by FDA Center for Drug Evaluation and Research (CDER) in June 2006. It is noteworthy that the above mentioned guidance documents reflect the FDA's current thinking on these topics and should be considered as recommendations. Changes may be expected when information from more HCV drug discovery projects and clinical trials become available in the future or due to the particular nature of certain antiviral drug studies.

The FDA Guidance covers the entire spectrum from the initial IND (investigational novel drug) to the NDA (novel drug application) and post-marketing phases of antiviral products.

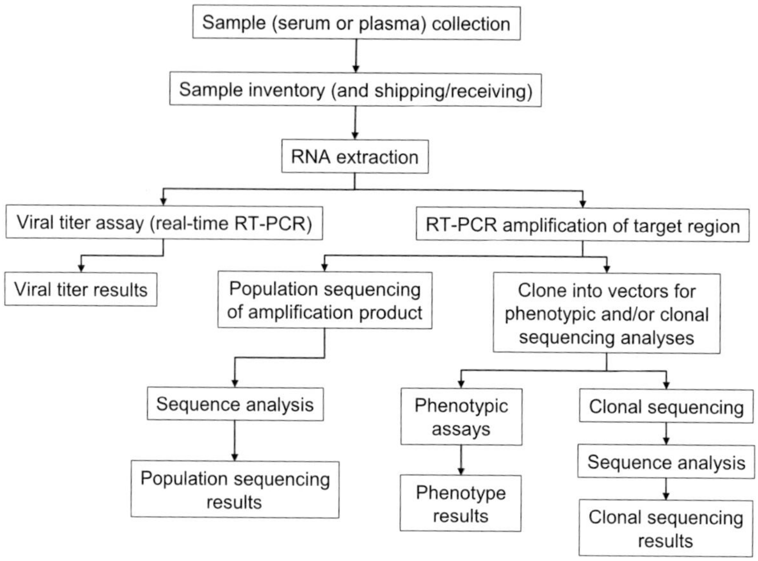

Figure 8.1 Typical workflow of a clinical virology study.

Although the FDA Guidance concerns both pre-/non-clinical and clinical virology study reports, we will discuss only the guidance text related to clinical virology studies in this chapter. Part IV of the FDA Guidance is the '*Proposal for Monitoring Resistance Development*' in clinical studies. Viral drug resistance analysis is described as a main focus and high priority during clinical trials, especially those targeting viruses with high mutations rates including HIV, HBV and HCV. It is recommended that a plan for monitoring viral drug resistance in clinical studies, as a part of the overall clinical development plan, should be developed and submitted to the FDA before initiation of clinical trials in virus-infected patients. The drug resistance monitoring plan should cover the following aspects:

- methods for sample collection, storage, handling and shipping;
- a description of viral load assay(s), assay protocols and performance characteristics;
- genotypic and phenotypic assays for analysing viral resistance, assay protocols and performance characteristics;
- a description of additional resistance analyses;
- time points when samples are collected for viral load, genotypic and phenotypic assays, and other resistance analyses.

The FDA Guidance recommends that clinical trial sponsors conduct 'comprehensive resistance testing' throughout the drug development process in a manner consistent with the future clinical use of the antiviral product. This recommendation can be met by having well-defined genotypic and phenotypic assays for viral resistance characterization. The genotypic and phenotypic viral resistance assays may serve the following purposes:

- detection and examination of drug-resistant viruses;
- identification of viral mutations that contribute to reduced drug susceptibility;
- establishment of relationship between drug resistance and clinical virological failures;
- define treatment options and predict treatment outcomes;
- guide designs of future clinical studies.

The importance of genotypic and phenotypic analyses of patients' baseline samples is specifically recognized in the FDA Guidance. Analyses of baseline samples may be used to predict and/or understand virological response and treatment outcomes, for example, by identifying pre-existing viral mutations and polymorphisms, and phenotypic drug susceptibility of baseline isolates. Comparison of genotypic assay results between baseline samples and viral breakthrough and rebound samples can identify drug-resistance mutations. Collection of genotypic and phenotypic data from all baseline samples, in addition to endpoint samples from all virological failures and discontinuations, is recommended for studies of both naïve and treatment-experienced patients. It is noteworthy that 'virological failure and discontinuation samples should be collected when the patient is still on the study product', emphasizing the need to collect and store baseline and on-treatment samples throughout clinical trials. Analyses of both baseline and post-treatment samples should be completed in a timely manner to allow characterization of an investigational drug's viral resistance profile and its potential for causing cross-resistance with other antiviral drugs.

The viral load, genotypic and phenotypic assays may be either conducted by the clinical trial sponsors themselves, or outsourced to contract research organizations. FDA-approved assays, which are well characterized and validated, are recommended to be used whenever possible, while sponsors using investigational assays are recommended to provide information on assay procedures and performance characteristics (accuracy, precision, limit of detection and quantification, specificity, linearity, range, stability, etc.). The same assay for any particular measurement or analysis should be consistently used, and the same assay should be consistently used for a particular patient. The Attachment to the FDA Guidance (FDA, 2006b) recommends that HCV RNA titres should be analysed with a 'sensitive and specific HCV RNA assay with lower limits of quantification in the range of less than 100 copies/ml'. Although commercial viral subtype and viral titre assays are available for HCV, there are still no certified assays for genotypic and phenotypic analyses. In particular, the FDA Guidance recognizes the challenge of phenotypic analysis of HCV, due to the lack of a robust infection system for most HCV strains. In addition, if the viral resistance pathway can be clearly defined by genotypic analysis alone for an antiviral product, phenotypic analysis may become optional. The various HCV clinical virology assays and methods are discussed in Part II of this chapter. The FDA Guidance recognizes that *in vitro* resistance studies 'can provide useful information for the design of *in vivo* studies and can help predict the development of resistant viruses *in vivo*'. Part III of this chapter discusses *in vitro* drug resistant studies on HCV drug candidates and the comparison of results from *in vitro* studies and clinical studies.

The Attachment to the Guidance (FDA, 2006b) specifies the particular clinical virology data to be submitted to the FDA:

- viral titres (HCV RNA $\log_{10}$ copies/ml) at baseline, at predefined time points during treatment, and at time of virological failure or discontinuation due to adverse events; $\log_{10}$ change in viral titres from baseline to endpoint and optional, additional time points; viral titre data of follow-up samples for determination of sustained virological response;
- HCV subtype information;
- genotype information (presented as amino acid sequence) of relevant viral genomic regions that highlights changes from baseline and polymorphisms;
- phenotypic data that include EC_{50} values of reference strains (usually publicly available standard laboratory strains), EC_{50} values of baseline samples, EC_{50} values of endpoints or virological failures, and EC_{50} values of follow-up samples, and fold change in EC_{50} values of drug-treated samples relative to those of baseline samples or reference strains;
- *if available*, EC_{50} values of reference strains, baseline, endpoint, virological failure samples, *for approved and other investigational anti-HCV products*, and fold changes in EC_{50} values of baseline samples relative to EC_{50} values of reference strains, of endpoint and virological failure samples relative to reference strains and baseline samples.

EMA guidelines

The EMA (European Medicines Agency) Committee for Medicinal Products for Human Use (CHMP) adopted on April 23, 2009 the *Guideline on the Clinical Evaluation of Direct Acting Antiviral Agents Intended for Treatment of Chronic Hepatitis C* (hereby referred to as the EMA Guideline; EMA, 2009), which became effective as of 1 November 2009. Again, revisions to the Guideline should be expected due to the dynamic nature of anti-HCV drug development. The EMA Guideline also emphasizes the concern with drug resistance development, and its impact on therapeutic failure, relapse, and future therapeutic interventions. In particular, the EMA Guideline considers the development of drug resistance potentially harmful, and suggests the use of clinical study designs and strategies that minimize the risk of resistance development.

The EMA Guideline specifically addresses the HCV subtyping and viral load assays to be used for clinical virology studies. For HCV subtype analysis, the EMA Guideline recommends the systemic use of direct sequencing assays of either commercial or validated in-house sources and a combination of sequence determination and phylogenetic analyses on 'a sufficiently large portion of the NS5B gene', while methods based on analysis of the 5′ UTR alone are not recommended due to their high rates of mis-subtyping. For HCV viral titres, the EMA Guideline *requires* the use of standardized quantitative assays-based on real-time PCR methods, and that the assays' lower limit of detection should be in the range of 10–15 IU/ml and their performance should be optimized for the subtypes in the study population to avoid underestimation. In addition, the same viral titre assay should be consistently used for at least a single study, and preferentially throughout an entire drug development programme.

The EMA Guideline also defines a series of endpoints for virological analyses. Sustained virological response (SVR, or undetectable HCV RNA 6 months after completion of therapy) is the primary endpoint for confirmatory studies. Other endpoints include rapid viral response (RVR, or undetectable HCV RNA after 4 weeks of treatment), early viral response (EVR, or undetectable HCV RNA or at least 2 $\log_{10}$ decline after 12 weeks of treatment), and end-of-treatment response (ETR, or undetectable HCV RNA at the planned end of treatment).

Regarding genotypic and phenotypic analyses, the EMA Guideline recommends sufficiently frequent sampling of patients to allow adequate monitoring of the evolution of drug resistance, in addition to viral kinetics/titre analyses, which is particularly important for patients who show virological failure or relapse. Moreover, patients who do not achieve SVR should be followed up for at least one more year with frequent sampling for viral titre and genotypic analyses. During the follow-up, either reversion to the baseline/wild-type viral population or persistence of drug-resistant viral variants should be recorded.

HCV DRAG recommendations

The HCV DRAG (Drug Resistance Advisory Group) is an *ad hoc* group that originated from a proposal made at the 1st International Workshop on Hepatitis C Resistance and New Compounds (October 2006, Boston, USA), and is supported by the Forum for Collaborative HIV Research (www.hivforum.org). It consists of representatives from academia and industry who are primarily scientists and clinicians interested in genotypic and phenotypic analysis of clinical samples for HCV drug development. The HCV DRAG's intent is to bring together relevant parties, to produce consensus recommendations of methods for HCV drug resistance analyses for both drug development and clinical practice, and to provide scientific guidance to facilitate communication between industry, clinicians, and regulatory agencies in this area. The HCV DRAG is organized into working groups that focus on specific topics such as genotype/sequence analysis, phenotype analysis, clinical trial design, and database development.

Neil Parkin, a member of the HCV DRAG organizing committee, presented a proposal on HCV resistance assessment on behalf of the DRAG group, at the 2008 EASL (European Association for the Study of Liver) Conference on Hepatitis B and C Virus Resistance to Antiviral Therapies (Parkin, 2008). The report contains recommendations from both the genotype and phenotype working groups of DRAG. Again, the DRAG recommendations are dynamic and will be updated periodically with new information.

The HCV DRAG genotype working group's recommendations include:

- standardized nomenclature for naming and reporting viral amino acid substitutions (Kuiken *et al.*, 2006);
- proposed reference strains for subtype 1a, 1b, and 2a (strain H77, Con1, and JFH-1, respectively);
- reporting of pre-existing mutations/polymorphisms in baseline samples relative to the relevant reference strains;
- the target detection sensitivity for clinical virology analyses (1000 IU/ml);
- validation of assay sensitivity and accuracy of mutation and minor species detection and the impact of viral titre on assay performance;
- viral regions (primarily direct drug targets) to be sequenced for genotypic analysis;
- the use of population vs. clonal sequencing analyses (population sequencing should be used for initial analysis; clonal sequencing may be optional if population sequencing explains clinical outcome, but is recommended for early clinical studies, linkage analysis, and minor species detection);
- the sensitivity of clonal sequencing analysis as a function of the number of clones sequenced and the desired confidence level (Table 8.1).

The HCV DRAG phenotype working group's recommendations include:

- types of phenotypic assays (recombinant replicon/'shuttle-vector' assays, cell-based non-replicon assays, and *in vitro* biochemical assays);
- validation of all phenotypic assays with known drug resistant variants;
- factors to consider for recombinant replicon-based assays (effects of replicon strains and adaptive mutations, compatibility between clinical isolates and replicon vectors, variation in cellular background, etc.);
- reference strains for phenotypic analysis (H77, Con 1, and JFH-1 for subtype 1a, 1b, and 2a, respectively);
- the use of relative drug susceptibility (fold change in EC_{50} values relative to the reference strains and baseline samples) and absolute drug susceptibility (actual EC_{50} values)
- the definition of resistance and phenotypic resistance cut-offs;
- the definition of replication capacity (RC);
- the use of population *vs.* clonal phenotypic analysis (use primarily population-based analysis; clonal analysis is recommended on a selected subset of samples when population analysis cannot explain clinical results).

Part II: HCV clinical virology assays and methods

In this section we discuss four different assay classes and their underlying technologies that are necessary for HCV clinical virology studies. They are (a) HCV subtype assays, (b) quantitative HCV RNA titre assays, (c) genotypic/sequencing assays, and (d) phenotypic assays for drug

Table 8.1 Sensitivity of clonal sequencing analysis as a function of the number of clones analysed and the desired confidence interval (CI)

		Number of clones analysed					
	Confidence	10	20	40	50	80	100
Detection sensitivity (%)	95% CI	25.9	13.9	7.2	5.8	3.7	3.0
	90% CI	20.6	10.9	5.6	4.5	2.8	2.3
	80% CI	14.9	7.7	3.9	3.2	2.0	1.6

Summarized from Chevaliez and Pawlotsky (2005); Forns and Costa (2006); Scott and Gretch (2007); Le Guillou-Guillemette and Lunel-Fabiani (2009).

resistance analyses. Serologic assays such as EIA (enzyme immunoassays), which are only used for initial diagnosis and blood screening, are beyond the scope of this chapter.

HCV genotype–subtype assays

Due to its high mutation rate and genetic heterogeneity, HCV is classified into six major genotypes (1 to 6) and numerous subtypes. Patients carrying different HCV genotypes show different response profiles to the current SOC (standard-of-care, or a combination of PEG-interferon and ribavirin). In addition, recent studies demonstrated that different HCV genotypes, even closely related subtypes such as 1a and 1b, have different susceptibility levels and resistance profiles to direct antiviral agents (DAA's) including NS3 protease inhibitors, NS5A inhibitors, and non-nucleoside inhibitors of NS5B polymerase (see Part III of this chapter). Therefore, determination of HCV genotype/subtype in patients is important not only for SOC treatment decisions and designs, but also for patient selection, choice and dose of DAA's, and interpretation of development of drug resistance, for clinical trials and future therapy with specific anti-HCV agents.

There are two types of assays for HCV genotype/subtype determination,-based on either molecular biology methods ('genotyping') or serological methods ('serotyping'). The serological methods use competitive ELISA technology and can only differentiate the major genotypes, but not subtypes. The molecular biology methods use one of the following three technologies: (i) direct sequencing analysis (of 5'UTR/E1/core/NS5B); (ii) reverse hybridization (of 5'UTR/core); and (iii) real-time PCR with genotype-specific primers/probes (of 5'UTR/NS5B). HCV genotype/subtype assays solely-based on the 5'UTR region are not recommended due to concerns with high-levels of mis-subtyping. As discussed earlier in this chapter, the EMA Guideline recommends direct sequencing and phylogenetic analysis of the NS5B region for genotype/subtype determination. Direct sequencing followed by phylogenetic analysis of certain regions of HCV genome is considered the gold standard for HCV genotyping/subtyping. See recent reviews for further information on HCV genotype/subtype assays (Chevaliez and Pawlotsky, 2005; Domiati-Saad and Scheuermann, 2006; Forns and Costa, 2006; Scott and Gretch, 2007). The following HCV genotype/subtype assays are commercial available:

- Abbott Real-Time Genotype Test (Abbott/Celera, real-time PCR of 5'UTR+NS5B);
- Inno-LiPA HCV 1.0 (Innogenetics, reverse hybridization of 5' UTR);
- Inno-LiPA HCV 2.0 (Innogenetics, reverse hybridization of 5' UTR +core);
- Murex HCV Serotyping 1–6 HC02 (Abbott, competitive EIA);
- Trugene 5'NC HCV Genotyping Kit (Bayer/Seimens, direct sequencing of 5' UTR);
- Versant HCV Genotyping Assay (Bayer, reverse hybridization of 5' UTR).

Quantitative HCV RNA titre ('viral load') assays

Quantitative HCV RNA assays are used to determine HCV RNA titre before, during, and after antiviral therapy to monitor the efficacy of the treatment. The HCV RNA titre analysis is also essential to monitoring viral kinetics and assessing virological response during clinical trials. The WHO Collaborative Study Group and the WHO Expert Committee on Biological Standardization established in 1997 the First International Standard for HCV RNA (96/790), and defined 'international units per milliliter' (IU/ml) as the unit for HCV RNA assays (Saldanha *et al.*, 1999). The Standard also specified that the minimum sensitivity of quantitative HCV RNA assays is 50 IU/ml. The 2nd International Standard for HCV RNA (96/798) was established in 2005 to replace the 1st International Standard (Saldanha *et al.*, 2005). The International Standards allow comparison and standardization of different HCV RNA quantification tests. Still, it is recommended that the same assay be consistently used for any given patient throughout therapy, to avoid assay-to-assay variability.

Quantitative HCV RNA assays are-based on a variety of technologies including bDNA (branched DNA), RT-PCR, and real-time

Table 8.2 Comparison of commercial HCV RNA tests

Classification	Assay name	Supplier	Method	Dynamic range[a] or lower limit of detection[b] (IU/ml)	Automation level
Quantitative	Abbott Real-time HCV Assay	Abbott	real-time RT-PCR	12 (or 30)[c] to 100,000,000	High
	Amplicor HCV Monitor v2.0	Roche	RT-PCR	600–500,000	Manual
	COBAS Amplicor HCV Monitor v2.0	Roche	RT-PCR	600–500,000	Semiautomated
	COBAS TaqMan HCV Test	Roche	real-time RT-PCR	43–69,000,000	High
	HCV SuperQuant	National Genetics Institute	RT-PCR	30–1,470,000	Semiautomated
	LCx HCV RNA Quantitative Assay	Abbott	RT-PCR	25–2,630,000	Average
	Versant HCV RNA 3.0 Assay	Siemens/Bayer	bDNA	615–7,700,000	Average
Qualitative	Amplicor HCV v2.0	Roche	RT-PCR	50	Manual
	Ampliscreen HCV v2.0	Roche	RT-PCR	50	?
	Cobas Amplicor HCV v 2.0	Roche	RT-PCR	50	Semiautomated
	UltraQual HCV	National Genetics Institute	RT-PCR	40	?
	Versant HCV Qualitative Assay	Siemens/Bayer	TMA	5~10	Manual

[a]For quantitative assays.

[b]For qualitative assays.

[c]Depending on the sample volume (0.2 ml or 0.5 ml).

Source: Jianling Li, Biostatistics, Abbott Laboratories.

RT-PCR (Table 8.2). Real-time RT-PCR assays are fast, reliable, cost-effective, fully automatable, and not prone to carryover contamination. Real-time RT-PCR assays also have wide dynamic ranges and low limits of detection. The performance characteristics of real-time RT-PCR assays allow them to be useful for all clinical applications of HCV RNA testing, both as a quantitative and a sensitive qualitative test. Several reviews provide analysis of the HCV tests available (Chevaliez and Pawlotsky, 2005; Forns and Costa, 2006; Le Guillou-Guillemette and Lunel-Fabiani, 2009; Scott and Gretch, 2007). Several commercial systems for HCV RNA quantification are currently available:

- Abbott Real-time HCV Assay (Abbott);
- Amplicor HCV Monitor v2.0 (Roche);
- COBAS Amplicor HCV Monitor v2.0 (Roche);
- COBAS TaqMan HCV Test (Roche);
- HCV SuperQuant (National Genetics Institute);
- LCx HCV RNA Quantitative Assay (Abbott);
- Versant HCV RNA 3.0 Assay (Siemens/Bayer).

Besides quantitative HCV RNA assays, there are also older, qualitative HCV RNA tests, which are-based on either PCR or TMA (transcription-mediated amplification) technologies. The qualitative tests can detect as low as 5–10 IU/ml of HCV RNA, and are useful for detecting very low levels of viraemia and confirming SVR after antiviral therapy. However, the conventional PCR-based assays are already being replaced by real-time RT-PCR assays due to the advantages of the latter. There are several commercially available qualitative HCV RNA tests:

- Amplicor HCV v2.0 (Roche);
- Ampliscreen HCV v2.0 (Roche);
- Cobas Amplicor HCV v 2.0 (Roche);
- UltraQual HCV (National Genetics Institute);
- Versant HCV Qualitative Assay (Siemens/Bayer).

Genotypic sequencing assays for HCV drug resistance analysis

The current standard method for genotypic analysis of HCV drug resistance is Sanger sequencing, a technology that has been available for decades and supported the Human Genome Project. Direct Sanger sequencing can be performed on an entire HCV quasispecies population isolated from a clinical sample ('population sequencing'), and the result is a reflection of the predominant HCV variants in that sample. Alternatively, sequencing can be conducted on individual clones isolated from a clinical sample ('clonal sequencing'), usually by picking individual colonies after a viral quasispecies is inserted into a vector and transformed into a bacterial host, with each sequence representing a unique quasispecies variant in the clinical isolate. Due to the quasispecies nature of HCV and the existence of variants with different frequencies in a clinical sample, and the non-quantitative nature and inherent signal-to-noise ratio of Sanger sequencing, the resulting population sequence reflects the dominant species in a sample, and has limited sensitivity for detection of the minor, low-frequency, species (< 20% of population). Since drug-resistant variants may constitute only a small portion of the entire viral quasispecies in baseline and early on-treatment isolates, population sequencing has limited use for analysing drug resistance in these samples. On the other hand, clonal sequencing may offer superior detection sensitivity and quantification for minor variants in clinical samples, dependent on the number of clones sequenced and the desired confidence level (Parkin, 2008; Table 8.1). Clonal sequencing is also ideal for analysing the linkage/association of multiple mutations within a single viral species. However, clonal sequencing is limited by tedious handling, low throughput, and high cost. In a study on HCV drug resistance analyses of *in vivo* samples (Chen *et al.*, 2007), both population and clonal sequencing methods were used on samples from two HCV-infected chimpanzees treated with a polymerase inhibitor to identify drug-resistant mutants and analyse their frequency and linkage, which provided a proof-of-concept for later studies on *in vivo* drug resistance analyses (also see discussion in Part III of this chapter).

Besides direct sequencing, HCV variants can be detected and quantified by allele-specific real-time PCR. Two different methods-based on this technology have been reported. In one method (Lu *et al.*, 2007), two distinctive allele-specific probes, one matching the wild-type allele and the other matching the M414T mutant allele of HCV polymerase, were used to detect and quantify the corresponding alleles in a sample. In a later report (Curry *et al.*, 2008), the TaqMAMA (TaqMan mismatch amplification mutation assay) method was used to detect various NS3 protease inhibitor resistant mutants. In this method, a single probe that carries one mismatch against the mutant allele and two mismatches against the wild-type allele was designed to differentiate between the two species. Both methods used predefined and prequantified synthetic DNA molecules to establish standard curves for quantification of different alleles, and their results were confirmed by clonal sequencing analysis. Both studies reported a detection sensitivity limit of 0.1% for minor mutant species in quasispecies samples, including those from HCV-infected patients. Despite the great sensitivity, low cost, and high-throughput of this technology, it is limited by other factors. First, the technology is designed to detect previously known variants/mutations only, and thus not useful for identifying novel mutations. In addition, due to the heterogeneity and quasispecies characteristics of HCV, it can be difficult to design probes suitable to cover all HCV clinical isolates or even different quasispecies from the same sample.

In recent years, commercial sequencing systems-based on 'next-generation' sequencing technologies, which allow much higher throughput and generate much more data than Sanger sequencing, have become available (Ansorge, 2009; MacLean *et al.*, 2009; Metzker, 2010). These commercial next-generation sequencing systems can be largely divided into two categories-based on reaction methodology, sequencing-by-synthesis (Roche/454, Illumina/Solexa, Qiagen/Biotage PyroMark, and Helicos Heliscope), and sequencing-by-ligation (ABI SOLiD). Among all these platforms, the Helicos Heliscope is the only commercial single-molecule-sequencing system, while all others require amplification of sequencing templates. The next-generation sequencing technologies have revolutionized genomics research in recent years and greatly facilitated genomic sequencing of more species and more individuals. In addition, they are also powerful tools for analyses of genetic variations and gene expression profiles ('transcriptomes'), and may contribute to 'personalized therapy' in the future.

Despite their short history, the next-generation sequencing methods have already been applied to research on human viruses, including HIV (Bushman *et al.*, 2008), HBV (Margeridon-Thermet *et al.*, 2009; Solmone *et al.*, 2009), and influenza virus (Deyde and Gubareva, 2009). The applications include viral genome sequencing, viral genotyping, quasispecies analysis, and detection of minor and/or drug-resistant variants. Very recently, the pyrosequencing technology has also been used for HCV sequence analysis and detection of minor, drug-resistant HCV NS3 mutants from clinical samples (Lenz *et al.*, 2009; Rodriguez-Frias *et al.*, 2009). The next-generation sequencing methods have demonstrated their ability to detect and quantify minor viral species with a frequency as low as 0.1%, at much higher throughput and sensitivity relative to conventional sequencing. However, all current commercial next-generation sequencing systems are limited by high instrument and reagent costs, short sequence reads (in the best case, a few hundred bases on average with Roche/454 pyrosequencing), and low raw date accuracy compared with Sanger sequencing. Although date accuracy can be compensated for by higher coverage of samples, the next-generation sequencing methods are not suitable for mutation linkage analysis across large regions of viral genomes, which are still best addressed by clonal sequencing. Moreover, their high throughput and large datasets demand good bioinformatics support such as data storage, raw data analysis, and sequence alignment/assembly. A potential breakthrough in the field of next-generation sequencing technologies is the SMRT (single-molecule, real-time) sequencing technology from Pacific BioSciences. This technology may allow long sequence reads from single DNA molecule-based templates, but has not yet been developed into a commercial platform.

Phenotypic assays for HCV drug resistance analysis

Since there is no efficient infection system for most HCV strains, *in vitro* phenotypic analyses (drug resistance, viral fitness, etc.) of clinical isolates using a bona fide viral infection model cannot be performed, and thus other technologies must be employed. As outlined in the HCV DRAG report (also see the chapter on HCV assays), three types of phenotypic assays for HCV samples are currently available: (a) recombinant replicon-based shuttle vector assays, (b) cell-based viral enzyme assays, and (c) biochemical enzyme assays.

Recombinant replicon-based shuttle vector assays

The HCV replicon is a cell-based subgenomic viral RNA replication model system that has been widely used for HCV research. The replicon-based shuttle vectors are replicon constructs that allow restriction site-dependent insertion of certain fragments of HCV genome derived from other sources, such as clinical samples, into a lab strain replicon backbone ('shuttle'), so that the inserted fragments can be tested in replicon-based phenotypic assays in cell culture. The clinical isolates are usually amplified and cloned into shuttle vectors as mixtures (pools) of different quasispecies from the original samples. In addition, the replicon-based shuttle vectors may be used for clonal sequencing or phenotyping of the inserted fragments if individual colonies are picked from the cloning products. So far the available HCV replicon systems and replicon-based shuttle vectors are limited to a few HCV subtypes, namely 1a, 1b, and 2a.

The first-wave of shuttle vectors were developed for the HCV NS5B polymerase (Ludmerer *et al.*, 2005; Middleton *et al.*, 2007), and were used to test NS5B genes derived from HCV-infected patient samples against HCV antiviral drugs, and to analyse viral samples from HCV-infected chimpanzees treated with a polymerase inhibitor (Chen *et al.*, 2007). These early studies showed that shuttle vectors are useful tools for phenotypic and genotypic analyses of HCV clinical samples, especially phenotypic drug-resistance analysis. A follow-up study (Tripathi *et al.*, 2007) demonstrated that shuttle vectors can be applied to longer HCV genome fragments, specifically the NS5A-NS5B region derived from patient samples, although in general 'shuttling' of the longer NS5A-NS5B fragment resulted in reduced replication levels relative to either NS5A or NS5B alone. Another study (Middleton *et al.*, 2006) tested the feasibility of shuttling even longer HCV genome fragments such NS3-NS5B. Recently, a shuttle vector designed for the HCV NS3 gene, either the NS3 protease domain alone or the full-length NS3 protease-helicase, became available (Chua and Najera, 2008; Pilot-Matias *et al.*, 2009; Qi *et al.*, 2009). These and later studies found that insertion of viral genes from different clinical isolates resulted in very different replication levels in the same shuttle vector system, indicating that not all clinical isolates are compatible with replicon-based shuttle vectors possibly due to the lack of co-evolution of viral genes from clinical isolates and lab strains. Moreover, Insertion of viral genes from other HCV genotypes into genotype 1-based shuttle vectors may also cause loss of replication. Shuttle vectors can be used for testing of target genes from clinical isolates against antiviral drugs *in vitro* and evaluation of phenotypic drug resistance by determining fold changes in EC_{50} values compared with reference or baseline samples. But their detection sensitivity for phenotypic drug resistance is limited by the abundance and replication capacity of resistant variants in quasispecies samples that contain both drug-sensitive and –resistant species. In addition, clinical samples that contain mixtures of mutant species with different levels of drug resistance may yield non-standard drug titration curves and pose a further challenge to shuttle vector-based phenotypic assays such as EC_{50} determination (Chen *et al.*, 2007; Middleton *et al.*, 2007). Another application of the shuttle vector system is selection of pre-existing drug-resistant variants from clinical samples by applying drug pressure during shuttle vector-based *in vitro* replication assays. This method has been used to demonstrate the existence of antiviral drug-resistant variants in drug-naïve HCV-infected patients (Middleton *et al.*, 2007).

Cell-based viral enzyme assays

In addition to shuttle vector-based assays, non-replicon, cell-based assays have also been developed for NS3 protease (Pellerin and Lamarre, 2007; Hong *et al.*, 2008, 2009; Ludmerer *et al.*, 2008). These cell-based NS3 protease assays are usually-based on vectors that contain a reporter gene such as luciferase or SEAP (secreted alkaline phosphatase), whose activation is dependent on NS3 protease activity. This type of assay has been used to analyse NS3 protease activity from ITMN-191 treated clinical samples and can be performed on both quasispecies and clonal samples (Hong *et al.*, 2008). Although the cell-based NS3 protease assays lack HCV replication, they are not limited by replication capacity and genotype compatibility issues associated with replicon-based shuttle vector assays. Moreover, by utilizing transient transfection, a secreted reporter, and a homogeneous assay, the cell-based NS3 protease assays can be fully automated to enable high throughput (Hong *et al.*, 2008). However, caution should be taken in interpretation of results from this assay since the effect of protease mutants on enzyme kinetics, drug resistance, and replication capacity in the context of a viral replication system is not evaluated.

Biochemical enzyme assays

Finally, the NS3 protease derived from clinical isolates can also be tested in enzymatic biochemical assays (Kieffer *et al.*, 2005). HCV NS3 genes derived from clinical samples are expressed in bacterial cells, and the resulting protease enzymes are purified, and used in enzymatic assays to determine IC_{50} values of protease inhibitors. This assay method appears to be a rather tedious approach due to numerous enzyme preparations, while it has the advantage of allowing enzymatic and protein structure studies.

Part III: *In vitro* drug resistance studies and their clinical implications

Pre-existing drug resistance and natural variants of HCV

HCV is a quasispecies RNA virus with high genetic variability. High rates of viral replication and low fidelity of the HCV NS5B RNA-dependent RNA polymerase are the basis of this phenomenon. HCV viral particles are estimated to have a half-life of 2–5 hours and the production and clearance of 10^{10} to 10^{12} virions per patient per day is expected (Neumann *et al.*, 1998). HCV NS5B polymerase lacks any proofreading function. Therefore, a large number of variants are produced and accumulated during the time course of HCV infections (Moradpour *et al.*, 2007). Many of the variants produced lack replication ability due to the loss of function of encoded proteins or intolerance by the host's immune system. Some variants with reduced susceptibility to small molecule compounds used in DAA-based therapy pre-exist as natural species in treatment-naïve patients (Tong *et al.*, 2006; Cubero *et al.*, 2008).

The proportion of drug-resistant quasispecies in a patient in the absence of drugs depends on the replicative fitness of each variant relative to drug-sensitive variants and other factors such as immune selection. Kuntzen T. *et al.* (2008) analysed HCV genome sequences from 507 isolates of HCV genotype 1-infected treatment-naïve patients for presence of previously reported resistance mutations in the NS3 gene. The prevalence of these mutations was found to be between 0.3% and 2.8% in the population (Table 8.3), and 8.6% of the subtype 1a and 1.4% of the subtype 1b patients carried at least one resistance mutation. In another study, Le Pogam *et al.* (2008) found pre-existing mutations conferring resistance to non-nucleoside polymerase inhibitors. However, none of the clinical isolates carried any known mutation conferring resistance to nucleoside inhibitors. Furthermore, they analysed 1110 clones isolated from 13 genotype 1 patients. Some previously described signature mutations conferring resistance to non-nucleoside polymerase inhibitors were found in low frequencies that ranged from 1% to 5% (Table 8.4). The existence of naturally occurring drug-resistant variants may limit the efficacy of drugs or allow rapid selection of resistant mutants in the presence of drug pressure, which may result in viral breakthroughs and treatment failures.

Table 8.3 Prevalence of naturally occurring NS3 drug-resistance mutations in 507 treatment-naïve genotype 1-infected patients

NS3 mutations	Affected inhibitors	Prevalence (%)
V36M/L	Teleprevir, Boceprevir	2.3
T54S	Teleprevir, Boceprevir	1.9
R155K	Teleprevir, Boceprevir, MK-7009, ITMN-191	0.8
D168E	MK-7009, ITMN-191	0.3

Summarized from Kuntzen *et al.* (2008).

Methods for preclinical characterization of drug resistance

The preclinical characterization of resistance mutations has largely depended on the use of subgenomic HCV replicons, due to the lack of efficient infection systems for most HCV strains. To identify mutations conferring resistance to an investigational drug, replicon cells are co-cultured with the compound for a period of time until resistant mutants emerge. Replicons that survive under the drug pressure are then isolated and their nucleotide sequences are compared with those of untreated replicons to reveal mutations conferring resistance. Two types of assays have been used to characterize the drug susceptibility and replication level of a resistant mutant *in vitro,* cell-based assays (Lu *et al.*, 2004; Mo *et al.*, 2005) and enzyme-based assays (Trozzi *et al.*, 2003; Lin *et al.*, 2004; Le Pogam *et al.*, 2006a; Howe *et al.*, 2008). In both assays, a specific mutation is introduced into a wild-type replicon vector by site-directed mutagenesis. In cell-based assays, the mutant replicon RNA templates are transfected into Huh-7 cells to generate a cell line carrying mutant replicons which can be used to examine activities of inhibitors against the mutant. The replication level of a mutant replicon, usually termed replication capacity (RC), is an indirect measure of viral fitness *in vitro*. RC reflects a mutant's ability to increase or decrease replicon copy numbers, relative to that of the wild-type control, in the absence of drugs (Lu *et al.*, 2004). RC from replicon-based assays must be interpreted with caution due to the lack of infection and limited choice of HCV strains in the assay system.

Resistant colony selection can also be used to estimate prevalence of pre-existing resistance mutations in HCV replicons (Mo *et al.*, 2005). In this assay, a fixed number of cells are plated on cell culture dishes in the presence of both a HCV inhibitor drug, which inhibits drug-sensitive replicons while selecting resistant ones, and G418, which kills cells lacking HCV replicons. After 2–3 weeks of continuous culture, survival cells carrying drug-resistant mutant replicons form colonies

Table 8.4 Prevalence of naturally occurring NS5B drug-resistance mutations in 1110 HCV isolates cloned from 13 genotype 1-infected patients

NS5B mutations	Affected inhibitors	Prevalence (%)
C316Y	HCV-796, A-837093	1.2
M414I/L	HCV-796	1.2–5
M423I/V/T	VCH916, Filibuvir	1.0–2.6
426T/V	Filibuvir	1
Y448C	A-837093	1.2
G554D	A-837093	1.2

Summarized from Le Pogam *et al.* (2008).

on the cell culture dishes. Prevalence of pre-existing resistance mutations are calculated as the percentage ratio of the resistant colony number over the total beginning cell number.

Comparison of drug-resistance mutations selected *in vivo* and *in vitro*

There are at least nine protease inhibitors and a dozen polymerase inhibitors currently in clinical trials (see the chapter on HCV drug pipeline and clinical trials). Most of these drug candidates have been thoroughly studied for their *in vitro* resistance profiles in the replicon system. The resistance mutations found *in vivo* have been reported for some of the compounds as they move into phase II and III clinical trials (reviewed by Kieffer *et al.*, 2010; Sarrazin and Zeuzem, 2010). Most of the resistance mutations found *in vivo* have also been found by *in vitro* methods (see Table 8.5 for comparison of resistance mutations from *in vitro* and *in vivo* studies for three classes of HCV inhibitors). The similar patterns of *in vitro* and *in vivo* resistance of many of these inhibitors indicate that the replicon system is a reliable tool for modelling resistance mutations that may be selected *in vivo*. However, some resistance mutations selected by an inhibitor *in vitro* were not found *in vivo*, and vice versa. Despite the fact that limited numbers of samples were analysed in some studies, different pathways of resistance development between HCV-infected patients and replicon cells should be expected due to the higher complexity of HCV replication and quasispecies *in vivo* than in the replicons. In addition, recent studies (Gaudieri *et al.*, 2009; Pawlotsky, 2009) suggest that the human immune response may target certain drug-resistant mutants while favouring others *in vivo*, showing the complex nature of mutation selection *in vivo*.

Subtype–genotype difference of drug-resistance mutations

Most DAA compounds have been designed to treat HCV genotype 1 infection due to its prevalence and relatively poor response to current standard of care. Indeed, different susceptibilities and resistance mutations are observed between different genotype and subtypes both *in vitro* and *in vivo*. BILN-2061, a protease inhibitor, was significantly less active against genotypes 2 and 3 than genotype 1 (Tong *et al.*, 2006). Substitutions at NS3 amino acid residue number 78–80, together with residues 122 and 132, accounted for most of the reduced sensitivity of genotype 2 against BILN-2061. It is well known that most HCV protease inhibitors under development select resistance mutations at amino acid 168 where the wild type is aspartic acid in genotype 1. However, glutamine is the wild type amino acid in this position in HCV genotype 3, which could reduce the sensitivity of HCV protease to the inhibitors and leads to different resistance profiles. In recent clinical trials, HCV protease inhibitor telaprevir (Vertex) has been used to treat patients infected with different HCV genotypes (Reesink *et al.*, 2006; Benhamou *et al.*, 2009; Foster *et al.*, 2009). HCV genotype 2, 3 and 4 patients were randomized to receive 15 days of telaprevir in combination with pegylated interferon (peg-IFN) and ribavirin (RBV), or only treated with pegylated interferon (peg-IFN) and ribavirin (RBV). The addition of telaprevir resulted in additional 1.3-, 0.2- and 1.4 $\log_{10}$ IU/ml mean plasma HCV RNA reductions in genotype 2, 3 and 4 patients, respectively, compared with treatments without telaprevir. Genotype 2 treatment-naive patients achieved a mean 4 log reduction in viral load after 6 days of monotherapy with telaprevir. But genotype 3 patients only achieved a 0.8 $\log_{10}$ reduction in viral load.

In a clinical trial with 14 days of ITMN-191 monotherapy, genotype 1b patients experienced breakthroughs less frequently than genotype 1a patients (4/24 or 17% vs. 10/16 or 63%, respectively; Sarrazin *et al.*, 2009). It was also clear that MK-7009 (Merck protease inhibitor) was overall more effective in HCV 1b-infected patients than in subtype 1a-infected patients as demonstrated by a phase Ib monotherapy (Barnard *et al.*, 2009).

In clinical studies of telaprevir alone or in combination with pegylated interferon (with or without ribavirin), 11 of the 12 patients who experienced a viral breakthrough were infected with HCV subtype 1a, and only one was infected with subtype 1b (McHutchison *et al.*, 2009). The mutation most frequently associated with viral

Table 8.5a PI resistance mutations

	V36	Q41	F43	T54	V55	Q80	V148	R155	A156	D168	V170
Telaprevir (VX-950)	**A/M**			**A**				**K/T**	S/V/T		**A**
	A/M			A				K/T	S/V/T		
Boceprevir (SCH503034)	**L/A/M**			**A/S**	**A**		**I**	**K/T**	S/T		**A/L**
	M	R	S/C	A/S					S/T/V/D		A
TMC435 (TMC435350)						**R**		**K**		**E/N/V**	
		R	S			R		K/Q	G/T/V	A/E/H/I/T/ V/Y	
MK-7009								**E/G/K/N/S/T**	V	**A/G/I/T/V**	
		R	S					K	T	Y	
RG7227 (ITMN-191, R-7227)								**K/G/Q/W**	T	**A/E/G/N/ T/V**	
		R	S					K	T/S/V	E/T/V/A	
Narlaprevir (SCH900518)	**M**							**K**	T		
				A/S					S/T/V		
BI201335								**K/W**		**V/E**	
	No data										

Bold: mutations *in vivo*; not bold: mutations *in vitro*.

Table 8.5b Nucleoside polymerase inhibitor-resistance mutations

Inhibitor	*In vivo* mutation	*In vitro* mutation
Valopicitabine (NM283)	None	S282T
R1626 (R1479)	None	S96T
RG7128 (R7128, PSI-6130)	None	S282T
IDX184	None	S282T

Table 8.5c Non-nucleoside polymerase inhibitor-resistance mutations

	C 316	I 363	S 365	S 368	L 392	A 395	N 411	M 414	L 419	M 423	M 426	N 444	C 445	Y 448	I 482	V 494	A 553	G 554	Y 555	S 556	D 559
VCH 916									**S/M**	**A/I/T/V**					**L**	**A**					
									M	T/V					L						
HCV 796	**Y/N**		**T/A**																		
	F/Y/N/S	V	A/T/L		F			I/V/T					F								
Filibuvir (PF-00868554)										**I/T/V/A**											
										I/T/V	T				T						
A-837093 (Chimp study)	**Y**													**H/C**				**D**			**G**
				A										H				D	C		G

Bold: mutations *in vivo*; not bold: mutations *in vitro*.

breakthroughs is R155K among these clinical trails. Substitution of arginine (R) with lysine (K) requires 1 nucleotide change in subtype 1a, whereas 2 nucleotide changes are needed by subtype 1b, which indicates that subtype 1b poses a higher resistance barrier than subtype 1a to compounds that select R155K as a major resistance mutation. Indeed, genotype 1b R155K mutations were selected by protease inhibitors in patients but not reported in *in vitro* replicon studies in which relatively short drug treatment and less genomic diversity might be the causal factors.

The binding site of nucleoside analogues on NS5B protein is highly conserved across genotypes and subtypes. Few resistance mutations were identified *in vitro* among subtype 1a and 1b HCV replicons (Table 8.5b). No resistance mutations were found in clinical studies against nucleoside analogues NM283, IDX184 (Idenix), R1626 and RG7128 (Roche/Pharmasset), and the inhibitors exhibited similar activities against 1a- or 1b-infected patients (Table 8.5b). Non-nucleoside polymerase inhibitors have a greater chance for variable antiviral activities and different resistance profiles in patients infected with different subtypes. McCown M. *et al.* (2009) compared resistance profiles of HCV-796 between subtype 1a and 1b in the replicon system, and NS5B mutation C316Y was found in both subtypes whereas L392F and M414T were only selected in 1a replicons. Administration of VCH-759 (ViroChem), a non-nucleoside polymerase inhibitor, reduced HCV RNA by 2.5 $\log_{10}$ IU/ml in genotype 1 infected patients, whereas no viral decline was observed in a genotype 6–infected patient in a 10-day therapy (Cooper *et al.*, 2009). Another NS5B inhibitor filibuvir (Pfizer, PF-00868554) showed a more than 50-fold loss of its activity against genotypes 2, 3 and 4 HCV isolates compared with genotype 1 *in vitro* (Shi *et al.*, 2009). Filibuvir selected different resistance mutations between patients infected with HCV subtypes 1a and 1b after 8 days of monotherapy (Troke *et al.*, 2009). In addition, a triple therapy with filibuvir, pegylated interferon alpha 2a and ribavirin resulted in a rapid virological response more frequently in patients infected with HCV subtype 1a (14 of 19) vs. 1b (3 of 7; Jacobson *et al.*, 2009).

Accumulation of multiple resistance mutations: effects on fitness and resistance

Unlike HIV, in which accumulation of multiple resistance mutations seems necessary to provide sufficient resistance to drug pressure, single amino acid substitutions in HCV genes can result in high-level resistance to drugs (Koev and Kati, 2008). Although these HCV single mutations confer reduced susceptibilities to DAA's, many of them are also associated with reduced viral replication abilities. It is known that the frequency of double mutants increases during a prolonged drug selection and with more stringent selective pressure with a protease inhibitor *in vitro* (Tong *et al.*, 2008). In a clinical trial with a protease inhibitor telaprevir (Sarrazin *et al.*, 2007), 13 patients experienced viral breakthroughs and 8 patients showed plateau of viral response among the total of 28 HCV-infected patients receiving the therapy. A double mutant V36M+R155K was detected in the viral breakthrough and plateau groups, but not in the continuous viral decline patients. Single mutations V36M and R155K conferred low-levels of resistance against telaprevir, with IC_{50} values 4- and 9-fold over that of the wild type, respectively. The double mutant V36M + R155K exhibited a 71-fold increase in its IC_{50} value compared with the wild type. Addition of V36M also increased *in vivo* fitness of the R155K mutant from 60% to more than 80% relative to the wild type. Therefore, the double mutants possess selective advantage over the single mutants as the result of their combined effect on drug resistance and viral fitness in the presence of drug pressures.

Preclinical studies using HCV replicons rarely select double mutants, probably due to the short treatment and less diversity of viral quasispecies in replicons compared with chronically infected patients or chimpanzees. Frequencies of pre-existing double mutants might be high in some hosts, as demonstrated by a preclinical study of a polymerase inhibitor (A-837093) in two HCV-infected chimpanzees (Chen *et al.*, 2007) as well telaprevir and boceprevir clinical studies. Variants carrying two mutations, C316Y and G554D, were detected in 2 out of 21 colonies sequenced from a HCV genotype 1b-infected chimpanzee after 2 days of treatment with A-837093. At day 28 post therapy, the double mutant reached its highest level, from 10% to 26% of total viral clones sequenced. On the other hand, a double mutant containing G554D and D559G was only detected at day 16 post therapy from 1 of 33 (3%) clones isolated from a genotype 1a-infected chimpanzee. Early appearance and high prevalence of double mutants in the subtype 1b chimpanzee indicate that the variants pre-existed with relatively high frequency and possibly better fitness in the animal.

Resistance mutations selected by combination treatment of STAT-C agents

Similar to what was learned from HIV regimens, and-based on information from the limited number of clinical trials on HCV-infected patients, an optimal anti-HCV therapy is likely to involve multiple drugs targeting different genes, or drugs targeting the same gene but without causing cross-resistance, to achieve maximal efficacy and minimal chance of viral breakthrough caused by resistance development. A combination study *in vitro* (Mo *et al.*, 2005) examined the activities of an NS5B polymerase inhibitor A-782759 and an NS3 protease inhibitor BILN-2061 alone, and in combination in HCV replicons. The combination of two inhibitors exhibited enhanced antiviral activities compared with either of the inhibitors alone. The frequency of mutants dually resistant to both inhibitors was significantly lower than that of variants with reduced susceptibility to either single compound. The dual resistance was due to mutants carrying mutations in both protease and polymerase genes. Mutations selected by an inhibitor alone or in combination with another inhibitor were not exactly matched. However, this may be partially due to the limited number of dually resistant mutants analysed.

In two independent *in vitro* studies, HCV protease inhibitor boceprevir was used in combination treatment of subtype 1b-con1 replicon cells with a non-nucleoside NS5B polymerase inhibitor HCV-796 (Flint *et al.*, 2009), or indole-N-acetamide, or nucleoside analogue NS5B inhibitor 2′-C-methyl-adenosine (Chase *et al.*, 2009). All the polymerase inhibitors selected their signature resistance mutations in combination treatments. Interestingly, boceprevir selected different resistance mutations in different combinations. NS3 mutation V170A was selected by the combination of boceprevir and HCV-796. M175L was identified as the major resistance mutation in replicons treated by boceprevir in combination with either indole-N-acetamide or 2′-C-methyl-adenosine. Both V170A and M175L conferred low-level resistance, with 5-fold and 2-fold reduced susceptibilities against boceprevir, respectively, and neither is the major resistance mutation identified by treatment with boceprevir alone. It is noteworthy that mutants carrying V170A or M175L displayed similar replication capacities as the wild type replicon. NS5B mutation C316Y was selected together with other NS5B mutations, C445F (Flint *et al.*, 2009), or L392F and M414T (McCown *et al.*, 2009), by the polymerase inhibitor HCV-796 in combination with a protease inhibitor (boceprevir or telaprevir, respectively). All of these combinations (C316Y+C445F, C316Y+L392F and C316Y+M414T) conferred significantly higher resistance than any single mutation. These observations indicate that development of resistance mutations in a combination treatment could be different from those in single drug treatments.

Combined treatment with two polymerase inhibitors targeting palm and thumb domains, respectively, also exhibited enhanced antiviral efficacy (Le Pogam *et al.*, 2006b). The frequency of resistant colonies selected by two inhibitors was significantly lower than those of single treatments. About 65% of the variants isolated from replicons selected by the double treatment possessed multiple amino acid substitutions conferring resistance to both inhibitors. Among those, the M414L+M423T double mutants were dominant. The double mutants also showed significantly reduced replication capacities compared with the wild type and single mutants carrying either M414L or M423T.

Conclusions and future directions

- HCV drug development, and consequently HCV clinical virology support, is a dynamic field. Current guidelines and recommendations from regulatory agencies and opinion groups reflect their current thinking on these topics, and updates and revisions of study designs and methods should be expected.

- Quantitative HCV RNA titre assays, especially those-based on real-time RT-PCR, are becoming the method of choice for monitoring virological response during anti-HCV therapy and clinical trials.
- The best method for HCV genotyping should be-based on direct sequencing and phylogenetic analysis of at least the NS5B gene, preferentially in combination with other viral genome regions.
- There are still no certified assays for genotypic and phenotypic analyses of HCV drug resistance. Exploratory assay methods have been developed and are currently used for clinical trial support. Direct Sanger sequencing, either for viral population or individual clones, is the current standard of genotypic analysis, but the next-generation sequencing methodologies may become a popular option in the near future. The most common phenotypic assays are-based on recombinant replicons or vectors expressing individual viral enzymes, which all lack viral infection and make it difficult to correlate *in vitro* replication capacity with *in vivo* fitness.
- Despite its shortcoming, the HCV replicon system has been proven to be a useful *in vitro* tool for characterizing and predicting drug resistance. Drug-resistance mutations may pre-exist in treatment-naïve patients as natural variants and may affect virological response and therapy outcome. Viral mutations may impact both drug susceptibility and viral fitness, and can be regulated by immune response *in vivo*. Although single mutations of HCV may cause high levels of drug resistance, and even retain wild-type fitness levels in certain cases, combined mutations have been observed from both *in vitro* and *in vivo* studies. The resistance profiles of certain classes of anti-HCV drugs may differ between viral genotypes/subtypes. A combination therapy involving multiple drugs targeting different HCV genes has the potential for a highly potent regimen with a high genetic barrier to viral drug resistance.

References

Ansorge, W.J. (2009). Next-generation DNA sequencing techniques. Nat. Biotechnol. *25*, 195–203.

Barnard, R.J.O., Feng, M., Ramos, G., Himmelberger, A.L., Miller, M.L., Graham, D.J., Ludmerer, S.W., Carroll, S.S., Olsen, D.B., Quirk, K., and Hazuda, D.J. (2009). Analysis of antiviral resistance to MK-7009 in protocol *004*, a phase 1b monotherapy study. Paper presented at: 4th International Workshop on Hepatitis C Resistance and New Compounds (Boston, MA).

Benhamou, Y., Moussalli, J., Ratziu, V., Lebray, P., Gysen, V., de Backer, K., Ghys, A., van Heeswijk, R., Vangeneugden, T., Picchio, G., and Beumont-Mauviel, M. (2009). Results of a proof of concept study (C210) of telaprevir monotherapy and in combination with peginterferon alfa-2a and ribavirin in treatment-naïve genotype 4 hcv patients. Paper presented at: EASL 44th Annual Meeting (Copenhagen, Denmark).

Bushman, F.D., Hoffmann, C., Ronen, K., Malani, N., Minkah, N., Rose, H.M., Tebas, P., and Wang, G.P. (2008). Massively parallel pyrosequencing in HIV research. Aids *22*, 1411–1415.

Chase, R., Skelton, A., Xia, E., Curry, S., Liu, S., McMonagle, P., Huang, C., and Tong, X. (2009). A novel HCV NS3 protease mutation selected by combination treatment of the protease inhibitor boceprevir and NS5B polymerase inhibitors. Antiviral Res. *84*, 178–184.

Chen, C.M., He, Y., Lu, L., Lim, H.B., Tripathi, R.L., Middleton, T., Hernandez, L.E., Beno, D.W., Long, M.A., Kati, W.M., *et al.* (2007). Activity of a potent hepatitis C virus polymerase inhibitor in the chimpanzee model. Antimicrob. Agents Chemother. *51*, 4290–4296.

Chevaliez, S., and Pawlotsky, J.M. (2005). Use of virologic assays in the diagnosis and management of hepatitis C virus infection. Clin. Liver Dis. 9, 371–*382*, v.

Chua, K., and Najera, I. (2008). HCV NS3 replicon shuttle vectors. WO/2008/148671.

Cooper, C., Lawitz, E.J., Ghali, P., Rodriguez-Torres, M., Anderson, F.H., Lee, S.S., Bedard, J., Chauret, N., Thibert, R., Boivin, I., *et al.* (2009). Evaluation of VCH-759 monotherapy in hepatitis C infection. J. Hepatol. *51*, 39–46.

Cubero, M., Esteban, J.I., Otero, T., Sauleda, S., Bes, M., Esteban, R., Guardia, J., and Quer, J. (2008). Naturally occurring NS3-protease-inhibitor resistant mutant A156T in the liver of an untreated chronic hepatitis C patient. Virology *370*, 237–245.

Curry, S., Qiu, P., and Tong, X. (2008). Analysis of HCV resistance mutations during combination therapy with protease inhibitor boceprevir and PEG-IFN alpha-2b using TaqMan mismatch amplification mutation assay. J. Virol. Methods *153*, 156–162.

Deyde, M., and Gubareva, L.V. (2009). Influenza genome analysis using pyrosequencing method: current applications for a moving target. Expert Rev. Mol. Diagn. *9*, 493–509.

Domiati-Saad, R., and Scheuermann, R.H. (2006). Nucleic acid testing for viral burden and viral genotyping. Clin. Chim. Acta *363*, 197–205.

E.M.A. (2009). Guideline on the Clinical Evaluation of Direct Acting Antiviral Agents Intended for Treatment of Chronic Hepatitis C. European Medicines Agency. Available from: http://www.ema.europa.eu/pdfs/human/ewp/3003908enfin.pdf

F.D.A. (2006a). Guidance for Industry Antiviral Product Development – Conducting and Submitting Virology Studies to the Agency. Food and Drug Administration. Available from: http://www.fda.gov/OHRMS/DOCKETS/98fr/05d-0183-gdl0002–01.pdf

F.D.A. (2006b). Guidance for Submitting HCV Resistance Data. Food and Drug Administration. Available from: http://www.fda.gov/OHRMS/DOCKETS/98fr/05d-0183-gdl0002–05-HCV.pdf

Flint, M., Mullen, S., Deatly, A.M., Chen, W., Miller, L.Z., Ralston, R., Broom, C., Emini, A., and Howe, A.Y. (2009). Selection and characterization of hepatitis C virus replicons dually resistant to the polymerase and protease inhibitors HCV-796 and boceprevir (SCH 503034). Antimicrob. Agents Chemother. *53*, 401–411.

Forns, X., and Costa, J. (2006). HCV virological assessment. J. Hepatol. *44*, S35–39.

Foster, G.R., Hezode, C., Bronowicki, J.-P., Carosi, G., Weiland, O., Verlinden, L., van Heeswijk, R., Vangeneugden, T., Picchio, G., and Beumont-Mauviel, M. (2009). Activity of telaprevir alone or in combination with peginterferon alfa-2a and ribavirin in treatment-naive genotype 2 and 3 hepatitis-C patients: interim results of study C209. Paper presented at: EASL 44th Annual Meeting (Copenhagen, Denmark).

Gane, E., Roberts, S., Stedman, C., Angus, P., Ritchie, B., Elston, R., Ipe, D., Morcos, P., Najera, I., Chu, T., *et al.* (2009). Combination therapy with A nucleoside polymerase (RG7128) and protease (RG7227/ITMN-191) Inhibitor In CV: Safety, harmacokinetics, And Virologic Results from INFORM-1. Paper presented at: 60th Annual Meeting of the American Association for the Study of Liver Diseases (Boston, MA).

Gaudieri, S., Rauch, A., Pfafferott, K., Barnes, E., Cheng, W., McCaughan, G., Shackel, N., Jeffrey, G.P., Mollison, L., Baker, R., *et al.* (2009). Hepatitis C virus drug resistance and immune-driven adaptations: relevance to new antiviral therapy. Hepatology *49*, 1069–1082.

Hong, J., Qin, X., Lim, B.H., Blatt, M., and Seiwert, S. (2008). A high throughput system for phenotypic analysis of NS3 sequences. Paper presented at: American Association for the Study of Liver Diseases (San Francisco, CA).

Hong, J., Seiwert, S., Lim, S., and Qin, X. (2009). HCV genotyping and phenotyping (Intermune, Inc.). WO/2009/059021.

Howe, A.Y., Cheng, H., Johann, S., Mullen, S., Chunduru, S.K., Young, D.C., Bard, J., Chopra, R., Krishnamurthy, G., Mansour, T., and O'Connell, J. (2008). Molecular mechanism of hepatitis C virus replicon variants with reduced susceptibility to a benzofuran inhibitor, HCV-796. Antimicrob. Agents Chemother. *52*, 3327–3338.

Jacobson, I., Pockros, P.J., Lalezari, J., Lawitz, E., Rodriguez-Torres, M., DeJesus, E., Haas, F., Martorell, C., Pruitt, R., Durham, K., *et al.* (2009). Antiviral Activity of Filibuvir in Combination with Pegylated Interferon Alfa-2a and Ribavirin for 28 Days in Treatment-Naive Patients Chronically Infected with HCV Genotype 1. Paper presented at: EASL 44th Annual Meeting (Copenhagen, Denmark).

Kieffer, T., Sarrazin, C., Bartels, D., Hanzelka, B., Muh, U., Welker, M., Wincheringer, D., Lin, C., Grossman, T., Weegink, C., *et al.* (2005). Genetic heterogeneity in the HCV NS3 protease of untreated genotype 1 patients has little effect on the sensitivity to VX-950. Paper presented at: 56th Annual Meeting of the American Association for the Study of Liver Diseases (San Francisco, CA).

Kieffer, T.L., Kwong, D., and Picchio, G.R. (2010). Viral resistance to specifically targeted antiviral therapies for hepatitis C (STAT-Cs). J. Antimicrob. Chemother. *65*, 202–212.

Koev, G. and Kati, W. (2008). The emerging field of HCV drug resistance. Expert Opin. Investig. Drugs *17*, 303–319.

Koev, G., Mondal, R., Beyer, J., Reisch, T., Masse, S., Kati, W., Hutchinson, D., Flentge, C., Randolph, J.T., Donner, P., *et al.* (2009). Characterization of Resistance Mutations Selected In Vitro by Non-Nucleoside HCV Polymerase Inhibitors ABT-333 and ABT-072. Paper presented at: 44th Annual Meeting of the European Association for the Study of the Liver (Copenhagen, Denmark).

Kuiken, C., Combet, C., Bukh, J., Shin, I.T., Deleage, G., Mizokami, M., Richardson, R., Sablon, E., Yusim, K., Pawlotsky, M., and Simmonds, P. (2006). A comprehensive system for consistent numbering of HCV sequences, proteins and epitopes. Hepatology *44*, 1355–1361.

Kuntzen, T., Timm, J., Berical, A., Lennon, N., Berlin, A.M., Young, S.K., Lee, B., Heckerman, D., Carlson, J., Reyor, L.L., *et al.* (2008). Naturally occurring dominant resistance mutations to hepatitis C virus protease and polymerase inhibitors in treatment-naive patients. Hepatology *48*, 1769–1778.

Le Guillou-Guillemette, H., and Lunel-Fabiani, F. (2009). Detection and quantification of serum or plasma HCV RNA: mini review of commercially available assays. Methods Mol. Biol. *510*, 3–14.

Le Pogam, S., Jiang, W.R., Leveque, V., Rajyaguru, S., Ma, H., Kang, H., Jiang, S., Singer, M., Ali, S., Klumpp, K., *et al.* (2006a). In vitro selected Con1 subgenomic replicons resistant to 2′-C-methyl-cytidine or to R1479 show lack of cross resistance. Virology *351*, 349–359.

Le Pogam, S., Kang, H., Harris, S.F., Leveque, V., Giannetti, A.M., Ali, S., Jiang, W.R., Rajyaguru, S., Tavares, G., Oshiro, C., *et al.* (2006b). Selection and characterization of replicon variants dually resistant to thumb- and palm-binding nonnucleoside polymerase inhibitors of the hepatitis C virus. J. Virol. *80*, 6146–6154.

Le Pogam, S., Seshaadri, A., Kosaka, A., Chiu, S., Kang, H., Hu, S., Rajyaguru, S., Symons, J., Cammack, N., and Najera, I. (2008). Existence of hepatitis C virus NS5B variants naturally resistant to non-nucleoside, but not to nucleoside, polymerase inhibitors among untreated patients. J. Antimicrob. Chemother. *61*, 1205–1216.

Lenz, O., Vijgen, L., Lin, T.I., Verloes, R., Verbinnen, T., Berke, J.M., van Marck, H., Vandenbroucke, I., Meyvisch, P., Simmen, K., and Fanning, G. (2009). Analysis of the NS3 region in HCV genotype 1-infected patients treated with 5-day monotherapy of TMC435 in a Phase 1 study. Paper presented at: 16th International Symposium on Hepatitis C Virus and Related Viruses (Nice, France).

Lin, C., Lin, K., Luong, Y.P., Rao, B.G., Wei, Y.Y., Brennan, D.L., Fulghum, J.R., Hsiao, H.M., Ma, S., Maxwell, J.P., *et al.* (2004). In vitro resistance studies of hepatitis C virus serine protease inhibitors, VX-950 and BILN 2061: structural analysis indicates different resistance mechanisms. J. Biol. Chem. *279*, 17508–17514.

Lin, C., Gates, C.A., Rao, B.G., Brennan, D.L., Fulghum, J.R., Luong, Y.P., Frantz, J.D., Lin, K., Ma, S., Wei, Y.Y., *et al.* (2005). In vitro studies of cross-resistance mutations against two hepatitis C virus serine protease inhibitors, VX-950 and BILN 2061. J. Biol. Chem. *280*, 36784–36791.

Lu, L., Pilot-Matias, T.J., Stewart, K.D., Randolph, J.T., Pithawalla, R., He, W., Huang, P.P., Klein, L.L., Mo, H., and Molla, A. (2004). Mutations conferring resistance to a potent hepatitis C virus serine protease inhibitor in vitro. Antimicrob. Agents Chemother. *48*, 2260–2266.

Lu, L., Mo, H., Pilot-Matias, J., and Molla, A. (2007). Evolution of resistant M414T mutants among hepatitis C virus replicon cells treated with polymerase inhibitor A-782759. Antimicrob. Agents Chemother. *51*, 1889–1896.

Ludmerer, S.W., Graham, D.J., Boots, E., Murray, E.M., Simcoe, A., Markel, E.J., Grobler, J.A., Flores, O.A., Olsen, D.B., Hazuda, J., and LaFemina, R.L. (2005). Replication fitness and NS5B drug sensitivity of diverse hepatitis C virus isolates characterized by using a transient replication assay. Antimicrob. Agents Chemother. *49*, 2059–2069.

Ludmerer, S.W., Graham, D.J., Patel, M., Gilbert, K., Stahlhut, M., and Olsen, D.B. (2008). A transient cell-based phenotype assay for hepatitis C NS3/4A protease: application to potency determinations of a novel macrocyclic inhibitor against diverse protease sequences isolated from plasma infected with HCV. J. Virol. Methods *151*, 301–307.

McCown, M.F., Rajyaguru, S., Kular, S., Cammack, N., and Najera, I. (2009). GT-1a or GT-1b subtype-specific resistance profiles for hepatitis C virus inhibitors telaprevir and HCV-796. Antimicrob. Agents Chemother. *53*, 2129–2132.

McHutchison, J.G., Everson, G.T., Gordon, S.C., Jacobson, I.M., Sulkowski, M., Kauffman, R., McNair, L., Alam, J., and Muir, A.J. (2009). Telaprevir with peginterferon and ribavirin for chronic HCV genotype 1 infection. N. Engl. J. Med. *360*, 1827–1838.

MacLean, D., Jones, D., and Studholme, D.J. (2009). Application of 'next-generation' sequencing technologies to microbial genetics. Nat. Rev. Microbiol. *7*, 287–296.

Margeridon-Thermet, S., Shulman, N.S., Ahmed, A., Shahriar, R., Liu, T., Wang, C., Holmes, S.P., Babrzadeh, F., Gharizadeh, B., Hanczaruk, B., *et al.* (2009). Ultra-deep pyrosequencing of hepatitis B virus quasispecies from nucleoside and nucleotide reverse-transcriptase inhibitor (NRTI)-treated patients and NRTI-naive patients. J. Infect. Dis. *199*, 1275–1285.

Metzker, M.L. (2010). Sequencing technologies – the next generation. Nat. Rev. Genet. *11*, 31–46.

Middleton, T., He, Y., Kati, W., and Molla, A. (2006). Phenotypic and genotypic characterization of the HCV NS3–5B region of isolates from HCV-infected patients. Paper presented at: American Society for Virology Annual Meeting (Madison, WI).

Middleton, T., He, Y., Pilot-Matias, T., Tripathi, R., Lim, B.H., Roth, A., Chen, C.M., Koev, G., Ng, T.I., Krishnan, P., *et al.* (2007). A replicon-based shuttle vector system for assessing the phenotype of HCV NS5B polymerase genes isolated from patient populations. J. Virol. Methods *145*, 137–145.

Mo, H., Lu, L., Pilot-Matias, T., Pithawalla, R., Mondal, R., Masse, S., Dekhtyar, T., Ng, T., Koev, G., Stoll, V., *et al.* (2005). Mutations conferring resistance to a hepatitis C virus (HCV) RNA-dependent RNA polymerase inhibitor alone or in combination with an HCV serine protease inhibitor in vitro. Antimicrob. Agents Chemother. *49*, 4305–4314.

Moradpour, D., Penin, F., and Rice, C.M. (2007). Replication of hepatitis C virus. Nat. Rev. Microbiol. *5*, 453–463.

Neumann, A.U., Lam, N.P., Dahari, H., Gretch, D.R., Wiley, T.E., Layden, J., and Perelson, A.S. (1998). Hepatitis C viral dynamics in vivo and the antiviral efficacy of interferon-alpha therapy. Science *282*, 103–107.

Parkin, N. (2008). HCV resistance assessment: a proposal from the HCV DRAG (Drug Resistance Advisory Group). Paper presented at: EASL-AASLD-APASL-ALEH-IASL Conference Hepatitis B and C Virus Resistance to Antiviral Therapies (Paris, France).

Pawlotsky, J.M. (2009). Hepatitis: HCV variability, the immune system and resistance to antiviral drugs. Nat Rev Gastroenterol Hepatol *6*, 383–385.

Pellerin, C., and Lamarre, D. (2007). Surrogate cell-based system and method for assaying the activity of hepatitis C virus NS3 protease [Boehringer Ingelheim (Canada) Ltd, Laval].

Pilot-Matias, T., Tripathi, R., Krishnan, P., Dekhtyar, T., Pithawalla, R., Ng, T.I., Lanford, R.E., Gai, Y., Or, Y.S., Kempf, D.J., *et al.* (2009). Development of an HCV Genotype 1a NS3 Shuttle Vector: Validation with Viral Isolates from Infected Patients and from a Chimpanzee Treated with a Protease Inhibitor. Paper presented at: 16th International Symposium on Hepatitis C Virus and Related Viruses (Nice, France).

Qi, X., Bae, A., Liu, S., Yang, H., Sun, S.C., Harris, J., Delaney, W., Miller, M., and Mo, H. (2009).

Development of a replicon-based phenotypic assay for assessing the drug susceptibilities of HCV NS3 protease genes from clinical isolates. Antiviral Res. *81*, 166–173.

Reesink, H.W., Zeuzem, S., Weegink, C.J., Forestier, N., van Vliet, A., van de Wetering de Rooij, J., McNair, L., Purdy, S., Kauffman, R., Alam, J., and Jansen, P.L. (2006). Rapid decline of viral RNA in hepatitis C patients treated with VX-950: a phase Ib, placebo-controlled, randomized study. Gastroenterology *131*, 997–1002.

Rodriguez-Frias, F., Cubero, M., Esteban, J.I., Buti, M., Tabernero, D., Ortega, I., Sanchez, A., Mosquera, J.L., Schaper, M., Homs, M., *et al.* (2009). detection of clinically relevant minor HCV and HBV antiviral-resistant mutants in treatment-naïve patients using ultra-deep pyrosequencing. Paper presented at: EASL (Copenhagen, Denmark).

Saldanha, J., Lelie, N., and Heath, A. (1999). Establishment of the first international standard for nucleic acid amplification technology (NAT) assays for HCV RNA. WHO Collaborative Study Group. Vox Sang. *76*, 149–158.

Saldanha, J., Heath, A., Aberham, C., Albrecht, J., Gentili, G., Gessner, M., and Pisani, G. (2005). World Health Organization collaborative study to establish a replacement WHO international standard for hepatitis C virus RNA nucleic acid amplification technology assays. Vox Sang *88*, 202–204.

Sarrazin, C., Kieffer, T.L., Bartels, D., Hanzelka, B., Muh, U., Welker, M., Wincheringer, D., Zhou, Y., Chu, H.M., Lin, C., *et al.* (2007). Dynamic hepatitis C virus genotypic and phenotypic changes in patients treated with the protease inhibitor telaprevir. Gastroenterology *132*, 1767–1777.

Sarrazin, C., Hong, J.R., Lim, S., Qin, X., Moy, C., Susser, S., Bradford, B., Porter, S., Zeuzem, S., and Seiwert, S. (2009). Incidence of virologic escape observed during ITMN-191 (r7227) monotherapy is genotype dependent, associated with a specific NS3 substitution, and suppressed upon combination with peginterferon alfa-2a/ribavirin. Paper presented at: AASLD 60th Annual Meeting of the American Association for the Study of Liver Diseases (Boston, MA).

Sarrazin, C., and Zeuzem, S. (2010). Resistance to direct antiviral agents in patients with hepatitis C virus infection. Gastroenterology *138*, 447–462.

Scott, D., and Gretch, D.R. (2007). Molecular diagnostics of hepatitis C virus infection: a systematic review. JAMA *297*, 724–732.

Shi, S.T., Herlihy, K.J., Graham, J.P., Nonomiya, J., Rahavendran, S.V., Skor, H., Irvine, R., Binford, S., Tatlock, J., Li, H., *et al.* (2009). Preclinical characterization of PF-00868554, a potent nonnucleoside inhibitor of the hepatitis C virus RNA-dependent RNA polymerase. Antimicrob. Agents Chemother. *53*, 2544–2552.

Solmone, M., Vincenti, D., Prosperi, M.C., Bruselles, A., Ippolito, G., and Capobianchi, M.R. (2009). Use of massively parallel ultradeep pyrosequencing to characterize the genetic diversity of hepatitis B virus in drug-resistant and drug-naive patients and to detect minor variants in reverse transcriptase and hepatitis B S antigen. J. Virol. *83*, 1718–1726.

Tong, X., Guo, Z., Wright-Minogue, J., Xia, E., Prongay, A., Madison, V., Qiu, P., Venkatraman, S., Velazquez, F., Njoroge, G., and Malcolm, B.A. (2006). Impact of naturally occurring variants of HCV protease on the binding of different classes of protease inhibitors. Biochemistry *45*, 1353–1361.

Tong, X., Bogen, S., Chase, R., Girijavallabhan, V., Guo, Z., Njoroge, F.G., Prongay, A., Saksena, A., Skelton, A., Xia, E., and Ralston, R. (2008). Characterization of resistance mutations against HCV ketoamide protease inhibitors. Antiviral Res. *77*, 177–185.

Tripathi, R.L., Krishnan, P., He, Y., Middleton, T., Pilot-Matias, T., Chen, C.M., Lau, D.T., Lemon, S.M., Mo, H., Kati, W., and Molla, A. (2007). Replication efficiency of chimeric replicon containing NS5A-5B genes derived from HCV-infected patient sera. Antiviral Res. *73*, 40–49.

Troke, P., Lewis, M., Simpson, P., van der Ryst, E., Hammond, J., Craig, C., Perros, M., and Westby, M. (2009). Genotypic characterisation of HCV NS5B following 8-day monotherapy with the polymerase inhibitor PF-00868554 in hcv-infected subjects. Paper presented at: EASL 44th Annual Meeting (Copenhagen, Denmark).

Trozzi, C., Bartholomew, L., Ceccacci, A., Biasiol, G., Pacini, L., Altamura, S., Narjes, F., Muraglia, E., Paonessa, G., Koch, U., *et al.* (2003). In vitro selection and characterization of hepatitis C virus serine protease variants resistant to an active-site peptide inhibitor. J. Virol. *77*, 3669–3679.

NS3 Protease Covalent Inhibitors

9

Kevin X. Chen and F. George Njoroge

Abstract

The hepatitis C virus (HCV) NS3 protease is essential for viral replication. It is one of the most attractive targets for developing novel antiviral therapies. Two distinct classes of NS3 protease inhibitors have emerged with different mechanisms of action: non-covalent classical inhibitors and covalent inhibitors. Several types of warheads for covalent bonding have been studied. Among them, the ketoamides have been the most successful class of covalent inhibitors. In fact, the two most clinically advanced HCV NS3 protease drug candidates are both ketoamides: boceprevir and telaprevir, in phase III clinical trials. This chapter reviews the evolution and development of covalent NS3 protease inhibitors from early structure–activity relationship investigations to late stage clinical candidates.

Introduction

Hepatitis C virus (HCV) infection represents a major global health-care problem. It is estimated that a minimum of 3% of the world's population, or 170 million people, and more than 4 million Americans are chronically infected with HCV (Cuthbert, 1994; Alter *et al.*, 1999; Cohen, 1999; Shepard *et al.*, 2005). In roughly 80% of cases, the virus leads to a chronic form of hepatitis: a condition that is incurable in many patients. Without therapeutic intervention, it can lead to morbidity or mortality in 10–20 years through either cirrhosis and hepatic failure or hepatocellular carcinoma (Alter *et al.*, 2000; Hoofnagle, 2002). It is anticipated that a significant percentage of those currently infected will develop cirrhosis and other associated hepatic sequelae. HCV infection is the most common cause of liver transplantation (Brown *et al.*, 2003). Since the discovery of the virus in 1989 (Choo *et al.*, 1989), significant progress has been made in the development of HCV therapy. With the initial introduction of interferon- monotherapy, and subsequent recommendation of pegylated-IFN-α and ribavirin combination therapy (Neumann *et al.*,1998; Fried *et al.*, 2002; Feld *et al.*, 2005), the proportion of patients achieving sustained virological response (SVR) has increased significantly. IFN-α is a protein that stimulates the immune system, while ribavirin is a nucleoside analogue that works in concert with IFN-α to control the infection. The efficacy of this combination therapy against the predominant genotype 1 affecting North America, Europe and Japan is moderate, with only about 40% of the patients demonstrating SVR. Some patients also experience significant side-effects related to the treatment. With the prediction that the morbidity and mortality rates associated with HCV will rise in the coming years, more efficacious and tolerable anti-HCV agents are urgently needed.

The HCV genome is a 9.6-kb, enveloped, positive-strand RNA, which encodes a long single polyprotein of approximately 3000 amino acids (Lindenbach *et al.*, 2005; Kolykhalov *et al.*, 2000). The polyprotein is processed proteolytically by cellular and viral proteases into at least 10 individual proteins, including four structural proteins (core–E1–E2–P7) and six non-structural (NS) proteins (NS2–NS3–NS4A–NS4B–NS5A–NS5B). The virally encoded protease responsible for processing the non-structural portion of the polyprotein is located in the N-terminal third of the NS3 protein (Fig. 9.1). Besides autoproteolysis of the NS3–NS4A junction, the protease also cleaves the polyprotein at the NS4A–NS4B, NS4B–NS5A, and NS5A–NS5B junctions to release the downstream NS functional proteins. The replicative complex, or replisome, is subsequently generated with the mature proteins through self-assembly on the endoplasmic reticulum. Using the viral genome as a template, the replisome generates negative-strand viral RNA intermediates, which are then used as templates to synthesize new positive-strand (genomic) RNAs. The new RNAs are either translated to yield more polyprotein or, later in the infection cycle, encapsulated to generate progeny virions. Inhibition of the proteolytic activities of the NS3 protease would therefore suppress replisome formation, RNA replication and ultimately generation of new virions. On the other hand, the RNA-dependent RNA polymerase (RdRp) contained within the NS5B protein is the catalytic component of the HCV RNA replication machinery (Behrens *et al.*, 1996; Lesburg *et al.*, 1999). This enzyme synthesizes RNA using the RNA template. This biochemical activity is not present in mammalian cells, offering the opportunity to identify very selective inhibitors of the viral enzyme.

The development of anti-HCV therapy has been hampered by several obstacles for many years after the virus has been discovered and characterized. The lack of a robust *in vitro* cell culture system capable of supporting viral replication has retarded traditional approaches to develop or evaluate antiviral compounds. In addition, there is no conventional small animal model to conveniently assess the *in vivo* efficacy. Most of our knowledge of HCV has been derived from surrogate experimental systems that approximate infection and often preclude definitive interpretation. Preclinical evaluation of potential anti-HCV agents became possible only after the development of the HCV autonomous subgenomic replicon system (Lohmann *et al.*, 1999). The chronically infected chimpanzee model (Walker *et al.*, 1997) and the severe combined immunodeficiency disease (SCID) mouse with chimeric human liver model (Mercer *et al.*, 2001) also proved effective in limited preclinical evaluation of anti-HCV therapies, although both animal models suffer from limitations that make them less than ideal for expanded studies. HCV infects only humans and chimpanzees. The chronically infected chimpanzee model (Walker *et al.*, 1997), the 'gold standard' for HCV studies, is challenging and expensive because one out of three chimpanzees spontaneously resolve their HCV infection. An immune deficient SCID mouse-human liver xenograft system was developed by researchers at the University of Alberta (Mercer *et al.*, 2001). In this model, the liver of neonate SCID beige mice are colonized with infused human hepatocytes which rescues them from a fatal transgene. Infection of these human liver grafts by several genotypes of HCV and the therapeutic effects of INF-α have been reported. Unfortunately, the animals are fragile and scale up of the colony has been slower than expected, thus limiting access to the system.

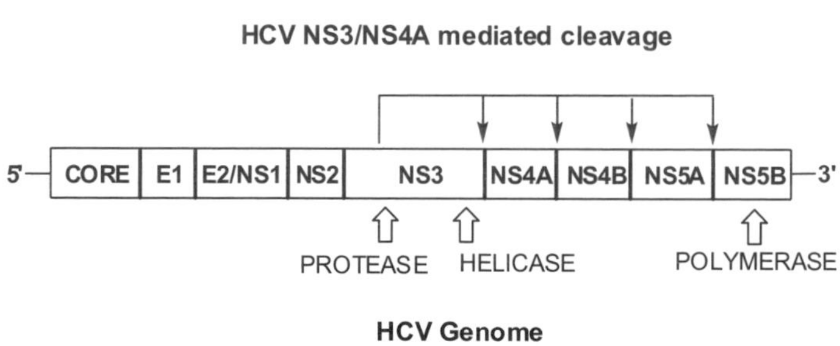

Figure 9.1 Schematic representation of the HCV genome and NS3 protease mediated cleavage (shown as arrows). NS3 is a bifunctional protein with protease and helicase activities.

Despite the daunting challenges in HCV antiviral drug development, extensive research efforts and progress have been made in discovering drugs that halt HCV replication through the inhibition of NS3 protease and other enzymes. A number of promising small-molecule inhibitors of the NS3/4A protease, the NS5B polymerase and the NS5A are in clinical development (Nettles *et al.*, 2008; Chen *et al.*, 2009a; Furman *et al.*, 2009; Liu *et al.*, 2009; Gao *et al.*, 2010). Early testings have demonstrated strong antiviral activity both *in vitro* and in patients. Inhibitors towards other potential targets such as IRES and replicase are in earlier stages of preclinical investigation (De Francesco *et al.*, 2005; White *et al.*, 2006; Manns *et al.*, 2007). Although efforts are ongoing to develop a vaccine, the unusually rapid genetic drift of HCV makes this a daunting task (Houghton *et al.*, 2005; Manns *et al.*, 2007). A major challenge for any successful direct acting anti-HCV therapy is the rapid emergence of the drug-resistant viruses under the selective pressure. The fast turnover rate and the intrinsic low fidelity of the HCV replication machinery endowed the virus with the ability to fully explore its genome space and quickly come up with mutations that rendered it resistant to antiviral drugs.

NS3 protease

The NS3 protease is in many ways a typical β-barrel serine protease, with a canonical Asp-His-Ser catalytic triad similar to the well-studied digestive enzymes, trypsin and chymotrypsin. The X-ray structures of NS3 protease have been resolved either as an isolated domain or as part of the full-length NS3 protein (Love *et al.*, 1996; Yan *et al.*, 1998). The data has provided detailed structural insights to facilitate rational inhibitor design. Histidine-57 and aspartic acid-81 of the catalytic triad are located in the N-terminal, whereas serine-139 forms part of the C-terminal subdomain. The active site, located near the surface in a cleft between two β-barrel subdomains, is shallow, featureless and highly solvent exposed. At other sites, cysteine is conserved in the P1 position of the natural substrate for the protease in all three *trans*-cleavage sites, and it is replaced by a threonine in the *cis*-cleavage event. The P1′ is a small hydrophobic amino acid, either a serine or an alanine. The P2, P3, and P4 sites are all amino acids with hydrophobic side chains, and P5 and P6 usually contain polar amino acids such as aspartic acid or glutamic acid. The NS3 protease uses an extended polydentate binding cleft, with several recognition subsites to ensure specificity. It forms a heterodimeric complex with the NS4A protein, an essential cofactor that activates the protease and assists in anchoring the heterodimer to the endoplasmic reticulum.

The critical role played by NS3 protease in HCV viral replication makes it an excellent target for the creation of new HCV therapy (Lamarre *et al.*, 2003; De Francesco *et al.*, 2005; Perni *et al.*, 2006; Chen *et al.*, 2009a). Development of small-molecule inhibitors for this enzyme would potentially arrest the processing of the aforementioned polyprotein required for viral replication (Malcolm *et al.*, 2006). This has been a field of intensive investigation by various groups worldwide (Chen *et al.*, 2009a). However, developing HCV protease inhibitors as drugs was not a trivial task. There were no viable small molecule lead structures from screening efforts of multi-million compounds to initiate a drug discovery effort. Early inhibitors were designed-based on the substrate-enzyme active site interactions (Llinas-Brunet *et al.*, 1998; Njoroge *et al.*, 2008). However, the HCV protease requires an extensive peptide substrate with multiple weak interactions. The requirement of large substrates was a major concern in the development of orally bioavailable small molecule drugs. It was feared that the enzyme may only be inhibited by molecules large enough to mimic the natural substrate. Indeed, early leads were long peptidic compounds which occupied much of the substrate-binding site to take advantage of multiple hydrogen bonding and hydrophobic interactions. The major challenge was to modify these large molecules to the less peptidic and lower molecular weight drug candidates with desirable pharmacokinetic profiles, while retaining or improving potency in the enzymatic and cellular assays. The lack of the cavities, holes and flaps, or so-called binding pockets at the vicinity of the active site, that had been exploited as anchor points to design potent and selective inhibitors of other proteases, made it a difficult task.

As a result of these challenges, no HCV protease inhibitors have been approved yet for the treatment of HCV disease in spite of the fact that the virus was fully characterized more than two decades ago. However, efforts within the pharmaceutical industry have resulted in a number of candidates in clinical development (Chen *et al.*, 2009a). Ciluprevir (BILN-2061) from Boehringer Ingelheim (Ingelheim, Germany) was the first HCV NS3 protease inhibitor that went into first-in-human clinical trial (Lamarre *et al.*, 2003). It demonstrated rapid viral load reduction in humans and established the proof of concept for HCV protease inhibitors. The development of ciluprevir was discontinued because the compound exhibited cardiac toxicity in monkeys at higher doses. Shortly after the success of the proof of concept by BILN-2061, two other novel protease inhibitors, boceprevir [SCH 503034, Schering-Plough (Whitehouse Station, NJ, USA); Venkatraman *et al.*, 2006; Njoroge *et al.*, 2008] and telaprevir [VX-950, Vertex (Cambridge, MA, USA); Perni *et al.*, 2006] were advanced into clinical studies in humans and demonstrated to be safe and efficacious. They are both currently in phase III clinical trials. A number of other candidates are also at various stage of clinical or preclinical development. Among them, ITMN-191 [also known as R-7227, InterMune (Brisbane, CA, USA)/Roche (Basel, Switzerland); Seiwert *et al.*, 2008], TMC-435350 [Tibotec (Beerse, Belgium)/Medivir (Huddinge, Sweden); Raboisson *et al.*, 2008; Lin *et al.*, 2009], vaniprevir (MK-7009, Merck; Liverton *et al.*, 2010; McCauley *et al.*, 2010) and narlaprevir (SCH 900518, Schering-Plough; Arasappan *et al.*, 2010) have entered phase II and appear to be more advanced than others.

Two classes of inhibitors

Two distinct mechanism-based classes of anti-HCV protease inhibitor molecules have been discovered: (i) non-covalent product-based inhibitors, such as ciluprevir (BILN-2061) and ITMN-191; and (ii) covalently bonded reversible inhibitors, also known as serine-trap inhibitors. The most clinically advanced NS3 protease inhibitors, telaprevir and boceprevir are both covalently bonded reversible inhibitors.

Non-covalent inhibitors

The discovery that HCV protease was susceptible to significant inhibition by the N-terminal peptide products released from substrate cleavage (Llinas-Brunet *et al.*, 1998; Steinkuhler *et al.*, 1998) has been exploited by several research groups using a structure-based drug design approach. Several classes of such substrate-based non-covalent inhibitors have been developed. The early product-derived hexapeptide P1-cysteine inhibitor such as compound *1* (Ingallinella *et al.*, 1998; Fig. 9.2), evolved into more elaborated hexapeptide inhibitor 2 (Pause *et al.*, 2003) with a crucial vinylcyclopropane carboxylic acid P1 residue and a 2-phenyl-4-oxoquinoline substituted proline P2 residue which resulted in a dramatic improvement in potency (K_i = 40 nM for *1*, and 0.074 nM for *2*). To overcome poor pharmacokinetic properties of the peptidic inhibitors, macrocyclization has been used as one of the depeptization methods. Upon optimization, researchers at Boehringer Ingelheim discovered BILN-2061 (*3*), the first HCV protease inhibitor progressed to human clinical trials and proven to be efficacious against HCV infection (Lamarre *et al.*, 2003; Llinas-Brunet *et al.*, 2004). A number of clinical candidates based on similar P1-P3 macrocyclic frame have been developed since the introduction of BILN-2061 (Chen *et al.*, 2009a). A unique class of P2-P3 cap macrocycles was discovered by scientists at Merck with the ultimate identification of the clinical candidate MK-7009 (*4*, McCauley *et al.*, 2010). The non-covalent HCV protease inhibitors will be the subject of an extensive review in the next chapter.

Covalent inhibitors

Besides traditional competitive inhibitors, another strategy that has been used successfully in the discovery of other serine, tyrosine, cysteine and threonine protease inhibitors (Sanderson *et al.*, 1998; Powers *et al.*, 2002; Turk, 2006) has been adapted to the development of HCV protease inhibitors. A covalent trap, or 'warhead', such as an electrophilic aldehyde, ketone, α-ketoamide, α-ketoacid, boronic acid/ester, or a phosphonate/phosphonic acid group was incorporated into a substrate-based inhibitor to react

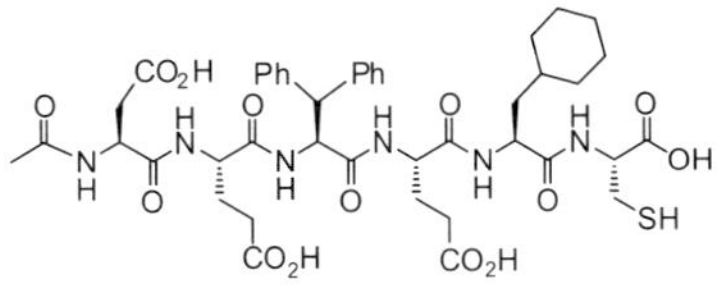

1

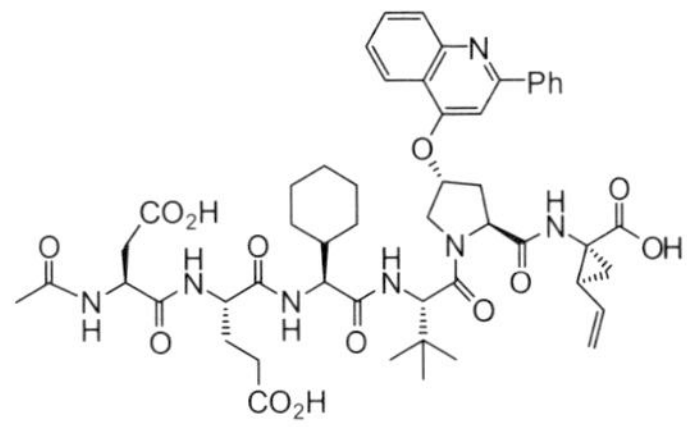

2

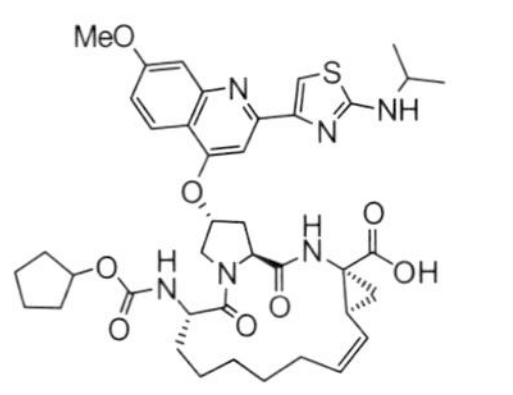

3
BILN-2061

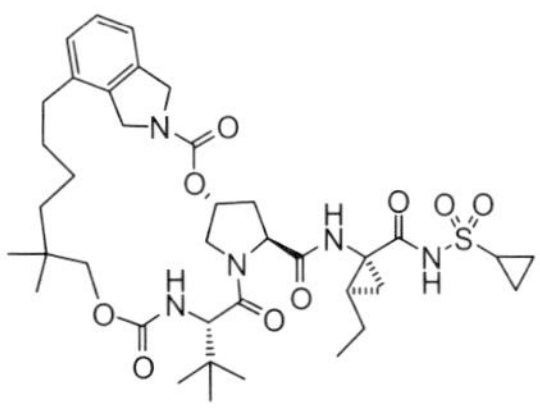

4
MK-7009

Figure 9.2 Non-covalent HCV NS3 protease inhibitors.

with the serine hydroxyl of NS3 protease catalytic site (Lin *et al.*, 2006). As illustrated in Fig. 9.3, reaction of the active site serine (Ser-139) with an activated carbonyl in the inhibitor (*5*) and subsequent trapping of the resulting transition-state analogues (*6*) by the active site triad (Ser-139, His-57 and Asp-81; Perni *et al.*, 2006) provided effective inhibition through a stable, covalent and reversible complex with the enzyme. An example of a boronic acid covalent inhibitor is bortezomib [marketed as Velcade for the treatment of multiple myeloma (*7*); Fig. 9.3]. It is a proteasome inhibitor which forms a covalent bond with the active site threonine (Groll *et al.*, 2006). The electrophilic 'warhead' was essential for the protease inhibitory activity of these compounds. The time required for stable covalent adduct formation is in the order of minutes, much longer than that required in traditional non-covalent binding.

Covalent HCV protease inhibitors

Several classes of HCV protease inhibitors with electrophilic functionality have been reported. The so-called serine trap 'warheads' form covalent reversible or irreversible bonds with hydroxyl of the active site catalytic serine residue of the protease. The approaches using aldehydes, ketones, ketoacids/ketoesters/ketoamides, boronic acids/boronates, phosphonates and Michael receptors are discussed in this review.

Aldehydes

Aldehyde is the smallest and simplest functionality that can serve as an electrophilic warhead to form a covalent bond with serine proteases. This class of covalent inhibitors have been used in the development of potential therapeutics for thrombin (Rebello *et al.*, 1997) and cathepsin K (Votta *et al.*, 1997) serine proteases (Sanderson *et al.*, 1998; Powers *et al.*, 2002; Turk, 2006). However, aldehydes, particularly aliphatic aldehydes, are susceptible to oxidation to their corresponding carboxylic acids. Although aldehydes are rarely developed into drug candidates due to their relative instability, they are useful tools in SAR investigations because of the ease of synthesis.

Several research groups have reported HCV NS3 protease inhibitors with an aldehyde warhead incorporated (Ede *et al.*, 2000; Tung *et al.*, 2001; Narjes *et al.*, 2002). Some examples (*9–16*) are shown in Fig. 9.4. As a substrate mimic hexapeptide, compound 9 had an IC_{50} of 5.5 μM against the NS3 protease (Ede *et al.*, 2000). When more elaborated amino acid residues were introduced into the inhibitors such as

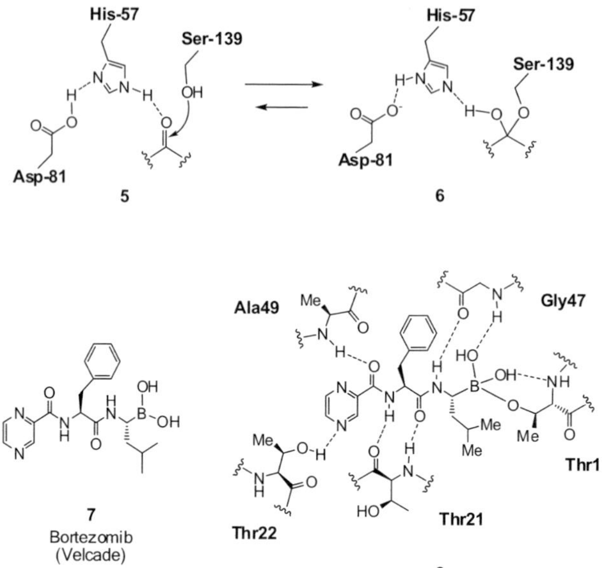

Figure 9.3 Nucleophilic attack of the carbonyl by Ser-139 of HCV NS3 protease (**5**) led to a covalent tetrahedron intermediate (**6**) stabilized by residues His-57 and Asp-81. Boronic acid covalent inhibitor Bortezomib (**7**), its covalent and hydrogen bonding network (**8**) in the active site of yeast 20S proteasome.

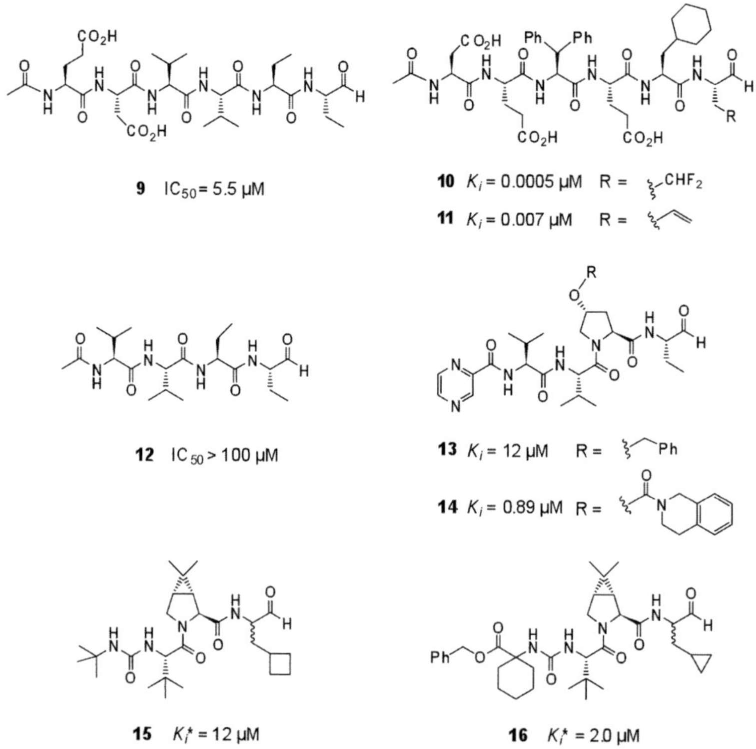

Figure 9.4 Aldehyde covalent HCV NS3 protease inhibitors.

10 and *11*, extraordinary potency (K_i of 0.0005 and 0.007 μM) was achieved (Narjes *et al.*, 2002). The discovery of Abu (*9*, *12–14*) as the replacement of reactive and unstable cysteine as the P1 residue was a great progress towards more stable and workable molecules. Subsequent findings of difluoro-Abu (*10*), allyl (*11*) and cyclic P1 (*15*, *16*) residues provided better alternatives for the future development of leads with desirable properties. However, the peptidic nature and the large molecular weight (MW) deprived the structures such as *10* and *11* of any desirable pharmacokinetic properties. Truncation efforts led to smaller molecules such as *12–16*. Removal of P6 glutamic acid and P5 aspartic acid from *9* resulted in a tetrapeptide *12*, which was inactive (IC_{50} > 100 μM). Incorporation of 4-benzyloxy substituted proline P2 and a pyrazine amide cap restored some activity to tetrapeptide *13* (IC_{50} = 12 μM; Perni *et al.*, 2003). The larger and extended hydrophobic groups from P2 proline enhanced the potency even further, the tetrahydroisoquinoline carbamate analogue *14* had a respectable K_i of 0.89 μM.

Scientists at Schering-Plough investigated aldehyde inhibitors incorporating the unique dimethyl-4,5-cyclopropylproline P2 residue and a urea P3 capping group. The *tert.*-butyl urea capped tripeptide *15* had a K_i^* (Morrison *et al.*, 1988) of 12 μM, while the analogue with a larger P3 capping (*16*) improved the potency to 2 μM.

Ketones

Activated carbonyl compounds have been studied extensively as covalent, reversible inhibitors of proteases. Simple ketones are not reactive enough to act as the warhead towards enzymatic nucleophiles. They can be activated by placing various electron-withdrawing groups adjacent to the ketone functionality. Inhibitors with an α-diketone, aromatic ketone, halogen-substituted alkyl ketone have been reported for HCV NS3 protease.

Several activated ketone NS3 protease inhibitors are shown in Fig. 9.5. Consistent with observations from aldehydes, the larger and more elaborated hexapeptide derivative with a aryl

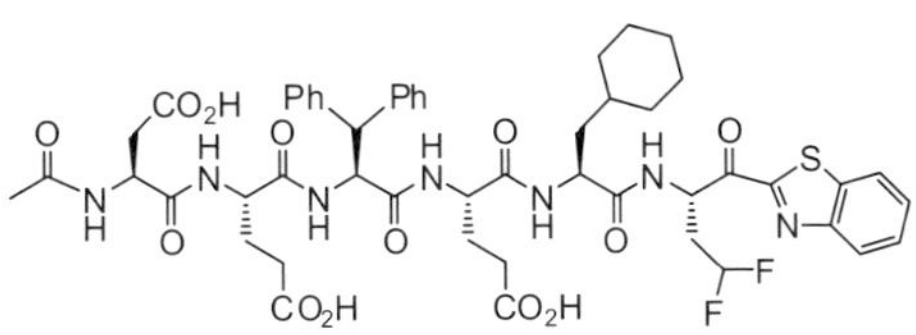

17 K_i = 0.15 μM

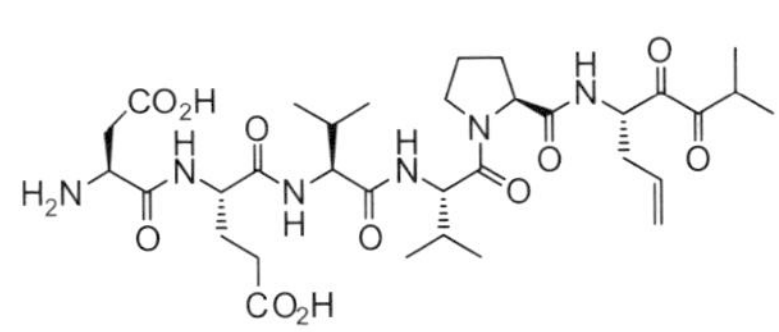

18 IC_{50} = 4.8 μM

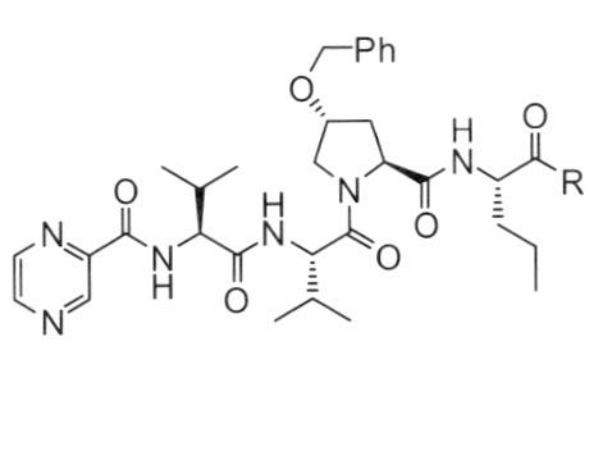

19 R = CF_3 K_i = >50 μM

20 R = CF_2CF_3 K_i = 18 μM

21 R = 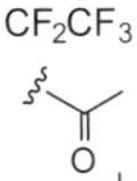 K_i = 50 μM

22 R = 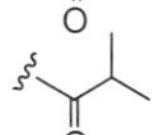 K_i = 2.7 μM

23 R = 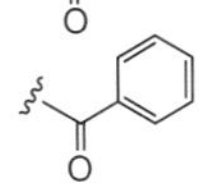 K_i = 10 μM

Figure 9.5 Ketone inhibitors of HCV NS3 protease.

ketone warhead (*17*) demonstrated good activity (K_i = 0.15 μM; Narjes *et al.*, 2002), while slightly smaller and less extended hexapeptide diketone *18* had only moderate IC_{50} of 4.8 μM (Han *et al.*, 2000). A series of tetrapeptide ketones (*19–23*) were reported by the Vertex researchers (Perni *et al.*, 2003). The fluoroalkyl ketones (*19* and *20*) were not very active against NS3 protease. Among the diketones, the smaller methyl diketone *21* was basically inactive, while larger isopropyl diketone *22* had some activity (K_i = 2.7 μM). The aromatic diketone *23* was somewhere in between (K_i = 10 μM). The ketone inhibitors have attracted less attention because they were generally less potent than their corresponding α-ketoamides. There are also selectivity concerns about the elevated reactivity of the ketone functionality.

Ketoacids, ketoesters and ketoamides

Ketoacids, ketoesters and ketoamides belong to the same group of activated carbonyl reversible covalent inhibitors (Fig. 9.6). The acid, ester or amide functionalities are electron-withdrawing enough to make the ketone an effective warhead for serine protease, yet not too strong to create selectivity concerns. α-Ketoacids provide both tight covalent bonding from the strongly electrophilic ketone moiety and additional non-covalent bindings through electrostatic interactions with the charged carboxylate terminus. As a result of optimized binding and covalent bond formation, large hexapeptide derivative *24* was discovered to possess superb potency against NS3 protease with K_i of 0.01 nM. (Narjes *et al.*, 2002). Kinetic investigation showed that the potency of ketoacid

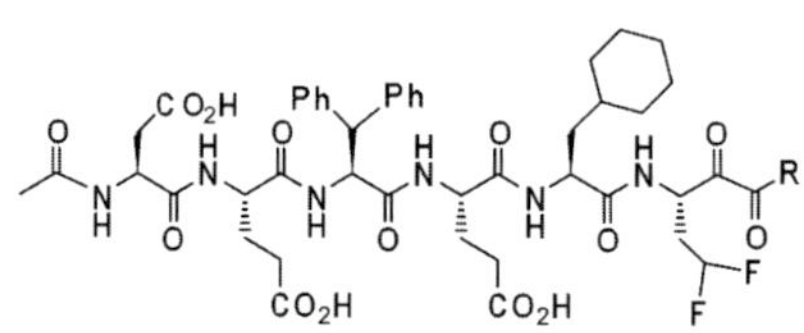

24 R = OH K_i = 0.00001 μM
25 R = OMe K_i = 0.002 μM
26 R = NHBn K_i = 0.0015 μM

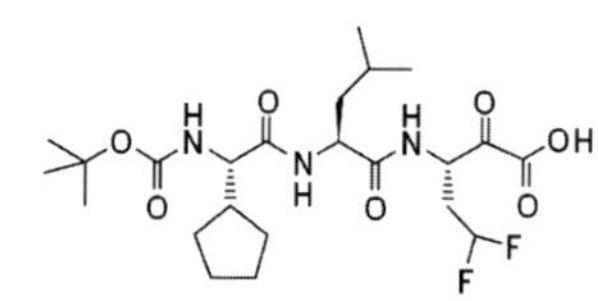

27 IC_{50} = 0.38 μM

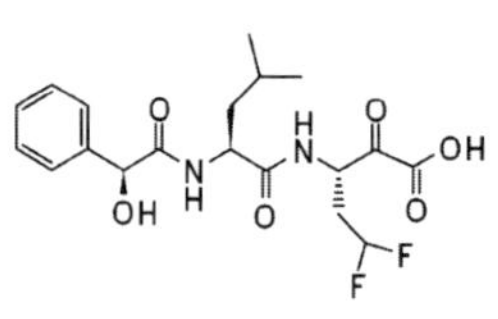

28 IC_{50} = 3 μM

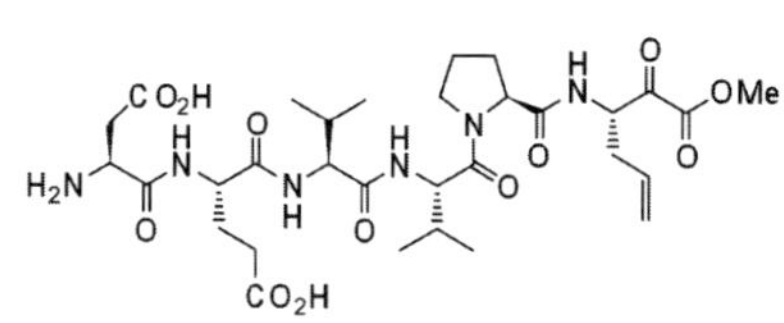

29 IC_{50} = 4.3 μM

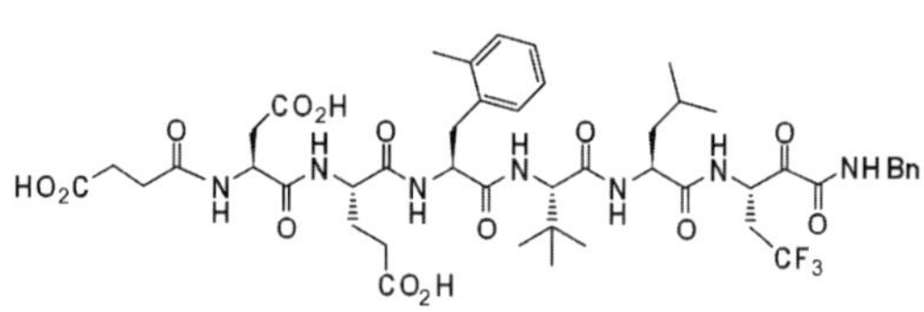

30 IC_{50} = 0.004 μM

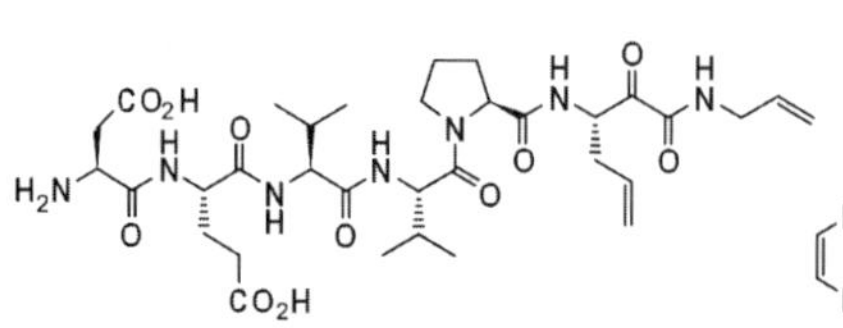

31 IC_{50} = 0.34 μM

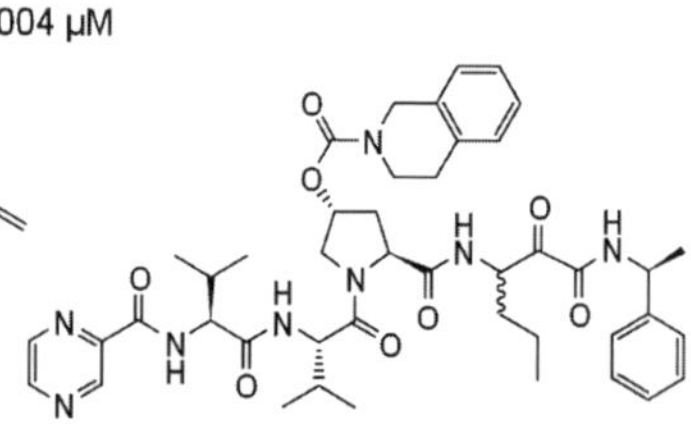

32 K_i = 0.10 μM, EC_{50} = 0.31 μM

Figure 9.6 α-Ketoacid, α-ketoester, α-ketoamide covalent HCV NS3 protease inhibitors.

24 was due to the extremely slow dissociation rate of the covalent adduct. Besides contribution from covalent binding, there was a strong electrostatic stabilization of the terminal carboxyl group by Lys-136 of the enzyme. However, compounds like *24* with a molecular weight approaching 1000 and multiple charged moieties rarely achieve desirable pharmacokinetic properties required for a drug candidate. Fortunately, much smaller truncated tripeptide such as *27* had good activity in enzyme assay (IC_{50} = 0.38 μM). Even dipeptide *28* had some decent potency against NS3 protease (IC_{50} = 3 μM; Nizi *et al.*, 2002, 2004). α-Ketoacids are charged molecules under neutral pH. They have difficulty getting into cells to be effective. In addition, the α-Ketoacid moiety is chemically not very stable. It can easily decarboxylate to give corresponding carboxylic acid, which usually is a less potent inhibitor.

α-Ketoesters are generally less potent than their corresponding α-ketoacids. This is evident for 25. Although still an excellent inhibitor, *25* was 20-fold less potent than ketoacid *24*. The less elaborated hexapeptide analogue *29* (K_i = 4.3 μM) was much less active than *25*, probably due to the loss of multiple non-covalent interactions (Han *et al.*, 2000). α-Ketoesters are susceptible to hydrolysis to give α-Ketoacids which, in turn, have their own liabilities.

As part of the search for the best warhead that is appropriate for HCV NS3 protease, several research groups also investigated inhibitors incorporating α-ketoamides moiety. Some of the early leads such as *26* and *30* were derived from product-based peptides (Fig. 9.6). Compound *26* is a hexapeptide with a large molecular weight (MW) of about 1050 (Narjes *et al.*, 2002). It had a very impressive potency (K_i) of 1.5 nM against NS3 enzyme. Similarly, hexapeptide ketoamide inhibitor *30*, with a MW of more than 1000, also exhibited excellent activity with an IC_{50} of 4 nM (Bennett *et al.*, 2000). Due to the size of the molecules and the presence of multiple charged species, these compounds were not likely to have good oral bioavailability. Nonetheless, they served as leads for the development of smaller and less peptidic drug candidates. Two examples of the truncated smaller inhibitors were pentapeptide *31* (Han *et al.*, 2000) and tetrapeptide *32* (Perni *et al.*, 2004a), both with a MW of around 700. With IC_{50} of 0.34 and 0.10 μM for *31* and *32*, respectively, significant loss of activity was observed. But more importantly, compound *32* demonstrated good activity in cell-based replicon assay with an EC_{50} of 0.31 μM. Although inhibitor *32* was still larger than ideal size, the fact that it contained only four amino acid residues and no charged groups rendered it a very meaningful progress towards desirable molecules, which was evidenced by the demonstration of potency in cellular replicon assay. The loss of potency compared with larger hexapeptides was discouraging, but it was a necessary exercise in the pursuit of viable candidates for further development. The evolution of two important α-ketoamide series that led to the discovery of the most advanced HCV NS3 protease inhibitor clinical candidates will be described in more details in the following sections.

Boronic acids and boronates

Analogous to the structures formed by active carbonyl warheads discussed above, boronic acids and boronates have been employed as electrophiles that could potentially react with the nucleophilic hydroxyl group of NS3 serine protease to form covalent tetrahedral boron complexes (Turk, 2006). This strategy has been used in the development of inhibitors of the thrombin receptor and other serine proteases (Deadman *et al.*, 1995). Potent boronic acid HCV NS3 protease inhibitors have been reported by several groups of researchers (Fig. 9.7). Bristol-Myers Squibb reported a series of potent (+)-pinanediol derived boronates exemplified by compound *33* (Priestley *et al.*, 2000, 2002). The hexapeptide analogue had a unusually large substituted phenethyl P1 residue and had an excellent K_i of 2 nM in enzyme assay. Of course, these molecules were too large and highly charged to be considered as a drug development candidate. The truncated and depeptized compounds such as *34* were much more interesting (Glunz *et al.*, 2003; Zhang *et al.*, 2003). The bicyclic pyrimidinone moiety was designed as β-sheet peptidomimetics. These boronates inhibitors were good inhibitors against NS3 protease with an IC_{50} of 20 nM, which was quite impressive considering the size and non-peptidic

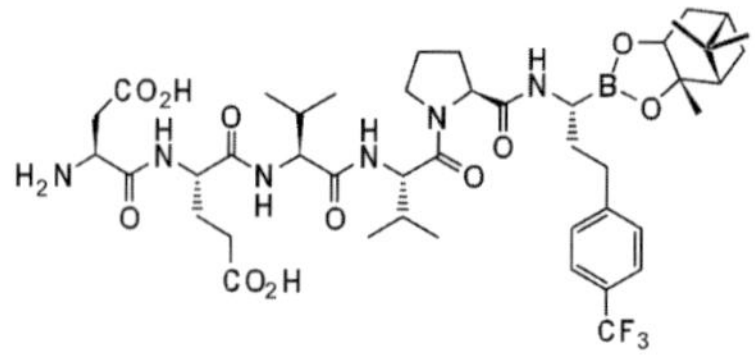

33 K_i = 0.002 μM

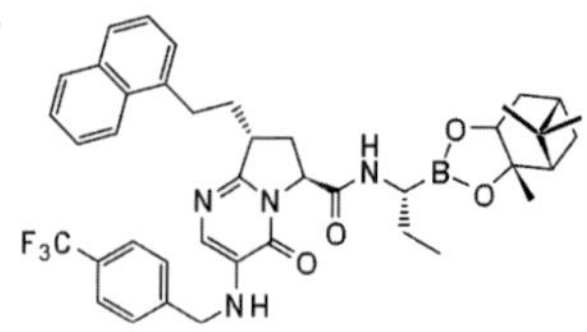

34 IC_{50} = 0.02 μM

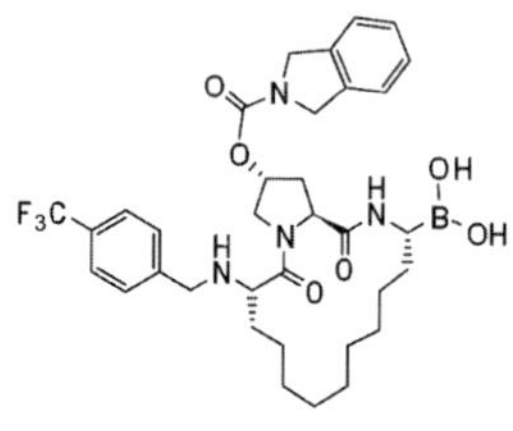

35

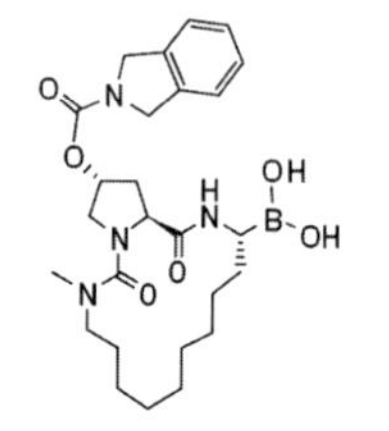

36

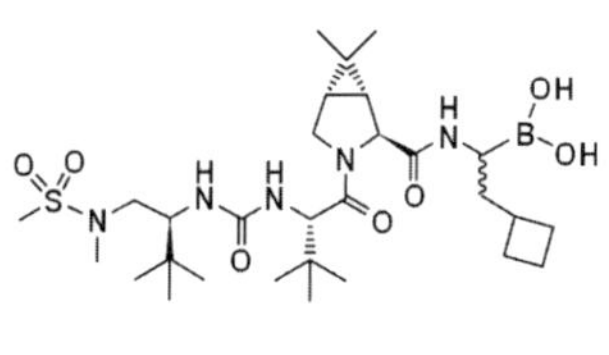

37 K_i = 0.0002 μM

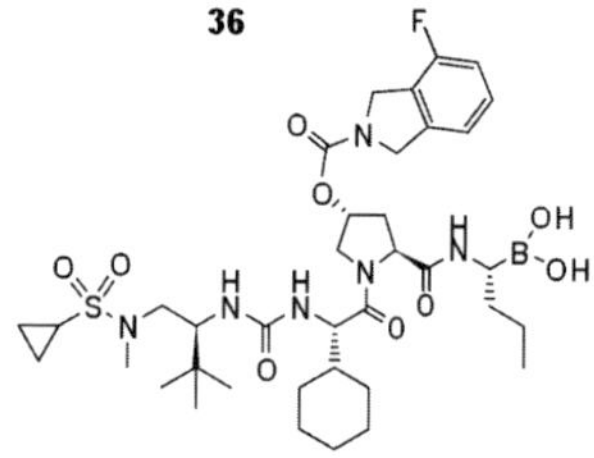

38

Figure 9.7 Boronic acid/boronate HCV NS3 protease inhibitors.

nature of the molecule. Unfortunately, there was no further report on the cellular potency and development of this series of compounds.

More recently, scientists at Phenomix published patents disclosing boronic acids and boronates incorporating structure elements reported by other research groups. The major series concentrated on P1–P3 macrocyclic inhibitors such as compounds 35 and 36 as P1 boronic acids (Campell *et al.*, 2008). Another series focused on linear peptidic structures (e.g. compound 38; Campell *et al.*, 2007). These boronic acids or boronates protease inhibitors have a fluoroisoindoline carbamate moiety attached to P2 proline, which was a signature scaffold discovered by Intermune scientists (Seiwert *et al.*, 2008). No potency data have been reported for these analogues. Researchers from Schering-Plough have converted the ketoamide moiety of some of their potent α-ketoamide protease inhibitors into boronates and corresponding boronic acids (e.g. compound 37; Venkatraman *et al.*, 2009a). These compounds were extraordinarily potent in enzymatic assays and had demonstrated subnanomolar K_i values of 0.2 nM for 37 and 0.5 nM for its corresponding (+)-pinene-based boronate. The more elaborated *tert.*-leucine amine-derived sulfonamide P3 capping group in compound 37 was necessary for the sub-nanomolar activity since simple *tert.*-butyl urea capped analogue had a K_i of 10 nM. The X-ray structure of inhibitor 37 bound to NS3 protease was solved (Venkatraman *et al.*, 2009a). Besides all the traditional hydrophobic and hydrogen-bonding interactions between the inhibitor and the enzyme, the oxygen atom of the Ser-139 residue at the active site was covalently bonded to boron to form a tetrahedral boronate derivative. Unfortunately, when 37 and similar analogues were evaluated in cell-based replicon assays, all compounds had EC_{90} of greater than 5 μM suggesting that the compounds had poor cell membrane permeability due to the presence of the charged groups under physiological conditions.

Other covalent inhibitors

Phosphonates

Phosphonic mono- or diesters and phosphonic acids have been used to mimic the tetrahedron transition state in the enzymatic cleavage of the substrates. Their potency can be greatly enhanced through the formation of a covalent bond with the serine hydroxyl group in proteases (Nishiyama *et al.*, 2002, 2004; Cihlar *et al.*, 2006).

Researchers at Gilead Sciences reported a number of phosphonic mono- or diesters and phosphonic acids (e.g. compounds *39* and *40*; Chaudhary *et al.*, 2006; Fig. 9.8) as HCV NS3 protease inhibitors. In these compounds, phosphonic acids or esters moieties were used as replacements for carboxylic acid which is a common feature in non-covalent NS3 inhibitors. Scientists at Merck and Co. Inc. and Istituto di Ricerca di Biologia Molecolare Pietro Angeletti have also developed a series of novel P2–P3 macrocyclic inhibitors, which contained a spirocyclopropyl phosphonic acid (e.g. compound *41*, R = H), phosphonate (e.g. compound *41*, R = Et) or an alkyl phosphinic acid (e.g. *42*) P1 residue (Liverton *et al.*, 2008). Potency data in either enzyme or cellular replicon assays for these phosphonate and phosphonic acid inhibitors *39–42* were not reported. Based on the activity trend of the boronic acids discussed above, it is not unreasonable to speculate that these compounds are quite potent in NS3 protease enzymatic assays. However, the results in cell-based replicon assay will probably be disappointing due to poor permeability encountered by boronic acids.

Halides

Activated α-carbonyl halides are, in theory, also capable of forming a covalent bond with a nucleophilic functionality from an amino acid residue in the protein backbone. There have been limited reports about this type of covalent inhibitors. A few large peptide analogues were tested by scientists at Schering-Plough (Zhang *et al.*, 2002) to explore activity of α-carbonyl chloride or α-carbonyl bromide. One such example was compound *43* (Fig. 9.8). Unfortunately, even large extended peptidic inhibitors did not demonstrate any activity towards NS3 protease. Recently, a few α-carbonyl chlorides were reported by Avila Therapeutics to have very good potency ($IC_{50} < 10$ nM) in enzyme assay (Petter *et al.*, 2009). The actual activity of these compounds against NS3 serine protease needs to be further clarified.

Michael receptors and β-lactams

A number of covalent irreversible inhibitors acting as Michael receptors were also reported by Avila scientists (Petter *et al.*, 2009). The α,β-unsaturated carbonyl or sulfonyl moieties act as Michael acceptors towards Cys-159 of the protease to form an irreversible covalent bond. Most of these compounds had excellent potency against not only wide-type NS3 protease but also several known mutants such as HCV A156S, A156T or D168A that were resistant to other protease inhibitors. For example, compound *44* had an impressive potency of less than 10 nM in both enzyme and cellular replicon assays. It was also very active towards mutant A156S and A156T ($IC_{50} < 10$ nM and 100 nM, respectively). The Michael acceptor analogue of boceprevir was also prepared (*45*; Niu *et al.*, 2010). In theory, this compound could form two independent covalent bonds with the NS3 protease, one reversible covalent bond from ketone and Ser-139, the other irreversible covalent bond from α,β-unsaturated ester and Cys-159. Unfortunately, the expected enhancement was not observed. With an IC_{50} of 53 nM, it was even less potent than its parent boceprevir. Several examples of β-lactams as covalent bond formation partner were also studied by Avila researchers (Petter *et al.*, 2009). Some of these β-lactam analogues exhibited excellent activity ($IC_{50} < 10$ nM) against NS3 protease and its mutants. Fine-tuning of reactivity vs selectivity has to be performed before any of these novel covalent NS3 protease inhibitors can be developed into drug candidates.

Boceprevir and narlaprevir

At the early stage of research for NS3 protease inhibitors, multi-million compound libraries were screened at Schering-Plough to generate potential leads to initiate a drug discovery effort. No small molecule lead structures were identified from the screening effort (Njoroge *et al.*, 2008).

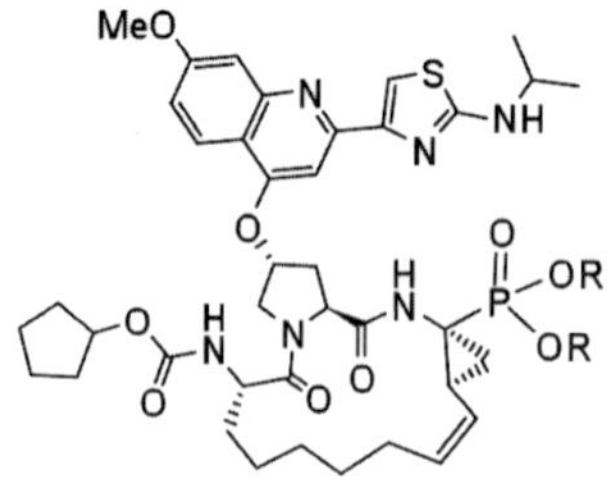

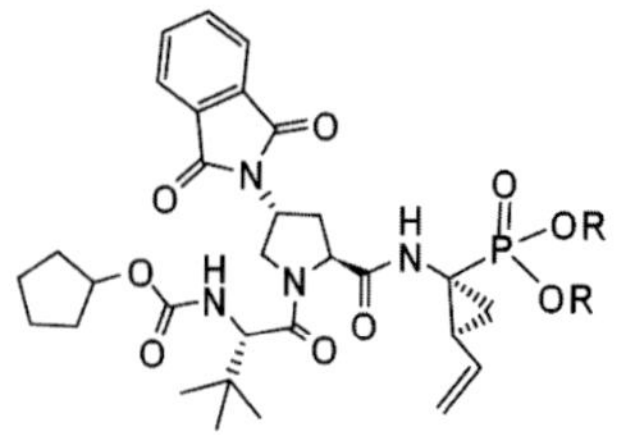

39 R = H, Et **40**

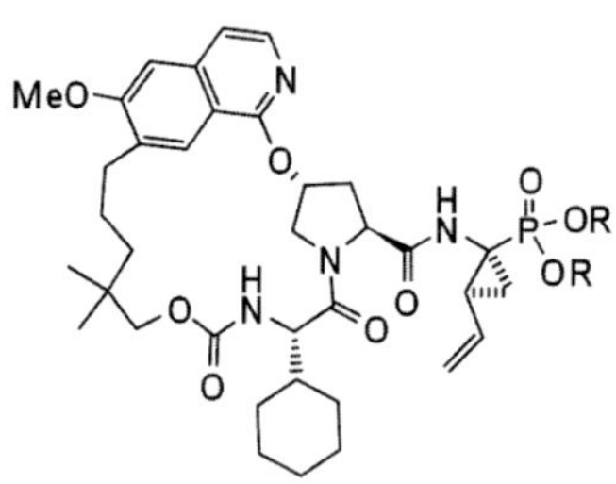

41 R = H, Et

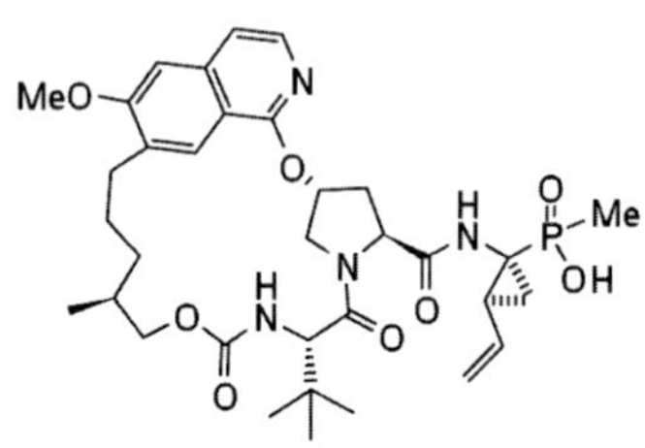

42

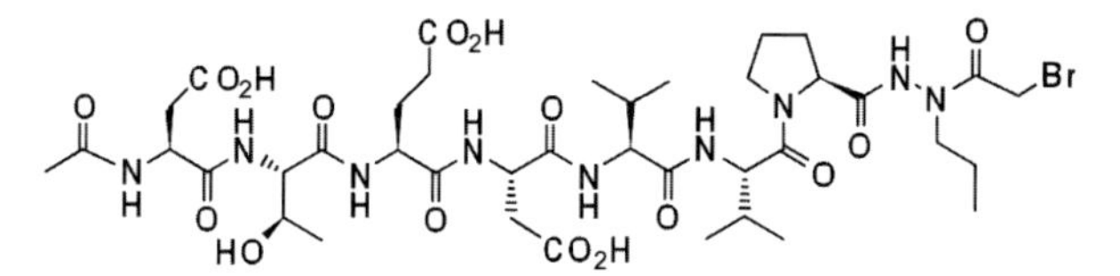

43 K_i = >100 μM

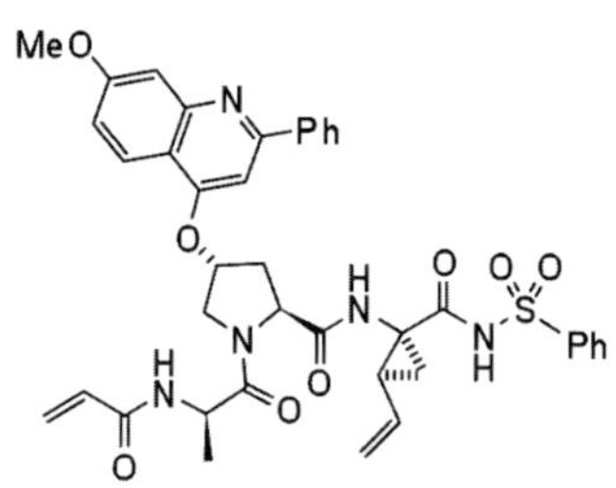

44 IC_{50} < 0.01 μM, EC_{50} < 0.01 μM

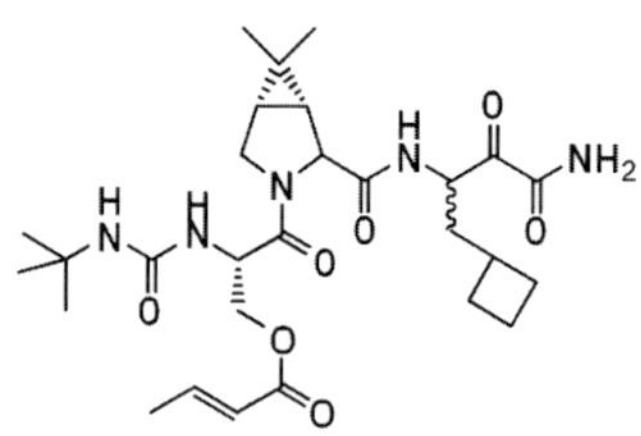

45 IC_{50} = 0.053 μM

Figure 9.8 Phosphonates, phosphonic acids, phosphinic acids, activated halide and Michael receptors as HCV NS3 protease inhibitors.

A structure-based drug design approach was then adopted. The fact that the enzyme is susceptible to marked inhibition by the N-terminal substrate cleavage products and that a electrophilic warhead can be employed in developing other potent serine-protease inhibitors (Sanderson *et al.*, 1998; Powers *et al.*, 2002; Turk, 2006) guided the early effort that eventually led to the discovery of the undecapeptide α-ketoamide inhibitor compound *46* (Venkatraman *et al.*, 2006; Fig. 9.9). Compound *46* exhibited excellent activity in continuous enzyme assay (K_i* = 1.9 nM; Morrison *et al*, 1988; Zhang *et al.*, 1999), but was not expected to have any desirable pharmacokinetic properties because of its size (MW = 1265). Nonetheless, it had served as a lead for further SAR studies. The efforts in truncating and optimizing this type of lead was greatly aided by X-ray

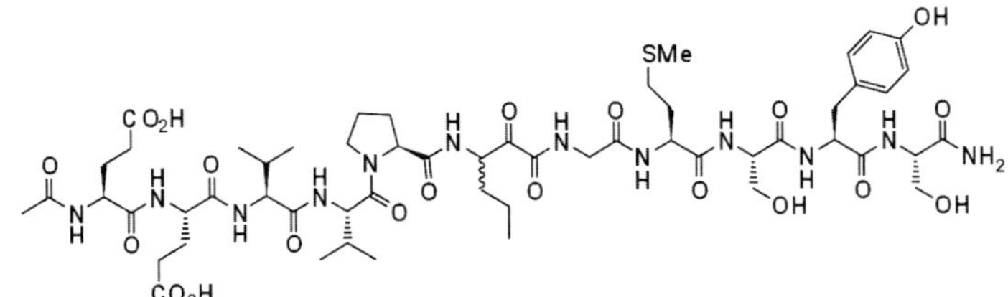

46 K_i^* = 0.0019 μM

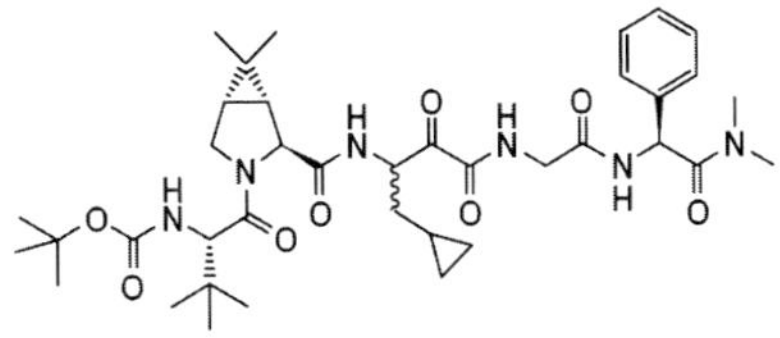

47 K_i^* = 0.005 μM, EC_{90} = 0.1 μM

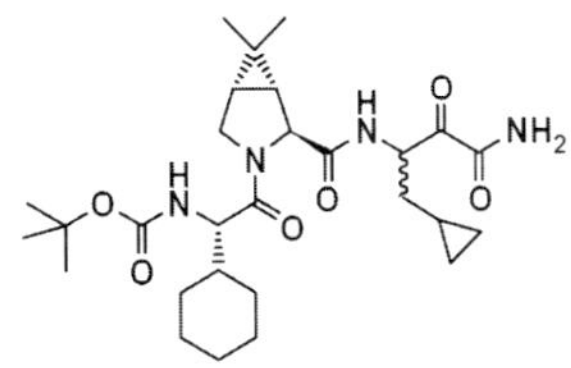

48 K_i^* = 0.025 μM, EC_{90} = 0.4 μM

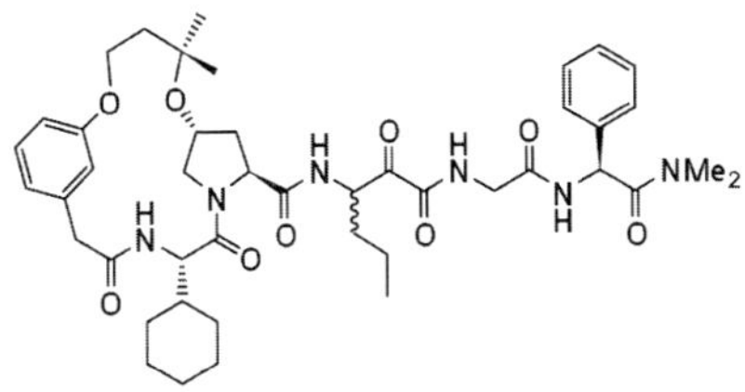

49 K_i^* = 0.006 μM, EC_{90} = 0.9 μM

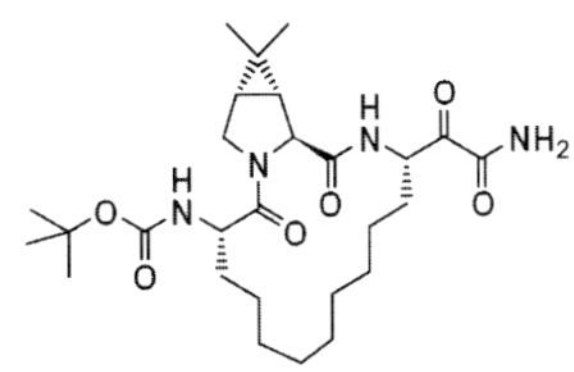

50 K_i^* = 0.03 μM, EC_{90} = 0.6 μM

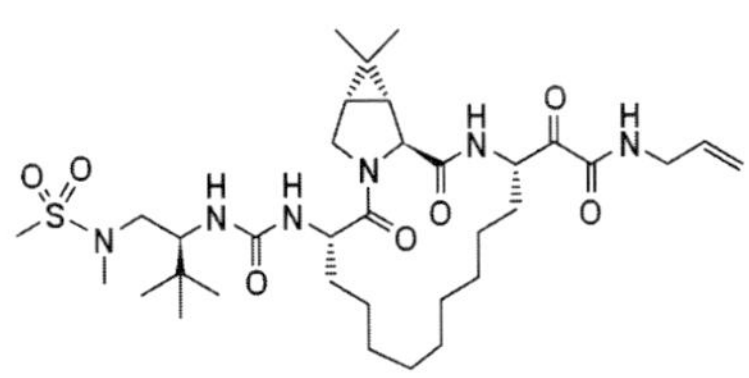

51 K_i^* = 0.003 μM, EC_{90} = 0.055 μM

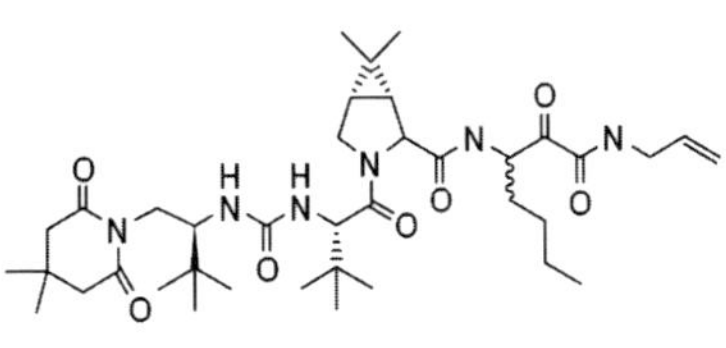

52 K_i^* = 0.002 μM, EC_{90} = 0.02 μM

Figure 9.9 From undecapeptide to tripeptide HCV NS3 protease inhibitors through truncations, cyclizations and capping modifications.

structures (Prongay *et al.*, 2009) of inhibitors bound to the HCV NS3 protease and continuous drug metabolism/pharmacokinetic evaluation.

To render it a drug-like molecule, compound *46* needed to be much smaller and possess appropriate structural features. A series of truncations were performed at both prime and non-prime sides. After trimming seven amino acid residues and optimize each moiety at remaining positions, a potent pentapeptide inhibitor was identified (*47*, Schering 6; Bogen *et al.*, 2006a). With the featured *gem*-dimethyl-3,4-cyclopropylproline as P2 residue, *tert.*-leucine as P3, cyclopropylalanine at P1, phenylglycine at P2′ and *tert.*-butyl urea as P3 capping group, this pentapeptide inhibitor was potent in both enzyme (K_i^* = 5 nM) and cellular assays (EC_{90} = 0.1 μM). The most important discovery during this round of truncation and optimization occurred at P2 position. The *gem*-dimethylcyclopropylproline was demonstrated to be a small but superior P2 residue compared with P2 moieties identified by other researchers. The X-ray structure of 47 bound to the HCV protease revealed hydrophobic interaction between the cyclopropyl ring and one of the methyl substituents and the methyl side chain of Ala-156. The reversible covalent bond, the P1–P2′ 'C'-shaped clamp around Lys-136 side chain and other interactions all contributed to the excellent potency of *47*. However, this compound was still larger than preferred size for an ideal drug candidate (Lipinski, 2000). Not surprisingly, the pharmacokinetic profile of *47* was poor in animals. Further modifications were clearly needed to achieve desirable pharmacokinetic properties. Additional truncation efforts were focused on the prime side. The P2′ phenylglycine and P1′ glycine residues were replaced by hydrogen or a small alkyl group such as methyl, ethyl or allyl. Unfortunately, all secondary and tertiary P1 ketoamides were significantly less potent than compound *47*. However, it was discovered that unsubstituted primary ketoamides had good potency in both enzyme and replicon assays. One representative tripeptide inhibitor *48* had a K_i^* of 25 nM and a replicon EC_{90} of 0.40 μM with a t-Boc P3 capping and cyclohexylglycine P3 residue. Although it was slightly less potent than *47*, its size was significantly smaller and was within the preferred range of a drug candidate.

The fact that it was difficult to acquire desirable pharmacokinetic properties from peptidic molecules because of their susceptibility to hydrolysis by an army of peptidases *in vivo* prompted us to devote substantial efforts in depeptizing the lead compound *47*. Replacement of regular amino acid residue and amide bond with a hydrazine urea-based aza-peptide or a ketone moiety resulted in substantial loss of activity (Bogen, *et al.*, 2006b). Cyclization or macrocyclization between different amino acid residues as a strategy of depeptization of NS3 protease inhibitors was also extensively investigated (Chen *et al.*, 2009a). The macrocyclization of P2 and P4 moieties resulted in a number of 15- to 18-membered macrocyclic inhibitors. The macrocycle that connected phenylacetamide P3 capping to C-4 hydroxyl of P2 proline through a *tert.*-alkyl ether linkage afforded the most potent series (e.g. *49*; Chen *et al*, 2005, 2006). The 16-membered macrocyclic ring in *49* formed a doughnut-shaped circle over the methyl group of Ala-156 as evidenced by X-ray crystallography. This compound exhibited excellent potency (K_i^*) of 0.006 μM in enzyme assay, but demonstrated only moderate activity (EC_{90} = 0.90 μM) in replicon assay. The close proximity of the P1 and P3 side chains revealed by the X-ray structures of the inhibitors bound to the protease had inspired the discovery of a number of P1–P3 macrocyclic inhibitors (Chen *et al.*, 2009a). The tripeptide analogue *50* with a 16-membered P1–P3 macrocycle exhibited decent potency in both enzyme and cell-based assays (K_i^* = 0.03 μM, EC_{90} = 0.6 μM).

Optimization at P1 position demonstrated that only small cycloalkyls or certain alkyl chains were tolerated in the S1 pocket with the optimum ring size being three- or four-membered and *n*-butyl the optimal chain length (Venkatraman *et al.*, 2006). The selectivity (HNE/HCV) of these compounds was associated with the size of the P1 side chain. The larger the size of the P1 side chain, the greater the selectivity against human neutrophil elastase (HNE). When both potency and selectivity were considered, cyclopropyl-, cyclobutyl-alanine or norleucine appeared to be the best residues for P1 position. The S3 pocket of HCV NS3 protease is shallow and hydrophobic which is occupied by isoleucine in the natural substrate. We explored the tolerance of this site by introducing several branched alkyl, and cycloalkyl P3 side chains. *tert.*-Butyl and cyclohexyl groups were found to be the best in both enzyme and replicon assays, inhibitors with the former were more selective against elastase while those with the later were slightly more potent. The P3 capping studies revealed that in general urea capped inhibitors were more potent and selective. Larger alkyl or cycloalkyl groups were somewhat more potent than smaller ones. Since the larger and more substituted cappings tended to possess poor pharmacokinetic properties, the *tert.*-butyl urea was determined to be the best choice because of the balance of good potency, selectivity and reasonable pharmacokinetic profiles. Ultimately, the most potent and selective inhibitor, boceprevir (*53*, SCH 503034; Fig. 9.10) came from the combination of *tert.*-butyl urea-capped inhibitor with *tert.*-butylglycine P3, cyclobutylalanine P1. With a K_i^* of 0.014 μM, EC_{90} of 0.35 μM and an extremely high selectivity (HNE/HCV) of 2200, compound *53* had the best overall profile among the molecules of similar size. The oral dosing area under the curve (AUC) and bioavailability of boceprevir were good in dogs and rats, but poor in monkeys. Boceprevir was highly concentrated

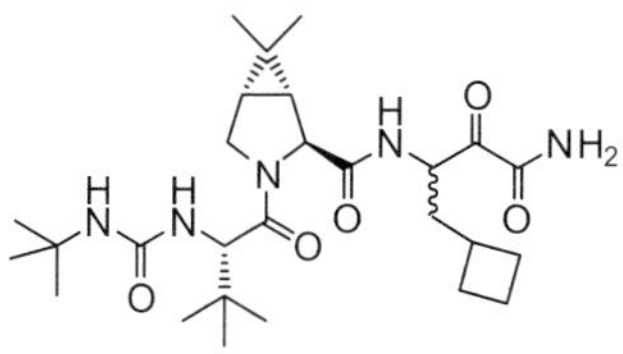

53
Boceprevir (SCH 503034)
Ki^* = 0.014 μM
EC_{90} = 0.35 μM
HNE/HCV = 2000

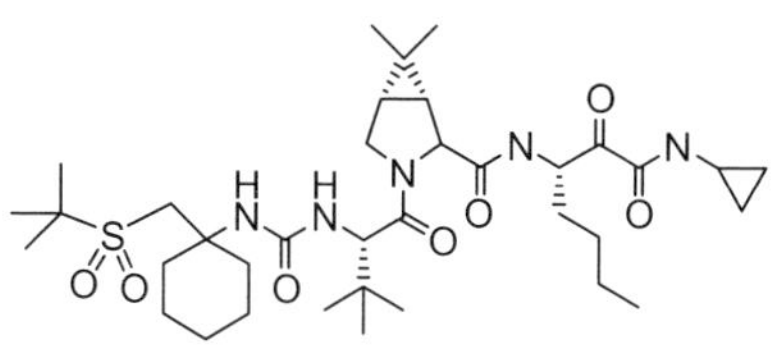

54
Narlaprevir (SCH 900518)
Ki^* = 0.006 μM
EC_{90} = 0.04 μM
HNE/HCV = 600

Figure 9.10 HCV protease inhibitor clinical candidates boceprevir and narlaprevir.

in the liver, with a liver/plasma ratio of approximately 30. The compound was selected for drug safety studies and then progressed to clinical trials.

Boceprevir has been reported as safe and well tolerated in phase I and phase II clinical trials. In a phase I proof-of-concept trial, the maximum mean viral load reduction in non-responding patients with HCV genotype 1 during 14 days of treatment was observed with the group receiving a dose of 400 mg three times daily. The reduction in viral load observed with the use of boceprevir in combination with PEG-Intron was 2.85 log_{10} (Mederacke *et al.*, 2009). In a phase II trial (SPRINT-1), patients treated with boceprevir (800 mg t.i.d.) for 48 weeks in combination with PEG-Intron and ribavirin regimens demonstrated SVR rates of 67–75% compared with SVR rates of 38% in the control arm (Kwo, 2009). Boceprevir is currently in late-stage phase III trials, and results are expected in 2010 (Clinicaltrials.gov identifiers: NCT00627926, NCT00708500). If these phase III trials proceed without any significant adverse findings, boceprevir could be approved as a drug in the near future.

Narlaprevir (SCH-900518)

With the success of the proof of concept of the first generation α-ketoamide covalent inhibitor boceprevir, the research was continued to identify a second generation candidate. The goal was to find a molecule which was 10-fold more potent in replicon assay and had at least comparable pharmacokinetic profile to boceprevir. X-ray structures revealed that larger moieties could be accommodated in S4 pocket for enhanced binding, and that additional hydrogen bond could be established with backbone Cys-159 through an extended functional group. In this extensive effort searching for better P3 capping groups, a variety of structures were investigated. Most of the new cappings were either derived from *tert.*-leucinol/*tert.*-leucine amine, or from 1-aminocyclohexyl methanol scaffold. Both classes succeeded in achieving desired higher potency. Carbamate (Arasappan *et al.*, 2009), sulfonamide (*51*; Venkatraman *et al.*, 2009b,c; Fig. 9.9), sultam (Chen *et al.*, 2009b) and cyclic imide (*52*; Chen *et al.*, 2009c; Fig. 9.9) cappings were derived from *tert.*-leucinol/*tert.*-leucine amine. A number of these inhibitors had excellent potency in cell-based assay with EC_{90} of at or below 50 nM range. Sulfonamide analogue *51* had a K_i^* of 3 nM and an EC_{90} of 55 nM, while glutarimide analogue *52* had an impressive K_i^* of 2 nM and EC_{90} of 20 nM. Ketone, ester, amide (Bogen *et al.*, 2009) and sulfone (Bennett *et al.*, 2010) cappings were developed from 1-aminocyclohexyl methyl group. Replicon EC_{90} values below 50 nM were also achieved with inhibitors having these caps. As evidenced by X-ray crystal structures, both chemotypes established an additional strong hydrogen bond between sulfonyl or carbonyl oxygen and N-H of Cys-159 residue. The *tert.*-butyl or cyclohexyl moieties were also extending into S4 pocket for enhanced interactions.

Among various new P3 cappings investigated, sulfones demonstrated better overall pharmacokinetic profiles than other functionalities, especially in monkeys. Optimization of moieties at different positions led to the discovery of compound 54 (narlaprevir, SCH-900518; Fig. 9.10), a second generation HCV NS3 protease inhibitor (Arasappan *et al.*, 2010). It has a *tert.*-butyl cyclohexylmethyl sulfone capping group, a *tert.*-butylglycine as P3, a norleucine as P1. The potency against NS3 protease in enzyme assay (K_i^*) was 6 nM and in replicon assay (EC_{90} genotype 1b) was 40 nM, which was about 10-fold more potent than boceprevir 53. Compound 54 was selective against HNE (HNE/HCV = 600), and had better overall pharmacokinetic profile than other similar molecules. The oral dosing AUC and bioavailability of 54 were good in rats (AUC = 6.5 μM h, F = 46%, 10 mg/kg) and monkeys (AUC = 1.1 μM h, F = 46%, 3 mg/kg), moderate in dogs (AUC = 0.9 μM h, F = 29%, 3 mg/kg). Narlaprevir was also highly concentrated in the target organ. The ratio of rat liver concentration at 8 hours after dosing over that of replicon EC_{90} was about 25. As a result of its overall profile, narlaprevir was selected for preclinical drug safety studies and then progressed into clinical trials. Phase I proof–of-concept trials of narlaprevir, both as monotherapy and in combination with PEG-Intron (without ribavirin), have demonstrated strong antiviral activity, with greater than 4 $\log_{10}$ reduction in circulating HCV RNA (Reesink *et al.*, 2009). Narlaprevir is currently in phase II clinical trials.

Telaprevir

Independently, researchers at Vertex and Eli Lilly were also working on α-ketoamide NS3 protease inhibitors. Truncation and optimization of peptidic product based inhibitors led to pentapeptide analogues such as 55 (Perni *et al.*, 2004a; Fig. 9.11) and 56 (Perni *et al.*, 2004b). With a large tetrahydroisoquinoline substituent at P2 proline, the pyrazine amide capped pentapeptide 55 was potent in enzyme assay (K_i = 20 nM). But it was inactive in replicon assay. To obtain desired pharmacokinetic property and gain activity in cellular assay, the size of the molecule needed to

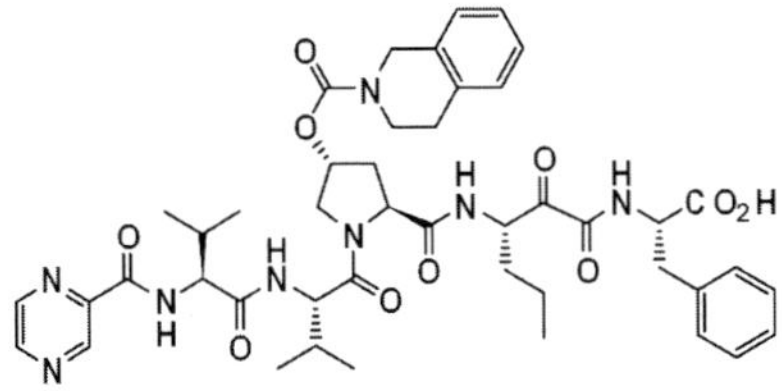

55 K_i = 0.02 μM, EC_{50} = > 10 μM

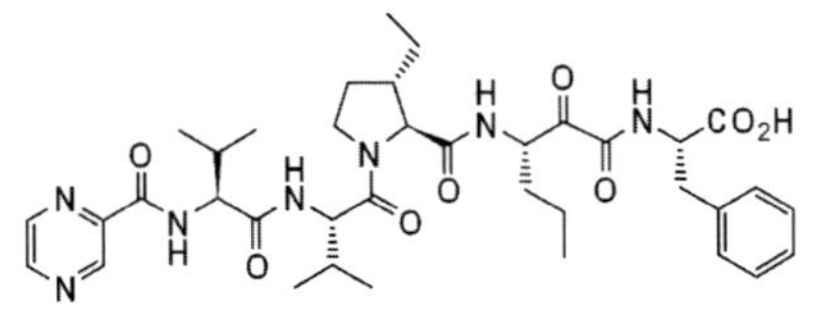

56 K_i = 0.22 μM, EC_{50} = > 10 μM

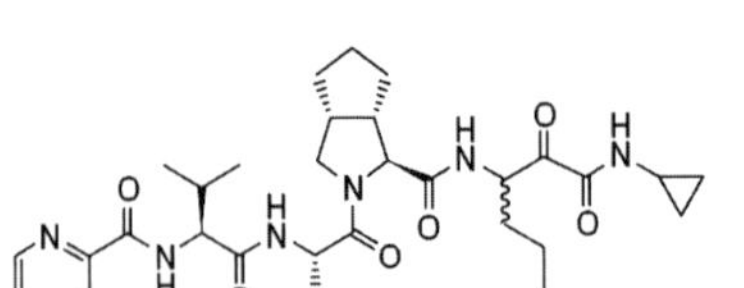

57 K_i = 0.12 μM, EC_{50} = 7.0 μM

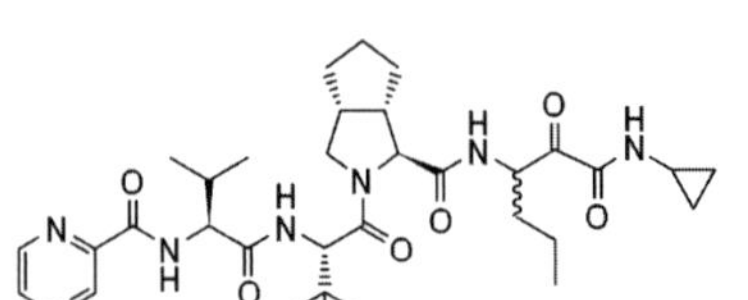

58 K_i = 0.084 μM, EC_{50} = 2.2 μM

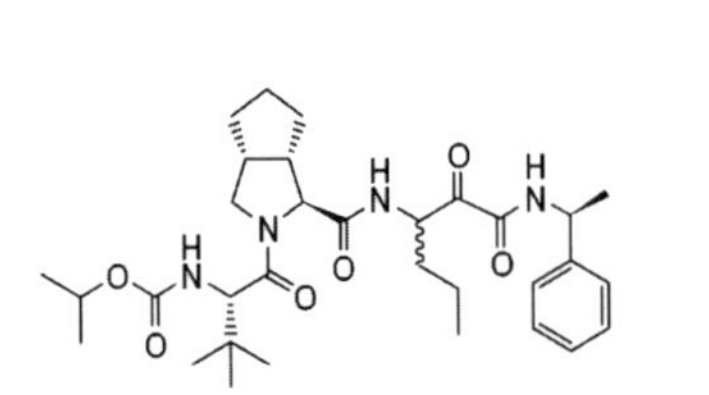

59 K_i = 1.8 μM, EC_{50} = 50 μM

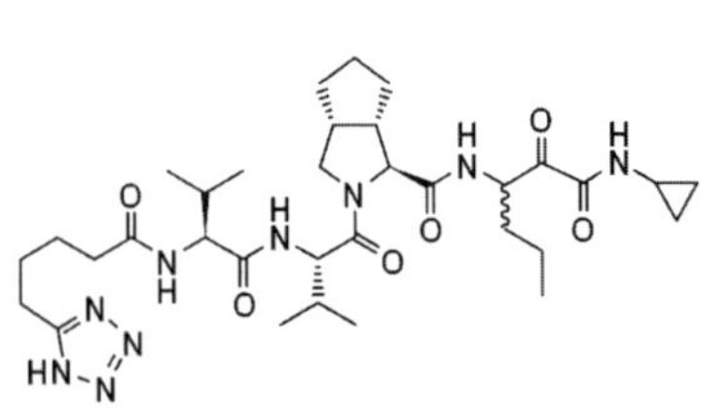

60 K_i = 0.17 μM, EC_{50} = >10 μM

Figure 9.11 α-Ketoamide NS3 protease inhibitors from Vertex and Eli Lily.

be further reduced. Thus, the large aromatic P2 proline C4-substituent was replaced with a small C3-ethyl substituent in 56. A loss of ten-fold in potency was observed (K_i = 220 nM). Cyclization of the ethyl group to proline C4 led to the discovery of the key bicyclic proline P2 moiety in 57 (Yip *et al.*, 2004). Even with the elimination of P1′ phenylalanine residue and replacing it with a small cyclopropyl group, compound 57 was more potent (K_i = 120 nM) than its larger analogue 56. More importantly, it exhibited activity in replicon assay with an EC_{50} of 7.0 μM. Subsequent optimization at P3 position identified *tert.*-butyl glycine as the best residue which led to compound 58 with further improvement of potency both in enzyme and cellular assays (K_i = 84 nM and EC_{50} = 2.2 μM; Victor *et al.*, 2004). Attempts in further truncation of P4 valine as in 59 (Lamar *et al.*, 2004) or modification of pyrazine moiety as in 60 (Sun *et al.*, 2004) resulted in significant loss of activity (EC_{50} > 10 μM).

Ultimately, SAR investigation at P4 position led to the identification of cyclohexylglycine as the optimal residue as shown in compound 61 (Fig. 9.12), which was selected as the clinical candidate telaprevir (VX-950; Yip *et al.*, 2004; Perni *et al.*, 2006). It has been shown that the reaction of the active site serine (Ser-139) with the α-ketoamides and subsequent trapping of the resulting transition-state analogues (6; Fig. 9.3) by the active site triad (Ser-139, His-57 and Asp-81) provided effective inhibition through a stable, covalent and reversible complex with the enzyme (Perni *et al.*, 2006). Telaprevir was very potent in enzyme assay with a K_i* of 7 nM and in genotype 1b replicon assay with an EC_{50} of 0.35 μM. Telaprevir exhibited a favourable pharmacokinetic profile in rats and dogs. At an oral dose of 40 mg/kg in rats, telaprevir achieved an AUC_{0-inf} of 3.3 μg h/ml and a bioavailability of 25%. In dogs at 9.6 mg/kg, the AUC_{0-inf} was 1.6 μg h/ml and the bioavailability was 41%. Telaprevir demonstrated potent inhibition of HCV protease activity in a mouse model (Lin *et al.*, 2006). In a phase Ib clinical trial in humans, telaprevir rapidly reduced the plasma viral load of patients with HCV genotype 1 by a mean of 3 $\log_{10}$ in 2 days (Reesink *et al.*, 2006). Telaprevir was generally safe and well tolerated. In a phase II clinical trial (PROVE-1), the observed rate of SVR was 35–67% with telaprevir (750 mg t.i.d.), compared with 41% in the control arm (Muir *et al.*, 2008). In the combination therapy phase II (PROVE-2) trial, twelve weeks of treatment with telaprevir in combination with pegylated interferon and ribavirin, followed by a further 12 weeks of treatment with pegylated interferon and ribavirin resulted in an SVR rate of 69% (Forns *et al.*, 2008). Although telaprevir demonstrated the potential to shorten the treatment period for standard HCV therapy from 12 to 6 months, the side-effects of this compound remain a concern (McHutchison *et al.*, 2009; Hézode *et al.*, 2009). Telaprevir is currently undergoing phase III clinical trials, and the results are expected in 2010 (clinicaltrials.gov identifier: NCT00627926). Telaprevir could be launched as

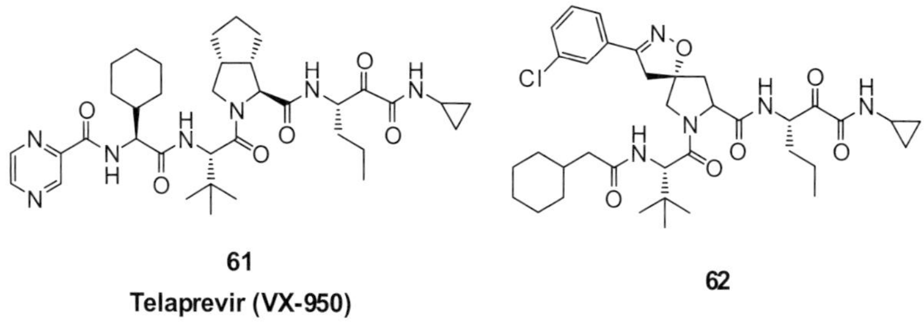

61

Telaprevir (VX-950)

K_i = 0.04 μM
K_i* = 0.007 μM
EC_{50} = 0.35 μM

62

Figure 9.12 HCV protease inhibitor clinical candidate telaprevir (VX-950) and an example of new structures.

a new drug for the treatment of HCV infection in the near future if no major adverse effects are identified during phase III trials.

Next-generation protease inhibitor from Vertex

Vertex scientists have continued the research in HCV NS3 protease inhibitors. The company recently announced that it has identified a couple of next generation inhibitors and has progressed one backup compound, VX-985, into phase I clinical trial (Vertex, 2009). The structure and profile of the second generation candidate have not been disclosed. A number of novel structures (e.g. compound *62*; Fig. 9.12) were published in a few recent patents (Cottrell *et al.*, 2007). Among the structures revealed were several different spiro-[5,5]-bicyclic P2 moieties, and the predominant ring system appeared to be the proline-spiroisoxazoline shown in *62*. The aromatic group attached to the spiroisoxazoline ring varied in size and substituents. A number of amide, carbamate and urea P3 cappings were incorporated. Hopefully, the detailed profile of VX-985 will be reported in the near future.

Conclusions

The small size of HCV NS3 protease, with short interstrand loops, presents a shallow and featureless active site with mostly open binding pockets. This has been a major challenge for the discovery of potent and orally bioavailable HCV NS3 protease inhibitor as a drug candidate for the treatment of hepatitis C virus infection. In addition, the lack of a robust HCV infection small animal model has forced scientists to reply on a combination of HCV replicon cellular assay measurement and pharmacokinetic profiling in animals as surrogate indicators of efficacy of the drug candidates before they can be evaluated in clinical trials. As a result of these challenges, no HCV protease inhibitors have been approved as drugs for the treatment of HCV infection to date. However, through the tremendous efforts from the pharmaceutical industry, impressive progress has been made which resulted in a number of experimental drugs in clinical development.

Without any viable small molecule leads, early research focused on substrate-based large peptide inhibitors. Guided by X-ray crystal structures of the inhibitor bound to the enzyme, a structure-based drug design approach was pursued by many research groups. Stepwise truncations and systematic depeptidizations on both prime and non-prime sites gave rise to smaller hexa- or pentapeptides that were potent inhibitors in enzyme assay, but did not possess any desirable cellular assay potency or pharmacokinetic properties due to their size and charged moieties. Further truncation and modifications on the prime side led to the discovery of two distinct classes of mechanism-based NS3 protease inhibitors: non-covalent competitive inhibitors, and covalently bonded reversible inhibitors, also known as serine-trap inhibitors. Both classes have yielded several advanced clinical candidates in phase II or above trials. The proof of concept of efficacy by a NS3 protease inhibitor was achieved by ciluprevir (BILN-2061), a traditional non-covalent inhibitor. However, the two most advanced NS3 protease inhibitors, telaprevir (VX-950) and boceprevir (SCH 503034), which are in phase III clinical trials, and both covalently bonded reversible inhibitors.

Several types of so-called electrophilic warheads were investigated in the development of NS3 protease inhibitors. They range from simple aldehydes, ketones, diketones, α-ketoacids/ketoesters, α-ketoamides to more elaborated boronic acid/boronates, phosphonates/phosphoric acids, α,β-unsaturated Michael acceptors and β-lactams. One class of warhead, namely, α-ketoamide has emerged as the best choice due to its potency, selectivity and stability. Extensive research into the truncation and SAR optimization from large peptidic early leads led to the identification of two late stage clinical candidates: boceprevir and telaprevir. They had good *in vitro* potency and cell-based replicon assay potency. They demonstrated moderate to good oral bioavailabilities in rats and dogs and were found to be highly concentrated in the target organ liver. When administered in combination with peginterferon and ribavirin in phase II clinical trials, they were proved to be highly efficacious with an SVR rate of about 65–70%. They are currently in phase III clinical trials as a third agent in triple combination therapy with peg-IFN- and ribavirin. Second generation α-ketoamide clinical candidates were also developed by Merck and possibly by Vertex. Narlaprevir, which was about 10-fold more potent in genotype 1 replicon assay than boceprevir, is currently in phase II clinical trials. With their proven clinical efficacy and most advanced status, the covalent inhibitors of HCV NS3 protease have been, and will always be a very important class of compounds for the development of future treatment of HCV infection.

Future perspectives

Despite daunting challenges presented by HCV NS3 protease, significant progress has been made in recent years to identify potent, selective and orally bioavailable small molecule inhibitors. Clinical proof-of-concept has been recently achieved by a number of candidates. With the expectation that the most advanced candidates gaining approval in the next couple of years, the future prospect of winning the war against HCV infection has never been brighter.

However, most of the studies with novel NS3 protease inhibitors conducted so far have focused on the response in patients infected with genotype 1. Studies of these agents in patients with other genotypes and in non-responder populations with refractory disease are not as intensive. Some early indications have highlighted that antiviral activity may be less pronounced in patients infected with genotypes 2 or 3 compared with those infected with genotype 1. In addition, since the active site of the NS3 protease for inhibitor binding is a long shallow groove, a single-point mutation in this enzyme might be sufficient to hinder the binding of an inhibitor. Several mutations have been identified in the patients treated with telaprevir and boceprevir. Therefore, combination therapy with peg-interferon-α and ribavirin is required to suppress the development of resistance.

The lessons learned from the development of treatment regimens for HIV will be valuable in future approaches towards the development of effective therapy for the treatment of chronic HCV infections. Analogous to HIV therapy, the success of future anti-HCV agents will be greatly affected by their resistance profiles. Combination therapies using multiple small molecules designed to attack different targets along the viral life cycle and perhaps, the hosts themselves, will almost certainly be required to eliminate the infection and prevent the emergence of drug-resistant viral mutants. For example, a protease inhibitor can be combined with a non-nucleoside polymerase inhibitor and a nucleoside polymerase inhibitor. Development of such new combination strategies other than current protease inhibitor/peg-interferon-α/ribavirin therapy will potentially provide higher SVR rate and shorten treatment duration while minimizing the potential for developing resistance to any single agent. The ultimately goal is to have an all oral combination therapy of two or more antiviral agents that is highly effective towards multiple genotypes, has shorter treatment time and less side-effects.

References

Alter, M.J., Kruszon-Moran, D., Nainan, O.V., McQuillan, G.M., Gao, F., Moyer, L.A., Kaslow, A., and Margolis, H.S. (1999). The prevalence of hepatitis C virus infection in the United States, 1988 through 1994. N. Engl. J. Med. *341*, 556–562.

Alter, H.J., and Seeff, L.B. (2000). Recovery, persistence, and sequelae in hepatitis C virus infection: A perspective on long-term outcome. Semin. Liver Dis. *20*, 17–35.

Arasappan, A., Padilla, A.I., Jao, E., Bennett, F., Bogen, S.L., Chen, K.X., Pike, R.E., Sannigrahi, M., Soares, J., Venkatraman, S., Vibulbhan, B., *et al.* (2009). Toward second generation hepatitis C virus NS3 serine protease inhibitors: Discovery of novel P4 modified analogues with improved potency and pharmacokinetic profile. J. Med. Chem. 52, 2806–2817.

Arasappan, A., Bennett, F., Bogen, S.L., Venkatraman, S., Blackman, M., Chen, K.X., Hendrata, S., Huang, Y., Huelgas, R.M., Nair, L., *et al.* (2010). Discovery of Narlaprevir (SCH 900518): a potent, second generation HCV NS3 Serine Protease Inhibitor. ACS Med. Chem. Lett. *1*, 64–69.

Behrens, S.E., Tomei, L., and De Francesco, R. (1996). Identification and properties of the RNA-dependent RNA polymerase of hepatitis C virus. EMBO J. *15*, 12–22.

Bennett, F., Raymond, L.G., Huang, Y., Hendrata, S., Bogen, S.L., Chen, K.X., Venkatraman, S., Sannigrahi, M., Velazquez, F., Arasappan, A., *et al.* (2010). The introduction of P4 substituted 1-methylcyclohexyl groups into Boceprevir: A change in direction in the search for a second generation HCV NS3 protease inhibitor. Bioorg. Med. Chem. Lett. *20*, 2617–2621.

Bennett J.M., Campbell, A.D., Campbell, A.J., Carr, M.G., Dunsdon, R.M., Greening, J.R., Hurst, D.N., Jennings, N.S., Jones, P.S., Jordan, S., *et al.* (2001). The identification of α-ketoamides as potent inhibitors of hepatitis c virus NS3–4A proteinase. Bioorg. Med. Chem. Lett. *11*, 355–357.

Bogen, S.L., Arasappan, A., Bennett, F., Chen, K., Jao, E., Liu, Y.-T., Lovey, R.G., Venkatraman, S., Pan, W., Parekh, T., *et al.* (2006a). Discovery of SCH 446211 (SCH6): A new ketoamide inhibitor of the HCV NS3 serine protease and HCV subgenomic RNA replication NS3 serine. J. Med. Chem. *49*, 2750–2757.

Bogen, S.L., Ruan, R., Saksena, A.K., Njoroge, F.G., Girijavallabhan, V., Agrawal, S., Liu, R., Pichardo, J., Baroudy, B., and Prongay, A. (2006b). Depeptidization efforts on P3-P2′ α-ketoamide inhibitors of HCV NS3–4A serine protease: Effect on HCV replicon activity. Bioorg. Med. Chem. Lett. *16*, 1621–1627.

Bogen, S.L., Pan, W., Ruan, S., Nair, L.G., Arasappan, A., Bennett, F., Chen, K.X., Jao, E., Venkatraman, S., Vibulbhan, B., *et al.* (2009). Toward the back-up of Boceprevir (SCH 503034): discovery of new extended P4-capped ketoamide inhibitors of hepatitis C virus NS3 serine protease with improved potency and pharmacokinetic profiles. J. Med. Chem. *52*, 3679–3688.

Brown, R.S., Jr., and Gaglio, P.J. (2003). Scope of worldwide hepatitis C problem. Liver Transpl. *9*, S10–S13.

Campbell, D.A., Hepperle, M.E., Winn, T., and Betancort, J.M. (2007). Preparation of peptidomimetics, particularly peptidyl boronic acid derivatives, as hepatitis C serine protease inhibitors for treating HCV infections. WO-2007-089618, p. 60, example 128.

Campbell, D.A., Hepperle, M.E., Winn, T., and Betancort, J.M. (2008). Preparation of macrocyclic hepatitis C serine protease inhibitors and uses therefore. WO-2008-070733, p. 68, compound 46.

Chaudhary, K., Fleury, M., Kim, C.U., Mcmurtrie, J., and Sheng, X.C. (2006). Preparation of tripeptides bearing a cyclopropyl moiety and phosphorous groups as antiviral compounds. WO-2006-020276, pp. 316 and 350, examples 83 and 104.

Chen, K.X., Njoroge, F.G., Vibulbhan, B., Prongay, A., Pichardo, J., Madison, V., Buevich, A., Chan, T.-M. (2005). Proline-based macrocyclic inhibitors of the hepatitis C virus: Stereoselective synthesis and biological activity. Angew. Chem. Int. Ed. *44*, 7024–7028.

Chen, K.X., Njoroge, F.G., Arasappan, A., Venkatraman, S., Vibulbhan, B., Yang, W., Parekh, T.N., Pichardo, J., Prongay, A., Cheng, K.-C., *et al.* (2006). Novel potent hepatitis C virus NS3 serine protease inhibitors derived from proline-based macrocycles. J. Med. Chem. *49*, 995–1005.

Chen, K.X., and Njoroge, F.G. (2009a). A review of HCV protease inhibitors. Curr. Opin. Invest. Drugs. 10, 821– 837.

Chen, K.X., Vibulbhan, B., Yang, W., Nair, L.G., Tong, X., Cheng, K.-C., and Njoroge, F.G. (2009b). Novel potent inhibitors of hepatitis C virus (HCV) NS3 protease with cyclic sulfonyl P3 cappings. Bioorg. Med. Chem. Lett. *19*, 1105–1109.

Chen, K.X., Nair, L., Vibulbhan, B., Yang, W., Arasappan, A., Bogen, S.L., Venkatraman, S., Bennett, F., Pan, W., Blackman, M.L., *et al.* (2009c). Second-generation highly potent and selective inhibitors of the hepatitis C virus NS3 serine protease. J. Med. Chem. *52*, 1370–1379.

Choo, Q.L., Kuo, G., Weiner, A.J., Overby, L.R., Bradley, D.W., and Houghton, M. (1989). Isolation of a cDNA clone derived from a blood-borne non-A, non-B viral hepatitis genome. Science (Washington DC) *244*, 359–362.

Cihlar, T., He, G.-X., Liu, X., Chen, J.M., Hatada, M., Swaminathan, S., McDermott, M.J., Yang, Z.-Y., Mulato, A.S., Chen, X., *et al.* (2006). Suppression of HIV-1 protease inhibitor resistance by phosphonate-mediated solvent anchoring. J. Mol. Biol. *363*, 635–647.

Cohen, J. (1999). The scientific challenge of hepatitis C. Science (New York, NY) *285*, 26–30.

Cottrell, K.M., Maxwell, J., Tang, Q., Grillot, A.-L., Le Tiran, A., and Perola, E. (2007). Preparation of spiroisoxazoline-based peptidomimetics as inhibitors of serine proteases, particularly HCV NS3-NS4A protease. WO-2007025307.

Cuthbert, J.A. (1994). Hepatitis C: progress and problems. Clin. Microbiol. Reviews 7, 505–532.

Deadman, J.J., Elgendy, S., Goodwin, C.A., Green, D., Baban, J.A., Patel, G., Skordalakes, E., Chino, N., and Claeson, G. (1995). Characterization of a class of peptide boronates with neutral P1 side chains as highly selective inhibitors of thrombin. J. Med. Chem. *38*, 1511–1522.

De Francesco, R., and Migliaccio, G. (2005). Challenges and successes in developing new therapies for hepatitis C. Nature (London) *436*, 953–960.

Ede, N.J., Eagle, S.N., Wickham, G., Bray, A.M., Warne, B., Shoemaker, K., and Rosenberg, S. (2000). Solid phase synthesis of peptide aldehyde protease inhibitors. Probing the proteolytic sites of hepatitis C virus polyprotein. J. Peptide Sci. *6*, 11–18.

Feld, J.J., and Hoofnagle, J.H. (2005). Mechanism of action of interferon and ribavirin in treatment of hepatitis C. Nature (London) *436*, 967–972.

Forns, X., Marcellin, P., Goeser, T., Ferenci, P., Nevens, F., Carosi, G., Drenth, J.P., De Backer, K., van Heeswijk, R., Vangeneugden, T.J., Picchio, G., and Beumont-Mauviel, M. (2008). Telaprevir in combination with peginterferon-α–2a with or without ribavirin in the treatment of chronic hepatitis C: final results of the PROVE 2 study. 59th Annual Meeting of the American Association for the Study of Liver Disease (AASLD; San Francisco, CA).

Fried, M.W., Shiffman, M.L., Reddy, D.K.R., Smith, C., Marinos, G., Gonçales, F.L., Häussinger, D., Diago, M., Carosi, G., Dhumeaux, D., Craxi, A., Lin, A.; Hoffman, J., and Yu, J. (2002). PEG interferon-α-2a plus ribavirin for chronic hepatitis C virus infection. New Engl. J. Med. *347*, 975 – 982.

Furman, P.A., Lam, A.M., and Murakami, E. (2009). Nucleoside analog inhibitors of hepatitis C viral replication: recent advances, challenges and trends. Fut. Med. Chem. *1*, 1429–1452.

Gao, M., Nettles, R.E., Belema, M., Snyder, L.B., Nguyen, V.N., Fridell, R.A., Serrano-Wu, M.H., Langley, D.R., Sun, J.-H., *et al.* (2010). Chemical genetics strategy identifies an HCV NS5A inhibitor with a potent clinical effect. Nature *465*, 96–100.

Glunz, P.W., Douty, D., and Decicco, C.P. (2003). Design and synthesis of bicyclic pyrimidinone-based HCV NS3 protease inhibitors. Bioorg. Med. Chem. Lett. *13*, 785–788.

Groll, M., Berkers, C.R., Ploegh, H.L., and Ovaa, H. (2006). Crystal structure of the boronic acid-based proteasome inhibitor Bortezomib in complex with the yeast 20S proteasome. Structure *14*, 451–456.

Han, W., Hu, Z., Jiang, X., and Decicco, C.P. (2000). α-Ketoamides, α-ketoesters and α-diketones as HCV NS3 protease inhibitors. Bioorg. Med. Chem. Lett. *10*, 711–713.

Hézode, C., Forestier, N., Dusheiko, G., Ferenci, P., Pol, S., Goeser, T., Bronowicki, J.P., Bourlière, M., Gharakhanian, S., Bengtsson, L., and McNair, L. (2009). Telaprevir and PEGinterferon with or without ribavirin for chronic HCV infection. N. Eng. J. Med. *360*, 1839–1850.

Hoofnagle, J.H. (2002). Course and outcome of hepatitis C. Hepatology *36*, S21–S29.

Houghton, M., and Abrignani, S. (2005). Prospects for a vaccine against the hepatitis C virus. Nature (London) *436*, 961–966.

Ingallinella, P., Altamura, S., Bianchi, E., Taliani, M., Ingenito, R., Cortese, R., De Francesco, R., Steinkühler, C., and Pessi, A. (1998). Potent peptide inhibitors of human hepatitis C virus NS3 protease are obtained by optimizing the cleavage products. Biochemistry *37*, 8906–8914.

Kolykhalov, A.A., Mihalik, K., Feinstone, S.M., and Rice, C.M. (2000). Hepatitis C virus-encoded enzymatic activities and conserved RNA elements in the 3′ non-translated region are essential for virus replication *in vivo*. J. Virol. *74*, 2046–2051.

Kwo, P. (2009). HCV SPRINT-1 final results: SVR 24 from a phase 2 study of boceprevir plus PEGIntron/ribavirin in treatment-naïve subjects with genotype-1 chronic hepatitis C. 44th European Association for the Study of the Liver (EASL) Annual Meeting (Copenhagen, Denmark).

Lamar, J., Victor, F., Snyder, N., Johnson, R.B., Wang, Q.M., Glass, J.I., Chen, S.-H. (2004). Novel P4 truncated tripeptidyl α-ketoamides as HCV protease inhibitors. Bioorg. Med. Chem. Lett. *14*, 263–266.

Lamarre, D., Anderson, P.C., Bailey, M., Beaulieu, P., Bolger, G., Bonneau, P., Bös, M., Cameron, D.R., Cartier, M., Cordingley, M.G., *et al.* (2003). An NS3 protease inhibitor with antiviral effects in humans infected with hepatitis C virus. Nature (London) *426*, 186–189.

Lesburg, C.A., Cable, M.B., Ferrari, E., Hong, Z., Mannarino, A.F., and Weber, P.C. (1999). Crystal structure of the RNA-dependent RNA polymerase from hepatitis C virus reveals a fully encircled active site. Nature Struct. Biol. *6*, 937 – 943.

Lin, C., Kwong, D., and Perni, R.B. (2006). Discovery and development of VX-*950*, a novel, covalent, and reversible inhibitor of hepatitis C virus NS34A serine protease. Infect. Dis. Drug Targets *6*, 3–16.

Lin, T.I., Lenz, O., Fanning, G., Verbinnen, T., Delouvroy, F., Scholliers, A., Vermeiren, K., Rosenquist, A., Edlund, M., Samuelsson, B., and Vrang, L. (2009). *In vitro* activity and preclinical profile of TMC435*350*, a potent hepatitis C virus protease inhibitor. Antimicrob. Agents Chemother. *53*, 1377–1385.

Lindenbach, B.D., and Rice, C.M. (2005). Unravelling hepatitis C virus replication from genome to function. Nature (London) *436*, 933–938.

Lipinski, C.A. (2000). Drug-like properties and the causes of poor solubility and poor permeability. J. Pharm. Tox. Meth. *44*, 235–249.

Liu, Y., He, Y. and Molla, A. (2009). Hepatitis C virus polymerase as a target for antiviral drug intervention: non-nucleoside inhibitors. In Antiviral Research: Strategies in Antiviral Drug Discovery, La Femina, RL ed. (ASM Press, Washington, DC), pp. 137–151.

Liverton, N.J., Summa, V., and Vacca, J.P. (2008). Preparation of macrocyclic peptides as HCV NS3 protease inhibitors. WO-2008-051475, p. 40, product from step 11.

Liverton, N.J., Carroll, S.S., Di Muzio, J., . Fandozzi, C., Graham, D.J., Hazuda, D., Holloway, M.K., Ludmerer, S.W., McCauley, J.A., McIntyre, C.J., *et al.* (2010). MK-7*009*, a potent and selective inhibitor of hepatitis C virus NS3/4A protease. Antimicrob. Agents Chemother. *54*, 305–311.

Llinas-Brunet, M., Bailey, M.D., Fazal, G., Goulet, S., Halmos, T., LePlante, S., Maurice, R., Poirier, M., Poupart, M.-A., Thibeault, D., *et al.* (1998). Peptide-based inhibitors of the hepatitis C virus serine protease. Bioorg. Med. Chem. Lett. *8*, 1713–1718.

Llinas-Brunet, M., Bailey, M.D., Bolger, G., Brochu, C., Faucher, A.-M., Ferland, J.M., Garneau, M., Ghiro, E., Gorys, V., Grand-Matre, C. *et al.* (2004). Structure–activity study on a novel series of macrocyclic inhibitors of the hepatitis C virus NS3 protease leading to the discovery of BILN 2061. J. Med. Chem. *47*, 1605–1608.

Lohmann, V., Korner, F., Koch, J.-O., Herian, U., Theilmann, L., and Bartenschlager, R. (1999). Replication of subgenomic hepatitis C virus RNAs in a hepatoma cell line. Science (Washington, DC) *285*, 110–113.

Love, R.A., Parge, H.E., Wichersham, J.A., Hostomsky, Z., Habuka, N., Moomaw, E.W., Adachi, T., and Hostomska, Z. (1996). The crystal structure of hepatitis C virus NS3 proteinase reveals a trypsin-like fold and a structural zinc binding site. Cell *87*, 331–342.

McCauley, J.A., McIntyre, C.J., Rudd, M.T., Nguyen, K.T., Romano, J.J., Butcher, J.W., Gilbert, K.F., Bush, K.J., Holloway, M.K., Swestock, J., *et al.* (2010). Discovery of Vaniprevir (MK-7009), a macrocyclic hepatitis C virus NS3/4a protease inhibitor. J. Med. Chem. *53*, 2443–2463.

McHutchison, J.G., Everson, G.T., Gordon, S.C., Jacobson, I.M., Sulkowski, M., Kauffman, R., McNair, L., Alam, J., and Muir, A.J. (2009). Telaprevir with PEGinterferon and ribavirin for chronic HCV genotype 1 infection. N. Eng. J. Med. *360*, 1827–1838.

Malcolm, B.A., Liu, R., Lahser, F., Agrawal, S., Belanger, B., Butkiewicz, N., Chase, R., Gheyas, F., Hart, A., Hesk, D.,*et al.* (2006). SCH 503034, a mechanism-based inhibitor of hepatitis C virus NS3 protease, suppresses polyprotein maturation and enhances the antiviral activity of alpha interferon in replicon cells. Antimicrob. Agents Chemother. *50*, 1013–1020.

Manns, M.P., Foster, G.R., Rockstroh, J.K., Zeuzem, S., Zoulim, F., and Houghton, M. (2007). The way forward in HCV treatment – finding the right path. Nat. Rev. Drug Discov. *6*, 991–1000.

Mederacke, I., Wedemeyer, H., and Manns, M.P. (2009). Boceprevir, an NS3 serine protease inhibitor of hepatitis C virus, for the treatment of HCV infection. Curr. Opin. Investig. Drugs *10*, 181–189.

Mercer, D.F., Schiller, D.E., Elliott, J.F., Douglas, D.N., Hao, C., Rinfret, A., Addison, W.R., Fischer, K.P., Churchill, T.A., Lakey, J.R.T. *et al.* (2001). Hepatitis C virus replication in mice with chimeric human livers. Nature Med. *7*, 927–933.

Morrison, J.F., and Walsh, C.T. (1988). The behavior and significance of slow-binding enzyme inhibitors. In Advances in Enzymolology, Meister, A., ed. (John Wiley and Sons, Chichester) vol. 61, pp. 201–301. For a definition of K_i* and discussions.

Muir, A.J., Lawitz, E.J., McHutchison, J.G., Gordon, S.C., Jacobson, I.M., Adiwijaya, B., Bengtsson, L., McNair, L., and Rodriguez-Torres, M. (2008). Viral responses in African-Americans, Latinos and Caucasians in the US phase 2 study (PROVE 1) of telaprevir with Peginterferon-α–2a and ribavirin in treatment-naïve genotype 1-infected subjects with hepatitis C. 59th Annual Meeting of the American Association for the Study of Liver Disease (AASLD; San Francisco, CA).

Narjes, F., Koehler, K.F., Koch, U., Gerlach, B., Colarusso, S., Steinkuhler, C., Brunetti, M., Altamura, S., De Francesco, R., and Matassa, V.G. (2002). A designed P1 cysteine mimetic for covalent and non-covalent inhibitors of HCV NS3 protease. Bioorg. Med. Chem. Lett. *12*, 701–704.

Nettles, R.E., Chien, C., Chung, E., Persson, A., Gao, M., Belema, M., Meanwell, N., DeMicco, M., Marbury, T., Goldwater, R., and Northup, P.G. (2008). BMS-790052 is a First-in-class Potent Hepatitis C Virus (HCV) NS5A Inhibitor for Patients with Chronic HCV Infection: results from a Proof-of-concept Study. 59th Annual Meeting of the American Association for the Study of Liver Diseases (AASLD; San Francisco, CA).

Neumann, A.U., Lam, N.P., Dahari, H., Gretch, D.R., Wiley, T.E., Layden, T.J., and Perelson, A.S. (1998). Hepatitis C viral dynamics *in vivo* and the antiviral efficacy of interferon-α therapy. Science (Washington, DC) *282*, 103–107.

Nishiyama, Y., Taguchi, H., Luo, J.-Q., Zhou, Y.-X., Burr, G., Karle and S., and Paul, S. (2002). Covalent reactivity of phosphonate monophenyl esters with serine proteinases: an overlooked feature of presumed transition state analogs. Archiv. Biochem. Biophys. *402*, 281–288.

Nishiyama, Y., Bhatia, G., Bangale, Y., Planque, S., Mitsuda, Y., Taguchi, H., Karle, S., and Paul, S. (2004). Toward selective covalent inactivation of pathogenic antibodies: a phosphonate diester analog of vasoactive intestinal peptide that inactivates catalytic autoantibodies. J. Biol. Chem. *279*, 7877–7883.

Niu, D., Petter, R., Qiao, L., and Singh, J. (2010). Preparation of peptide derivatives as HCV protease inhibitors. US 2010041674.

Nizi, E., Koch, U., Ponzi, S., Matassa, V.G., and Gardelli, C. (2002). Capped dipeptide α-ketoacid inhibitors of the HCV NS3 protease. Bioorg. Med. Chem. Lett. *12*, 3325–3328.

Nizi, E., Koch, U., Ontoria, J.M., Marchetti, A., Narjes, F., Malancona, S., Matassa, V.G., and Gardelli, C. (2004). Capped dipeptide phenethylamide inhibitors of the HCV NS3 protease. Bioorg. Med. Chem. Lett. *14*, 2151–2154.

Njoroge, F.G., Chen, K.X., Shih, N.-Y., and Piwinski, J.J. (2008). Challenges in modern drug discovery: a case study of boceprevir, an HCV protease inhibitor for the treatment of hepatitis C virus infection. Acc. Chem. Res. *41*, 50–59.

Pause, A., Kukolj, G., Bailey, M., Brault, M., Dô, F., Halmos, T., Lagacé, L., Maurice, R., Marquis, M., McKercher, *et al.* (2003). An NS3 serine protease inhibitor abrogates replication of subgenomic hepatitis C virus RNA. J. Biol. Chem. *278, 20374–20380.*

Perni, R.B., Britt, S.D., Court, J.C., Courtney, L.F., Deininger, D.D., Farmer, L.J., Gates, C.A., Harbeson, S.L., Kim, J.L., Landro, J.A. *et al.* (2003). Inhibitors of hepatitis C virus NS3·4A protease 1. Non-charged tetrapeptide variants. Bioorg. Med. Chem. Lett. *13*, 4059–4063.

Perni, R.B., Pitlik, J., Britt, S.D., Court, J.J., Courtney, L.F., Deininger, D.D., Farmer, L.J., Gates, C.A., Harbeson, S.L., Levin, R.B., *et al.* (2004a). Inhibitors of hepatitis C virus NS3–4A protease 2. Warhead SAR and optimization. Bioorg. Med. Chem. Lett. *14*, 1441–1446.

Perni, R.B., Farmer, L.J., Cottrell, K.M., Court, J.J., Courtney, L.F., Deininger, D.D., Gates, C.A., Harbeson, Scott, L.,Kim, J.L. *et al.* (2004b). Inhibitors of hepatitis C virus NS3·4A protease. Part 3: P 2 proline variants. Bioorg. Med. Chem. Lett. *14*, 1939–1942.

Perni, R.B., Almquist, S.J., Byrn, R.A., Chandorkar, G., Chaturvedi, P.R., Courtney, L.F., Decker, C.J., Dinehart, K., Gates, C.A., Harbeson, S.L., *et al.* (2006). Preclinical profile of VX-950, a potent, selective, and orally bioavailable inhibitor of hepatitis C virus NS3–4A serine protease. Antimicrob. Agents Chemother. *50*, 899–909.

Petter, R., Singh, J., Kluge, A.F., Niu, D., Qiao, L., and Ghosh, S. (2009). Preparation of peptides as HCV protease inhibitors. WO-2009-082697, p. 78 example I-46.

Powers, J.C., Asgian, J.L., Ekici, D., and James, K.E. (2002). Irreversible inhibitors of serine, cysteine, and threonine proteases. Chem. Rev. *102*, 4639–4750.

Priestley, E.S., and Decicco, C.P. (2000). 1-Aminocyclopropaneboronic acid: synthesis and incorporation into an inhibitor of hepatitis C virus NS3 protease. Org. Lett. *2*, 3095–3097.

Priestley, E.S., De Lucca, I., Ghavimi, B., Erickson-Viitanen, S., and Decicco, C.P. (2002). P1 Phenethyl peptide boronic acid inhibitors of Hepatitis C virus NS3 protease. Bioorg. Med. Chem. Lett. *12*, 3199–3202.

Prongay, A.J., Guo, Z., Yao, N., Pichardo, J., Fischmann, T., Strickland, C., Myers, J., Jr., Weber, P.C., Beyer, B.M., Ingram, R., *et al.* (2007). Discovery of the HCV NS3/4A protease inhibitor (1R, 5S)-N-[3-Amino-1-(cyclobutylmethyl)–2, 3-dioxopropyl]–3- [2(S)-[[[(1, 1-dimethylethyl)amino] carbonyl]amino]–3, 3-dimethyl-1-oxobutyl]- 6, 6-dimethyl-3-azabicyclo[3.1.0]hexan-2(S)-carboxamide (Sch 503034). Key steps in structure-based optimization. *J. Med. Chem. 50*, 2310–2318.

Raboisson, P., de Kock, H., Rosenquist, A., Nilsson, M., Salvador-Oden, L., Lin, T.I., Roue, N., Ivanov, V., Wähling, H., Wickström, K., and Hamelink, E. (2008). Structure-activity relationship study on a novel series of cyclopentane-containing macrocyclic inhibitors of the hepatitis C virus NS3/4A protease leading to the discovery of TMC435350. Bioorg. Med. Chem. Lett. *18*, 4853–4858.

Rebello, S.S., Miller, B.V., Basler, G.C., and Lucchesi, B.R. (1997). CVS-1*123*, a direct thrombin inhibitor, prevents occlusive arterial and venous thrombosis in a canine model of vascular injury. J. Cardiovasc. Pharmacol. *29*, 240–249.

Reesink, H.W. (2009). Safety and antiviral activity of SCH 900518 administered as monotherapy and in combination with peginterferon-α-2b to naïve and treatment-experienced HCV-1 infected patients. 44th European Association for the Study of the Liver (EASL) Annual Meeting (Copenhagen, Denmark). Abstract No. 86.

Reesink, H.W., Zeuzem, S., Weegink, C.J., Forestier, N., van Vliet, A., van de Wetering de Rooij, J., McNair, L., Purdy, S., Kauffman, R., Alam, J., and Jansen, P.L. (2006). Rapid decline of viral RNA in hepatitis C patients treated with VX-950: a phase Ib, placebo-controlled, randomized study. Gastroenterology *131*, 997–1002.

Saha, A.K., Young, C., Del Vecchio, A.M., Bailey, T.A., Reinhardt, J.A., Kulkarni, B.A., Faitg, T.H., Feng, H., Rippin, S.R., Blackledge, C.W., *et al.* (2007). Discovery of HCV–796: A potent and orally bioavailable hepatitis-C polymerase inhibitor under clinical development. Abstracts of Papers, 233rd ACS National Meeting, 25–29 March 2007 (Chicago, IL). MEDI-238.

Sanderson, P.E.J., and Naylor-Olsen, A.M. (1998). Thrombin inhibitor design. Curr. Med. Chem. *5*, 289–304.

Seiwert, S.D., Andrews, S.W., Jiang, Y., Serebryany, V., Tan, H., Kossen, K., Rajagopalan, P.T., Misialek, S., Stevens, S.K., Stoycheva, A., *et al.* (2008). Preclinical characteristics of the hepatitis C virus NS3/4A protease inhibitor ITMN-191 (R7227). Antimicrob. Agents Chemother. *52*, 4432–4441.

Shepard, C.W., Finelli, L., and Alter, M.J. (2005). Global epidemiology of hepatitis C virus infection. Lancet Infect. Dis. *5*, 558–567.

Steinkühler, C., Biasiol, G., Brunetti, M., Urbani, A., Koch, U., Cortese, R., Pessi, A., and De Francesco, R. (1998). Product inhibition of the hepatitis C virus NS3 protease. Biochemistry *37*, 8899–8905.

Sun, D.X., Liu, L., Heinz, B., Kolykhalov, A., Lamar, J., Johnson, R.B., Wang, Q.M., Yip, Y., and Chen, S.-H. (2004). P4 cap modified tetrapeptidyl α-ketoamides as potent HCV NS3 protease inhibitors. Bioorg. Med. Chem. Lett. *14*, 4333–4338.

Tung, R.D., Harbeson, S.L., Deininger, D.D., Mercko, M.A., Bhisetti, G.R., and Farmer, L.J. (2001). Preparation of peptide analogs as inhibitors of serine proteases, particularly hepatitis C virus NS3 protease. US Patent 6,265,380.

Turk, B. (2006). Targeting proteases: successes, failures and future prospects. Nat. Rev. Drug Discov. *5*, 785–799.

Venkatraman, S., Bogen, S.L., Arasappan, A., Bennett, F., Chen, K., Jao, J., Liu, Y.-T., Lovey, R., Hendrata, S., Huang, Y., *et al.* (2006). Discovery of (1R, 5S)-*N*-[3-amino-1-(cyclobutylmethyl)–2, 3-dioxopropyl]–3-[2(S)-[[[(1, 1-dimethylethyl)amino]carbonyl]amino]–3, 3-dimethyl-1-oxobutyl]–6, 6-dimethyl-3-azabicyclo[3.1.0] hexan-(S)-carboxamide (SCH 503034), a selective, potent, orally bioavailable hepatitis C virus NS3 protease inhibitor: A potential therapeutic agent for the treatment of hepatitis C infection. J. Med. Chem. *49*, 6074–6086.

Venkatraman, S., Wu, W., Prongay, A., Girijavallabhan, V., and Njoroge, F.G. (2009a). Potent inhibitors of HCV-NS3 protease derived from boronic acids. Bioorg. Med. Chem. Lett. *19*, 180–183.

Venkatraman, S., Blackman, M., Wu, W., Nair, L., Arasappan, A., Padilla, A., Bogen, S., Bennett, F., Chen, K., Pichardo, J., *et al.* (2009b). Discovery of novel P3 sulfonamide-capped inhibitors of HCV NS3 protease. Inhibitors with improved cellular potencies. Bioorg. Med. Chem. *17*, 4486–4495.

Venkatraman, S., Velazquez, F., Wu, W., Blackman, M., Chen, K.X., Bogen, S., Nair, L., Tong, X., Chase, R., Hart, A., *et al.* (2009c). Discovery and structure-activity relationship of P1-P3 ketoamide derived macrocyclic inhibitors of hepatitis C virus NS3 protease. J. Med. Chem. *52*, 336–346.

Vertex Pharmaceuticals. (2009). http://www.vrtx.com/current-projects/drug-candidates/vx-985.html.

Victor, F., Lamar, J., Snyder, N., Yip, Y., Guo, D., Yumibe, N., Johnson, R.B., Wang, Q.M., Glass, J.I., and Chen, S.-H. (2004). P1 and P3 optimization of novel bicycloproline P2 bearing tetrapeptidyl α-ketoamide based HCV protease inhibitors. Bioorg. Med. Chem. Lett. *14*, 257–261.

Votta, B.J., Levy, M.A., Badger, A., Bradbeer, J., Dodds, R.A., James, I.E., Thompson, S., Bossard, M.J., Carr, T., Connor, J.R., Thomaszek, T.A., Szewczuk, L., Drake, F.H., Veber, D.F., and Gowen, M. (1997). Peptide aldehyde inhibitors of cathepsin K inhibit bone resorption both in vitro and in vivo. J. Bone. Miner. Res., *12*, 1396–1406.

Walker, C.M. (1997). Comparative features of hepatitis C virus infection in humans and chimpanzees. Springer Semin. Immunopathol. *19*, 85 – 98.

White, P.W., Llinas-Brunet, M., and Bos, M. (2006). Blunting the Swiss army knife of hepatitis C virus: inhibitors of NS3/4A protease. In King FD, Lawton G, eds, Progress in Medicinal Chemistry (Elsevier, New York), vol. *44*, chapter 2, pp. 65–107.

Yan, Y., Li, Y., Munshi, S., Sardana, V., Cole, J.L., Sardana, M., Steinkuehler, C., Tomei, L., De Francesco, R., Kuo, L.C., and Chen, Z. (1998). Complex of NS3 protease and NS4A peptide of BK strain hepatitis C virus: A 2.2 Å resolution structure in a hexagonal crystal form. Protein Sci. 7, 837–847.

Yip, Y., Victor, F., Lamar, J., Johnson, R., Wang, Q.M., Barket, D., Glass, J.I., Jin, L., Liu, L., Venable, D., Wakulchik, M., Xie, C., Heinz, B., Villarreal, E., Colacino, J., Yumibe, N., Tebbe, M., Munroe, J., and Chen, S.-H. (2004). Discovery of a novel bicycloproline P2 bearing peptidyl α-ketoamide LY514962 as HCV protease inhibitor. Bioorg. Med. Chem. Lett. *14*, 251–256.

Zhang, R., Beyer, B.M., Durkin, J., Ingram, R., Njoroge, F.G., Windsor, W.T., and Malcolm, B.A. (1999). A continuous spectrophotometric assay for the hepatitis C virus serine protease. Anal. Biochem. 270, 268–275. The substrate Ac-DTEDVVP(Nva)-O-PAP was used in the present study.

Zhang R., Durkin, J.P., and Windsor, W.T. (2002). Azapeptides as inhibitors of the Hepatitis C virus NS3 serine protease. Bioorg. Med. Chem. Lett. *12*, 1005–1008.

Zhang X., Schmitt, A.C., Jiang, W., Wasserman, Z., and Decicco, C.P. (2003). Design and synthesis of potent, non-peptide inhibitors of HCV NS3 protease. Bioorg. Med. Chem. Lett. *13*, 1157–1160.

NS3 Protease Non-covalent Inhibitors 10

Brad O. Buckman, Karl Kossen, John B. Nicholas and Scott D. Seiwert

Abstract

Progress is rapidly being made to improve the standard of care for chronic hepatitis C (HCV), which currently requires long durations of weekly pegylated interferon (peg-IFN) injections in combination with twice daily orally administered ribavirin (RBV). More effective future treatments are likely to include direct acting antiviral agents (DAAs). Several small, orally active non-covalent peptidomimetic inhibitors of NS3/4A protease have demonstrated potent reduction of HCV RNA and promising safety profiles in HCV infected patients. These agents may represent important components of future therapies. Here, the discovery, preclinical and clinical characteristics of these inhibitors are described.

Background and introduction

Chronic hepatitis C virus (HCV) infection afflicts more than 170 million people worldwide and is the leading cause of liver transplantation in the United States (2002; Shepard *et al.*, 2005). Standard of care (SOC) for this disease is weekly injection of pegylated interferon alpha and twice-daily oral administration of ribavirin. This regimen has significant side-effects and achieves the clinically relevant endpoint of sustained virological response (SVR), defined as viral negativity 6 months following cessation of therapy, in only half of treated patients (Manns *et al.*, 2001; Fried *et al.*, 2002). Poorer response rates are observed in certain subpopulations, including individuals harbouring genotype 1 virus or a high viral load, cirrhotic patients, and African Americans (Manns *et al.*, 2001; Fried *et al.*, 2002). Thus, novel therapeutic approaches that have milder side-effects and that enhance SVR rates compared with SOC are needed to better treat this prevalent and serious disease.

Direct antiviral agents (DAA) target essential HCV-encoded enzymes, including the protease activity of non-structural protein 3/4A (NS3/4A) and the RNA-directed RNA polymerase activity of NS5B (Pawlotsky, 2004; Lemon *et al.*, 2005; Neukam *et al.*, 2009). NS3 is a chymotrypsin-like serine protease that is activated by association with NS4A. Following translation of the HCV RNA genome into a polyprotein, the NS3/4A protease cleaves four sites to liberate five viral proteins essential for HCV replication. In addition, the proteolytic activity of NS3/4A recently has been shown to dampen cellular sensing of viral components, and in doing so reduce type 1 interferon production (Gale and Foy, 2005; Thimme *et al.*, 2006). Thus, inhibitors of NS3/4A may disrupt two separate processes relevant for replication of HCV.

NS3/4A protease inhibitors (PIs) have shown potent antiviral activity in early clinical trials, highlighting the significant potential of this class of compounds. In landmark studies ciluprevir (BILN-2061) was found to reduce the average plasma concentration of genotype 1 HCV by approximately 3.0 $\log_{10}$ following twice-daily dosing of 200 mg for 2 days (Lamarre *et al.*, 2003; Hinrichsen *et al.*, 2004). Despite robust and rapid antiviral activity that at the time was unprecedented, further clinical development of ciluprevir was discontinued due to severe cardiac toxicity in non-clinical species (Reiser *et al.*, 2005).

Subsequently, several other NS3/4A PIs have demonstrated robust antiviral activity as monotherapy in chronic HCV (Forestier *et al.*, 2008; Lawitz *et al.*, 2008; Manns *et al.*, 2008a,b; Reesink *et al.*, 2009a). In combination with SOC for 14–28 days these compounds drive viral loads below the limit of quantification or detection in a majority of patients (Manns *et al.*, 2008a,b, 2009b; Forestier *et al.*, 2009). Two linear/non-macrocyclic mechanism-based covalent inhibitors administered three times a day, boceprevir (SCH 503034; Njoroge *et al.*, 2008; Mederacke *et al.*, 2009) and telaprevir (VX-950; Perni *et al.*, 2006; Hezode *et al.*, 2009; McHutchison *et al.*, 2009), are considered first generation NS3/4A PIs. Both boceprevir and telaprevir substantially improve SVR when added to SOC and; therefore, demonstrate the significant clinical utility of NS3/4A PIs (Hezode *et al.*, 2009; Kwo *et al.*, 2009; McHutchison *et al.*, 2009). However, both compounds demonstrate that side-effects associated with triple combination therapy can compromise clinical utility due to treatment discontinuation (Hezode *et al.*, 2009; Kwo *et al.*, 2009; McHutchison *et al.*, 2009). Moreover, the requirement for three times daily dosing with first generation PIs may lead to patient non-compliance, which in turn may allow drug resistant viral variants to develop. Second generation PIs have attempted to address these issues in order to further improve treatment outcomes. All of the second generation inhibitors with publicly disclosed structures inhibit NS3/4A in a non-covalent fashion.

Covalent and non-covalent inhibitors

Multiple structures of the HCV NS3 protease domain in complex with the activating protein cofactor NS4A have been reported (Yan *et al.*, 1998; Yao *et al.*, 1999; Di Marco *et al.*, 2000; Ontoria *et al.*, 2004). The protein consists of two β-barrel subdomains similar to other chymotrypsin-like proteases. The NS4A cofactor contributes one beta strand to the N-terminal β-barrel and forms extensive interactions with NS3. A comparison of structures in the presence and absence of NS4A demonstrates that cofactor binding orders the active site residues which sit at the interface of the two domains.

The catalytic triad of NS3 contains conserved aspartic acid (Asp-81), histidine (His-57), and serine (Ser-139) moieties. In the classical serine protease mechanism, nucleophilic attack of the catalytic triad serine on the carbonyl moiety of the scissile peptide bond forms a tetrahedral enzyme–substrate adduct (Fig. 10.1). The nascent negative charge on this intermediate is stabilized by interactions with backbone amides in the oxyanion hole of the protease (Gly-137 and Ser-139). Collapse of this intermediate results in the expulsion of the first cleavage product. Subsequent hydrolysis of the remaining acyl-enzyme intermediate completes product release and regenerates the active enzyme.

The substrate cleft of NS3 protease lacks structural features that form a defined substrate binding surface in other serine proteases (Lin *et al.*, 2006). The resulting binding surface of the NS3 protease is flat, solvent exposed, and has frequently been described as featureless. A common nomenclature for peptide substrates describes the amino acid positions relative to the cleavage site. For example, NH_2-P6-P5-P4-P3-P2-P1-P1′-P2′-P3′-P4′-OH denotes a decapeptide with six amino acids N-terminal to the cleavage site, which is between P1 and P1′, and four amino acids C-terminal to this site. NS3 cleavage sites in the HCV polyprotein have an acidic residue at P6, a cysteine or threonine at P1, and a small amino acid at P1′ (Lin *et al.*, 2006). However, many of the interactions between NS3 and a bound peptide are sequence-independent contacts to the peptide backbone.

Both covalent and non-covalent NS3 protease inhibitors have been developed to bind in this flat and featureless active site. The covalent inhibitors are linear peptides that bind the active site and adjacent N-terminal substrate binding site in the P4 to P1′ region. The most advanced inhibitors of this type, boceprevir (SCH503034) and telaprevir (VX-950), incorporate a di-keto amide moiety in place of the scissile bond (Fig. 10.2). Upon nucleophilic attack by the active site serine, these mechanism-based inhibitors trap the protease in a reversible covalent adduct (Fig.

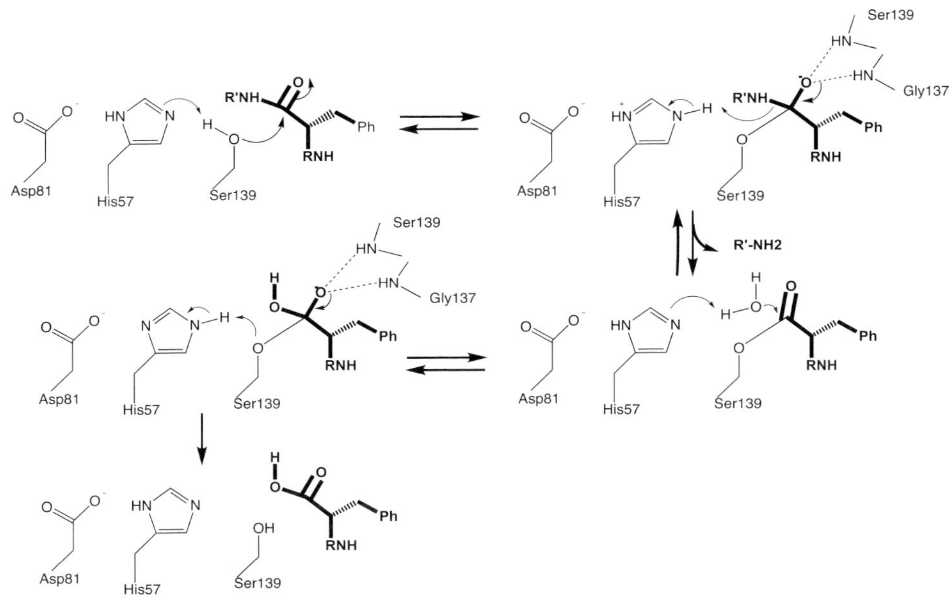

Figure 10.1 Generalized reaction mechanism of a serine protease. Substrate shown in bold.

10.3A) which increases the affinity of the protease inhibitor complex by >100-fold (Flores *et al.*, 2009). Taking covalent inhibitor boceprevir as an example, the X-ray structure of the α-ketoamide boceprevir is shown bound to the NS3–4A protease active site (Prongay *et al.*, 2007; Fig. 10.2A). As shown, the γ-oxygen of Ser139 is covalently bound to the P1 keto carbon of the ligand. The negatively charged keto oxygen hydrogen bonds to the δ-nitrogen atom of His57. The amide carbonyl of the ligand is situated in the oxyanion hole and is involved in hydrogen bonds with the backbone amide nitrogens of both Ser139 and Gly137. The amide nitrogen of the ketoamide does not appear to have any hydrogen bonding partners.

Additional classes of covalent inhibitors including those that form a reversible covalent adduct with a boron containing inhibitor (Li *et al.*, 2010), and a class that forms an irreversible covalent bond to NS3 (Hagel *et al.*, 2010; Niu *et al.*, 2010) are in earlier stages of discovery/development and are not described further.

Non-covalent NS3 protease inhibitors rely on an extensive network of hydrogen bonding interactions and polar interactions to obtain binding energy. Despite the lack of a covalent bond to the protease, the most potent non-covalent macrocyclic inhibitors display sub-nanomolar potencies in biochemical protease assays. The earliest in this class, ciluprevir (BILN-2061), was the first direct acting antiviral compound to demonstrate significant viral reduction in patients (Fig. 10.2). Ciluprevir contains a P1 carboxylic acid that mimics the products of NS3 protease cleavage of the HCV polypeptide. An oxygen of the P1 carboxylate forms a weak hydrogen bond to the δ-nitrogen atom of His57. The other carboxylate oxygen is located in the oxyanion hole, and forms a hydrogen bond with the backbone amide of Gly137 (Condroski *et al.*, 2006; Fig. 10.3B).

The most advanced non-covalent NS3 protease inhibitors currently in clinical development are danoprevir (ITMN-191/RG7227), TMC435350, vaniprevir (MK-7009), BI 201335, and BMS-650032 (Fig. 10.2). All of these non-covalent inhibitors bind the active site and adjacent N-terminal substrate binding site although they may lack a P1′ moiety (ciluprevir) or a P4 moiety (TMC435350). Danoprevir and TMC435350 are macrocyclic compounds in which P1 and P3 moieties serve as the site of cyclization. In vaniprevir, P2 and P4 groups serve as the site of cyclization. The macrocycle

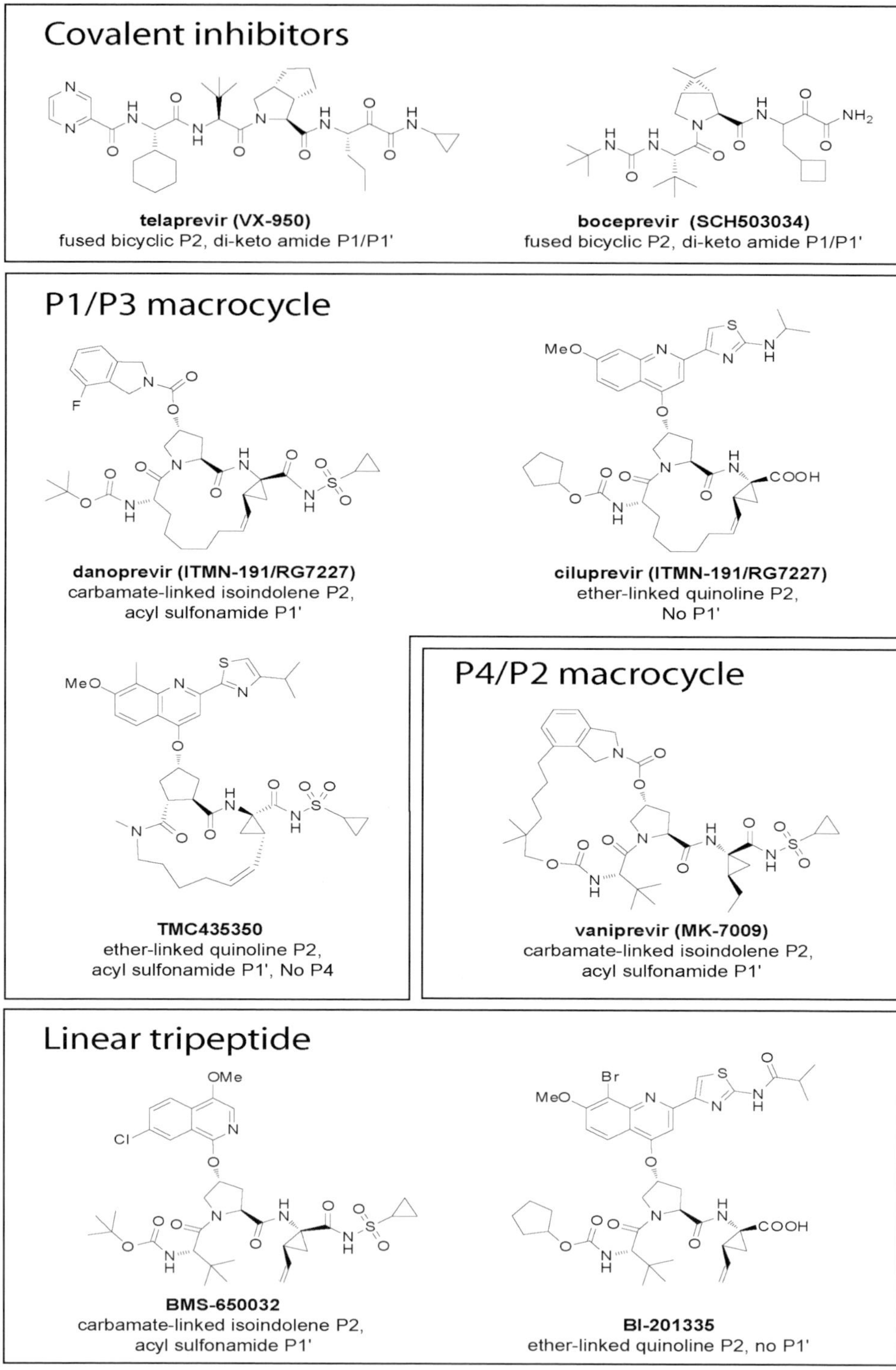

Figure 10.2 NS3/4A Protease inhibitors in clinical development with known structures.

in these compounds rigidifies the unbound conformation and is designed to reduce the entropic cost of inhibitor binding (LaPlante and Llinàs-Brunet, 2005). BI 201335 and BMS-650032

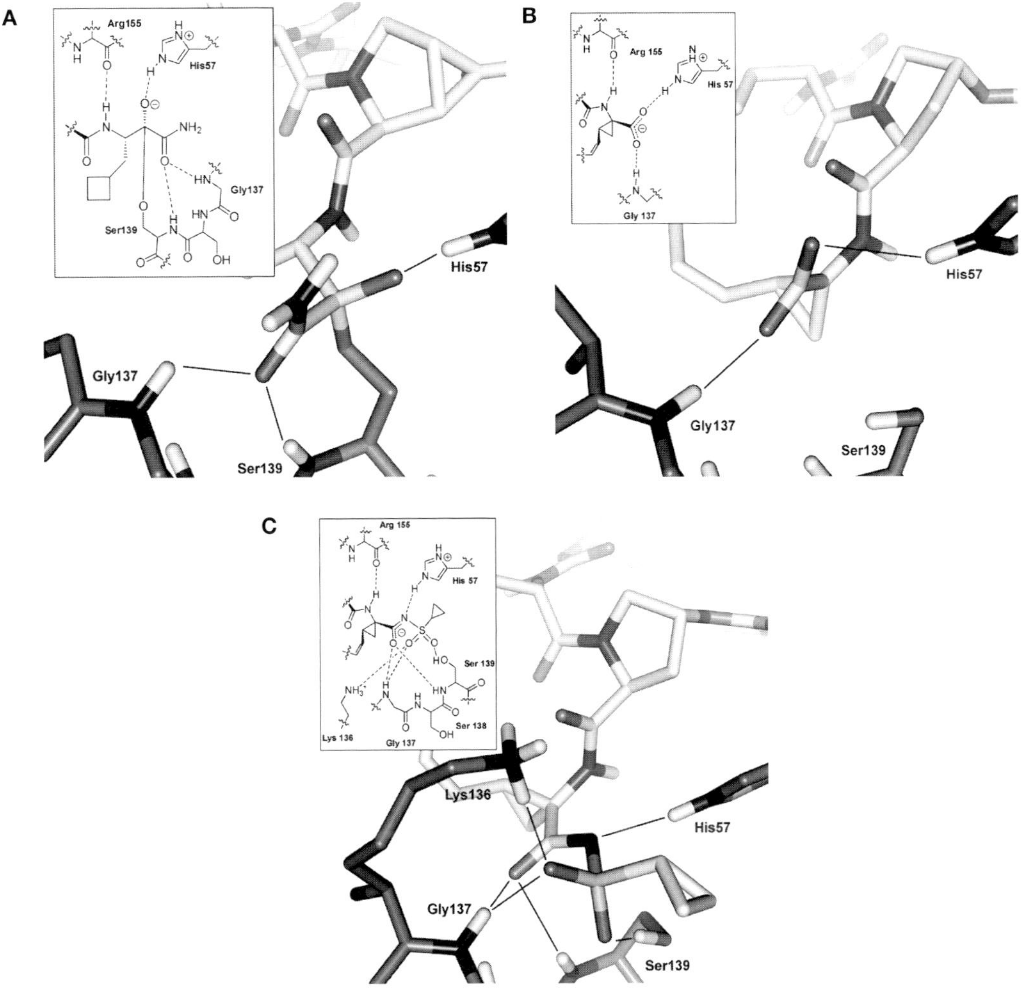

Figure 10.3 Interactions of NS3/4A with various inhibitor classes as determined by X-ray crystallography. (A) 2.66 Å resolution X-ray structure of the α-ketoamide SCH503034 (Boceprevir, EC_{50} = 200 nM, PDBID 2OC8) bound to the active site of NS3/4A (Prongay *et al.*, 2007). As shown, the γ-oxygen of Ser139 is covalently bound to the P1 keto carbon of the ligand, with a C–γO bond distance of 1.5 Å. The negatively charged keto oxygen hydrogen bonds to the δ-nitrogen atom of His57, which we assume is protonated. The δN–O hydrogen bond distance is 2.7 Å. The amide carbonyl of the ligand is situated in the oxyanion hole and is involved in hydrogen bonds with the backbone amide nitrogens of both Ser-139 (2.8 Å) and Gly137 (3.0 Å). The amide nitrogen of the ketoamide does not appear to have any hydrogen bonding partners. Only polar hydrogens are shown, and some atoms have been removed for clarity. (B) 2.95 Å resolution X-ray structure of the carboxylic acid ciluprevir (BILN-2061, EC_{50} = 2.5 nM) bound to the active site of NS3/4A (Condroski *et al.*, 2006). One of the oxygens of the P1 carboxylate forms a weak hydrogen bond to the δ-nitrogen atom of His57, shown protonated in the figure. The δN–O hydrogen bond distance is long at 3.6 Å. The other carboxylate oxygen is located in the oxyanion hole, and forms a hydrogen bond with the backbone amide of Gly-137. The O–N hydrogen bond distance is 3.2 Å. (C) 1.70 Å resolution X-ray structure of the acylsulfonamide-containing compound danoprevir (ITMN-191/RG7227, EC_{50} = 1.8 nM) bound to the active site of NS3/4A (Condroski *et al.*, 2006). A complex arrangement of hydrogen bonds is evident in the P1 P1′ region, in contrast to di-keto amide and carboxylic acid containing molecules. The P1 acyl group of the ligand is assumed to bear a full negative charge, shared between the acyl nitrogen and oxygen. The acyl nitrogen hydrogen bonds with the δ-nitrogen of His57, at a N–N distance of 3.3 Å. The acyl oxygen is involved in hydrogen bonds to the backbone NH groups of both Gly137 and Ser139. One of the P1′ sulfonamide oxygens hydrogen bonds to the γ-oxygen of Ser-139. The other sulfonamide oxygen forms hydrogen bonds to the backbone NH of Gly137 and the ζ-nitrogen of Lys-136.

are linear tetrapetides, but nevertheless display potent activity in subgenomic replicon systems. Ciluprevir and BI 201335 carry a carboxylic acid in the P1–P1′ region; all other non-covalent inhibitors of NS3/4A described herein contain bioisosteric acylsulfonamides as replacements of this group. In contrast to boceprevir, telaprevir and non-covalent inhibitors that carry carboxylic acids at P1, the NS3/4A co-crystal structure of acylsulfonamide-containing compounds such as danoprevir shows a complex arrangement of hydrogen bonds (Condroski *et al.*, 2006; Fig. 10.3C). The acyl nitrogen hydrogen bonds with the δ-nitrogen of His57. The acyl oxygen is involved with hydrogen bonds to the backbone NH groups of both Gly137 and Ser139. One of the P1′ sulfonamide oxygens hydrogen bonds to the γ-oxygen of Ser139. The other sulfonamide oxygen forms hydrogen bonds to the backbone NH of Gly137 and the ζ-nitrogen of Lys136.

Several NS3 protease inhibitors bind in a two-step (slow binding) mechanism with formation of an initial enzyme–inhibitor complex (EI) that subsequently isomerizes into a highly stable complex (EI*; Rajagopalan, 2009; Flores, 2009). The biochemical potencies of these compounds are described in terms of the inhibition constant K_i, which considers only the first equilibrium between unbound enzyme and inhibitor and EI, and K_i^* which accounts for the complete binding equilibrium. In the case of the di-keto amide-based covalent inhibitors, the transition from EI to EI* is associated with formation of a covalent adduct. The mechanism leading to slow binding inhibition in the non-covalent inhibitors is not well defined and appears to reflect the aggregate effect of optimized interactions throughout the molecule (Rajagopalan, 2009). As a result, the presence of slow binding and its contribution to potency varies significantly among non-covalent inhibitors. In the case of danoprevir, the conversion of EI to EI* is associated with a 1600-fold increase in potency and the resultant complex affords persistent inhibition of protease activity with a half-life of at least 5 hours (Rajagopalan *et al.*, 2009).

Non-covalent protease inhibitors in clinical trials

Ciluprevir (BILN-2061)

Ciluprevir was the first NS3/4A protease inhibitor tested for antiviral effect in humans. The robust, multiple $\log_{10}$ decrease in HCV RNA promoted by ciluprevir in short duration clinical studies was at the time unprecedented (Lamarre *et al.*, 2003). Clinical study of this compound was terminated in 2004 due to significant mitochondrial toxicity in cardiac myocytes of several preclinical species (Vanwolleghem *et al.*, 2007). The structural features of ciluprevir responsible for this toxicity were never elucidated. Nevertheless, this compound generated excitement and interest in both clinical and drug discovery teams worldwide and served as a template for drug discovery efforts for numerous groups. Several features of ciluprevir discovered during its pioneering medicinal chemistry optimization can be seen in more recent inhibitors.

The structure-based discovery of ciluprevir originated with the finding that NS3/4A serine protease activity could be inhibited by the N-terminal product peptides that contain a P1 carboxylic acid (Llinàs-Brunet *et al.*, 1998a,b; Steinkuhler *et al.*, 1998). The initial peptides were based on the product peptides of the natural cleavage sites of the NS3/4A, NS4A/NS4B, NS4B/NS5A, and NS5A/NS5B junctions. Progress in increasing potency was achieved primarily by substitution of the P2 proline with a napthylmethanol and P1 with aminocyclopropyl carboxylic acid. Further optimization of the P2 prolinol ether with substituted quinoline analogues and P1 with vinyl cyclopropyl carboxylic acid afforded potent tetrapeptides (IC_{50} = 30 nM). A strategy of macrocyclization was applied in the next stage of optimization. Macrocyclization has been employed and shown to improve stability and permeability by depeptidization, and improve potency and selectivity by decreasing the flexibility and confining the molecule to the preferred target-binding conformation (Driggers *et al.*, 2008). Introduction of the P1-P3 macrocycle was guided by X-ray and NMR data that showed the proximity of the S1 and S3 binding sites and the P1 and P3 side chains of the peptide inhibitors, respectively. Ciluprevir (Fig. 10.2) was selected as the clinical candidate from a series of analogues discovered by optimization that mainly focused on the P4 capping group and P2 heterocyclic ethers to improve cell-based potency and pharmacokinetics (Llinas-Brunet *et al.*, 2004).

Ciluprevir is a potent, reversible inhibitor of NS3/4A protease from HCV genotype 1a (K_i = 0.3 nM; EC_{50} = 4 nM) and 1b (K_i = 0.66 nM; EC_{50} = 3 nM) with a CC_{50} of 3 μM (Lamarre *et al.*, 2003). A one-step slow binding mechanism for ciluprevir was recently proposed (Flores *et al.*, 2009). This report suggested a 40 minute half-life for the enzyme inhibitor complex, which is longer than that observed in surface plasmon resonance ($t_{½}$ = 8 minutes) and rapid dilution experiments (Rajagopalan, 2009; Lamarre, 2003). Ciluprevir shows good selectivity for NS3 protease against representative serine and cysteine proteases human leucocyte elastase and cathepsin B (IC_{50} > 30 μM) and only minimal effect in a small number of receptor binding and enzyme inhibition assays at 10–30 μM concentrations (Lamarre *et al.*, 2003). The selectivity obtained is ascribed to the P1 carboxylic acid which affords product-based inhibition of the NS3 protease.

In vitro metabolism in rat showed BILN-2061 to be stable, with dog, monkey and human half-lives ($t_{½}$) of 105, 89 and >385 min, respectively. The compound showed high human plasma protein binding (>99.1%), but only a modest 10-fold shift in potency in the HCV replicon system in the presence of 50% human serum (Lamarre *et al.*, 2003). Ciluprevir showed an IC_{50} > 7 μM when evaluated against a panel of cytochrome P450 isozymes (Lamarre *et al.*, 2003).

Preclinical pharmacokinetic parameters after an i.v. dose of 2 mg/kg showed clearance rates (Cl) in rat, dog and monkey of 14.3 ml/min/kg, 2.3 ml/min/kg, and 5.5 ml/min/kg, respectively (Lamarre *et al.*, 2003). In the same species, terminal half-life ($t_{½}$) was 0.8 h, 2.2 h, and 7 h, respectively. Upon oral dosing (5 mg/kg), exposure parameters were similar in rat and monkey, and superior in dog. In these three species, C_{max} was 250 ng/ml, 300 ng/ml and 2100 ng/ml, respectively. Total exposure (AUC_{0-inf}) was 800 ng h/ml, 900 ng.h/ml, and 35,000 ng h/ml, while bioavailability (%*F*) was 13%, 7% and 38%, respectively. In rat, ciluprevir showed liver to plasma exposure ratio (LPR) of 5 to 39 depending on the time point examined. Higher exposure in the liver compared with plasma compartment is now known to be a general feature of NS3/4A protease inhibitors, though the LPR is both compound- and species-specific.

In humans, a single 200 mg oral dose delivered in 80% PEG 400, 20% ethanol to healthy male volunteers showed C_{max}, AUC_{0-inf} and $t_{½}$ of 800 ng/ml, 3300 ng h/ml, and ~4 h, respectively. At the same dose, exposure in HCV patients was approximately 2-fold higher and increased a further 2-fold after a third administration on a q12 schedule (Hinrichsen *et al.*, 2004). Increased exposure in HCV patients compared with healthy volunteers is a common feature of NS3/4A inhibitors, and may reflect impaired hepatic clearance in HCV patients. A 200 mg dose administered twice daily was predicted to provide a 12 hours post dose plasma concentration of 42 nM, representing 14-fold the EC_{50} of ciluprevir against a genotype 1b HCV replicon (Lamarre *et al.*, 2003).

Three publications have described the antiviral activity of ciluprevir (Lamarre *et al.*, 2003; Hinrichsen *et al.*, 2004; Reiser *et al.*, 2005). All of these studies relied on twice daily dosing, a schedule consistent with the terminal half-life of the compound in humans (Lamarre *et al.*, 2003). In patients carrying genotype 1 with minimal liver fibrosis, a twice-daily dose of 25 mg, 200 mg, or 500 mg resulted in progressively more robust antiviral activity, with the highest dose providing uniformly rapid and ≥3 log_{10} IU/ml reduction in HCV RNA in nearly all patients. Nearly all patients treated with the 25 mg b.i.d. dose showed a decrease of >2 log_{10} in HCV RNA (7 of 9, 78%). These findings were truly remarkable at the time against a backdrop where 2 log_{10} IU/ml HCV RNA decline in 2 weeks with SOC was considered encouraging.

In patients with advanced fibrosis (Ishak score of 3 to 4 or a Metavir score of 2 to 3) or in patients with compensated liver cirrhosis (Ishak score of 5 to 6 or a Metavir score of 3 to 4), a dose of 200 mg provided antiviral activity similar to that observed in patients with minimal fibrosis described above (Ishak score <3 or Metavir score <2). Undiminished activity regardless of the degree of liver fibrosis may be a general property of NS3/4A inhibitors and is distinct from pegylated interferon, whose activity is markedly decreased by increased fibrosis (Bruno *et al.*, 2010).

The antiviral activity of ciluprevir was also examined in patients harbouring genotype 2 or 3 at a twice daily dose of 500 mg (Reiser *et al.*, 2005). Antiviral activity in these patients was less pronounced than that observed in patients harbouring genotype 1 virus, likely reflecting the significantly lower affinity of ciluprevir for NS3/4A derived from genotype 3 (Seiwert *et al.*, 2008), the predominant HCV genotype included in this study. Importantly, patients who responded most strongly to ciluprevir displayed the highest exposure to drug, suggesting that a lower affinity to drug may be overcome by increased exposure. This same principle may have utility in the context of drug-resistant viral variant suppression.

Danoprevir (ITMN-191/RG7227)

The campaign that led to danoprevir was a rational, structure guided approach. Virtual tetrapeptide compound libraries were generated and scored in a docking model developed from NS3 X-ray crystal structures and binding motifs. These libraries carried a P1 carboxylic acid group and diverse collection of P2 linkers and P2 substituents to probe the lipophilic S2 protein surface region. P2 tetrahydroisoquinoline carbamates emerged as promising leads. Two strategies were pursued to increase cellular potency: (1) the P4 amino acid residue was truncated, and (2) macrocyclization was used to rigidify the molecule. By linking P1 and P3 portions and truncating the P4 one residue at the same time, the 15-membered macrocyclic tripeptide mimetic was obtained. Compounds were further optimized using a strategy to improve cellular permeability by replacing the P1 carboxylic acid with a bioisostere such as an acylsulfonamide (Johansson *et al.*, 2003). Profiling of the macrocyclic protease analogues revealed that a select group of these compounds dissociate very slowly from NS3/4A protease leading to prolonged enzyme inhibition (Rajagopalan *et al.*, 2007). This characteristic was unexpected in non-covalent inhibitors since they lack the catalytic warhead found in first generation protease inhibitors which forms a covalent bond to the active site serine and results in slowly reversible enzyme inhibition (Perni *et al.*, 2006). Additional medicinal chemistry effort, relying in part on screening potent, slowly dissociating compounds for liver exposure in rodent, led to the discovery of danoprevir. Danoprevir features a 4-fluoro-substituted tetrahydroisoindoline P2 and an acyl sulfonamide in P1′ (Fig. 10.2).

Danoprevir inhibited a reference genotype 1 NS3/4A protein in a two-step binding mechanism with a slow dissociation rate. Under pre-equilibrium conditions, 290 pM danoprevir half-maximally inhibited the reference 1b NS3/4A protease. Sub-nanomolar biochemical potency was maintained against NS3/4A derived from HCV genotypes 4, 5 and 6, while single-digit nanomolar potency was observed against NS3/4A from genotypes 2b and 3a. A detailed kinetic analysis using genotype 1b protein defined a K_i 100 nM and a true biochemical potency (K_i*) of ~62 pM (Rajagopalan, 2009). The latter value accounts for the contribution of slow binding to inhibitor potency. None of a panel of 53 proteases or a panel of 23 ion channels and receptors is inhibited by more than 50% with 10 μM danoprevir, indicating an IC_{50} higher than 10 μM against all of these proteins (Seiwert *et al.*, 2008). Against a genotype 1b (con1) subgenomic HCV replicon, EC_{50} was 1.8 nM while median danoprevir potency against multiple clinical isolates of subtype 1a NS3 (n = 16) and subtype 1b NS3 (n = 24) was 3.5 nM and 4.5 nM respectively. CC_{50} values range from 75 μM to 340 μM in Huh7 cells and primary cell cultures, indicating that specificity in cellular assays is greater than 41,000-fold.

Preclinical pharmacokinetic parameters following an oral dose of 30 mg/kg showed plasma C_{max}, AUC_{0-inf} and C_{12h} of 1.1 μg/ml, 9.5 μg h/ml, and 0.16 μg/ml in rat and 0.02 μg/ml, 0.06 μg h/ml, and 0.001 μg/ml in cynomolgus monkey. In both species, liver concentrations were significantly above plasma concentrations. In rat, the LPR based on AUC_{0-inf} is 10-fold, while in monkeys it is 127-fold. Thus, the lower plasma exposure in monkey compared with rat is somewhat compensated for by the difference in LPR between the species.

Initial clinical experience with danoprevir was obtained in a single escalating dose study in healthy volunteers. Doses of 2 mg, 100 mg, 200 mg, 400 mg, 800 mg, and 1600 mg were administered under fasted conditions. The pharmacokinetics of danoprevir was linear from 100 to 800 mg; a more than dose-proportional exposure increase was observed at 1600 mg. Food consumption increased both AUC and C_{min}, while slightly lowering C_{max}. As with preclinical species, plasma exposure was low; a 400 mg dose under fed condition provided mean C_{max}, C_{12hr} and AUC_{0-inf} of 75 ng/ml, 0.3 ng/ml and 122 ng h/ml, respectively. The exposure and *t* of danoprevir (1.5 hours to 2.1 hours) suggested two or three times daily administration in HCV patients.

To initially investigate the antiviral effects of danoprevir, treatment naïve, chronic HCV patients harbouring genotype 1 were administered 100 mg or 200 mg of danoprevir either twice daily or three times daily for 14 days (Forestier *et al.*, 2008). Danoprevir reduced HCV RNA in a dose-dependent manner with median HCV RNA reductions at EOT of 3.1 and 3.8 $\log_{10}$ IU/ml when 200 mg was dosed twice daily or three times daily, respectively. The lower antiviral effect with twice daily administration suggested that to achieve equivalent antiviral effects a twice daily dose would need to be higher than a three times daily dose.

To gauge antiviral effects in treatment-experienced patients, prior non-responders to SOC (defined as $\leq 2 \log_{10}$ IU/ml decrease within 12 weeks) or partial responders to interferon (defined as detectable HCV RNA at week 24) were administered 300 mg twice daily. Median EOT viral decline in antiviral effect was 2.1 $\log_{10}$ IU/ml, less than that in treatment-naïve patients, despite that fact that that exposure was higher in this cohort than in treatment-naïve patients receiving 200 mg. These data may suggest that different doses of an NS3/4A inhibitor may be required in these two patient populations. Danoprevir was safe and well tolerated in both patient populations in this study (Forestier *et al.*, 2008).

Virological rebound was observed in a subset of primarily genotype 1a patients treated with danoprevir monotherapy and was primarily associated with R155K substitution (Sarrazin *et al.*, 2007). All patients experiencing virological rebound carried R155K. One rebound patient and two plateau patients also carried substitutions at D168. Interestingly, the virological rebound was associated with R155K in genotype 1b despite the requirement for 2 nucleotide substitutions. Virological rebound in subtype 1b with other P1/P3 macrocyclic inhibitors such as TMC435350 is exclusively associated with D168 substitutions (Reesink *et al.*, 2009b), which require only a single nucleotide substitution in subtype 1b. The differing reliance on R155K for virological rebound in subtype 1b with different NS3 protease inhibitors is curious, and may reflect the level of resistance provided by D168 substitutions for different compounds (He *et al.*, 2008).

Antiviral effects of danoprevir in combination with SOC were first examined in a 14 day study in treatment-naïve patients harbouring HCV genotype 1 (Forestier *et al.*, 2009). Twice daily doses in this study (400 mg, 600 mg and 900 mg) were higher than three times daily doses (100 mg, 200 mg and 300 mg). In this study, the AE profile was similar to that observed historically for SOC, and no SAEs or Grade 4 AEs were observed. The best performing twice daily (600 mg) and three times daily (200 mg) cohorts provided median HCV RNA reductions at EOT of 5.4 $\log_{10}$ IU/ml and 5.7 $\log_{10}$ IU/ml. At EOT, HCV RNA was below the limit of detection (10 IU/ml) in 1/7 and 4/8 patients at doses of 200 mg t.i.d. and 600 mg b.i.d., respectively. No viral breakthrough was observed in either twice daily or three times daily cohorts.

Danoprevir was also studied in short duration combination with the nucleoside inhibitor RG7128. *In vitro* studies indicate the active moiety of R7128 has a high barrier to resistance and suppresses the emergence of danoprevir-resistant variants (Le Pogam *et al.*, 2008, 2009; Tan *et al.*, 2008, 2009). The first of its kind INFORM-1 study examined the combined antiviral effect of danoprevir and R7128 in the absence of SOC components. Twice-daily doses of danoprevir (either 600 mg or 900 mg) and RG7128 (1000 mg) resulted in HCV RNA reductions approaching or exceeding 5 $\log_{10}$ IU/ml. Responses were nearly identical in treatmen-naïve patients and prior null responders to SOC. HCV RNA was undetectable in the majority of patients at the end of treatment and no viral breakthrough occurred in treatment-naïve or prior null responders to SOC.

Interestingly, the magnitude and kinetics of viral suppression was similar to that observed when a single direct antiviral agent is used in triple combination with SOC. While it remains to be determined if SVR can be achieved in the absence of pegylated interferon, INFORM-1 demonstrated that an all oral regimen combining a NS5B NI and an NS3/4A PI can drive viral loads below the limit of detection in short duration studies at the same frequency as reported for triple combination cocktails comprising a NS3/4A PI, pegylated interferon and ribavirin (Manns *et al.*, 2008a,b, 2009b; Forestier *et al.*, 2009).

Danoprevir is currently in phase IIb clinical development in combination with SOC at unboosted doses of 300 mg t.i.d., and 600 mg b.i.d. Treatment duration is 12 weeks followed by 12 weeks of SOC. A higher b.i.d. dose in this study (900 mg) was discontinued due to grade 4 liver enzyme elevations in three patients, and an increase in bilirubin in one of these patients who was also receiving the known hepatotoxic agent allopurinol. Interim efficacy analysis of this study indicated 73%, 86% and 86% of patients in the above cohorts were HCV RNA negative by week 4 of treatment, and 88%, 89% and 92% were HCV RNA negative by week 12 of treatment.

Danoprevir displays a short $t_{1/2}$ in humans due to rapid cytochrome P450-mediated metabolism. Ritonavir 'boosting' is commonly employed in HIV treatment to reduce the cytochrome-mediated clearance of co-administered protease inhibitors (van Heeswijk *et al.*, 2001). Therefore, co-administration of danoprevir with ritonavir was examined. In healthy volunteers, ritonavir co-administration increased danoprevir concentration 12 hours post dose by 18 times in healthy volunteers, indicating that it dramatically reduced the clearance rate of danoprevir (Haznedar *et al.*, 2010). Antiviral effects of ritonavir-boosted danoprevir subsequently were examined in combination with SOC in genotype 1, treatment-naïve patients for 15 days. In this regimen, danoprevir doses of 100 mg b.i.d., 200 mg q.d., or 200 mg b.i.d. resulted in 50–100% of patients HCV RNA negative with 15 days of treatment (Gane *et al.*, 2010). These antiviral effects were superior to those provided by higher danoprevir doses in unboosted t.i.d. or b.i.d. regimens, despite the lower overall exposure of danoprevir they provide. Subsequent development of this compound will employ low b.i.d. and/or q.d. ritonavir-boosted doses given these data.

Vaniprevir (MK-7009)

Merck has reported the discovery and development of vaniprevir (Liverton, 2008, 2010; McCauley *et al.*, 2008, 2010). Vaniprevir employs a P2–P4 linker to generate a macrocyclic structure instead of a P1–P3 linker (Fig. 10.2). The discovery was aided by computational analysis of known macrocyclic inhibitors docked in the published full-length NS3/4A protein containing the helicase domain that suggested a P2–P4 linker could be accommodated. Analogues containing various P2 quinolyl ethers and P2 carbamates were generated by ring closing metathesis of appropriately substituted P2 and P4 dialkenes. Vaniprevir was selected from a series that contains P2 isoindoline carbamates identical to danoprevir. Other P2–P4 analogues described in this series contained 2-amino-1,3-diazoles as a P4 carbamate replacement and compounds with basic amine functionality at P3–P4 (Harper *et al.*, 2009).

Vaniprevir is a low nanomolar inhibitor of protease activity of genotypes 1a (K_i = 0.07 nM) and 1b (K_i = 0.06 nM) and, like other inhibitors in this class, is less active against genotype 2 enzymes (K_i = 1.0 and 1.4 nM; Liverton *et al.*, 2010). Vaniprevir shows high selectivity against a panel of 169 receptors, enzymes and ion channels tested ($IC_{50} > 10$ μM), except for inhibitory activity against chymotrypsin ($IC_{50} = 520$ nM). In a cell-based replicon assay, vaniprevir showed potent inhibition against genotype 1b ($EC_{50} = 3$ nM), and a modest ~6-fold loss of potency with addition of 50% human serum.

Preclinical pharmacokinetics after i.v. dosing in rat, dog and rhesus monkey showed moderate to high plasma clearance (74, 11, 18 ml/min/kg, respectively) and a short $t_{1/2}$ (0.9, 1.2, 1.3 h, respectively). Following oral dosing, exposure (AUC) was very low in rat and monkey (0–0.1, 0.05 μM h, respectively) but somewhat higher in dog (1.2 μM h). However, liver concentration at 24 hours in both rat and dog was high relative to the compound's EC_{50} (133-fold and 166-fold in rat and dog, respectively; Liverton *et al.*, 2010).

Human pharmacokinetics of vaniprevir was first examined in healthy volunteers. Following oral administration, AUC_{0-inf} and C_{max} values increased greater than dose proportionally. The apparent terminal half-life averaged ~ 4–5 h. Food did not have a meaningful effect on plasma pharmacokinetics of vaniprevir. Multiple dose administration indicated modest accumulation over 14 days (both AUC and C_{max}), but no change to terminal half-life (Wright *et al.*, 2008). As in preclinical species, plasma exposure was low.

The antiviral effects of vaniprevir were first examined in monotherapy for 8 days duration in genotype 1 patients (Lawitz *et al.*, 2008). Both treatment-naïve and treatment-experienced patients were included in this study. Patients were predominantly genotype 1a. The study examined twice daily (25 mg, 75 mg, 250 mg, 500 mg, 700 mg) and once daily (125 mg and 600 mg) doses. No serious adverse events were observed. The most common AEs were headache, nausea and diarrhoea. Dose-dependent HCV RNA declines were observed, with b.i.d. doses providing more robust antiviral responses than q.d. doses. A dose of 700 mg b.i.d. provided a continuous decline in HCV RNA and the greatest HCV RNA decline, 4.6 $\log_{10}$ IU/ml at end of treatment (EOT). HCV RNA rebounded in all other cohorts with the greatest level of rebound observed in the low dose q.d. cohort. NS3 variants were only observed in genotype 1a patients. Population-based NS3/4A sequence analysis demonstrated the presence of R155K/T/N/S/E/G, D168V/G/A/N/S, and A156T in a single patient. In several patients NS3/4A variants were identified by day 3 of treatment, but only R155K persisted 13 days following treatment. Interestingly, an allele-specific RT-qPCR analysis for select patients indicated that, for patients harbouring R155K at low levels at baseline, vaniprevir effectively prevented viral load rebound at higher doses (500 mg and 700 mg b.i.d.) but not lower doses (250 mg b.i.d.). This analysis clearly shows that the ability of a substitution in NS3/4A to promote viral rebound is dependent on drug dose, and by extension drug exposure.

A 4 week study in combination with SOC examined b.i.d. (300 mg and 600 mg) and q.d. (600 mg and 800 mg) doses in treatment-naïve genotype 1 patients (Manns *et al.*, 2009b). After 2 weeks of treatment, rates of HCV RNA negativity in the various cohorts ranged from <10% to >40%, but after 4 weeks of treatment the rate of HCV RNA negativity in the cohorts was tightly grouped, ranging from 69% to 82%. Viral breakthrough was observed in ~12% of vaniprevir-treated patients (8/63). While the rate of RVR provided by vaniprevir is comparable to other non-covalent inhibitors, the elevated incidence of virological rebound in this patient population is a rather unique finding.

Of the four rebound patients with amplifiable NS3 sequence, three were genotype 1a and one was genotype 1b. Population analysis indicated that all three genotype 1a patients carried R155K. One of these patients additionally carried D168V. The genotype 1b rebound patient carried D168T/I/A/V. More detailed clonal analysis indicated that the two dominant variants, R155K (genotype 1a) and D168T (genotype 1b), persisted in the absence of selective pressure by vaniprevir at least to day 84.

The most common adverse events were headache, nausea, vomiting, fatigue and flu-like illness. A dose of 600 mg b.i.d. resulted in vomiting in 40% of patients. In this cohort, vomiting duration was up to 23 days and was coincident with vaniprevir administration. Importantly, vomiting did not require anti-emetic treatment and did not lead to discontinuation, or dose reduction. Further studies will be required to understand how this side-effect can be best managed.

Vaniprevir is currently in phase IIb development in combination with SOC in treatment-experienced patients. Doses of 300 mg and 600 mg b.i.d. are being examined. Treatment durations include 24 weeks (600 mg), 48 weeks

(300 mg and 600 mg), and 24 weeks of triple therapy followed by 24 weeks of SOC (600 mg).

TMC435350

Researchers at Tibotec and Medivir developed a class of related of NS3 inhibitors that differed from previously disclosed structures in two ways: a cyclopentane replacement for the proline and the P3 amino and its P4 capping group have been modified or eliminated. Initial reports described the synthesis and potency of a series of reversible non-covalent tetrapeptide-like inhibitors in which the usual P2 hydroxyproline was replaced by a hydroxy-substituted cyclopentane dicarboxylic acid (Johansson *et al.*, 2006). Optimized compounds featuring P3–P4 L-amino acids and the P2 aryl 2-phenyl-7-methoxy-4-quinoline substituent originally reported by Boehringer Ingelheim researchers (Tsantrizos *et al.*, 2003) provide potent inhibitors in the low nM range (K_i = 16 nM). Further work on these cyclopentane inhibitors involved attempts at macrocyclization and optimization of P2 quinoline substituents for pharmacokinetic improvement. An iso-propyl thiazole substituent at position 2 and a methyl group at position 8 provided the most promising potency and *in vitro* and *in vivo* characteristics, and was selected as the clinical candidate TMC435350 (Fig. 10.2).

TMC435350 displayed potent activity against genotype 1a and 1b NS3/4A enzyme (K_i = 0.5 and 0.4 nM, respectively; Lin *et al.*, 2009). The $t_{½}$ for a complex of TMC435350 with NS3/4A derived from genotype 1a and 1b was ~5.7 min and 2.2 min, respectively (Lin *et al.*, 2008), indicating that in and of themselves a P1/P3 macrocyclic core and acyl sulfonamide P1′ could not be responsible for slow dissociative behaviour. Selectivity tests against a panel of 20 human proteases, including leucocyte elastase, trypsin and chymotrypsin provided IC_{50} values >1000 times that against HCV NS3/4A protease, with only cathepsin S showing an IC_{50} = 0.8 ±M. However, in a relevant cellular assay for cathepsin S activity, TMC435350 was found to be inactive at the highest dose tested (10 μM). Against genotype 1a and 1b HCV replicon constructs, EC_{50} values of 28.4 and 8.1 nM were obtained, respectively, and increase only slightly (1.4- to 1.8-fold) in the presence of plasma proteins and human serum. CC_{50}s against a panel of cell lines derived from different tissues and primary PBMC were >10 μM, indicating a selectivity index of at least 344-fold.

TMC435350 was metabolically stable in human liver microsomes (<6 μl/min/mg) and displayed moderate permeability in a Caco-2 assay (P_{app} = 8.4 × 10^{-6} cm/s). A 4 mg/kg i.v. administration in rats provided AUC_{0-inf} = 1750 ng h/ml, Cl = 2.31/h/kg and $t_{½}$ = 4 hours. A 40 mg/kg oral administration in rat provided a C_{max} = 1430 ng/ml with a T_{max} = 2 hours. The % *F* in rats was calculated to be 44%. LPR of 39 was observed, and liver concentrations remained above the EC_{99} for up to 31 hours after dosing. Other tissue concentrations were similar or below plasma AUC levels. Pharmacokinetics in Beagle dog following oral dosing (6.5 mg/kg) showed excellent bioavailability (100%*F*), with C_{max} = 4.72 μM, AUC_{0-inf} = 14,986 ng h/ml and $t_{½}$ = 5.1 hours.

The exposure and terminal half-life of TMC435350 in humans support once daily dosing of this compound (van 't Klooster *et al.*, 2008). Initial pharmacokinetics with TMC435350 suggested slow or prolonged absorption of TMC435350 (delayed T_{max}), possibly reflecting solubility limited absorption of this compound in the gastric compartment. Plasma exposure increase in TMC435350 was more than dose-proportional, and exposure parameters were not modulated by food consumption. Steady-state drug exposure was achieved within 5 days. In HCV-infected individuals, exposure was 3-fold higher than that in healthy volunteers and was associated with a longer elimination half-life (41 ± 33 h vs. 16 ± 5 h; van 't Klooster *et al.*, 2008).

Antiviral activity of TMC435350 was initially examined in a 5 day open label monotherapy study in genotype 1 patients who were non-responders or relapsers to prior treatment with pegylated interferon alpha and ribavirin (Reesink *et al.*, 2008). A dose of 200 mg provided a median reduction in HCV RNA of 3.9 log_{10} IU/ml on day 6. Rapid decline in HCV viral load was observed in both genotype 1a and 1b patients. NS3 substitutions known to confer reduced sensitivity to NS3/4A PIs emerged during or shortly after treatment. In genotype 1a patients these substitutions include T54A, Q80R, R155K, and D168A/V/E. In genotype 1b patients, Q80R and D168V/E were observed. Viral rebound was not associated with the appearance of these substitutions, but the duration of treatment may have been too short for viral rebound to become apparent.

OPERA-1 examined TMC435350 in combination with the standard of care (SOC) for up to 28 days in treatment-naïve and treatment-experienced patients infected with genotype 1 hepatitis C virus (Manns *et al.*, 2008a, 2009a; Marcellin *et al.*, 2009). TMC435350 was generally safe and well tolerated in this study.

OPERA-1 consisted of three dosing panels. In panel A, treatment-naive patients received TMC435350 (25, 75, 200 mg q.d.) or placebo as monotherapy for 7 days, followed by triple therapy consisting of TMC-435 (25, 75, 200 mg q.d.) or placebo, plus SOC for a further 21 days. In panel B, treatment-naive patients received TMC435350 (25, 75, 200 mg q.d.) plus SOC for 28 days. In panel C relapsers/non-responders to pegylated interferon received SOC in combination with higher doses of TMC (75, 150, 200 mg q.d.) for 28 days.

In panel A, 7-day monotherapy resulted in HCV RNA reductions of 2.6, 3.4, and 4.1 $\log_{10}$ IU/ml for doses of 25 mg, 75 mg, and 200 mg q.d., respectively. The viral load reduction at 200 mg in this study was in close agreement to that observed after 6 days of dosing in treatment-experienced patients in the open label study described above.

In panel B, TMC435350 doses of 25 mg, 75 mg, or 200 mg q.d. resulted in mean EOT reductions in HCV RNA of 4.7, 5.5 and 5.4 $\log_{10}$ IU/ml, respectively (Manns *et al.*, 2009a). At EOT, these same doses resulted, respectively, in 3/9, 8/9 and 7/10 patients with undetectable HCV RNA (<10 IU/ml). No viral breakthrough was observed in treatment naive patients. The lower two doses were reported to provide steady-state $C_{24\ h}$ concentrations in plasma that were 12-fold and 44-fold the compound's EC_{50} against genotype 1b HCV replicon.

In panel C, at EOT doses of 75 mg, 150 mg, or 200 mg q.d. resulted in mean HCV RNA reductions of 4.3, 5.5 and 5.3 $\log_{10}$ IU/ml, respectively, and viral negativity in 2/9, 5/9 and 3/10 treatment-experienced patients, respectively. Three genotype 1b patients experienced virological breakthrough, and all three were prior null responders to interferon (≤ 2 $\log_{10}$ IU/ml decrease within 12 weeks). Two of these patients were in the 75 mg group and one was in the 150 mg group. D168V was detected in all three patients by population sequencing (Marcellin *et al.*, 2009). These data suggest that the severely attenuated antiviral effects of interferon in these patients may have provided a lower barrier to drug resistance. ALT and AST decreased to normal ranges in patients treated with 75 mg and 150 mg TMC435350, but reductions were not significant in patients receiving 200 mg TMC435350.

OPERA-1 provided a well controlled assessment of triple therapy responses in treatment-naïve patients compared with treatment-experienced patients. In comparing 75 mg and 200 mg in treatment-naïve patients (panel B) and treatment-experienced patients (panel C), it is clear that both viral kinetic responses and the fraction of patients HCV RNA negative at EOT were reduced in patients with compromised interferon responses. For example, the viral response in treatment-naïve patients was more than a 1 $\log_{10}$ IU/ml greater than treatment-experienced patients at a dose of 75 mg. Moreover, at this dose 8/9 treatment-naïve patients were HCV RNA negative at EOT compared with 2/9 treatment-experienced patients. Even within panel C, responses in prior null responders (defined as <2 $\log_{10}$ IU/ml RNA reduction after 12 weeks of prior IFN-based therapy) were found to be lower than responses in prior relapsers (detectable HCV RNA after achieving undetectable HCV RNA at the end of treatment). This study supports observations with the covalent NS3/4A inhibitor telaprevir (Poordad *et al.*, 2008; Berg *et al.*, 2010) suggesting that patients with compromised interferon responses respond differently to SOC-based triple therapy regimens containing a single direct acting antiviral agent.

TMC435350 is currently in phase IIb clinical development in treatment-naïve and interferon experienced patients in combination with SOC. TMC435350 doses to be explored in treatment-naïve patients will be 75 mg and 150 mg q.d.

Both doses will be explored for 12 weeks in triple combination followed by 12 weeks of SOC, and for 24 weeks in triple combination. Treatment experienced patients will be administered 100 mg or 150 mg q.d. TMC435350. In this patient class, triple therapy durations of 12, 24 and 48 weeks will be explored with the former two durations followed by 36 and 24 weeks of SOC, respectively.

BI 201335

BI 201335 is the next generation protease inhibitor from Boehringer Ingelheim. The most interesting structural feature is that it is not a macrocycle but a linear tripeptide (Fig. 10.2). Similar to ciluprevir, it retains a carboxylic acid in the P1 position and a cyclopentane carbamate at P4. The large P2 group appears to be optimized from ciluprevir by addition of a bromine to the quinoline moiety and modification of the thiazole substituent. A recent report of the structure of BI 201335 bound to NS3/4A highlights three interesting features: a tightly bound water molecule forms a hydrogen bond to the P1 carboxylic acid, a salt bridge between R155 and D168 provides an appropriate surface for accommodation of P2 through hydrophobic interactions, and the presence of a halogen bond between the bromine on the P2 substituent and an oxygen from Asp 79 (Lemke *et al.*, 2010). The later interaction produces a 3.6-fold reduction in K_i relative to an analogue lacking the bromine.

The preclinical profile of BI 201335 has been described in outline form only (White *et al.*, 2010). In biochemical assays, BI 201335 displays a K_i of 2.6 and 2.0 nM against genotype 1a and 1b NS3, respectively. Similar potency is displayed against NS3 derived from genotypes 4, 5, and 6, but reduced potency is observed against genotypes 2 and 3. BI 201335 appears to bind in a one-step reaction mechanism although a two-step mechanism with a very loosely bound initial complex cannot be ruled out (Lemke *et al.*, 2010). The off-rate determined from a global kinetic analysis of compound binding suggests that the half-life of the bound complex is approximately 30 minutes. In subgenomic replicon systems, potency against genotype 1a and 1b was 6.5 and 3.1 nM, respectively, and CC_{50} in Huh7 cells was reported as 30 μM. *In vitro* resistance selections suggested D168 substitutions as predominant mutations for genotype 1b, although A156 substitutions were also observed. Selections using genotype 1a replicon associated both D168 and R155K substitutions with drug resistance (Kukolj *et al.*, 2010).

In the Caco-2 system, BI 201335 demonstrated similar apical-to-basolateral and basolateral-to-apical permeability suggesting it is not subject to extensive efflux transport. *In vitro* metabolic stability was investigated in microsomes derived from rat, dog and monkey. Expressed as percentage of hepatic blood flow in each species, clearance was <5.6%, <10.9%, and 18.6%, respectively. Upon oral dosing at 5 mg/kg (0.5% Methocel, 0.3% Tween-80) AUC_{0-inf} was highest in dog (14 μM h), poorest in rat (1.6 μM h), and intermediate in monkey (8.3 μM h). In rats, LPR was 42 indicating that, like other NS3/4A inhibitors, compound exposure in the plasma compartment underpredicts liver exposure.

BI 201335 is being developed for q.d. administration. The antiviral activity of this compound was initially investigated in an interesting phase I programme in both treatment-naïve patients and treatment-experienced patients (Manns *et al.*, 2008b). In treatment naive patients, 14 days of BI 201335 monotherapy were followed by 14 days of SOC-containing triple therapy. In monotherapy, maximal HCV RNA reductions were 2.9, 3.5, 3.7, and 4.0 log_{10} IU/ml in 20, 48, 120, 240 mg q.d. dose groups, respectively. Maximal HCV RNA suppression occurred between days 2 and 4 days, with viral rebound seen in most patients from all dose groups. Subsequent 14-day triple combination therapy failed to reduce viral loads in a pronounced manner. Characterization of the viral isolates showed that NS3 variants selected during BI 201335 monotherapy were R155K (genotype 1a) and D168V substitutions (genotype 1b; Kukolj *et al.*, 2009). Thus, in this instance drug resistant viral variants existing prior to triple combination therapy could not be suppressed by combination treatment.

A phase Ib 28-day triple therapy trial with treatment-experienced patients demonstrated a median maximal decline in VL during 28-day combination therapy of 4.8, 5.2, and 5.3 log_{10} IU/ml for the 48 mg, 120 mg, and 240 mg q.d. dose cohorts, respectively. No rebound during treatment was seen in the 240 mg q.d. dose cohort in the first 28 days. Viral rebound was observed in 2 patients in the 48 mg dose group and 1 patient in the 120 mg dose group. Sequencing of the NS3/4A protease at baseline and at viral rebound revealed selection of variants in the NS3 protease domain shown to confer *in vitro* resistance to BI 201335 (Manns *et al.*, 2008b).

In the above studies, BI 201335 was generally safe and well tolerated. Similar to TMC435350, dose-dependent hyperbilirubinaemia (indirect bilirubin) was observed at doses of 48 mg and above. Increased gastrointestinal effects were also observed relative to placebo treated patients.

BI 201335 has also been characterized in genotype 1 patients with compensated liver cirrhosis who were non-responsive to previous SOC. Patients were treated with 240 mg once or twice daily in combination with SOC for 28 days. Mean viral load declines on Day 28 in the 240 mg q.d. and b.i.d. groups were −4.8 and −5.0 $\log_{10}$, respectively. Similar to patients with less severe liver disease, mild-to-moderate jaundice due to unconjugated hyperbilirubinaemia was reported in 2/6 and 1/7 patients at 240 mg q.d. and b.i.d., respectively. There were no serious AEs (SAEs) in the 240 mg q.d. group and 2 SAEs in the 240 mg b.i.d. group leading to discontinuations. Since similar antiviral activity was observed for the q.d. as the b.i.d. dose group, the q.d. dose was suggested to provide a better risk profile for these patients.

Phase II trials (SILEN-C1) are ongoing and interim data from week 12 have been disclosed (Sulkowski *et al.*, 2009). HCV GT1 treatment-naïve patients were randomized 1:2:2:1 to placebo, 240 mg BI 201335, 240 mg BI 201335 with a 3-day lead-in phase of SOC, and 120 mg BI 201335 with a 3-day lead-in phase of SOC. In each group, triple therapy is administered for 24 weeks, and followed by SOC for an additional 24 or 48 weeks, dependent on whether a patient is HCV RNA negative after 4 and 12 weeks of treatment. At week 4, HCV RNA decline was observed to be below the limit of quantification for 92%, 84%, and 90% of patients in the 240 mg q.d., 240 mg q.d. lead-in and 120 mg lead-in groups, respectively. At week 12, HCV RNA decline below the limit of detection was observed for 91%, 82%, and 84% of patients in the 240 mg q.d., 240 mg q.d. lead-in and 120 mg lead-in groups, respectively. Virological rebound was seen in <3% of patients in the BI 201335 dose groups. These impressive virological results support a once daily schedule for this agent. However, 16 patients reported drug-related SAEs. Rash was observed more frequently with BI 201335. Eighteen (5.0%) patients discontinued due to AEs of which 1.7% were due to rash. Severe rashes occurred in 2.5% and 1.4% of patients treated with BI 201335 and placebo, respectively. As discussed above, BI 201335 leads to a dose-dependent unconjugated hyperbilirubinaemia associated with jaundice, and in this study this was observed in 15.9% of patients in the BI 201335 dose groups versus 1.4% of patients receiving placebo.

The SILEN-C2 programme examined 240 mg in treatment-experienced patients in combination with SOC (Sulkowski *et al.*, 2010). BI 201335 was administered q.d. without or with a 3-day SOC lead in, or 240 mg was administered twice daily with a 3-day lead in of SOC. No control group was included in this study. After 4 weeks of treatment, 62%, 64% and 69% of patients had HCV RNA below the limit of quantification, respectively. After 12 weeks of treatment, 59%, 59%, and 54% of patients has HCV RNA below the limit of detection, respectively. Viral breakthrough was observed in ~22% of patients treated with once daily BI 201335, and in ~14% of patients treated twice daily. Rash was very frequent in this trial, occurring in 33% to 59% of patients depending on dose. As in previous trials, jaundice was observed in roughly 10% of patients when BI 201335 was administered once daily and 34% when dosed twice daily. Nausea, diarrhoea and vomiting were also common. SAEs and treatment discontinuations were more common with twice daily dosing.

BMS-650032

Like BI 201335, BMS-650032 is a linear tripeptide (Fig. 10.2). In contrast to BI 201335 and similar to the macrocyclic inhibitors currently in clinical development, it contains a cyclopropyl acylsulfonamide in P1/P1′. The P2 group is a largely undecorated 2-quinoline moiety and P4 is a carbamate-linked butyl.

The preclinical characteristics of BMS-650032 have been described in outline form only (McPhee and al., 2010). In biochemical assays, BMS-650032 displays a K_i of 0.7 nM and 0.3 nM against NS3/4A derived from genotype 1a and 1b, respectively. In genotype 1a and 1b subgenomic replicon systems in the absence of human serum, potencies of 4 nM and 1.2 nM, respectively, are observed. *In vitro* resistance selections with BMS-650032 identified R155K, various D168 substitutions, and I170T as NS3 variants associated with drug resistance.

Publicly disclosed clinical data for BMS-650032 are limited to a single dose study in genotype 1 HCV patients that were either naïve or experienced with interferon therapy, and a 3-day multiple dose study in genotype 1 in treatment-naïve patients (Pasquinelli *et al.*, 2009). Single doses of 10 mg, 60 mg, 200 mg, and 600 mg resulted in a proportional increase in exposure. Similar to other NS3 PIs, exposure at all doses was at least several fold higher than in healthy volunteers. T_{max} was ~2 to 4 h, terminal half-life was ~15 to 20 h, and oral clearance ranged from 302 to 491 l/h. Exposure in the plasma compartment appeared low.

Following administration of a single dose of 10 mg, 60 mg, 200 mg or 600 mg in HCV genotype 1 patients, maximal mean declines in HCV RNA of ~0, ~0.5, ~2.0, ~2.5 $\log_{10}$ IU/ml, respectively. A pre-existing non-consensus amino acid at position 80 that is common in genotype 1a (K80 vs. Q80) reduced viral kinetic responses and conferred reduced sensitivity to BMS-650032 (McPhee *et al.*, 2009). A three day, twice daily course of 200 mg, 400 mg, or 600 mg in treatment-naïve patients resulted in mean declines of ~2.5, ~3.5 and ~3.5 IU/ml at EOT. The mean viral kinetic profile of all three cohorts was that of continual decline, but a longer duration would likely be required for viral rebound to be observed. NS3/4A sequence and phenotype have not been reported.

A two part study in treatment naïve, genotype 1 and 4 patients is currently underway to examine the combined antiviral effects of BMS-650032 in combination with SOC. In part A of this study, described as phase IIa, the antiviral effects and safety of 200 mg b.i.d., 600 mg b.i.d., and 600 mg q.d. will be investigated with SOC for a 48 weeks. The second part of this study, described as phase IIb, will examine the optimal dose in treatment durations of 24 and 48 weeks.

In addition, BMS-650032 is being examined in an innovative programme in combination with the highly potent NS5A inhibitor BMS-790052 in prior null responders to SOC. Three regimens are being examined; all employ a 600 mg b.i.d. dose of BMS-650032 and a 60 mg q.d. dose of BMS-790052. The two agents will be examined in 24 week and 48 week durations. The third cohort will be for a 24 week duration and include SOC. Results of this innovative programme are eagerly anticipated.

ACH-1625

ACH-1625 is an open chain, non-covalent, reversible inhibitor of NS3 protease but its structure has not been disclosed. Non-clinical data describing ACH-1625 has been disclosed in outline form only (Huang *et al.*, 2010; Stauber *et al.*, 2010).

Similar to danoprevir, ACH-1625 shows slow/tight binding, with K_i >47 nM, K_i^* = 61 pM, and a $t_{½}$ for NS3 residence >5.9 hours against NS3 derived from genotype 1b. Potency is reduced approximately 3-fold against genotype 1a, and reduced more significantly against NS3 derived from genotype 3. Selectivity against a panel of 16 proteases was high (<50% inhibition at 10 μM), except that an IC_{50} of 4.5 μM was obtained against cathepsin S. Cathepsin S has been found to be inhibited by protease inhibitors from multiple classes.

A potency of 11 nM was observed against a reference NS3 sequence (con1) in an HCV replicon system, and the potency showed a modest 3-fold shift in the presence of 40% human serum. Genotype 1a or 1b NS3 derived from patients was sensitive to ACH-1625 when inserted into a replicon system; the greatest loss of potency was 7-fold. In all primary cells tested, ACH-1625 displayed CC_{50} > 32 μM, indicating a selectivity of 2900-fold relative to inhibition of con1 NS3 sequence in the replicon system.

ACH-1625 is stable in microsomes from human, rat and dog ($t_{½}$ of 106, >90 and 115 min., respectively) and less stable in monkey ($t_{½}$ = 10 min) This compound was rapidly eliminated from the plasma compartment in rat following i.v. dosing, but showed high liver exposure. Similar to danoprevir and vaniprevir, upon oral dosing plasma exposure in rat and dog was low, with liver concentrations significantly higher. In the case of rat, a 50 mg/kg oral dose provided liver $C_{24\,h}$ = 339 ng/g, representing 33-fold the EC_{50} and a LPR of 505. In dogs, a 20 mg/kg oral dose provided liver C_{24h} = 134 ng/g, representing 12-fold the EC_{50} and a LPR = 55. Tissue distribution studies in the rat with ^{14}C-ACH-1625 confirmed prolonged hepatic distribution, with distribution into most other tissues (excluding the GI tract) being modest. Specific transport into cell lines containing expressed human organic anion-transporting polypeptide (OATP) uptake transporters OATP1B1, OATP1B3 and OATP2B1 and rat rOatp1a1 indicated a role for active transport in liver partitioning of ACH-1625.

The pharmacokinetics of ACH-1625 in healthy volunteers and HCV patients has been reported (Detishin *et al.*, 2010). In single dose escalation, a greater than dose proportional increase in exposure was seen at doses of 50 mg to 1000 mg. Administration of compound in the fasted state greatly reduced exposure (~10-fold reduction). As in preclinical species, the compound was rapidly eliminated from the plasma compartment but displayed a terminal $t_{1/2}$ of 11–14 h. Following twice daily administration of either 500 mg or 600 mg for 4.5 days to healthy volunteers, accumulation was more than anticipated from the single dose study (>6-fold) and, in contrast to other NS3 PIs, exposure in HCV patients was comparable to healthy volunteers. Both doses provided robust virological effect, with HCV RNA declines of 4.25 and 3.94 $\log_{10}$ IU/ml reported for the 500 mg and 600 mg twice daily doses, respectively. Based on the strength of these virological data, once daily administration of this compound has been initiated. ACH-1625 was generally safe and well tolerated in this study, with no SAE or discontinuations with single or multiple dose administration.

GS-9256

Public information concerning GS-9256 is derived from a single publication (Lawitz *et al.*, 2010). This compound displays a K_i = 90 pM, and an EC_{50} against HCV replicon of 20 nM against an undisclosed genotype of HCV. Selectivity against human proteases is 1000- to 28,500-fold. Single dose administration to healthy subjects and HCV patients showed, as nearly all other NS3/4A inhibitors, that exposure was slightly higher in HCV infected patients. In HCV patients, a single dose of 150 mg, 300 mg, and 450 mg resulted in median maximal HCV RNA declines of 1.8, 2.4 and 2.8 $\log_{10}$ IU/ml, respectively.

A 3-day dose ranging study was conducted to evaluate safety, pharmacokinetics and antiviral effects in predominantly genotype 1a HCV patients using both solution (75 mg b.i.d., 200 mg b.i.d., 300 mg q.d.) and capsule (25 mg b.i.d., 75 mg b.i.d.) dosage forms. Across all dosage groups, exposure increased at least 2-fold from the first to last dose of the study. Solution doses of 75 mg b.i.d., 200 mg b.i.d. and 300 mg q.d. provided day 3 $AUC_{0\text{-trough}}$ of 38, 285, and 205 μ g h/ml. These plasma exposures are several orders of magnitude greater than the plasma exposures provided by danoprevir, vaniprevir and ACH-1625. Terminal $t_{1/2}$ ranged from 6.9 to 13.7 hours in the three cohorts employing solution dosing. Similar pharmacokinetic performance was provided when 75 mg was delivered in capsule or solution form. Viral load reductions were not provided, but from graphical representation appear to be 2.8 to 3.0 $\log_{10}$ IU/ml for dose groups above 25 mg b.i.d., with the most robust effects observed with a dose of 200 mg b.i.d. Maximal RNA decline could be fit to an E_{max} model using $AUC_{0\text{-trough}}$, and provided E_{max} = 3.1 $\log_{10}$ IU/ml with an EC_{50} = 4.6 μg h/ml. Interestingly, and in contrast to several other NS3/4A inhibitors, drug resistant viral variants were detected by population sequencing in this 3-day monotherapy study. Variants included R155K, various D168 substitutions, and A156V.

Like danoprevir, BMS-650032, and the covalent NS3/4A inhibitor telaprevir, the antiviral activity and safety of GS-9256 is currently being studied in combination with a second direct acting antiviral. In this case, an open label study will investigate GS-9256 (75 mg b.i.d.) with the non-nucleoside inhibitor GS-9190 (40 mg b.i.d.) alone and in combination with weight-based ribavirin, which is also dosed twice daily. Treatment naïve patients will be administered these regimens for 28 days and will then initiate SOC.

PHX-1766 and ABT-450

PHX-1766 is a reversible inhibitor of NS3 protease replicon assay but its structure has not been disclosed. A phase Ia randomized, placebo-controlled study of PHX-1766 in both healthy volunteers and genotype 1 HCV patients has been reported (Hotho *et al.*, 2009). As with other NS3/4A PIs, exposure was higher in HCV patients compared with healthy volunteers. A single dose of 500 mg in HCV patients resulted in maximal HCV RNA decline of ~1 $\log_{10}$ IU/ml in 24 hours. In a 6- to 9-day multiple dose study, the average maximal HCV RNA reduction was 1.2 and 1.8 $\log_{10}$ IU/ml in the 400 mg and 800 mg b.i.d. groups, respectively.

Abbott is developing ABT-450, an acylsulfonamide NS3/4A protease inhibitor of undisclosed structure. No information for this compound is publicly available, though it is thought to be in development as a ritonavir boosted product.

Summary

A structurally diverse collection of non covalent inhibitors of the NS3/4A protease active site is currently in mid-stage clinical trials. These inhibitors may offer advantages in terms of reduced dose, less frequent administration, and more favourable side-effect profile compared with the first generation, covalent inhibitors telaprevir and boceprevir. Nearly all of the compounds described herein promote greater than a 3 $\log_{10}$ IU/ml reduction in HCV RNA in short duration studies at their optimal schedule, and when reported, provide rates of HCV RNA negativity after 4 weeks of treatment with SOC that can approach 90%. While sustained virological response rates have not yet been reported for any of these compounds, and the full safety profiles for many are yet to be elucidated, the data to date suggest that non-covalent inhibitors of NS3/4A protease may meaningfully increase cure rates in chronic HCV when combined with SOC. Additionally, no fewer than three of these agents (danoprevir, BMS-650032, and GS-9256) are under study in combination with a second experimental direct acting antiviral agent in various all oral regimens, suggesting that non-covalent inhibitors of NS3/4A are poised to play a transformative role in HCV treatment.

References

Berg, T., McHutchison, J.G., Adda, N., Poordad, F., Shiffman, M.L., Ferenci, P., Heathcote, J., Pawlotsky, J.-M., Zeuzem, S., Reesink, H., *et al.* (2010). SVR with telaprevir, peginterferon alfa-2a and ribavirin in HCV patients with well characterized prior null response, partial response, viral breakthrough or relapse after PR. J. Hepatol. *52*, S2.

Bruno, S., Shiffman, M.L., Roberts, S.K., Gane, E.J., Messinger, D., Hadziyannis, J., and Marcellin, P. (2010). Efficacy and safety of peginterferon alfa-2a (40KD) plus ribavirin in hepatitis C patients with advanced fibrosis and cirrhosis. Hepatology *51*, 388–397.

Condroski, K., Zhang, D.E., Seiwert, S.D., Ballard, J.A., Bernat, B.A., Brandhuber, B., Colwell, H., Smith, C., Vigers, G., Andrews, S.W., *et al.* (2006). Structure-based design of novel isoindolene inhibitors of HCV NS3/4A protease and binding mode analysis of ITMN 191 by x-ray crystallography. Gastroenterology *130*, A835.

Detishin, V., Haazen, W., Robison, H., Robarge, L., and Olek, E. (2010). Virological response, safety, and pharmacokinetic profile following single- and multiple dose administration of ACH-0141625 protease inhibitor to healthy volunteers and HCV genotype-1 patients. J. Hepatol. *52*, S468.

Di Marco, S., Rizzi, M., Volpari, C., Walsh, M.A., Narjes, F., Colarusso, S., De Francesco, R., Matassa, G., and Sollazzo, M. (2000). Inhibition of the hepatitis C virus NS3/4A protease. The crystal structures of two protease-inhibitor complexes. J. Biol. Chem. *275*, 7152–7157.

Driggers, E.M., Hale, S.P., Lee, J., and Terrett, N.K. (2008). The exploration of macrocycles for drug discovery – an underexploited structural class. Nat. Rev. Drug Discov. *7*, 608–624.

Flores, M.V., Strawbridge, J., Ciaramella, G., and Corbau, R. (2009). HCV-NS3 inhibitors: determination of their kinetic parameters and mechanism. Biochim. Biophys. Acta *1794*, 1441–1448.

Forestier, N., Larrey, D., Guyader, D., Marcellin, P., Rouzier, R., Patat, A., Bradford, B., Porter, S., Blatt, M., and Zeuzem, S. (2008). Treatment of chronic hepatitis C virus (HCV) genotype 1 patients with the NS3/4A protease inhibitor ITMN-191 leads to rapid reductions in plasma HCV RNA: results of a phase 1b multiple ascending dose (MAD) study. Hepatology *48*, 1132A.

Forestier, N., Larrey, D., Marcellin, P., Benhamou, Y., Guyader, D., Bradford, W., Porter, S., Patat, A., Rouzier, R., and Zeuzem, S. (2009). Antiviral activity and safety of ITMN-191 in combination with peginterferon Alfa-2a and ribavirin in patients with chronic Hepatitis C virus (HCV). J. Hepatol. *50*, S35.

Fried, M.W., Shiffman, M.L., Reddy, K.R., Smith, C., Marinos, G., Goncales, F.L., Jr., Haussinger, D., Diago, M., Carosi, G., Dhumeaux, D., *et al.* (2002). Peginterferon alfa-2a plus ribavirin for chronic hepatitis C virus infection. N. Engl. J. Med. *347*, 975–982.

Gale, M., Jr. and Foy, E.M. (2005). Evasion of intracellular host defense by hepatitis C virus. Nature *436*, 939–945.

Gane, E., Rouzier, R., Stedman, C., Wiercinska-Drapalo, A., Horban, A., Chang, L., Zhang, Y., Sampeur, P., Najera, I., Shulman, N., *et al.* (2010). Ritonavir boosting of low dose RG7227/ITMN-*191*, HCV NS3/4A protease inhibitor, results in robust reduction in HCV RNA at lower exposures than provided by unboosted regimens. J. Hepatol. *52*, S16–S17.

Hagel, M., Niu, D., St. Martin, T., Sheets, M., Qiao, L., Chaturvedi, P., Labenski, M., Nacht, M., Petter, R., Singh, J., *et al.* (2010). AVL-192: potency against HCV NS3 mutants allows for replicon clearance as a monotherapy and in combination studies. J. Hepatol. *52*, S292–S293.

Harper, S., Ferrara, M., Crescenzi, B., Pompei, M., Palumbi, M.C., Dimuzio, J.M., Donghi, M., Fiore, F., Koch, U., Liverton, N.J., *et al.* (2009). Inhibitors of the hepatitis C virus NS3 protease with basic amine functionality at the P3-amino acid N-terminus: discovery and optimization of a new series of P2–P4 macrocycles. J. Med. Chem. *52*, 4820–4837.

Haznedar, J.Ö., Fretland, J., Leong, G., Blotner, S., Hill, T., Smith, P., and Tran, J.Q. (2010). Impact of low dose ritonavir on the pharmacokinetics of RG7227 (ITMN-191), a highly potent and selective inhibitor of the HCV NS3/4A protease. J. Hepatol. *52*, S293.

He, Y., King, M.S., Kempf, D.J., Lu, L., Lim, H.B., Krishnan, P., Kati, W., Middleton, T., and Molla, A. (2008). Relative replication capacity and selective advantage profiles of protease inhibitor-resistant hepatitis C virus (HCV) NS3 protease mutants in the HCV genotype 1b replicon system. Antimicrob. Agents. Chemother. *52*, 1101–1110.

van Heeswijk, R.P., Veldkamp, A., Mulder, J.W., Meenhorst, P.L., Lange, J.M., Beijnen, H., and Hoetelmans, R.M. (2001). Combination of protease inhibitors for the treatment of HIV-1-infected patients: a review of pharmacokinetics and clinical experience. Antivir. Ther. *6*, 201–229.

Hezode, C., Forestier, N., Dusheiko, G., Ferenci, P., Pol, S., Goeser, T., Bronowicki, J.P., Bourliere, M., Gharakhanian, S., Bengtsson, L., *et al.* (2009). Telaprevir and peginterferon with or without ribavirin for chronic HCV infection. N. Engl. J. Med. *360*, 1839–1850.

Hinrichsen, H., Benhamou, Y., Wedemeyer, H., Reiser, M., Sentjens, R.E., Calleja, J.L., Forns, X., Erhardt, A., Cronlein, J., Chaves, R.L., *et al.* (2004). Short-term antiviral efficacy of BILN *2061*, a hepatitis C virus serine protease inhibitor, in hepatitis C genotype 1 patients. Gastroenterology *127*, 1347–1355.

Hotho, D., Bruijne, J.D., O'Farrell, A.-M., Boyea, T., Li, J., Weegink, C.J., Schinkel, J., Molenkamp, R., Wetering, J.V.D., Vliet, A.A.V., *et al.* (2009). Accelerated clinical trial design to assess the safety, tolerability and antiviral activity of PHX1*766*, a novel HCV NS3/4A protease inhibitor, in healthy volunteers and chronic hepatitis C patients. Hepatology *50*, 1031A–1032A.

Huang, M., Fabrycki, J., Patel, D., Podos, S., Yang, G., Zhao, Y., Marlor, C.W., Stauber, K., Agarwal, A., Wang, X., *et al.* (2010). Antiviral activity, combination and resistance of ACH-1*625*, a potent HCV NS3 protease inhibitor. J. Hepatol. *52*, S254.

Johansson, A., Poliakov, A., Akerblom, E., Wiklund, K., Lindeberg, G., Winiwarter, S., Danielson, U.H., Samuelsson, B., and Hallberg, A. (2003). Acyl sulfonamides as potent protease inhibitors of the hepatitis C virus full-Length NS3 (protease-helicase/NTPase): a comparative study of different C-terminals. Bioorg. Med. Chem. *11*, 2551–2568.

Johansson, P.O., Back, M., Kvarnstrom, I., Jansson, K., Vrang, L., Hamelink, E., Hallberg, A., Rosenquist, A., and Samuelsson, B. (2006). Potent inhibitors of the hepatitis C virus NS3 protease: use of a novel P2 cyclopentane-derived template. Bioorg. Med. Chem. *14*, 5136–5151.

van't Klooster, G.A.E., Vanwelkenhuysen, I., Hooijmaijers, R., Bol, K., Voets, M., Van Houdt, J., Verloes, R.A.L., Aharchi, F., Marien, K., Van Remoortere, P., *et al.* (2008). Once-daily regimens of the HCV NS3/4A-protease inhibitor TMC435350 are predicted to provide therapeutic exposure in plasma and liver. J. Hepatol. *48*, S321.

Kukolj, G., Benhamou, Y., Manns, M.P., Bourliere, M., Pol, S., Schuchmann, M., Cartier, M., Huang, D., Lagace, L., Steinmann, G., *et al.* (2009). BI 201335, a potent HCV NS3 protease inhibitor, in treatment-naive and -experienced chronic HCV genotype-1 infection: genotypic and phenotypic analysis of the NS3 protease domain. J. Hepatol. *50*, S347.

Kukolj, G., Bousquet, C., Dansereau, N.,F., Dô, F., Lagacé, L., Llinàs-Brunet, M., Marquis, M., Massariol, M.-J., Maurice, R., Spickler, C., *et al.* (2010). Characterization of resistance mutations selected in vitro by the NS3/4A protease inhibitor BI 201335 J. Hepatol. *52*, S295.

Kwo, P., Lawitz, E.J., McCone, J., Schiff, E.R., Vierling, J.M., Pound, D., Davis, M., Galati, J.S., Gordon, S.C., Ravendhran, N., *et al.* (2009). Final results SVR 24 NS3 protease inhibitor boceprevir plus PegIFN alpha-2b/ribavirin HCV 1 treatment naïve patients. J. Hepatol. *50*, S4.

Lamarre, D., Anderson, P.C., Bailey, M., Beaulieu, P., Bolger, G., Bonneau, P., Bos, M., Cameron, D.R., Cartier, M., Cordingley, M.G., *et al.* (2003). An NS3 protease inhibitor with antiviral effects in humans infected with hepatitis C virus. Nature *426*, 186–189.

LaPlante, S.R. and Llinàs-Brunet, M. (2005). Dynamics and structure-based design of drugs targeting the critical serine protease of the hepatitis C virus – from a peptidic substrate to BILN 2061. Curr. Med. Chem. Anti-Infect. Agents *4*, 22.

Lawitz, E., Sulkowski, M., Jacobson, I., Faruqui, S., Kraft, W., Maliakkal, B., Al-Ibrahim, M., Ghalib, R., Gordon, S.C., Kwo, P., *et al.* (2008). Safety, tolerability and antiviral activity of MK-7009, a novel inhibitor of the hepatitis C virus NS3/4A protease, in patients with chronic HCV genotype 1 infection. Hepatology *48*, 403A.

Lawitz, E.J., Marbury, T.C., Vince, B.D., Grunenberg, N., Rodriguez-Torres, M., De Micco, M.P., Tarro, J.N., Shelton, M.J., West, S., Zong, J., *et al.* (2010). Dose-ranging, three-day monotherapy study of the NS3 protease inhibitor GS-9256. J. Hepatol. *52*, S466–S467.

Le Pogam, S., Seshaadri, A., Kosaka, A., Chiu, S., Kang, H., Hu, S., Rajyaguru, S., Symons, J., Cammack, N., and Najera, I. (2008). Existence of hepatitis C virus NS5B variants naturally resistant to non-nucleoside, but not to nucleoside, polymerase inhibitors among untreated patients. J. Antimicrob. Chemother. *61*, 1205–1216.

Le Pogam, S., Seshaadri, A., Kosaka, A., Hu1, S., Beard, A., Symmons, J., Cammack, N., and Najera, I. (2009). No evidence of R7128 drug resistance after up to 4 weeks treatment of genotype 1, 2, and 3 hepatitis C virus infected individuals. J. Hepatol. *50*, S348.

Lemke, C., Zhao, S., Goudreau, N., Hucke, O., Thiebeault, D., Llinas-Brunet, M., and White, P. (2010). Bound Structure and Biochemical Mechanism of Action of Non-Covalent HCV NS3/4A Protease Inhibitor BI 201335. The 23rd International Conference on Antiviral Research (ICAR) (San Francisco, CA).

Lemon, S.M., Yi, M., and Li, K. (2005). "Strong reasons make strong actions"–The antiviral efficacy of NS3/4A protease inhibitors. Hepatology *41*, 671–674.

Li, X., Zhang, Y.-K., Liu, Y., Ding, C.Z., Zhou, Y., Li, Q., Plattner, J.L., Baker, S.J., Zhang, S., Kazmierski, W.M., *et al.* (2010). Novel macrocyclic HCV NS3 protease inhibitors derived from α-amino cyclic baronates. Bioorg. Med. Chem. Lett. *20*, 5695–5700.

Lin, C., Kwong, D., and Perni, R.B. (2006). Discovery and development of VX-950, a novel, covalent, and reversible inhibitor of hepatitis C virus NS3.4A serine protease. Infect. Disord. Drug Targets *6*, 3–16.

Lin, T., Devogelaere, B., Lenz, O., Nyanguile, O., van der Helm, E., Vermeiren, K., Vandercruyssen, G., Cleiren, E., Lindberg, J., Edlund, M., *et al.* (2008). Inhibitory activity of TMC435350 on HCV NS3/4A proteases from genotypes 1 to 6. Hepatology *48*, 1166A.

Lin, T.I., Lenz, O., Fanning, G., Verbinnen, T., Delouvroy, F., Scholliers, A., Vermeiren, K., Rosenquist, A., Edlund, M., Samuelsson, B., *et al.* (2009). In vitro activity and preclinical profile of TMC435350, a potent hepatitis C virus protease inhibitor. Antimicrob. Agents Chemother. *53*, 1377–1385.

Liverton, N.J., Holloway, M.K., McCauley, J.A., Rudd, M.T., Butcher, J.W., Carroll, S.S., DiMuzio, J., Fandozzi, C., Gilbert, K.F., Mao, S.S., *et al.* (2008). Molecular modeling based approach to potent P2–P4 macrocyclic inhibitors of hepatitis C NS3/4A protease. J. Am. Chem. Soc. *130*, 4607–4609.

Liverton, N.J., Carroll, S.S., Dimuzio, J., Fandozzi, C., Graham, D.J., Hazuda, D., Holloway, M.K., Ludmerer, S.W., McCauley, J.A., McIntyre, C.J., *et al.* (2010). MK-7009, a potent and selective inhibitor of hepatitis C virus NS3/4A protease. Antimicrob. Agents Chemother. *54*, 305–311.

Llinàs-Brunet, M., Bailey, M., Déziel, R., Fazal, G., Gorys, V., Goulet, S., Halmos, T., Maurice, R., Poirier, M., Poupart, M.-A., *et al.* (1998a). Studies on the C-terminal of hexapeptide inhibitors of the hepatitis C virus serine protease. Bioorg. Med. Chem. Lett. *8*, 2719–2724.

Llinàs-Brunet, M., Bailey, M., Fazal, G., Goulet, S., Halmos, T., Laplante, S., Maurice, R., Poirier, M., Poupart, M.-A., Thibeault, D., *et al.* (1998b). Peptide-based inhibitors of the hepatitis C virus serine protease. Bioorg. Med. Chem. Lett. *8*, 1713–1718.

Llinas-Brunet, M., Bailey, M.D., Bolger, G., Brochu, C., Faucher, A.M., Ferland, J.M., Garneau, M., Ghiro, E., Gorys, V., Grand-Maitre, C., *et al.* (2004). Structure-activity study on a novel series of macrocyclic inhibitors of the hepatitis C virus NS3 protease leading to the discovery of BILN 2061. J. Med. Chem. *47*, 1605–1608.

McCauley, J.A., Rudd, M.T., Nguyen, K.T., McIntyre, C.J., Romano, J.J., Bush, K.J., Varga, S.L., Ross, C.W., 3rd, Carroll, S.S., DiMuzio, J., *et al.* (2008). Bismacrocyclic inhibitors of hepatitis C NS3/4a protease. Angew. Chem. Int. Ed. Engl. *47*, 9104–9107.

McCauley, J.A., McIntyre, C.J., Rudd, M.T., Nguyen, K.T., Romano, J.J., Butcher, J.W., Gilbert, K.F., Bush, K.J., Holloway, M.K., Swestock, J., *et al.* (2010). Discovery of vaniprevir (MK-7009), a macrocyclic hepatitis C virus NS3/4a protease inhibitor. J. Med. Chem. *53*, 21.

McHutchison, J.G., Everson, G.T., Gordon, S.C., Jacobson, I.M., Sulkowski, M., Kauffman, R., McNair, L., Alam, J., and Muir, A.J. (2009). Telaprevir with peginterferon and ribavirin for chronic HCV genotype 1 infection. N. Engl. J. Med. *360*, 1827–1838.

McPhee, F. (2010). Identification and preclinical profile of the novel NS3 protease inhibitor BMS-650032. J. Hepatol. *52*, S296.

McPhee, F., Hernandez, D., Sheaffer, A., Lee, M., Falk, P., Yu, F., Zhai, G., Chaniewski, S., and Pasquinelli, C. (2009). Genotypic and phenotypic analysis of samples from HCV-infected subjects treated with BMS-650032 in a single ascending dose study. Hepatology *50*, 1048A.

Manns, M.P., McHutchison, J.G., Gordon, S.C., Rustgi, V.K., Shiffman, M., Reindollar, R., Goodman, Z.D., Koury, K., Ling, M., and Albrecht, J.K. (2001). Peginterferon alfa-2b plus ribavirin compared with interferon alfa-2b plus ribavirin for initial treatment of chronic hepatitis C: a randomised trial. Lancet *358*, 958–965.

Manns, M., Reesink, H., Moreno, C., Berg, T., Benhamou, Y., Horsmans, Y., Dusheiko, G., Flisiak, R., Meyvisch, P., Lenz, O., *et al.* (2008a). Safety and antiviral activity of TMC436350 in treatment-naive genotype 1 HCV-infected patients. Hepatology *48*, 1023A.

Manns, M.P., Bourliere, M., Benhamou, Y., Schuchmann, M., D., H., Pol, S., Bonacini, M., Steinmann, G., Huang, D.B., Mikl, J., *et al.* (2008b). Safety and antiviral activity of BI 201335, a new HCV NS3 protease inhibitor, on combination therapy with peginterferon alfa-2a (P) and ribavirin (R) for 28 days in P+R treatment-experienced patients with chronic hepatitis C genotype 1 infection treatment-experienced patients with chronic hepatitis C genotype 1 infection. Hepatology *48*, 1151A–1152A.

Manns, M., Reesink, H., Moreno, C., Berg, T., Benhamou, Y., Horsmans, Y., Dusheiko, G., Flisiak, R., Meyvisch, P., Lenz, O., *et al.* (2009a). Opera-1 Trial: interim analysis of safety and antiviral activity of TMC435 in treatment-naive genotype 1 HCV patients. J. Hepatol. *50*, S7.

Manns, M.P., Gane, E., Rodriguez-Torres, M., Stoehr, A., Yeh, C.T., Wiedmann, R., Hwang, P., Quirk, E., Silber, J., and Lee, A. (2009b). MK-7009 significantly improves rapid viral response (RVR) in combination with pegylated interferon alfa-2a and ribavirin in patients with chronic hepatitis C (CHC) genotype 1 infection. J. Hepatol. *50*, S384–S384.

Marcellin, P., Reesink, H., Berg, T., Cramp, M., Flisiak, R., Van Vlierberghe, H., Verloes, R., Lenz, O., Peeters, M., Sekar, V., *et al.* (2009). Antiviral activity and safety of TMC435 combined with peginterferon alpha-2a and ribavirin in patients with genotype-1 hepatitis C infection who failed previous IFN-based therapy. J. Hepatol. *50*, S385.

Mederacke, I., Wedemeyer, H., and Manns, M.P. (2009). Boceprevir, an NS3 serine protease inhibitor of hepatitis C virus, for the treatment of HCV infection. Curr. Opin. Investig. Drugs *10*, 181–189.

Neukam, K., Macias, J., Mira, A., and Pineda, J.A. (2009). A review of current anti-HCV treatment regimens and possible future strategies. Expert Opin. Pharmacother. *10*, 417–433.

Niu, D., Hagel, M., Qiao, L., St. Martin, T., Sheets, M., Chaturvedi, P., Labenski, M., Nacht, M., Westlin, W., Petter, R.C., *et al.* (2010). AVL-*181*, A potent and selective irreversible HCV protease inhibitor, requires a helper amino acid in the binding microenvironment to form a covalent bond with CYS-159. J. Hepatol. *52*, S297–S297.

Njoroge, F.G., Chen, K.X., Shih, Y., and Piwinski, J.J. (2008). Challenges in modern drug discovery: a case study of boceprevir, an HCV protease inhibitor for the treatment of hepatitis C virus infection. Acc. Chem. Res. *41*, 50–59.

Ontoria, J.M., Di Marco, S., Conte, I., Di Francesco, M.E., Gardelli, C., Koch, U., Matassa, V.G., Poma, M., Steinkuhler, C., Volpari, C., *et al.* (2004). The design and enzyme-bound crystal structure of indoline based peptidomimetic inhibitors of hepatitis C virus NS3 protease. J. Med. Chem. *47*, 6443–6446.

Pasquinelli, C., Eley, T., Villegas, C., Sandy, K., Mathias, E., Wendelburg, P., Liao, S., McPhee, F., Scola, P.M., Sun, L.-Q., *et al.* (2009). Safety, tolerability, pharmacokinetics and antiviral activity following single- and multiple- dose administration of BMS-650*032*, a novel HCV NS3 inhibitor, in subjects with chronic genotype 1 HCV infection. Hepatology *50*, 411A–412A.

Pawlotsky, J.M. (2004). Hepatitis C: it's a long way to new therapy, it's a long way to go. Gastroenterology *127*, 1629–1632.

Perni, R.B., Almquist, S.J., Byrn, R.A., Chandorkar, G., Chaturvedi, P.R., Courtney, L.F., Decker, C.J., Dinehart, K., Gates, C.A., Harbeson, S.L., *et al.* (2006). Preclinical profile of VX-*950*, a potent, selective, and orally bioavailable inhibitor of hepatitis C virus NS3–4A serine protease. Antimicrob. Agents Chemother. *50*, 899–909.

Poordad, F., Shiffman, M., Sherman, K.E., Smith, J.B., Yao, M., George, S., Adda, N., and McHutchison, J.G. (2008). A Study of Telaprevir (TVR) with Peginterferon Alfa-2a (P) and Ribavirin (R) in Subjects with Well-Documented Prior PR Null Response, Non-Response or Relapse: Preliminary Results. J. Hepatol. *48*, S374–S375.

Prongay, A.J., Guo, Z., Yao, N., Pichardo, J., Fischmann, T., Strickland, C., Joseph Myers, J., Weber, P.C., Beyer, B.M., Ingram, R., *et al.* (2007). Discovery of the HCV NS3/4A Protease Inhibitor (1R, 5S)-N-[3-Amino-1-(cyclobutylmethyl)–2, 3-dioxo-propyl]–3- [2(S)-[[[(1, 1-dimethylethyl)amino] carbonyl]amino]–3, 3-dimethyl-1-oxobutyl]- 6, 6-dimethyl-3-azabicyclo[3.1.0]hexan-2(S)-carboxamide (Sch 503034) II. Key Steps in Structure-Based Optimization. J. Med. Chem. *50*, 2310–2318.

Rajagopalan, P.T.R., Misialek, S.S., Blatt, L.M., Seiwert, S.D.K., and Kossen, K. (2007). Characterization of HCV NA3/4a protease inhibition by ITMN 191 reveals picomolar potency and slow dissociation: implications for the use of ITMN 191 in chronic HCV treatment. Gastroenterology *132*, A–782.

Rajagopalan, R., Misialek, S., Stevens, S., Myszka, D., Brandhuber, B., Ballard, J., Andrews, S., Seiwert, S., and Kossen, K. (2009). Inhibition and binding kinetics of the hepatitis C virus NS3 protease inhibitor ITMN-191 reveals tight binding and slow dissociative behavior. Biochemistry *48*, 2559–2568.

Reesink, H., Verloes, R., Abou Farha, K., Van Vliet, A., Weegink, C., van 't Klooster, G., Aharchi, F., Marien, K., Van Remoortere, P., de Kock, H., *et al.* (2008). Safety of the HCV protease inhibitor TMC435350 in healthy volunteers and safety and activity in chronic hepatitis C infected individuals: a phase 1 study. J. Hepatol. *48*, S28–S29.

Reesink, H., Bergmann, J., de Bruijne, J., Weegink, C., van Lier, J., van Vliet, A., Keung, A., Li, J., O'Mara, E., Treitel, M., *et al.* (2009a). Safety and antiviral activity of SCH900518 administered as monotherapy and in combination with peginterferon alfa-2b to naive and treatment-experienced HCV-1 infected patients. J. Hepatol. *50*, S35.

Reesink, H., de Bruijne, J., Weegink, C., Lenz, O., Simmen, K., Peeters, M., De Smedt, G., Sekar, V., and Verloes, R. (2009b). Antiviral activity and safety of TMC435 combined with pegylated interferon and ribavirin in hepatitis C patients with genotype 1 who had previous exposure to TMC435. Hepatology *50*, 1024A.

Reiser, M., Hinrichsen, H., Benhamou, Y., Reesink, H.W., Wedemeyer, H., Avendano, C., Riba, N., Yong, C.L., Nehmiz, G., and Steinmann, G.G. (2005). Antiviral efficacy of NS3-serine protease inhibitor BILN-2061 in patients with chronic genotype 2 and 3 hepatitis C. Hepatology *41*, 832–835.

Sarrazin, C., Kieffer, T.L., Bartels, D., Hanzelka, B., Muh, U., Welker, M., Wincheringer, D., Zhou, Y., Chu, H.M., Lin, C., *et al.* (2007). Dynamic hepatitis C virus genotypic and phenotypic changes in patients treated with the protease inhibitor telaprevir. Gastroenterology *132*, 1767–1777.

Seiwert, S.D., Andrews, S.W., Jiang, Y., Serebryany, V., Tan, H., Kossen, K., Rajagopalan, P.T., Misialek, S., Stevens, S.K., Stoycheva, A., *et al.* (2008). Preclinical characteristics of the hepatitis C virus NS3/4A protease inhibitor ITMN-191 (R7227). Antimicrob. Agents Chemother. *52*, 4432–4441.

Stauber, K., Marlor, C.W., Mushtaq, G., Gadhachanda, V., Pais, G., Hashimoto, A., Phadke, A., Huang, M., and Deshpande, M. (2010). Characterization of the Hepatoselective distribution of ACH-1625, a potent, clinical stage HCV NS3 protease inhibitor. J. Hepatol. *52*, S431.

Steinkuhler, C., Biasiol, G., Brunetti, M., Urbani, A., Koch, U., Cortese, R., Pessi, A., De Francesco, R. (1998). Product inhibition of the hepatitis C virus NS3 protease. Biochemistry *37*, 8899–8905.

Sulkowski, M., Ferenci, P., Emanoil, C., Asselah, T., Alexandru, C.F., Lalezari, J., Bourlière, M., Mauss, S., Grange, J.-D., Berg, T., *et al.* (2009). SILEN-C1: Early Antiviral Activity and Safety of BI 201335 Combined with Peginterferon alfa-2a and Ribavirin in Treatment-naïve Patients with Chronic Genotype 1 HCV infection. Hepatology *50*, 90A.

Sulkowski, M., Bourliere, M., Bronowicki, J.-P., Streinu-Cercel, A., Preotescu, L., Asselah, T., Pawlotsky, J.-M., Shafran, S., Pol, S., Caruntu, F.A., *et al.* (2010). SILEN-C2: Early antiviral activity and safety of BI201335 combined with pegylated interferon alfa-2a and ribavirin (PEGIFN/RBV) in chronic HCV genotype-1 patients with non-response to PEGIFN/RBV. J. Hepatol. *52*, S462.

Tan, H., Rajyaguru, S., Wu, K., McCown, M.F., Ali, S., Jiang, W., Otto, M.J., Furman, P.A., Najera, I., Klumpp, K., *et al.* (2008). Combination of the NS3/4A protease inhibitor ITMN-191 (R7227) with the active moiety of the NS5B inhibitors R1626 or R7128 enhances replicon clearance and reduces the emergence of drug resistant variants. Hepatology *48*, 1153A.

Tan, H., Wang, G., Moy, C., Kossen, K., Rajagopalan, P.T., Misialek, S., Ruhrmund, D., Hooi, L., Snarskaya, N., Stoycheva, A., *et al.* (2009). Combination of the NS3/4A Protease Inhibitor ITMN-191 with the Allosteric NS5B Polymerase Inhibitor ITMN-8020 Enhances Replicon Clearance and Reduces the Emergence of Drug Resistant Variants. J. Hepatol. *50*, S340–S341.

Thimme, R., Lohmann, V., and Weber, F. (2006). A target on the move: innate and adaptive immune escape strategies of hepatitis C virus. Antiviral Res. *69*, 129–141.

Tsantrizos, Y.S., Bolger, G., Bonneau, P., Cameron, D.R., Goudreau, N., Kukolj, G., LaPlante, S.R., Llinas-Brunet, M., Nar, H., and Lamarre, D. (2003). Macrocyclic inhibitors of the NS3 protease as potential therapeutic agents of hepatitis C virus infection. Angew. Chem. Int. Ed. Engl. *42*, 1356–1360.

Vanwolleghem, T., Meuleman, P., Libbrecht, L., Roskams, T., De Vos, R., and Leroux-Roels, G. (2007). Ultra-rapid cardiotoxicity of the hepatitis C virus protease inhibitor BILN 2061 in the urokinase-type plasminogen activator mouse. Gastroenterology *133*, 1144–1155.

White, P.W., Llinas-Brunet, M., Amad, M., Bolger, G., Cordingley, M.G., Duan, J., Garneau, M., Lagace, L., Thibeault, D., and Kukolj, G. (2010). Preclinical characterization of non-covalent HCV NS3/4A protease inhibitor BI 201335. J. Hepatol. *52*, S302.

Wright, D., Miller, J.L., Verlinden, I., Cilissen, C., Valentine, J., Sun, P., Smet, M.D., Hoon, J.D., Depré, M., Cavens, L., *et al.* (2008). Safety, tolerability, and pharmacokinetic data following single- and multiple-dose administration of MK-7009, a hepatitis C virus non-structural 3/4a protease inhibitor, to healthy male subjects. Hepatology *48*, 1165A.

Yan, Y., Li, Y., Munshi, S., Sardana, V., Cole, J.L., Sardana, M., Steinkuehler, C., Tomei, L., De Francesco, R., Kuo, L.C., *et al.* (1998). Complex of NS3 protease and NS4A peptide of BK strain hepatitis C virus: a 2.2 A resolution structure in a hexagonal crystal form. Protein Sci. *7*, 837–847.

Yao, N., Reichert, P., Taremi, S.S., Prosise, W., and Weber, P.C. (1999). Molecular views of viral polyprotein processing revealed by the crystal structure of the hepatitis C virus bifunctional protease-helicase. Structure *7*, 1353–1363.

Discovery of Hepatitis C Virus NS3–NS4A Complex Inhibitors

11

Mingjun Huang, Kathe Stauber, Atul Agarwal, Milind Deshpande and Avinash Phadke

Abstract

Hepatitis C virus (HCV) infection is a serious global health problem because of its high prevalence, chronic nature and significant morbidity of the resulting diseases. The current standard of care (peginterferon-α2a/b and ribavirin combination) is limited by numerous side-effects and suboptimal efficacy in genotype-1 HCV-infected patients underscoring the unmet medical need for new therapies. Here we describe our efforts in discovery and optimization of a novel series of specific inhibitors leading to the nomination of ACH-806 (or GS-9132) for clinical development. These inhibitors interrupt the functions of HCV NS3–NS4A complex likely by binding to NS4A. We have summarized our studies on antiviral activity, resistance induction, mechanism of action, and pharmacokinetics with selected inhibitors including ACH-806. Finally, we present the data on a proof-of-concept clinical trial with ACH-806 that validates this novel mechanism of action.

Introduction

Conventional target-based drug-discovery relies on the use of biochemical or biophysical *in vitro* assays for lead identification which is exemplified by the successful identification of multiple classes of drugs to treat human immunodeficiency virus (HIV) infection. Using this approach, anti-HCV compounds have also been discovered that inhibit HCV-NS3 protease or HCV-NS5B polymerase by binding at the active or allosteric site of the enzyme A handful of these compounds have already been shown to reduce viral load in the HCV-infected patients alone and in combination with peginterferon alpha-2a/b and ribavirin (Sarrazin and Zeuzem, 2010). However, emergence of the resistant HCV variants to these agents is deemed inevitable because of the error prone nature of HCV NS5B polymerase and the high replication rate of HCV in patients (Martell *et al.*, 1992; Neumann *et al.*, 1998). In fact, the appearance of resistant viruses to anti-HCV drug candidates has already been observed during clinical trials (Sarrazin and Zeuzem, 2010). Learning from HIV, combination therapies employing multiple classes of antiviral agents that act through distinct mechanisms are believed to be necessary for sustained suppression of the HCV replication. Therefore, the cell-based subgenomic HCV replicons (Lohmann *et al.*, 1999) are being used in identifying anti-HCV compounds that inhibit HCV by novel mechanisms. Using this approach we have discovered a class of compounds by inhibiting the HCV NS3–4A complex.

Hit to lead discovery

Prior to the availability of HCV replicons carrying the luciferase reporter gene, anti-replicon activity of compounds was determined by quantifying the HCV RNA by using real-time PCR and hybridization dot blot assay. These methods of HCV RNA quantitation were expensive and their throughput was low. Therefore we started our HCV drug discovery efforts by developing a cost effective high throughput screening assay.

Cell lines stably harbouring HCV replicons were established by transfecting the HCV replicon molecules that encode a selectable marker. The neomycin phosphotransferase II (NPTII)

encoded by the *neo* gene is the most used selection maker for this purpose. Since NPTII is expressed only by the replicon molecules, the quantity of NPTII is directly correlated with the quantity of replicon molecules. We developed an *in situ* ELISA assay to quantify the amount of this NPTII in each well of 96-well plates. Our *in situ* ELISA assay has a throughput of 8000 compounds per week. The reported *in situ* ELISA assays use one of the HCV proteins as a surrogate marker. Our assay uses commercially available anti-NPTII antibody and overcomes the need of HCV specific antibody. It also eliminates the need of changing the antibodies for different genotypes and strains of HCV replicons.

Subsequently we screened a set of about 70,000 compounds that included about 35,000 compounds purified from plant cell cultures or insects and an additional 35,000 synthetically prepared small molecules by using our novel high-throughput assay. Our effort resulted in the discovery of 30 hits belonging to four distinct chemical classes. After confirmation of anti-replicon activity and cellular toxicity and demonstration of antiviral specificity, an acylthiourea compound (ACH-392) was chosen as a lead for optimization (Fig. 11.1).

Lead optimization

The structure of ACH-392 (compound 1) was divided into three parts for the purpose of studying the SAR and optimizing its potency. These three parts included an acyl group on the left hand side (LHS) of ACH-392, an acylthiourea functional group in the centre and an aniline on the right hand side (RHS) as shown in Fig. 11.1B. We derived a dataset by modifying each part independently that showed a clear structure activity relationship and helped in optimizing the potency of ACH-392.

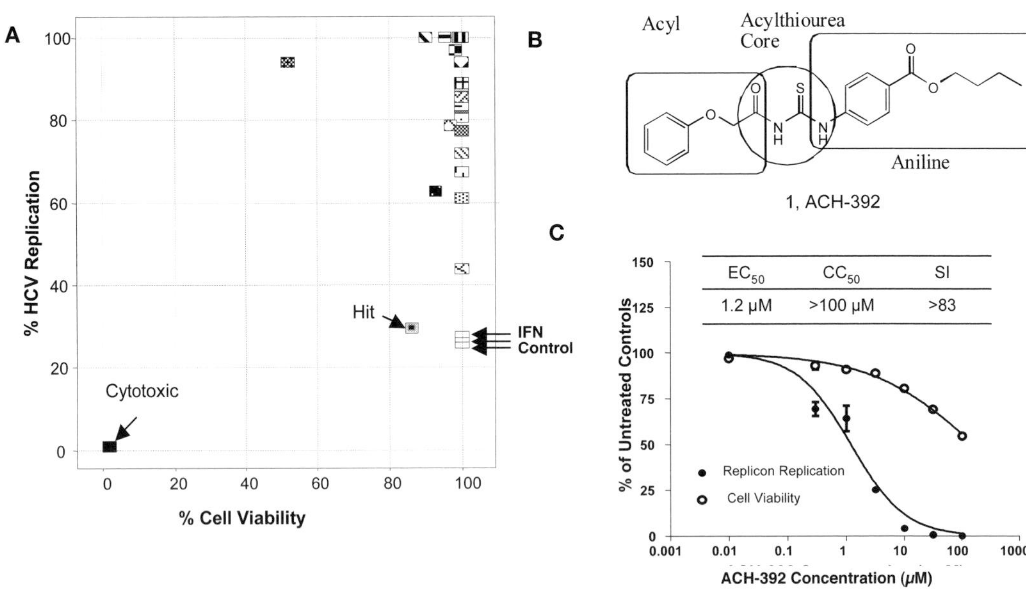

Figure 11.1 Representative screening and confirmative data. (A) Huh-9–13, a prototype HCV replicon-containing cell line (Lohmann *et al.*, 1999), was seeded in a 96-well plate and treated with 10 μM of each compound for 3 days. The cell viability was determined with MTS assay that was followed by in situ ELISA to detect the amount of NPTII in each well. The percentages of the cell viability and HCV replicon replication were calculated relative to untreated controls (defined as 100%) and plotted at the x-axis and y-axis, respectively. (B) Chemical structure of the hit (ACH-392). (C) To confirm the screening results, Huh-9-13 cells were seeded in a 96-well plate and treated with ACH-392 at six concentrations, each in triplicates, for 3 days. The cell viability and HCV replicon replication were determined as above and 100% was defined as the values determined for untreated controls. Non-linear regression analysis was performed with Prism from GraphPad Software (La Jolla, CA). SI, selective index, the ratio CC_{50} over EC_{50}.

We started by modifying the acylthiourea function where C=S was replaced by C=O, (compound 2) or the C=O group was removed (compound 3) or moved (compound 4). All of these changes diminished the compound activity indicating the necessity of the acylthiourea functional group as shown in Fig. 11.2A.

Next we explored the need of NH protons by synthesizing *N*-methyl (compounds 5 and 7) and *S*-methyl (compound 6) analogues. These

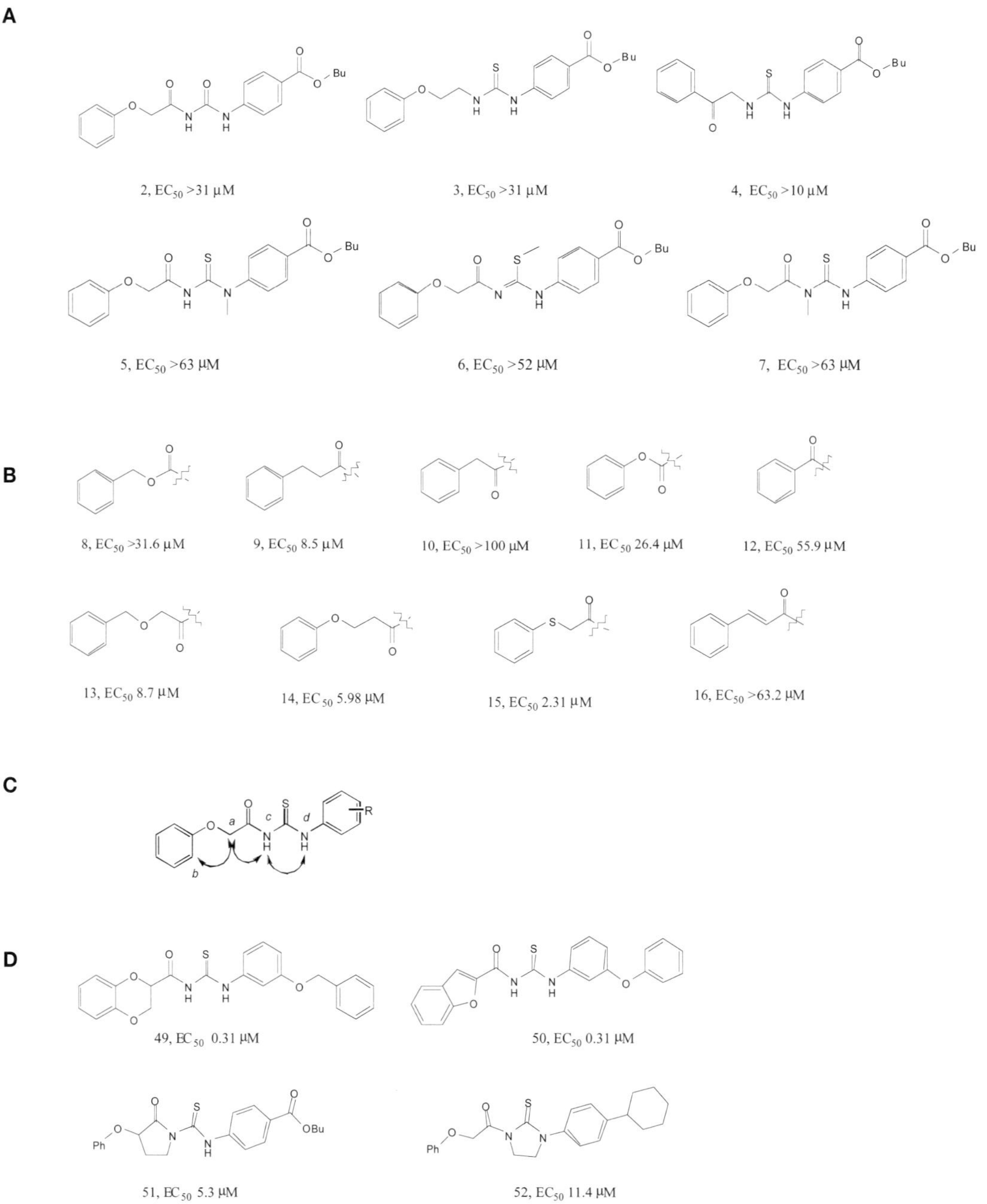

Figure 11.2 Optimization of acylthioureas. (A) Acylthiourea core modification. (B) Linker modification. (C and D) Reduction of the number of rotatable bonds.

changes resulted in a complete loss of activity (Fig. 11.2A) indicating that the NH protons are essential for the activity. Further exploring the SAR of ACH-392, the acyl group was modified by changing the linker that joined C=O with the aryl group and the presence and position of the 'O' while keeping the remaining structure constant. The resulting compounds 8, 10, 11, 12 and 16 were less potent than ACH-392. Increasing the spacer length and moving the oxygen away from the carbonyl or replacing the oxygen with $-CH_2$- or sulfur (compounds 13, 14, 9 and 15) reduced the potency less significantly as compared to the aforementioned structural modifications (Fig. 11.2B).

Next we explored the effect of substitutions on the phenoxyacetyl and assessed their effect on the compound potency. As can be seen from the data in Table 11.1A, introducing an electron donating or withdrawing group at the ortho, meta or the para position did not alter the potency of these acylthioureas significantly.

The structural modifications thus far yielded inactive or equipotent compounds but we noticed a trend in our SAR dataset that was derived by modifying the aniline on the right hand side of ACH-392. A small aliphatic, electron withdrawing or donating group at the ortho, meta or para position of the aniline nitrogen afforded nearly inactive analogues (compounds 32–38; Table 11.1B). However, the potency improved significantly when large hydrophobic groups were introduced at the meta or para position of the aniline (compounds 39–48; Table 11.1B). Furthermore, there was also a clear preference for aromatic substituents at the meta position

Table 11.1 Modification of phenoxyacetyl (left-hand side, A) and analine (right-hand side, B)

A

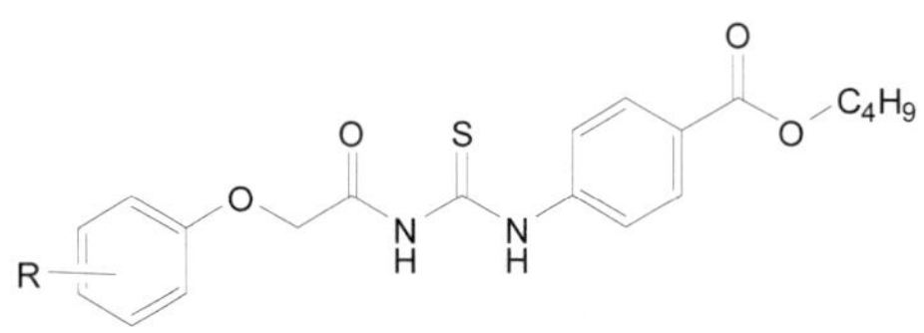

B

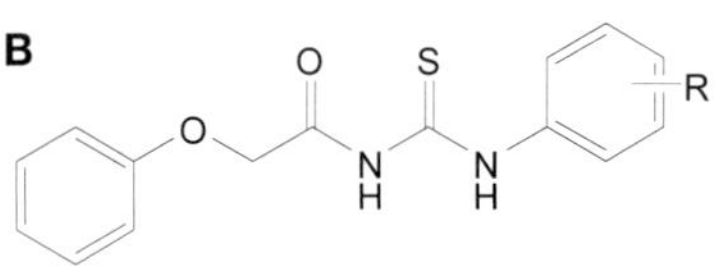

No.	R	EC_{50} (μM)	No.	R	EC_{50} (μM) a, Ortho	b, Meta	c, Para
17	H	2	32	Cl	>31	26	36
18	4-Cl	7.3	33	F	>63	>63	>63
19	3-Cl	3.0	34	CH_3	>31	>31	>31
20	4-Me	1.8	35	OMe	>31	>31	>31
21	2-Me	4.3	36	CN	>31	>31	12
22	2-OMe	4.1	37	COOMe	>31	>31	>31
23	3-OMe	5.4	38	COOEt	–	38	2
24	4-OMe	1.6	39	OPhenyl	>31	1	2
25	4-F	3.4	40	Phenyl	41	0.9	1.6
26	2,4-DiCl	2.2	41	Benzyl	>63	0.33	10
27	3,4-DiCl	2.7	42	*n*-Hexyl	–	–	<0.3
28	4-tBu	2.0	43	*n*-Pentyl	–	–	0.3
29	2,4,6-TriCl	5.2	44	*n*-Butyl	–	–	3.5
30	2,6-DiCl	5.5	45	*O*-*n*-Pentyl	–	–	0.5
31	4-OBn	5.2	46	*O*-*n*-Butyl	–	–	0.3
			47	*O*-Benzyl	–	0.19	2
			48	c-Hexyl	–	–	0.7

(compare 39b, 40b and 41b with 39c, 40c and 41c) and larger aliphatic substituents at the para position (compounds 42c-48c).

It has been shown that a reduction in the number of rotatable bonds and flexibility can improve a compound's potency and pharmacokinetic properties (Hann and Oprea, 2004). After having improved the potency by more than 10-fold, next we tried to reduce the number of rotatable bonds and rigidify these structures. As shown in Fig. 11.2C, we explored several relatively rigid analogues. Cyclization between atoms a and c (compound 51) or between c and d (compound 52) afforded compounds with significantly reduced potency whereas cyclization through atoms a and b (compounds 49, 50) maintained or slightly improved the potency (Fig. 11.2D).

Several more hydrophobic substituents were tried on the anilines with the benzofuran acyl on the RHS because aromatic/heteroaromatic acyls were synthesized with high yields compared with the aliphatic acyl groups (Table 11.2). The aforementioned trend was reconfirmed that aliphatic groups were preferred at the para position (compounds 60 and 61) while the aromatic groups were preferred at the meta position (compounds 53 and 56, 54 and 57, 55 and 58). An increase in the aliphatic chain length improved the potency where 4-*n*-pentyl (compound 62) and 4-*n*-pentyloxy (compound 66) were the optimum alkyl and alkyloxy length, respectively. Subsequently, the aniline ring was disubstituted by introducing halogens and haloalkyl groups at the meta position while keeping 4-*n*-pentyloxy at the para position.

Table 11.2 Benzofuranyl thioureas

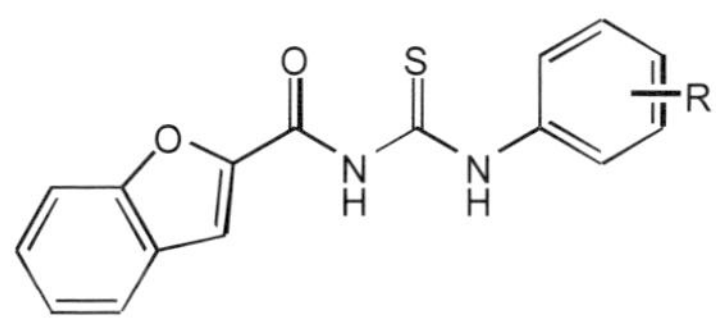

	R	EC_{50} (μM)
53	4-Benzyl	5.67
54	4-Phenyl	9.65
55	4-OPhenyl	2.4
56	3-Benzyl	0.09
57	3-Phenyl	0.55
58	3-OPhenyl	0.31
59	3-OBenzyl	6.51
60	3-O-n-Butyl	6.56
61	4-O-n-Butyl	0.81
62	4-n-Pentyl	0.31
63	4-n-Butyl	4.12
64	4-n-Propyl	6.95
65	4-Ethyl	>10
66	4-O-*n*-Pentyl	0.32
67	4-O-*n*-Hexyl	0.81
68	4-O-*n*-Octyl	7.15
69	3-Chloro, 4-OPentyl	0.23
70	3-Fluoro, 4-OPentyl	0.26
71	3-Trifluoromethyl, 4-OPentyl	0.22

These disubstituted anilines yielded more potent analogues (compounds 69, 70 and 71 compared with 66).

More heterocycles were explored on the left hand side to identify six membered nitrogen heteroaryls like pyridine (compound 74, Table 11.3A) that showed very good potency. The most potent analogues (compounds 90, 91, 92, 94; Table 11.3B) were obtained by combining the best anilines (Table 11.2) with 3-pyridyl on the left hand side (Table 11.3A). The most potent acylthiourea, compound 92 (ACH-806) was recommended for clinical development based on its other properties described in the following sections.

Virology studies on selected inhibitors

Antiviral activity

Anti-replicon activity of the compounds was determined in a Huh-7 cell line stably harbouring a subgenomic replicon derived from a genotype 1b strain (Con-1). Their specificity was also determined during lead optimization by testing them against a panel of mammalian viruses. Results for representative compounds in Table 11.4 show that these acylthioureas specifically inhibit HCV over the other mammalian viruses.

The anti-HCV spectrum of our development candidate (ACH-806) was determined by assessing its activity in the following systems as well: H/SG-neo, a Huh-7 cell line stably carrying a subgenomic replicon derived from a genotype 1a strain (H77; Blight *et al.*, 2003) and four Huh-7 cell lines stably harbouring subgenomic chimeric replicons. These four chimeric replicons were made by replacing both NS3 and NS4A genes in the *neo*-encoding subgenomic Con-1 replicon with a cassette amplified from 'quasispecies' pools of HCV-infected patient sera. A control chimeric replicon was also made by replacing both NS3 and NS4A genes in the *neo*-encoding subgenomic Con-1 replicon with that from H77. Following transfection of 'quasispecies' replicon RNA molecules into Huh-7 cells and a short selection under G418 of the transfected cells (a process to enrich replicon-containing cells), surviving colonies

Table 11.3 Optimized acylthioureas

A

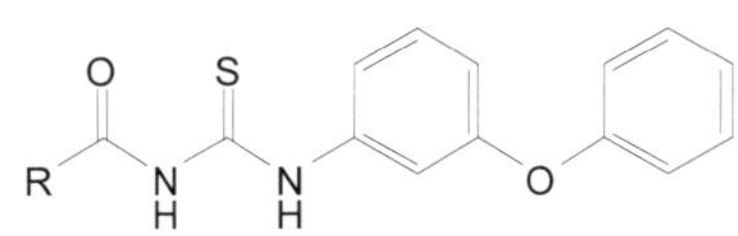

B

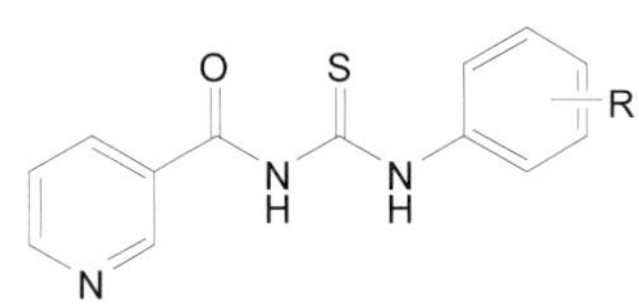

No.	R	EC_{50} (µM)	No.	R	EC_{50} (µM)
72	2-Benzofuranyl	0.09	85	3-OBenzyl	0.35
73	3- Benzofuranyl	>10	86	3-Benzyl	0.49
74	3- Pyridyl	0.15	87	4-OPentyl	0.3
75	2-Pyrazine	2.7	88	4-Octyl	0.064
76	1-Methylimidazol-2-Yl	6.06	89	4-OOctyl	0.072
77	2-Quinoxalinyl	0.6	90	4-OHexyl	0.059
78	5-Isoxazolyl	0.64	91	3-Fluoro-4-OPentyl	0.053
79	3-Indolyl	>10	92	3-Trifluromethyl-4-OPentyl	0.014
80	2-Benzothiophenyl	5.1	93	4-OButyl	1.2
81	2-Indolyl	>10	94	3-Trifluoromethyl-4-OHeptyl	0.049
82	6-Quinolinyl	0.93	95	3-Cyano-4-OPentyl	0.92
83	5-Azabenzofuran-2-Yl	0.7			
84	5-Pyrimidinyl	0.20			

Table 11.4 Activity of lead compounds against selected mammalian viruses

ACH/compound number	HCV (μM)[a]		HIV (μM)[b]		BVDV (μM)[b]		RSV(μM)[b]		HSV-1 (μM)[b]	
	EC_{50}	CC_{50}	EC_{50}	CC_{50}	EC_{50}	CC_{50}	EC_{50}	CC_{50}	EC_{50}	CC_{50}
392/1	1.2	>100	>93	93	>100	>100	>100	>100	>100	>100
659/47[b]	0.19	>100	>100	>100	25	>32	>100	>100	>100	>100
209/56	0.09	>100	>96	96	>100	>100	>100	>100	>100	>100
806/92	0.014	>10	>50	50	>50	>50	>40	>40	>40	>40

[a]EC_{50} values were obtained either in Huh-9-13 with NPT-II assay or in Huh-luc/neo (Vrolijk *et al.*, 2003) with luciferase assay. CC_{50} values were obtained simultaneously with MTS assay.

[b]HIV-1: human immunodeficiency virus type 1; BVDV: bovine viral diarrhoea virus; RSV: respiratory syncytial virus; HSV-1: herpes simplex virus type 1. The host cells used in the assays: CEM-SS (HIV-1), MDBK (BVDV), Hep-2 (RSV) and Vero (HSV-1).

were pooled to test their susceptibility to the inhibitors. The activity of ACH-806 and the control inhibitor (VX-950, an NS3 protease inhibitor currently in phase III development; Sarrazin and Zeuzem, 2010) against these chimeric replicons and against H77 and Con-1 replicons is summarized in Table 11.5. The average EC_{50} values of ACH-806 against genotype 1a and 1b replicons are 0.042 and 0.033 μM, respectively which is 11- to 17-fold greater potency than VX-950. In addition, the difference in the EC_{50} values of ACH-806 against different replicons used is less than 4-fold which is comparable to that of VX-950. Our results indicated that ACH-806 is a potent inhibitor against both genotype 1a and 1b HCV.

The EC_{50} value of ACH-806 increases by 10-fold in the presence of a physiological concentration of human serum albumin (45 mg/ml) whereas it is unchanged in the presence of a physiological concentration of α-acidic glycoprotein (1 mg/ml).

Combination with other HCV inhibitors

It is expected that clinical combination therapy with inhibitors acting by different mechanisms will be necessary to achieve sustained virological response. Therefore, anti-HCV activity of the two drug combinations was evaluated where ACH-806 was combined with interferon-α (IFN-α), VX-950, NM-107 (an NS5B polymerase nucleoside inhibitor, 2′-C-methyl-cytosine; Golitsina *et al.,* 2009) or HCV-796 (an NS5B polymerase non-nucleoside inhibitor; Kneteman *et al.,* 2009). Following the conventional method (Prichard *et al.,* 1993; Prichard and Shipman, 1990), HCV replicon-containing cells were treated with ACH-806 or one of the other inhibitors or ACH-806 in combination with one of the inhibitors in a checker board manner. The inhibition of HCV replicon by each treatment was analysed using a method developed by Prichard and Shipman (Prichard and Shipman, 1990; Prichard *et al.,* 1993). Consistent with the previous report (Wyles *et al.,* 2008), ACH-806 is not antagonistic with the above mentioned inhibitors and more importantly, it is synergetic with VX-950, NM-107 and HCV-796 (data not shown).

The ultimate goal of HCV therapy is to eradicate the virus or *cure* when this effect is studied *in vitro*. Therefore, we evaluated the *curing* effect of ACH-806 in combination with VX-950, NM-107 or HCV-796 in a *curing* assay as follows: HCV replicon-containing cells were incubated with one or two inhibitors at concentrations about 3 to 5× EC_{50} in the absence of G418 for 9 days, during which the cells were split twice a week

Table 11.5 Potency (EC_{50} in μM) of ACH-806 against various replicons[a]

Compound	H/SG-Neo (1a/H77)[b]	Huh-9-13 (1b/Con-1)[b]	Chimeric replicon[c]					Average ±SD	
			H77 (1a)	P1(1a)	P2 (1a)	P3 (1b)	P4 (1b)	1a	1b
ACH-806	0.052 ± 0.042	0.022 ± 0.009	0.022	0.024	0.068	0.030	0.048	0.042 ± 0.017	0.033 ± 0.013
VX-950	0.348 ± 0.125	0.236 ± 0.011	0.402	0.692	0.402	0.844	0.563	0.461 ± 0.185	0.548 ± 0.304

[a]EC_{50} values were obtained using hybridization dot blot assay as described previously (Zhao *et al.*, 2008).

[b]Average of two independent experiments.

[c]Chimeric replicons were established in the backbone of genotype1b/Con-1 replicon. NS3 and NS4A genes in H77 chimeric replicon were derived from one HCV c-DNA genome; NS3 and NS4A genes in P1, P2, P3 and P4 chimeric replicon were derived from a quasispecies of HCV c-DNA genomes amplified from HCV-infected patients.

and inhibitor(s) was replenished. The possible cytotoxic or cytostatic effect was precluded by re-seeding the same number of cells at each passage. At the end of the treatment with the inhibitor(s), the cells were re-plated at the same density and selected in the presence of G418. As a result, the cells that were rid of replicon molecules (or *cured*) during the 9-day treatment with the inhibitor(s) died under G418 selection whereas the cells that were not rid of replicon molecules (or not *cured*) survived and formed colonies in about three weeks. Hence, this assay measured the *curing* effect of inhibitor(s).

The *curing* effect of ACH-806 in combination with IFN-α, VX-950 and NM-107 or HCV-796 is summarized in Fig. 11.3. A modest reduction in the number of colonies was observed after the replicon cells were treated alone with ACH-806, IFN-α, VX-950, HCV-796 or NM-107 for 9 days. In contrast, a significant reduction in the number of colonies was observed after the replicon cells were treated with ACH-806 in combination with any of the other inhibitors for 9 days. Owing to the nature of the assay, it was difficult to quantify whether the combination effect was additive or synergistic. Nevertheless, the results clearly showed that combination of ACH-806 with any of the inhibitors will facilitate the cells ridding of the replicons.

Resistance studies

Extensive resistance studies were conducted with select compounds including ACH-392 and ACH-806. For selection of the replicon variants resistant to the class, HCV replicon-containing cells were exposed to one of the leads at a concentration at least 10-fold higher than its EC_{50} in the presence of G418. With time, either cellular or viral adaptation occurred leading to loss of inhibitory activity of the compound. Consequently, the cells survived and formed colonies in the presence of G418. The duration to obtain colonies in the size > 1 mm is between 4 and 6 weeks and the frequency to form the colonies is between 0.015% and 0.05% with our leads. These numbers are similar to those observed with NS3 protease inhibitors, NS5B nucleoside inhibitors and some non-nucleoside polymerase inhibitors.

The replicon cellular clones obtained with the first lead (ACH-392) were tested for their susceptibility to its analogues, an NS3 protease inhibitor (PI-1), an NS5B non-nucleoside inhibitor (NNI-1) and an NS5B nucleoside inhibitor (NI-1). As shown in Table 11.6 with the example of the resistant cellular clone #5, ACH-392 is cross-resistant to its analogues [ACH-659 (compound 47) and ACH-209 (compound 56) but not to the other classes of inhibitors tested.

As the lead compound was discovered through screening the HCV replicon-containing cells,

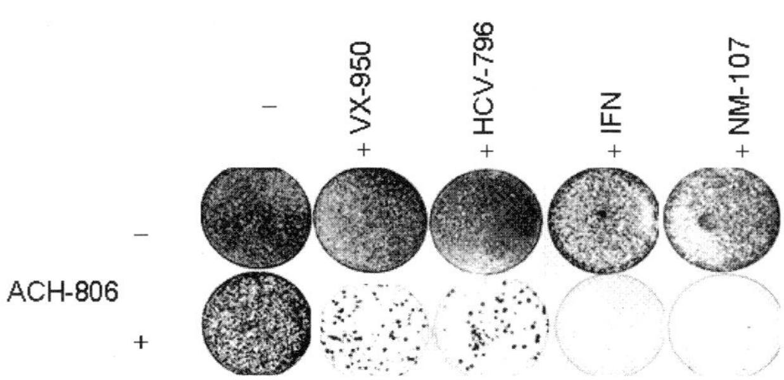

Figure 11.3 Combination of ACH-806 with other inhibitors. Huh-9-13 cells were treated with the inhibitors indicated either alone or in combination with ACH-806 as described in the text. The concentration used was approximately 5× of the EC_{50} (0.1 μM) for ACH-806, 4× EC_{50} (1 μM) for VX-950, 2.5× EC_{50} (0.05 μM) for HCV-796, 5× EC_{50} for NM-107 (6 μM) and 10× EC_{50} (15 IU/ml) for IFN. The EC_{50} values in Huh-9-13 cells were obtained by measuring HCV RNA using hybridization dot blot assay (Zhao *et al.*, 2008).

Table 11.6 Cross resistance among ACH-392 series and between ACH-392 series and other classes of inhibitors

Compound on cell clone[b]	EC_{50} (μM)[a] Against naïve replicon cell clone[b]	Against ACH-392[R] replicon cellular clone[b]
ACH-392	1.5	20
ACH-659	0.28	8.9
ACH-209	0.09	3.1
PI-1[c]	1.4	0.94
NNI-1[d]	1.2	1.1
NI-1[e]	0.58	0.81

[a]EC_{50} was determined by measuring HCV RNA with hybridization dot blot assay (Zhao *et al.*, 2008).

[b]The naïve or parent replicon cell line is Huh-9–13; the resistant cellular clone in the study is #5.

[c]PI-1 is an NS3 protease inhibitor and its structure has been reported previously (Yang *et al.*, 2008).

[d]NN-1 is a non-nucleoside NS5B polymerase inhibitor (a benzothiadiazine derivative) and its structure has been reported previously (Yang *et al.*, 2008).

[e]NI-1 is a nucleoside NS5B polymerase inhibitor (2′-C-methyl-adenosine).

it was important to confirm that the resistant cellular clones resulted from adaptation in the viral genome rather than in the host factors (Fig. 11.4A). We performed two studies with the resistant cellular clones obtained under the selection of ACH-392. In the first study, we cured one of the ACH-392 resistant cellular clones (#5) with interferon-α in the presence of ACH-392 at the concentration used for selection. The purpose of adding ACH-392 during the curing was to prevent reversal of mutation(s). Following this curing (4 weeks' co-treatment with interferon-α and ACH-392), the cells were transfected with the naïve transient HCV replicon and were subjected for susceptibility tests to ACH-659 (ACH-392 analogue). If the adaptation was due to the change in host factor(s), the naïve HCV replicon in the ACH-392 resistant cured cells will be less susceptible to ACH-659 (Fig. 11.4B). However, we did not observe any reduction in the susceptibility to ACH-659 of the cured ACH-392 resistant cells transfected with the naïve replicon compared with the cured naïve cells (Fig. 11.4D). This result suggested that the adaptation was not derived from the host factor(s).

In the second study, we isolated the replicon RNA (along with host RNA) from both the naïve and one of ACH-392 resistant replicon cellular clones (#5) and transferred the RNA into naïve Huh-7 cells by transfection. After treating the cells with G418 for 10 days (a process to enrich replicon-containing cells), all of the cells were pooled and re-plated into 96-well plates to evaluate the EC_{50} values of ACH-392 and ACH-659. If the transferred replicon RNA carried the genetic information which conferred the replicon resistance to ACH-392 series, the fold shift in the EC_{50} values obtained with these two sets of newly transfected cells should be retained (Fig. 11.4C). As shown in Fig. 11.4E, EC_{50} values of ACH-392 and ACH-659 against the cells transfected with the RNA isolated from the naïve replicon cellular clone are 1.1 and 0.18 μM, respectively whereas those values against the cells transfected with the RNA isolated from the ACH-392 resistant cellular clone are 15 and 4.0 μM, respectively. The fold increase in EC_{50} values between two sets of transfected cells is similar to that observed with two cellular clones (the naïve and the ACH-392 resistant) from which RNA was isolated (compare Table 11.6 and Fig. 11.4E), demonstrating that the replicon RNA residing in the ACH-392 resistant cellular clone carries sufficient genetic information to confer the replicon resistance to ACH-392 series. In conclusion, resistance to ACH-392 and its analogues developed in the replicon cellular clones resulted from viral adaptation.

Confirmation of the viral adaptation was followed by genotypic analyses to map the location of resistance mutations in the resistant replicon molecules. Replicon RNA molecules were isolated from a few replicon cellular clones obtained under selection of ACH-392 and its analogues and were reverse amplified with PCR. The PCR products were sequenced and the mutations identified in ≥ 50% clones are summarized in Fig. 11.5A. In addition, we carried out genotypic analyses of replicon RNA molecules isolated from PI-1, NNI-1 and NI-1 resistant cellular clones and reported their unique resistant mutations (Migliaccio *et al.*, 2003; Nguyen *et al.*, 2003; Trozzi *et al.*, 2003) along with those found in ≥ 50% of ACH-392 resistant cellular clones in Fig. 11.5A. The mutation present exclusively in the acylthioureas resistant clones is a substitution of

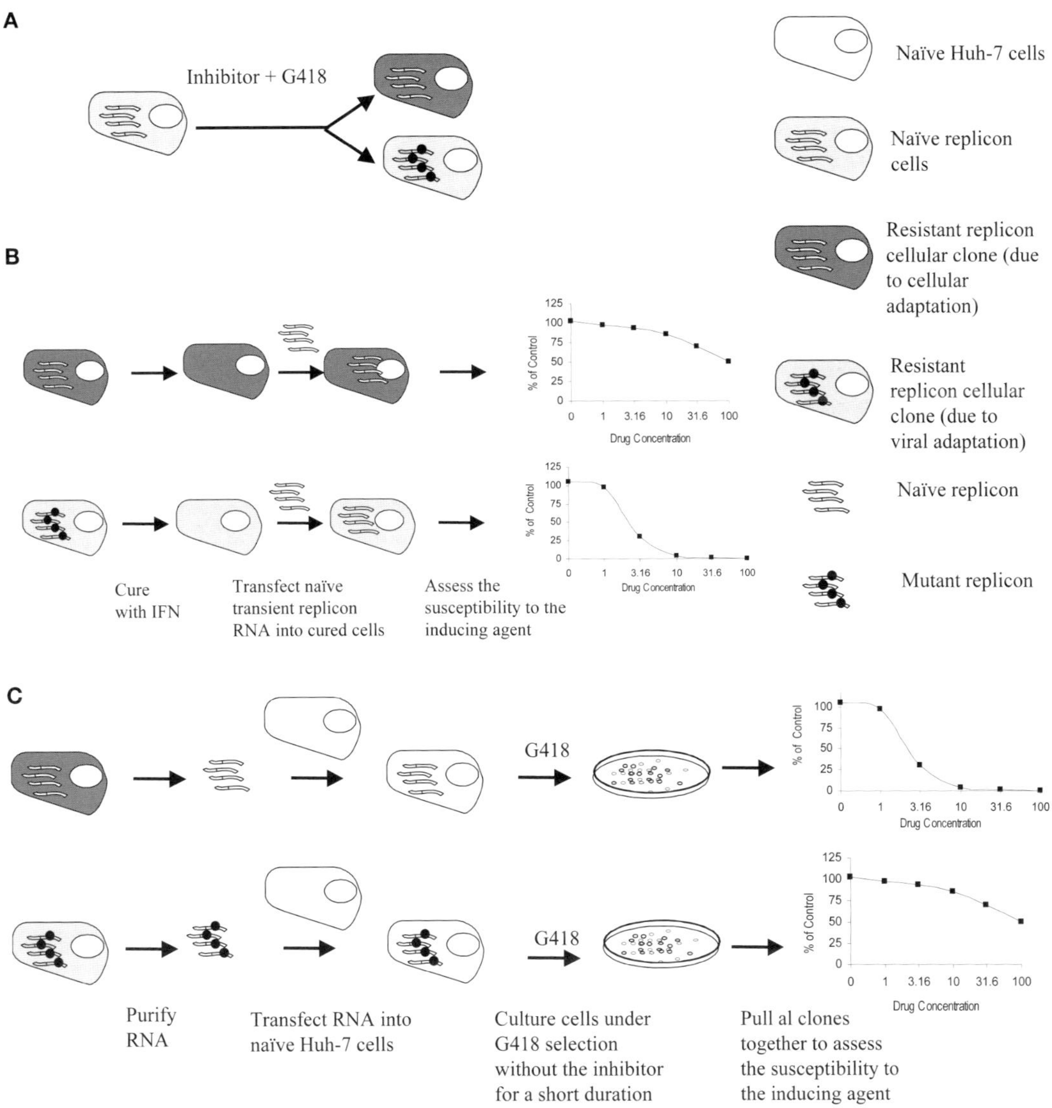

Figure 11.4A–C Viral vs. cellular adaptation. (A) Emergence of the resistant replicon cellular clones under the selection of an inhibitor due to either the viral adaptation or the cellular adaptation. (B) Approach to exclude the cellular adaptation. (C) Approach to confirm the viral adaptation.

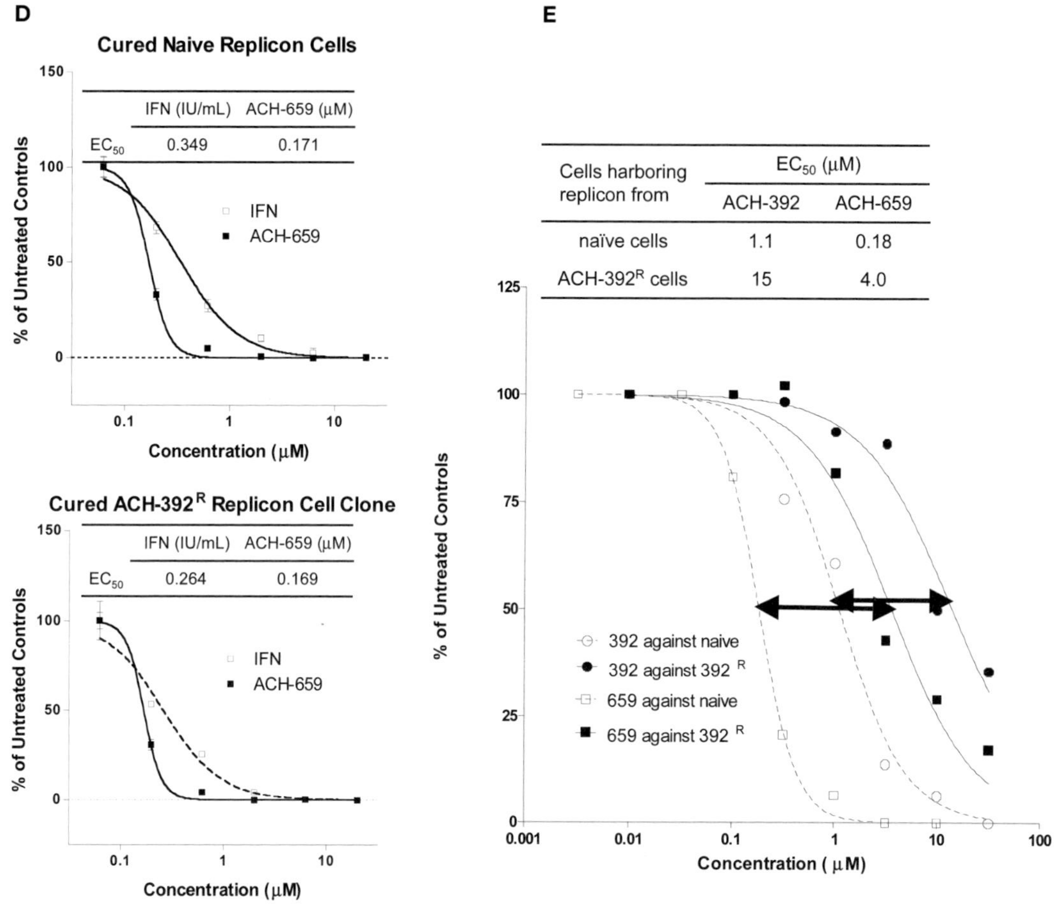

Figure 11.4D–E (D) Comparison of the susceptibility of the naïve replicon after its transfection into the cured naïve or the resistant cellular clone (#5). Similar EC_{50} values were obtained for the ACH-392 analogue (ACH-659) as well as for the control inhibitor, IFN- between two sets of transfected cells. EC50 values were determined with luciferase assay and non-linear regression analysis was performed with Prism from GraphPad Software (La Jolla, CA). (E) Comparison of susceptibility of the cells harbouring the replicon RNA from the naïve replicon cellular clone or the resistant cellular clone (#5). The EC_{50} values for both ACH-392 and its analogue (ACH-659) are >10-fold higher in the cells transfected with the replicon RNA isolated from the ACH-392 resistant cellular clone than in the cells transfected with the replicon RNA isolated from the naïve cells. EC50 values were determined with hybridization dot blot assay (Zhao *et al.*, 2008) and non-linear regression analysis was performed with Prism from GraphPad Software (La Jolla, CA).

alanine (ala or A) at position 39 of NS3 to valine (val or V).

Next, we carried out several reverse genetic studies to confirm the role of A39V mutation (Fig. 11.5B):

1. In order to exclude the role of mutations in regions other than NS3 and NS4A proteins, we replaced the NS3-NS4A genes in the naïve replicon with that from one of the resistant replicons and compared the susceptibility of the parent (naïve) and resulting chimeric replicons to select acylthioureas.
2. Subsequently, to rule out the role of non-consensus mutations in the NS3 and NS4A proteins, we reversed valine at position 39 of NS3 to alanine in the aforesaid chimeric replicon and determined any change in the susceptibility of the derived replicon to select acylthioureas.

A

Amino Acid				Naïve	392^{R}			659^{R}	209^{R}		PI^{R}	NNI^{R}	NI^{R}
Change		Protein	Position		#5	#7	#12	A	N3	M			
A	V	NS3	39	–	+	+	+	+	+	+	–	–	–
D	V	NS3	168	–	–	–	–	–	–	–	+	–	–
E	G	NS3	176	–	–	–	–	+	–	–	+	–	–
K	E	NS3	583	–	+	+	NA	–	+	+	+	–	+
S	I	NS5A	232	+	+	+	NA	NA	+	+	NA	+	+
D	E	NS5A	282	+	+	+	NA	NA	+	+	NA	+	+
I	V	NS5A	352	+	+	+	NA	NA	+	+	NA	+	+
S	T	NS5B	282	–	–	–	NA	NA	–	–	NA	–	+
S	T	NS5B	414	–	–	–	–	NA	–	–	NA	+	–

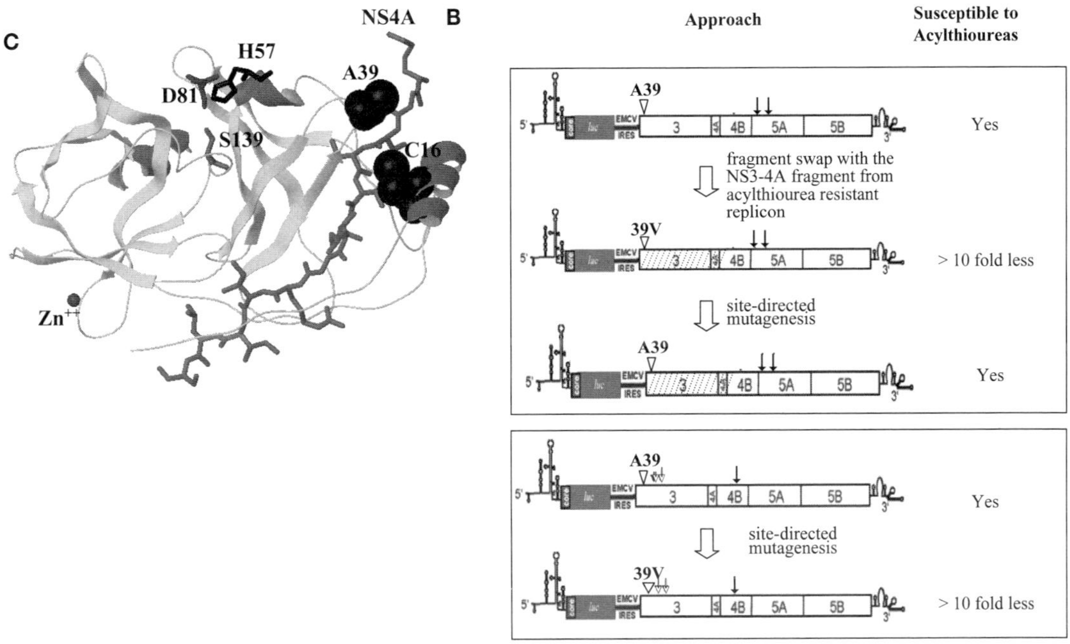

Figure 11.5 Genotyping and reverse genetic studies. (A) Representative genotyping results are shown. The '+' in bold represents the mutations present exclusively in a class of the inhibitors, which is A39V in NS3 for acylthioureas (ACH-392, 659 and 209), D168V in NS3 for NS3 protease inhibitor (PI-1), M414T in NS5B for NS5B non-nucleoside inhibitor (NNI-1) and S282T in NS5B for NS5B nucleoside inhibitor (NI-1). (B) Reverse genetic studies. A bicistronic non-selectable (transient) replicon with adaptive mutations (depicted as arrows) either in NS3 and NS4B (Lohmann *et al.*, 2003) or in NS5A (Guo *et al.*, 2001) was used as the parent to construct the mutant replicons. The susceptibility of the parent and the mutant replicons to acylthioureas was compared following transfection them into Huh-7 cells and the results were summarized on the right. The open triangle represents the amino acid residue at position 39 of NS3, which is either A (alanine) or V (valine). (C) Location of C16 and A39 of NS3. NS3 protease domain (pdb code: 1NS3) is shown with light grey colour where b-strands and -helices are shown as ribbons; catalytic site residues are shown with thick lines (D81, H57, S139); structural zinc ion is displayed as a sphere; NS4A co-factor is shown with thick lines; and the resistant mutants (A39 and C16) are shown with CPK representation. The distance between C16 (Cb) of NS3 and V26 (CG1) of NS4A and A39 (Cb) of NS3 and G21 (C) of NS4A is 4.1 Å and 7.3 Å, respectively.

3 Lastly, to confirm that A39V was sufficient, we introduced this mutation in the naïve replicon and determined the susceptibility of this replicon to select acylthioureas. Results of these studies are schematically summarized in Fig. 11.5B, demonstrating that the A39V mutation is sufficient to confer HCV replicon resistance to this class of inhibitors.

The aforementioned genotyping, phenotyping and reverse genetic studies were repeated by using ACH-806. In addition to A39V, a cysteine (cys or C) to serine (ser or S) substitution at position 16 of NS3 was also found to confer HCV replicon resistance to ACH-806 (Yang *et al.*, 2008). The location of C16 and A39 are shown in Fig. 11.5C, along with the binding position of the NS4A cofactor, both in the proximity of NS4A cofactor.

Mechanism of action studies

Detailed mechanism of action studies will be published elsewhere. Briefly, these acylthioureas displayed no significant activity against NS3 protease and NS5B polymerase in the enzymatic assays. These inhibitors induced formation of a dimeric form of NS4A in the replicon cells, designated as p14 (Fig. 11.6), the amount of which increased dose-dependently (Fig. 11.6A). Expression of NS4A alone was observed to be sufficient for the induction of p14 formation and this effect was correlated with the anti-replicon activity (Fig. 11.6B). A photoaffinity-labelled analogue of ACH-806 was found to bind selectively to NS4A both *in vitro* and *in vivo*. The interaction between NS3 and NS4A was reduced in the replicon cells upon treatment with acylthioureas leading to a reduction of the amount of NS3 because of the reduced stability of NS3. In addition, the internal cleavage of NS3 was blocked in the replicon cells, presumably due to its stringent dependency on the NS3–NS4A complex (see discussion below). These results, together with the location of resistance mutations that are in the NS3 domain that directly interacts with NS4A strongly suggest that this class of inhibitors interrupt the functions of the NS3–NS4A complex likely by binding to NS4A.

Blocking the interaction between NS3 and NS4A was once proposed as an approach to inhibit NS3 protease because NS4A enhances the activity of NS3 protease. However, the success of this approach was thought unlikely because of the very tight binding between NS3 and NS4A (Kd is estimated to be <1 nM; Gallinari *et al.*, 1999). Our results are consistent with this predication that despite interfering with the interaction

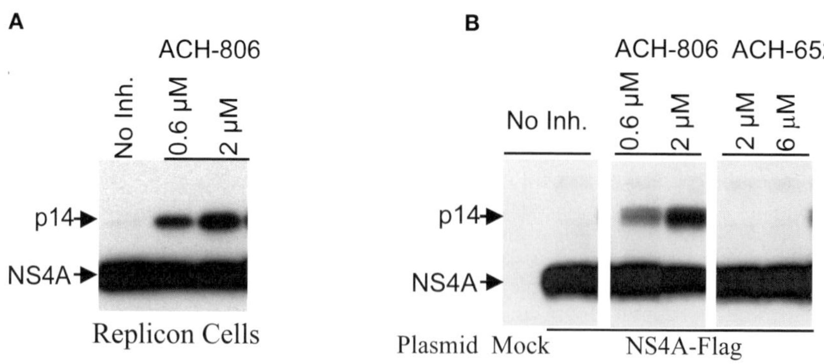

Figure 11.6 Induction of a dimeric form of NS4A (p14). (A) HCV replicon cells (Huh-9–13) were treated overnight and were subjected to cell lysis, immunoprecipitation, SDS-PAGE and Western blot with anti-NS4A antibody. (B) Huh-7 cells were transfected with a plasmid expressing NS4A fused at its C-terminal with a flag epitope. After 40 hours' exposure to ACH-806 or its urea derivative (ACH-652, EC_{50}>31 µM), the cells were lysed and the cell lysates were subjected to SDS-PAGE and Western blot analysis with anti-flag antibody. Positions of viral proteins are indicated on the left.

between NS3 and NS4A, acylthioureas do not effectively block the viral polyprotein processing. Still, these acylthioureas display a potent activity against HCV RNA replication which could not have been predicted *a priori*.

NS4A is the smallest non-structural protein of HCV and plays multiple roles in the HCV life cycle. Most importantly, NS4A interacts with NS3 to enhance its intracellular stability, (Wolk *et al.*, 2000), to facilitate its protease activity (Failla *et al.*, 1994; Tanji *et al.*, 1995; Butkiewicz *et al.*, 1996; Koch *et al.*, 1996) and to regulate its helicase activity (Pang *et al.*, 2002; Frick *et al.*, 2004; Kuang *et al.*, 2004). NS4A also directs the membrane association of NS3 (Hijikata *et al.*, 1993; Lin and Rice, 1995; Tanji *et al.*, 1995; Wolk *et al.*, 2000) and forms a complex with other HCV proteins, such as NS4B and NS5A (Asabe *et al.*, 1997; Lin *et al.*, 1997), which suggest its role in assisting the viral RNA replication complex formation. NS4A modulates NS5A phosphorylation either by itself or together with NS3 or NS4B (Tanji *et al.*, 1995; Asabe *et al.*, 1997; Koch and Bartenschlager, 1999; Lindenbach *et al.*, 2007). Lastly, point mutations in the central domain of NS4A reduce the amount of the NS3–4A complex and result in the blockage of NS3 internal cleavage but have no apparent effect on the viral polyprotein processing, which demonstrates more stringent dependence of the NS3 internal cleavage on the NS3–NS4A complex than the viral polyprotein processing (Kou *et al.*, 2007). Moreover, the internal cleavage of NS3 appears to be needed for HCV RNA replication because when we eliminated some of the internal cleavage products by introducing the mutations at their cleavage sites, we observed no growth of the replicon (data not shown). All together, these diverse functions suggest a critical role of NS4A in viral replication and also highlight the importance of the interaction between NS3 and NS4A for viral replication. Hence, it is not surprising that potent inhibition of HCV RNA replication can be achieved with the optimized acylthioureas that bind to NS4A and reduce the interaction between NS3 and NS4A even though this class has no apparent effect on viral polyprotein processing.

Pharmacokinetic studies on ACH-806

Incorporation of absorption, distribution, metabolism and elimination (ADME) profiling into the drug discovery process is standard practice in the pharmaceutical industry (Tsaioun and Jacewicz, 2009). While *in vitro* physicochemical (solubility and permeability) and biological (metabolic stability) screens provide valuable information to develop and optimize compounds, pharmacokinetic characterization of key compounds in a discovery target series is an integral part of the drug optimization and selection process (Singh, 2006). Besides being a pivotal bridge between *in vitro* potency and efficacy *in vivo,* pharmacokinetic investigations early in the discovery process provide unique information on ADME characteristics that may not be discerned from *in vitro* screens alone. In addition, these early pharmacokinetic studies can confirm the reliability of *in vitro* screens as part of the iterative drug design process.

For this series of HCV NS3–4A complex inhibitors, the use of *in vitro* ADME studies combined with antiviral potency, guided the chemical synthesis forward to the nomination of ACH-806 as a development candidate. Pertinent to the selection of ACH-806 was its high potency (at least 10-fold greater potency than others in the series) and its high metabolic stability (165 min) in human liver microsomes (Table 11.7).

In this section, we describe the pharmacokinetic profile of ACH-806. The pharmacokinetic parameters after an intravenous dose (1 mg/kg) are listed on Table 11.8. While the half-life was short in rat, a notably longer half-life and lower clearance values were observed in the dog. The lower clearance in the dog was consistent with the rank order of metabolic stability provided from the *in vitro* assays.

Typically, initial oral preclinical pharmacokinetic studies in rodents assess the potential for lead compounds to reach efficacious plasma concentrations. While allometric scaling and human exposure simulations are performed throughout development, these initial pharmacokinetic studies, combined with *in vitro*

Table 11.7 Summary of metabolic stability of ACH-806 in liver microsomes[a]

	Half-life (min)			
Compound	Rat	Dog	Monkey	Human
ACH-806	53	100	82	165

[a]Liver microsomes (1 mg/ml) were incubated with 1 μM ACH-806 for 60 min at 37°C in 50 mM phosphate buffer containing 1 mM NADPH. The metabolism of ACH-806 was monitored by disappearance of parent drug during 0 to 60 minutes of incubation. Half-life values were calculated from the per cent ACH-806 remaining at each time point.

Table 11.8 Summary of pharmacokinetic parameters after intravenous administration[a]

	Pharmacokinetic parameter (mean ± SD)			
Compound	Clearance (l/h/kg)	Volume of distribution (l/kg)	AUC[b] (ng h/ml)	$t_{½}$ (h)
Dog	0.7 ± 0.11	10.63 ± 0.78	1519 ± 254	16.9 ± 4.36
Rat	2.52 ± 0.12	2.23 ± 0.26	385 ± 20	1.7 ± 0.43

[a]ACH-806 was formulated as 1 mg/ml oral dose and administered intravenously as 1 mg/kg to Sprague-Dawley rats and Beagle dogs (*n* = 3). Serial blood samples were obtained from each animal at selected time points post dose. ACH-806 plasma concentrations were measured by high performance liquid chromatography with tandem mass spectrometric detection (LC-MS/MS).

[b]AUC: area under the plasma concentration–time curve from time zero to infinity.

assessments of metabolic stability, are useful to flag compounds that might not be suitable for oral administration. Our early pharmacokinetic studies with potent lead molecules assessed the likelihood of achieving clinical efficacy by comparison of plasma concentrations with the antiviral EC_{50} after an oral dose. As a measure of efficacy, we chose to target a trough plasma concentration of ACH-806 equal to the antiviral EC_{50} (0.014 nM). Oral bioavailability studies with ACH-806 utilized a single 10 mg/kg dose to the rat and the dog. When formulated in PBS containing 10% gelucire and 1% dimethylacetamide, showed an oral bioavailability of about 24% in the rat and 11% in the dog (Table 11.9). In both species, however, plasma concentrations remained above the antiviral EC_{50} through to 6 to >24 hour post dose (Fig. 11.7). In the dog, the low oral bioavailability was offset by the long half-life such that efficacious plasma concentrations were maintained through at least 24-hours post dose. The plasma concentrations after an oral dose were similar for many of the lead compounds, however, the antiviral potency of ACH-806 was higher.

Obtaining efficacious plasma concentrations in these studies, particularly in the dog, indicated that reasonable oral doses levels of ACH-806 could be expected to provide therapeutic drug levels in humans. ACH-806 showed limited aqueous solubility and therefore a series of studies was initiated to explore potential vehicles and dose levels such that sufficiently high plasma concentrations of ACH-806 could be attained in toxicology studies. As shown in Table 11.9, attempts to formulate ACH-806 as a simple suspension in 0.5% Tylose, reduced oral bioavailability to about 8% compared with 25% when the drug was formulated with a lipid component. A series of studies (data not shown) in the rat were conducted to assess the oral bioavailability of ACH-806 when formulated in complex formulation using mixtures of lipophilic excipients, such as Labrasol, Capryol, Solutol, and Gelucire. These formulations were tested for increased plasma levels of ACH-806 across a range of dose levels and also for tolerability to the vehicle under conditions of repeated administration. A formulation, PATG, consisting of Peg400:Akoline

Table 11.9 Summary of oral pharmacokinetic parameters of ACH-806 in rat and dog

			Pharmacokinetic parameter (mean) or (mean ± SD)			
Species	Dose (mg/kg)	Formulation	AUC (ng h/ml)	C_{max} (ng/ml)	t_{max} (h)	F (%)
Rat[a]	10	Dimethylacetamide:Gelucire:PBS (1:10:89)	949 ± 414	528 ± 233	1 ± 0	25
	10	0.5% Tylose	288 ± 139	74 ± 24	2.7 ± 1.2	7.5
	10	0.5% Tylose, micronized	334 ± 143	100 ± 56	1.7 ± 0.6	8.7
	100	0.5% Tylose	908 ± 347	195 ± 49	2.3 ± 1.5	2.3
	100	0.5% Tylose, micronized	995 ± 412	253 ± 79	2.0 ± 1.0	2.6
	5[b]	PATG[d]	1030	34.4	1	54
	150[b]	PATG[d]	14500	2140	0.5	25
Dog[c]	10	DMA:Gelucire:PBS (1:10:89)	1680 ± 520	218 ± 44	2.7 ± 1.7	11
	5	PATG[d]	3810	787	2	50
	75	PATG[d]	55000	6410	2	48

[a]ACH-806 was formulated as above and administered via oral gavage to male Sprague-Dawley rats (n = 3). Serial blood samples were obtained from each animal through a jugular vein catheter at selected time points post dose. ACH-806 plasma concentrations were measured by LC-MS/MS.

[b]Non-serial blood samples obtained from individual animals via orbital sinus venipuncture (n = 3 rats/time point).

[c]ACH-806 was formulated as above and administered via oral gavage to male Beagle dogs (n = 3). Serial blood samples were obtained from each animal through venipuncture at selected time points post dose. ACH-806 plasma concentrations were measured by LC-MS/MS.

[d]PATG consists of Peg400:Akoline MCM:Tetraglycol:Gelucire (50:20:20:10).

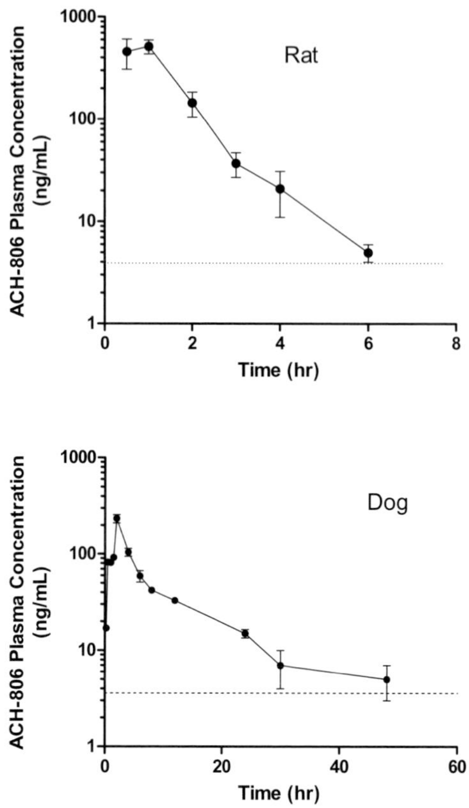

Figure 11.7 Comparison of ACH-806 plasma concentration in the rat and dog with antiviral EC_{50} following a single oral dose (10 mg/kg). ACH-806 was formulated in PBS containing 1% dimethylacetamide and 10% gelucire and administered orally to the rat and dog (*n* = 3). Figures represent the plasma concentration of ACH-806 across time following dose administration compared with the antiviral EC_{50} (dotted line).

MCM:Tetraglycol:Gelucire (50:20:20:10) showed a notable increase in oral bioavailability in both rats and dogs when compared with the initial, predominantly aqueous formulation (Table 11.9). In the rat, oral bioavailability at the lower doses increased from 25% in a predominantly aqueous formulation to 54% using PATG. Similarly, the use of PATG to improve oral exposure in the dog resulted in an approximate 4-fold increase of oral bioavailability over that in the initial formulation. This vehicle was well tolerated and successfully implemented in the toxicology programme.

These early pharmacokinetic studies highlight the challenge to attain high plasma levels for potent molecules with low solubility that are moving forward to multiple dose toxicology studies. In our pharmacokinetic studies with ACH-806, we were able to achieve high drug levels using PATG. The vehicle was utilized for 28-day IND-enabling studies where levels of exposure attained were sufficient to support initiation of phase I clinical trials.

Clinical proof-of-concept study with ACH-806

Based on the data collected from preclinical studies, ACH-806 was selected for IND-enabling studies and finally was evaluated in a proof-of concept clinical trial. The trial was randomized, double-blind, placebo-controlled and conducted in full compliance with the guidelines of Good Clinical Practice and of the World Medical Assembly Declaration of Helsinki. Prior to participation in the trial all patients provided written informed consent. Ten HCV genotype-1 infected patients (8 active: 2 placebo) were enrolled in the first cohort. These patients had compensated and non-cirrhotic liver disease. Except for one patient (in the group treated with ACH-806), all had undergone prior interferon-based treatment for HCV which had ended at least 6 months prior to entry into this trial. Each subject received 300 mg b.i.d. of either ACH-806 or a placebo for 4 days (days 1 through 4) with a single dose on Day 5. Plasma samples were drawn at various times and plasma HCV RNA levels were determined.

The mean change in HCV RNA ($\log_{10}$) from baseline is shown in Fig. 11.8. At day 5, the mean change was −0.91 for ACH-806 treated subjects vs. +0.05 for placebo subjects. Four of the ACH-806 treated subjects vs. none of the placebo treated subjects achieved HCV RNA reductions of > 1.0 $\log_{10}$ (maximum reduction of 2.38 $\log_{10}$). No serious adverse reactions occurred during this trial. ACH-806 was generally well tolerated clinically, but small increases in serum creatinine were observed in subjects treated with ACH-806. Mean increases in serum creatinine at day 3 and day 5 were 0.43 and 0.49 (mg/dl) vs. 0.1 and 0.05 in placebo-treated patients. Creatinine levels returned to baseline by 1 week post dosing. Based upon review of the laboratory data, it was decided that ACH-806 was associated with renal tubular dysfunction and hence dose escalation to the next cohort was not initiated. Nevertheless, this is the first demonstration of anti-HCV activity of an NS3–NS4A complex inhibitor in patients with chronic HCV infection and warrants continued evaluation and development of the class belonging to a different chemical series.

Conclusion

In this chapter, we described a series of novel compounds discovered by screening against the HCV replicon. Our mechanism of action and resistance induction studies helped in identifying the NS3–NS4A complex as the target of acylthioureas. Demonstration of antiviral activity with a representative acylthiourea, ACH-806, in HCV-infected patients validated this mechanism of HCV inhibition and warrants further discovery and development of the next generation of compounds acting through this mechanism.

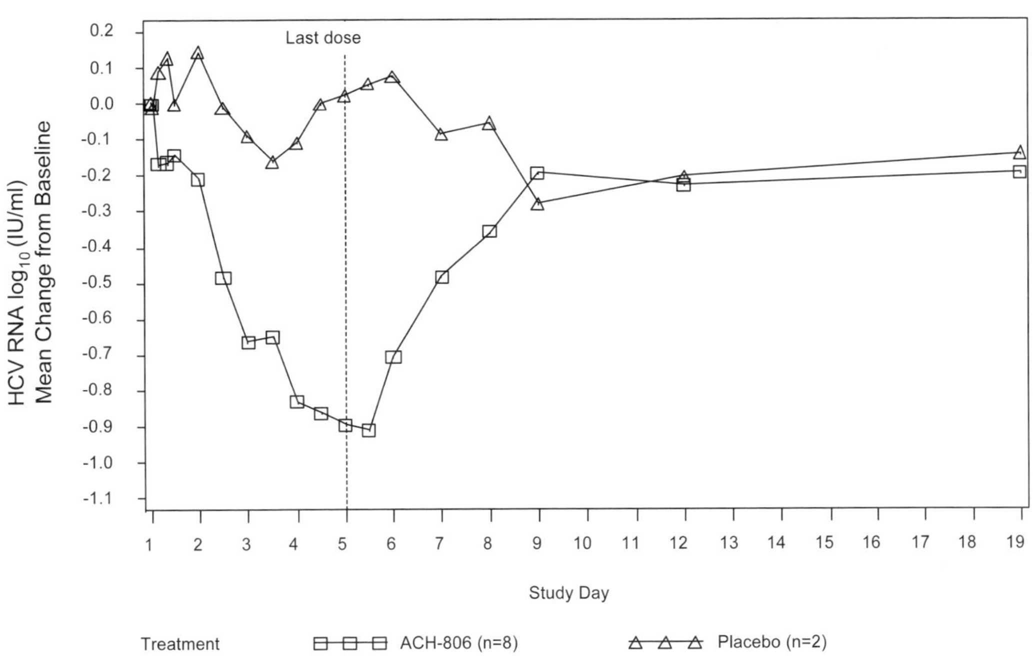

Figure 11.8 Antiviral activity of ACH-806 in HCV (genotype-1)-infected patients. The mean change in HCV RNA ($\log_{10}$) from baseline is shown. HCV RNA levels were determined using the Cobas Amplicor HCV Monitor V2.0 Assay (Roche Diagnostics). All samples were analysed by a central laboratory.

Future trends

Combination therapies consisting of specific anti-HCV agents are expected to be more effective in treating chronic hepatitis C infection. Due to the error prone nature of HCV RNA polymerase, variants resistant to one or more inhibitors almost always exist in any patient prior to the therapy. Hence, three or more HCV-specific inhibitors will be necessary to achieve sustained virological response. Including NS3–NS4A complex inhibitors into combination therapies will be advantageous due to their unique mechanism of action and lack of cross-resistance with other classes of inhibitors.

Acknowledgements

We thank our colleagues in the discovery group for their contribution towards hit identification, lead optimization, virology and preclinical studies. We thank Achillion development team led by John Pottage and also clinical investigators Nezam Afdhal (Beth Israel Deaconess Medical Center, Liver Research Center, Boston, MA, USA), Reem Ghalib (The Liver Institute at Methodist Transplant Physicians, Dallas, TX, USA), Eric Lawitz (Alamo Medical Research, San Antonio, TX, USA), Paul Martin (Mt. Sinai Medical Center, Division of Liver Diseases, New York, NY, USA), Dago Mazur (Parexel International GmbH, Berlin, BE, Germany), Marion Peters (University of California San Francisco, San Francisco, CA, USA), Hugo Vargas (Mayo Clinic Hospital, Phoenix, AZ, USA), and David Wyles (University of California, San Diego, La Jolla, CA, USA). We acknowledge all members of the ACH-806 study team at Gilead Sciences, Inc. We also thank our colleagues in Regulatory Affairs, Quality Assurance and Pharmaceutical Technology. Finally, we thank the patients who participated in the clinical trial.

References

Asabe, S.I., Tanji, Y., Satoh, S., Kaneko, T., Kimura, K., and Shimotohno, K. (1997). The N-terminal region of hepatitis C virus-encoded NS5A is important for NS4A-dependent phosphorylation. J. Virol. *71*, 790–796.

Blight, K.J., McKeating, J.A., Marcotrigiano, J., and Rice, C.M. (2003). Efficient replication of hepatitis C virus genotype 1a RNAs in cell culture. J. Virol. *77*, 3181–3190.

Butkiewicz, N.J., Wendel, M., Zhang, R., Jubin, R., Pichardo, J., Smith, E.B., Hart, A.M., Ingram, R., Durkin, J., Mui, P.W., *et al.* (1996). Enhancement of hepatitis C virus NS3 proteinase activity by association with NS4A-specific synthetic peptides: identification of sequence and critical residues of NS4A for the cofactor activity. Virology *225*, 328–338.

Failla, C., Tomei, L., De Francesco, R. (1994). Both NS3 and NS4A are required for proteolytic processing of hepatitis C virus nonstructural proteins. J. Virol. *68*, 3753–3760.

Frick, D.N., Rypma, R.S., Lam, M., and Gu, B. (2004). The nonstructural protein 3 protease/helicase requires an intact protease domain to unwind duplex RNA efficiently. J. Biol. Chem. *279*, 1269–1280.

Gallinari, P., Paolini, C., Brennan, D., Nardi, C., Steinkuhler, C., De Francesco, R. (1999). Modulation of hepatitis C virus NS3 protease and helicase activities through the interaction with NS4A. Biochemistry *38*, 5620–5632.

Golitsina, N.L., Danehy, F.T., Jr., Fellows, R., Cretton-Scott, E., Standring, D.N. (2009). Evaluation of the role of three candidate human kinases in the conversion of the hepatitis C virus inhibitor 2′-C-methyl-cytidine to its 5′-monophosphate metabolite. Antiviral Res *85*, 470–481.

Guo, J.T., Bichko, V., and Seeger, C. (2001). Effect of alpha interferon on the hepatitis C virus replicon. J. Virol. *75*, 8516–8523.

Hann, M., and Oprea, T.I. (2004). Pursuing the leadlikeness concept in pharmaceutical research. Curr Opin Chem Biol *8*, 255–263.

Hijikata, M., Mizushima, H., Tanji, Y., Komoda, Y., Hirowatari, Y., Akagi, T., Kato, N., Kimura, K., and Shimotohno, K. (1993). Proteolytic processing and membrane association of putative nonstructural proteins of hepatitis C virus. Proc. Natl. Acad. Sci. U.S.A. *90*, 10773–10777.

Kneteman, N.M., Howe, A.Y., Gao, T., Lewis, J., Pevear, D., Lund, G., Douglas, D., Mercer, D.F., Tyrrell, D.L., Immermann, F., *et al.* (2009). HCV796: a selective nonstructural protein 5B polymerase inhibitor with potent anti-hepatitis C virus activity in vitro, in mice with chimeric human livers, and in humans infected with hepatitis C virus. Hepatology *49*, 745–752.

Koch, O., and Bartenschlager, R. (1999). Modulation of hepatitis C virus NS5A hyperphosphorylation by nonstructural proteins NS3, NS4A, and NS4B. J. Virol. 73, 7138–7146.

Koch, J.O., Lohmann, V., Herian, U., and Bartenschlager, R. (1996). In vitro studies on the activation of the hepatitis C virus NS3 proteinase by the NS4A cofactor. Virology *221*, 54–66.

Kou, Y.H., Chang, M.F., Wang, Y.M., Hong, M., and Chang, S.C. (2007). Differential requirements of NS4A for the internal NS3 cleavage and polyprotein processing of hepatitis C virus. J. Virol. 23, 23.

Kuang, W.F., Lin, Y.C., Jean, F., Huang, Y.W., Tai, C.L., Chen, D.S., Chen, J., and Hwang, L.H. (2004).

Hepatitis C virus NS3 RNA helicase activity is modulated by the two domains of NS3 and NS4A. Biochem. Biophys. Res. Commun. *317*, 211–217.

Lin, C., and Rice, C.M. (1995). The hepatitis C virus NS3 serine proteinase and NS4A cofactor: establishment of a cell-free trans-processing assay. Proc. Natl. Acad. Sci. U.S.A. *92*, 7622–7626.

Lin, C., Wu, J.W., Hsiao, K., and Su, M.S. (1997). The hepatitis C virus NS4A protein: interactions with the NS4B and NS5A proteins. J. Virol. *71*, 6465–6471.

Lindenbach, B.D., Pragai, B.M., Montserret, R., Beran, R.K., Pyle, A.M., Penin, F., and Rice, C.M. (2007). The C-terminus of hepatitis C virus NS4A encodes an electrostatic switch that regulates NS5A hyperphosphorylation and viral replication. J. Virol. *20*, 20.

Lohmann, V., Korner, F., Koch, J., Herian, U., Theilmann, L., and Bartenschlager, R. (1999). Replication of subgenomic hepatitis C virus RNAs in a hepatoma cell line. Science *285*, 110–113.

Lohmann, V., Hoffmann, S., Herian, U., Penin, F., and Bartenschlager, R. (2003). Viral and cellular determinants of hepatitis C virus RNA replication in cell culture. J. Virol. *77*, 3007–3019.

Martell, M., Esteban, J.I., Quer, J., Genesca, J., Weiner, A., Esteban, R., Guardia, J., and Gomez, J. (1992). Hepatitis C virus (HCV) circulates as a population of different but closely related genomes: quasispecies nature of HCV genome distribution. J. Virol. *66*, 3225–3229.

Migliaccio, G., Tomassini, J.E., Carroll, S.S., Tomei, L., Altamura, S., Bhat, B., Bartholomew, L., Bosserman, M.R., Ceccacci, A., Colwell, L.F., *et al.* (2003). Characterization of resistance to non-obligate chain-terminating ribonucleoside analogs that inhibit hepatitis C virus replication in vitro. J. Biol. Chem. *278*, 49164–49170.

Neumann, A.U., Lam, N.P., Dahari, H., Gretch, D.R., Wiley, T.E., Layden, J., and Perelson, A.S. (1998). Hepatitis C viral dynamics in vivo and the antiviral efficacy of interferon-alpha therapy. Science *282*, 103–107.

Nguyen, T.T., Gates, A.T., Gutshall, L.L., Johnston, V.K., Gu, B., Duffy, J., and Sarisky, R.T. (2003). Resistance profile of a hepatitis C virus RNA-dependent RNA polymerase benzothiadiazine inhibitor. Antimicrob. Agents Chemother. *47*, 3525–3530.

Pang, P.S., Jankowsky, E., Planet, J., and Pyle, A.M. (2002). The hepatitis C viral NS3 protein is a processive DNA helicase with cofactor enhanced RNA unwinding. EMBO J. *21*, 1168–1176.

Prichard, M.N. and Shipman, C. Jr. (1990). A three-dimensional model to analyze drug–drug interactions. Antiviral Res *14*, 181–205.

Prichard, M.N., Prichard, L.E., Shipman, C. Jr. (1993). Strategic design and three-dimensional analysis of antiviral drug combinations. Antimicrob. Agents Chemother. *37*, 540–545.

Sarrazin, C. and Zeuzem, S. (2010). Resistance to direct antiviral agents in patients with hepatitis C virus infection. Gastroenterology *138*, 447–462.

Singh, S.S. (2006). Preclinical pharmacokinetics: an approach towards safer and efficacious drugs. Curr. Drug Metab. *7*, 165–182.

Tanji, Y., Hijikata, M., Satoh, S., Kaneko, T., and Shimotohno, K. (1995). Hepatitis C virus-encoded nonstructural protein NS4A has versatile functions in viral protein processing. J. Virol. *69*, 1575–1581.

Trozzi, C., Bartholomew, L., Ceccacci, A., Biasiol, G., Pacini, L., Altamura, S., Narjes, F., Muraglia, E., Paonessa, G., Koch, U., *et al.* (2003). In vitro selection and characterization of hepatitis C virus serine protease variants resistant to an active-site peptide inhibitor. J. Virol. *77*, 3669–3679.

Tsaioun, K., and Jacewicz, M. (2009). De-risking drug discovery with ADDME -- avoiding drug development mistakes early. Altern Lab Anim 37 (Suppl. 1), 47–55.

Vrolijk, J.M., Kaul, A., Hansen, B.E., Lohmann, V., Haagmans, B.L., Schalm, W., and Bartenschlager, R. (2003). A replicon-based bioassay for the measurement of interferons in patients with chronic hepatitis C. J. Virol. Methods *110*, 201–209.

Wolk, B., Sansonno, D., Krausslich, H.G., Dammacco, F., Rice, C.M., Blum, E., and Moradpour, D. (2000). Subcellular localization, stability, and trans-cleavage competence of the hepatitis C virus NS3-NS4A complex expressed in tetracycline-regulated cell lines. J. Virol. *74*, 2293–2304.

Wyles, D.L., Kaihara, A., and Schooley, R.T. (2008). Synergy of a hepatitis C virus (HCV) NS4A antagonist in combination with HCV protease and polymerase inhibitors. Antimicrob. Agents Chemother. *52*, 1862–1864.

Yang, W., Zhao, Y., Fabrycki, J., Hou, X., Nie, X., Sanchez, A., Phadke, A., Deshpande, M., Agarwal, A., and Huang, M. (2008). Selection of replicon variants resistant to ACH-*806*, a novel hepatitis C virus inhibitor with no cross-resistance to NS3 protease and NS5B polymerase inhibitors. Antimicrob. Agents Chemother. *52*, 2043–2052.

Zhao, Y., Sanchez, A., Nie, X., Liu, D., Hou, X., Fabrycki, J., Phadke, A., Deshpande, M., Huang, M., and Yang, W. (2008). A highly sensitive and reproducible HCV RNA hybridization method valuable for antiviral drug discovery. J. Virol. Methods *151*, 154–157.

Hepatitis C Virus NS3 Helicase Inhibitors

12

Craig A. Belon and David N. Frick

Abstract

The hepatitis C virus (HCV) non-structural protein 3 (NS3) is a complex multifunctional enzyme. In addition to processing the viral polyprotein, NS3 functions as a viral helicase capable of separating duplex RNA and DNA in reactions fuelled by ATP hydrolysis. A functioning helicase is necessary for HCV replication, showing that the NS3 helicase could be an antiviral drug target. Although early screens for HCV helicase inhibitors yielded few if any antiviral leads, more recent studies have found potent helicase inhibitors that are active against the HCV replicon. Noteworthy HCV helicase inhibitors that are relatively non-toxic to cells and inhibit the HCV replicon include triphenylmethanes, acridones, amidinoanthracyclines, and a rationally designed substituted pyrrole. Also discussed here are assay systems available for discovering and analysing HCV helicase inhibitors, which can broadly be grouped into two categories: those that measure helicase-catalysed hydrolysis of ATP, and those that measure helicase-catalysed separation of double-stranded nucleic acid substrates.

Current HCV therapy is toxic, expensive, and not effective against all HCV genotypes, in part, because interferon and ribavirin were not developed to specifically target HCV. Direct acting antivirals targeting proteins made by HCV and key events in the viral lifecycle might therefore be better than current treatments. Direct acting HCV antivirals, which are sometimes called specifically targeted antiviral therapy for HCV (STAT-C), have been discovered and some have shown marked success in clinical trials. Present STAT-C agents, like the polymerase and protease inhibitors, will need to be supplemented with other compounds if they are to replace current treatment because viruses resistant to most polymerase and protease inhibitors have already been identified. One possible STAT-C target is the hepatitis C virus NS3 helicase, which is needed for HCV replicon replication (Lam and Frick, 2006) and for HCV to replicate in chimpanzees (Kolykhalov *et al.*, 2000).

The HCV helicase was first intensively studied as a drug target starting around 1993, but interest in developing helicase inhibitors as antivirals waned after early attempts failed to yield any promising lead compounds. In the intervening years new assays have been developed for screening, the molecular helicase mechanism has been unveiled and, most excitingly, a few compounds inhibiting HCV NS3 helicase-catalysed reactions have been shown to prevent viral RNA replication in the HCV replicon system. Even if these new inhibitors might not be valuable as lead compounds, they could be used to finally understand why HCV and similar positive sense RNA viruses need to encode a helicase to replicate in cells.

This chapter focuses mainly on assay systems, screening protocols and publicly disclosed HCV helicase inhibitors. We start, however, by briefly reviewing HCV helicase structure and function, highlighting the few studies that were published since we last reviewed the field (Frick, 2007; Belon and Frick, 2009a). If a reader is interested in how the compounds mentioned here might be further developed or used to study HCV biology, there are methods discussed in other chapters of this book and in the recent book *Hepatitis C: Methods and Protocols* (Tang, 2008). The recent book *Helicases: Methods and Protocols* (Abdelhaleem, 2010), is also recommended if a reader desires to employ or develop alternate assays or use compounds discussed here as mechanistic probes.

The HCV helicase

HCV non-structural protein 3 (NS3) is a multifunctional protein possessing two major functional domains. The N-terminal one third of the NS3 protein joins with NS4A to produce the mature NS3-NS4A protease, which is responsible for cleaving several locations in the HCV polyprotein to release mature proteins. The C-terminal two thirds of NS3 forms the helicase core, which catalyses the hydrolysis of ATP and uses the free energy to separate double stranded nucleic acids. Recombinant 'full-length' NS3 protein containing both the protease and helicase functional domains can be expressed in *E. coli* and purified. Recombinant NS3 retains its ATP hydrolysis and nucleic acid strand separating helicase activities.

A recombinant protein in which NS3 is truncated to remove the protease portion entirely, termed 'NS3h' for NS3-helicase fragment, can be more easily expressed to higher levels in *E. coli*. NS3h retains an ability to hydrolyse ATP and unwind nucleic acids, although its ability to unwind RNA is somewhat diminished (Frick *et al.*, 2004; Beran *et al.*, 2007). It should be remembered, also, that in most assays not all NS3h proteins are equal. Depending on where the protein is truncated and the location and nature of any attached fusion proteins, various NS3h constructs have remarkably different activities in almost all assays (Frick *et al.*, 2004).

Role of the NS3 helicase in HCV replication

DNA helicases are clearly needed to separate the double helix so that DNA can be copied and transcribed into RNA. The biological roles of RNA helicases are more elusive, but they are probably needed for RNA folding or to modulate secondary structures (Cordin *et al.*, 2006). The NS3 helicase likely separates local regions of double stranded RNA produced during viral replication, but it might also assist in the packaging of viral RNA into virions as has been seen with the closely related Kunjin virus (Liu *et al.*, 2002). Genetic evidence hints at a similar relationship in HCV (Ma *et al.*, 2008), and HCV helicase is most active at a somewhat acidic pH as is typical of the Golgi apparatus, where assembly likely starts (Lam *et al.*, 2004). Other clues to a biological role might come from the fact that in cells, the helicase portion of NS3 protein is methylated (Duong *et al.*, 2005), acetylated, and phosphorylated (Liefhebber *et al.*, 2010). The HCV helicase's promiscuous substrate specificity also hints at possible role in modulating host gene expression. HCV is an RNA virus that is thought to replicate outside the nucleus yet, in addition to being capable of unwinding double stranded RNA, NS3 helicase unwinds double stranded DNA and RNA/DNA hybrids (Tai *et al.*, 1996). The biological significance of HCV helicase action on DNA remains enigmatic, but NS3 has been spotted in the nucleus (Ishido *et al.*, 1997; Muramatsu *et al.*, 1997; Errington *et al.*, 1999) and has the capacity to transform cells (Sakamuro *et al.*, 1995).

Within the HCV replication complex, HCV helicase clearly interacts with itself and other proteins. In such a context, the helicase may function either as a monomer, dimer, or higher order oligomer (Jennings *et al.*, 2009). The NS3 protease/helicase modulates the activity of the NS5B polymerase (Piccininni *et al.*, 2002) and vice versa (Zhang *et al.*, 2005; Jennings *et al.*, 2008). The NS4A protease co-factor enhances the NS3–RNA interaction (Beran *et al.*, 2009) and the presence of an intact NS3 protease domain enhances the ability of NS3 helicase to unwind RNA (Frick *et al.*, 2004; Beran *et al.*, 2007).

HCV helicase: structure and mechanics

In order to understand how compounds might inhibit HCV helicase, it is helpful to understand how HCV helicase is evolutionarily related to other ATP hydrolysing enzymes, its structure, and its functions. This information can be used to design assays, characterize compound specificity, and to rationally design a next generation of derivatives.

Precisely how HCV helicase is related to other viral and cellular helicases has been reviewed in detail previously (Frick and Lam, 2006), but it is important to mention some key points here because if a compound inhibits any particular helicase, it is possible that it will inhibit related helicases as well, sometimes with even greater potency (Zhang *et al.*, 2003; Belon *et al.*, 2010). Helicases are similar to each other and cellular motor proteins, proteasomes, chaperones, kinases, and small GTPases because all these proteins share a common ancestor. As evidence, each retains a 'Walker-type' nucleotide binding site formed by two sequence motifs called the Walker A and B sites (Walker *et al.*, 1982). Helicase amino acid sequences also share several additional sequence motifs, which are used to classify helicases into families and these families into superfamilies (Hall and Matson, 1999; Gorbalenya and Koonin, 2003). HCV helicase is part of the viral DEx/CH box family, which in turn is part of helicase superfamily 2 (SF2). Closely related helicases typically share many structural and functional features; for example HCV NS3 helicase and Dengue virus NS3 helicase share well-conserved sequences and consequently have three-dimensional structures that superimpose nicely.

Structure homology is not limited to closely related helicases. All members of the helicase class of motor proteins possess a similar three-dimensional fold that resembles one first seen in the bacterial recombination protein RecA (Story *et al.*, 1992). The first crystal structure of HCV helicase showed a Y-shaped protein composed of three domains, two of which resemble RecA (Yao *et al.*, 1997). All helicase structures reported to date possess at least one 'RecA-like' motor domain, which function as the core of the helicase molecular motor. HCV helicase, like other SF1 and SF2 helicases has two Rec-A like motor domains linked in tandem on the same polypeptide.

To date, over ten atomic structures of HCV helicase have been deposited in the Protein Data Bank (PDB, www.pdb.org). The ATP and RNA binding sites on NS3 have been resolved (Kim *et al.*, 1998; Mackintosh *et al.*, 2006; Gu and Rice, 2010), and the various structures reveal the precise positions of various amino acid side chains when the protein assumes a variety of conformations. Two PDB files show NS3h from either HCV genotype 1a (PDB file 1HEI) or 1b (PDB file 8OHM) with its motor domains in either a closed or open conformation (Yao *et al.*, 1997; Cho *et al.*, 1998). PDB file 1CU1 shows full-length NS3 with a portion of NS4A tethered to its N-terminus (Yao *et al.*, 1999). There are several views of NS3h bound to DNA oligonucleotides (Kim *et al.*, 1998; Mackintosh *et al.*, 2006; Gu and Rice, 2010). File 1A1V shows a single NS3h with a bound 8-nucleotide long polyU DNA oligonucleotide (Kim *et al.*, 1998) and file 2F55 shows how two NS3h protomers might assemble in tandem on a 13-nuclotide long stretch of DNA (Mackintosh *et al.*, 2006). File 3KQM and file 3KQK show NS3h bound to 6-nt long polymers of (d)A and (d)T, respectively. Precisely how the protein changes conformation upon ATP binding can be visualized by comparing file 3KQK with 3KQN, in which the ground state ATP-mimic ADP(BeF_3) is bound between the motor domains. Changes induced by ATP hydrolysis can be seen by comparing file 3KQK (or 3KQN) with 3KQL, in which ADP(AlF_4)$^-$ mimics the transition state formed between the motor domains upon ATP hydrolysis. File 3KQU shows another ADP(BeF_3) complex bound to a longer 19-nt long (d)T oligonucleotide (Gu and Rice, 2010). Such a wealth of structural data is a goldmine for rational drug design.

Although all helicases are homologues, they have each evolved to fill a different biological niche. This divergent evolution is reflected in their different enzymatic activities, which vary with regard to substrate specificity, fuel specificity, and direction of movement. Each of these properties affects the design and selection of appropriate assay systems, and how the protein might interact with inhibitors. Because helicases are motor proteins, it is often helpful to classify them based on which direction they move along a nucleic acid strand. Helicase directionality can be probed by creating a partially duplex nucleic acid substrate with a short single-stranded 'tail' of about 20 nucleotides on either the 3′ or 5′ end. If the helicase requires a 3′ tail to catalyse strand separation, it likely moves in 3′–5′ direction, and, inversely, if a 5′ tail is required, the helicase likely moves in 5′–3′ direction. Since HCV NS3 helicase will only unwind nucleic acid that has a 3′ single-stranded region, it classified as a 3′–5′ helicase. Proving directionality is difficult, but two commonly used methods are to assess either ability of helicase to displace proteins bound to either the 3′ end or 5′ end of oligonucleotides (Morris and Raney, 1999) or a helicase's ability to quench end-labelled fluorescent oligonucleotides in a time-dependent manner (Fischer *et al.*, 2004). Both of these experiments have been performed with HCV helicase and each suggests that the protein moves in a 3′ to 5′ direction (Morris *et al.*, 2002; Matlock *et al.*, 2010).

Exactly how HCV helicase unwinds RNA or DNA is still hotly debated and has been the subject of other reviews (Frick, 2007; Pyle, 2008). Most models assume HCV helicase acts like a molecular inchworm, crawling along nucleic acids by rotating the second motor domain relative to the rest of the protein in a reaction modulated by ATP binding and hydrolysis. Nucleotide modulated Rec-A like motor domain rotation, and how it might move the protein on DNA, was recently visualized in the above series of structures of NS3h bound to nucleotide analogues. In those structures, nucleotide driven conformational changes appear to transfer individual backbone phosphates between two nucleic acid binding sites, located along the nucleic acid binding surface of the two RecA-like motor domains (Gu and Rice, 2010). Other key recent developments include new mathematical models that can be used to analyse single molecule studies (Arunajadai, 2009) like those previous published with NS3 (Dumont *et al.*, 2006; Myong *et al.*, 2007), a plausible explanation for the very different kinetic step sizes that have been reported in various studies (Serebrov *et al.*, 2009), and a clear demonstration that phosphate release limits how rapidly ATP hydrolysis can fuel RNA unwinding (Wang *et al.*, 2010).

Methods to identify and analyse helicase inhibitors

Helicases are enzymes that use the energy derived from NTP hydrolysis to separate regions of double stranded nucleic acid. In this reaction, ATP is considered an activator and the nucleic acid is the substrate. As such, it is possible to measure helicase activity in at least two ways: by measuring the cleavage of ATP or by measuring the conversion of double stranded nucleic acid into single strands. Many methods have been developed to accomplish either of these goals, and each method has advantages and disadvantages, but they all can provide valuable mechanistic insights.

ATPase assays

To fuel its movement on nucleic acids, HCV helicase catalyses water-mediated cleavage of ATP into ADP and inorganic phosphate (P_i). The rate of ATP hydrolysis can be determined by measuring the loss of ATP, the appearance of ADP, or the appearance of P_i.

The most direct method to monitor helicase-catalysed ATP hydrolysis is to measure amounts of ATP and ADP in helicase reactions using high performance liquid chromatography (HPLC) or thin layer chromatography (TLC). In either case, the substrate and product can be detected based on the fact that nucleotides absorb ultraviolet (u.v.) light (Du *et al.*, 2002). More sensitive assays use ATP labelled with a radioisotope. For example, hydrolysis of ATP can be directly measured by combining helicase, buffer, divalent metal, nucleic acid, and [α^{32}P]ATP. After stopping the reaction with acid or a metal chelator, products can be separated using TLC, and quantified by exposing the dried plate to X-ray film (Suzich *et al.*, 1993) or a phosphorimager screen (Howe *et al.*, 1999). Such an assay measures both [α^{32}P] ATP and [α^{32}P]ADP, but not the amount of P_i liberated. By substituting [γ^{32}P]ATP, the phosphate product can be measured either by using HPLC, TLC, or by simply adding activated charcoal (i.e. Norit; Lam *et al.*, 2003a). Norit absorbs nucleotides and can be rapidly removed by centrifugation or filtration, leaving only phosphate in the supernatant. Assays using radiolabelled nucleotides are remarkably sensitive. Because only the specific activity of ATP limits sensitivity, such assays are routinely performed with sub-nanomolar substrate concentrations in a few microlitres. Although [^{32}P]ATP is typically used for this type of assay, ATP labelled with other nuclides is commercially available as are other radiolabelled nucleoside triphosphates and some ATP analogues. The disadvantages to this procedure are that use of radioisotopes can be relatively hazardous, and the specialized equipment needed to safely handle and measure radioisotopes is not always available.

Although not as sensitive as the above radioisotope-based assays, there are numerous high-throughput methods available to monitor ATP hydrolysis by measuring the substrate or products using a spectrophotometer (Table 12.1). Phosphate can be measured with one of the many colorimetric assays that use the Fiske-Subbarow ammonium molybdate reagent (Fiske and Subbarow, 1925). For example, an acidic ammonium molybdate solution can be added to a HCV helicase-catalysed reaction after nucleotides are removed using Norit (Lam *et al.*, 2003a). Alternatively, phosphate can be determined without removing ATP using a modified procedure in which colour is amplified with the dye malachite green molybdate. In such an assay, a phosphate complex forms that absorbs light at 630 nm (Lanzetta *et al.*, 1979). Helicase, metal, and ATP are combined in reaction buffer, incubated for a period of time, then quenched with an acidic solution of malachite green and molybdate. A citrate solution is then quickly added, absorbance is measured, and the amount of phosphate generated determined by comparison to a standard curve. For 30 μl helicase reactions performed in microplates, such an assay has a linear range between 10 μM and 200 μM phosphate (Frick *et al.*, 2007; Belon and Frick, 2009b).

Phosphate production can also be measured continuously by coupling its production to other reactions that can be directly monitored. For example, the enzyme purine nucleoside phosphorylase (PNPase), which normally functions for purine salvage, also acts on nucleoside analogues whose absorbance spectrums change upon phosphorolysis, such 7-methylthioguanosine (MESG). MESG absorbs light at 330 nm

Table 12.1 High-throughput screening assays useful for HCV helicase inhibitor discovery and analysis

Assay	Reaction monitored	Equipment needed	Protocol
Luciferin	ATP hydrolysis	Luminescence spectrophotometer	Koresawa and Okabe (2004)[a]
NADH	ADP production	UV–vis spectrophotometer (A_{340})	Suzich *et al.* (1993)
Malachite Green	P_i production	UV–vis spectrophotometer (A_{630})	Frick *et al.* (2007), Belon and Frick (2009b)
MESG	P_i production	UV–vis spectrophotometer (A_{360})	Funk *et al.* (2004)[a]
MDCC-PBP	P_i production	Fluorescence spectrophotometer	Wang *et al.* (2010)
GlassMilk: Filter plates	Oligo:M13 unwinding	Membrane filter plates, plate washer, microplate scintillation counter	Sivaraja *et al.* (1998)[a]
SPA	DNA unwinding	SPA beads, microplate scintillation counter	Kyono *et al.* (1998)
FlashPlate	DNA unwinding	FlashPlates, microplate scintillation counter	Hicham Alaoui-Ismaili *et al.* (2000)
ECL	DNA unwinding	Magnetic flow cell, ECL instrument	Zhang *et al.* (2001)[a]
ELISA	RNA unwinding	UV–vis spectrophotometer (A_{450})	Hsu *et al.* (1998)
FRET	DNA or RNA unwinding	Fluorescence spectrophotometer	Porter *et al.* (1998), Boguszewska-Chachulska *et al.* (2004), Tani *et al.* (2010)
TR-FRET	DNA unwinding	Fluorescence spectrophotometer	Earnshaw *et al.* (1999)[a]
MBHA	DNA or RNA unwinding	Fluorescence spectrophotometer	Belon and Frick (2008)

[a]Protocol refers to use with a protein other than HCV helicase.

and when phosphate is present, PNPase converts MESG to ribose-1-phosphate and a base (2-amino-6-mercapto-7-methylpurine) that absorbs at a longer wavelength (Webb, 1992). Since the hydrolysis of ATP by the HCV NS3 helicase produces ADP and P_i, the activity of the enzyme is linked to the appearance of an absorbance at 360 nm. This assay is sensitive to ~2 μM phosphate. One concern when using this assay with HCV helicase is that the k_{cat} of the PNPase (~40 s^{-1}) is slower than the k_{cat} seen with HV NS3h under optimal conditions (Frick *et al.*, 2007). The PNPase reaction is most useful, however, because it is essentially irreversible, and therefore can be used to remove phosphate contaminates from ATP or other reagents. Such a 'phosphate mop' is necessary when one desires to precisely measure very small amounts of phosphate produced when a helicase cleaves ATP.

Helicase-catysed phosphate release can be most precisely measured using a fluorescently labelled *E. coli* phosphate binding protein (PBP). This assay uses a site directed mutant of the *E. Coli* PBP that contains a cysteine to which the fluorescent molecule *N*-[2-(l-maleimidyl) ethyl]-7-diethylamino)coumarin-3-carboxamide (MDCC) is attached. In the presence of P_i, the emission maxima of MDCC-PBP changes from 474 nm to 464 nm (when excited at 425 nm) and fluorescence changes by approximately 5-fold (Brune *et al.*, 1994). To use this very sensitive assay, P_i contaminating ATP or any other reagents must first be removed using MESG and PNPase to mop P_i contamination to less than 0.1 μM (Brune *et al.*, 1994). MDCC-PBP is now commercially available (e.g. the 'phosphate sensor' from Invitrogen), making this assay relatively easy to set up if a fluorescence spectrophotometer is available (Wang *et al.*, 2010).

Rather than measuring P_i, HCV helicase-catalysed ATP hydrolysis can also be monitored by coupling the production of ADP with other enzymes that use ADP as a substrate. For example, when pyruvate kinase (PK) is include in a HCV helicase reaction, along with lactate dehydrogenase (LDH), phosphoenolpyruvate (PEP) and reduced nicotinamide adenine dinucleotide (NADH), ATP hydrolysis will be proportional to NADH oxidation, which can be followed with a spectrophotometer (Suzich *et al.*, 1993). The ADP produced will react with PEP via PK to produce ATP and pyruvate. LDH then reduces pyruvate using NADH to produce lactate and NAD^+. Since NADH absorbs at 340 nm while NAD^+ does not, this assay can be continuously monitored so that rates can be directly measured. Sensitivity is limited, though, by the extinction coefficient of NADH (ε= 6.22/ mM/cm), and any compounds that absorb light at 340 nm or that inhibit PK or LDH will interfere with this assay.

For screening purposes, probably the best method to identify compounds inhibiting the NS3 ATPase is to use one of the many available luminescent ATP assays (Funk *et al.*, 2004). As is routinely done with kinases (Koresawa and Okabe, 2004) and related proteins, ATP remaining after a helicase reaction is terminated can be measured using the enzyme luciferase to react residual ATP with luciferin, producing a luminescent signal. By comparing the amount of luminescence at the end of a helicase reaction to a control reaction, the amount of ATP consumed can be calculated; the amount of luminescence produced is directly proportional to the amount of ATP remaining in the reaction mixture and thus inversely proportional to ATPase activity. These assays are commercially available (e.g. Promega's Kinase Glo Kits), and are amenable to screening since only compounds that inhibit luciferase should interfere with the reaction (Koresawa and Okabe, 2004). They are also linear over a wide range of ATP concentrations (1 μM to 500 μM). Such an assay is less useful for precise 'steady-state' kinetic analysis because an accurate measurement requires consumption of more than 20% of the ATP substrate.

The above ATPase assays are useful not only for studying HCV NS3 helicase function, but also for analysing the mechanism of action of helicase inhibitors. Any of the above assays could be used for screening, but if they are, hit compounds need to also be tested for their ability to inhibit any number of other similar ATPase enzymes. Non-specific compounds are more likely to prove toxic in cell-based replicon assays. From the point of view of inhibitor discovery, simply screening for ATPase inhibitors may miss compounds functioning by another mechanism such as interaction at the RNA binding site or a possible allosteric site.

Unwinding assays

The ability of a helicase to separate a double helix was first demonstrated using a radiolabelled DNA substrate and S1 nuclease, which only degrades duplex DNA. In such a rudimentary helicase assay, single-stranded DNA products can be estimated by measuring the amount of acid soluble radioactivity produced in the presence of helicase (Abdel-Monem *et al.*, 1976; Hotta and Stern, 1978). Today, the prototype helicase assay measures the conversion of double stranded to single-stranded nucleic acids using isotope-labelled oligonucleotides.

The substrates for these assays are typically made of a labelled shorter strand annealed to a longer complementary strand. Helicase, buffer, divalent metal ions (e.g. Mg^{2+}) and ATP are then combined, the reaction allowed to proceed for some time, and metal chelating agents are added to terminate the reaction. Products are then separated using non-denaturing polyacrylamide gel electrophoresis, and detected with autoradiography or a similar method (Matson and Kaiser-Rogers, 1990). Since single-stranded nucleic acids run faster in such a gel, the amount of single-stranded product formed can be used to calculate reaction velocity. With HCV helicase, the shorter oligonucleotide needs to be annealed to a complementary longer strand to leave a 3′ tail necessary for HCV helicase to function. During the reaction (or shortly after the reaction is stopped) a third oligonucleotide must be added that is complementary to one strand to 'trap' reaction products and prevent the substrate from re-forming. This process is extremely sensitive, limited only by the specific activity of the labelled DNA probe, and is still the gold standard in the field today. The helicase assays outlined below illustrate how this basic assay can be modified to provide a higher throughput or so that more data points can be obtained to follow reaction kinetics in real time (Table 12.1).

The above assay can be directly used for high throughput screening by employing DNA binding beads (e.g. GeneClean Glassmilk, Qbiogene) and filter plates, as was done to identify an inhibitor of the helicase encoded by herpes simplex virus (HSV; Sivaraja *et al.*, 1998). In this high throughput helicase assay, the substrate is made of a labelled-oligonucleotide annealed to M13 viral DNA. Reactions are performed in well plates with hydrophobic membranes on the bottom, and the beads are added to bind the viral DNA (and any remaining annealed oligonucleotide) along with a detergent. When the plates are then washed, the labelled single-stranded product of the helicase reaction is removed, and only substrate remains in the wells, which can be measured using a scintillation counter.

While the above bead-based assay requires a long DNA substrate, shorter helicase substrates can be monitored in a high throughput format using oligonucleotides and a scintillation proximity assay (SPA). Kyono *et al.* developed a SPA for HCV helicase, which uses a substrate made of a ^{3}H-labelled oligonucleotide annealed to an unlabelled RNA oligonucleotide. When the helicase reaction is run, the ^{3}H-labelled strand is released, and it is captured with a complementary biotinylated oligonucleotide that is present in high concentration in the reaction mixture. This biotin-labelled [^{3}H]DNA duplex then binds to the scintillant bead via biotin/streptavidin, allowing for detection (Kyono *et al.*, 1998). SPA beads emit photons only when they directly complex radioisotopes. A similar assay using a duplex RNA substrate is now commercially available (GE healthcare).

A FlashPlate (NEN Lifesciences/Perkin Elmer) assay is a modified scintillation proximity assay where the well plate itself is coated with a scintillant rather than using a bead scintillant. A DNA oligonucleotide directly bound to the plate by a biotin linker is annealed to a complementary [^{33}P]DNA (Earnshaw and Pope, 2001). When a helicase and ATP is added, the radiolabelled DNA is released into solution, away from the scintillant coating of the well, and lower counts are recorded. Such an assay is advantageous in that the plates can be read at any point without the need to quench the reaction (Earnshaw and Pope, 2001). Such a FlashPlate assay was used by Hicham Alaoui-Ismaili *et al.* (2000) to measure HCV helicase-catysed unwinding of DNA in a high-throughput format.

A HCV helicase HTS-capable unwinding assay that eliminates the need for radioisotopes is based on electrochemiluminescence (ECL) and was tested by Zhang *et al.* (2001) using the DnaB helicase. The ECL substrate is made by attaching a DNA nucleotide to a ruthenium chelate which is in turn annealed to an unlabelled DNA (Zhang *et al.*, 2001). A helicase reaction is run in the presence of a trapping strand that is conjugated to biotin, and the trapped reaction product complex is captured on streptavidin-coated magnetic beads. Application of an electric potential causes excitation of the ruthenium chelate at the electrode, resulting in the chemiluminescent emission of photons in proportion to the amount of single-stranded DNA produced in the helicase reaction.

HCV helicase-catalysed RNA unwinding can also be measured in a high throughput format using an enzyme linked immunosorbent assay (ELISA; Hsu *et al.*, 1998). The helicase ELISA substrate is composed of a digoxigenin (DIG) labelled RNA annealed to a biotin-labelled RNA. The biotin of the ELISA substrate is bound to a streptavidin coated well plate, then helicase and ATP are added to initiate RNA unwinding (Hsu *et al.*, 1998). After incubating for an hour, plates are washed to remove free DIG-labelled RNA and anti-DIG antibody is added and detected using standard techniques (Hsu *et al.*, 1998). Both biotin- and DIG- labelled oligonucleotides are commercially available.

Most fluorescent helicase assays rely on Förster resonance energy transfer (FRET), which involves two chromophores referred to as the 'donor' and the 'acceptor.' For FRET to occur, the emission wavelength of the donor must overlap the absorption wavelength of the acceptor. Light of a defined wavelength is used to excite the donor, and either fluorescence emission of the acceptor or the donor is monitored. If the accepter chromophore is not fluorescent, it will simply act to quench the fluorescence of the donor. In most fluorescent helicase assays the donor and acceptor chromophores are tethered to complementary strands of DNA. As the name FRET implies, a photon is not involved in the transfer of energy from donor to acceptor: it is a dipole–dipole interaction and the efficiency of this transfer falls off rapidly as distance between donor and acceptor increases, as it would if the donor and accepter are bound to two complementary DNA strands that are separated by a helicase (Lakowicz, 1999).

A wide variety of chromophores have been used in FRET-based unwinding assays. The original helicase FRET assays used coumarin as a donor and fluorescein as an acceptor (Houston and Kodadek, 1994), or a fluorescein/hexachlorofluorescein donor/acceptor pair (Bjornson *et al.*, 1994). The later assay was used by Porter *et al.* to study HCV NS3h in early mechanistic studies (Porter *et al.*, 1998). Other FRET assays used for HCV helicase high throughput screens have employed a Cy3/Black hole Quencher donor/acceptor pair (Boguszewska-Chachulska *et al.*, 2004), or an Alexa Fluor 448/Black Hole Quencher pair (Tani *et al.*, 2010). Similar fluorescence assays can be used to monitor unwinding with the DNA itself acting as the acceptor chromophore. For example, fluorescence of a DNA oligonucleotide containing the fluorescent base 2-aminopurine in place of adenine decreases when it is annealed to a complementary strand (Raney *et al.*, 1994). Similarly, some fluorescent moieties, such as BODIPY FL, can be quenched through photoinduced electron transfer by nearby guanines in complementary strands (Tani *et al.*, 2009).

The above fluorescent helicase assays are all set up like the prototype radioisotope-based helicase assay except that the trapping oligonucleotide must always be added at the start of the reaction. Otherwise, no reaction will be observed when the acceptor-bearing strand and the donor-bearing strand re-anneal. How additional DNA strands might interact with HCV helicase is not precisely known, although a recent report suggest that HCV helicase might simultaneously interact with three complementary strands of DNA to facilitate strand exchange in a reaction reminiscent of those catalysed by similar helicases needed for DNA recombination (Rypma *et al.*, 2010). Spurred by the idea that HCV helicase might function to primarily rearrange nucleic acids in such a manner, we recently developed a FRET-based helicase assay that utilizes dual-labelled oligonucleotides, called molecular beacons, which are capable of forming both hetero-duplexes and hairpin structures (Tyagi and Kramer, 1996). Such a molecular beacon-based helicase assay (MBHA) has the advantage over other of FRET-based assays in that no trap is required in the reaction (Belon and Frick, 2008). Upon strand separation by HCV NS3, the beacon forms a hairpin, bringing the donor and acceptor next to each other and fluorescence is quenched. By utilizing a molecular beacon, each strand acts as its own trapping agent, making the reaction essentially irreversible. Also, since the donor and acceptor modifications are located on the same strand, and opposite the strand on which the helicase likely translocates, possible protein–fluorophore interactions are minimized. The MBHA can also be performed in the same vessel as fluorescent NS3 protease assays so that both activates of NS3 can be simultaneously monitored (Frick *et al.*, 2010).

The most common problem seen when fluorescence-based assays are used for high throughput screens is that compounds that absorb or emit light at either the excitation or emission wavelengths will interfere with the assay, either by fluorescing themselves or by quenching the fluorophore being monitored. The two most common methods used to attempt to circumvent compound interference involve monitoring either changes in fluorescence polarization (or anisotropy) or time-resolved fluorescence, instead of fluorescence intensity. Polarization changes when a fluorescence probe-labelled oligonucleotide is displaced from its complement, and such assays have been extensively been used to monitor helicase action (Xu *et al.*, 2003). A time resolved fluorescence energy transfer (TR-FRET) assay, featuring a Eu^{3+}/tetramethyl rhodamine donor/acceptor, has also been used in high throughput screens to identified several novel bacterial and viral helicase inhibitors (Earnshaw *et al.*, 1999).

The above list of assays that could be used to screen for inhibitors of HCV helicase-catalysed DNA or RNA unwinding is by no means exhaustive. There are a seemingly endless variety of assays that monitor the action of helicases on DNA or RNA, and numerous HCV helicase assays have been reported that might also be useful for screening and for analysing the effects of the inhibitors listed below. Notable other assays include one that can be used to track NS3 progression along duplex RNA (Serebrov and Pyle, 2004), a procedure that monitors helicase action on an RNA thread held with optical tweezers (Dumont *et al.*, 2006), and a procedure that uses single molecule fluorescence to monitor HCV helicase action on DNA (Myong *et al.*, 2007). More detailed reviews and protocols for these and other helicase can be found in the recent book *Helicases: Methods and Protocols* (Abdelhaleem, 2010). The selection of a helicase assay depends on experimental goals as well as the time, resources, and instruments available.

Compounds that inhibit HCV helicase

Helicases are difficult targets as evidenced by the fact that no helicase inhibitors are approved for clinical use. There has been some preclinical success, however, in developing inhibitors of the herpes simplex virus (HSV) helicase-primase as antiviral drugs (Crute *et al.*, 1995; Crute *et al.*, 2002; Kleymann *et al.*, 2002) and, for this reason alone, all HCV helicase inhibitors should be critically examined for those that might have antiviral potential. Below follows a list of known HCV helicase inhibitors grouped based on where they might interact with NS3, along with summaries of their known potency. For each compound, up to three values are listed that describe its ability to inhibit HCV helicase in helicase assays by 50% (IC_{50}), its ability to inhibit the HCV replicon by 50% (EC_{50}), and its cellular toxicity reported as the concentration of compound necessary to kill half the cells (CC_{50}). When discussing compounds that were tested in replicons, we also report a 'selectivity index', (SI), defied here as the simple ratio EC_{50}/CC_{50}. The SI is a measure of how effective a compound is at inhibiting cellular HCV RNA replication compared with its toxicity in cells. A high SI indicates that the compound is relatively non-toxic. In addition to small molecules, we also briefly discuss small peptide inhibitors, which might be useful in designing peptidomimetic compounds. Other inhibition strategies have been investigated as well, such the use of antibodies inhibiting NS3 helicase (Tessmann *et al.*, 2002; Prabhu *et al.*, 2004) and RNA aptamers (Hwang *et al.*, 2004), but these are not discussed here. These alternate approaches have been reviewed previously in detail (Frick and Lam, 2006; Frick, 2007).

Compounds that likely displace ATP

Most nucleoside monophosphates (NMPs), nucleoside diphosphates (NDPs), and nucleoside triphosphates (NTPs) bind HCV helicase weakly with dissociation constants that mimic what is seen with ATP. In the presence of RNA, HCV helicase binds ATP with an apparent K_d of 7 mM and in the absence of RNA the protein binds ATP with K_d of 0.9 mM (Wang *et al.*, 2010). In our lab, we have found that all NTPs tested, including ribavirin triphosphate and the non-obligate chain terminators used as NS5B inhibitors, can substitute for ATP and fuel unwinding assays. Guanine derivatives are typically hydrolysed more slowly and support the slowest unwinding rates. As such, under certain conditions, high concentrations of GTP can inhibit ATP hydrolysis and ATP-fuelled unwinding (Heck *et al.*, 2008; Belon and Frick, 2009b). Others have also observed that NTPs with inverted sugars such as L-β-dATP, L-β-dTTP, L-β-dCTP and L-β-ddCTP are also very poor fuels for the unwinding reaction (Locatelli *et al.*, 2001). Non-hydrolysable NTP analogues, such as α, β-methylene-ATP and β, γ-imido-ATP, are poor inhibitors of HCV catalysed DNA unwinding with IC_{50} values of 140 μM and 1.8 mM, respectively (Belon and Frick, 2008).

Some modified nucleosides appear to interact somewhat more tightly with HCV helicase than the above NTPs. For example, some potent compounds were found in screens of ring-expanded, or so-called 'fat nucleosides' (Fig. 12.1A). Fat nucleosides were tested for their ability to inhibit the ATPase and helicase action of NS3 helicases from HCV, West Nile virus, (WNV), and Japanese encephalitis virus (JEV; Zhang *et al.*, 2003). Fat nucleoside compound *24* (Fig. 12.1A) inhibits HCV helicase-catalysed DNA and RNA unwinding with an IC_{50} of 11 μM and 5.5 μM,

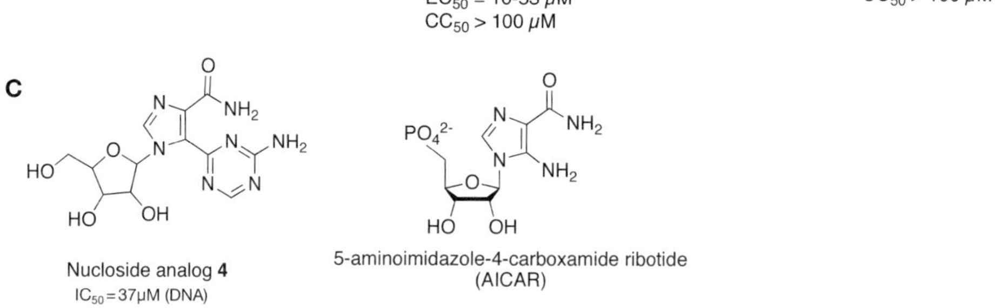

Figure 12.1 Nucleoside and base analogue helicase inhibitors presumed to bind to ATP binding site. (A) 'Fat' nucleoside analogue **24** (Zhang *et al.*, 2003). This compound inhibits helicase-catalysed DNA unwinding with an IC_{50} of 11 μM, and inhibits RNA unwinding with an IC_{50} of 5.5 μM. This compound also inhibits WNV helicase (DNA IC_{50} = 5.7 μM, RNA IC_{50} = 3.3 μM), JEV helicase (DNA IC_{50} = 150 μM, RNA IC_{50} > 500 μM) and human Suv3 helicase (DNA IC_{50} = 5 μM, RNA IC_{50} = 6 μM). (B) Halogenated nucleoside or base analogues. 5,6-dichloro(ribofuranosyl)benzotriazole (DRBT) inhibits helicase-catysed DNA unwinding with an IC_{50} of 1.5 μM but the closely related 5,6-dichloro(ribofuranosyl)benzimidazole (DRB) inhibits unwinding with an IC_{50} of 450 μM (Borowski *et al.*, 2003). Tetrabromobenzotriazole (TBBT) inhibits HCV NS3 catalysed DNA unwinding with an IC_{50} of 20 μM (Borowski *et al.*, 2003). Both DRBT and TBBT are active in various different replicons with EC_{50} *of ~10–65* μM and no observed toxicity up to 100 μM (Paeshuyse *et al.*, 2008). (C) Compound **4** is an aminotriazine modified analogue of 5-aminoimidazole-4-carboxamide ribotide (AICAR), an intermediate in the purine biosynthesis pathway. It inhibits helicase-catysed DNA unwinding with an IC_{50} of 37 μM but does not inhibit RNA unwinding (Ujjinamatada *et al.*, 2007).

respectively. Although this series of analogues would seem likely to compete for the NS3 ATP binding site, none of the ring-expanded nucleosides inhibit the ATPase activity of JEV, HCV or WNV helicases. The efficacy of fat nucleosides is not isolated to flaviviral helicases. The human cellular DEAD box helicase, DDX3, which has been shown to be necessary for HIV replication (Yedavalli *et al.*, 2004), is also inhibited by fat nucleosides, and this effect is sufficient to halt HIV replication in cell culture (Yedavalli *et al.*, 2008).

Other nucleoside analogues that inhibit HCV helicase are built on benzazole base scaffolds. Halogenated benzotriazoles (Fig. 12.1B), such as 5,6-dichloro(ribofuranosyl)benzotriazole (DRBT) have been tested on HCV, WNV and JEV NS3 helicases, and they inhibit DNA unwinding by HCV helicase selectively with an IC_{50} of 1.5 μM. They do not inhibit WNV or JEV helicases at concentrations up to 500 μM (Borowski *et al.*, 2003). The importance of the benzotriazole ring was demonstrated by replacing it with a benzimidazole (Fig. 12.1B). The benzimidazole nucleosides are much less potent with an IC_{50} of 450 μM (Borowski *et al.*, 2003). A halogenated benzotriazole lacking the ribose moiety, tetrabromobenzotriazole (TBBT; Fig. 12.1B), is also somewhat effective in inhibiting HCV catalysed DNA unwinding with an IC_{50} of 20 μM, but it is more potent against WNV NS3 helicase with a 1.7 μM IC_{50} (Borowski *et al.*, 2003). Both DRBT and TBBT have been tested in four different replicon cell lines and they both inhibit HCV replication (DRBT EC_{50} values of 10–53 μM, TBBT EC_{50} values of 40–65 μM). Neither is toxic at concentrations up to 100 μM (Paeshuyse *et al.*, 2008).

Nucleoside analogue *4*, an analogue of an intermediate in the purine biosynthesis pathway (Fig. 12.1C), was tested in helicase assays against WNV, HCV, JEV and DV helicases. The compound inhibited DNA unwinding by the HCV helicase with an IC_{50} of 37 μM, and was slightly more potent against the WNV helicase with an IC_{50} of 23 μM, but was ineffective against JEV and DV helicases. Surprisingly, when the substrate was switched from DNA to RNA, no inhibition of unwinding was observed for any of the helicases (Ujjinamatada *et al.*, 2007).

Only two classes of HCV inhibitors that do not resemble nucleosides are believed to target the HCV helicase ATP binding site. The first class in made of pyrophosphate analogues. The pyrophosphate analogue imidodiphosphate, (Fig. 12.2A) has a K_i of 12 μM in assays measuring the HCV NS3 catalysed ATP hydrolysis, and pyrophosphate itself appears in such assays to bind HCV NS3 significantly tighter than any nucleoside analogue with an K_i of 20 μM (Mukovnya *et al.*, 2008). The other class comprises compounds derived for the dye soluble blue HT. Hu *et al.* used molecular modelling to identify soluble blue HT (Fig. 12.2B) as a potential NS3 helicase inhibitor (Hu *et al.*, 2003). They also solved a co-crystal structure (PDB file 2ZJO) that shows soluble blue HT bound to NS3 with one of its three benzene sulfonate moieties directly adjacent to the catalytic Walker A site (SF2 helicase motif I). Such an arrangement would prevent the enzyme from binding and hydrolysing ATP. Blue HT inhibits HCV NS3 catalysed DNA unwinding with an IC_{50} of 40 μM (Chen *et al.*, 2009). Blue HT was later subjected to several rounds of structural refinement by creating derivatives and assaying compounds for efficacy, eventually resulting in the discovery of compound *12* (Fig. 12.2B), which inhibits HCV NS3-catalysed DNA unwinding with an IC_{50} of 10.1 μM. Compound *12* inhibits an HCV subgenomic replicon with an EC_{50} of only 2.7 μM and a CC_{50} of 10.5 μM, a significant increase in selectivity index (i.e. EC_{50}/CC_{50}) compared with other active compounds in the series that had specificities of approximately 1.4–1.5 (Chen *et al.*, 2009).

Compounds acting at the DNA/RNA binding site

Even though the atomic details of how HCV helicase interacts with nucleic acids have been known for over ten years (Kim *et al.*, 1998), only a few classes of compounds are known to bind this site. Probably the largest class is the set of symmetrical benzimidazole-phenyl-carboximides that were first reported in a series of patents by Viropharma (Diana and Bailey, 1997) and were later the

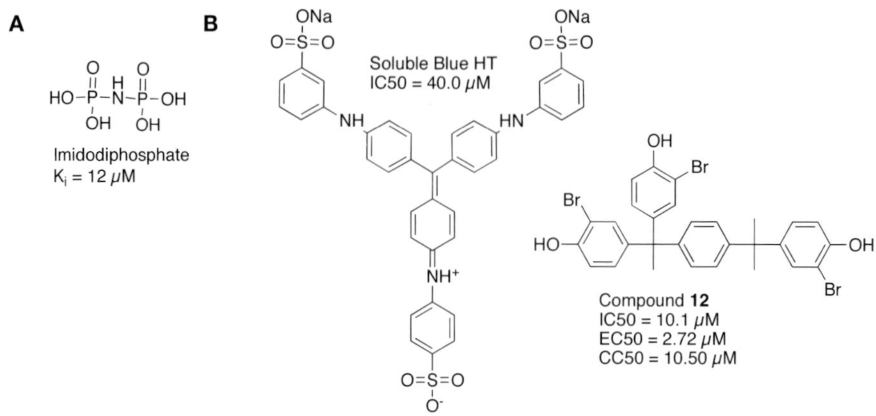

Figure 12.2 Non-nucleoside inhibitors presumed to bind the ATP binding site. (A) Pyrophosphate analogue helicase inhibitor imidodiphosphate. Imidodiphosphate inhibits helicase-catysed ATP hydrolysis, and has a K_i of 12 μM (Mukovnya *et al.*, 2008). (B) Triphenylmethane derivatives soluble blue HT (Hu et al., 2003) and compound **12** (Chen *et al.*, 2009). Soluble blue HT inhibits helicase-catysed DNA unwinding with an IC_{50} *of 40* μM. Compound **12** inhibits DNA unwinding with an IC_{50} of 10.1 μM and has an EC_{50} of 2.7 μM and a CC_{50} of 10 μM in an HCV subgenomic replicon.

subject of a detailed structure activity relationship study (Phoon *et al.*, 2001). We have recently studied the mechanism of action of the prototype compound in this series: a head-to-head dimer composed of benzimidazole-phenyl-carboxamide subunits linked to benzene we term $(BIP)_2B$ (Belon *et al.*, 2010; Fig. 12.3A). $(BIP)_2B$ inhibits DNA and RNA unwinding with IC_{50}'s of 0.7 to 5 μM, and binds the HCV NS3h protein in place of nucleic acid, preventing the action of the helicase on its substrate (Belon *et al.*, 2010). The compound also inhibits related helicases from Dengue virus (DV), Japanese encephalitis virus (JEV) as well human DDX3 helicase, and binds DV and JEV helicases substantially tighter than either HCV helicase or DDX3, suggesting $(BIP)_2B$ may be an more effective inhibitor of the flaviviral helicases. $(BIP)_2B$ does not inhibit the HCV replicon.

Maga *et al.* describe a series of quinoline and quinoxalines that mimic the electronic characteristic of purines and pyrimidines and compete with nucleic acid but do not affect ATPase activity (Maga *et al.*, 2005). Through progressive rounds of synthesis, molecular modelling and inhibition assays, they created QU663 (Fig. 12.3B), a quinoline-based compound that inhibits DNA unwinding with a K_i of only 750 nM. QU663 and is not toxic to 3T3, NSO, Daudi, or normal human lymphocyte cells (Maga *et al.*, 2005), but was not reportedly tested in a HCV replicon system. Molecular modelling supports the idea that QU663 binds the RNA binding groove. Such models predict QU663 interacts strongly with W501 as well as other key residues located in the RNA binding site including Phe557, Arg393, Thr298, Gln434, and Glu493 (Maga *et al.*, 2005).

Kandil *et al.* used molecular modelling to *de novo* design an NS3 helicase inhibitor that binds in place of RNA (Kandil *et al.*, 2009). First, using the crystal structure of NS3h bound to a DNA oligonucleotide (Kim *et al.*, 1998) they sought to target a region surrounding Arg393, a residue that is critical for RNA unwinding (Lam *et al.*, 2003b) because it interacts with the DNA/RNA backbone. Because Arg393 is entirely solvent exposed, the group hypothesized that a molecule interacting with the protein near Arg393 would be significantly stabilized if it formed a covalent bond with an adjacent cysteine, Cys431 (Kandil *et al.*, 2009). The parameters of the initial docking and model building were tailored to produce molecules of comparatively small size with few chiral centres. The necessity of the covalent bond-forming Michael acceptor was verified by the target compound, compound 4 (Fig. 12.3C), inhibiting HCV NS3 catalysed DNA unwinding with an IC_{50} of only 0.26 μM (Kandil *et al.*, 2009).

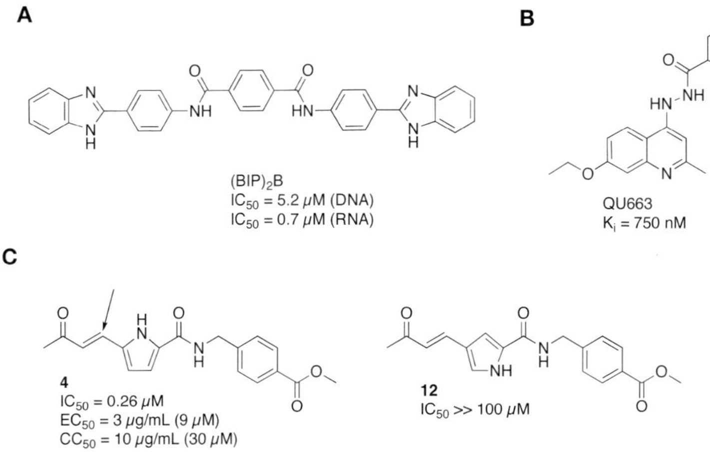

Figure 12.3 Compounds acting at the RNA binding site. (A) 1-*N*,4-*N*-bis[4-(1H-benzimidazol-2-yl)phenyl] benzene-1,4- dicarboxamide, (BIP)$_2$B (Phoon *et al*., 2001), inhibits helicase-catysed DNA unwinding with an IC_{50} *of 5.4* μM and RNA unwinding with an IC_{50} of 0.7 μM (Belon *et al.,* 2010). (B) QU663, a quinoline derivative, inhibits helicase-catysed DNA unwinding with a K_i of 750 nM (Maga *et al*., 2005). (C) Substituted pyrrole helicase inhibitors. Compound **4** was designed to place a Michael acceptor (arrow) near active site Cys431, thereby forming a covalent helicase–inhibitor complex. Compound **4** inhibits helicase-catysed DNA unwinding with an IC_{50} of 0.26 μM, and inhibits a HCV subgenomic replicon with an EC_{50} of 3 μg/ml (9 μM) and a CC_{50} of 10 μg/ml (30 μM; Kandil *et al*., 2009). On the basis of computer docking, regioisomer **12** is predicted to bind with the Michael acceptor close to Cys431; it has no *in vitro* inhibitory activity at up to 100 μM (Kandil *et* al., 2009).

In docking simulations, regioisomer *12* (Fig. 12.3C) bound in a conformation where the reactive Michael acceptor was distant to Cys431 and showed no *in vitro* activity at all, up to the 100 μM tested. Compound *4* was then tested in an HCV subgenomic replicon and demonstrated an EC_{50} of 3 μg/ml (9 μM) but killed cells with a CC_{50} of 10 μg/ml (30 μM; Kandil *et al.,* 2009).

Compounds acting at allosteric or unknown sites

Five other classes of compounds have been reported to be potent HCV helicase inhibitors, but less is known about where these compounds might bind the enzyme and how they might exert their actions. Two of these classes, the acridone derivatives and the anthracyclines, are known to also intercalate into the strands of double stranded nucleic acids and this activity might contribute to their potency. The other classes are composed of tropolones, thiadiazoles, and substituted pyrimidines.

The acridone derivatives are particularly intriguing because they were discovered to be helicase inhibitors independently by two different groups using different approaches. The first group to characterize acridone-based helicase inhibitors initially screened a library of acridone derivatives with unwinding assays. Two of the compounds, *20* and *27* (Fig. 12.4A) inhibit DNA unwinding with IC_{50}'s of 8.9 and 3.8 μM, respectively (Stankiewicz-Drogon *et al.,* 2008). Acridone derivatives *20* and *27* also inhibit RNA replication in a subgenomic replicon with EC_{50}'s of 9.0 and 10.2 μM while being relatively non toxic with CC_{50} values of 174.8 and 411.9 μM, respectively (Stankiewicz-Drogon *et al.,* 2008). Contrary to the methodology used by Stankiewicz-Drogon *et al.,* Manfroni *et al.* first screened drugs for activity against a subgenomic replicon, identified active compounds, and then investigated their mechanism of action. By starting with a drug that showed activity against bovine diarrhoeal virus (BVDV), the group synthesized over 40 derivative compounds and tested them for activity in an HCV subgenomic replicon; 5 of these compounds, derivatives *12, 14, 19, 23,* and *35* were active (Manfroni *et al.,* 2009; Fig. 12.4B). Of these 5 compounds, none inhibited NS5B polymerase or cellular topoisomerase II, but compound *23*

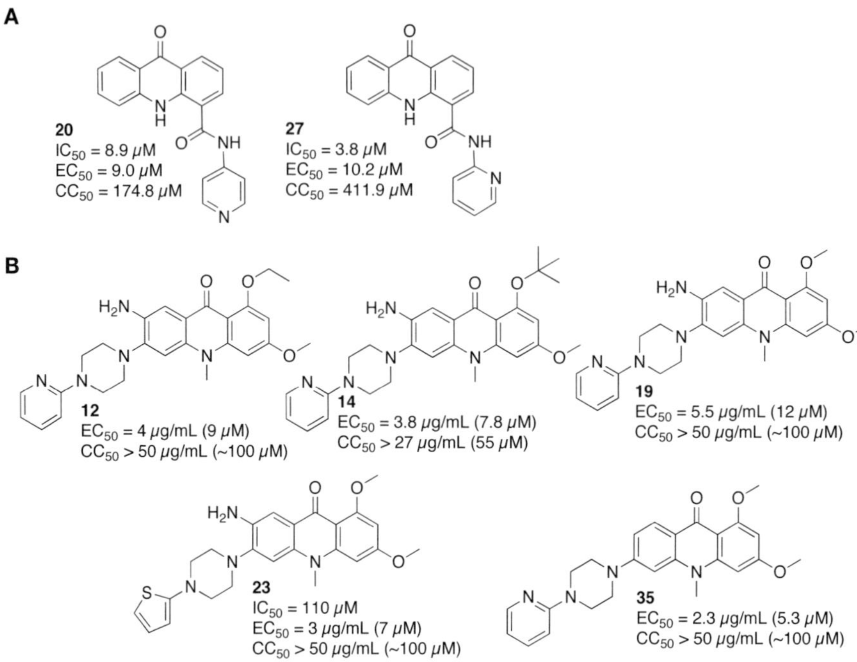

Figure 12.4 Acridone-based helicase inhibitors (A) Acridone derivative helicase inhibitors **20** and **27** inhibit helicase-catysed DNA unwinding and RNA replication in a subgenomic replicon at low micromolar concentrations (Stankiewicz-Drogon *et al.*, 2008). (B) Acridone compounds **12**, **14**, **19**, **23**, and **35** were identified by screening for activity in an HCV subgenomic replicon, and all inhibit RNA replication with an EC_{50} of 2.3–5.5 μg/ml (5–12 μM; Manfroni *et al.*, 2009). Only compound **23** inhibited helicase-catysed DNA unwinding, with an IC_{50} of 50 μg/ml (110 μM; Manfroni *et al.*, 2009).

inhibited HCV helicase activity with an IC_{50} of 110 μM.

By employing an initial high throughput enzymatic screen of about 70 compounds, followed by replicon testing, Krawczyk *et al.* discovered several high potency, low toxicity compounds based on the known anti-cancer drug doxorubicin (Fig. 12.5A). For example, compound *3* had and very low EC_{50} of 0.057 μM in cell culture, and compound *21* had an EC_{50} of 0.129 μM, both of which had a selectivity index (i.e. EC_{50}/CC_{50}) of 20 or greater. The anthracyclines are known DNA intercalating agents, and the group demonstrated clear intercalation of compound into DNA with competition studies, direct measurement of substrate–inhibitor complex formation, and ATPase inhibition experiments. These experiments imply a mechanism where helicase is unable to interact productively with the DNA substrate, and may form a stable ternary enzyme–inhibitor DNA complex (Krawczyk *et al.*, 2009).

Tropolones are compounds known to have antiviral, anticancer and other therapeutic uses (Zhao, 2007), and several such compounds are potent HCV helicase inhibitors. For example, dibromo-morpholinomethyltropolone (DBMTr; Fig. 12.5B) inhibits HCV helicase-catalysed DNA unwinding with an IC_{50} of 18 μM and shows little cytotoxicity in yeast (Boguszewska-Chachulska *et al.*, 2006). Tropolone effects on the HCV NS3 ATPase and HCV helicase-catysed RNA unwinding, however, are generally small; some even stimulate ATP hydrolysis. For example DMBTr stimulates the NS3 helicase-catalysed ATP hydrolysis rate by 8% at 100 μM (Boguszewska-Chachulska *et al.*, 2006) and does not inhibit helicase-catysed RNA unwinding (Borowski *et al.*, 2007). The tropolone derivatives do not appear to compete with ATP, DNA or RNA in any assays.

Little has been reported about the final two classes of HCV helicase inhibitors, which have

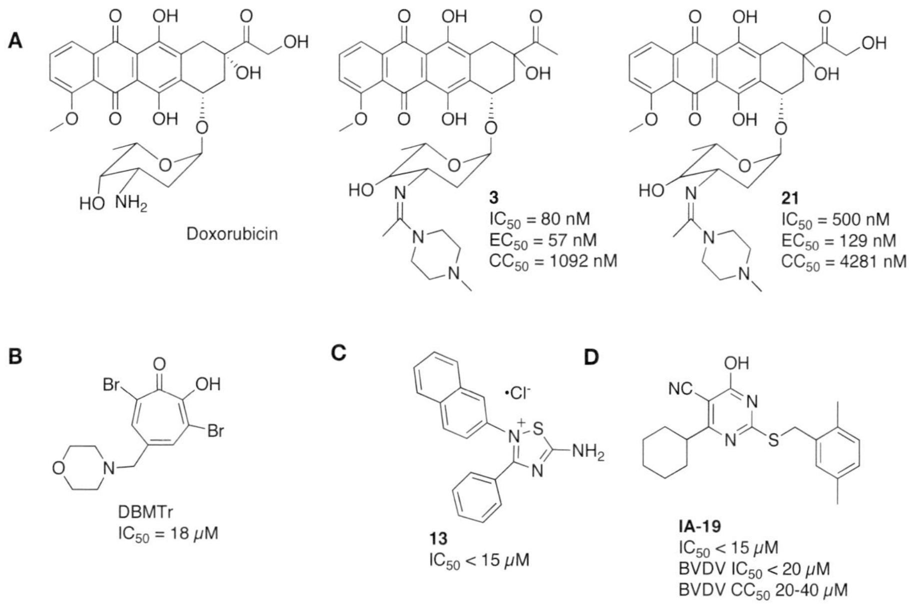

Figure 12.5 Helicase inhibits acting at unknown sites. (A) Selected amidinoanthracycline helicase inhibitors. These compounds are analogues of the anti-cancer drug doxorubicin. Compounds **3** and **21** are shown due to their high potency in both DNA unwinding assays (IC_{50}=80, 500 nM) and subgenomic replicons (EC_{50}=57, 129 nM) as well as for their high selectivity index (i.e. EC_{50}/CC_{50}) of 19 and 33 for compounds **3** and **21,** respectively (Krawczyk *et al*., 2009). (B) Dibromo-morpholinomethyltropolone (DBMTr) inhibits helicase-catysed DNA unwinding with an IC_{50} of 18 μM (Boguszewska-Chachulska *et a*l., 2006). (C and D) Thiadiazole (Janetka *et al*., 2000) and substituted pyrimidine (Hale *et al*., 2001) helicase inhibitors reported in patents assigned to Vertex Pharmaceuticals, Inc. Compound **13** and **IA-19** have IC_{50}'s of less than 15 μM in a helicase-catysed DNA unwinding assay; compound **IA-19** also inhibits viral replication in a bovine viral diarrhoea virus (BVDV) assay with an IC_{50} of < 20 μM, although the CC_{50} is 20–40 μM.

only been descried in patents. Vertex Inc. found 56 substituted thiadiazole compounds (Janetka *et al.*, 2000) that all had the ability to inhibit HCV NS3 catalysed ATP hydrolysis with IC_{50} values of less than 15 μM. Of the five most potent compounds in this set that were tested for an ability to inhibit helicase-catysed DNA unwinding only compound *13* (Fig. 12.5C), had an IC_{50} of less than 25 μM (Janetka *et al.*, 2000). Vertex also reported testing 17 substituted pyrimidine compounds for a their ability to inhibit HCV NS3 catalysed RNA unwinding (Hale *et al.*, 2001). Nine of the 17 compounds inhibit RNA unwinding with an IC_{50} of less than 15 μM and of those, one compound, *IA-19* (Fig. 12.5D), also inhibits replication of bovine viral diarrhoea virus in a cell culture system with an IC_{50} of less than 20 μM. The compound was, however, fairly toxic with a CC_{50} between 20 and 40 μM (Hale *et al.*, 2001).

Peptide-based inhibitors

Two small peptides have been reported to inhibit HCV helicase, each by a different mechanism, and they each also inhibit the HCV replicon in cells. A 2008 patent by Raney *et al.* describes a short peptide with the sequence HIDAHFLSQTK-GGG-YARAAARQARA that inhibits HCV replicons. The peptide mimics a portion of NS3 that is suspected to be involved in the formation of NS3 dimers and assembly of higher order oligomers (Raney *et al.*, 2008). The other known small peptide HCV helicase inhibitor appears to bind both ATP and RNA binding sites. This 14 amino acid peptide termed 'p14' has the sequence RRGRTGRGRRGIYR and was designed to mimic arginine-rich conserved helicase motif VI (Gozdek *et al.*, 2008). In DNA unwinding assays, p14 inhibited with an IC_{50} of 725 nM while not affecting ATP hydrolysis at all (Gozdek *et al.*,

2008). In other studies, p14 proved to be slightly more potent than initially described, inhibiting DNA unwinding with an IC_{50} of 200 nM (Borowski *et al.*, 2008). NMR and computer docking studies suggest that p14 occupies the nucleic acid binding groove, preventing the enzyme from acting on its nucleic acid substrate, while curving into the region near the P-loop, preventing interaction between helicase domains 1 and 2 (Gozdek *et al.*, 2008). p14 inhibits HCV replicons with an EC_{50} of 83 μM (Gozdek *et al.*, 2008).

Conclusion and future directions

To date, no inhibitors of any helicase are in clinical use, and much more work must be done if HCV helicase inhibitors are ever to enter the clinic. Recent major progress in the HCV inhibitor field centres on the fact that several relatively non-toxic compounds have been shown to halt HCV RNA replication in cell culture-based replicon assay systems. However, to our knowledge, none of these compounds have been tested in HCV strains capable of replicating in cell culture (e.g. HCVcc), and no HCV strains resistant to any of these compounds have been identified. Likewise, only a couple studies have made use of the information in the many NS3 structural or mechanistic studies to target key regions or rationally design compound derivatives. The next steps towards drug development for HCV helicase inhibitors should therefore include validating inhibitors in the HCVcc system. Resistant strains will then need to be isolated and sequenced to understand how rapidly resistance develops and if amino acid combinations conferring resistance already exist in nature. Compounds could also be soaked into NS3 crystals or the binding sites could be determined with NMR so that modelling could be used to facilitate the synthesis of more potent analogues.

At this point the utility of compounds that have shown activity in the replicon system is not limited strictly to drug design. Compounds could be used as chemical probes to understand why HCV needs a helicase using one of the subgenomic, genomic, or HCVcc culture systems. Similarly, helicase inhibitors can be used to study the HCV helicase itself, a prototype helicase and molecular motor. Inhibitors functioning by various mechanisms of action could be used to elucidate steps in the ATP hydrolysis cycle and the mechanism of helicase-catalysed nucleic acid unwinding. Although HCV NS3 helicase inhibitors may not represent the next generation of anti-HCV drugs, they are currently useful for studying HCV and represent another promising way to fight this potentially serious disease in the future.

Acknowledgement

The work was supported by National Institute of Health grant AI088001.

References

Abdel-Monem, M., Durwald, H., and Hoffmann-Berling, H. (1976). Enzymic unwinding of DNA. 2. Chain separation by an ATP-dependent DNA unwinding enzyme. Eur. J. Biochem. *65*, 441–449.

Abdelhaleem, M.M. (2010). Helicases: Methods and Protocols. In Walker, J.M., ed. series (Humana Press, New York).

Arunajadai, S.G. (2009). RNA unwinding by NS3 helicase: a statistical approach. PLoS One *4*, e6937.

Belon, A., and Frick, D.N. (2008). Monitoring helicase activity with molecular beacons. BioTechniques 45, 433–40, 442.

Belon, A., and Frick, D.N. (2009a). Helicase inhibitors as specifically targeted antiviral therapy for hepatitis C. Future Virol. *4*, 277–293.

Belon, A., and Frick, D.N. (2009b). Fuel specificity of the hepatitis C virus NS3 helicase. J. Mol. Biol. 388, 851–864.

Belon, C.A., High, Y.D., Lin, T.I., Pauwels, F., and Frick, D.N. (2010). Mechanism and specificity of a symmetrical benzimidazolephenylcarboxamide helicase inhibitor. Biochemistry *49*, 1822–1832.

Beran, R.K., Serebrov, V., and Pyle, A.M. (2007). The serine protease domain of hepatitis C viral NS3 activates RNA helicase activity by promoting the binding of RNA substrate. J. Biol. Chem. *282*, 34913–34920.

Beran, R.K., Lindenbach, D., and Pyle, A.M. (2009). The NS4A protein of hepatitis C virus promotes RNA-coupled ATP hydrolysis by the NS3 helicase. J. Virol. *83*, 3268–3275.

Bjornson, K.P., Amaratunga, M., Moore, J., and Lohman, T.M. (1994). Single-turnover kinetics of helicase-catalyzed DNA unwinding monitored continuously by fluorescence energy transfer. Biochemistry 33, 14306–14316.

Boguszewska-Chachulska, A.M., Krawczyk, M., Stankiewicz, A., Gozdek, A., Haenni, L., and Strokovskaya, L. (2004). Direct fluorometric measurement of hepatitis C virus helicase activity. FEBS Lett. *567*, 253–258.

Boguszewska-Chachulska, A.M., Krawczyk, M., Najda, A., Kopanska, K., Stankiewicz-Drogon, A., Zagorski-Ostoja, W., and Bretner, M. (2006). Searching for a new anti-HCV therapy: synthesis and properties of tropolone derivatives. Biochem. Biophys. Res. Commun. *341*, 641–647.

Borowski, P., Deinert, J., Schalinski, S., Bretner, M., Ginalski, K., Kulikowski, T., and Shugar, D. (2003). Halogenated benzimidazoles and benzotriazoles as inhibitors of the NTPase/helicase activities of hepatitis C and related viruses. Eur. J. Biochem. *270*, 1645–1653.

Borowski, P., Lang, M., Haag, A., and Baier, A. (2007). Tropolone and its derivatives as inhibitors of the helicase activity of hepatitis C virus nucleotide triphosphatase/helicase. Antivir. Chem. Chemother. *18*, 103–109.

Borowski, P., Heising, M.V., Miranda, I.B., Liao, C.L., Choe, J., and Baier, A. (2008). Viral NS3 helicase activity is inhibited by peptides reproducing the Arg-rich conserved motif of the enzyme (motif VI). Biochem. Pharmacol. *76*, 28–38.

Brune, M., Hunter, J.L., Corrie, E., and Webb, M.R. (1994). Direct, real-time measurement of rapid inorganic phosphate release using a novel fluorescent probe and its application to actomyosin subfragment 1 ATPase. Biochemistry 33, 8262–8271.

Chen, C.S., Chiou, C.T., Chen, G.S., Chen, S.C., Hu, C.Y., Chi, W.K., Chu, Y.D., Hwang, L.H., Chen, P.J., Chen, D.S., *et al.* (2009). Structure-based discovery of triphenylmethane derivatives as inhibitors of hepatitis C virus helicase. J. Med. Chem. *52*, 2716–2723.

Cho, H.S., Ha, N.C., Kang, L.W., Chung, K.M., Back, S.H., Jang, K., and Oh, B.H. (1998). Crystal structure of RNA helicase from genotype 1b hepatitis C virus. A feasible mechanism of unwinding duplex RNA. J. Biol. Chem. *273*, 15045–15052.

Cordin, O., Banroques, J., Tanner, K., and Linder, P. (2006). The DEAD-box protein family of RNA helicases. Gene *367*, 17–37.

Crute, J.J., Lehman, I.R., Gambino, J., Yang, T.F., Medveczky, P., Medveczky, M., Khan, N.N., Mulder, C., Monroe, J., and Wright, G.E. (1995). Inhibition of herpes simplex virus type 1 helicase-primase by (dichloroanilino)purines and -pyrimidines. J. Med. Chem. *38*, 1820–1825.

Crute, J.J., Grygon, C.A., Hargrave, K.D., Simoneau, B., Faucher, A.M., Bolger, G., Kibler, P., Liuzzi, M., and Cordingley, M.G. (2002). Herpes simplex virus helicase-primase inhibitors are active in animal models of human disease. Nat. Med. *8*, 386–391.

Diana, D., and Bailey, T.R. (1997). Compounds, compositions and methods for treatment of hepatitis C. Viropharma Incorporated USA, *5, 633*, 388.

Du, M.X., Johnson, R.B., Sun, X.L., Staschke, K.A., Colacino, J., and Wang, Q.M. (2002). Comparative characterization of two DEAD-box RNA helicases in superfamily II: human translation-initiation factor 4A and hepatitis C virus non-structural protein 3 (NS3) helicase. Biochem. J. *363*, 147–155.

Dumont, S., Cheng, W., Serebrov, V., Beran, R.K., Tinoco, I.J., Pyle, M., and Bustamante, C. (2006). RNA translocation and unwinding mechanism of HCV NS3 helicase and its coordination by ATP. Nature *439*, 105–108.

Duong, F.H., Christen, V., Berke, J.M., Penna, S.H., Moradpour, D., and Heim, M.H. (2005). Upregulation of protein phosphatase 2Ac by hepatitis C virus modulates NS3 helicase activity through inhibition of protein arginine methyltransferase 1. J. Virol. *79*, 15342–15350.

Earnshaw, D.L., Moore, K.J., Greenwood, C.J., Djaballah, H., Jurewicz, A.J., Murray, J., and Pope, A.J. (1999). Time-Resolved Fluorescence Energy Transfer DNA Helicase Assays for High Throughput Screening. J. Biomol. Screen. *4*, 239–248.

Earnshaw, L., and Pope, A.J. (2001). FlashPlate scintillation proximity assays for characterization and screening of DNA polymerase, primase, and helicase activities. J. Biomol. Screen. *6*, 39–46.

Errington, W., Wardell, A.D., McDonald, S., Goldin, D., and McGarvey, M.J. (1999). Subcellular localisation of NS3 in HCV-infected hepatocytes. J. Med. Virol. *59*, 456–462.

Fischer, C.J., Maluf, K., and Lohman, T.M. (2004). Mechanism of ATP-dependent translocation of *E. coli* UvrD monomers along single-stranded DNA. J. Mol. Biol. *344*, 1287–1309.

Fiske, H., and Subbarow, Y. (1925). The colorimetric determination of phosphorus. J. Biol. Chem. *66*, 675–400.

Frick, D.N. (2007). The hepatitis C virus NS3 protein: a model RNA helicase and potential drug target. Curr. Issues Mol. Biol. *9*, 1–20.

Frick, D.N. and Lam, A.M. (2006). Understanding helicases as a means of virus control. Curr. Pharm. Des. *12*, 1315–1338.

Frick, D.N., Rypma, R.S., Lam, M., and Gu, B. (2004). The nonstructural protein 3 protease/helicase requires an intact protease domain to unwind duplex RNA efficiently. J. Biol. Chem. *279*, 1269–1280.

Frick, D.N., Banik, S., and Rypma, R.S. (2007). Role of divalent metal cations in ATP hydrolysis catalyzed by the hepatitis C virus NS3 helicase: magnesium provides a bridge for ATP to fuel unwinding. J. Mol. Biol. *365*, 1017–1032.

Frick, D.N., Ginzburg, O., and Lam, A.M. (2010). A method to simultaneously monitor hepatitis C virus NS3 helicase and protease activities. Meth. Mol. Bio. *587*, 223–233.

Funk, C.J., Davis, A.S., Hopkins, A., and Middleton, K.M. (2004). Development of high-throughput screens for discovery of kinesin adenosine triphosphatase modulators. Anal. Biochem. *329*, 68–76.

Gorbalenya, E., and Koonin, E.V. (2003). Helicases: amino acid sequence comparisons and structure–function relationships. Curr. Opin. Struct. Biol. *3*, 419–429.

Gozdek, A., Zhukov, I., Polkowska, A., Poznanski, J., Stankiewicz-Drogon, A., Pawlowicz, J.M., Zagorski-Ostoja, W., Borowski, P., and

Boguszewska-Chachulska, A.M. (2008). NS3 Peptide, a novel potent hepatitis C virus NS3 helicase inhibitor: its mechanism of action and antiviral activity in the replicon system. Antimicrob. Agents Chemother. *52*, 393–401.

Gu, M., and Rice, C.M. (2010). Three conformational snapshots of the hepatitis C virus NS3 helicase reveal a ratchet translocation mechanism. Proc. Natl. Acad. Sci. U.S.A. *107*, 521–528.

Hale, M., Maltais, F., Baker, C., Janetka, J.W., Moon, C., and Saunders, J. (2001). Inhibitors of viral helicase. Vertex Pharmaceuticals Inc. International, WO 01/07027 A2.

Hall, C., and Matson, S.W. (1999). Helicase motifs: the engine that powers DNA unwinding. Mol. Microbiol. *34*, 867–877.

Heck, J.A., Lam, A.M., Narayanan, N., and Frick, D.N. (2008). Effects of mutagenic and chain-terminating nucleotide analogs on enzymes isolated from hepatitis C virus strains of various genotypes. Antimicrob. Agents Chemother. *52*, 1901–1911.

Hicham Alaoui-Ismaili, M., Gervais, C., Brunette, S., Gouin, G., Hamel, M., Rando, F., and Bedard, J. (2000). A novel high throughput screening assay for HCV NS3 helicase activity. Antiviral Res. *46*, 181–193.

Hotta, Y., and Stern, H. (1978). DNA unwinding protein from meiotic cells of Lilium. Biochemistry *17*, 1872–1880.

Houston, P., and Kodadek, T. (1994). Spectrophotometric assay for enzyme-mediated unwinding of double-stranded DNA. Proc. Natl. Acad. Sci. U.S.A. *91*, 5471–5474.

Howe, A.Y., Chase, R., Taremi, S.S., Risano, C., Beyer, B., Malcolm, B., and Lau, J.Y. (1999). A novel recombinant single-chain hepatitis C virus NS3-NS4A protein with improved helicase activity. Protein Sci. *8*, 1332–1341.

Hsu, C.C., Hwang, L.H., Huang, Y.W., Chi, W.K., Chu, D., and Chen, D.S. (1998). An ELISA for RNA helicase activity: application as an assay of the NS3 helicase of hepatitis C virus. Biochem. Biophys. Res. Commun. *253*, 594–599.

Hu, C.-Y., Chen, S.-J., Liaw, S.-H. (2003). Rational drug designs based on crystal structures of the hepatitis C virus NS3 helicase-inhibitor complexes. KEK Progress Report *2002–2*, 183.

Hwang, B., Cho, J.S., Yeo, H.J., Kim, J.H., Chung, K.M., Han, K., Jang, K., and Lee, S.W. (2004). Isolation of specific and high-affinity RNA aptamers against NS3 helicase domain of hepatitis C virus. RNA *10*, 1277–1290.

Ishido, S., Muramatsu, S., Fujita, T., Iwanaga, Y., Tong, W.Y., Katayama, Y., Itoh, M., and Hotta, H. (1997). Wild-type, but not mutant-type, p53 enhances nuclear accumulation of the NS3 protein of hepatitis C virus. Biochem. Biophys. Res. Commun. *230*, 431–436.

Janetka, J.W., Ledford, E., and Mullican, M.D. (2000). Pentacyclic compounds useful as inhibitors of hepatitis C virus NS3 helicase. Vertex Pharmaceuticals Inc. International, WO 00/24725.

Jennings, T.A., Chen, Y., Sikora, D., Harrison, M.K., Sikora, B., Huang, L., Jankowsky, E., Fairman, M.E., Cameron, E., and Raney, K.D. (2008). RNA Unwinding Activity of the Hepatitis C Virus NS3 Helicase Is Modulated by the NS5B Polymerase. Biochemistry *47*, 1126–1135.

Jennings, T.A., Mackintosh, S.G., Harrison, M.K., Sikora, D., Sikora, B., Dave, B., Tackett, A.J., Cameron, E., and Raney, K.D. (2009). NS3 helicase from the hepatitis C Virus can function as a monomer or oligomer depending on enzyme and substrate concentrations. J. Biol. Chem. *284*, 4806–4814.

Kandil, S., Biondaro, S., Vlachakis, D., Cummins, A.C., Coluccia, A., Berry, C., Leyssen, P., Neyts, J., and Brancale, A. (2009). Discovery of a novel HCV helicase inhibitor by a de novo drug design approach. Bioorg. Med. Chem. Lett. *19*, 2935–2937.

Kim, J.L., Morgenstern, K.A., Griffith, J.P., Dwyer, M.D., Thomson, J.A., Murcko, M.A., Lin, C., and Caron, P.R. (1998). Hepatitis C virus NS3 RNA helicase domain with a bound oligonucleotide: the crystal structure provides insights into the mode of unwinding. Structure *6*, 89–100.

Kleymann, G., Fischer, R., Betz, U.A., Hendrix, M., Bender, W., Schneider, U., Handke, G., Eckenberg, P., Hewlett, G., Pevzner, V., *et al.* (2002). New helicase-primase inhibitors as drug candidates for the treatment of herpes simplex disease. Nat. Med. *8*, 392–398.

Kolykhalov, A.A., Mihalik, K., Feinstone, M., and Rice, C.M. (2000). Hepatitis C virus-encoded enzymatic activities and conserved RNA elements in the 3′ nontranslated region are essential for virus replication in vivo. J. Virol. *74*, 2046–2051.

Koresawa, M., and Okabe, T. (2004). High-throughput screening with quantitation of ATP consumption: a universal non-radioisotope, homogeneous assay for protein kinase. Assay Drug Dev. Technol. *2*, 153–160.

Krawczyk, M., Wasowska-Lukawska, M., Oszczapowicz, I., and Boguszewska-Chachulska, A.M. (2009). Amidinoanthracyclines – a new group of potential anti-hepatitis C virus compounds. Biol. Chem. *390*, 351–360.

Kyono, K., Miyashiro, M., and Taguchi, I. (1998). Detection of hepatitis C virus helicase activity using the scintillation proximity assay system. Anal. Biochem. *257*, 120–126.

Lakowicz, J.R. (1999). Principles of Fluorescence Spectroscopy, Chapter 12: Energy Transfer (Kluwer Academic/Plenum Publishers, New York), pp. 367–394.

Lam, A.M., and Frick, D.N. (2006). Hepatitis C virus subgenomic replicon requires an active NS3 RNA helicase. J. Virol. *80*, 404–411.

Lam, A.M., Keeney, D., Eckert, Q., and Frick, D.N. (2003a). Hepatitis C virus NS3 ATPases/helicases from different genotypes exhibit variations in enzymatic properties. J. Virol. *77*, 3950–3961.

Lam, A.M., Keeney, D., and Frick, D.N. (2003b). Two novel conserved motifs in the hepatitis C virus NS3 protein critical for helicase action. J. Biol. Chem. *278*, 44514–44524.

Lam, A.M., Rypma, S., and Frick, D.N. (2004). Enhanced nucleic acid binding to ATP-bound hepatitis C virus NS3 helicase at low pH activates RNA unwinding. Nucleic Acids Res. *32*, 4060–4070.

Lanzetta, P.A., Alvarez, L.J., Reinach, S., and Candia, O.A. (1979). An improved assay for nanomole amounts of inorganic phosphate. Anal. Biochem. *100*, 95–97.

Liefhebber, J.M., Hensbergen, P.J., Deelder, A.M., Spaan, W.J., and van Leeuwen, H.C. (2010). Characterization of hepatitis C virus NS3 modifications in the context of replication. J. Gen. Virol. *91*, 1013–1018.

Liu, W.J., Sedlak, P.L., Kondratieva, N., and Khromykh, A.A. (2002). Complementation analysis of the flavivirus Kunjin NS3 and NS5 proteins defines the minimal regions essential for formation of a replication complex and shows a requirement of NS3 in cis for virus assembly. J. Virol. *76*, 10766–10775.

Locatelli, G.A., Gosselin, G., Spadari, S., and Maga, G. (2001). Hepatitis C virus NS3 NTPase/helicase: different stereoselectivity in nucleoside triphosphate utilisation suggests that NTPase and helicase activities are coupled by a nucleotide-dependent rate limiting step. J. Mol. Biol. *313*, 683–694.

Ma, Y., Yates, J., Liang, Y., Lemon, M., and Yi, M. (2008). NS3 helicase domains involved in infectious intracellular hepatitis C virus particle assembly. J. Virol. *82*, 7624–7639.

Mackintosh, S.G., Lu, J.Z., Jordan, J.B., Harrison, M.K., Sikora, B., Sharma, S.D., Cameron, C.E., Raney, D., and Sakon, J. (2006). Structural and biological identification of residues on the surface of NS3 helicase required for optimal replication of the hepatitis C virus. J. Biol. Chem. *281*, 3528–3535.

Maga, G., Gemma, S., Fattorusso, C., Locatelli, G.A., Butini, S., Persico, M., Kukreja, G., Romano, M.P., Chiasserini, L., Savini, L., Novellino, E., Nacci, V., Spadari, S., and Campiani, G. (2005). Specific targeting of hepatitis C virus NS3 RNA helicase. Discovery of the potent and selective competitive nucleotide-mimicking inhibitor QU663. Biochemistry *44*, 9637–9644.

Manfroni, G., Paeshuyse, J., Massari, S., Zanoli, S., Gatto, B., Maga, G., Tabarrini, O., Cecchetti, V., Fravolini, A., and Neyts, J. (2009). Inhibition of Subgenomic Hepatitis C Virus RNA Replication by Acridone Derivatives: Identification of an NS3 Helicase Inhibitor. J. Med. Chem. *52*, 3354–3365.

Matlock, D.L., Yeruva, L., Byrd, A.K., Mackintosh, S.G., Langston, C., Brown, C., Cameron, C.E., Fischer, J., and Raney, K.D. (2010). Investigation of translocation, DNA unwinding, and protein displacement by NS3h, the helicase domain from the hepatitis C virus helicase. Biochemistry *49*, 2097–2109.

Matson, S.W., and Kaiser-Rogers, K.A. (1990). DNA helicases. Annu. Rev. Biochem. *59*, 289–329.

Morris, P.D., and Raney, K.D. (1999). DNA helicases displace streptavidin from biotin-labeled oligonucleotides. Biochemistry *38*, 5164–5171.

Morris, P.D., Byrd, A.K., Tackett, A.J., Cameron, C.E., Tanega, P., Ott, R., Fanning, E., and Raney, K.D. (2002). Hepatitis C virus NS3 and simian virus 40 T antigen helicases displace streptavidin from 5′-biotinylated oligonucleotides but not from 3′-biotinylated oligonucleotides: evidence for directional bias in translocation on single-stranded DNA. Biochemistry *41*, 2372–2378.

Mukovnya, A.V., Tunitskaya, V.L., Khandazhinskaya, A.L., Golubeva, N.A., Zakirova, N.F., Ivanov, A.V., Kukhanova, K., and Kochetkov, S.N. (2008). Hepatitis C virus helicase/NTPase: an efficient expression system and new inhibitors. Biochemistry (Mosc) *73*, 660–668.

Muramatsu, S., Ishido, S., Fujita, T., Itoh, M., and Hotta, H. (1997). Nuclear localization of the NS3 protein of hepatitis C virus and factors affecting the localization. J. Virol. *71*, 4954–4961.

Myong, S., Bruno, M.M., Pyle, M., and Ha, T. (2007). Spring-loaded mechanism of DNA unwinding by hepatitis C virus NS3 helicase. Science *317*, 513–516.

Paeshuyse, J., Vliegen, I., Coelmont, L., Leyssen, P., Tabarrini, O., Herdewijn, P., Mittendorfer, H., Easmon, J., Cecchetti, V., Bartenschlager, R., Puerstinger, G., and Neyts, J. (2008). Comparative in vitro anti-hepatitis C virus activities of a selected series of polymerase, protease, and helicase inhibitors. Antimicrob. Agents Chemother. *52*, 3433–3437.

Phoon, C.W., Ng, P.Y., Ting, A.E., Yeo, L., and Sim, M.M. (2001). Biological evaluation of hepatitis C virus helicase inhibitors. Bioorg. Med. Chem. Lett. *11*, 1647–1650.

Piccininni, S., Varaklioti, A., Nardelli, M., Dave, B., Raney, D., and McCarthy, J.E. (2002). Modulation of the hepatitis C virus RNA-dependent RNA polymerase activity by the non-structural (NS) 3 helicase and the NS4B membrane protein. J. Biol. Chem. *277*, 45670–45679.

Porter, D.J., Short, S.A., Hanlon, M.H., Preugschat, F., Wilson, J.E., Willard, J., and Consler, T.G. (1998). Product release is the major contributor to kcat for the hepatitis C virus helicase-catalyzed strand separation of short duplex DNA. J. Biol. Chem. *273*, 18906–18914.

Prabhu, R., Khalap, N., Burioni, R., Clementi, M., Garry, F., and Dash, S. (2004). Inhibition of hepatitis C virus nonstructural protein, helicase activity, and viral replication by a recombinant human antibody clone. Am. J. Pathol. *165*, 1163–1173.

Pyle, A.M. (2008). Translocation and unwinding mechanisms of RNA and DNA helicases. Annu. Rev. Biophys. *37*, 317–336.

Raney, K.D., Sowers, L.C., Millar, P., and Benkovic, S.J. (1994). A fluorescence-based assay for monitoring helicase activity. Proc. Natl. Acad. Sci. U.S.A. *91*, 6644–6648.

Raney, K.D., Cameron, C.E., Dave, B., Sakon, J., Lu, J.Z., Mackintosh, G., and Jennings, T.A. (2008). Compounds and methods for inhibiting hepatitis C virus replication. The Board of Trustees of the University of Arkansas *USA*, US 7, 465, 537 B2.

Rypma, R.S., Lam, M., and Frick, D. (2010). Effect of substrate traps on hepatitis C virus NS3 helicase catalyzed DNA unwinding: evidence for enzyme catalyzed strand exchange. In Bacterial DNA, DNA Polymerases and DNA Helicases, Knudsen, W.D., Bruns, S.S., eds (Nova Publishers, New York), pp. 389–407.

Sakamuro, D., Furukawa, T., and Takegami, T. (1995). Hepatitis C virus nonstructural protein NS3 transforms NIH 3T3 cells. J. Virol. *69*, 3893–3896.

Serebrov, V., and Pyle, A.M. (2004). Periodic cycles of RNA unwinding and pausing by hepatitis C virus NS3 helicase. Nature *430*, 476–480.

Serebrov, V., Beran, K., and Pyle, A.M. (2009). Establishing a mechanistic basis for the large kinetic steps of the NS3 helicase. J. Biol. Chem. *284*, 2512–2521.

Sivaraja, M., Giordano, H., and Peterson, M.G. (1998). High-throughput screening assay for helicase enzymes. Anal. Biochem. *265*, 22–27.

Stankiewicz-Drogon, A., Palchykovska, L.G., Kostina, V.G., Alexeeva, I.V., Shved, A.D., and Boguszewska-Chachulska, A.M. (2008). New acridone-4-carboxylic acid derivatives as potential inhibitors of hepatitis C virus infection. Bioorg. Med. Chem. *16*, 8846–8852.

Story, R.M., Weber, T., and Steitz, T.A. (1992). The structure of the *E. coli* recA protein monomer and polymer. Nature *355*, 318–325.

Suzich, J.A., Tamura, J.K., Palmer-Hill, F., Warrener, P., Grakoui, A., Rice, C.M., Feinstone, M., and Collett, M.S. (1993). Hepatitis C virus NS3 protein polynucleotide-stimulated nucleoside triphosphatase and comparison with the related pestivirus and flavivirus enzymes. J. Virol. *67*, 6152–6158.

Tai, C.L., Chi, W.K., Chen, S., and Hwang, L.H. (1996). The helicase activity associated with hepatitis C virus nonstructural protein 3 (NS3). J. Virol. *70*, 8477–8484.

Tang, H. (2008). Hepatitis C: Methods and Protocols, Second Edition. In Walker, J.M. ed. series (Humana Press, New York).

Tani, H., Akimitsu, N., Fujita, O., Matsuda, Y., Miyata, R., Tsuneda, S., Igarashi, M., Sekiguchi, Y., and Noda, N. (2009). High-throughput screening assay of hepatitis C virus helicase inhibitors using fluorescence-quenching phenomenon. Biochem. Biophys. Res. Commun. *379*, 1054–1059.

Tani, H., Fujita, O., Furuta, A., Matsuda, Y., Miyata, R., Akimitsu, N., Tanaka, J., Tsuneda, S., Sekiguchi, Y., and Noda, N. (2010). Real-time monitoring of RNA helicase activity using fluorescence resonance energy transfer in vitro. Biochem. Biophys. Res. Commun. *393*, 131–136.

Tessmann, K., Erhardt, A., Haussinger, D., and Heintges, T. (2002). Cloning and molecular characterization of human high affinity antibody fragments against Hepatitis C virus NS3 helicase. J. Virol. Methods *103*, 75–88.

Tyagi, S., and Kramer, F.R. (1996). Molecular beacons: probes that fluoresce upon hybridization. Nat. Biotechnol. *14*, 303–308.

Ujjinamatada, R.K., Baier, A., Borowski, P., and Hosmane, R.S. (2007). An analogue of AICAR with dual inhibitory activity against WNV and HCV NTPase/helicase: synthesis and in vitro screening of 4-carbamoyl-5-(4, 6-diamino-2, 5-dihydro-1, 3, 5-triazin-2-yl)imidazole-1-beta-D-ribofuranoside. Bioorg. Med. Chem. Lett. *17*, 2285–2288.

Walker, J.E., Saraste, M., Runswick, J., and Gay, N.J. (1982). Distantly related sequences in the alpha- and beta-subunits of ATP synthase, myosin, kinases and other ATP-requiring enzymes and a common nucleotide binding fold. EMBO J. *1*, 945–951.

Wang, Q., Arnold, J.J., Uchida, A., Raney, D., and Cameron, C.E. (2010). Phosphate release contributes to the rate-limiting step for unwinding by an RNA helicase. Nucleic Acids Res. *38*, 1312–1324.

Webb, M.R. (1992). A continuous spectrophotometric assay for inorganic phosphate and for measuring phosphate release kinetics in biological systems. Proc. Natl. Acad. Sci. U.S.A. *89*, 4884–4887.

Xu, H.Q., Zhang, A.H., Auclair, C., and Xi, X.G. (2003). Simultaneously monitoring DNA binding and helicase-catalyzed DNA unwinding by fluorescence polarization. Nucleic Acids Res. *31*, e70.

Yao, N., Hesson, T., Cable, M., Hong, Z., Kwong, A.D., Le, V., and Weber, P.C. (1997). Structure of the hepatitis C virus RNA helicase domain. Nat. Struct. Biol. *4*, 463–467.

Yao, N., Reichert, P., Taremi, S.S., Prosise, W., and Weber, P.C. (1999). Molecular views of viral polyprotein processing revealed by the crystal structure of the hepatitis C virus bifunctional protease-helicase. Structure *7*, 1353–1363.

Yedavalli, V.S., Neuveut, C., Chi, Y.H., Kleiman, L., and Jeang, K.T. (2004). Requirement of DDX3 DEAD box RNA helicase for HIV-1 Rev-RRE export function. Cell *119*, 381–392.

Yedavalli, V.S., Zhang, N., Cai, H., Zhang, P., Starost, M.F., Hosmane, S., and Jeang, K.T. (2008). Ring Expanded Nucleoside Analogues Inhibit RNA Helicase and Intracellular Human Immunodeficiency Virus Type 1 Replication. J. Med. Chem. *51*, 5043–5051.

Zhang, L., Schwartz, G., O'Donnell, M., and Harrison, R.K. (2001). Development of a novel helicase assay using electrochemiluminescence. Anal. Biochem. *293*, 31–37.

Zhang, N., Chen, H.M., Koch, V., Schmitz, H., Liao, C.L., Bretner, M., Bhadti, V.S., Fattom, A.I., Naso, R.B., Hosmane, S., and Borowski, P. (2003). Ring-expanded ("fat") nucleoside and nucleotide analogues exhibit potent in vitro activity against flaviviridae NTPases/helicases, including those of the West Nile virus, hepatitis C virus, and Japanese encephalitis virus. J. Med. Chem. *46*, 4149–4164.

Zhang, C., Cai, Z., Kim, Y.C., Kumar, R., Yuan, F., Shi, P.Y., Kao, C., and Luo, G. (2005). Stimulation of hepatitis C virus (HCV) nonstructural protein 3 (NS3) helicase activity by the NS3 protease domain and by HCV RNA-dependent RNA polymerase. J. Virol. *79*, 8687–8697.

Zhao, J. (2007). Plant troponoids: chemistry, biological activity, and biosynthesis. Curr. Med. Chem. *14*, 2597–2621.

NS4B Targets and Inhibitors

13

Menashe Elazar and Jeffrey S. Glenn

Abstract

Hepatitis C virus (HCV) non-structural protein 4B (NS4B) is emerging as a major target for antiviral intervention. NS4B's crucial role in different aspects of HCV's life cycle and newly identified functions of this protein enabled development of specific assays that were utilized for high-throughput small molecule screens, which in turn lead to identification of small molecule inhibitors of NS4B functions and HCV replication, some of which are undergoing clinical evaluation and preclinical development. In this review, we highlight the recent identification of small molecule inhibitors that target NS4B's RNA binding activity and an amphipathic helix-mediated membrane interaction. We further discuss the potential of other candidate activities or domains within NS4B that may serve as targets for antiviral intervention.

Introduction

Hepatitis C virus (HCV) is the leading cause of liver transplantation and hepatocellular carcinoma in the US and is a significant cause of morbidity and mortality, estimated to affect more then 150 million individuals worldwide (Shepard *et al.*, 2005). Although treatment options and outcome have progressed significantly over the past decade, current therapies that include combination of pegylated interferon alpha (pegIFN) with ribavirin is of benefit for only ~50% of treated patients (Manns *et al.*, 2007). To address this problem, an effort has been made to better understand the molecular virology of HCV in order to identify new potential viral targets for antiviral intervention. The viral protease/helicase (NS3) and RNA-dependent RNA polymerase (RdRp-NS5B) functions were identified early on following the isolation of the virus based on sequence homology with known homologues in other RNA viruses. The protease and polymerase activities thus immediately stood out as candidate targets for antiviral intervention (Bouffard *et al.*, 1995; Behrens *et al.*, 1996; Carroll *et al.*, 2003). This set the stage for important drug discovery efforts even prior to the establishment of efficient HCV viral replication systems. These efforts have resulted in the development of specific inhibitors targeting the functions of both NS3 and NS5B (Thompson and McHutchison, 2009). HCV infection, however, is a highly dynamic process with a high replication activity. This, combined with NS5B's lack of a proofreading function helps result in a high genetic variability of the virus. This variability, in turn, represents the basis whereby HCV can rapidly develop resistance to new drugs, especially when the latter are used as monotherapies. Such is indeed the case for several compounds developed to target NS3 and NS5B (Tong *et al.*, 2006; Lu *et al.*, 2007; Koev and Kati, 2008; Flint *et al.*, 2009; Thompson and McHutchison, 2009). Thus, similar to HIV and tuberculosis, it is believed that an effective pharmacological control of HCV will require a cocktail of multiple agents, each targeting an independent virus-specific function (Tomei *et al.*, 2005; Thompson *et al.*, 2009a). Hence the identification of novel classes of drugs to include in such future cocktails would be of great benefit.

It was not until the establishment of the HCV replicon system (Lohmann *et al.*, 1999; Blight *et al.*, 2000) that the roles and functions of other non-structural (NS) proteins began to be unravelled at an increased pace, providing insights into the roles of these proteins in the viral life cycle and leading to the identification of new targets that might form the basis for novel antiviral strategies. In this review, we will explore the progress towards targeting NS4B that has resulted in some exciting new discoveries of targets that possess high potential for pharmacological intervention.

NS4B: general characteristics

The existence of NS4B as one of the protein products of HCV infection was demonstrated early following the cloning of HCV (Grakoui *et al.*, 1993). However in the subsequent HCV literature, NS4B was mostly referred to as 'a protein of unknown function'. The establishment of the HCV replicon system (Lohmann *et al.*, 1999; Blight *et al.*, 2000) and the identification of the infectious genome that gives rise to productive infection in cell culture (Lindenbach *et al.*, 2005; Wakita *et al.*, 2005; Zhong *et al.*, 2005) led to a better appreciation of the crucial role NS4B plays in HCV's viral life cycle and facilitated the identification of several different functions in NS4B (for review see Sklan and Glenn, 2006), many of which possess the potential to serve as antiviral targets.

NS4B is a 27-kDa integral membrane protein that is associated with the endoplasmic reticulum (ER) membranes or ER-derived membranes (Selby *et al.*, 1993; Kim *et al.*, 1999; Hugle *et al.*, 2001; Lundin *et al.*, 2003). NS4B is believed to possess at least four transmembrane domains (TMDs; Fig. 13.1) as predicted by transmembrane prediction tools (Lundin *et al.*, 2003) and its association with the membrane was shown to occur cotranslationally (Hugle *et al.*, 2001). Neither the crystal structure nor the definitive topology of NS4B has been determined yet; however, Lundin *et al.* demonstrated that the loop around amino acid 161 is lumenal, suggesting that TMDs 3 and 4 traverse the membrane. Using their results to constrain the TMD prediction analysis of NS4B, Lundin *et al.* confirmed that the protein contains 4 TMDs (Lundin *et al.*, 2003). Furthermore, it was shown that a portion of NS4B's N-terminus becomes lumenal post-translationally, predicting a fifth TMD designated TMX (Lundin *et al.*, 2003). Interestingly, NS4B from other viruses in the Flaviviridae family also orient such that their N-terminus is located in the lumen, although this occurs via a different mechanism suggesting that NS4B across the Flaviviridae family might possess a similar topology and functions (Cahour *et al.*, 1992; Lin *et al.*, 1993).

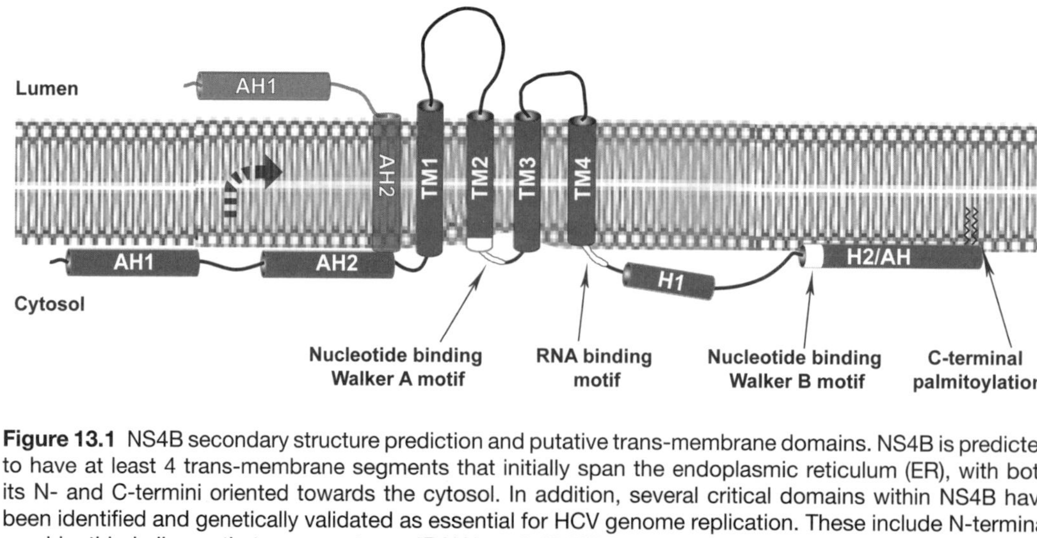

Figure 13.1 NS4B secondary structure prediction and putative trans-membrane domains. NS4B is predicted to have at least 4 trans-membrane segments that initially span the endoplasmic reticulum (ER), with both its N- and C-termini oriented towards the cytosol. In addition, several critical domains within NS4B have been identified and genetically validated as essential for HCV genome replication. These include N-terminal amphipathic helices—that we now term 4BAH1 and 4BAH2, a nucleotide-binding motif (NBM), an RNA binding domain, a C-terminal palmitoylation motif, and two C-terminal predicted helices.

The TMDs of NS4B are sufficient to confer the protein's membrane association (Elazar *et al.*, 2004; Gouttenoire *et al.*, 2009a), however this is not the only domain of the protein that associates with membranes. The extreme N-terminus of NS4B possesses an amphipathic helix (AH1; Fig. 13.1) that was shown to mediate the correct localization of the replication complex components as well as HCV RNA replication (Elazar *et al.*, 2004).

Recently, Cho *et al.* and Gouttenoire *et al.* identified, immediately following AH1, a second AH (AH2; Fig. 13.1) that is sufficient to confer tight membrane association and is required for viral replication (Cho *et al.*, 2009; Gouttenoire *et al.*, 2009a). AH2 was also shown to adopt a transmembrane orientation, confirming the observation by Lundin *et al.* (Lundin *et al.*, 2003; Gouttenoire *et al.*, 2009a). Interestingly, Cho *et al.* demonstrated that AH2 mediates oligomerization and acts to cause lipid vesicle aggregation (Cho *et al.*, 2009), a function that might be relevant for the role of NS4B in forming the membranous-web replication platform (Egger *et al.*, 2002; see below).

Amphipathic helices are not exclusive to the N-terminal portion of NS4B. A few reports demonstrated that the C-terminus of NS4B possesses an AH (cAH; Fig. 13.1) that can mediate membrane association (Aligo *et al.*, 2009; Gouttenoire *et al.*, 2009b; Guillen *et al.*, 2009). The cAH was shown to be essential for viral replication and for formation of the viral-mediated membrane foci (Aligo *et al.*, 2009; Gouttenoire *et al.*, 2009b). Interestingly, individual residues within this helix were shown to mediate localization of Rab5 to NS4B induced foci (Aligo *et al.*, 2009). Further studies are necessary to determine whether this interaction/colocalization facilitates formation of NS4B-induced foci or is a consequence of their formation.

A second helix in the C-terminal domain of NS4B (although not amphipathic in nature) was identified (Fig. 13.1) and residues around this helix were shown to be involved in viral particle production while others are involved in RNA replication without affecting the NS4B-mediated foci formation (Jones *et al.*, 2009).

An important feature of NS4B's role in HCV replication is its involvement in host cell membrane rearrangements. Egger *et al.* demonstrated that expression of some of HCV's proteins results in host cell membrane alterations (Egger *et al.*, 2002). However expression of NS4B alone was sufficient to generate distinct membrane structures termed membranous webs. The membranous webs harboured all of the non-structural proteins when these were expressed along with NS4B in the context of the entire non-structural coding region of the HCV genome, suggesting that these membrane structures represent the scaffold upon which HCV replication takes place (Egger *et al.*, 2002). Subsequently it was confirmed that HCV's non-structural proteins as well as nascent HCV RNA are associated with the membranous webs in cells supporting active HCV replication, strengthening this notion (Gosert *et al.*, 2003).

These replication-supporting membrane structures also appear to have a particular biochemical composition. For example, both HCV proteins and RNA were detected in low-density membrane fractions that are resistant to solubilization with 1% NP40 (Shi *et al.*, 2003). Although all of the HCV non-structural proteins can be found associated with lipid rafts upon expression of the polyprotein (Shi *et al.*, 2003), only NS4B was shown to do so when expressed alone (Gao *et al.*, 2004). This suggests that the viral replication complex is associated with cholesterol rich lipid rafts and that NS4B plays a role in recruiting the viral replication complex components to these unique membrane subdomains. Interestingly, HCV's core protein was also shown to co-localize with lipid raft-like domains (Matto *et al.*, 2004) yet it is not clear if the latter are common to those with which NS4B associates. Matto *et al.* speculated that association of core with lipid rafts might be related to its role in virion morphogenesis. Recently it was reported that residues within the C-terminal part of NS4B may be involved in viral particle production and release (Jones *et al.*, 2009). It is temping to speculate that NS4B and core associate with the same lipid rafts manifesting the contribution of NS4B to virus assembly and release.

A nucleotide-binding motif associated with GTP hydrolysis that is essential for HCV replication was described for NS4B (Einav *et al.*, 2004). Thompson *et al.* confirmed this finding recently and also demonstrated that NS4B harbours an adenylate kinase activity (Thompson *et al.*, 2009b). Interestingly the NS4B nucleotide-binding motif was shown to mediate NS4B's tumorigenic potential *in vitro* and *in vivo* (Einav *et al.*, 2008b).

Multiple protein–protein interactions between HCV non-structural proteins have been demonstrated (Dimitrova *et al.*, 2003). NS4B was shown to interact with itself suggesting that it can form a homo-multimer (Dimitrova *et al.*, 2003). Yu *et al.* further demonstrated that the N-terminus of NS4B is important for oligomerization and suggested that this process is facilitated by palmitoylation on two C-terminal cysteines of NS4B (Yu *et al.*, 2006). NS4B was also shown to interact with CREB-RP/ATF6beta suggesting a role in modulation of host response to ER-stress (Tong *et al.*, 2002), however the consequences of this interaction deserve further clarification.

In the next part of this review we will discuss the potential of some of the above-described features of NS4B as antiviral targets and demonstrate recent advancements towards pharmacologically inhibiting these features.

NS4B as an antiviral target

Like most known drug targets, NS4B is a membrane protein (Overington *et al.*, 2006) and as such its stable natural conformation is formed jointly by the protein and the membrane in which it resides. This provides added difficulty to determine the protein's structure experimentally and to use this structure to design small molecules that target the protein. In order to target membrane proteins, one needs to identify specific functions or structural domains within the protein to focus on for drug screening.

The establishment of the HCV replicon system (Lohmann *et al.*, 1999; Blight *et al.*, 2000) and later on the tissue culture infectious clone (Lindenbach *et al.*, 2005; Wakita *et al.*, 2005; Zhong *et al.*, 2005) were key milestones in studying the complete life cycle of HCV and the involvement and function of various viral proteins in this cycle. As a result, the many functions of NS4B are being unravelled, giving us both a better understanding of this protein and the identification of several potential targets for antiviral intervention. This in turn has led to several interesting small molecule hits, leads, and even clinical candidates, some of which are entering IND-enabling studies and clinical trials.

NS4B RNA binding

NS4B shares several structural and functional characteristics with the 2C protein in the picornaviruses including an N-terminal amphipathic helix and an apparent role in rearranging intracellular membranes (Rodriguez and Carrasco, 1995; Kusov *et al.*, 1998). Based on the similarities between 2C and NS4B, Einav *et al.* hypothesized that the RNA binding activity of the 2C protein might also be a feature of NS4B. To overcome some of the limitations associated with the expression and purification of a membrane protein, Einav *et al.* used a mechanical trapping of molecular interactions (MITOMI) in a microfluidics device to develop an in-vitro RNA binding assay (Einav *et al.*, 2008a; Fig. 13.2). Advantages of this system include nanolitre protein consumption and expression of the protein in the presence of microsomal membranes, which enables membrane proteins to adopt a more natural folding. Using that system, it was shown that NS4B is an RNA binding protein with preference for the 3′ terminus of the HCV negative strand RNA, an important region for the production of new viral genome molecules (Einav *et al.*, 2008a). Furthermore the RNA-binding motif within NS4B was identified and shown to be essential for viral replication, suggesting that efficient RNA binding is necessary for HCV replication (Einav *et al.*, 2008a).

Using a microfluidics device combined with MITOMI enables simultaneous manipulation of multiple individual cells (Maerkl, 2009), a feature that allows performing a high throughput screen (HTS). Einav *et al.* utilized the combination of microfluidics and MITOMI to identify several small molecules that specifically inhibit the interaction of NS4B with RNA (Fig. 13.3A and B). The identified molecules were further shown to inhibit HCV replication in the HCV replicon

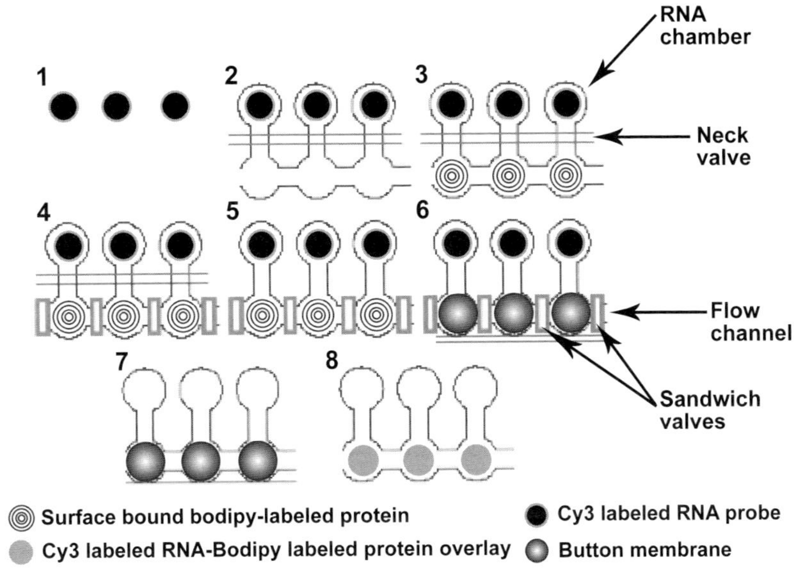

Figure 13.2 MITOMI and microfluidics-based RNA binding assay for NS4B–RNA interaction. (1) Target RNA sequences labelled with Cy3 were spotted onto an epoxy-coated slide as a microarray; (2) The microfluidic device was aligned and bonded to the slide allowing contact of each spot in the array with a unit cell in the device. The device was subjected to surface patterning that resulted in a circular area coated with biotinylated anti-histidine antibodies within each unit cell. The 'neck valve' separating the flow channels from the RNA chamber remained closed during this process; (3) The device was then loaded with *in vitro* transcription/translation mixture containing DNA template encoding a his-labelled protein. Bodipy-labelled tRNALys was added for protein labelling (surface-bound bodipy-labelled protein); (4) Each unit cell was then isolated by the control of two 'sandwich' micromechanical valves; (5) the 'neck valve' was opened forming a single compartment by combining the flow channel compartment with the RNA chamber. This was followed by an incubation to allow protein synthesis, binding of the synthesized protein to the surface biotinylated anti-his antibodies, solvation of target RNA, and equilibration of proteins and target RNA; (6) MITOMI was then performed by actuation of a 'button' membrane trapping surface-bound complexes while expelling any solution phase molecules; (7) The 'sandwich' valves were opened followed by a brief wash to remove untrapped unbound material; (8) the 'button' membrane was opened and the device was scanned by an array scanner. Trapped RNA molecules and expressed protein were detected. The ratio of bound RNA to expressed protein was calculated for each data point by measuring the median signal of Cy3 to median signal of bodipy (Cy3-labelled RNA-bodipy labelled protein overlay). Reproduced from Einav *et al.* (2008a).

and infectious clone models (Einav *et al.*, 2008a). One of the identified molecules, clemizole hydrochloride (1-*p*-chlorobenzyl-2-(1-pyrrolidinyl) methylbenzimidazole hydrochloride; Fig. 13.3E), an H1 histamine receptor antagonist, was found to have a strong inhibitory effect on the NS4B–RNA interaction and moderate HCV replication inhibitory activity with no toxicity at the tested concentrations ($IC_{50} = 24 \pm 1$ nM and $EC_{50} = 8$ μM; Einav *et al.*, 2008a; Fig. 13.3C and D). The 400-fold difference in the potency of the drug between the RNA binding assay and the viral replication assay is not surprising and can be accounted for by a variety of factors including cellular uptake and metabolism.

Selection for resistant mutants using the HCV subgenomic replicon resulted in the emergence of resistant colonies over several months. The identified mutations conferring resistance in these colonies were mapped to either the NS4B coding region, the 3′-terminus of the HCV negative strand, or both (Einav *et al.*, 2008a), further substantiating clemizole's mechanism of action.

The antiviral activity is not due to clemizole's antihistamine activity and structure–activity studies have yielded more potent derivatives (Dvory-Sobol *et al.*, unpublished). In spite of its modest *in vitro* EC_{50}, because clemizole was widely used in humans with excellent safety and tolerability in the past, it represents

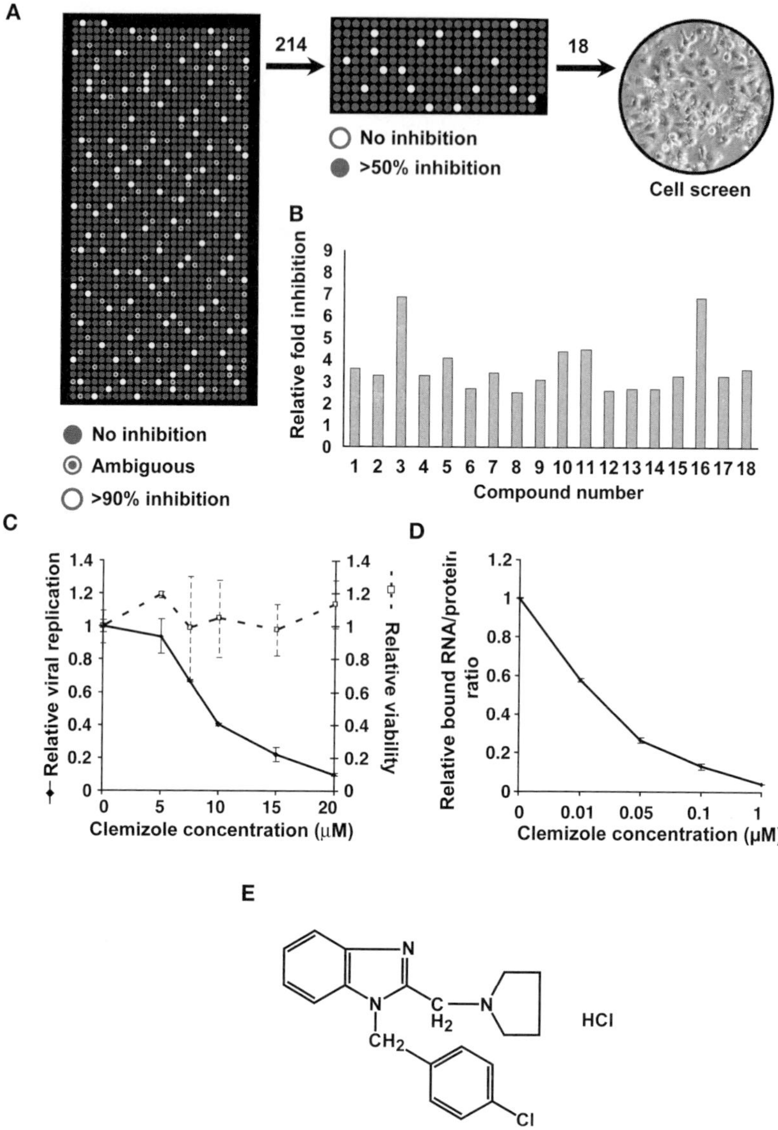

Figure 13.3 Small-molecule screen reveals that clemizole hydrochloride inhibits RNA binding by NS4B and HCV RNA replication in cell culture. (A) The first screen represented a low-stringency measurement of inhibition of 1280 compounds where the latter were categorized as having high (white), ambiguous (white/grey) or no (grey) inhibition. Based on the initial screen, 214 compounds were then measured again with higher stringency and with a greater number of replicates and the best 18 inhibitors were tested for their ability to inhibit HCV replication via a cellular assay. (B) *In vitro* inhibition of NS4B-RNA binding by the top 18 small molecules. (C) HCV luciferase reporter–linked cellular assay showing that clemizole inhibits HCV replication (solid line) with no measurable toxicity to the cell as measured by Alamar Blue (dashed line). (D) *In vitro* NS4B-RNA binding: inhibition curve of clemizole. Each data point represents the mean of 10–20 replicates and the bars represent the standard error. (E) Chemical structure of clemizole hydrochloride. Reproduced from Einav *et al.*, 2008a.

an exciting opportunity for clinical evaluation in hepatitis C, especially because metabolism and pharmacokinetic parameters may significantly enhance a compound's antiviral activity *in vivo*. Indeed, clemizole is currently in clinical development (for example, see www.clinicaltrials.gov) and some of the features described below may make it ideally suited for both additions to current standard of care agents and as a component of future interferon-free oral anti-HCV cocktails.

One possible outcome when using a cocktail of antiviral compounds that acts against different viral targets is that the antiviral effect would be cumulative or at best synergistic. Recent studies have demonstrated that combination of an NS4A inhibitor (Yang *et al.*, 2008) with interferon has an additive effect *in vitro*, while combination with protease inhibitors (PIs) or polymerase inhibitors resulted in a mildly synergistic effect (Wyles *et al.*, 2008). Similarly combinations of PIs with polymerase inhibitors generally have an additive effect, with a mild synergistic effect at the IC_{90} combinations tested (Wyles *et al.*, 2007; Grunberger *et al.*, 2008). While combination of clemizole with interferon, ribavirin or polymerase inhibitors resulted in an additive effect, when combined with PIs clemizole's effect is highly synergistic (Einav *et al.*, 2010). Furthermore, cotreatment of clemizole with PIs resulted in a decrease in the emergence of resistant mutants compared with treatment with either drug alone, and no cross-resistance between the tested drugs was observed (Einav *et al.*, 2010).

The inhibition of NS4B-RNA binding is a new and exciting antiviral target. It is clear that we are at an early stage of development for this target and we expect that promising new compounds will be developed against this target, either as clemizole derivatives or new small molecules that inhibit NS4B-RNA binding.

NS4B's membrane association

One of the hallmarks of all positive-strand RNA viruses is the close association of their genome replication with host intracellular membranes and HCV is no exception to that. NS4B is associated with membranes and was shown to be the sole protein necessary for inducing host cell membrane rearrangements that results in the formation of the membranous-webs (Egger *et al.*, 2002; Gosert *et al.*, 2003). The membrane association of NS4B was attributed early on to the predicted *trans*-membrane domains (TMDs) of the protein. In recent years however, the membrane association of NS4B has proven to be more complex then simply TMDs that span the membrane. At least three amphipathic helices (AHs) were identified in NS4B's N- (Elazar *et al.*, 2004; Cho *et al.*, 2009; Gouttenoire *et al.*, 2009a) and C-terminal domains (Aligo *et al.*, 2009; Gouttenoire *et al.*, 2009b; Guillen *et al.*, 2009; Boleti *et al.*, 2010). The amphipathic nature of the three AHs that were identified in NS4B is essential for viral replication and the ability of the replication complex components to adopt their correct localization or the formation of the membranous-webs (Elazar *et al.*, 2004; Aligo *et al.*, 2009; Cho *et al.*, 2009; Gouttenoire *et al.*, 2009a,b; Guillen *et al.*, 2009; Boleti *et al.*, 2010). The crucial role of these structural domains for HCV replication suggests that pharmacologic interference with these domains might be a novel means of inhibiting HCV replication.

NS4B AH2

As described above, the second N-terminal AH of NS4B (amino acids 43–62 of NS4B), termed 4BAH2, mediates membrane association and viral RNA replication (Cho *et al.*, 2009; Gouttenoire *et al.*, 2009a). Cho *et al.* studied the interaction of a peptide comprising the 4BAH2 sequence with lipid vesicles in an attempt to characterize its interaction with membranes. Dynamic light scattering (DLS) measurements of uniformly sized lipid vesicles demonstrated that the average particle size increased more then a 1000-fold following incubation with 4BAH2 (Cho *et al.*, 2009). This effect was specific to 4BAH2, since a peptide corresponding to 4BAH1 (the first AH in NS4B's N-terminus) had no such effect. Transmission electron microscopy and atomic force microscopy studies showed that the increase in average particle size is a result of 4BAH2-mediated vesicle aggregation, as opposed to vesicle fusion (Cho *et al.*, 2009). The vesicle-aggregation activity of 4BAH2 was dependant on the amphipathic nature of the peptide. Since 4BAH2 is crucial for HCV RNA replication it was hypothesized by Cho *et al.* that pharmacological inhibition of 4BAH2's lipid vesicle-aggregating activity would result in abrogation of HCV replication. To that end, Cho *et al.* utilized fluorescently labelled small unilamellar lipid vesicles to develop an assay for small molecule inhibitors of 4BAH2's vesicle aggregating activity. Upon treatment of the labelled vesicles with a peptide corresponding to 4BAH2, a large-scale aggregation of vesicles was observed that could be monitored by fluorescence microscopy or DLS. This concept was adapted for a high throughput screen using automated high-content microscopic imaging coupled with pattern recognition software analysis trained to quantify the amount of fluorescence contained within the 4BAH2-induced aggregates (Fig. 13.4, top). Screening a small molecule library to identify inhibitors of the 4BAH2-mediated aggregation was performed and resulted in several candidate inhibitors. Two of the latter, termed A2 and C4 (Fig. 13.4, bottom), were further characterized (Cho *et al.*, 2009). The inhibitory effect of these two compounds on vesicle aggregation was confirmed by DLS measurements and their effect

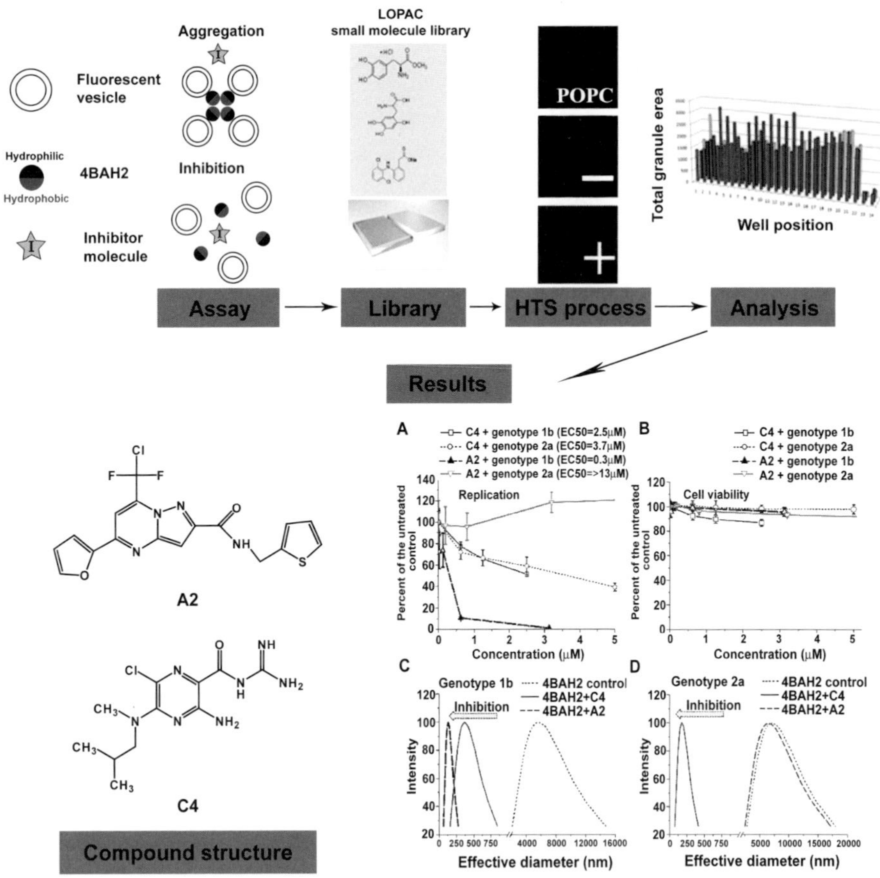

Figure 13.4 Schematic flow of high-throughput screen (HTS) against 4BAH2 and screen results. Schematic of the imaging-based method used to quantitatively analyse the high throughput screening measurements of 4BAH2-induced aggregation of lipid vesicles containing a fluorescent lipid probe. Inhibition of this mode of vesicle aggregation by small molecule compounds was monitored. The assay was performed in a 384 well format and analysed by automated high-content microscopic imaging coupled with pattern recognition software trained to quantify the fluorescence contained within granule size aggregates of vesicles in order to identify small molecule inhibitor candidates. Left bottom, chemical structure of 4BAH2 aggregation activity inhibitors A2 and C4. Right bottom, HCV genome replication is inhibited by small-molecule inhibitors of 4BAH2 function in a genotype specific manner. (A and B) Huh7.5 cells harbouring genotype 1b replicon RNA (Bart79I-luc) or full-length genotype 2a HCV RNA (J6/JFH-luc) were treated with various concentrations of C4 or A2 and assayed for (A) viral replication by luciferase activity (n = 5) and (B) cell viability (n = 5). (C and D) DLS assays of 4BAH2-induced vesicle aggregation were performed with genotype 1b 4BAH2 (C) or genotype 2a 4BAH2 (D) in the absence (control) or presence of either C4 or A2. Note that the x-axis scale is separated into two linear size ranges to directly compare the average vesicle size distribution in the presence or absence of compounds. Reproduced from Cho *et al.*, 2010.

on HCV replication was tested (Fig. 13.4A–D). Although HCV variance has no effect on the vesicle aggregation activity (i.e. 4BAH2 from all genotypes induces vesicle aggregation) genotype specificity for the two compounds was observed. In particular, both C4 and A2 were inhibitory in vesicle aggregation assays with a peptide derived from genotype 1b 4BAH2, but only C4 inhibited genotype 2a 4BAH2-induced vesicle aggregation (Fig. 13.4C,D). Analogous genotype specificity was observed in HCV replication assays with both compounds inhibiting genotype 1b replication, but only C4 inhibiting genotype 2a replication (Cho *et al.*, 2009; Fig. 13.4A).

As suggested by AFM studies, 4BAH2 self-aggregates after adsorption onto the substrate while C4 abrogates this self-oligomerization (Cho *et al.*, 2009). In contrast, A2 inhibits interaction of 4BAH2 with the membranes, as was revealed by quartz crystal microbalance with dissipation (QCM-D; Cho *et al.*, 2009). These results show that 4BAH2 activity can be inhibited in two distinct fashions, with each of the above two compounds exemplifying one of the modes of inhibition: interfering with 4BAH2:4BAH2 interactions (C4), or disrupting 4BAH2:membrane interactions (A2).

Available resistance data to date indicates that resistance mutations, as expected, map to the NS4B coding region(Cho *et al.*, 2009; Bryson *et al.*, 2010). Interestingly these included mutations beyond the AH2 region including some at the cytoplasmic C-terminal of the protein. The mechanisms of resistance conferred by such residues distinct from the primary target site deserve further investigation. Although this likely simply reflects their effect on the folding of NS4B, it is also tempting to speculate that intra molecular interactions might be in play in the overall activity of NS4B associated with membrane rearrangements. It is yet unclear if cross-resistance between A2 and C4 can develop, but since they appear to disrupt functions that may be mediated by opposite faces of the 4BAH2 amphipathic helix, resistance to one type of compound may have a minimal effect on the other's antiviral efficacy.

Other amphipathic helices within NS4B are involved in mediating membrane association, including AH1 at the extreme N-terminus and cAH at the cytoplasmic C-terminus. The finding that AH2 can be specifically targeted via two distinct mechanisms suggests that these other AHs of NS4B might be targetable as well.

NS4B AH1

AH1 was shown to mediate the correct localization of the HCV replication complex components and to be essential for HCV replication. (Elazar *et al.*, 2004). Similar to AH2, AH1 can be targeted to result in interference of its interaction with the membrane. The most appropriate specific assays with the capacity for HTS for small molecule inhibitors are currently being explored.

AH1 and PIP2 binding

AH1 was shown to possess a recently described novel domain, termed basic amino acid PIP2 pincer (BAAPP) domain (Lee *et al.*, 2008), which is present in amphipathic helices from different pathogens including the HCV NS5A AH and NS4B AH1 (Fig. 13.5). In the case of NS5A, the BAAPP domain was shown to mediate specific interaction of the NS5A AH with phosphatidylinositol 4,5-bisphosphate (PIP2) lipids, as well as HCV replication. Replication was also amenable to pharmacological inhibition with neomycin that is known to bind and neutralize the anionic headgroups of PIP2 (Wang *et al.*, 1984). Thus, the BAAPP domain in the NS4B AH1 represents yet another potential antiviral target that might be inhibited by small molecules and should be further explored.

NS4B cAH

The C-terminal AH of NS4B can mediate membrane association (Aligo *et al.*, 2009; Gouttenoire *et al.*, 2009b; Guillen *et al.*, 2009) and was shown to be essential for viral replication and formation of the viral-mediated membrane foci (Aligo *et al.*, 2009; Gouttenoire *et al.*, 2009b). Similar to 4BAH2, one can hypothesize that this helix might be targetable with small molecules. Aligo *et al.* (2009) suggested that the cAH is responsible for recruitment of Rab-5 into the replication complex sites. Specifically, Valine 233 and Leucine 237 within the cAH of NS4B mediated this activity, although it is not clear whether cAH or more specifically V233 and L237 are involved in

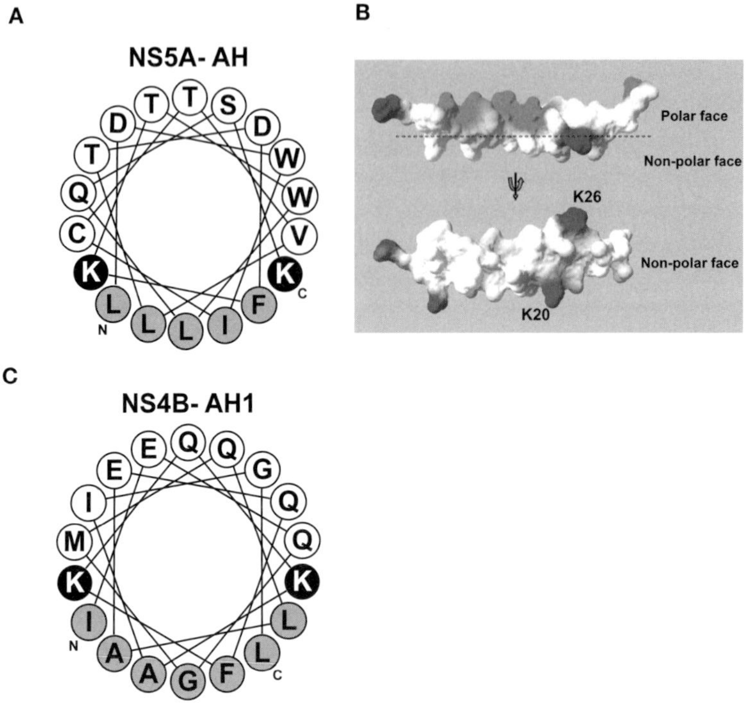

Figure 13.5 BAAPP domain in HCV NS5A and NS4B amphipathic helices. Helix wheel projection of NS5A AH (A) and NS4B AH1 (C) showing the amphipathic nature of the helix flanked by positive charges. Wheel projections were generated using HELIQUEST web-based server (Gautier *et al*., 2008). Grey, amino acids located on the hydrophobic face of the helix; Black, BAAPP domain lysines; White, amino acids outside the hydrophobic face of the helix. (B) A molecular surface model of NS5A AH created based on the NMR structure of NS5A AH (Protein Data Bank ID 1R7C) using the Swiss-PDB viewer (Guex and Peitsch, 1997). Surface electrostatic potential was calculated using charged residues. Dark grey, negative charges; light grey, positive charges; white, non-charged residues. (We thank Dr Pang for help with generating this figure.)

direct protein–protein interactions with Rab-5. If a direct protein–protein interaction is at play, we can envision an assay for identifying specific inhibitors of the Rab-5:NS4B interaction.

NS4B's nucleotide hydrolysis and adenylate kinase activity

The only enzymatic activity attributed to NS4B was its ability to bind and hydrolyse nucleotides (Einav *et al.*, 2004; Thompson *et al.*, 2009a). A nucleotide-binding motif (NBM), consisting of a set of highly conserved amino-acid sequences that are common to the GTP-binding members of the G-protein superfamily of proteins, as well as several viral proteins known to bind nucleotides, was identified in NS4B (Einav *et al.*, 2004). Binding and hydrolysis of nucleotides mediates a variety of signalling, membrane trafficking and membrane fusion events suggesting that this activity is essential for NS4B's function in HCV replication. Indeed, the NS4B NBM is essential for NTP hydrolysis (Einav *et al.*, 2004; Thompson *et al.*, 2009a) and HCV replication (Einav *et al.*, 2004). In addition, as mentioned earlier, the NS4B NBM also mediates cellular transformation and tumor formation in nude mice (Einav *et al.*, 2008b). Recently, an adenylate kinase activity, catalysing the conversion of ADP to ATP, has been demonstrated for NS4B, an activity that is also mediated via the NBM (Thompson *et al.*, 2009b). The exact role of these enzymatic activities of NS4B and their function in the HCV life cycle still await clarification, but it is clear that such activities appear to be essential for viral replication.

The ability of NS4B to bind and hydrolyse different nucleotides (ATP and GTP) suggests that nucleoside analogues may be considered for targeting NS4B. Since GTPases are abundant proteins in the host cell one might expect that inhibitors of NS4B's NTPase activity might result in host toxicity, however the amino-acid residues outside the NBM are highly conserved across all HCV genotypes yet highly different from those in known host cell GTPases. Thus, highly selective inhibitors can be envisaged.

NS4B's specificity for ATP or GTP is somewhat controversial. Einav *et al.* (2004) suggested specificity for GTP while Thompson *et al.* (2009b) suggested that ATP is the preferred substrate. It was suggested that the slow hydrolysis of GTP as reported by Thompson *et al.* represents the intrinsic hydrolysis of the enzyme and that a yet unknown NS4B-GTPase activating protein (GAP; Thompson *et al.*, 2009b) might stimulate catalysis. Considering the different NS4B preparations in these two reports (bacterial expression vs. mammalian expression) this is a possible scenario. If there is such a host GAP that mediates NS4B's NTPase activity, additional antiviral strategies may be contemplated such as inhibitors of the interaction with this putative GAP or direct inhibitors of its activity.

Conclusions and future perspectives

NS4B is involved in many functions, many of which are still not fully characterized. It appears, however, that NS4B serves as a hub in HCV replication and possibly other stages of the viral life cycle. This positions NS4B as a key focal point for antiviral intervention and the multiple targets within the protein make NS4B a rich source of potential new classes of anti-HCV drugs.

The screens described in this review were performed on small libraries with a good success rate. It is reasonable to hypothesize that screening larger libraries will result in more candidate hits with similar or completely unrelated chemical entities that can contribute to the arsenal of NS4B and HCV inhibitors. These, as well as those compounds already identified against NS4B's RNA binding and vesicle aggregation can in turn be the basis of focused structure–function studies designed to yield more potent derivatives and eventual clinical candidates.

We expect that NS4B will become a major target for anti-HCV intervention. Moreover, chemical compounds that inhibit NS4B function may be used as leads for development of antivirals against homologous proteins in related viruses.

References

Aligo, J., Jia, S., Manna, D., and Konan, K.V. (2009). Formation and function of hepatitis C virus replication complexes require residues in the carboxy-terminal domain of NS4B protein. Virology *393*, 68–83.

Behrens, S.E., Tomei, L., De Francesco, R. (1996). Identification and properties of the RNA-dependent RNA polymerase of hepatitis C virus. EMBO J. *15*, 12–22.

Blight, K.J., Kolykhalov, A., and Rice, C.M. (2000). Efficient initiation of HCV RNA replication in cell culture. Science *290*, 1972–1974.

Boleti, H., Smirlis, D., Dalagiorgou, G., Meurs, E.F., Christoforidis, S., and Mavromara, P. (2010). ER targeting and retention of the HCV NS4B protein relies on the concerted action of multiple structural features including its transmembrane domains. Mol. Membr. Biol. *27*, 50–74.

Bouffard, P., Bartenschlager, R., Ahlborn-Laake, L., Mous, J., Roberts, N., and Jacobsen, H. (1995). An in vitro assay for hepatitis C virus NS3 serine proteinase. Virology *209*, 52–59.

Bryson, P.D., Cho, N.J., Einav, S., Lee, C., Tai, V., Bechtel, J., Sivaraja, M., Roberts, C., Schmitz, U., and Glenn, J.S. (2010). A small molecule inhibits HCV replication and alters NS4B's subcellular distribution. Antiviral Res. *87*, 1–8.

Cahour, A., Falgout, B., and Lai, C.J. (1992). Cleavage of the Dengue virus polyprotein at the NS3/NS4A and NS4B/NS5 junctions is mediated by viral protease NS2B-NS3, whereas NS4A/NS4B may be processed by a cellular protease. J. Virol. *66*, 1535–1542.

Carroll, S.S., Tomassini, J.E., Bosserman, M., Getty, K., Stahlhut, M.W., Eldrup, A.B., Bhat, B., Hall, D., Simcoe, A.L., LaFemina, R., *et al.* (2003). Inhibition of hepatitis C virus RNA replication by 2'-modified nucleoside analogs. J. Biol. Chem. *278*, 11979–11984.

Cho, N.-J., Dvory-Sobol, H., Lee, C., Cho, S.-J., Bryson, P., Masek, M., Elazar, M., Frank, W., and Glenn, J.S. (2009). Identification of a class of HCV inhibitors directed against the nonstructural protein NS4B. Science Translational Medicine 2, 15ra16–15ra16.

Dimitrova, M., Imbert, I., Kieny, P., and Schuster, C. (2003). Protein-protein interactions between hepatitis C virus nonstructural proteins. J. Virol. *77*, 5401–5414.

Egger, D., Wolk, B., Gosert, R., Bianchi, L., Blum, H.E., Moradpour, D., and Bienz, K. (2002). Expression of hepatitis C virus proteins induces distinct membrane alterations including a candidate viral replication complex. J. Virol. *76*, 5974–5984.

Einav, S., Dvory-Sobol, H., Gehrig, E., and Glenn, J.S. (2010). The Hepatitis C virus (HCV) NS4B RNA binding inhibitor, clemizole, is highly synergistic with HCV protease inhibitors. J. Infect. Dis. *202*, 65–74.

Einav, S., Elazar, M., Danieli, T., and Glenn, J.S. (2004). A nucleotide binding motif in hepatitis C virus (HCV) NS4B mediates HCV RNA replication. J. Virol. *78*, 11288–11295.

Einav, S., Gerber, D., Bryson, P.D., Sklan, E.H., Elazar, M., Maerkl, S.J., Glenn, S., and Quake, S.R. (2008a). Discovery of a hepatitis C target and its pharmacological inhibitors by microfluidic affinity analysis. Nat. Biotechnol. *26*, 1019–1027.

Einav, S., Sklan, E.H., Moon, H.M., Gehrig, E., Liu, P., Hao, Y., Lowe, W., and Glenn, J.S. (2008b). The nucleotide binding motif of hepatitis C virus NS4B can mediate cellular transformation and tumor formation without Ha-ras co-transfection. Hepatology *47*, 827–835.

Elazar, M., Liu, P., Rice, M., and Glenn, J.S. (2004). An N-terminal amphipathic helix in hepatitis C virus (HCV) NS4B mediates membrane association, correct localization of replication complex proteins, and HCV RNA replication. J. Virol. *78*, 11393–11400.

Flint, M., Mullen, S., Deatly, A.M., Chen, W., Miller, L.Z., Ralston, R., Broom, C., Emini, A., and Howe, A.Y. (2009). Selection and characterization of hepatitis C virus replicons dually resistant to the polymerase and protease inhibitors HCV-796 and boceprevir (SCH 503034). Antimicrob. Agents Chemother. *53*, 401–411.

Gao, L., Aizaki, H., He, W., and Lai, M.M. (2004). Interactions between viral nonstructural proteins and host protein hVAP-33 mediate the formation of hepatitis C virus RNA replication complex on lipid raft. J. Virol. *78*, 3480–3488.

Gautier, R., Douguet, D., Antonny, B., and Drin, G. (2008). HELIQUEST: a web server to screen sequences with specific alpha-helical properties. Bioinformatics *24*, 2101–2102.

Gosert, R., Egger, D., Lohmann, V., Bartenschlager, R., Blum, H.E., Bienz, K., and Moradpour, D. (2003). Identification of the hepatitis C virus RNA replication complex in huh-7 cells harboring subgenomic replicons. J. Virol. *77*, 5487–5492.

Gouttenoire, J., Castet, V., Montserret, R., Arora, N., Raussens, V., Ruysschaert, J.M., Diesis, E., Blum, H.E., Penin, F., and Moradpour, D. (2009a). Identification of a novel determinant for membrane association in hepatitis C virus nonstructural protein 4B. J. Virol. *83*, 6257–6268.

Gouttenoire, J., Montserret, R., Kennel, A., Penin, F., and Moradpour, D. (2009b). An amphipathic alpha-helix at the C terminus of hepatitis C virus nonstructural protein 4B mediates membrane association. J. Virol. *83*, 11378–11384.

Grakoui, A., Wychowski, C., Lin, C., Feinstone, M., and Rice, C.M. (1993). Expression and identification of hepatitis C virus polyprotein cleavage products. J. Virol. *67*, 1385–1395.

Grunberger, C., Wyles, D.L., Kaihara, A., and Schooley, R.T. (2008). 3-drug synergistic interactions of small molecular inhibitors of hepatitis C virus replication. J. Infect. Dis. *197*, 42–45.

Guex, N., and Peitsch, M.C. (1997). SWISS-MODEL and the Swiss-PdbViewer: an environment for comparative protein modeling. Electrophoresis *18*, 2714–2723.

Guillen, J., Gonzalez-Alvarez, A., and Villalain, J. (2009). A membranotropic region in the C-terminal domain of Hepatitis C virus protein NS4B Interaction with membranes. Biochim. Biophys. Acta. *1798*, 327–337.

Hugle, T., Fehrmann, F., Bieck, E., Kohara, M., Krausslich, H.G., Rice, C.M., Blum, E., and Moradpour, D. (2001). The hepatitis C virus nonstructural protein 4B is an integral endoplasmic reticulum membrane protein. Virology *284*, 70–81.

Jones, D.M., Patel, A.H., Targett-Adams, P., and McLauchlan, J. (2009). The hepatitis C virus NS4B protein can trans-complement viral RNA replication and modulates production of infectious virus. J. Virol. *83*, 2163–2177.

Kim, J.E., Song, W.K., Chung, K.M., Back, H., and Jang, S.K. (1999). Subcellular localization of hepatitis C viral proteins in mammalian cells. Arch. Virol. *144*, 329–343.

Koev, G., and Kati, W. (2008). The emerging field of HCV drug resistance. Expert Opin. Investig. Drugs *17*, 303–319.

Kusov, Y.Y., Probst, C., Jecht, M., Jost, P.D., and Gauss-Muller, V. (1998). Membrane association and RNA binding of recombinant hepatitis A virus protein 2C. Arch. Virol. *143*, 931–944.

Lee, C., Cho, N.J., Pang, S.P., Frank, W., and Glenn, J.S. (2008). Phosphatidylinositol 4, 5-biphosphate is a ligand of HCV NS5A and mediates HCV genome replication. 15th International Symposium on Hepatitis C Virus and Related Viruses (San Antonio, Texas), pp. 223.

Lin, C., Amberg, S.M., Chambers, J., and Rice, C.M. (1993). Cleavage at a novel site in the NS4A region by the yellow fever virus NS2B-3 proteinase is a prerequisite for processing at the downstream 4A/4B signalase site. J. Virol. *67*, 2327–2335.

Lindenbach, B.D., Evans, M.J., Syder, A.J., Wolk, B., Tellinghuisen, T.L., Liu, C.C., Maruyama, T., Hynes, R.O., Burton, D.R., McKeating, J.A., *et al.* (2005). Complete replication of hepatitis C virus in cell culture. Science *309*, 623–626.

Lohmann, V., Korner, F., Koch, J., Herian, U., Theilmann, L., and Bartenschlager, R. (1999). Replication of subgenomic hepatitis C virus RNAs in a hepatoma cell line. Science *285*, 110–113.

Lu, L., Dekhtyar, T., Masse, S., Pithawalla, R., Krishnan, P., He, W., Ng, T., Koev, G., Stewart, K., Larson, D., *et al.* (2007). Identification and characterization of mutations conferring resistance to an HCV RNA-dependent RNA polymerase inhibitor in vitro. Antiviral Res. *76*, 93–97.

Lundin, M., Monne, M., Widell, A., Von Heijne, G., and Persson, M.A. (2003). Topology of the

membrane-associated hepatitis C virus protein NS4B. J. Virol. 77, 5428–5438.
Maerkl, S.J. (2009). Integration column: Microfluidic high-throughput screening. Integr. Biol. (Camb.) *1*, 19–29.
Manns, M.P., Foster, G.R., Rockstroh, J.K., Zeuzem, S., Zoulim, F., and Houghton, M. (2007). The way forward in HCV treatment--finding the right path. Nat. Rev. Drug Discov. *6*, 991–1000.
Matto, M., Rice, C.M., Aroeti, B., and Glenn, J.S. (2004). Hepatitis C virus core protein associates with detergent-resistant membranes distinct from classical plasma membrane rafts. J. Virol. *78*, 12047–12053.
Overington, J.P., Al-Lazikani, B., and Hopkins, A.L. (2006). How many drug targets are there? Nat. Rev. Drug Discov. *5*, 993–996.
Rodriguez, L., and Carrasco, L. (1995). Poliovirus protein 2C contains two regions involved in RNA binding activity. J. Biol. Chem. *270*, 10105–10112.
Selby, M.J., Choo, Q.L., Berger, K., Kuo, G., Glazer, E., Eckart, M., Lee, C., Chien, D., Kuo, C., and Houghton, M. (1993). Expression, identification and subcellular localization of the proteins encoded by the hepatitis C viral genome. J. Gen. Virol. *74*, 1103–1113.
Shepard, C.W., Finelli, L., and Alter, M.J. (2005). Global epidemiology of hepatitis C virus infection. Lancet Infect. Dis. *5*, 558–567.
Shi, S.T., Lee, K.J., Aizaki, H., Hwang, B., and Lai, M.M. (2003). Hepatitis C virus RNA replication occurs on a detergent-resistant membrane that cofractionates with caveolin-2. J. Virol. *77*, 4160–4168.
Sklan, H., and Glenn, J.S. (2006). The hepatitis C virus non-structural protein NS4B: from obscurity to central stage. In Hepatitis C Viruses: Genomes and Molecular Biology, Tan, S.L., ed. (Horizon Scientific Press, Norwich), pp. 245–266.
Thompson, A., Patel, K., Tillman, H., and McHutchison, J.G. (2009a). Directly acting antivirals for the treatment of patients with hepatitis C infection: a clinical development update addressing key future challenges. J. Hepatol. *50*, 184–194.
Thompson, A.A., Zou, A., Yan, J., Duggal, R., Hao, W., Molina, D., Cronin, N., and Wells, P.A. (2009b). Biochemical characterization of recombinant hepatitis C virus nonstructural protein 4B: evidence for ATP/GTP hydrolysis and adenylate kinase activity. Biochemistry *48*, 906–916.
Thompson, J. and McHutchison, J.G. (2009). Review article: investigational agents for chronic hepatitis C. Aliment. Pharmacol. Ther. *29*, 689–705.
Tomei, L., Altamura, S., Paonessa, G., De Francesco, R., and Migliaccio, G. (2005). HCV antiviral resistance: the impact of in vitro studies on the development of antiviral agents targeting the viral NS5B polymerase. Antivir. Chem. Chemother. *16*, 225–245.
Tong, W.Y., Nagano-Fujii, M., Hidajat, R., Deng, L., Takigawa, Y., and Hotta, H. (2002). Physical interaction between hepatitis C virus NS4B protein and CREB-RP/ATF6beta. Biochem. Biophys. Res. Commun. *299*, 366–372.
Tong, X., Chase, R., Skelton, A., Chen, T., Wright-Minogue, J., and Malcolm, B.A. (2006). Identification and analysis of fitness of resistance mutations against the HCV protease inhibitor SCH 503034. Antiviral Res. *70*, 28–38.
Wakita, T., Pietschmann, T., Kato, T., Date, T., Miyamoto, M., Zhao, Z., Murthy, K., Habermann, A., Krausslich, H.G., Mizokami, M., *et al.* (2005). Production of infectious hepatitis C virus in tissue culture from a cloned viral genome. Nat. Med. *11*, 791–796.
Wang, B.M., Weiner, N.D., Takada, A., and Schacht, J. (1984). Characterization of aminoglycoside-lipid interactions and development of a refined model for ototoxicity testing. Biochem. Pharmacol. *33*, 3257–3262.
Wyles, D.L., Kaihara, K.A., Vaida, F., and Schooley, R.T. (2007). Synergy of small molecular inhibitors of hepatitis C virus replication directed at multiple viral targets. J. Virol. *81*, 3005–3008.
Wyles, D.L., Kaihara, A., and Schooley, R.T. (2008). Synergy of a hepatitis C virus (HCV) NS4A antagonist in combination with HCV protease and polymerase inhibitors. Antimicrob. Agents Chemother. *52*, 1862–1864.
Yang, W., Zhao, Y., Fabrycki, J., Hou, X., Nie, X., Sanchez, A., Phadke, A., Deshpande, M., Agarwal, A., and Huang, M. (2008). Selection of replicon variants resistant to ACH-*806*, a novel hepatitis C virus inhibitor with no cross-resistance to NS3 protease and NS5B polymerase inhibitors. Antimicrob. Agents Chemother. *52*, 2043–2052.
Yu, G.Y., Lee, K.J., Gao, L., and Lai, M.M. (2006). Palmitoylation and polymerization of hepatitis C virus NS4B protein. J. Virol. *80*, 6013–6023.
Zhong, J., Gastaminza, P., Cheng, G., Kapadia, S., Kato, T., Burton, D.R., Wieland, S.F., Uprichard, S.L., Wakita, T., and Chisari, F.V. (2005). Robust hepatitis C virus infection in vitro. Proc. Natl. Acad. Sci. U.S.A. *102*, 9294–9299.

NS5A Inhibitors

14

Pilar Najarro, Neil Mathews and Stuart Cockerill

Abstract

The HCV NS5A protein plays critical roles in the life cycle of the virus and has been associated with a multitude of host–pathogen events. As part of the replication complex, it is essential for RNA replication. In addition, its propensity to engage in many protein–protein interactions enables this protein to facilitate the assembly of viral particles and to counteract the host immune response. However, the inherent intractability of NS5A as a drug target has caused it to be of low priority in the 'rational' drug hunter's portfolio of antiviral targets. This would change with the development of HCV subgenomic replicon technology, which provided an opportunity to identify inhibitors of HCV with mechanisms distinct from the established viral targets, i.e. the NS5B polymerase and NS3-4A protease. Whole cell screening using the replicon technology and subsequent generation of replicon mutants has pointed to NS5A as being the likely target of a number of hit compounds. A number of companies have now developed, through lead optimization, highly selective and potent inhibitors of NS5A. Recently these efforts have culminated in clinical proof of concept and established NS5A as a valid target for the development of small-molecule therapies.

Introduction

The non-structural 5A (NS5A) protein of HCV has become the subject of intensive research over the last decade due to its critical role in the life cycle of the virus. Along with the rest of the HCV proteins, NS5A is translated initially as part of a polyprotein structure that is subsequently cleaved into its different individual components by cellular and viral proteases. Once processed NS5A associates post-translationally with cytoplasmic membranes by means of its amphipathic helix (AH) and it is in this context that it contributes to its many host–virus interactions (Macdonald and Harris, 2004; He *et al.*, 2006). NS5A is essential for RNA replication and a key factor in the regulation of virus production and the host immune response.

Amongst those looking for novel treatments of HCV, the prosecution of NS5A as a drug target has been of low priority. It has been pursued by only a handful of pharmaceutical companies. The majority of publications describing small molecule inhibitors of NS5A have been confined to patent disclosures and only in recent times have NS5A emerged as a valid drug target with proof of concept in the clinic (Nettles *et al.*, 2008). A comprehensive review that examined the state of the art for NS5A drug discovery was published two years ago (Schmitz and Tan, 2008) highlighting its excellent potential as a target for HCV therapy. Opportunities for drug discovery with this target are based on (i) the key role of NS5A within the replication complex that amalgamates several non-structural proteins of the virus; (ii) its intimate association with ER-derived membranes, its protein phosphorylation and interactions with numerous host factors; and (iii) its ability to bind to RNA. The protein has no known sequence homologues within any eukaryotic genome, suggesting the potential for exquisite selectivity for drugs targeting NS5A.

The numerous ways by which NS5A is able to subvert the innate immune response underscores its importance in HCV immune evasion and further enhances its appeal as a target. Over the years, several lines of evidence have supported a role for NS5A in promoting viral persistence, pathogenesis and possibly carcinogenesis (Macdonald and Harris, 2004). Studies conducted in NS5A transgenic mice (Kanda *et al.*, 2009; Wang *et al.*, 2009) indicate that NS5A, independent of other HCV viral proteins, could play an important role in the development of liver damage, including steatosis and hepatocellular carcinoma. Moreover, the intrahepatic accumulation of IFN-γ is reduced in NS5A transgenic mice (Kanda *et al.*, 2009) which indicates a role for this protein in the establishment of chronic HCV infection and links it directly to the observed impairment of the immune response in patients chronically infected with HCV.

The association of NS5A with the response of infected subjects to IFN-α therapy has been very controversial and the predictive value of the so-called 'IFN sensitivity-determining region' (ISDR) of NS5A is contradictory (Torres-Puente *et al.*, 2008a). The discrepancies probably underscore the complex interactions behind the indirect action of IFN on HCV-infected cells. A closer look at the data generated from various studies indicates no direct correlation between NS5A ISDR and IFN response (Torres-Puente *et al.*, 2008b).

The development of an infectious HCV cell culture system (HCVcc) based on an HCV genotype 2a isolate (JFH1; Wakita, *et al*, 2005) has further facilitated the study of NS5A and its contribution to the viral life cycle (Miyanari *et al.*, 2007). Hitherto the function of this protein could only be partially studied in the HCV replicon system. Key interactions subsequently described for NS5A include those with non-structural proteins, lipid droplets (LD) and lipoproteins. There is now substantial evidence for the essential function of NS5A in viral morphogenesis, particle formation and egress based on studies with HCVcc. Mutations in the N-terminal domain of NS5A block its localization to LD whereas the C-terminus has been shown to interact with core proteins (McLauchlan, 2009). In addition preliminary data suggest that NS5A is packed in the virion (Jiang *et al.*, 2008) and if so NS5A inhibitors could potentially access the target extracellularly–intravirion, with a systemic antiviral effect. Moreover, the presence of NS5A in the infectious particles could also imply a role for the protein in the initial steps of infection.

The study of NS5A was biochemically challenging. It is difficult to express and purify both the full-length and soluble versions in which the AH has been removed. The NS5A protein is organized into three distinct domains (1, 2 and 3) that are connected by low-complexity sequences. Crystallography data have been reported for domain 1 only (Tellinghuisen *et al.*, 2004; Love *et al.*, 2009). Additional *in silico* sequence motif analysis along with 3D structure prediction (El Hefnawi *et al.*, 2009) results in further subdivisions. But, it is difficult to ascribe *bona fide* functionalities to the different protein partitions. Domain 1 is most conserved and includes a zinc-binding domain with RNA binding activity (Huang *et al.*, 2005). The rest of the protein primary sequence is significantly less conserved across the different HCV genotypes. Structural predictions suggest high disorder in domains 2 and 3. Domain 2 is needed for HCV replication and was shown to be natively unfolded (Liang *et al.*, 2007). Domain 3 is dispensable for HCV RNA replication but plays an essential role in viral particle production and assembly. The plasticity of the C-terminus of the molecule may facilitate its interaction with cellular and/or viral proteins.

NS5A is a phosphoprotein and there are numerous reports on the relevance of its hypo- and hyper-phosphorylation states to influence virus and RNA replication *in vitro*. When HCV subgenomic replicons were first established it was noticed that several cell culture adaptive mutations mapped to NS5A (Blight *et al.*, 2000) and that a decrease in NS5A hyperphosphorylation by selective inhibition of cellular kinases increased HCV replicon RNA replication (Neddermann *et al.*, 1999). Recent studies in the infectious system indicate that phosphorylation of serine 457 on NS5A by casein kinase II (CKII) could be acting as a switch between the RNA replication mode and the particle assembly mode (Tellinghuisen *et al.*, 2008). These and many other kinase-associated events are likely to be critical during HCV chronic infection (Gretton *et al.*, 2010). One of the most interesting areas of research focuses on how the different phospho-variants may guide the virus through its lifecycle and how they may be of relevance to viral replication and host–virus interaction (Tellinghuisen, 2009).

Here we focus on the NS5A replicon inhibitors reported thus far and recent advances in the field. We review the properties of these drugs in terms of potency, genotype and resistant replicon profiles, their clinical development, and their potential for therapy combination that will likely become the future standard of care (SOC) for HCV.

The discovery of compounds targeting NS5A

Screening and chemical series evolution

During the early years, HCV drug discovery was marked by a lack of tools to facilitate the study of the replicative cycle of the virus. It took scientists over 10 years since the discovery of HCV before Lohmann and colleagues (Lohmann *et al.*, 1999) obtained and made available a robust cell culture system that allowed broad screening of compounds. Since then the replicon system has been widely used by numerous research groups to look for inhibitors of RNA replication. For a detailed look at the HCV tool box see Chapter 2. In the case of conventional viral targets, e.g. the protease and its cofactor (NS3/NS4A) and the RNA-dependent RNA polymerase (NS5B), the replicon system has been used in conjunction with enzyme systems and X-ray or NMR structural analysis to drive drug discovery programmes in a tried and trusted manner. The absence of functional assays and crystallography data for NS5A required a different approach to identify and progress inhibitors.

A pragmatic, 'cell-to-mutant-to-sequence' method emerged, essentially a chemical genetics approach where inhibitors uncovered by this method probes the essentiality of the HCV genome. The compounds that target NS5A have all been discovered using this screening approach.

Structural definition

Specifically, Huh.7 cells containing the Con1 (1b)-derived subgenomic replicon have been used to identify HCV replication inhibitors. Several of the active compounds were characterized by their ability to induce changes in the domain I of NS5A during mutant generation *in vitro*. The most recurrent change corresponds to variants of tyrosine at position 93 (Y93) (Schmitz and Tan, 2008).

BMS790052 and AZD7295 lead the field, both having demonstrated clinical efficacy in phase Ib studies in patients (Nettles *et al.*, 2008; Gane *et al.*, 2010). Both compounds (and other NS5A inhibitors) are characterized by structural complexity and a high molecular weight in comparison with conventional 'drug-like' small molecules. The structures of the lead compounds are shown in Fig. 14.1. Their common structural components can be summarized by the core 'C' and the peripheral 'P' regions, used as a schematic shown in the same figure. Each of these peripheral or core components can possess high levels of identity with each other. This schematic will be used to rationalize and organize the understanding of the structure of compounds throughout this chapter.

Specifically, NS5A inhibitors possess two central core elements; often these are identical and inherently hydrophobic. Then there are two peripheral elements, again often identical. In the case of the BMS type structure, these peripheral elements are dipeptide in nature whereas this is not the case with AZD7295 as seen in Fig. 14.1. However, this periphery is the source of the structural complexity of this class of inhibitors. There have been a range of approaches to designing inhibitors of this general structure, and fragments associated with these approaches are shown in Fig. 14.1B.

A characteristic of these series is their great efficacy in suppressing viral RNA replication in the replicon assay. This level of potency may be viewed as a prerequisite for progression of compounds of this type into clinical evaluation; the extreme potency should help to mitigate formulation and potential toxicology issues inherent in such large and relatively insoluble molecules.

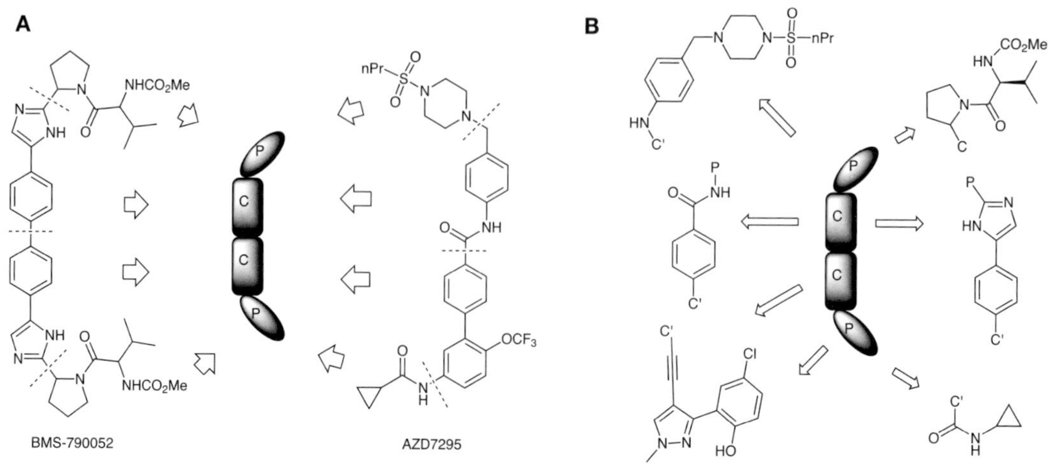

Figure 14.1 Peripheral and core components of clinically efficacious NS5A inhibitors.

Biphenyl imidazole core inhibitors
The most prominent inhibitors of this type described thus far are those from the Bristol Myers Squibb (BMS) group in Massachusetts. Their initial approach was described in a patent in 2004 (Colonno *et al.*, 2004) and revolved around optimization of a tri-functionalized thiazolidine imine system (Fig. 14.2A). The evolution of this series appears to be a process of simplification of the central core, elimination of unstable functionality, and potency improvement via manipulation of the periphery of the molecule. Initial progress was documented by the 2006 claim on symmetrical and pseudo symmetrical stilbenes (Serrano-Wu *et al.*, 2006; Fig. 14.2B). The retention of the proline moiety later proved to be significant. Subsequently, the imidazole was incorporated as a presumed bioisosteric replacement for the potentially metabolically labile amido groups linking the proline to the central core (C), (Bachand *et al.*, 2008b). Elaboration with a number of amino acids was then published in 2008 (Bachand *et al.*, 2008a,c) with the identification of the clinical compound BMS-790052, presumed to be the methyl carbonyl protected bis-valinate as shown by process and form patents which specifically described this compound (Kim *et al.*, 2009; Pack *et al.*, 2009) and subsequently confirmed by BMS (Gao *et al.*, 2010).

More details have emerged recently (Gao *et al.*, 2008, 2009; Bachand *et al.*, 2009d) on the discovery of this pseudo-dimeric series from the origins of the thiazolidine imine (Fig. 14.2). A potent example of the thiazolidine imine series, BMS-824 (D) was found to be unstable. Mass spectra identified potentially dimeric contaminants. When synthesized, these dimeric species (E) were found to be potent. Subsequently simplified dimeric structures (F) were also found to be potent. This fits into the identification of stilbene derivatives (B) described earlier. This undoubtedly can be taken as one of the examples that demonstrate the role that serendipity plays in drug discovery and how cell screening systems broaden the opportunities for finding new potent antivirals.

Subsequent patents were reported by BMS from late 2008 to 2009 (Fig. 14.3). These describe a series of modifications to the lipophilic central core of the molecules, usually through the addition of further rigidifying ring systems. These filings would appear defensive in nature and perhaps add little to the ongoing evolution of NS5A inhibitors *per se* (Fig. 14.3G–J, L; Bachand *et al.*, 2009b,d,e).

Perhaps a publication of more interest, Bachand *et al.* (2009a) addressed the introduction of a carboxylic acid functionality into the lipophilic core of the molecule (K). This is most likely a modification with the presumed intent of improving the solubility profile of this series. This incorporation of a hydrophilic acid appears

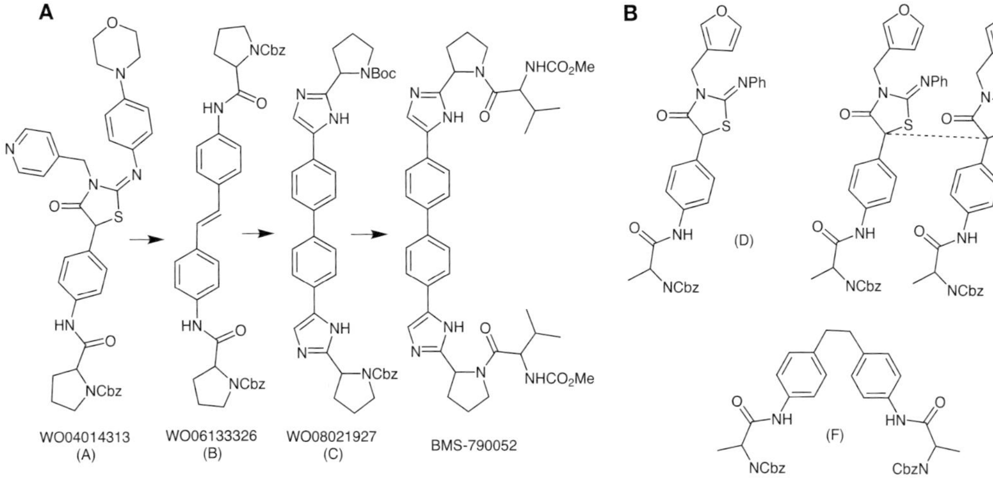

Figure 14.2 Evolution of BMS NS5A targeting inhibitors and discovery of dimeric structure activity.

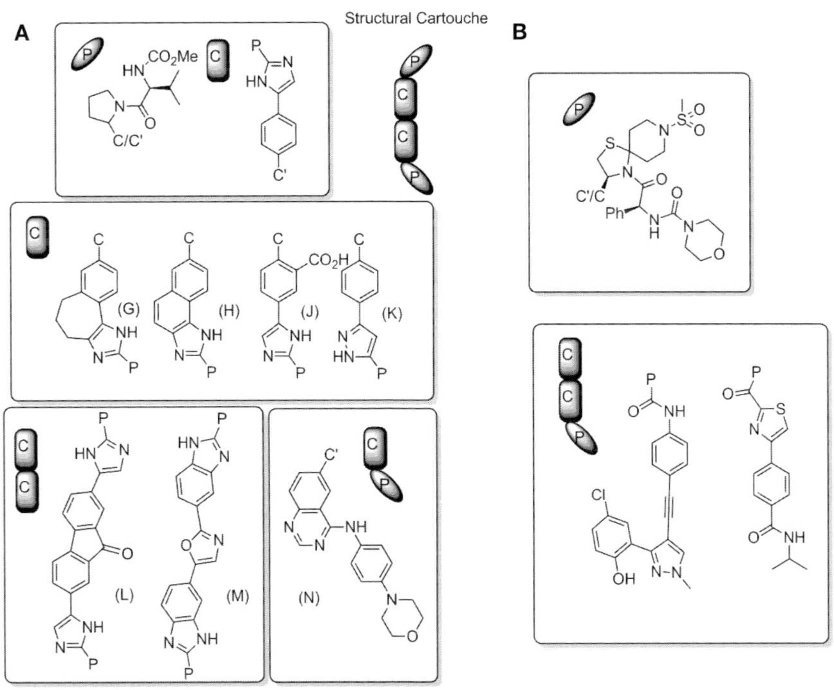

Figure 14.3 Structural cartouches from BMS and Genelabs.

to contradict a consistent picture of the lipophilic spacer group as a key element of this active series.

A further patent from BMS described the anti-HCV activity of a series of 6-phenyl quinazolines (N) (Bachand *et al.*, 2009c), which is ostensibly a hybrid of capped phenyl imidazole proline and an earlier series of 6-phenyl quinazolines described by the Arrow group. These hybrid compounds are an interesting concept given the disparity in mutation patterns and therefore the presumed modes of binding to NS5A that have been observed by these distinct core and structural elements. Further progress in this area is awaited with interest.

Most recently, an additional patent disclosure by BMS describes further hydrophobic linker modifications. This patent describes bis-benzimidazole modifications in tandem with acetylene, phenyl and heterocyclic centrepieces. From these structures, there are non-obvious regiochemistries that are presented as potent examples. See compound M in Fig. 14.3A (Belema *et al.*, 2010).

Related inhibitor types

The complex structural nature of this class of antivirals has been supported by the publication of patents from both Presidio (published under the name of XTL) and Genelabs. Little structural information has been revealed by Presidio, with only earlier structures currently available for comments. The phenolic pyrazoles and other classes described in 2008 have presumably undergone significant evolution, from both druggability and IP standpoints, as lead compounds have progressed into preclinical toxicological evaluation. The characteristics of this potent series of compounds are discussed elsewhere in the chapter.

Genelabs, now part of GSK, published patents most recently containing a spirocyclic oxazolidine fragment (Fig. 14.3B). Little is known about the status or properties of these increasingly complex molecules and further data are indeed eagerly awaited.

Diphenyl carboxamides (DPCs)

In contrast to the peptoidal complexity associated with the BMS and Genelabs series, the Arrow series of Y93-targeted inhibitors are relatively simple structurally. They were derived from a library enrichment programme which looked at cores amenable to further functionalization and possessing a drug-like provenance (Gaster and Mulholland, 1996). The central core in this series had been associated with clinical compounds in both CNS and anti-inflammatory therapeutic space. The amide functionality allowed for 2-dimensional diversification and initially moderately sized libraries were prepared around this core for screening against a range of viral targets (Fig. 14.4).

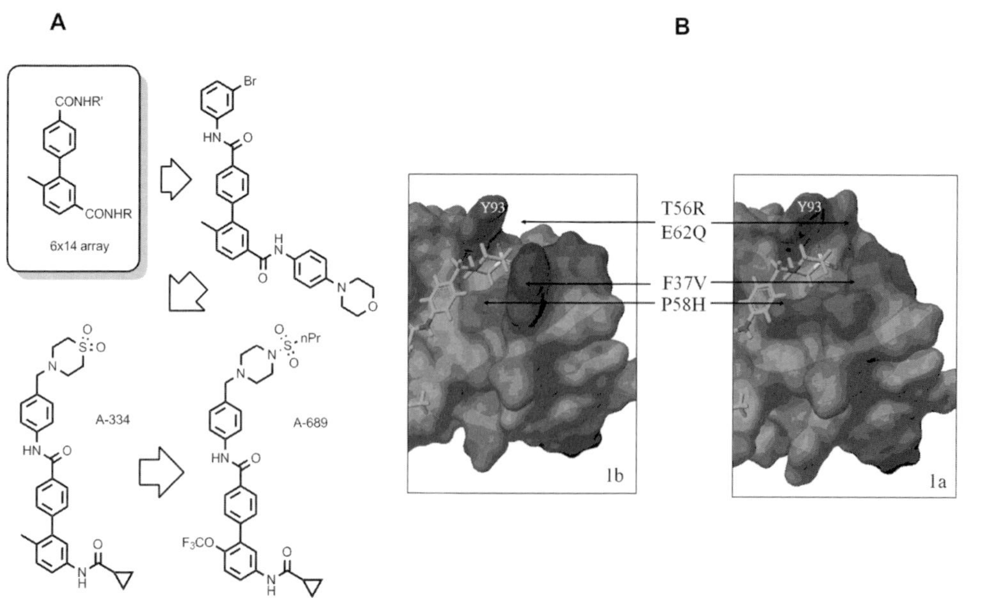

Figure 14.4 Diphenyl Carboxamide Optimization. (A) Central core and key compounds on the series. (B) Partial view of domain 1 of NS5A showing a 3D reconstruction of a putative compound-protein surface interaction based in 1b (left) and 1a (right) genotypes sequences.

Screening this library against the HCV Con-1-derived 1b replicon from ReBLikon GmbH (Mainz, Germany) identified bis-anilino amides with modest micromolar activity (A-266). Subsequent rounds of optimization (Wheelhouse *et al.*, 2007), focusing on potency and pharmacokinetic issues, allowed a progression through 3 orders of magnitude in potency and the identification of A-689 (later renamed AZD7295) as a lead and potential clinical compound (Carter *et al.*, 2010).

Despite the apparent physicochemical issues associated with this compound (solubility, molecular weight, lipophilicity) AZD7295 showed acceptable pharmacokinetics in multiple species, a clean toxicology profile and a short and scalable synthesis to allow the compound to progress into studies in human (Gane *et al.*, 2010).

Other approaches

A parallel approach adopted by Arrow concerned the discovery of anilino-quinazolines as NS5A inhibitors. The anilino-quinazoline core is a well documented core in drug design in the 21st century. Several examples exist for successful progression of this drug type to the market (Tarceva, Iressa, Lapatinib) with numerous further examples in development. The Arrow approach was to view this quinazoline ATP-competitive template as a generic nucleoside base binder. Structures distinct from nucleosides which may bind to similar sites may potentially target nucleoside dependant functions associated with viral replication. Although this approach may incur issues associated with the inhibition of cellular function through kinase effects, it served as a valid basis for an alternate screening set selection.

Rapid optimization work at Arrow, allowed the identification of leads (Mathews *et al.*, 2007) like A-831 (later renamed AZD2836) suitable for advancement towards the clinic (Fig. 14.5). Development of this compound is described later in the chapter.

Other HCV inhibitors reported to select mutations in NS5A in the replicon assay screen are the cyclophilins inhibitors (DEBIO-025, SCY635, and NIM811). Although they target cellular factors, NS5A seems to be most critical for their action as changes in NS5A have the greatest impact on resistance to this class of compounds (Puyang *et al.*, 2010) These antiviral agents are reviewed in detail in Chapter 18.

Mode of action studies

In order to characterize these compounds, several approaches were investigated, ranging from target specificity, protein–compound interactions, mutant generation, reverse genetics, and microscopy, with some approaches more successful than others.

Molecular and cellular biology

Following the discovery of the series as inhibitors of the HCV replicon in cell culture, their specificity for the virus has been confirmed by us and others (Lemm *et al.*, 2010). NS5A compounds are inactive when tested *in vitro* against unrelated viruses such as HIV, HSV and respiratory viruses and have shown only marginal activity against related Flaviviruses. The potency of BMS-790052 against bovine viral diarrhoea virus (BVDV) is reduced by six orders of magnitude compared with HCV.

Most efforts in the HCV drug discovery field have been directed towards classical enzyme targets expressed by the virus, namely the protease and its cofactor (NS3/NS4A) and the RNA-dependent RNA polymerase (NS5B). The initial hits that were derived from the replicon screening described earlier do not show activity in these biochemical assays and compound-resistant mutant generation was an obvious pathway to determine the molecular target of the hits.

Several experimental approaches have been used in order to generate replicon variants with a drug-resistant phenotype. In the case of NS5A mutant selection, the selective pressure of the drug can be either maintained in a constant manner or increased progressively during the selection period. This approach depends on the potency of the compound, its solubility and toxicity. Ultimately, the conditions of the experiment are optimized to induce changes in the replicon-containing cells that will allow cellular growth in the presence of both the compound of interest and the antibiotic selection (contained within the replicon). These conditions ensure that dividing cells will still harbour the subgenomic

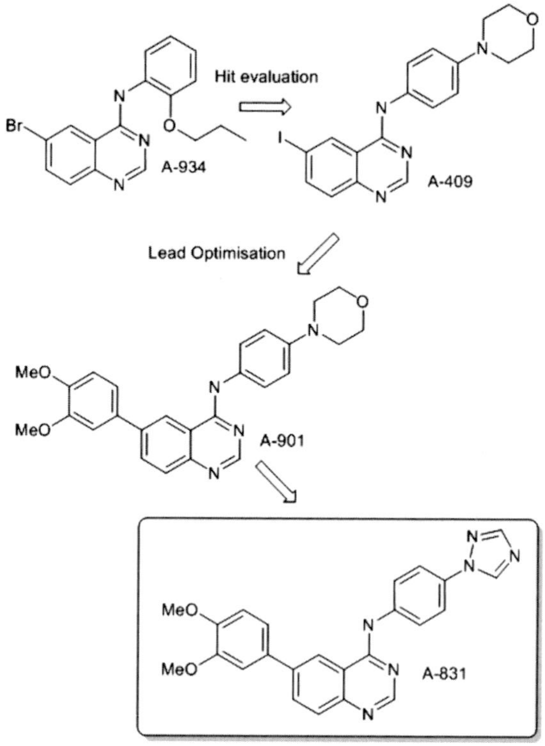

Figure 14.5 Quinazoline optimization to AZD2836.

viral RNA at the end of the process. Several compounds with unknown mode of action have been use to interrogate the replicon. Early work in mutant generation for these series was carried out on the 1b replicon (ReBLikon GmbH, Mainz, Germany); mostly due to the limited availability of other replicon genotypes. More recently, variant analysis has been conducted in 1a (H77) and 2a replicon backgrounds. Although some overlapping changes are observed, there is a distinct pattern of amino acid variations in the resistant clones in NS5A domain 1 from 1b, 1a and 2a replicons. For the sake of simplicity, we will focus on variants from the Con1 sequence unless specified otherwise.

Early studies to characterize the mode of action of these compounds were aimed to ensure that reduced sensitivity to a compound was due to viral and not to cellular changes. One way to address this is to transfer total RNA from the resistant cells into naïve cells and determine if there are any differences between wild type (wt) and mutant-derived RNA material. In this approach, after the initial assessment for resistance, the transfected cells were exposed to the antibiotic selection. Viable clones were then selected, expanded and the drugs tested for their potency against the clones to prove that the reduced sensitivity was not transmitted with the cellular RNA (Lemm *et al.*, 2010).

To confirm virus-based resistance the replicon RNA is extracted from the selected clones and sequenced. The variants obtained in cell culture for most chemical series described in this chapter (Figs. 14.1 to 14.4) map into NS5A. A summary of the most prominent changes and their frequency is depicted in Fig. 14.6B. The mutation profile of BMS790052 and AZD7295 *in vitro* appears restricted to NS5A and no other significant changes have been observed after sequencing the entire replicon from the resistant variants by clonal analysis. A cumulative scheme of the variations in amino acid sequence observed in 1b during mutant generation for the dicarboxamide series, exemplified by AZD7295, shows that the most prominent changes in NS5A localized to domain 1 with the hallmark mutation involving Y93 (Fig. 14.6B). The modifications observed in this amino acid occur by single nucleotide substitutions that translate into a histidine or cysteine residue (Y93H or Y93C). Domain 2, including the ISDR, does not vary above background levels. In domain 3, two 'hot-spot' areas were observed. The first one localized to what has been described as variable region 3 (V3) that has, intrinsically, high capacity to accumulate mutations (Jardim *et al.*, 2009) and a good degree of variation was seen in both control samples and resistant clones (Fig. 14.6B). The second one is right at the C-terminus of NS5A, in a region involved in viral morphogenesis that is postulated to interact with cellular proteins involved in lipid droplet formation. In the infectious system, domain 3 phosphorylation by casein kinase II (CKII) and other cell factors is important for particle formation and replication (Tellinghuisen *et al.*, 2008; Tellinghuisen, 2009). In the case of BMS-790052, only changes in the first 100 amino acid of the protein have been reported, with slightly different patterns in 1b (L31V, Y93H) and 1a (M28T, Q30H/R, L31M/V, Y93C) backgrounds. Research groups at XTL, GeneLabs, IRBM, BMS and Arrow demonstrated using reverse genetics that Y93 changes alone can recapitulate the lack of sensitivity to their NS5A compound series (De Francesco, 2005; Bechtel *et al.*, 2007; Wobbe *et al.*, 2007;

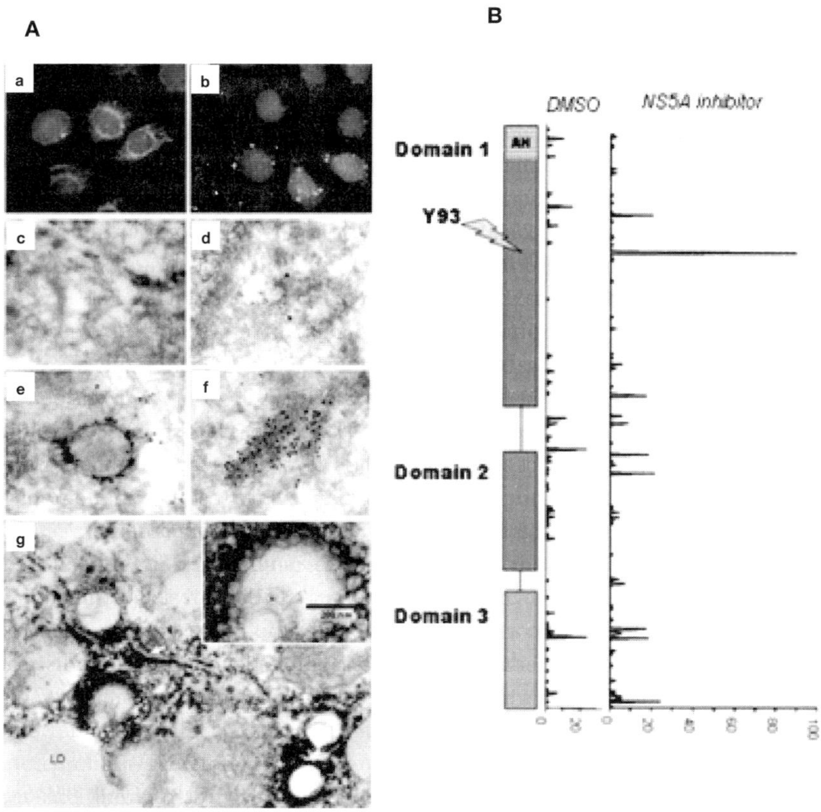

Figure 14.6 Molecular and cell biology studies on the mode of action of DPCs. (A) NS5A redistribution after incubation with NS5A inhibitors. Top panels: indirect immunofluorescence after DMSO control (a) or NS5A inhibitor treatment (b). Middle panels: cryosections and immunogold electron microscopy of replicon-containing cells using anti-NS5A mAb after DMSO (c) or NS5A inhibitor treatment (d). Closer view of aggregation effect (e) and tangential section through NS5A-enriched membranous structures following NS5A inhibitor treatment (f). Bottom panel: electron microscopy on epon-sections labelled with anti-NS5A followed by HRP and DAB precipitation reaction. Black deposits correspond with NS5A localization (g). Inset shows a detail of the NS5A-loaded spherule-like structures surrounding lipid droplets. (B) Schematic of the distribution pattern of amino acid substitution that occur in the 1b replicon during selection of resistant variants to DPC. The DMSO background variation is shown as a reference.

Deneka *et al.*, 2009; Lemm *et al.*, 2010). It will be important to determine whether some of the other variations observed during clonal selection are genotype-specific (such as L31V for BMS-790052 in genotype 1a) or how many of the background variants represent compensatory mutations that would counteract the poor replication fitness of some of the single point variants.

Along with the Y93 changes, another salient feature of the early compounds targeting NS5A series was the lack of, or a severely reduced, activity against 1a replicon (Bechtel *et al.*, 2007; Wobbe *et al.*, 2007; Najarro *et al.*, 2008). In many cases the compound potency against the 1b genotype decreased by at least two orders of magnitude when tested against the 1a, H77 derived replicon (Figs. 14.2 and 14.3). This genotypic selectivity was very striking and lacked a clear explanation.

NS5A is a temperamental protein. It is not easy to express and purify and the intrinsic disordered nature of the molecule has been a challenge for all researchers involved in its study. Domain 1 was initially published as a dimeric structure (Marcotrigiano and Tellinghuisen, 2009) and it was not until 4 years later when the structural biology group at Pfizer (Love *et al.*, 2009) succeeded in crystallizing, once more, albeit the N-terminus of NS5A this time starting with a slightly shorter construct. The diffraction data derived from these crystals indicated protein subunits paired in a dimeric fashion, yet, in the Pfizer structure, the monomer's orientation to one another was diametrically opposed to the earlier NS5A structure. The NS5A (33–202) monomer–monomer interface pointed to interactions of the most-highly conserved face of each subunit. These somehow opposing configurations of the protein cast some degree of scepticism in the NS5A drug discovery field about using a protein-based approach to try to elucidate the 1b/1a dichotomy of the series. Therefore, attempts to rationalize mode-of-action findings using the protein modelling were abandoned and instead it seemed more useful to rely on a ligand-based approach such that structural analysis was focused on the pharmacophore as illustrated in the descriptive chemistry section above. Indirect evidence on interacting key surfaces was derived from (i) the mutations isolated after selecting resistant variants *in vitro* and (ii) cross-resistance studies on genetically engineered replicons. If compounds that target Y93 bind in the same site, it is plausible to look for similarities between their chemical structures and build up a 3D alignment that shows their common features and differences and, by doing so, design changes to obtain pan-genotypic compounds. To undertake these studies and to overcome the limited availability of reagents to investigate *in vitro* HCV genotypes other than 1b, it was necessary to resort to genetic engineering of chimeric replicons (Lanford *et al.*, 2006; Bell *et al.*, 2009; Lemm *et al.*, 2010) as well as replicons with point mutations for reverse mutagenesis analysis and compound cross-screening. To characterize the genotypic selectivity of NS5A compounds, we have conducted studies in which the NS5A derived from H77 (1a) has been shuttled into the Con1 (1b) replicon background and have been able to demonstrate that NS5A is directly related to the genotypic shift observed (Bell *et al.*, 2009). In an effort to study the mode of action of the early compounds that were characterized by their lack of potency against 1a (Lemm *et al.*, 2010), BMS scientists swapped different regions of domain 1 and narrowed the genotypic specificity of the series to the first 76 amino acids of NS5A (Gao, 2010).

The overall amino acid sequence homology of NS5A from different genotypes is high and is over 80% conserved between 1a and 1b. The N-terminus is also conserved and tyrosine in position 93 is the consensus residue for both subtypes. There are, however, several amino acid variations that greatly modify the exposed surface in the published structure and that might have an effect on the ability of these compounds to interact with their target. As seen in Fig. 14.6B, residues T56, F37, E62 and P58 are positioned in the vicinity of Y93 and the surrounding surface contours may have an impact on the 1a and 1b potency difference. Indeed, when amino acid mutations were engineered to mimic the structural conformations of 1a into 1b, the compound activity against the former became identical to the latter (Chapman *et al*, in preparation). The potency changes observed with the four-variant construct (T56R, E62Q, F37V and P58H) were specific to the Y93-NS5A compounds since the efficacy of HCV protease inhibitors (PIs) and non-nucleoside inhibitors (NNIs) against these constructs was not affected. Conversely, introducing the reverse changes in NS5A into 1a appeared to increase the potency of the compounds. A similar modelling-based rational was used by XTL to explain the relative potencies observed by their chemical series as well as BMS and IRBM compounds in the 1b, 1a and 2a genotypes (Wobbe *et al*, 2007). Amino acid changes in V37 and H58 for 1a and E152 and R157 in 2a were postulated to be responsible for reduced compound affinity although no bio-physical data were demonstrated.

The fact that Y93 and other amino acids involved in resistance are distinct in genotypes 1a and 1b, together with the data on selective 1b/1a compounds, suggest that the genotype selectivity and resistance are not directly associated. Further chemistry work on the series and additional studies using a variety of variants have been used to target these structures with the aim to obtain broad spectrum coverage as illustrated by the range of activities displayed by BMS-790052 against six different HCV genotypes (Gao *et al.*, 2010).

Some tangential evidence of compound-NS5A dimer interaction can be derived from the evolution of the chemical series described previously. A crude comparison of the data available in the different genotypes and subtypes suggest a great leap in efficacy that coincides when simple (single) structures are linked together resulting in quasi symmetrical chemophores and so it is tempting to speculate that these drugs might be acting on the NS5A dimer. The IRBM group released data on a series of substituted piprazinyl-N-(aryl) benzamides that are suggested to have anti HCV activity via modulation of the dimerization of NS5A (Conte *et al.*, 2009). There has been some speculation as well on how other dimeric compounds such as BMS-790052 could be interfering either directly or in an allosteric fashion with the function of the protein by association with the dimer (Gao *et al.*, 2010). Alternative explanations for the extremely high potency of the NS5A compounds could, perhaps, rely on the synergistic inhibition of two or more activities of the protein.

The cellular distribution of NS5A has been analysed by differential centrifugation, electron microscopy and immunofluorescence. Like the rest of the positive-stranded RNA viruses, HCV replication is associated with cell membranes, with its replicase's proteins inducing membrane invaginations or vesicles, which could function as protective environments for RNA replication (Salonen *et al.*, 2005). In replicon-containing cells, NS5A co-purifies with the rest of the non-structural proteins in the microsomal fraction. By microscopy, the localization appears to be restricted to the perinuclear membrane and ER derived membranes (Penin *et al.*, 2004; Najarro *et al.*, 2008). The way NS5A inserts into the membrane, its topology, is believed to be determined by direct contact of the hydrophobic side of the AH with the cytoplasmic phospholipid leaflet. This would yield an orientation parallel to the lipid bilayer in which polar residues of the AH are exposed to the cytoplasm defining a unique platform in which protein–protein interactions with itself, other HCV proteins, and cellular factors can occur. It is also likely that this monotopic insertion mode could allow NS5A to transfer without anchoring restraints from double, virus-modified membranes onto the single phospholipid sheet on lipid droplets and thus directly contribute to the morphogenesis of viral particles.

NS5A in replicon-containing Huh.7 cells displays a punctuated, reticular pattern of expression in the cytoplasm. The replicon-driven overexpression of HCV non-structural proteins modifies the microsomal architecture of the secretory organelles and associated vesicles and NS5A is intimately linked to them. When Y93 mutant-inducing compounds are added to replicon cells there is a clear shift in the distribution of the protein such that a strong concentration of NS5A epitopes are seen grouping the signal in punctuated aggregations (Fig. 14.6A). This effect is obvious both by indirect immunofluorescence (IF; Fig. 14.6A, upper panels, A and B) as well as by direct observations using a GFP–NS5A construct. Importantly in the context of the whole virus, a similar redistribution pattern is observed. Ultra-structural studies on cryosections by electron microscopy after addition of DPCs to replicon cells show anti-NS5A antibodies conjugated to gold particles 'clumping' in membrane-rich areas, confirming the effect on the distribution pattern of NS5A seen by IF (Fig. 14.6A, middle panels, C, D, E and F). In epon sections, where the cell morphology is well preserved during sample processing, the most intriguing picture emerged. In these samples, NS5A localization was revealed by the DAB (black) deposits, localizing the replication complexes (Fig. 14.6A, lower panel). Most of the electron-dense signal associated with NS5A was distributed along the ER tracks and was dispersed among the cytoplasm in untreated cells. After incubation with DPCs a strong accumulation of NS5A coating electron-lucent areas is evident. A close look into these NS5A-enriched areas showed spherules of an average diameter of 50 nm covering the outside layer of lipid droplet-like structures. Live imaging of this effect in HCV infected cells will need to be conducted to closely monitor whether this effect is mediated by a blockage of NS5A movement or by an active redirection of the protein within the cell.

NS5A protein is presented in eukaryotic cells in both unphosphorylated (Huang *et al.*, 2005) and phosphorylated forms (basal-phosphorylated p56 and hyperphosphorylated p58). There have been numerous studies related to NS5A phosphorylation, which have been reviewed recently (Huang *et al.*, 2007). Integration of the findings related to the differential expression of these forms and the roles that each species might play in the viral life cycle is still not fully understood. It is clear though that NS5A has key functions that are likely to be mediated selectively by p56 or p58. Most of the data published so far have been in the context of replicon cells and more recently in the HCVcc system that is limited to the JFH1 strain. Interpretation of the *in vivo* relevance of some of these findings has to be treated with caution. Among the possible functions proposed for the phospho-NS5A isoforms are: the differential interaction with host factors; the packaging of viral RNA, its dimerization, interactions with other NS proteins and binding to viral untranslated regions, etc (Huang *et al.*, 2007). The balance between p56 and p58 has been shown to be a critical determinant of HCV replication and infectivity. *In vitro* several cell-culture adaptive mutations map to NS5A and impair NS5A hyperphosphorylatio. Yet some level of p58 is essential for HCV RNA replication and *in vivo* NS5A phosphorylation variant S225P (S2197P in the polyprotein) prevented productive replication in chimpanzees (Bukh *et al.*, 2002). Amino acid changes that enhance RNA replication, those commonly referred to as cell culture adaptive mutations, are different from those affecting virus titres. Recent studies with genotype 1b viruses indicate that the underlying mechanism interfering with virus assembly probably involves a cross-talk between structural and non-structural proteins during the assembly process (Pietschmann *et al.*, 2009).

The effect of NS5A replicon active compounds on the level of p58 was first reported by the IRBM group (Migliaccio, 2004). It was related to the ability of diarylcarboxamides to modulate HCV replication in the cell culture adapted and non-adapted replicons, inhibiting and activating them respectively. The partial mode of action was characterized and linked to their ability to alter NS5A biogenesis resulting in a reduction of p56 to p58 conversion and the resistance mutations identified mapped to NS5A, A92V, Y93H and R157W (De Francesco, 2005). Patent publications on the series followed (Migliaccio, 2004; Conte *et al.*, 2006) and the structure–activity relationship has been discussed (Conte *et al.*, 2009).

The same group reported a series of 2,4,5-trisubstituted imidazole inhibitors which selectively block NS5A hyperphosphorylation in the replicon system. The mode of action of this series implicated conserved serine residues and underscore the importance of NS5A associated kinases (Neddermann *et al.*, 2004; Quintavalle *et al.*, 2007).

Chemical series from both Arrow and BMS have been shown to reduce the levels of p58. (Gao *et al.*, 2008; Najarro *et al.*, 2008). It is unclear how certain compounds reduce the levels of NS5A p58 and why the loss of genotype coverage and the effect on p58 correlate. In principle, the effect could be mediated directly, by inhibiting a kinase or activating a phosphatase acting on NS5A as well as indirectly, by blocking or enhancing the kinase or phosphatase action responsible for p56 to p58 conversion. Interestingly, HCV NS3 protease inhibitors can also reduce p58 levels, which shows inhibitors with no kinase activity can also block p58 formation and highlights the importance of the protein–protein interactions that occur within the replication complexes (Lemm *et al.*, 2010).

It will be important to find out the number of basally phosphorylated residues that constitute p56, and how many more phosphorylated residues contribute to p58. It is also not known whether p56 and p58 are homogeneous or heterogeneous in nature and which proteins are responsible for this modulation. Intriguingly NS5A hyperphosphorylation could be modulated independently of kinase inhibition or mutations within non-structural HCV proteins, supporting the idea of a 'built-in' control of HCV replication orchestrated by NS5A (McCormick *et al.*, 2006). Phosphorylation of the NS5A C-terminus is required for virus nucleocapsid formation and it is during this early step of virus particle assembly that phosphorylated NS5A might facilitate its core to bind to HCV RNA, whereas unphosphorylated protein does not (Masaki *et al.*, 2008). In a different approach studying the *in vitro* effects of unphosphorylated and phosphorylated NS5A protein on its binding to RNA (Ivanov *et al.*, 2009), it was shown that when viral RNA was used as a template, CKII-phosphorylated NS5A completely blocked the RNA-dependent RNA polymerase activity of NS5B, whereas the unphosphorylated NS5A had no effect. Most probably NS5A basal phosphorylation is crucial for other stages in the virus life cycle. Recently, quantification and comparison of the different binding affinities of NS5A domains to RNA show that the stronger associations occur via domain 2 (Foster *et al.*, 2009), which harbours several phosphorylation sites.

Taken together, NS5A-targeted inhibitors are likely to disrupt the balance towards a particular phosphorylated isoform that in turn perturbs the equilibrium of the different tasks orchestrated by NS5A. However, it remains to be determined whether these effects will hold true in a *bona fide* infection system.

Biochemistry and binding studies

In order to address the potential ways in which NS5A targeting inhibitors are affecting replication, several binding studies have been tried by us and others (Gao *et al.*, 2008). Attempts to measure low molecular weight compounds binding to NS5A by surface plasmon resonance is challenging because low molecular weight compounds are poor analytes and they tend to suffer from low solubility and thus present technical difficulties.

NS5A is associated with cellular membranes by means of its N-terminal AH. The association occurs post-transcriptionally and the interaction of the hydrophobic side of the helix occurs by its insertion into the cytosolic lipid leaflet (Tellinghuisen *et al.*, 2004) A peptide derived from this region has potent antiviral activity against HCV (and HIV) and probably the ability to bind and modify the membrane structure is responsible for the virocidal activity observed (Cheng *et al.*, 2008). Recently, Palomares-Jerez and colleagues have reported an assay (Glenn *et al.*, 2010; Palomares-Jerez *et al.*, 2010) by which the relationship between the AH of NS5A and the membrane could be exploited as a tool to screen anti-HCV molecules. Indeed, genetic disruption of the AH abrogated viral RNA replication (Elazar *et al.*, 2003; Tellinghuisen *et al.*, 2004) and thus validates this region as a potential antiviral target. It is possible that the surface area of the protein encompassing Y93 and its vicinity interacts with the exposed residues on the polar, cytoplasmic side of the AH and the compounds discussed above could be disrupting such an interaction. Interestingly, some compounds have been shown to induce mutations very close to the AH (Bechtel *et al.*, 2007; Gao *et al.*, 2010) and within it (Bell *et al.*, 2009). However, attempts to set up binding assays in which fluorescent labelled peptides mimicking this region were incubated with purified, soluble NS5A were inconclusive.

Arrow has investigated compounds tagged with fluorescent labels *in vitro* with purified NS5A (or domain 1 only), but assays have been unable to detect significant binding over background. Similar frustrating results were obtained for equilibrium dialysis assays, NMR, co-crystallization studies and affinity chromatography experiments.

The only convincing evidence presented of a direct interaction between NS5A inhibitors and their target comes from the BMS group. Gao and colleagues produced a biotin-tagged HCV NS5A inhibitor that was incubated with replicon cells followed by pull down with streptavidin-coated beads. The proteins associated with the drug included NS5A but not NS3 or NS5B (Gao *et al.*, 2008; Gao *et al.*, 2010)

Several studies using different methodology have been undertaken to unravel what are the cellular partners of NS5A and a plethora of host proteins have been suggested to have functional interactions with it. Different approaches so far include the use of inhibitor affinity chromatography (Quintavalle *et al.*, 2006) and the yeast two-hybrid system. The latter identified the SH3 domain binding and the proline-rich motif in the C-terminal region of NS5A as potential sites for therapeutic intervention (Nanda *et al.*, 2006). Other studies include the analysis of proteins that co-purify differentially when full-length NS5A or NS5A deletions were used (Tellinghuisen, 2009). Another possible approach may be the probing of protein platforms such as the ProtoArray® system with unphosphorylated or CKII phosphorylated recombinant NS5A.

The development of compounds targeting NS5A

HCV has high mutation rates owing to lack of proofreading of the RNA-dependent RNA polymerase (RdRp) and this can create non-viable strains as well as a pool of potential 'good' mutations. This allows HCV genome to circulate as a diverse set of variants that are different but closely related strains within each genotype, commonly called quasispecies. This natural heterogeneity facilitates virus evolution and adaptation to new environments. Thus, the drugs that target any HCV protein will face the challenge of acting on a range of distinct structural conformations of the protein within an infected individual. There is general consensus that the future therapy will be made of a combination of IFN-α and direct acting antivirals (DAA) (Miller, 2009). For more details on HCV clinical trial design, see Chapter 7.

During treatment with IFN-α there have been several regions in the NS5A protein that tend to accumulate variation and have been proposed to be linked to therapy outcome. The association between NS5A heterogeneity in the sequences attributed to influence virulence (ISDR, V3 and IRRDR) and the response to current standard of care remains controversial (Aus dem Siepen *et al.*, 2005; Tsai *et al.*, 2008; Chen *et al.*, 2010). The implications of these changes on the clinical efficacy of NS5A targeting inhibitors will remain to be seen.

In vitro studies with IFN and other direct-acting antivirals

There are at least three established ways *in vitro* to study the behaviour of two or more HCV antiviral compounds in combination. The first one is to investigate if their potencies together are additive, antagonistic or synergistic. The second is to determine their combined ability to 'cure', i.e. eliminate viral RNA from replicon cells. The third one is to analyse the genetic barrier to resistance of one or more compounds together and define the threshold to generate resistant variants that are viable with drugs targeting different sites or functions.

NS5A targeting compounds have been shown to be additive to synergistic with IFN-α *in vitro* (Najarro *et al.*, 2008). Combination with protease inhibitors (PIs) and non-nucleoside inhibiotrs (NNIs) have proven to be more potent than each individual input (Gao *et al.*, 2010). Regarding their curing ability, NS5A compounds can eliminate the replicon RNA more efficiently when dosed together with either a PI or IFN-α (Deneka *et al.*, 2009). In addition, combination studies looking at the genetic barrier to resistance, as determined by the ability to generate drug-resistant clones after prolonged treatment with NS5A compounds either alone or simultaneously with PIs or NNIs, showed combinations of two compounds at their IC_{90} concentrations significantly reduced viable clones (Deneka *et al.*, 2009). So far, the replicon system has been a good predictor for therapy outcomes in the clinic. Thus, it is not unreasonable to forecast that the curing potential of NS5A-targeted inhibitors will be greatly enhanced when used in combination with IFN and other anti-HCV agents.

Genotype selectivity: clinical quasi-species

Based on the population genetic studies reported so far (Bouzgarrou *et al.*, 2009; Jardim *et al.*, 2009), it is difficult to establish any correlation between NS5A sequences at baseline and the outcome of current therapies. It is possible that deeper and broader molecular analysis of the NS5A region will help to delineate the role of baseline mutations following HCV therapy (Gane *et al.*, 2010; Gao *et al.*, 2010). For combination of NS5A with the current standard of care, and in future drug cocktails in which NS5A inhibitors are included, it will be important to assess the prevalence of variants in the population (Y93, L31, etc), their level of resistance to particular compounds and their response to other direct acting antivirals and IFN.

Looking at the available clinical sequences that are public (see web sources), it is estimated that ≥3% of the genotype 1 patients carry either H or C at the Y93 position. Little is known about the origin of these sequences; whether they represent a major quasispecies in that percentage of infected patients or if they account for a small proportion of all the variants within an infected subject. From the limited data on hand, it will appear that the calculated 3% reflects the amount of subjects in the data base with Y93C/H data derived from an 8-fold coverage sequencing of the so called dominant sequence. It is important, however, to notice that the fitness of the variants resistant to NS5A targeting compounds reported so far (from *in vitro* studies) is lower than the parental, wild type variant. In the case of NS3 and NS5B inhibitors, replicon fitness extrapolation to *in vivo* settings has been shown yet it is still unknown whether the Y93H/C are less fit *in vivo* or if compensatory mutations in different part of the HCV genome could play a role in generation of resistance.

Efficacy *in vitro* and extrapolation to clinical data

As clinical data is published, an improved understanding is emerging of the relationship between drug pharmacokinetic properties and *in vitro* anti-replicon activity, which allows improved dose efficacy predictions in the clinic. Several groups have attempted to model the dynamic behaviour of the virus during therapy using the results from IFN-α treatment (Neumann *et al.*, 1998). This particular compartmental model assumes the viral load and number of infected cells to be at steady state before treatment is initiated. At the start of treatment, *de novo* infection and viral production drop, giving rise to a biphasic decline in serum viral titre. This model continuously calculates changes in both infected cell and free virion numbers throughout the pharmacokinetic profile and represents an improvement over predictions based solely on discreet pharmacokinetic parameters (e.g. trough plasma levels, average AUC etc).

Application of this model to the clinical dataset available for BMS790052 can be used to estimate the factor linking *in vitro* potency with measured clinical effect (Arrow, unpublished). This approach could facilitate the early clinical development of future compounds with regard to dose selection and robust predictions of outcome (Adiwijaya *et al.*, 2010; Rong and Perelson, 2010).

The quinazoline series from Arrow, specifically the 6-phenyl variant A-831 (later termed AZD2836) was the first NS5A targeting inhibitor to enter the clinic. One of the most attractive aspects of this series was the potential to achieve pan-genotypic antiviral activity. Mutants were generated against several quinazolines and scattered mutations were localized in NS5A in triples that occurred either all in NS5A or appeared concomitant with changes in the C-terminus of NS4B or the N-terminus of NS5B (Najarro *et al.*, 2006). Extensive reverse genetics studies in which these mutations were reintroduced in the replicon (single, double and triple combinations) by transient transfections did not reproduce the resistance profile observed in the original clones. The exact nature of their target is uncertain, but it is possible that quinazolines interfere with host–viral interactions. In the search for potential kinase interference mechanism, Arrow undertook a screening of a broad kinase panel but it did not yield any significant hits. Despite the lack of a clear mode of action for this series, its safe preclinical profile allowed Arrow to advance the compound to full clinical development in 2007.

Publication of the clinical results for AZD2836 is still pending. Astra Zeneca listed the compound as 'on hold' for further efforts in formulation at the end of 2008, and announced its discontinuation in the pipeline early in 2009 (AZ Annual Report 2009). The lack of adequate exposure was a great challenge and subsequent work on formulation did not result in improved parameters that could have provided plasma concentrations sufficient to inhibit virus replication in the liver. Perhaps additional members of this chemotype could take advantage of the latest advances in HCV small animal models, where it should be possible now to use the humanized mouse model in order to measure *in vivo* activity and obtain proof of concept for this series.

BMS-790052 has been demonstrated to be a potent inhibitor of HCV replication in HCV genotype 1 patients with viral load declines of 1.8, 3.2, and 3.3 logs at 24 hours after single 1 mg, 10 mg, and 100 mg doses respectively (Nettles *et al.*, 2008; Gane *et al.*, 2010; Gao *et al.*, 2010). Viral load declines were maintained for 144 hours after dosing in the 100 mg dose group. Additional insight into the individual data show that the maximum response was observed in the 1b patients included in the cohort. It is important to note that the genotypic analyses in this study using serum samples at baseline and at 24 and 144 hours showed both pre-existing variants and emergence of resistance in some of the subjects. This is no surprise as other direct-acting antivirals (DAAs) have experienced similar challenges when entering the clinic and highlights the importance of early combination trials rather than initial monotherapy. In fact, this compound is now in clinical trials in combination with peg-IFN and ribavirin (Pol *et al*, 2010) and also in combination with an NS3 inhibitor, BMS-650032 (see web reference). Interim 12 week treatment data on BMS-790052 in combination with peg-IFN and ribavirin have recently been reported (Pol *et al.*, 2010). Treatment-naive HCV genotype 1 patients were randomized to receive 3 mg, 10 mg, 60 mg, or placebo once daily in combination with peg-IFN-2α and ribavirin for 48 weeks. eRVR (HCV RNA undetectable at weeks 4 and 12) was achieved in 42%, 83%, and 75% in the 3 mg, 10 mg, and 60 mg groups, respectively, compared with 8% in the placebo group. Viral breakthrough was observed in two patients in the 3 mg dosing group prior to the 12 week time point. BMS-790052 was well tolerated with a safety profile comparable to that observed in the placebo/peg-IFN-2α/ribavirin group.

Astra Zeneca/Arrow have also reported clinical data for AZD7295 (Gane *et al*, 2010). This selective inhibitor of HCV NS5A is very potent *in vitro* against 1b replicon (7 nM) but much less potent against the 1a replicon (IC_{50} ~1.24 μM, (Carter *et al.*, 2010) and in a NS5A genotype 3 chimera (IC_{50} ~1.3 μM). The study investigated safety, tolerability, pharmacokinetics, and antiviral activity of AZD7295 in patients with chronic HCV infection with genotypes 1a, 1b, or 3. AZD7295 was administered for 5 days to 4 groups of patients who were either treatment-naïve or treatment-experienced using different doses and dosing regimens. The results are summarized in Fig. 14.7. Other than for GI effects at higher doses, which may have been due to the high volumes of excipients used at higher doses, the study medication was well tolerated. No clinically significant vital sign, ECG, or clinical laboratory abnormalities were reported. HCV RNA declined in a dose-dependent manner in HCV genotype 1b patients with mean log declines from baseline of −1.2 to −2.4 within 24 hours (intention-to-treat analysis), which were maintained through to day 6. Plasma levels achieved with a suboptimal formulation were 2- to 9.4-fold in excess of the *in vitro* genotype 1b IC90 but below the IC_{90} for genotypes 1a and 3. Although preclinical liver concentration data suggested that efficacious concentrations may be achievable in the liver for genotypes 1a and 3, no antiviral response was demonstrated in patients infected with these genotypes. Further studies of a new formulation of AZD7295 combined with pegylated IFN plus ribavirin in HCV genotype 1b patients are planned.

Presidio Pharmaceuticals has recently announced that it has been authorized by the Medicines and Healthcare Products Regulatory Agency to initiate a phase I first-in-human clinical trial of PPI-461, their NS5A inhibitor, in healthy volunteers. With a preclinical profile similar to BMS-790052 (Colonno *et al.*, 2010), trials in patients have followed swiftly.

It is clear that efforts developing drugs targeting NS5A have resulted in safe and well tolerated regimens with a clear relationship between the *in vitro* potency data and pharmacodynamic effects in the clinic. Clinical data from progression of these compounds in combination studies with IFN and ribavirin show an advantage of adding NS5A inhibitors to the current SoC over SoC alone (Pol *et al.*, 2010). Further data from clinical studies in which NS5A is combined with other DAAs are eagerly awaited.

Conclusion

Screening in the replicon system has been applied as practical approach to broaden target opportunities in the discovery of new potent DAAs.

After a long and sustained effort, initial hits derived from HCV replicon screening have been explored and optimized as anti-HCV direct acting antivirals. An element of serendipity, as well as careful investigation, has focused the area on large, atypical drug molecules. These compounds with extreme potency against RNA replication highlight the use of cell-based assays in antiviral drug discovery and illustrate the role that observation and perseverance play in drug discovery.

Compounds inhibiting NS5A have the ability to interfere with the phosphorylation and cellular distribution of NS5A. All publicly disclosed NS5A inhibitors appear sensitive to Y93 mutations but other residues in domain 1 are key in defining the genotypic specificity of the drugs. As for NS3 and NS5B inhibitors, pre-existing variants in the population exist (Y93H, Y93C, L31) and whether it will be needed to determine how these variants respond to drug pressure. Also, it remains to be seen if the lack of sensitivity *in vitro* translates to the *in vivo* situation and if the fitness of the virus is altered by these mutations in the HCV genome.

Combination of NS5A inhibitors and NS3 inhibitors are currently being explored in the clinic and other combinations are likely to enter the clinic in the near future.

Future trends

The NS5A area is an exciting new area in HCV research and development, one where potent antiviral activity has already been demonstrated in the clinic with two different NS5A inhibitors. Given the successful nature of clinical trial data published thus far from both the BMS and Arrow groups, NS5A inhibitors can now be viewed as a potential future cornerstone of HCV combination therapy. In some ways the position of this area is not unlike that observed in the field of

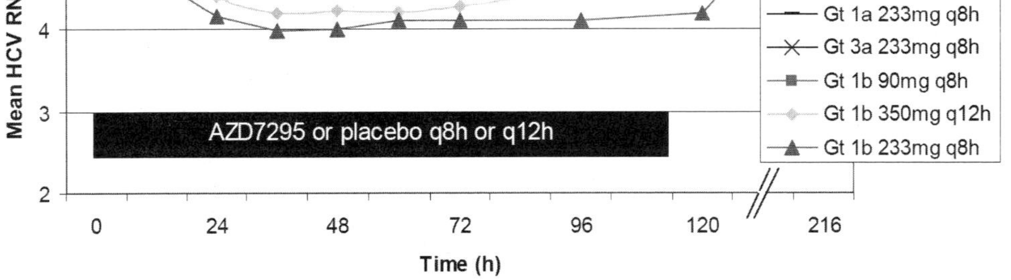

Figure 14.7 Clinical efficacy of NS5A inhibitors. AZD7295 oral solution or placebo (4:1) was administered for 5 days to treatment-naïve and treatment-experienced patients without decompensated liver disease or histological or clinical evidence of cirrhosis. Patients were randomized to the dose levels shown on the right. At each active dose level there were 4 patients per genotype. Six patients received placebo.

HCV protease inhibitor development back in 2003–2005. We saw Boehringer-Ingelheim and Vertex starting to release data on the progress of their protease inhibitors (Lamarre *et al.*, 2003; Lin *et al.*, 2004) and demonstrating improved efficacies over the nucleoside polymerase inhibitors described at that time.

NS5A as a target is also an increasingly competitive area for drug discovery. Multiple companies are entering the space and searching for clinical compounds of similar structure and complexity to those currently under development. However, the future may lie in the requirement to find simpler, more conventional molecules than those currently identified in order to demonstrate a clear advantage over the clinical leads. On the other hand, the HCV market may have less time to reach peak value than in the early days of protease inhibitor development. The commercial window allowing development of second or third generation NS5A inhibitors may already be closing. Perhaps the exciting clinical efficacies of the leading molecules currently in development demonstrating strong viral load reductions will be the best we see. This will leave the commercial returns to those early companies that invested in the area despite the obvious risks associated at the outset with taming such a seemingly intractable target.

In addition, the pace of HCV drug development has increased. It took more than 5 years from the time Telaprevir was first dosed in humans before it was combined with another DAA. For NS5A inhibitors, PoC for such combination therapy studies occured in a matter of months. Triple combinations, with or without IFN, with other directly acting antivirals are currently in clinical trials and a similar acceleration of these trials containing an NS5A inhibitor are expected.

Because of the exciting clinical potencies observed thus far, we expect to see the NS5A basic research area to keep expanding and eventually to capitalize on the knowledge of the functions of this fascinating protein in HCV-mediated pathogenesis, which will allow the discovery of new and better drugs.

Web sources

Clinical trials
http://ctr.bms.com/OneBmsCtd/InitTrialAction

Sequences databases
http://www.broadinstitute.org/annotation/viral/HCV
http://www.hcvdb.org/sequence.asp
http://hcv.lanl.gov/content/sequence/HCV/ToolsOutline.html

Acknowledgements

The authors would like to thank all their colleagues, students and collaborators who contributed to the NS5A work at Arrow Therapeutics. The opinions expressed in this publication are those of the authors and do not necessarily state or reflect the views held by Arrow Therapeutics or their parent company, Astra Zeneca.

References

Adiwijaya, B.S., Herrmann, E., Hare, B., Kieffer, T., Lin, C., Kwong, A.D., Garg, V., Randle, J.C., Sarrazin, C., Zeuzem, S., and Caron, P.R. (2010). A multi-variant, viral dynamic model of genotype 1 HCV to assess the in vivo evolution of protease-inhibitor resistant variants. PLoS. Comput. Biol. *6*, e1000745.

Aus dem Siepen M., Lohmann, V., Wiese, M., Ross, S., Roggendorf, M., and Viazov, S. (2005). Nonstructural protein 5A does not contribute to the resistance of hepatitis C virus replication to interferon alpha in cell culture. Virology *336*, 131–136.

Bachand, C., Belema, M., Deon, D.H., Good, A.C., Goodrich, J., James, C.A., Lavoie, R., Lopez, O.D., Martel, A., Meanwell, *et al.* (2008a). WO 2008/021936. 2008a.

Bachand, C., Belema, M., Deon, D.H., Good, A.C., Goodrich, J., James, C.A., Lavoie, R., Lopez, O.D., Martel, A., *et al.* (2008b). WO 2008/021927 A2. 2008b.

Bachand, C., Belema, M., Deon, D.H., Good, A.C., Goodrich, J., James ,C.A., Lavoie, R., Lopez, O.D., Martel, A., Meanwell, N.A., *et al.* (2008c). WO 2008/021928. 2008c.

Bachand, C., Belema, M., Good, A.C., Goodrich, J., Lavoie, R., Lopez, O.D., Martel, A., Meanwell, N.A., Nguyen, V.N., Romine, J.L., *et al.* (2008d). WO 2008/144380. 2008d.

Bachand, C., Belema, M., Deon, D.H., Good, A.C., Goodrich, J., James, C.A., Lavoie, R., *et al.* (2009a). US 2009/068140. 2009a.

Bachand, C., Belema, M., Deon, D.H., Good, A.C., Goodrich, J., James, C.A., Lavoie, R., Lopez, O.D.,

Martel, A., Meanwell, N.A., *et al.* (2009b). WO 2009/102633. 2009b.

Bachand, C., Belema, M., Deon, D.H., Good, A.C., Goodrich, J., James, C.A., Lavoie, R., Lopez, O.D., Martel, A., Meanwell, N.A., *et al.* (2009c). WO 2009/102694. 2009c.

Bachand, C., Deon, D.H., Good, A.C., Goodrich, J., James, C.A., Lavoie, R., *et al.* (2009d). WO 2009/102318. 2009d.

Bachand, C., Deon, D.H., Good, A.C., Goodrich, J., James, C.A., Lavoie, R., *et al.* (2009e). WO 2009/102568. 2009e.

Bechtel, J., Arfsten, A., Koo-Mc-Coy, S., Zhang, J., Fung, K., Sivaraja, M., Leivers, M., Dehghani, A., Chan, S., Slobodov, S., *et al.* (2007). Novel inhibitor of HCV replication functioning via NS5A. 2007. 14th International Symposium on Hepatitis C Virus and Related Viruses, 25–28 September 2007 (Glasgow, UK).

Belema, M., Good, A.C., Goodrich, J., Karkala, R., Guo Lopez, O.D., Van Hguyen, N., Kapur, J., Qiu, Y., Romine, J.L., St. Laurent, D.R., *et al.* (2010). WO 2010/017401. 2010.

Bell, T.J., Granycome, C., Bone, P., Wilson, L., Wheelhouse, C., and Chapman, J. (2009). Generation and Characterisation of Genotype 1a and 1b HCV Replicons with Reduced Sensitivity to a Novel NS5A Inhibitor. HepDart 2009, 6–10 December 2009 (Hawaii, USA).

Blight, K.J., Kolykhalov, A., and Rice, C.M. (2000). Efficient initiation of HCV RNA replication in cell culture. Science *290*, 1972–1974.

Bouzgarrou, N., Hassen, E., Mahfoudh, W., Gabbouj, S., Schvoerer, E., Ben, Y.A., Ben, M.N., Triki, H., and Chouchane, L. (2009). NS5A(ISDR-V3) region genetic variability of Tunisian HCV-1b strains: Correlation with the response to the combined interferon/ribavirin therapy 1. J. Med. Virol. *81*, 2021–2028.

Bukh, J., Pietschmann, T., Lohmann, V., Krieger, N., Faulk, K., Engle, R.E., Govindarajan, S., Shapiro, M., St, M., and Bartenschlager, R. (2002). Mutations that permit efficient replication of hepatitis C virus RNA in Huh-7 cells prevent productive replication in chimpanzees. Proc. Natl. Acad. Sci. U.S.A. *99*, 14416–14421.

Carter, M., Alber. D., Baxter, B., Borger, J., Chapman, J., Fram, S., Goulding, E., Hallot, A., Lockyer, M., Mathews, N., *et al.* (2010). Diphenylcarboxamides As Inhibitors of HCV non-structural protein NS5A. 23rd International Conference of Antiviral Research, 25–28 April 2010 (San Francisco, USA).

Chen, T.M., Huang, P.T., Wen, C.F., Tung, J.N., Chow, C., and Chen, Y.P. (2010). Reappraisal of the importance of mutations in the NS5A-PKR-binding domain of hepatitis C-1b virus in the era of optimally individualized therapy 1. J. Viral Hepatis doi:10.1111/j.1365-2893.2010.01287.x.

Cheng, G., Montero, A., Gastaminza, P., Whitten-Bauer, C., Wieland, S.F., Isogawa, M., Fredericksen, B., Selvarajah, S., Gallay, P.A., Ghadiri, R., and Chisari, F.V. (2008). A virocidal amphipathic {alpha}-helical peptide that inhibits hepatitis C virus infection in vitro. Proc. Natl. Acad. Sci. U.S.A. *105*, 3088–3093.

Colonno, R., Lemm, J., O'Boyle, D., Gao, M., Romine, J.L., Martin, S.W., Snyder, L.B., Serrano-Wu, M., Deshpande, M., and Whitehouse, D. (2004). WO 2004/014313 A2. 2004.

Colonno, R., Peng, E., Bencsik, M., Huang, N., Zhong, M., Huq, A., Huang, Q., Williams, J., and Li, L. (2010). 33 Identification and characterization of PPI-*461*, A Potent And Selective HCV NS5A Inhibitor With Activity Against All Hcv Genotypes. J. Hepatol. *52*, S14–S15.

Conte, I., Ercolani, C., Guiliano, C., Migliaccio, G., and Stansfield, I. (2006). WO2006/038039 A1. 2006.

Conte, I., Giuliano, C., Ercolani, C., Narjes, F., Koch, U., Rowley, M., Altamura, S., De, F.R., Neddermann, P., Migliaccio, G., and Stansfield, I. (2009). Synthesis and SAR of piperazinyl-N-phenylbenzamides as inhibitors of hepatitis C virus RNA replication in cell culture 93. Bioorg. Med. Chem. Lett. *19*, 1779–1783.

De Francesco, R. (2005). Hepatitis C virus drug resistance: the impact of *in vitro* studies on the development of novel antiviral agents targeting viral or host factors. HepDart 2005, 6–10 December 2005 (Hawaii, USA).

Deneka, M., Quinkert, D., Rupassara, D., Thomas, E., Bone, P., Granycome, C., Bell, T., Chapman, J., Salter, J., Lumley, J., *et al.* (2009). Characterisation of small molecule inhibitors of HCV replication that target NS5A. 16th International Symposium on Hepatitis C Virus and Related Viruses, 3–7 October 2009 (Nice, France).

El Hefnawi, M.M., El Behaidy, W.H., Youssif, A.A., Ghalwash, A.Z., El Housseiny, A., and Zada, S. (2009). Natural genetic engineering of hepatitis C virus NS5A for immune system counterattack. Ann. N. Y. Acad. Sci. *1178*, 173–185.

Elazar, M., Cheong, K.H., Liu, P., Greenberg, H.B., Rice, M., and Glenn, J.S. (2003). Amphipathic helix-dependent localization of NS5A mediates hepatitis C virus RNA replication. J. Virol. *77*, 6055–6061.

Foster, T., Kalverda, A., Homans, S., Pearson, A., Harris, M. (2009). Analysis of the relative contribution of the three domains of NS5A to RNA binding. 2009. 16th International symposium on Hepatitis C virus and related viruses, 3–7 October 2009 (Nice, Fance).

Gane, E., Foster, G.R., Cianciara, C., Stedman, S., Ryder, M., Buti, E., Clark, E., and Tait, D. (2010). Antiviral Activity, Pharmacokinetics, Safety, and Tolerability of AZD7295, a Novel NS5A Inhibitor, Following Multiple Ascending Doses in HCV Genotype 1 and 3 Patients. 45th Annual meeting of the European Association for the Study of the Liver, 14–18 April 2010 (Vienna, Austria).

Gao, M., Fridell, R., O'Boyle, D., Qui, D., Sun, J.H., Lemm, J., Nower, P., Valera, L., Voss, S., Lui, M., *et al.* (2008). HCV NS5A Inhibitor: From Screen Hit to Clinic. 15th International Symposium on Hepatitis C virus and related viruses, 5–9 October 2008 (San Antonio, USA).

Gao, M., Lemm, J., Leet, J., O'Boyle, D., Fridell, R., Sun, J., Huang, S., Serrano-Wu, M., Meanwell, N., Romine, J., *et al.* (2009). Discovery of HCV NS5A inhibitors. In 16th International Symposium on Hepatitis C Virus and Related Viruses, 3 October 2009 (Nice, France).

Gao, M., Nettles, R.E., Belema, M., Snyder, L.B., Nguyen, V.N., Fridell, R.A., Serrano-Wu, M.H., Langley, D.R., Sun, J.H., O'Boyle Ii, D.R., *et al.* (2010). Chemical genetics strategy identifies an HCV NS5A inhibitor with a potent clinical effect 3. Nature *465*, 96–100.

Gaster, L.M. and Mulholland, K.R. (1996). Preparation of N-phenylbiphenylcarboxamide derivatives as 5HT1D antagonists. P.C.T Int. Appl. WO 9631508 A1.

Glenn, J.S., Cho, N.-J., and Yang, W. (2010). WO2010/039195 Screening for inhibitors of amphipathic helix (AH) function. 8 April 2010.

Gretton, S., Hughes, M., and Harris, M. (2010). Hepatitis C virus RNA replication is regulated by Ras-Erk signalling 1. J. Gen. Virol. *91*, 671–680.

He, Y., Staschke, K., and Tan, S. (2006). HCV NS5A: A Multifunctional Regulator of Cellular Pathways and Virus Replication. In Seng-Lai Tan, ed., Hepatitis C Viruses: Genomes and Molecular Biology (Horizon Bioscience, Norwich), pp. 267–292.

Huang, L., Hwang, J., Sharma, S.D., Hargittai, M.R., Chen, Y., Arnold, J.J., Raney, D., and Cameron, C.E. (2005). Hepatitis C virus nonstructural protein 5A (NS5A) is an RNA-binding protein. J. Biol. Chem. *280*, 36417–36428.

Huang, Y., Staschke, K., De, R., and Tan, S.L. (2007). Phosphorylation of hepatitis C virus NS5A nonstructural protein: a new paradigm for phosphorylation-dependent viral RNA replication? Virology *364*, 1–9.

Ivanov, A.V., Tunitskaya, V.L., Ivanova, O.N., Mitkevich, V.A., Prassolov, V.S., Makarov, A.A., Kukhanova, K., and Kochetkov, S.N. (2009). Hepatitis C virus NS5A protein modulates template selection by the RNA polymerase in *in vitro* system 1. FEBS Lett. *583*, 277–280.

Jardim, A.C., Yamasaki, L.H., de Queiroz, A.T., Bittar, C., Pinho, J.R., Carareto, C.M., Rahal, P., and Mello, I.M. (2009). Quasispecies of hepatitis C virus genotype 1 and treatment outcome with peginterferon and ribavirin. Infect. Genet. Evol. *9*, 689–698.

Jiang, J., Lui, F., and Luo, G.G. (2008). Role of HCV Nonstructural proteins in virion assembly. 15th International Symposium on HCV and related viruses, 5 October 2008 (San Antonio, USA).

Kanda, T., Steele, R., Ray, R., and Ray, R.B. (2009). Inhibition of intrahepatic gamma interferon production by hepatitis C virus nonstructural protein 5A in transgenic mice. J. Virol. *83*, 8463–8469.

Kim, S., Gao, Q., and Yang, F. (2009). WO 2009/020828.

Lamarre, D., Anderson, P.C., Bailey, M., Beaulieu, P., Bolger, G., Bonneau, P., Bos, M., Cameron, D.R., Cartier, M., Cordingley, M.G., *et al.* (2003). An NS3 protease inhibitor with antiviral effects in humans infected with hepatitis C virus 1. Nature *426*, 186–189.

Lanford, R.E., Guerra, B., and Lee, H. (2006). Hepatitis C virus genotype 1b chimeric replicon containing genotype 3 NS5A domain. Virology *355*, 192–202.

Lemm, J.A., O'Boyle, D., Liu, M., Nower, P.T., Colonno, R., Deshpande, M.S., Snyder, L.B., Martin, S.W., St Laurent, D.R., Serrano-Wu, M.H., *et al.* (2010). Identification of hepatitis C virus NS5A inhibitors. J. Virol. *84*, 482–491.

Liang, Y., Ye, H., Kang, B., and Yoon, H.S. (2007). Domain 2 of nonstructural protein 5A (NS5A) of hepatitis C virus is natively unfolded. Biochemistry *46*, 11550–11558.

Lin, K., Kwong, D., and Lin, C. (2004). Combination of a hepatitis C virus NS3-NS4A protease inhibitor and alpha interferon synergistically inhibits viral RNA replication and facilitates viral RNA clearance in replicon cells. Antimicrob. Agents Chemother. *48*, 4784–4792.

Lohmann, V., Korner, F., Koch, J., Herian, U., Theilmann, L., and Bartenschlager, R. (1999). Replication of subgenomic hepatitis C virus RNAs in a hepatoma cell line. Science *285*, 110–113.

Love, R.A., Brodsky, O., Hickey, M.J., Wells, A., and Cronin, C.N. (2009). Crystal structure of a novel dimeric form of NS5A domain I protein from hepatitis C virus 2. J. Virol. *83*, 4395–4403.

McCormick, C.J., Brown, D., Griffin, S., Challinor, L., Rowlands, J., and Harris, M. (2006). A link between translation of the hepatitis C virus polyprotein and polymerase function; possible consequences for hyperphosphorylation of NS5A. J. Gen. Virol. *87*, 93–102.

Macdonald, A., and Harris, M. (2004). Hepatitis C virus NS5A: tales of a promiscuous protein. J. Gen. Virol. *85*, 2485–2502.

McLauchlan, J. (2009). Hepatitis C virus: viral proteins on the move 11. Biochem. Soc. Trans. *37*, 986–990.

Marcotrigiano, J., and Tellinghuisen, T. (2009). Purification and crystallization of NS5A domain I of hepatitis C virus. Methods Mol. Biol. *510*, 85–94.

Masaki, T., Suzuki, R., Murakami, K., Aizaki, H., Ishii, K., Murayama, A., Date, T., Matsuura, Y., Miyamura, T., Wakita, T., and Suzuki, T. (2008). Interaction of hepatitis C virus nonstructural protein 5A with core protein is critical for the production of infectious virus particles 4. J. Virol. *82*, 7964–7976.

Mathews, N., Thomas, A., Spencer, K., Tiberghien, N., Pilkington, C., Jennens, L., Chana, S., and Fraser, I. (2007). WO 2007/080401 A1. Triazoloanilinopyrimidine derivatives for use as Antiviral Agents.

Migliaccio, G. (2004). Modulation of HCV replication by small molecules. 111th International Symposium on Hepatitis C and Related Viruses, 3–7 October 2004 (Heidelberg, Germany).

Miller, V. (2009). Anticipating a new clinical paradigm for hepatitis C virus treatment: overcoming resistance. The business of Antiviral agents: 2010 and beyond. HepDart. 6–9 December 2009. (Hawaii, USA).

Miyanari, Y., Atsuzawa, K., Usuda, N., Watashi, K., Hishiki, T., Zayas, M., Bartenschlager, R., Wakita, T., Hijikata, M., and Shimotohno, K. (2007). The lipid droplet is an important organelle for hepatitis C virus production 8. Nat. Cell Biol. *9*, 1089–1097.

Najarro, P., Powell, K., Budworth, J., Hallott, A., Griffiths, S., Borger, J., Baxter, B., Mathews, N., Tiberghien., Flack, S., and Cockerill, S. (2006). A-*831*, a novel

HCV inhibitor targeting NS5A, 25–26 October 2006 (Boston, USA).

Najarro, P., Chapman, J., Hallot, A., Baxter, B., Granycome, G., Rupassara, D., Quintet, D., Dixon, K., Salter, J., Bushnell, D., Wheelhouse, C., Carter, M., and Powell, K. (2008). Potent HCV inhibitors targeting NS5A. 2008. 15th International Symposium on Hepatitis C Virus and Related Viruses, 5–9 October 2008 (San Antonio, USA).

Nanda, S.K., Herion, D., and Liang, T.J. (2006). The SH3 binding motif of HCV [corrected] NS5A protein interacts with Bin1 and is important for apoptosis and infectivity 3. Gastroenterology *130*, 794–809.

Neddermann, P., Clementi, A., and De, F.R. (1999). Hyperphosphorylation of the hepatitis C virus NS5A protein requires an active NS3 protease, NS4A, NS4B, and NS5A encoded on the same polyprotein. J. Virol. *73*, 9984–9991.

Neddermann, P., Quintavalle, M., Di, P.C., Clementi, A., Cerretani, M., Altamura, S., Bartholomew, L., and De, F.R. (2004). Reduction of hepatitis C virus NS5A hyperphosphorylation by selective inhibition of cellular kinases activates viral RNA replication in cell culture 14. J. Virol. *78*, 13306–13314.

Nettles, R., Chien, C., Chung, E., Persson, A., Gao, M., Belema, M., Meanwell, N.A., DeMicco, M., Marbury, T.C., Goldwafer, R., *et al.* (2008). BMS-790052 is a First-in-class Potent Hepatitis C Virus (HCV) NS5A Inhibitor for Patients with Chronic HCV Infection: Results from a Proof-of-concept Study. 59th Annual Meeting of the American Association for the Study of Liver Disease, 31 October to 4 November 2008 (San Francisco, USA).

Neumann, A.U., Lam, N.P., Dahari, H., Gretch, D.R., Wiley, T.E., Layden, J., and Perelson, A.S. (1998). Hepatitis C viral dynamics in vivo and the antiviral efficacy of interferon-alpha therapy. Science *282*, 103–107.

Pack, S.K., Geng, P., Smith, M.J., and Hamm, J. (2009). WO 2009/020825.

Palomares-Jerez, M.F., Guillen, J., and Villalain, J. (2010). Interaction of the N-terminal segment of HCV protein NS5A with model membranes. Biochim. Biophys. Acta. doi: 10.1016/j.bbamem.2010.02.007.

Penin, F., Brass, V., Appel, N., Ramboarina, S., Montserret, R., Ficheux, D., Blum, H.E., Bartenschlager, R., and Moradpour, D. (2004). Structure and function of the membrane anchor domain of hepatitis C virus nonstructural protein 5A 5. J. Biol. Chem. *279*, 40835–40843.

Pietschmann, T., Zayas, M., Meuleman, P., Long, G., Appel, N., Koutsoudakis, G., Kallis, S., Leroux-Roels, G., Lohmann, V., and Bartenschlager, R. (2009). Production of infectious genotype 1b virus particles in cell culture and impairment by replication enhancing mutations. PLoS. Pathog. *5*, e1000475.

Pol S., Everson G., Ghalib R., Rustgi V., Martorell C., Tatum H.A., Lim J., Hezode C., Diva U., Yin P.D., and Hindes, R. (2010). Once-daily NS5A inhibitor (BMS-790052) plus peginterferon-alpha-2a and ribavirin produces high rates of extended rapid virological response in treatment-naive HCV-genotype 1 subjects: Phase 2a trial. J. Hepatol. *52*, S462.

Puyang, X., Poulin, D.L., Mathy, J.E., Anderson, L.J., Ma, S., Fang, Z., Zhu, S., Lin, K., Fujimoto, R., Compton, T., and Wiedmann, B. (2010). Mechanism of Resistance of HCV Replicons to Structurally Distinct Cyclophilin Inhibitors 1. Antimicrob. Agents Chemother. *54*, 1981–1987.

Quintavalle, M., Sambucini, S., Di, P.C., De, R., and Neddermann, P. (2006). The alpha isoform of protein kinase CKI is responsible for hepatitis C virus NS5A hyperphosphorylation. J. Virol. *80*, 11305–11312.

Quintavalle, M., Sambucini, S., Summa, V., Orsatti, L., Talamo, F., De, R., and Neddermann, P. (2007). Hepatitis C virus NS5A is a direct substrate of casein kinase I-alpha, a cellular kinase identified by inhibitor affinity chromatography using specific NS5A hyperphosphorylation inhibitors. J. Biol. Chem. *282*, 5536–5544.

Rong, L., and Perelson, A.S. (2010). Treatment of hepatitis C virus infection with interferon and small molecule direct antivirals: viral kinetics and modeling. Crit Rev. Immunol. *30*, 131–148.

Salonen, A., Ahola, T., and Kaariainen, L. (2005). Viral RNA replication in association with cellular membranes. Curr. Top. Microbiol. Immunol. *285*, 139–173.

Schmitz, U., and Tan, S.L. (2008). NS5A--from obscurity to new target for HCV therapy. Recent Pat. Antiinfect. Drug Discov. *3*, 77–92.

Serrano-Wu, M., Belema, M., Snyder, L.B., Meanwell, N., St.Laurent, D., Kakaria, R., Van Nguyen, N., Qui, Y., Yang, X., Leet, J.E., *et al.* (2006). WO 2006/13326 A1.

Tellinghuisen, T.L. (2009). Insights into NS5A: a New Emerging Drug Target. HepDart 2009: Frontiers in Drug Development for viral Hepatitis, 6–9 December 2009 (Hawaii, USA).

Tellinghuisen, T.L., Marcotrigiano, J., Gorbalenya, E., and Rice, C.M. (2004). The NS5A protein of hepatitis C virus is a zinc metalloprotein. J. Biol. Chem. *279*, 48576–48587.

Tellinghuisen, T.L., Foss, L., and Treadaway, J. (2008). Regulation of hepatitis C virion production via phosphorylation of the NS5A protein. PLoS. Pathog. *4*, e1000032.

Torres-Puente, M., Cuevas, J.M., Jimenez-Hernandez, N., Bracho, M.A., Garcia-Robles, I., Carnicer, F., del, O.J., Ortega, E., Moya, A., and Gonzalez-Candelas, F. (2008a). Hepatitis C virus and the controversial role of the interferon sensitivity determining region in the response to interferon treatment 11. J. Med. Virol. *80*, 247–253.

Torres-Puente, M., Cuevas, J.M., Jimenez-Hernandez, N., Bracho, M.A., Garcia-Robles, I., Wrobel, B., Carnicer, F., del, O.J., Ortega, E., Moya, A., and Gonzalez-Candelas, F. (2008b). Genetic variability in hepatitis C virus and its role in antiviral treatment response. J. Viral Hepat. *15*, 188–199.

Tsai, Y.H., Kuang, W.F., Lu, T.Y., Kao, J.H., Lai, M.Y., Liu, C.J., Chen, J., and Hwang, L.H. (2008). The nonstructural 5A protein of hepatitis C virus exhibits genotypic differences in interferon antagonism 100. J. Hepatol. *49*, 899–907.

Wakita, T., Pietschmann, T., Kato, T., Date, T., Miyamoto, M. Zhao, Z., Murthy, K., Habermann, A. Kräusslich, H.G., Mizokami, M., *et al.* (2005). Production of infectious hepatitis C virus in tissue culture from a cloned viral genome. Nat. Med. *11*, 791–796.

Wang, A.G., Lee, D.S., Moon, H.B., Kim, J.M., Cho, K.H., Choi, S.H., Ha, H.L., Han, Y.H., Kim, D.G., Hwang, B., and Yu, D.Y. (2009). Non-structural 5A protein of hepatitis C virus induces a range of liver pathology in transgenic mice. J. Pathol. *219*, 253–262.

Wheelhouse, C., Thomas, J., Bushnell, D., Lumley, J., Salter, J., Carter, M., Mathews, N., Pilkington, C., and Angell, R. (2007). WO 2007/031791 A1. Biphenyl Derivatives and their use in treating Hepatitis C.

Wobbe, C.R., Cao, J., Cho, H.-J., Fathi, R., Lam, A., Liao, Y., Nawoschik, K., Sandrasagra, A., Westby, G., Wu, R., Yanng, Z., and Zhu, Q. (2007). Potent small molecule HCV inhibitors affecting NS5A-dependent functions. 14th International Symposium on Hepatitis C Virus and Related Viruses, 9–13 September 2007(Glasgow, UK).

Nucleoside Inhibitors of Hepatitis C Virus

15

Klaus Klumpp and Mark Smith

Abstract

Nucleoside analogues have transformed our ability to treat viral diseases and have contributed to a significant reduction in morbidity, mortality and suffering worldwide caused by viral infections, including human immunodeficiency virus (HIV), hepatitis B virus and herpesvirus infections. Similarly, nucleoside analogues hold great promise to also become the backbone of the future standard of care for the treatment of hepatitis C virus infection, because the currently available data suggest that nucleoside analogues can provide high antiviral potency, a high barrier to resistance selection, antiviral potency across the whole spectrum of hepatitis C virus genotypes and favourable properties for combination with other antiviral agents. This chapter summarizes efforts in antiviral research that have led to the discovery of valopicitabine (NM283), balapiravir (R1626) and R7128 (RG7128). Current efforts aim to further improve efficacy by optimizing nucleoside intrinsic potency, phosphorylation efficiency and delivery of antiviral nucleoside analogues.

Introduction

Nucleoside analogues have had a major impact on the treatment viral diseases and have become standard of care for the treatment of human immunodeficiency virus (HIV), hepatitis B virus (HBV), and herpes virus infections. Nucleoside analogues typically target the viral polymerase enzymes and inhibit the viral replication process. Most nucleoside analogues are metabolized by cellular enzymes to their 5′-triphosphate derivatives, which are selective substrates for the viral polymerases and cause chain termination after incorporation (Fig. 15.1). A number of nucleoside analogues have also been explored for their potential to provide safe and efficacious new treatment options for hepatitis C virus (HCV) infection. Most of these HCV targeted nucleosides have been based on a sugar ribose backbone, in an attempt to facilitate recognition by the HCV RNA polymerase. In contrast, nucleosides targeting HIV, HBV or herpesvirus polymerases were mostly based on deoxyribose sugar backbone or acyclic ribose replacements, to allow recognition by the viral DNA polymerases. Selective nucleoside analogues that target RNA synthesis by the HCV RNA-dependent polymerase can provide a unique profile of combined features of both direct antiviral potency and high barrier to resistance. The combination of these features is expected to be ideal for achieving effective suppression of HCV replication for an extended duration of treatment required to achieve sustained viral response (SVR) without breakthrough of drug resistant virus.

New antiviral agents targeting HCV replication face the challenge of a remarkably high genetic diversity of HCV. Virus isolates from different genotypes differ in approximately 30–35% of nucleotides on average across the genome, whereas those from different subtypes differ approximately 20–25%. The level of sequence difference between genotypes is therefore almost similar to that of serologically and genetically distinct viruses within the genus *Flavivirus* (Bukh *et al.*, 1995; Simmonds, 2004). This remarkable

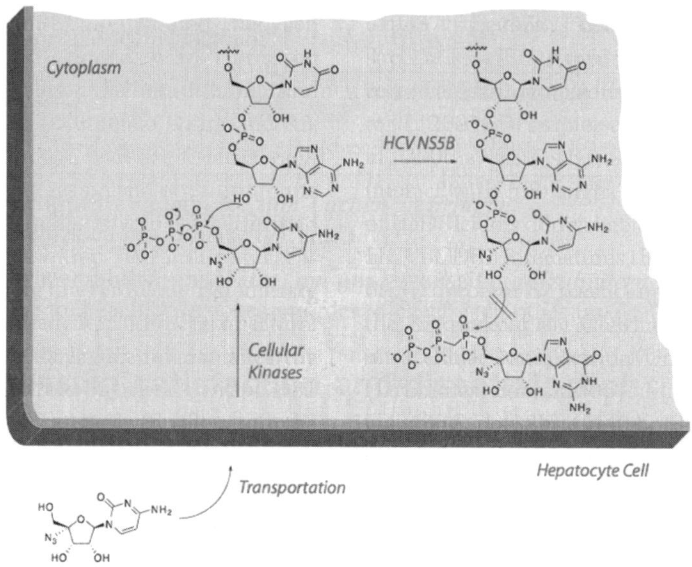

Figure 15.1 Uptake, intracellular metabolism, incorporation and chain termination are key steps in the process of typical nucleoside analogue mediated inhibition of HCV replication. Balapiravir (4′-azido-cytidine) is shown as an example. HCV polymerase catalyses the nucleophilic substitution at the -phosphate of the nucleoside triphosphate substrate with the 3′-hydroxyl-group of the growing RNA chain as the nucleophile and the β-γ-pyrophosphate as a leaving group. If a modified nucleoside analogue, such as 4′-azidocytidine, is incorporated into new RNA, further RNA chain elongation is blocked, most likely because of geometrical misalignment of the 3′-hydroxyl-group in the polymerase active site.

sequence diversity represents a true challenge to the design of small molecule replication inhibitors that could bind with high affinity to viral targets across the different HCV genotypes and subtypes. Also within infected subjects the sequence variability of HCV is substantial. The RNA-dependent RNA polymerase NS5B introduces nucleotide sequence changes at a rate of approximately 1 change per 1–10 genomes of ~10^4 nucleotides length, and HCV infected subjects generate an estimated 10^{12} virus particles per day (Neumann *et al.*, 1998). Therefore, all possible single and double amino acid mutations across the viral proteins can be expected to be generated daily at low frequency. The virus populations in HCV infected subjects are thus highly heterogeneous in sequence, forming a 'quasispecies', where at any moment in time one or a few sequences will be dominant. The sequence diversity among individual virus particles generated daily in this quasispecies allows rapid adaptation to changes in selection pressure and rapid drug resistance selection and thus constitutes a considerable challenge to antiviral drug discovery.

The high sequence diversity of the HCV quasispecies allows the virus to adapt to the presence of antiviral drugs. Exposure to HCV protease inhibitors or non-nucleoside polymerase inhibitors in monotherapy leads to the rapid selection of virus variants that can replicate in the presence of the drugs (Bartels *et al.*, 2008; Curry *et al.*, 2008; Kuntzen *et al.*, 2008; Le Pogam *et al.*, 2008; Kim *et al.*, 2009; Thompson and McHutchison, 2009). Most resistant virus variants have been detected as minority species at baseline. The presence of the antiviral drug effectively inhibits replication of drug sensitive virus variants within the quasispecies, such that drug-resistant virus variants become dominant in the virus population. For potent antiviral drugs therefore, a typical clinical resistance profile is characterized by an initial steep viral load reduction as the replication of the dominant sensitive virus is abolished, followed by a rebound in viral load as the drug resistant virus variants continue replicating and multiply from a minority species to eventually become the dominant virus in the quasispecies. Drug-resistant virus variants can also be generated de novo under drug pressure. This process requires ongoing virus replication in the presence of the drug and can occur, when the antiviral drug exposure is suboptimal and does not achieve complete inhibition of viral replication. The clinical barrier to resistance describes the likelihood for a drug resistant virus variant to emerge during antiviral treatment. The barrier to resistance is a function of the frequency of naturally occurring drug resistant virus variants at baseline, the replication capacity of resistant virus variants, the resistance fold shift associated with the resistance mutation and the exposure to the active drug at the end of the dosing interval relative to the intrinsic antiviral potency of the drug. The latter is often expressed as the inhibitory quotient. Thus, antiviral drugs with lower intrinsic antiviral potency (EC_{50}) can achieve high clinical efficacy if they reach high exposure, whereas antiviral agents with high intrinsic antiviral potency *in vitro* may be associated with low clinical efficacy and low barrier to resistance if they only achieve suboptimal exposure (Sweeney and Klumpp, 2008). Even highly potent antiviral agents with high exposure may still have a low barrier to resistance, if naturally resistant virus variants exist at baseline, which can still replicate in the presence of such drug concentrations. In addition to intrinsic potency, exposure, and absence of naturally resistant virus variants at baseline, other factors can also contribute to the barrier to resistance: most importantly, these are the replication capacity of drug resistant variants (fitness) and the relative impact of resistance mutations on the intrinsic inhibitory potency of the drug (resistance fold shift). Compounds with high intrinsic potency, that achieve high exposure relative to their antiviral potency, and for which naturally occurring drug resistant variants are rare in the quasispecies therefore have an increased likelihood of achieving a high barrier to resistance, especially, if resistance mutations substantially reduce virus replication fitness. Also, if the fold-shift of resistance mutations on antiviral potency is low, drugs at standard dose or slightly higher dose may achieve sufficiently high exposure to effectively prevent replication of such virus variants. Nucleoside analogues that inhibit HCV replication can achieve a high barrier to resistance through a combination of properties, including favourable pharmacokinetic/pharmacodynamic properties (ability to achieve a high inhibition quotient), very low prevalence of naturally resistant virus variants, low impact of resistance mutations on inhibition potency (low antiviral susceptibility fold-shift) and low replicative fitness of resistant mutant variants.

This chapter summarizes key information on nucleoside analogues that have been evaluated as inhibitors of HCV replication.

4′-Substituted nucleosides

The antiviral potential of 4′-substituted nucleoside analogues was first described for inhibition of HIV replication. Researchers at Syntex, Palo Alto, established that 4′-modified deoxyribonucleoside analogues could be efficient substrates for DNA synthesis by HIV reverse transcriptase and could induce chain termination despite the presence of a 3′-hydroxyl group (Chen *et al.*, 1992; Maag *et al.*, 1992). A large effort during the 1990s was placed on establishing cell-based systems to study HCV replication, with eventual first success described by Ralf Bartenschlager at the University of Mainz who established the HCV replicon system, based on a self-replicating bicistronic RNA that expressed the neomycin resistance gene as a selection marker from the HCV internal ribosomal entry site (IRES) and expressed the HCV non-structural genes from a downstream EMCV-IRES (Lohmann *et al.*, 1999). This breakthrough was followed more recently by the discovery of a HCV genotype 2 isolate capable of efficient replication and virion production in cell culture, the HCVcc system (Wakita *et al.*, 2005). These systems have since then been widely used to screen compound libraries, including nucleoside analogue libraries to identify inhibitors of HCV replication (Klumpp and Smith, 2010).

4′-Azidocytidine 2 (R1479) was discovered within the group of nucleoside analogues that demonstrated antiviral potency without apparent cytotoxic or cytostatic properties. **2** is a potent inhibitor of HCV replication in the HCV replicon system and mediates inhibition of RNA chain extension after incorporation by HCV polymerase into nascent RNA molecules (Klumpp *et al.*, 2006). A comparison of a number of 4′-substituted uridine and cytidine analogues was also published, and a summary is shown in Table 15.1 (Smith *et al.*, 2007). Most substitutions at the 4′-position resulted in nucleoside analogues that were inactive and non-toxic in the cell-based HCV replicon system. 4′-Propynyl uridine showed weak antiviral potency in the replicon system (76% inhibition at a concentration of 100 μM), with low cytotoxicity, suggesting that alkynyl substitutions may be further optimized to provide selective HCV replication inhibitors. Interestingly, 4′-azido uridine **1** and 4′-ethynylcytidine **7** displayed no antiviral potency in the replicon system. This was in contrast to their 2′-deoxy derivatives (2′-deoxy-4′-azidouridine and 2′-deoxy-4′-ethynylcytidine), which were potent inhibitors of HIV replication, targeting HIV reverse transcriptase and inducing DNA chain termination (Maag *et al.*, 1992; Ohrui *et al.*, 2000). The lack of activity displayed by many 4′-substituted nucleosides was investigated further. The triphosphates of 4′-azidouridine **1** and 4′-ethynylcytidine **7** were prepared and evaluated in the NS5b assay. Both compounds were shown to be inhibitors of RNA synthesis (4′-azidouridine-TP $IC_{50}=0.30$ μM; 4′-ethynylcytidine-TP $IC_{50}=2.7$ μM). These results suggested a block in either cellular uptake or phosphorylation in HCV replicon cells for 4′-azidouridine and 4′-ethynylcytidine (Smith *et al.*, 2007).

4′-Cyanocytidine **5** showed cytotoxicity in the HCV replicon system ($CC_{50}=20$ μM), a feature reminiscent of other 4′-cyano nucleosides that had previously been evaluated as inhibitors of HIV replication (Oyang *et al.*, 1992). 4′-Allyluridine **11** also displayed cytotoxicity in the replicon

Table 15.1 Evaluation of 4′-substituted nucleosides (Smith *et al.*, 2007)

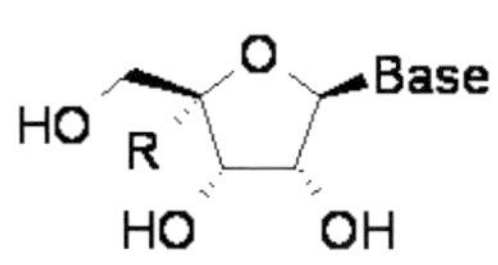

Entry	R	Base	Replicon inhibition at 20 μM[a]	Cytotoxicity at 20 μM[b]
1	N_3	U	13%	17%
2	N_3	C	97% ($EC_{50}=1.28$ μM)	17% (CC_{50} = >2000 μM
3	Et	C	20%[c]	20%[c]
4	EtO	C	11%	0%
5	C≡N	C	99%	100%
6	C≡N	U	0%	0%
7	C≡CH	C	3%	0%
8	C≡CH	U	0%	0%
9	C≡CMe	U	76%[c]	19%[c]
10	C≡CMe	C	11%	0%
11	Allyl	U	22	29
12	Allyl	C	0%[c]	0%[c]
13	Vinyl	C	53%[c]	0%[c]
14	Vinyl	U	50%[c]	0%[c]

[a]GT-1b replicon (2290–23 cells).

[b]Reduction of 2209–23 cell viability, determined using WST-1 reagent.

[c]Activity at 100 μM.

system (CC_{50} = 39 μM). It is interesting that the uridine analogue of **5**, 4′-cyanouridine **6** did not show any cytotoxicity, while the cytidine analogue **5** was cytotoxic. In contrast, among the 4′-allyl pair, it was the uridine analogue (4′-allyluridine **11**) that was cytotoxic, with no apparent cytotoxicity associated with 4′-allylcytidine **12**. Further evaluation of the intracellular metabolism of these compounds and their potential cellular targets would be required to elucidate the very compound specific observations of antiviral potency and selectivity.

The 5′-triphosphate of 4′-azidocytidine (**2**-TP) was prepared and shown to inhibit viral RNA synthesis by HCV NS5b polymerase, in a template base-dependent manner, consistent with competitive inhibition of cytidine monophosphate incorporation (Klumpp 2006). **2**-TP inhibited RNA synthesis by HCV polymerase with an IC_{50} of 0.29 μM, while the parent nucleoside **2** did not inhibit RNA synthesis up to a concentration of 1 mM, consistent with the requirement of the phosphorylated derivative as a substrate for incorporation by HCV NS5b. 4′-Azidocytidine-5′-triphosphate is therefore a substrate of HCV polymerase NS5b and is incorporated into HCV RNA in a template sequence-dependent manner by base-pairing to guanosine residues on HCV genomic RNA. Incorporation into nascent HCV RNA molecules by HCV polymerase leads to inhibition of RNA chain extension (Klumpp *et al.*, 2006).

4′-Azidocytidine was selected for development for the treatment of HCV infection and designated R1479. Oral bioavailability of R1479 could be significantly enhanced by administering a 2′,3′,5′-triisobutyrate prodrug of R1479 (Balapiravir, R1626). This tri-ester derivative of R1479 reduced the polarity of the nucleoside analogue and therefore improved uptake of the R1479 in the gastrointestinal tract. A comprehensive summary of R1479 prodrug optimization, pharmacokinetic/pharmacodynamic and clinical development of balapiravir is currently being published (Klumpp and Smith, 2010). Balapiravir showed dose and time-dependent reductions in plasma HCV RNA, when dosed as monotherapy for 14 days in treatment naive, HCV genotype 1 infected persons (Roberts *et al.*, 2006, 2008). No difference in response in persons infected with HCV subtypes 1a and 1b was observed, consistent with broad *in vitro* potency of R1479-triphosphate against all HCV genotypes (Leveque *et al.*, 2008). Dose-dependent antiviral effects of balapiravir in individual HCV infected persons were measured as reduction in plasma viral load and after 14 days ranged between 0 and 4.4 log IU/ml with mean viral load reduction values of 0.32, 1.2, 2.6 and 3.7 log IU/ml at doses of 500, 1500, 3000 or 4500 mg b.i.d. respectively. 5 of 9 patients in the 4500 mg b.i.d. dose group achieved more than 4 log viral load reduction within 14 days of balapiravir monotherapy.

Interestingly, a larger than expected enhancement of antiviral potency was observed in HCV infected persons who received balapiravir in combination with pegylated interferon alpha 2a (PEG) and ribavirin (RBV) (Pockros *et al.*, 2008b; Table 15.2). A significant contribution to overall antiviral potency at week 2 of treatment was observed for all three components, PEG, RBV and balapiravir (Table 15.2). 52% of subjects in the balapiravir 1500 mg b.i.d. + PEG + RBV TRIPLE1500 arm achieved HCV RNA levels <50 IU/ml and 42% <15 IU/ml at week 2, while 81% and 71% achieved these levels at week 4 (RVR; Pockros *et al.*, 2008b; Klumpp and Smith, 2010). Mean viral load reduction in the TRIPLE1500 arm at week 2 was 4.6 log IU/ml, compared with 1.6 log IU/ml in the PEG + RBV arm, 1.2 log IU/ml in the balapiravir 1500 mg b.i.d. monotherapy arm and 3.1 log IU/ml in the balapiravir 1500 mg b.i.d. + PEG dual combination arm. The observed efficacy synergy of the three components in the TRIPLE1500 dose group was consistent with *in vitro* analysis in the HCV replicon system, which showed antiviral synergy when each of the three components were combined separately (Jiang *et al.*, 2006). The molecular mechanism associated with this apparent antiviral synergy is unresolved, but its elucidation may contribute to the design of more efficacious combination therapies for the treatment of HCV infection.

The combination of balapiravir at the 1500 mg b.i.d. dose level with standard PEG and RBV was unfortunately associated with significant hematologic changes in HCV infected persons. Ninety per cent of subjects in the balapiravir + PEG +

Table 15.2 Viral load reduction after 14 days of monotherapy with balapiravir (R1626) or combination therapy of balapiravir with pegylated interferon-α2a (PEG) and/or ribavirin (RBV) in treatment-naive, genotype 1-infected persons[a]

	R1626 1500 mg b.i.d. monotherapy	R1626 1500 mg b.i.d. + PEG	R1626 1500 mg b.i.d. + PEG +RBV	PEG	PEG + RBV (SOC)
Mean viral load reduction at week 2 ($\log_{10}$ IU/ml)	**1.2**	**3.1**	**4.6**	**0.5[b]**	**1.6**
Median (range)	0.8 (0.49–2.46)	3.1 (0.0–5.5)	4.7 (2.5–6.6)		1.1 (0.1–3.8)
Additive effect[c]	1.2	1.7	2.8	0.5	0.5[d]

[a]Pockros *et al.* (2008b).

[b]Estimated, based on Herrmann E *et al.* (2003).

[c]adding monotherapy effects from the combination therapy components observed at week 2; monotherapy of PEG based on Herrmann *et al.* (2003).

[d]Lack of effect of RBV on plasma HCV RNA at week 2 (Lee *et al.*, 1998; Pawlotsky *et al.*, 2004).

RBV triple arm had dose modifications and 39% discontinued balapiravir before the end of the 4 week dosing period, in the majority of cases as a result of grade 4 neutropenia (Pockros *et al.*, 2008a,b). A further dose finding phase IIB study exploring different doses of balapiravir and PEG in combination with RBV also unveiled a progressive reduction in lymphocytes that became apparent with longer term dosing. The development of balapiravir for the treatment of HCV infection was discontinued.

Analogues of 4′-azidocytidine

Pharmacokinetic/pharmacodynamic analysis of balapiravir suggested that the maximal antiviral effect was not achieved because of dose limiting haematological toxicity. Further studies therefore focused on improving intrinsic antiviral potency of related nucleoside analogues. Very high antiviral potency was obtained with 4′-modified arabino nucleosides (Klumpp *et al.*, 2008). This finding was unexpected, because these nucleosides are 2′-deoxy nucleoside analogues and were therefore not expected to be accepted as substrates for RNA polymerases, such as HCV polymerase. Further structure–activity relationships in this series were also explored (Smith *et al.*, 2009a,b). The impact of 4′-substitution on selectivity of these nucleoside analogues was remarkable. The prototype arabino nucleoside, AraC **15** (arabinocytidine or cytarabine), is a potent inhibitor of cell proliferation. AraC inhibited tritiated thymidine incorporation by HCV replicon cells with an IC_{50} value of 29 nM and thus induced non-specific reduction of HCV replicon RNA in this system, as HCV replicon maintenance in the HCV replicon system is dependent on cell proliferation (Klumpp *et al.*, 2008; Smith *et al.*, 2009b). AraC **15** is a well characterized cytostatic agent and is approved for the treatment of acute myelogenous lymphoma. The introduction of a 4′-azido group converted AraC **15** to 4′-azido-aracytidine **16**, a potent compound in this series. **16** inhibited HCV replication with an EC_{50} value of 171 nM in 2209–23 cells, but did not inhibit cell proliferation at concentrations up to 100 μM. Therefore, the 4′-azido moiety conferred more than 3000-fold increased selectivity when comparing **16** to AraC, and at the same time increased antiviral potency 7.5-fold relative to balapiravir (prodrug of R1479, **2**). As mentioned earlier, the high potency of **16** as an inhibitor of HCV replication was unexpected as it lacks a 2′-α-hydroxy group which is generally involved in specific hydrogen bonding interactions with RNA polymerases and necessary for ribonucleoside recognition and efficient incorporation by RNA polymerases. The 5′-triphosphate of **16** was demonstrated to be an efficient substrate of HCV polymerase despite the absence of a 2′-hydroxyl group and incorporation into viral RNA resulted in chain termination despite the presence of a 3′-hydroxyl group (Klumpp *et al.*, 2008). This was similar to results obtained with 4′-azidocytidine **2**, suggesting that the 4′-azido group may generally lead to misalignment of the 3′-hydroxyl group in the HCV polymerase active site. It is unknown if chain termination will occur under all cellular and/or *in vitro* conditions, or if mutations within HCV polymerase may affect chain termination efficiency of nucleoside analogues. It was therefore of interest to investigate if obligatory chain terminators that lack the 3′-hydroxyl group may be converted into specific inhibitors of HCV replication. The prototype compound 3′-deoxy-CTP has been described as an efficient substrate of HCV polymerase and obligatory chain terminator, but lacks selectivity against human polymerases. Also, 3′-deoxy-cytidine is not converted to its 5′-triphosphate derivative by human kinases in HCV replicon cells (Shim *et al.*, 2003; Ma *et al.*, 2005). 4′-Azido-3′-deoxycytidine **18** was synthesized to evaluate the possibility of deriving an obligatory chain terminator directly from 4′-azidocytidine. Interestingly, **18** was completely inactive in the HCV replicon system (Table 15.3). Similar to what is known for 3′-deoxycytidine, **18** may therefore not be phosphorylated by cellular kinases to the 5′-triphosphate derivative. In contrast to **18**, **17** (4′-azidoxylocytidine) did show measurable antiviral potency in the replicon

Table 15.3 Structure activity relationship of 4′-azidocytidine analogues (Smith *et al.*, 2009a,b)

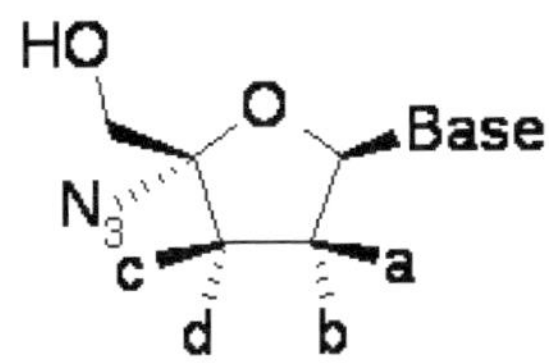

Entry	a	b	c	d	e	EC_{50} (μM)[a]	TP data, K_i (μM)[b]	CC_{50} (μM)[c]	Proliferation, IC_{50} (μM)[d]
2	H	OH	H	OH	N_3	1.28	0.04	>100	>100
15	OH	H	H	OH	H	0.020	0.09	>100	0.029
16	OH	H	H	OH	N_3	0.171	0.011	>1000	>100
17	H	OH	OH	H	N_3	27	nd	>100	nd
18	H	H	H	OH	N_3	>100	nd	nd	nd
19	Me	OH	H	OH	N_3	>100	nd	>100	nd
20	F	H	H	OH	H	0.114	nd	>100	0.174
21	F	H	H	OH	N_3	0.024	0.033	>100	>100
22	H	F	H	OH	N_3	4.02	nd	>100	>100
23	H	OH	F	H	N_3	>100	nd	>100	nd
24	H	OH	H	F	N_3	>100	nd	>100	nd
25	F	F	H	OH	N_3	0.066	0.19	>100	nd
Gemcitabine[e]	F	F	H	OH	H	0.0035	nd	~0.1	0.0075

[a]GT-1b replicon (2290–23 cells).

[b]HCV NS5B-570-BK-1b (Klumpp *et al.*, 2006, 2008).

[c]2209–23 cells viability, determined using WST-1 reagent.

[d]Inhibition of tritiated thymidine incorporation into cellular DNA of 2209–23 cells.

[e]Unpublished data.

nd: not determined.

system (EC_{50} = 27.8 μM), suggesting that cellular potency with obligatory chain terminators may be achievable. However, this potency was more than 20-fold lower than that of **2** (4′-azidocytidine), consistent with an important role for the 3′-hydroxyl group in the intracellular formation of active nucleoside triphosphate analogues (Smith *et al.*, 2009b).

4′-Azido-2′-C-methylcytidine **19** was generated as a hybrid of **2** and 2′-C-methylcytidine (**27**), both compounds that have progressed into clinical development. Surprisingly, **19** was found to be inactive in the replicon system, most likely due to lack of phosphorylation to the 5-triphosphate. This was interesting, as both **2** and **27** are phosphorylated to form their respective 5′-triphosphate derivatives in HCV replicon cells. However, **2** and **27** are phosphorylated by different cellular enzymes. Whereas **2** is a substrate for deoxycytidine kinase (dCK), **27** is phosphorylated by either uridine-cytidine kinase 1 or 2 (UCK-1, UCK-2; Golitsina *et al.*, 2010; Klumpp *et al.*, 2008). The combination of the structural features of 2′-hydroxy-2′-methyl and 4′-azido moieties on a single molecule **19** therefore resulted in the apparent exclusion of **19** at both the dCK and UCK-1/UCK-2 active sites.

The addition of 4′-modification to 2′-fluoro nucleoside analogues also had a dramatic effect on compound selectivity. Gemcitabine (2′-deoxy-2′,2′-difluorocytidine) is a highly cytotoxic cytidine analogue and approved for the treatment of several cancer indications including non-small cell lung cancer and pancreatic cancer. CC_{50} values for gemcitabine in human hepatoma cells have been reported to be in the nM range (Sabini *et al.*, 2003; Veltkamp *et al.*, 2008). The related monofluoro compound 2′-deoxy-2′-fluoro arabinocytidine (dFAraC, **20**) was also a highly potent inhibitor of cell proliferation (IC_{50} = 174 nM; Klumpp *et al.*, 2008). Introduction of the 4′-azido moiety into dFAraC **20** resulted in 4′-azido-2′-deoxy-2′-fluoroaracytidine **21**, which did not show any measurable inhibition of cell proliferation in HCV replicon cells. The introduction of the 4′-azido moiety therefore increased the selectivity by over 4000-fold when comparing **21** to **20**. In addition, **21** showed exceptionally high antiviral potency in the HCV replicon system and is the most potent nucleoside inhibitor of HCV that is currently known. With an antiviral EC_{50} value of 24 nM the potency of **21** was improved more than 50-fold compared with 4′-azidocytidine and 7-fold compared with 4′-azidoaracytidine **16**. Unexpectedly therefore, the lack of a 2′-hydroxyl group on nucleoside analogues was not only acceptable for binding and incorporation by the HCV RNA polymerase, but such 2′-deoxy-nucleoside analogues could become highly potent without apparent effects on cell viability and proliferation in the HCV replicon system. Interestingly, inversion of the 2′-fluoro group, to form 4′-azido-2′-deoxy-2′-fluorocytidine **22**, reduced potency by >150-fold compared with **21**. It has not yet been determined if this reduction in potency is due to reduced phosphorylation efficiency of **22** or whether incorporation efficiencies by HCV polymerase are different between **21** and **22** triphosphates. The related compound 2′-deoxy-2-fluorocytidine (FdC) **33** has also been characterized as an inhibitor of HCV replication. FdC was found to be a potent inhibitor of cell proliferation with an IC_{50} of 0.8 μM (Clark *et al.*, 2005). Therefore the introduction of the 4′-azido moiety into FdC to form **22** also abolished cell proliferation inhibition and increased selectivity by more than 125-fold.

As the 2′-difluoro motif of gemcitabine is accepted by human kinases and polymerases, it was of interest to evaluate the effect of 4′-substitutions on the antiviral and cytotoxicity profile of such gemcitabine derived compounds. As shown in Table 15.3, the introduction of the 4′-azido moiety (4′-azido-2′-dideoxy-2′,2′-difluorocytidine **25**) resulted in a very interesting compound with high antiviral potency in the HCV replicon system (EC_{50} = 66 nM) and no apparent cytotoxicity, suggesting a substantial increase in selectivity compared with gemcitabine (Smith *et al.*, 2009a). The effect of the 4′-azido moiety to confer high selectivity has therefore been consistent through **2**, **16**, **21**, **22** and **25**.

3′-Fluoro analogues of 4′-azidocytidine, **23** and **24**, were completely inactive as well as non-toxic in the replicon system (Smith *et al.*, 2009a). The 3′-hydroxyl group was suggested from structural studies with deoxycytidine kinase to be important for productive binding to the kinase, as hydrogen bonds are formed with E197 and Y86 in the human deoxycytidine kinase active site (Alexander *et al.*, 2003; Sabini *et al.*, 2003, 2008). Removal of the 3′-hydroxy group in **23** and **24** may prevent these compounds from binding productively to human dCK and thus from being phosphorylated to their triphosphate derivatives.

Historically, it has been hypothesized that the conformation a nucleoside adopts in solution may be related to its antiviral activity in cell and may therefore be used as a predictive measure in nucleoside design. Specifically, anti-HIV activity was observed in 2′-deoxy-2′-fluoro nucleosides that preferentially assume the Southern conformation in solution (Marquez *et al.*, 1993). It was therefore of interest to determine whether the addition of a 4′-azido moiety conferred any conformational bias on the parent nucleoside, potentially locking it in either a Northern or Southern conformation. Previously, it was suggested that 4′-azido-2′-deoxy nucleosides prefer to adopt the Northern conformation in solution (Prisbe *et al.*, 1993). It should be noted that cytidine itself forms a 3:2 mixture of Northern (C3′-*endo*) and Southern (C2′-*endo*) isomers in solution whereas 2′-deoxycytidine preferentially adopts the Southern conformation, in common with the three other natural deoxyribonucleosides (Altona and Sundaralingam, 1972).

4′-Azidocytidine **2** adopted the Northern conformation in deuterated water, similar to what had been observed with 4′-azido-2′-deoxy-nucleoside analogues (Table 15.4). This conformation remained the same when the experiment was performed in the presence of deoxycytidine kinase (dCK), to test a possible influence of kinase active site binding on nucleoside conformation. Similar results were obtained with the 2′,2′-difluroro analogue **25**. In contrast, the compounds **16** (4′-azido aracytidine) and **21** (4′-azido-2′-deoxy-2′-fluoro aracytidine) did appear to have a preference for the Southern conformation in D_2O, which can be attributed to the gauche effect caused by the 2′-substituent prevailing over the anomeric-like effect of the azide moiety at the 4′-position. Interestingly, both these compounds **16** and **21** adopted the Northern conformation in the presence of dCK, suggesting active site binding to the kinase in the northern conformation, similar to **2**. In fact, all the 4′-azido nucleosides that have been determined to be phosphorylated by dCK adopted the northern conformation in the presence of dCK, consistent with a requirement for an active site structural fit in the northern conformation. The 4′-azido-3′-deoxy-3′-fluorocytidine analogues **23** and **24** both adopted the southern conformation in the presence of dCK. Therefore,

Table 15.4 Conformational analysis of 4′-azido nucleosides (Maltseva *et al.*, 2008)

Entry	Conformation in D_2O solution	GT-1b replicon (2209–23) EC_{50} (μM)	Conformation in presence of dCK based on nOe
2	Northern	1.28	Northern
16	Southern (N:S 40:60)	0.171	Northern
21	Southern (N:S 20:80)	0.024	Northern
22	Northern (N:S 90:10)	4.02	Northern
23	(N:S 50:50)	>100	Southern
24	Southern (N:S 20:80)	>100	Southern
25	Northern	0.066	Northern

within the 4′-azidocytidine nucleoside series of compounds, it is not possible to predict antiviral activity based upon the preferred solution conformation (Maltseva *et al.*, 2008). However, the ability of the target human kinase to induce an appropriate conformation for phosphorylation deserves further investigation as a predictive measure for the ability to form 5′-phosphorylated derivatives in human cells.

2′-Modified nucleosides

2′-Modified nucleosides have also shown significant promise as selective inhibitors of HCV replication. To date, 2′-methyl nucleosides have been the best characterized HCV replication inhibitors. These compounds were initially discovered as promising antiviral agents using bovine viral diarrhoea virus (BVDV) as a viral test system. Compounds of this class were subsequently shown to also have potential as inhibitors of HCV replication (Carroll and Olsen, 2006; Sorbera and Castaner, 2006).

Analogous to studies in the 4′-azido nucleoside series, uridine analogues with substitution at the 2′-position, such as **26**, showed no antiviral properties in the replicon system because of inefficient phosphorylation by cellular enzymes (Furman *et al.*, 2009). 2′-*C*-methylcytidine (NM107) **27** shows similar potency compared with 4′-azidocytidine with an EC_{50} of 1.23 μM, high selectivity in cell culture and was evaluated as a clinical candidate for the treatment of HCV infection (Klumpp *et al.*, 2008; Sorbera and Castaner, 2006). The 3′-valine prodrug (NM283) of **27** (Fig. 15.2) provided significantly improved oral bioavailability of NM107 compared with that achieved by dosing of NM107 itself, and NM283 was therefore selected for full clinical development (Pierra *et al.*, 2005, 2006; Sorbera and Castaner, 2006). In a monotherapy study of NM283 in HCV infected persons, mean HCV viral load reductions of 0.15, 0.39, 0.3, 0.67 and 1.21 log IU/ml were observed at 2 weeks with 50, 100, 200, 400 and 800 mg/day NM283 (Sorbera and Castaner, 2006). Viral load reductions at the highest dose ranged from 0.41 to 2.37 log in individual patients. NM283 was associated with dose limiting gastrointestinal disturbances, nausea and vomiting, in particular in persons receiving doses of 400 mg/day or above. Eventually, the development of NM283 was discontinued, because no significant benefit beyond standard of care PEG + RBV could be demonstrated in NM283-based triple therapy at sufficiently well tolerated dose levels.

2′-*C*-Methyladenosine **29** also displayed high antiviral potency in the replicon assay with an EC_{50} value of 0.26 μM (Carroll *et al.*, 2003; Migliaccio *et al.*, 2003; Eldrup *et al.*, 2004b; Table 15.5). **29** was however found to be a substrate for adenosine deaminase and purine nucleoside phosphorylase, resulting in low metabolic stability and low oral bioavailability. The potency of 2′-*C*-methylguanosine **28** in the HCV replicon (EC_{50} = 3.5 μM) was lower than expected from the intrinsic polymerase inhibition potency of the 5′-triphosphate derivative of **28**, as **28** triphosphate was a more potent inhibitor of HCV polymerase compared with the triphosphate of **29** (Eldrup *et al.*, 2004a,b). Inefficient phosphorylation of **28** appears to limit the antiviral potential of **28**, which was not further pursued as a candidate. Instead, efforts have focused on designing 5′-monophosphate prodrugs of **28**, which may

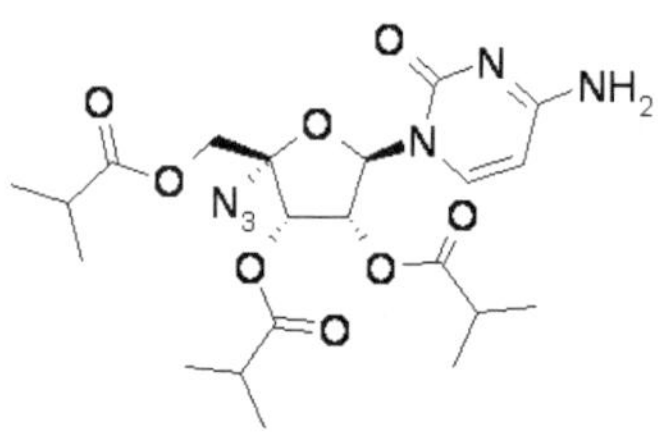

R1626

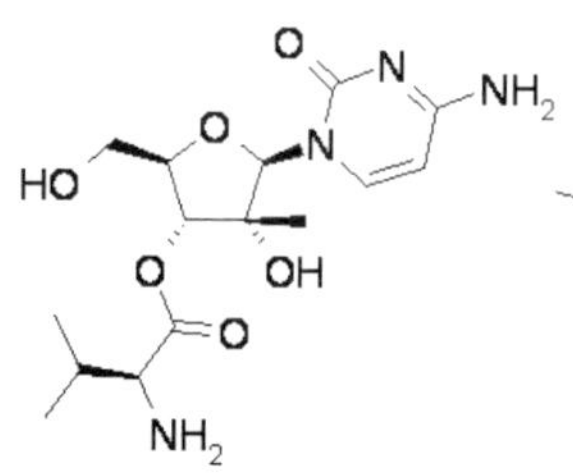

NM283

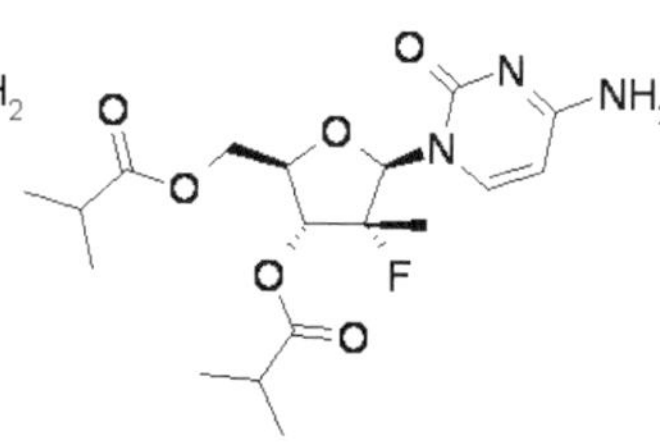

R7128

Figure 15.2 Nucleoside prodrugs that have shown efficacy for the treatment of HCV infection in clinical trials.

Table 15.5 Structure activity relationship of 2′-modified nucleosides

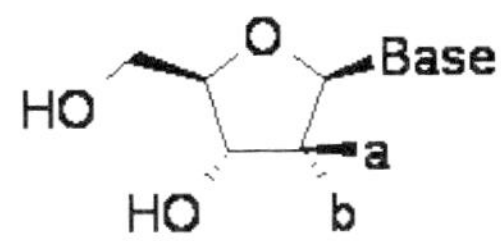

Entry	a	b	Base	Replicon EC_{50}	NS5b IC_{50}
26[a]	Me	OH	U	>100	Not reported
27[a] (NM107)	Me	OH	C	1.23	0.09
28[b]	Me	OH	G	3.5	0.13
29[b]	Me	OH	A	0.26	1.9
30[c] (MK-0608)	Me	OH	7-deaza-A	0.25	0.12
31[d]	Me	OH	7-F-7-deaza-A	0.07	Not reported
32[e]	H	OMe	C	21	2.7
33[f]	Me	F	U	>100	0.52
34[g] (PSI-6130)	Me	F	C	0.61	0.11
35[h]	H	F	C	5.0	Not reported

[a]Klumpp *et al.* (2006), Ma *et al.* (2007), Furman *et al.* (2009).

[b]Eldrup *et al.* (2004a).

[c]Olsen *et al.* (2004).

[d]Eldrup *et al.* (2004b).

[e]Carroll *et al.* (2003).

[f]Ma *et al.* (2007).

[g]Ali *et al.* (2008), McCown *et al.* (2008).

[h]Stuyver *et al.* (2004).

Cytostatic inhibition of cell proliferation IC_{50} = 0.8 μM.

deliver higher concentrations of **28**-triphosphate to target cells (Lalezari *et al.*, 2009; McGuigan *et al.*, 2009; Patti *et al.*, 2009). Further attention turned to modification of the adenine base moiety to increase metabolic stability relative to **29**. 2′-*C*-methyl-7-deazaadenosine **30** achieved that goal, was not a substrate for adenosine deaminase or purine nucleoside phosphorylase any more and retained antiviral potency in the replicon system similar to **29** (Eldrup *et al.*, 2004b; Olsen *et al.*, 2004). Interestingly, the 7-deaza modification appeared to improve the intrinsic inhibition potency of the triphosphate as an inhibitor of HCV polymerase compared with the triphosphate of **29**. Further exploration at the 7-position of the adenine base revealed that addition of a fluorine group, to give **31**, further increased potency in the replicon system by 3- to 4-fold (Eldrup *et al.*, 2004b). These modifications also improved the physicochemical properties of the compounds and substantially improved pharmacokinetic properties were achieved with **30** compared with **29** (Olsen *et al.*, 2004). Compound **30** (MK-0608) was chosen for further development and demonstrated promising antiviral efficacy in HCV infected chimpanzees at a relatively low oral dose of 1 mg/kg/day (Carroll *et al.*, 2009). MK-0608 was discontinued for undisclosed reasons after entering phase I clinical studies.

2′-*O*-Methyl cytidine **32** was reported as an inhibitor of HCV replication with a modest EC_{50}

value of 21 μM in the HCV replicon system. The corresponding nucleoside triphosphate was also only modestly active when tested in the *in vitro* enzyme assay with an IC_{50} of 2.7 μM (Carroll *et al.*, 2003; Tomassini *et al.*, 2005). Further studies with radiolabelled **32** revealed low intracellular stability and extensive deamination, demethylation or base transfer through nucleoside phosphorylase, leading to the formation of mainly CTP and UTP and only low concentrations of **32**-triphosphate in Huh-7 cells. In contrast, 2′-*C*-methylcytidine **26** was stable under the same conditions. Based on the different predicted solution conformations of **26** and **32** it was hypothesized that **32** may be a preferred substrate for cytidine deaminase due to it preferentially adopting the Southern conformation in solution (Tomassini *et al.*, 2005).

2′-Fluoro-2′-C-methyl nucleosides

As described above, the 5′-triphosphate of 2′-deoxy-2′-fluoro cytidine is a substrate for HCV polymerase, but the nucleoside is associated with high cell proliferation inhibition potency in the HCV replicon system (Stuyver *et al.*, 2004; Clark *et al.*, 2005). The addition of a methyl group to 2′-deoxy-2′-fluoro cytidine resulted in a highly selective inhibitor of HCV replication (2′-deoxy-2′-C-methyl-2′-fluoro cytidine, PSI-6130, **34**) with an EC_{50} of 0.61 μM and no apparent cytotoxicity in HCV replicon cells (Stuyver *et al.*, 2004, 2006; Ali *et al.*, 2008). The exchange of the 2′-hydroxyl group of **27** (NM107) with a fluorine results in **34** (PSI-6130). Interestingly, this small structural change has a significant impact on the selectivity of these compounds. Whereas NM107 exhibits activity against HCV and BVDV, PSI-6130 was selective for HCV and did not inhibit BVDV replication in cell culture (Stuyver *et al.*, 2006). Similar to the observations with most other ribonucleoside analogue pairs, the uridine analogue directly related to PSI-6130, 2′-deoxy-2′-C-methyl-2′-fluorouridine, **33**, was not active in the HCV replicon system due to inefficient phosphorylation (Ma *et al.*, 2007; Murakami *et al.*, 2008). Therefore, no potent uridine analogue inhibitor of HCV replication has been identified to date. However, based on the intrinsic potency of their corresponding 5′-triphosphates uridine analogues and other inefficiently phosphorylated nucleosides may be further developed as nucleoside monophosphate prodrugs, thus circumventing the limiting first phosphorylation step in the target cells (Hecker *et al.*, 2007; Perrone *et al.*, 2007a,b; Furman *et al.*, 2009). Interestingly, it was found that PSI-6130 formed significant levels of 2′-deoxy-2′-C-methyl-2′-fluorouridine triphosphate in human hepatocytes, in addition to 2′-deoxy-2′-C-methyl-2′-fluorocytidine triphosphate (Ma *et al.*, 2007). Exposure to PSI-6130 therefore results in the simultaneous formation of two triphosphate species with antiviral activity. Deamination was determined to occur at the level of 2′-deoxy-2′-C-methyl-2′-fluorocytidine monophosphate by cytidine deaminase, while PSI-6130 was only a very poor substrate of cytidine deaminase (Ma *et al.*, 2007; Murakami *et al.*, 2008). The triphosphate half-life of the uridine triphosphate analogue was 8-fold increased relative to PSI-6130-triphosphate (38 hours and 4.7 hours respectively). In contrast to **29**, where deamination resulted in an inactive metabolite, the deamination of PSI-6130-monophosphate in human hepatocytes resulted in the formation of two potent nucleoside triphosphate derivatives (Table 15.5) and therefore contributes in this case to overall antiviral potency of this compound.

The 3′,5′-diisobutyrate prodrug of **34** (R7128, RG7128) was selected for clinical development for the treatment of HCV infection. The diester prodrug of PSI-6130 improved oral bioavailability by increasing absorption from the gastrointestinal tract. RG7128 is currently in phase IIb clinical development and has been associated with a very benign safety and tolerability profile to date. In a 14-day monotherapy study in HCV infected persons, RG7128 showed high antiviral efficacy, achieving mean viral load decreases of 0.9, 1.5, 2.1 and 2.7 log IU/ml at the end of the dosing period in the dose groups receiving 750 mg once daily, 1500 mg once daily, 750 mg twice daily or 1500 mg twice daily, respectively (Reddy, 2007). Similar to the balapiravir monotherapy study, there was no evidence for resistance selection during 2 weeks of monotherapy in any of the dose groups (Le Pogam *et al.*, 2009b). A 4-week phase IIA study compared the combination of either 750 mg or 1500 mg RG7128 twice daily in combination with PEG and RBV and the PEG + RBV standard of care control. 30% (750 mg b.i.d.), 85% (1500 mg b.i.d.) and 10% (SOC) of treatment naive patients infected with HCV genotype 1 achieved undetectable viral load (<15 IU/ml) by week 4. There were no substantial differences in safety and tolerability between the dose groups, indicating that RG7128 was well tolerated in this combination and timeframe (Lalezari, 2008). RG7128 also demonstrated high efficacy in HCV genotype 2 or 3 infected patients. 90% of patients infected with HCV genotype 2 or 3 and failing prior SOC therapy (50% non-responders, 50% relapsers) achieved undetectable viral load after 4 weeks of 1500 mg twice daily RG7128 in combination with PEG + RBV (n = 20: 8 genotype 2, 12 genotype 3; Gane, 2008). In an improved oral formulation, 1000 mg b.i.d. RG7128 achieved antiviral efficacy similar to 1500 mg b.i.d. RG7128 in the previous formulation, with 88% of treatment naive genotype 1 infected patients achieving undetectable viral load after 4 weeks of combination treatment with PEG and RBV (Rodriguez-Torres *et al.*, 2008).

RG7128 has also shown high antiviral potency in combination with the HCV protease inhibitor RG7227 (ITMN-191) in interferon and ribavirin sparing combination therapy. After 14 days of dual combination of RG7128 1000 mg b.i.d. + RG7227 900 mg b.i.d., treatment naive patients infected with HCV genotype 1 achieved a mean viral load reduction of 5.1 log IU/ml (range 3.0–5.9) and 63% (5/8) achieved undetectable viral load as early as week 2 (Gane *et al.*, 2009, 2010). The dual combination of RG7128 and RG7227 therefore showed similar antiviral potency compared with highly potent triple combinations of PEG+RBV with an additional potent antiviral agent. In addition, there was no evidence for resistance selection in this dual combination, suggesting efficient suppression of the selection of protease inhibitor resistance by the combination with RG7128 (Le Pogam *et al.*, 2009a). RG7128 and RG7227 are currently in phase II clinical development.

Conclusions

The urgent clinical need for improved therapies for the treatment of HCV infection has sparked intensive research into the design of potent and selective ribonucleoside analogues as inhibitors of HCV replication. Optimization of nucleoside analogues is complicated by the requirement of maintaining the ability for interaction with multiple cellular enzymes that are required for the formation of intracellular triphosphate species, while selectivity is required with respect to cellular nucleoside metabolizing enzymes, cellular polymerases and signalling pathways. Modifications at the 2′- and 4′-modified nucleoside analogues have shown most potential to achieve antiviral potency and selectivity with modified nucleoside analogue. Importantly, it has been demonstrated that a hydrogen bonding interaction of the HCV RNA-dependent RNA polymerase with the 2′-hydroxyl group of ribonucleosides is not required to achieve high antiviral potency, thus increasing the scope of structural modifications on nucleosides that might have potential for inhibition of HCV replication. The most advanced candidate, RG7128, a prodrug of 2′-C-methyl-2′-fluorocytidine, is currently in phase II clinical development and has shown a highly promising safety, tolerability and efficacy profile. 88% of treatment naive, genotype 1 infected patients and 90% of treatment-experienced genotype 2 or 3 infected patients achieved undetectable viral load after 4 weeks of treatment (RVR)with a combination of RG7128, pegylated interferon alpha and ribavirin. In addition, the interferon free dual combination of RG7128 and the HCV protease inhibitor RG7227 demonstrated similar antiviral potency after 2 weeks of treatment compared with interferon based highly potent triple combinations. There has been no evidence of resistance selection in studies of balapiravir or RG7128 up to 2 weeks of monotherapy or up to 4 weeks (RG7128) or 12 weeks (balapiravir) of triple therapy, or in 2 weeks of interferon free combination of RG7128 and RG7227. These results suggest a high barrier to resistance of nucleoside analogues as inhibitors of HCV replication in HCV infected patients, high potency across genotypes and ability to suppress resistance selection against a co-administered HCV protease inhibitor.

Future trends

Future trends in nucleoside research will be around new structural classes of inhibitors that will provide increased intrinsic potency and an ability to reduce daily doses or dosing frequency. Also, a better understanding of specific nucleoside toxicity and mechanisms related to nucleoside interference with cellular functions will drive the design of better tolerated and safer nucleoside analogues. Improved ways to deliver nucleoside analogues to target cells or to deliver nucleoside monophosphate analogues will be explored and may result in overall increased efficacy or reduced dosing requirements. More and longer durations of combinations of antiviral agents will determine safety, tolerability and efficacy of interferon free therapy options and may mark a step change in how HCV infections will be cured in the future.

Acknowledgements

The authors wish to thank Silvia Patrick and Rebecca Bernat for their assistance with this manuscript.

References

Alexander, R.L., Morris-Natschke, S.L., Ishaq, K.S., Fleming, A., and Kucera, G.L. (2003). Synthesis and cytotoxic activity of two novel 1-dodecylthio-2-decyloxypropyl-3-phosphatidic acid conjugates with gemcitabine and cytosine arabinoside. J. Med. Chem. *46*, 4205–4208.

Ali, S., Leveque, V., Le Pogam, S., Ma, H., Philipp, F., Inocencio, N., Smith, M., Alker, A., Kang, H., Najera, I., Klumpp, K., Symons, J., Cammack, N., and Jiang, W.R. (2008). Selected replicon variants with low-level in vitro resistance to the hepatitis c virus NS5B polymerase inhibitor PSI-6130 lack cross-resistance with R1479. Antimicrob. Agents Chemother. *52*, 4356–4369.

Altona, C., and Sundaralingam, M. (1972). Conformational analysis of the sugar ring in nucleosides and nucleotides. A new description using the concept of pseudorotation. J. Am. Chem. Sci. *94*, 8205–8212.

Bartels, D.J., Zhou, Y., Zhang, E.Z., Marcial, M., Byrn, R.A., Pfeiffer, T., Tigges, A.M., Adiwijaya, B.S., Lin, C., Kwong, D., and Kieffer, T.L. (2008). Natural prevalence of hepatitis C virus variants with decreased sensitivity to NS3.4A protease inhibitors in treatment-naive subjects. J. Infect. Dis. *198*, 800–807.

Bukh, J., Miller, H., and Purcell, R.H. (1995). Genetic heterogeneity of hepatitis C virus: quasispecies and genotypes. Semin. Liver Dis. *15*, 41–63.

Carroll, S.S., and Olsen, D.B. (2006). Nucleoside analog inhibitors of hepatitis C virus replication. Infect. Disord. Drug Targets *6*, 17–29.

Carroll, S.S., Tomassini, J.E., Bosserman, M., Getty, K., Stahlhut, M.W., Eldrup, A.B., Bhat, B., Hall, D., Simcoe, A.L., LaFemina, R., *et al.* (2003). Inhibition of hepatitis C virus RNA replication by 2′-modified nucleoside analogs. J. Biol. Chem. *278*, 11979–11984.

Carroll, S.S., Ludmerer, S., Handt, L., Koeplinger, K., Zhang, N.R., Graham, D., Davies, M.E., MacCoss, M., Hazuda, D., and Olsen, D.B. (2009). Robust antiviral efficacy upon administration of a nucleoside analog to hepatitis C virus-infected chimpanzees. Antimicrob. Agents Chemother. *53*, 926–934.

Chen, M.S., Suttmann, R.T., Wu, C., and Prisbe, E.J. (1992). Metabolism of 4′-azidothymidine. A compound with potent and selective activity against the human immunodeficiency virus. J. Biol. Chem. *267*, 257–260.

Clark, J.L., Hollecker, L., Mason, J.C., Stuyver, L.J., Tharnish, P.M., Lostia, S., McBrayer, T.R., Schinazi, R.F., Watanabe, K.A., Otto, M.J., *et al.* (2005). Design, synthesis, and antiviral activity of 2′-deoxy-2′-fluoro-2′-C-methylcytidine, a potent inhibitor of hepatitis C virus replication. J. Med. Chem. *48*, 5504–5508.

Curry, S., Qiu, P., and Tong, X. (2008). Analysis of HCV resistance mutations during combination therapy with protease inhibitor boceprevir and PEG-IFN alpha-2b using TaqMan mismatch amplification mutation assay. J. Virol. Methods *153*, 156–162.

Eldrup, A.B., Allerson, C.R., Bennett, C.F., Bera, S., Bhat, B., Bhat, N., Bosserman, M.R., Brooks, J., Burlein, C., Carroll, S.S., *et al.* (2004a). Structure-activity relationship of purine ribonucleosides for inhibition of hepatitis C virus RNA-dependent RNA polymerase. J. Med. Chem. *47*, 2283–2295.

Eldrup, A.B., Prhavc, M., Brooks, J., Bhat, B., Prakash, T.P., Song, Q., Bera, S., Bhat, N., Dande, P., Cook, P.D., *et al.* (2004b). Structure-activity relationship of heterobase-modified 2′-C-methyl ribonucleosides as inhibitors of hepatitis C virus RNA replication. J. Med. Chem. *47*, 5284–5297.

Furman, P.A., Lam, M., and Murakami, E. (2009). Nucleoside analog inhibitors of hepatitis C viral replication: recent advances, challenges and trends. Future Med. Chem. *1*, 1429–1452.

Gane, E.J., Rodriguez-Torres, M., Nelson, D., Everson, G., Jacobson, I., McHutchison, J.G., Jeffers, L., Reddy, R., Beard, A., Walker, S., *et al.* (2008). Antiviral activity of the HCV nucleoside polymerase inhibitor R7128 in HCV genotype 2 and 3 prior non-responders: results of R7128, 1500mg BID with PEG-IFN and ribavirin For 28 days. Hepatology. *48*, LB10.

Gane, E.J., Roberts, S.K., Stedman, C.A., Angus, P.W., Ritchie, B., Elston, R., Ipe, D., Morcos, P.N., Najera, I., Chu, T., Berrey, M.M., *et al.* (2009). Combination therapy with a nucleoside polymerase (RG7128) and protease (RG7227/ITMN-191) inhibitor in

HCV: safety, pharmacokinetics, and virologic results from INFORM-1. The 60th Annual Meeting of the American Association for the Study of Liver Diseases (AASLD), 30 October – 3 November 2009 (Boston, MA).

Gane, E.J., Roberts, S.K., Stedman, C.A., Angus, P.W., Ritchie, B. Elston, R., Ipe, D., Morcaos, P.N., Baher, L., Najera, I., *et al.* (2010). Oral combination therapy with a nucleoside polymerase inhibitor (RG7128) and danoprevir for chronic hepatitis C genotype 1 infection (INFORM-1): a randomised, double-blind, placebo-controlled, dose escalation trial. Lancet *376*, 1467–1475.

Golitsina, N.L., Danehy, F.T., Jr., Fellows, R., Cretton-Scott, E., and Standring, D.N. (2010). Evaluation of the role of three candidate human kinases in the conversion of the hepatitis C virus inhibitor 2′-C-methyl-cytidine to its 5′-monophosphate metabolite. Antiviral Res. *85*, 470–481.

Hecker, S.J., Reddy, K.R., van Poelje, P.D., Sun, Z., Huang, W., Varkhedkar, V., Reddy, M.V., Fujitaki, J.M., Olsen, D.B., Koeplinger, K.A., *et al.* (2007). Liver-targeted prodrugs of 2′-C-methyladenosine for therapy of hepatitis C virus infection. J. Med. Chem. *50*, 3891–3896.

Herrmann, E., Lee, J.H., Marinos, G., Modi, M., and Zeuzem, S. (2003). Effect of ribavirin on hepatitis C viral kinetics in patients treated with pegylated interferon. Hepatology *37*, 1351–1358.

Jiang, W.-R., Chiu, S., Ali, S., Chapman, M., Daniel, C., Kretz, T., Cammack, N., Symons, J., and Klumpp, K. (2006). In vitro antiviral interactions of a novel HCV inhibitor R1479 with interferon alpha-2A, ribavirin and other HCV inhibitors. Hepatology *44*, 533A.

Kim, A.Y., Timm, J., Nolan, B.E., Reyor, L.L., Kane, K., Berical, A.C., Zachary, K.C., Lauer, G.M., Kuntzen, T., and Allen, T.M. (2009). Temporal dynamics of a predominant protease inhibitor-resistance mutation in a treatment-naive, hepatitis C virus-infected individual. J. Infect. Dis. *199*, 737–741.

Klumpp, K., Leveque, V., Le Pogam, S., Ma, H., Jiang, W., Kang, H., Granycome, C., Singer, M., Laxton, C., Hang, J.Q., *et al.* (2006). The novel nucleoside analog R1479 (4′-azidocytidine) is a potent inhibitor of NS5B-dependent RNA synthesis and hepatitis C virus replication in cell culture. J. Biol. Chem. *281*, 3793–3799.

Klumpp, K., Kalayanov, G., Ma, H., Le Pogam, S., Leveque, V., Jiang, W.R., Inocencio, N., De Witte, A., Rajyaguru, S., Tai, E., *et al.* (2008). 2′-deoxy-4′-azido nucleoside analogs are highly potent inhibitors of hepatitis C virus replication despite the lack of 2′-alpha-hydroxyl groups. J. Biol. Chem. *283*, 2167–2175.

Klumpp, K., and Smith, D.B. (2010). Discovery and clinical evaluation of the nucleoside analog Balapiravir (R1626) for the treatment of HCV infection. In Antiviral Drugs: Biology, Chemistry, Clinic, Kazmierski, W., ed. (John Wiley and Sons, Inc., Hoboken, NJ), in press.

Kuntzen, T., Timm, J., Berical, A., Lennon, N., Berlin, A.M., Young, S.K., Lee, B., Heckerman, D., Carlson, J., Reyor, L.L., *et al.* (2008). Naturally occurring dominant resistance mutations to hepatitis C virus protease and polymerase inhibitors in treatment-naive patients. Hepatology *48*, 1769–1778.

Lalezari, E.G., Rodriguez-Torres, M., De Jesus, E., Nelson, D., Everson, G., Jacobson, I., Reddy, R., Hill, G.Z., Beard, A., Symonds, W.T., *et al.* (2008). Potent antiviral activity of the HCV nucleoside polymerase inhibitor R7128 with PEG-IFN and Ribavirin: interim results of R7128, 500 mg BID for 28 days. J. Hepatol. *48*, S29.

Lalezari, E., Asmuth, D., Casiro, A., Vargas, H., Dubuc Patrick, G., Liu, W., Pietropaolo, K., Zhou, X.J., Sullivan-Bolyai, J., and Mayers, D. (2009). Antiviral activity, safety and pharmacokinetics of IDX-*184*, a liver-targeted nucleotide HCV polymerase inhibitor, in patients with chronic Hepatitis C. The 60th Annual meeting of the American Association for the Study of Liver Diseases (AASLD), 30 October – 3 November 2009 (Boston, MA).

Le Pogam, S., Seshaadri, A., Kosaka, A., Chiu, S., Kang, H., Hu, S., Rajyaguru, S., Symons, J., Cammack, N., and Najera, I. (2008). Existence of hepatitis C virus NS5B variants naturally resistant to non-nucleoside, but not to nucleoside, polymerase inhibitors among untreated patients. J. Antimicrob. Chemother. *61*, 1205–1216.

Le Pogam, S., Chhabra, M., Ali, S., Yan, J.M., Ilnicka, M., Kang, H., Wong, J., Kosaka, A., Ewing, A., Seshaadri, A., *et al.* (2009a). Combination therapy with nucleoside polymerase (RG7128) and protease (RG7227) inhibitors in genotype 1 HCV infected patients: interim resistance analysis of INFORM-1 cohorts A–D. The 60th Annual Meeting of the American Association for the Study of Liver Diseases (AASLD), 30 October – 3 November 2009 (Boston, MA).

Le Pogam, S., Kang, H., Seshaadri, A., Kosaka, A., Hu, S., Ewing, A., Yan, J.-M., Beard, A., Symons, J., Cammack, N., and Najera, I. (2009b). No evidence of R7128 drug resistance after up to 4 weeks treatment of GT1, 2 and 3 Hepatitis C Virus infected individuals. 44th Annual Meeting of the European Association for the Study of the Liver (EASL), 22–26 April 2009 (Copenhagen, Denmark).

Lee, J.H., von Wagner, M., Roth, W.K., Teuber, G., Sarrazin, C., and Zeuzem, S. (1998). Effect of ribavirin on virus load and quasispecies distribution in patients infected with hepatitis C virus. J. Hepatol. *29*, 29–35.

Leveque, V., Le Pogam, S., Ao-leong, G., Harris, S.F., Kang, H., Kosaka, A., Seshaadri, A., Chiu, S., Symons, J., Cammack, N., Klumpp, K., and Najera, I. (2008). R1626 (prodrug of R1479) demonstrates potent antiviral activity across HCV genotypes 1–6 in vitro. Paper presented at: 43rd Annual Meeting of the European Association for the Study of the Liver (EASL) (Milan, Italy).

Lohmann, V., Korner, F., Koch, J., Herian, U., Theilmann, L., and Bartenschlager, R. (1999). Replication of subgenomic hepatitis C virus RNAs in a hepatoma cell line. Science *285*, 110–113.

Ma, H., Leveque, V., De Witte, A., Li, W., Hendricks, T., Clausen, S.M., Cammack, N., and Klumpp, K. (2005).

Inhibition of native hepatitis C virus replicase by nucleotide and non-nucleoside inhibitors. Virology *332*, 8–15.

Ma, H., Jiang, W.R., Robledo, N., Leveque, V., Ali, S., Lara-Jaime, T., Masjedizadeh, M., Smith, D.B., Cammack, N., Klumpp, K., and Symons, J. (2007). Characterization of the metabolic activation of hepatitis C virus nucleoside inhibitor beta-D-2′-Deoxy-2′-fluoro-2′-C-methylcytidine (PSI-6130) and identification of a novel active 5′-triphosphate species. J. Biol. Chem. *282*, 29812–29820.

Maag, H., Rydzewski, R.M., McRoberts, M.J., Crawford-Ruth, D., Verheyden, P., and Prisbe, E.J. (1992). Synthesis and anti-HIV activity of 4′-azido- and 4′-methoxynucleosides. J. Med. Chem. *35*, 1440–1451.

McCown, M.F., Rajyaguru, S., Le Pogam, S., Ali, S., Jiang, W.R., Kang, H., Symons, J., Cammack, N., and Najera, I. (2008). The hepatitis C virus replicon presents a higher barrier to resistance to nucleoside analogs than to nonnucleoside polymerase or protease inhibitors. Antimicrob. Agents. Chemother. *52*, 1604–1612.

McGuigan, C., Perrone, P., Madela, K., and Neyts, J. (2009). The phosphoramidate ProTide approach greatly enhances the activity of beta-2′-C-methylguanosine against hepatitis C virus. Bioorg. Med. Chem. Lett. *19*, 4316–4320.

Maltseva, T., Usova, E., Johansson, N.G., Kalayanov, G., Eriksson, S., Sund, C., Winqvist, A., Zhou, X.X., Smith, D.B., Klumpp, K., Smith, M., Maag, H., and Martin, J.A. (2008). Structure-activity studies of deoxycytidine kinase with novel 4′-azido-2′-deoxy-nucleosides using NMR analysis. Small molecule NMR Conference (SMASH), 7–10 September (Santa Fe, NM).

Marquez, V.E., Lim, B.B., Barchi, J., and Nicklaus, M.C. (1993). Conformational Studies and Anti-HIV Activity of Mono- and Difluorodideoxynucleosides. In Nucleosides and Nucleotides as Antitumour and Antiviral Agents, Chu, C.K., and Baker, D., eds (Plenum Press, New York), pp. 256–284.

Migliaccio, G., Tomassini, J.E., Carroll, S.S., Tomei, L., Altamura, S., Bhat, B., Bartholomew, L., Bosserman, M.R., Ceccacci, A., Colwell, L.F., *et al.* (2003). Characterization of resistance to non-obligate chain-terminating ribonucleoside analogs that inhibit hepatitis C virus replication in vitro. J. Biol. Chem. *278*, 49164–49170.

Murakami, E., Niu, C., Bao, H., Micolochick Steuer, H.M., Whitaker, T., Nachman, T., Sofia, M.A., Wang, P., Otto, J., and Furman, P.A. (2008). The mechanism of action of beta-D-2′-deoxy-2′-fluoro-2′-C-methylcytidine involves a second metabolic pathway leading to beta-D-2′-deoxy-2′-fluoro-2′-C-methyluridine 5′-triphosphate, a potent inhibitor of the hepatitis C virus RNA-dependent RNA polymerase. Antimicrob. Agents Chemother. *52*, 458–464.

Neumann, A.U., Lam, N.P., Dahari, H., Gretch, D.R., Wiley, T.E., Layden, J., and Perelson, A.S. (1998). Hepatitis C viral dynamics in vivo and the antiviral efficacy of interferon-alpha therapy. Science. *282*, 103–107.

Ohrui, H., Kohgo, S., Kitano, K., Sakata, S., Kodama, E., Yoshimura, K., Matsuoka, M., Shigeta, S., and Mitsuya, H. (2000). Syntheses of 4′-C-ethynyl-beta-D-arabino- and 4′-C-ethynyl-2′-deoxy-beta-D-ribo-pentofuranosylpyrimidines and -purines and evaluation of their anti-HIV activity. J. Med. Chem. *43*, 4516–4525.

Olsen, D.B., Eldrup, A.B., Bartholomew, L., Bhat, B., Bosserman, M.R., Ceccacci, A., Colwell, L.F., Fay, J.F., Flores, O.A., Getty, K.L., *et al.* (2004). A 7-deaza-adenosine analog is a potent and selective inhibitor of hepatitis C virus replication with excellent pharmacokinetic properties. Antimicrob. Agents Chemother. *48*, 3944–3953.

Oyang, C., Wu, H.Y., Fraser-Smith, B., and Walker, K.A.M. (1992). Synthesis of 4′-cyanothymidine and analogs as potent inhibitors of HIV. Tetrahedron Letters *33*, 37–40.

Patti, J., Ames, B., Bryant, D., Chamberlain, S.D., Ganguly, B.S., Gorovits, E., Hall, A., Hunley, D.S., Hutchins, J.T., Kolykhalov, A., *et al.* (2009). Pharmacological Properties and in vitro characterization of INX-*189*, a liver-targeted phosphoramidate nucleoside analogue inhibitor of NS5b. Annual meeting of the American Association for the Study of Liver Diseases (AASLD), 30 October – 3 November 2009 (Boston, MA).

Pawlotsky, J.M., Dahari, H., Neumann, A.U., Hezode, C., Germanidis, G., Lonjon, I., Castera, L., and Dhumeaux, D. (2004). Antiviral action of ribavirin in chronic hepatitis C. Gastroenterology *126*, 703–714.

Perrone, P., Daverio, F., Valente, R., Rajyaguru, S., Martin, J.A., Leveque, V., Le Pogam, S., Najera, I., Klumpp, K., Smith, B., and McGuigan, C. (2007a). First example of phosphoramidate approach applied to a 4′-substituted purine nucleoside (4′-azidoadenosine): conversion of an inactive nucleoside to a submicromolar compound versus hepatitis C virus. J. Med. Chem. *50*, 5463–5470.

Perrone, P., Luoni, G.M., Kelleher, M.R., Daverio, F., Angell, A., Mulready, S., Congiatu, C., Rajyaguru, S., Martin, J.A., Leveque, V., *et al.* (2007b). Application of the phosphoramidate ProTide approach to 4′-azi-douridine confers sub-micromolar potency versus hepatitis C virus on an inactive nucleoside. J. Med. Chem. *50*, 1840–1849.

Pierra, C., Benzaria, S., Amador, A., Moussa, A., Mathieu, S., and Gosselin, G. (2005). NM283, an efficient prodrug of the potent anti-HCV agent 2′-C-Methylcytidine. Nucleosides, Nucleotides and Nucleic Acids *24*, 767–770.

Pierra, C., Amador, A., Benzaria, S., Cretton-Scott, E., D'Amours, M., Mao, J., Mathieu, S., Moussa, A., Bridges, E.G., Standring, D.N., *et al.* (2006). Synthesis and pharmacokinetics of valopicitabine (NM283), an efficient prodrug of the potent anti-HCV agent 2′-C-methylcytidine. J. Med. Chem. *49*, 6614–6620.

Pockros, P., Nelson, D., Godofsky, E., Rodriguez-Torres, M., Everson, G.T., Fried, M.W., Ghalib, R., Harrison, S., Nyberg, L., Shiffman, M., *et al.* (2008a). High relapse rate seen at week 72 for patients treated with R1626 combination therapy. Hepatology *48*, 1349–1350.

Pockros, P.J., Nelson, D., Godofsky, E., Rodriguez-Torres, M., Everson, G.T., Fried, M.W., Ghalib, R., Harrison, S., Nyberg, L., Shiffman, M.L., Najera, I., Chan, A., and Hill, G. (2008b). R1626 plus peginterferon Alfa-2a provides potent suppression of hepatitis C virus RNA and significant antiviral synergy in combination with ribavirin. Hepatology *48*, 385–397.

Prisbe, E.J., Maag, H., Verheyden, H., and Rydzewski, R.M. (1993). Structure–Activity Relationships Among HIV Inhibitory 4′-Substituted Nucleosides. In Nucleotides and nucleosides as Antitumor and Antiviral Agents Chu, C.K., and Baker, D., eds (Plenum Press, New York), pp. 101–113.

Reddy, R., Rodriguez-Torres, M., Gane, E., Robson, R., Lalezari, J., Everson, G., De Jesus, E., McHutchison, J.G., Vargas, H.E., Beard, A., *et al.* (2007). Antiviral activity, pharmacokinetics, safety, and tolerability of R7128, a novel nucleoside HCV RNA polymerase inhibitor, following multiple, ascending, oral doses in patients with HCV genotype 1 infection who have failed prior interferon therapy. Hepatology *46*, 232A.

Roberts, S., Cooksley, G., Dore, G., Robson, R., Shaw, D., Berns, H., Brandl, M., Fettner, S., Hill, G., Ipe, D., *et al.* (2006). Results of a Phase 1B, Multiple Dose Study of R1626, a Novel Nucleoside Analog Targeting HCV Polymerase in Chronic HCV Genotype 1 Patients. Hepatology *44*, 692A.

Roberts, S.K., Cooksley, G., Dore, G.J., Robson, R., Shaw, D., Berns, H., Hill, G., Klumpp, K., Najera, I., and Washington, C. (2008). Robust antiviral activity of R1626, a novel nucleoside analog: a randomized, placebo-controlled study in patients with chronic hepatitis C. Hepatology *48*, 398–406.

Rodriguez-Torres, M., Lalezari, J., Gane, E.J., DeJesus, E., Nelson, D.R., Everson, G.T., Jacobson, I.M., Reddy, K.R., McHutchison, J.G., Beard, A., *et al.* (2008). Potent antiviral response to the HCV nucleoside polymerase inhibitor R7128 for 28 days with Peg-IFN and ribavirin: subanalysis by race/ethnicity, weight and HCV genotype. The 59th Annual Meeting of the American Association for the Study of Liver Diseases (AASLD), 31 October – 4 November (San Francisco, CA).

Sabini, E., Ort, S., Monnerjahn, C., Konrad, M., and Lavie, A. (2003). Structure of human dCK suggests strategies to improve anticancer and antiviral therapy. Nat. Struct. Biol. *10*, 513–519.

Sabini, E., Hazra, S., Ort, S., Konrad, M., and Lavie, A. (2008). Structural basis for substrate promiscuity of dCK. J. Mol. Biol. *378*, 607–621.

Shim, J., Larson, G., Lai, V., Naim, S., and Wu, J.Z. (2003). Canonical 3′-deoxyribonucleotides as a chain terminator for HCV NS5B RNA-dependent RNA polymerase. Antiviral Res. *58*, 243–251.

Simmonds, P. (2004). Genetic diversity and evolution of hepatitis C virus – 15 years on. The J. Gen. Virol. *85*, 3173–3188.

Smith, D.B., Martin, J.A., Klumpp, K., Baker, S.J., Blomgren, P.A., Devos, R., Granycome, C., Hang, J., Hobbs, C.J., Jiang, W.R. *et al.* (2007). Design, synthesis, and antiviral properties of 4′-substituted ribonucleosides as inhibitors of hepatitis C virus replication: the discovery of R1479. Bioorg. Med. Chem. Lett. *17*, 2570–2576.

Smith, D.B., Kalayanov, G., Sund, C., Winqvist, A., Maltseva, T., Leveque, V.J., Rajyaguru, S., Le Pogam, S., Najera, I., Benkestock, K. *et al.* (2009a). The design, synthesis, and antiviral activity of monofluoro and difluoro analogues of 4′-azidocytidine against hepatitis C virus replication: the discovery of 4′-azido-2′-deoxy-2′-fluorocytidine and 4′-azido-2′-dideoxy-2′,2′-difluorocytidine. J. Med. Chem. *52*, 2971–2978.

Smith, D.B., Kalayanov, G., Sund, C., Winqvist, A., Pinho, P., Maltseva, T., Morisson, V., Leveque, V., Rajyaguru, S., Le Pogam, S. *et al.* (2009b). The design, synthesis, and antiviral activity of 4′-azidocytidine analogues against hepatitis C virus replication: the discovery of 4′-azidoarabinocytidine. J. Med. Chem. *52*, 219–223.

Sorbera, A., and Castaner, P.A. (2006). Valopicitabine. Drugs Fut. *31*, 320–324.

Stuyver, L.J., McBrayer, T.R., Whitaker, T., Tharnish, P.M., Ramesh, M., Lostia, S., Cartee, L., Shi, J., Hobbs, A., Schinazi, R.F., Watanabe, A., and Otto, M.J. (2004). Inhibition of the subgenomic hepatitis C virus replicon in huh-7 cells by 2′-deoxy-2′-fluorocytidine. Antimicrob. Agents Chemother. *48*, 651–654.

Stuyver, L.J., McBrayer, T.R., Tharnish, P.M., Clark, J., Hollecker, L., Lostia, S., Nachman, T., Grier, J., Bennett, M.A., Xie, M.Y., *et al.* (2006). Inhibition of hepatitis C replicon RNA synthesis by beta-D-2′-deoxy-2′-fluoro-2′-C-methylcytidine: a specific inhibitor of hepatitis C virus replication. Antivir. Chem. Chemother. *17*, 79–87.

Sweeney, K., and Klumpp, K. (2008). Improving non-nucleoside reverse transcriptase inhibitors for first-line treatment of HIV infection: the development pipeline and recent clinical data. Curr. Opin. Drug. Discov. Devel. *11*, 458–470.

Thompson, J., and McHutchison, J.G. (2009). Antiviral resistance and specifically targeted therapy for HCV (STAT-C). J. Viral. Hepat. *16*, 377–387.

Tomassini, J.E., Getty, K., Stahlhut, M.W., Shim, S., Bhat, B., Eldrup, A.B., Prakash, T.P., Carroll, S.S., Flores, O., MacCoss, M., *et al.* (2005). Inhibitory effect of 2′-substituted nucleosides on hepatitis C virus replication correlates with metabolic properties in replicon cells. Antimicrob. Agents Chemother. *49*, 2050–2058.

Veltkamp, S.A., Pluim, D., van Eijndhoven, M.A., Bolijn, M.J., Ong, F.H., Govindarajan, R., Unadkat, J.D., Beijnen, H., and Schellens, J.H. (2008). New insights into the pharmacology and cytotoxicity of gemcitabine and 2′, 2′-difluorodeoxyuridine. Mol. Cancer Ther. *7*, 2415–2425.

Wakita, T., Pietschmann, T., Kato, T., Date, T., Miyamoto, M., Zhao, Z., Murthy, K., Habermann, A., Krausslich, H.G., Mizokami, M., Bartenschlager, R., and Liang, T.J. (2005). Production of infectious hepatitis C virus in tissue culture from a cloned viral genome. Nat. Med. *11*, 791–796.

NS5B Polymerase Non-nucleoside Inhibitors

16

Martijn Fenaux and Hongmei Mo

Abstract

The inhibitors of HCV NS5B polymerase consist of two classes: nucleoside inhibitors (NIs) and non-nucleoside inhibitors (NNIs). In contrast to the nucleoside inhibitors which bind to the active Site of the polymerase, the HCV non-nucleoside polymerase inhibitors binds to one of the four allosteric binding Sites within the NS5B polymerase including: Site 1 (Thumb I) for JTK-109, Site II (Thumb II) for PF-868554, VCH-759, VCH-916 and VCH-222, Site III (Palm I) for ANA-598, A-848837 and ABT-333, and Site IV (Palm II) for HCV-796. Among these non-nucleoside inhibitors, HCV-796 was the first inhibitor to show an antiviral effect in HCV-infected patients. However, the development of this compound was discontinued due to the hepatic toxicity. Subsequently, PF-868554, VCH-759, VCH-916, VCH-222, ANA-598, and ABT-333 demonstrated antiviral activity in early clinical trials and some have been advanced to phase II. Owing to their distinctive binding sites, non-nucleoside polymerase inhibitors selected different NS5B mutations which exhibited non-cross-resistance profiles. Therefore, combination of non-nucleoside inhibitors may reduce the development of resistance. Given the importance of the NS5B non-nucleoside polymerase inhibitors for the treatment of HCV, this chapter will describes the preclinical resistance profiles as well as the clinical results of the above inhibitors.

Introduction

The HCV NS5B RNA-dependent RNA polymerase (RdRp) protein is the catalytic component of the HCV replication complex. The HCV replication complex is made up of multiple HCV non-structural proteins and host factors and is responsible for the transcription of HCV genomic RNA. NS5B transcribes the HCV RNA for protein translation and progeny genome production (Lindenbach and Rice 2005). The NS5B polymerase recombinant protein has been successfully expressed for crystallographic analysis (Ago *et al.*, 1999; Bressanelli *et al.*, 1999; Lesburg *et al.*, 1999)and *in vitro* enzymatic characterization of RNA-dependent RNA polymerase (RdRp) activity. Since mammalian cells lack a RNA-dependent RNA polymerase equivalent, targeting HCV NS5B with small molecule inhibitors has the potential to be a selective inhibitor for viral replication.

The structure of HCV NS5B polymerase, solved in 1999, showed that, similar to most polynucleotide polymerases, there are three domains of NS5B: the palm domain (green in Fig. 16.1), the finger domain (blue in Fig. 16.1) and the thumb domain (red in Fig. 16.1). However, the HCV NS5B structures demonstrated the unique architecture of the viral RdRp class, which includes collected figure and thumb domains. For example, one of the unusual features of this polymerase is the extent to which the fingers and thumb subdomains interact; a helix at the tip of the fingers binds to a lipophilic pocket in the back of the thumb domain. The palm domain serves as the foundation for the active site and the thumb and finger domains envelop the active site. The finger domain loops reach over the active site and connect with the top of the thumb domain

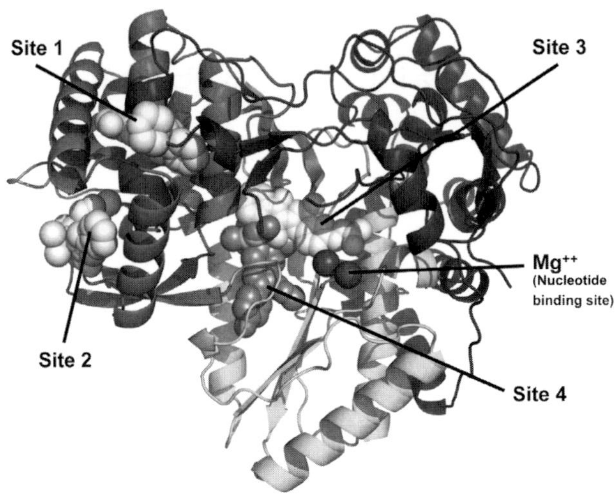

Figure 16.1 Crystal structure of NS5B and NNI binding sites (Cooper *et al.*, 2009). A colour version of this figure can be located in the plate section at the back of the book.

to completely envelope the active site. NS5B crystallographic structures so far have only been able to visualize this closed form of the protein and although it is suspected the finger loops can release the thumb domain to open the protein for catalytic activity this form has never been captured in crystal form. Within the active site is the β-loop attached to the thumb domain, which is involved in regulation of RNA synthesis (Bressanelli *et al.,* 1999). Because NS5B is structurally distinct from related mammalian DNA and RNA polymerase enzymes, it represents an attractive target for the development of novel anti-HCV agents.

Summary of the HCV non-nucleoside polymerase inhibitors under clinical investigation

The HCV NS5B polymerase inhibitors currently in development are divided into two groups, nucleoside analogues (NAs) and NNIs. Nucleoside analogues are converted by the host cell machinery to triphosphates and are then preferentially and erroneously incorporated into the nascent HCV RNA by the NS5B polymerase, resulting in chain termination and premature termination of HCV RNA synthesis. In contrast to NIs, NNIs achieve NS5B inhibition by binding to the allosteric enzyme sites, resulting in conformational changes of the protein-inhibiting catalytic activity of polymerase. This chapter describes the NNIs while NIs are covered in a different chapter.

NNIs in general were identified through high throughput screens. Since several companies pursued the same HTS approach, multiple different chemotypes were identified with potent anti-HCV activity *in vitro*. Currently, a number of HCV NNIs have entered the clinical trials and some demonstrated potent anti-HCV activity in HCV infected patients (Table 16.1). Interestingly, the crystal structure of NS5B complexed with the small molecule inhibitors and resistance analysis studies reveal at least four pockets as binding sites for the allosteric inhibitors (Site 1–4 in Fig. 16.1). These NNI binding sites include: 1) Site I, located at thumb domain near the fingertips where the finger residues interact with the thumb domain 2) Site II, further down the thumb domain and forms a cleft 35 angstroms away from the active site; 3) Site III, palm domain near the active site; 4) Site IV, partially overlapping with palm I and towards the active site. First, JTK-109 and other benzimidazoles and indoles analogues bind to the Site I, where the fingers and thumb subdomains interact; a helix at the tip of the fingers binds to a lipophilic pocket in the back of the thumb domain. Reduced susceptibility to these class of compound was associated with the selection of

Table 16.1 HCV NS5B non-nucleoside Inhibitors under clinical investigation

Inhibitors	Binding site	Company	Study phase
JTK-109	Site I	Japan-Tobac	Withdraw
PF-868554	Site II	Pfizer	Phase II
VCH-759	Site II	ViroChem Pharma	Phase I
VCH-916	Site II	ViroChem Pharma	Phase I
VCH-222	Site II	ViroChem Pharma	Phase I-II
ANA-598	Site III	Anadys Pharmaceuticals, Inc	Phase II
ABT-333	Site III	Abbott Laboratories	Phase II
ABT-072	Site III	Abbott Laboratories	Phase II
HCV-796	Site IV	Wyeth/Viropharma	Withdraw

mutations at resides 495 and 496 of the NS5B polymerase. Second, VCH-796, VCH-916 and VCH-222, the thiophene carboxylic acids, and PF-868554 bind to the Site II pocket (Biswal *et al.*, 2005; Li *et al.*, 2009), a remote region of the thumb subdomain and are presumed to affect the necessary flexibility. The most frequent mutations observed *in vitro* and *in vivo* for Site II inhibitors are located at the amino acid position 423 (M423T/V/I) followed by positions 419 (L419M) and 482 (I482L)(Nicolas 2009; Troke 2009). Third, the ABT-333, ABT-072, and ANA-598 benzothiadiazines analogues bind to Site III, consisting of elements of the β-hairpin and the palm subdomain simultaneously. Resistance selection in the presence of benzothiadiazines revealed amino acid substitutions at the residues 411, 414, 448, 451, 553, 554, 555, 556, and 559 (Thompson *et al.*, 2008, 2009; Wagner *et al.*, 2009). Lastly, HCV-796, benzofuran (Wyeth/Viropharma) binds to the Site IV (Howe *et al.*, 2008), a binding pocket overlaps partially with Site III, do not interact directly with the β-hairpin but occlude the region that would be occupied by the incoming nucleotide prior to ligation. HCV-796 has selected a major resistant mutation at amino acid residues 316 *in vitro* in replicon system and in early clinical trials.

Site I inhibitor: JTK-109

Site I or the finger loop thumb domain pocket is targeted by the benzimidazole JTK-109. JTK-109 binds to the site I pocket and is thought to disrupt the finger and thumb domain interaction required for NS5B RNA polymerase active conformation and forces the enzyme into the open confirmation. JTK-109 inhibits NS5B genotype 1b enzyme activity *in vitro* with an IC_{50} of 17nM (Hirashima *et al.*, 2007). It shows antiviral activity in replicon cells with EC_{50} values of 320 nM in genotype 1a replicon, 62 nM in genotype 1b replicon and 6.4 uM in genotype 2a replicon, with a 25 uM CC_{50} in Huh-5–2 cells. The resistance mutation associated with JTK-109 is P495L. JTK-109 development was terminated after initial clinical trials for undisclosed reasons.

Site II inhibitor

Currently two Site II inhibitor chemotypes are in development, the Pfizer compounds based on the dihydropyranone scaffold, and the second by ViroChem Pharma, now Vertex, based on the thiophene scaffold. Site II inhibition mechanism of action is thought to hinder conformational changes that are required for the RNA polymerase-mediated processive elongation of the RNA product.

The Pfizer Site II inhibitor PF-868554, a dihydropyranone, forms a hydrogen bond with Ser 476 and Tyr 477 (Li *et al.*, 2009). PF-868554 inhibits the 1b (BK) NS5B enzyme *in vitro* with an IC_{50} of 75nM. PF-868554 has antiviral activity in genotype 1a replicon with an EC_{50} of 59nM and no detectable CC_{50} in multiple cell lines (Li *et al.*, 2009). PF-868554 performed well in safety and tolerability clinical trials, including 33 HCV negative individuals given either 50 mg, 100 mg or 300 mg b.i.d., or 300 mg t.i.d. daily for 14 days. No participant withdrew from the safety trial due to SAEs and reported AE were mild including flatulence and headaches (Hammond, 2008a). In phase Ib clinical trials (Hammond, 2008b) PF-868554 was tested in a dose-escalating study of 100 mg, 300 mg or 450 mg b.i.d., or 300 mg t.i.d. as mono-therapy over 8 days in genotype 1 treatment-naïve patients with chronic infection. Within 48 hours of initial treatment all patients demonstrated antiviral response with a mean maximal reduction of plasma HCV RNA from 0.97 $\log_{10}$ drop (100 mg b.i.d.) to 2.3 $\log_{10}$ drop (300 mg t.i.d.). For all doses PF-868554 had a 10–12 hour half-life supporting a b.i.d. treatment schedule. However, virological rebound was detected and associated with resistance mutations in 46% of patients receiving PF-868554 in the Site II (M423A/T/I/V), pocket indicating the need for combined treatment. phase IIa clinical trials (Jacobson, 2009) included 35 treatment- naïve genotype 1 patients and were treated with either 200 mg, 300 mg or 500 mg PF-868554 b.i.d. or placebo in combination with standard of care (SOC; pegylated interferon-2a and ribavirin) for 4 weeks and an additional 44 weeks thereafter with SOC alone. By week 4 HCV RNA plasma levels dropped 4.29 $\log_{10}$ in the 200 mg b.i.d. group, 4.43 $\log_{10}$ in the 300 mg b.i.d. group and 3.44 log 10 in the 500 mg group compared with a drop of 0.98 $\log_{10}$ in the placebo + SOC group. All doses significantly increased the proportion of patients with undetectable HCV RNA. After 4 weeks of treatment 0%, 60%, 75% and 63% of patients receiving placebo, 200 mg, 300 mg or 500 mg b.i.d. had undetectable HCV RNA. After 12 weeks of treatment 50%, 80%, 88% and 63% of patients receiving placebo, 200 mg, 300 mg or 500 mg b.i.d. had undetectable HCV RNA. All patients with undetectable HCV RNA at week 4 remained HCV RNA negative on week 12. Creatinine elevation was reported in a 58-year-old patient with diabetes receiving 300 mg of PF-868554. At day 3 all treatment was halted as the patient experienced chills, nausea and vomiting. Serum creatinine was normalized with aggressive i.v. hydration. Other adverse events were limited to headache, fatigue and insomnia. AE severity or reporting frequency did not increase with increasing PF-868554 dose.

VCH-916 binds the Site II pocket and forms a hydrogen bond with Ser 476 and Tyr 477. VCH-916 (Nicolas, 2009) inhibits the NS5B RNA-dependent RNA polymerase (RdRp) enzymatic activity with a low μM IC_{50} and a genotype 1a and 1b EC_{50} of 79 and 100 nM, respectively, and a 400-fold CC_{50}/EC_{50} selectivity index. VCH-916 has a high protein-binding affinity for human plasma protein and by *in vitro* equilibrium dialysis 98% of the molecule was bound. In phase I safety and tolerability study healthy patients were given a single oral dose ranging from 50 mg to 600 mg, and all doses were well tolerated. In phase Ib clinical study (Proulx, 2008) four treatment-naïve patients with chronic HCV genotype 1 infections were given a single escalating dose of 100 mg or 200 mg t.i.d., or 300 mg or 400 mg b.i.d. mono-therapy over 3 days Within 24 hours of initial treatment all patients demonstrated rapid and significant antiviral response with HCV plasma RNA load drops of 0.6 $\log_{10}$ for 100 mg t.i.d., 1.5 $\log_{10}$ for 200 mg t.i.d., 1.5 $\log_{10}$ for 300 mg b.i.d. and 1.5 $\log_{10}$ for 400 mg b.i.d. Virological rebound was detected during the 3-day treatment period and was associated with polymerase mutation which persisted up to 140 days after treatment. Viral breakthrough was detected in the 100 mg and 200 mg treatment groups. Resistance mutations (Nicolas, 2009) in HCV RNA isolated from patients in these groups were found at L419M/S, R422K, M423T/V/I, I482L and V494A. At 140 days post treatment two patients had resistance variants L419M and M423I. Although no breakthrough was detected in the 300–400 mg treatment groups, mutations in Site II residues were observed, including M423I and V494A/I. No significant AE resulting in discontinuation were observed. Adverse effects were mild to moderate and included nausea, headache and dyspnoea.

VCH-759 binds to the Site II pocket and forms a hydrogen bond with Ser 476 and Tyr 477. VCH-759 (Cooper *et al.*, 2009; Nicolas, 2008) inhibits the NS5B enzyme *in vitro* at a low μM IC_{50}. Its antiviral activity in genotypes 1a and 1b replicons resulted in EC_{50} values of 410nM and 380nM, respectively, with a CC_{50}/EC_{50} selectivity index

>600-fold. In clinical trials VCH-759 was given as a monotherapy over 10 days to three cohorts of treatment-naïve genotype 1 infected patients who received either placebo, 400 mg t.i.d., 800 mg t.i.d. or 800 mg b.i.d. Mean maximal reduction of plasma HCV RNA was 1.9 $\log_{10}$, 2.3 $\log_{10}$ and 2.5 $\log_{10}$ for 400 mg t.i.d., 800 mg b.i.d. and 800 mg t.i.d., respectively. Virological rebound (Nicolas, 2008) was detected in a subset of patients in all three cohorts receiving VCH-759. Analysis of the HCV RNA of these patients revealed resistance mutations M423T/V and L419M. No serious AE resulting in discontinuation of treatment were observed. Reported adverse effects were fatigue, flatulence, dyspepsia, diarrhoea and headache. Neither severity nor frequency of reported AEs were VCH-759 dose dependent.

VCH-222 binds to the Site II pocket and forms a hydrogen bond with Ser-476 and Tyr-477. VCH-222 (Bedard, 2009) inhibits NS5B RDRP enzymatic activity with an IC_{50} of 1.2 μM in 1b (BK strain) and 0.94 μM in 1a (H77 strain). Its antiviral activity in HCV replicons representing genotypes 1a, 1b and 2a resulted in EC_{50} values of 22.3 nM, 11.2 nM and 4.6 μM, respectively. VCH-222 has a high binding affinity for human plasma protein and *in vitro* equilibrium dialysis showed that 99.6% of the compound was bound. VCH-222 has limited toxicity with a CC_{50} of 37.7 μM in HepG2 cells, resulting in a ~4000-fold therapeutic index. In a phase I study (Cooper, 2009) a single oral dose ranging from 250 mg to 1500 mg of VCH-222 was given to healthy patient and all doses were well tolerated. Phase Ib clinical study (Cooper, 2009) used a single dose of 750 mg b.i.d. monotherapy over 3 days in four genotype 1a and one genotype 1b treatment-naïve patients with chronic infection. Within 24 hours of initial treatment all patient demonstrated rapid and significant antiviral response with >3 $\log_{10}$ drop in plasma HCV RNA in both genotypes 1a and 1b patients. Mean maximal reduction of plasma HCV RNA, over the 3 days of treatment, was 3.7 $\log_{10}$. No virological rebound was detected during this 4-day treatment period. No SAE events or discontinuations were observed. Adverse effects for HCV patients were nausea, paraesthesia, abdominal pain, diarrhoea, headache and dizziness. ViroChem was acquired by Vertex in 2009 and telaprevir (NS3 protease inhibitor) + VCH-222 combo.

Site III inhibitors: ANA598, ABT-333 and ABT-072

ANA598

ANA 598 is an NNI of the HCV RNA polymerase discovered by Anadys and exhibits nanomolar potency against genotype 1 HCV replicons [EC_{50}: 1b-con = 3 nM, 1a H77 = 50 nM; EC_{95}: 1b-con = 30 nM (17 ng/ml), 1a H77 = 500 nM (280 ng/ml); Thompson *et al.*, 2008]. Similar potency was observed with genotype 1 treatment-naïve clinical isolates with a mean EC_{50} of 2.8 nM for genotype 1b (n = 10, 0.13 to 14.5 nM) and 29.2 nM for genotype 1a (n = 9, 8–52 nM; Showalter and Thompson, 2009). The difference in activity observed for ANA-598 against 1a and 1b can be attributed in part by the variation at position 415 in the NS5B gene sequence (Y415F in 1a). In addition, no change in susceptibility was observed when ANA598 was tested with replicons containing the primary mutations known to confer resistance to other direct antivirals including A156T (NS3 protease active site), S282T (NS5B active site) and M423T (NS5B non-nucleoside thumb binding site). Genotypic mutations resistant to ANA598 (M414T, M414L, G554D, D559G and M414T/G554D) were identified *in vitro* (Thompson *et al.*, 2008, 2009). Reduced potency was observed for ANA598 against all mutations with EC_{50} values of 0.47, 0.30, 0.63, 0.085 and 20 μM determined for M414T, M414L, G554D, D559G and M414T/G554D, respectively. Replicons containing mutations that confer resistance to ANA598 remained fully susceptible to IFN-a, Telaprevir (NS3 protease inhibitor), and PSI-6130 (NI). ANA598 was shown to be synergistic and/or additive with IFN-a (in both wt and M414T mutant replicons), telaprevir, PSI-6130, and ANA773 (Thompson *et al.*, 2009).

A phase I, randomized, double-blind, ascending, single-dose trial was conducted to evaluate the safety, tolerability and pharmacokinetics of ANA598 capsules (Anadys Pharmaceuticals. 2009). Healthy subjects (18 to 55 years of age) were randomized into treatment cohorts consisting of six subjects receiving single ascending oral doses of ANA598 (400, 800, 1400, 2000 mg or 2000 mg-fed) and two subjects receiving placebo. An additional cohort received 800 mg twice daily (b.i.d.) for 2 doses. In this study, ANA598 was well tolerated at all doses administered thus far. Preliminary data analyses indicate that systemic exposure generally increases with increasing dose of ANA598. Both C_{max} and AUC increased in subjects who received ANA598 after consuming a high-fat meal.

The antiviral activity of ANA598 was first evaluated in two sets of chimpanzees infected with HCV genotypes 1a and 1b (Steffy *et al.*, 2008). Two HCV 1a-infected chimpanzees were given a single oral dose at 30 mg/kg. This single dose of ANA598 produced 0.5 and 1.4 $\log_{10}$ reductions in viral load in the respective animals at 48 hours post-dose. ANA598 was administered to two HCV-1b infected chimpanzees as a once daily dose of 30 mg/kg for 4 consecutive days. Maximum viral load reductions of 2.1 and 2.6 $\log_{10}$ copies RNA/ml for each 1b-infected chimpanzee were observed within 24–48 hours after the initiation of treatment. After the initial drop in viral load, one chimpanzee maintained suppression of viral replication that lasted throughout the treatment period. In the other chimpanzee, a gradual rise of plasma HCV RNA levels was observed beginning at day 2 and continuing through dosing (0.9 log increase from the nadir to 24 hours after last dose). Viral titres in both animals returned to within 2-fold of the baseline value by day 16.

Subsequently, a randomized, double-blind, placebo-controlled, multiple ascending dose trial was conducted to evaluate the safety, tolerability and antiviral activity of orally administered ANA598 in treatment-naive patients with chronic HCV genotype 1 infection (Lawitz *et al.*, 2009). Patients were treated with ANA598 capsules (or matching placebo) at doses of 200 mg, 400 mg or 800 mg b.i.d. for three days. Thirty-five patients participated in the study, with 11 receiving 200 mg b.i.d., eight receiving 400 mg b.i.d., eight receiving 800 mg b.i.d. and eight receiving placebo (with none of the patients receiving placebo showing an end-of-treatment response greater than 0.2 $\log_{10}$ reduction in viral load). In the study, ANA598 treatment resulted in rapid and sustained reductions in HCV RNA with median reductions at end of treatment (day 4) exceeding 2 $\log_{10}$ (>99%) at all dose levels. At 200 mg b.i.d. (twice daily), the median viral load reduction was 2.4 $\log_{10}$ (range of 0.4 to 3.4); at 400 mg b.i.d., 2.3 $\log_{10}$ (range of 1.6 to 3.5); and at 800 mg b.i.d., 2.9 $\log_{10}$ (range of 2.2 to 3.4). Genotype 1a patients demonstrated median reductions of 1.4 $\log_{10}$, 1.8 $\log_{10}$, and 2.5 $\log_{10}$ at 200, 400 and 800 mg b.i.d., respectively. In 10 of the 12 genotype 1a patients who received ANA598, viral load was still declining at the end of the three days of treatment. Genotype 1b patients demonstrated median reductions of 2.6 $\log_{10}$, 2.5 $\log_{10}$, and 3.2 $\log_{10}$, at 200, 400 and 800 mg b.i.d., respectively. No patient showed evidence of viral rebound while on ANA598. ANA598 was well tolerated in this short-term study and there were no serious adverse events.

In addition, two long-term chronic toxicology studies of ANA598 (26 weeks duration in rats and 39 weeks duration in monkeys) were completed (Anadys Pharmaceuticals. 2009). The no observed adverse effect level (NOAEL), is 1000 mg/kg, the highest dose tested, in both the rat and monkey. The completed toxicology studies support the ongoing phase II clinical study as well as future clinical studies of longer duration. In the ongoing phase II study, treatment-naïve genotype 1 patients are to receive ANA598 or placebo in combination with Pegasys® (peginterferon alpha-2a) and Copegus® (ribavirin, USP) for 12 weeks at dose levels of 200 mg or 400 mg b.i.d., each with a loading dose of 800 mg b.i.d. on day one (Anadys Pharmaceuticals. 2009). After week 12, patients will continue receiving SOC alone. Patients who achieve undetectable levels of virus at weeks 4 and 12 will be randomized to stop all treatment at week 24 or 48. The primary endpoint of the study is the proportion of patients who achieve undetectable levels of virus at week 12 (cEVR). Additional endpoints include safety and tolerability as well as the proportion of patients with undetectable levels of virus at week 4 (RVR). Patients will be followed for 24 weeks after stopping therapy to determine the rate of sustained virological response, or SVR. Approximately 90 patients were to be and have been enrolled in this study – with approximately 30 patients receiving ANA598 and 15 receiving placebo at each dose level. Preliminary results from this ongoing phase II study were reported recently. It demonstrated that 73% of hepatitis C patients treated with 200 mg ANA598 twice daily in combination with pegylated interferon and ribavirin (SOC) achieved undetectable levels of virus (<15 IU/ml) at week 12, known as complete early virological response or cEVR. No patient experienced viral rebound on ANA598. ANA598 was well tolerated through twelve weeks, with no serious adverse events reported and a profile of adverse events in the ANA598 group comparable to the group receiving SOC alone.

ABT-333

ABT-333 is a novel NNI with potent inhibition of HCV genotype 1a and 1b polymerases (IC_{50} 2.2–10.7 nM in a biochemical enzymatic assay), but significantly less (> 400-fold) active against genotypes 2a, 2b, 3a and 4a (Wagner *et al.*, 2009). It also displays potent activity against genotype 1a (EC_{50} 7.7 nM) and 1b (EC_{50} 1.8 nM) HCV replicons. The replicon EC_{50} values increased ~13-fold in the presence of human plasma. *In vitro* resistance selection studies with ABT-333 in subgenomic replicons revealed the following common resistance mutations: C316Y, S368T, M414T/I/V, Y448H/C, A553V or S556G. Levels of potency loss by the compound against the mutants ranged from less than 10-fold to more than 1000-fold (C316Y, Y448H/C). Most mutants exhibited reduced replication capacity compared with the wild-type replicon. In short-term replicon combination assays, ABT-333 exhibited additive to synergistic interactions with IFN. In long-term replicon assays, combinations of ABT-333 and IFN exhibited ~1.5 log_{10} greater RNA inhibitory effect than ABT-333 or IFN alone.

The pharmacokinetic profile of ABT-333 in rat, monkey and dog was characterized by a wide range of plasma clearance values with high volumes of distribution in all species. Oral doses of ABT-333 distribute preferentially to the liver in rat and monkey at significant multiples over that observed in plasma and at relatively lower multiples in dog. The CYP metabolism and inhibition profile of ABT-333 together with a low potential for CYP induction suggest a low potential for clinically significant drug–drug interactions.

In a two-day monotherapy with ABT-333, 18 HCV genotype-1-infected subjects were randomized to receive ABT-333 at 100 mg q.d. (n = 2), 100 mg b.i.d. (n = 4), 600 mg q.d. (n = 2) or 600 mg b.i.d. (n = 4) or placebo (n = 6) for two days (Lawitz *et al.*, 2009). Mean Cmax values ranged from 257 to 1126 ng/ml and mean AUC(0–12) values ranged from 1465 to 6184 ng h/ml. The median maximum changes of HCV viral load from baseline were −0.02 log_{10} IU/ml for placebo, −0.6 log 10 IU/ml for 100 mg q.d., −0.46 log_{10} IU/ml for 100 mg b.i.d., −0.58 log_{10} IU/ml for 600 mg q.d. and -1.68 log_{10} IU/ml for 600 mg b.i.d., respectively. Population sequencing did not reveal any resistance mutation.

Subsequently, the antiviral activity of ABT-333 in combination with PEG-IFN/ribavirin (SOC) was evaluated (Klein *et al.*, 2009). In this study, 30 HCV genotype 1-infeced, treatment-naïve subjects were randomized to ABT-333 300 mg b.i.d. (n = 8), 600 mg b.i.d. (n = 8), 1200 mg q.d. (n = 8) or placebo (n = 6) for 2 days of monotherapy followed by 26 days of combination therapy with PEG-IFN/ribavirin. The mean maximum HCV RNA change from baseline was −3.7, −4.0, and −3.5 log_{10} IU/ml for 300 mg b.i.d., 600 b.i.d. and 1200 mg q.d. ABT-333 plus PEG-IFN/ribavirin, respectively, compared with −1.4 log_{10} IU/ml placebo/PEG-IFN/ribavirin. In all the groups who were treated with ABT-333 plus PEG-IFN/ribavirin, at Day 28 41.7% (10 out of 24 patients) had less than 25 IU/ml HCV RNA compared with 0% (0 out of 6 patients) in the placebo group (without ABT-333). The side-effects were reported to be mild in severity (85%) with 63% believed to be from PEG-IFN/ribavirin. There were no serious AEs or discontinuations due to AEs. Lab abnormalities were similar in subjects receiving ABT-333 and placebo.

ABT-072

ABT-072 is another NNI discovered by Abbott Laboratories. In HCV replicons (Wagner *et al.*, 2009), the mean EC_{50} values of ABT-072 were 1 and 0.3 nM against genotype 1a and 1b, respectively. The resistance profile for ABT-072 *in vitro* was similar to ABT-333. Following 2 days of dosing in treatment-naïve HCV genotype-1-infected subjects, the median (range) maximal $\log_{10}$ change in viral load from baseline was −1.4 (−2.3 to −0.8) in 4 subjects receiving ABT-072 160 mg q.d. compared with −0.3 (−0.4 to −0.20) for two placebo-treated subjects (Klein *et al.,* 2009). Further clinical studies with this compound are ongoing, but have not yet been reported.

HCV-796

HCV-796 is a benzofuran-C3-carboxamide inhibitor of HCV RdRp from collaborative discovery efforts by Wyeth Research and ViroPharma Incorporated.

The median inhibitory concentration (IC_{50}) values of HCV-796 were 0.01 to 0.02 μM for genotype 1, and 0.22 to 1.7 for genotypes 2, 3, and 4 in biochemical assays of polymerase inhibition (Howe *et al.,* 2008; Kneteman *et al.,* 2009; Reich *et al.,* 2010). In the subgenomic replicon assay, HCV-796 demonstrated reduction of steady-state levels of HCV RNA with an EC_{50} of 5 nM against genotype 1a and 9 nM against genotype 1b. In the presence of human serum, the EC_{50} values of HCV-796 increased by 2- to 4-fold. Multiple treatments with HCV-796 in replicon cells resulted in a 3 to 4 $\log_{10}$ reduction in HCV RNA levels, compared with 2 to 3 $\log_{10}$ reductions with interferon. Replicon resistance mutations, C316S/N/F/Y, L363V, S365A/T and M414I/V in NS5B polymerase, were associated with 5- to 212-fold reduced susceptibility to HCV-796 (Howe *et al.,* 2008; Flint *et al.,* 2009).

Prior to evaluation of HCV-796 in patients, the antiviral activity of HCV-796 was assessed in chimeric transgenic SCID mice (Kneteman *et al.,* 2009). Treatment with HCV-796 at 50 mg/kg three times daily for 5 days resulted in a 2.02–0.55 $\log_{10}$ decrease in HCV titre with one mouse below the level of detection, whereas HCV levels in the control mice were relatively stable (0.26–0.16 $\log_{10}$ decline) in the chimeric SCID mice. The plasma levels after acute and chronic dosing were 18-fold and 45-fold, respectively, above the EC_{50} in the replicon system, and 5-fold and 12.5-fold above the human serum adjusted EC_{50} of 8 ng/ml relative to the 1a replicon.

In a randomized, double-blind, placebo-controlled study, 102 treatment-naïve adult patients with HCV (72% genotype 1) received 50, 100, 250, 500, 1000, or 1500 mg oral doses of HCV-796 or placebo given as monotherapy twice daily (q12h) for 14 days (Villano *et al.,* 2006). It demonstrated dose-related antiviral activity across multiple HCV genotypes, with maximal effects observed at approximately day 4 of treatment. Mean reductions of HCV RNA at day 4 in the highest dose groups were 1.4–1.5 $\log_{10}$ below baseline. In all dose groups, the predominant virological response pattern demonstrated an initial decline of plasma HCV RNA levels followed by an increase of viral RNA levels that were not explained by changes in plasma concentrations of HCV-796. Among patients on therapy who experienced an initial decline in HCV RNA levels followed by an increase (n = 21), 11 subjects (52%) exhibited emergence of a viral variant carrying a C316Y amino acid substitution in NS5B.

To further evaluate HCV-796 when administered with pegylated interferon alpha-2a (PEG2a), adult patients with chronic HCV infection who were naïve to treatment were randomized to receive oral HCV-796 or placebo every 12 h for 14 days plus PEG (Villano SA and 2007). The mean baseline HCV RNA level was 6.4–6.5 $\log_{10}$ in each group and 71% of patients were infected with HCV genotype 1. In genotype 1 patients, mean reduction from baseline at Day 7 ranged from 1.5 to 2.3 $\log_{10}$ in the combination group versus 0.9 $\log_{10}$ for PEG-IFN alone. At Day 14, the viral titre reduction was 2.6 to 3.2 $\log_{10}$ (combination therapy) versus 1.3 $\log_{10}$. Non-genotype 1 patients demonstrated reductions from baseline of 2.8 to 3.5 $\log_{10}$ and 3.5 to 4.8 $\log_{10}$ on days 7 and 14, respectively, in comparison to 1.5 and 2.6 $\log_{10}$ for PEG-IFN alone. One-third of combination-treated patients dosed with HCV-796 at 250 mg twice daily or higher achieved undetectable viral levels at day 14 (limit of detection, 50 IU/ml HCV RNA). Despite the positive antiviral activity of HCV-796, further development of HCV-796 was withdrawn due to hepatic toxicity.

Conclusion

HCV NS5B polymerase is only one of the targets of the overall HCV antiviral discovery effort. Currently small molecules targeting the NS3 protease activity are in late clinical phases and novel NS5A compounds are undergoing clinical trials. Initial clinical trials with either NS5B NNIs or NS3 protease inhibitors have revealed the selection of drug resistance in several days of monotherapy. It appears that the genetic resistance barrier differ among various classes of antivirals, with NIs > protease inhibitors > NNIs. The low genetic barrier for NNIs will likely require their combination initially with pegylated interferon and ribavirin, later with protease inhibitors and other polymerase inhibitors, as well as NS5A inhibitors when these direct antivirals become available. Given the distinct binding sites and resistance profiles among different NNIs, it is likely that two to three NNIs could be used in combination. The ability of various NNIs to be used in combination and their potent antiviral effects make HCV NS5B NNIs a potential major contribution to the future of HCV treatment. Despite the promising antiviral activity and tolerability of the HCV NNIs demonstrated in days to weeks of treatment, numerous questions regarding safety, tolerability, efficacy, and resistance profiles during longer treatment remain to be answered by ongoing and future clinical trials.

References

American Association for the Study of Liver Diseases (AASLD), Boston, 30 October – 3 November. 3, Abstract# 1586. Hepatology. *50*, 1037A.

Anadys Pharmaceuticals. (2009). Accessed 10 April 2010, Available at: www.anadyspharma.com.

Ago, H., Adachi, T., Yoshida, A., Yamamoto, M., Habuka, N., Yatsunami, K., and Miyano M. (1999). Crystal structure of the RNA-dependent RNA polymerase of hepatitis C virus. Structure 7, 1417–1426.

Bedard, J. (2009). Identification and characterization of VCH-222, a novel potent and selective non-nucleoside HCV polymerase inhibitor. 44th annual EASL meeting (Copenhagen, Denmark).

Biswal, B.K., Cherney, M.M., Wang, M., Chan, L., Yannopoulos, C.G., Bilimoria, D., Nicolas, O., Bedard, J., and James M.N. (2005). Crystal structures of the RNA-dependent RNA polymerase genotype 2a of hepatitis C virus reveal two conformations and suggest mechanisms of inhibition by non-nucleoside inhibitors. J. Biol. Chem. *280*, 18202–18210.

Bressanelli, S., L. Tomei, Tomei, L., Roussel, A., Incitti, I., Vitale, R.L., Mathieu, M., De Francesco, R., and Rey F.A. (1999). Crystal structure of the RNA-dependent RNA polymerase of hepatitis C virus. Proc. Natl. Acad. Sci. U.S.A. *96*, 13034–13039.

Cooper, C. (2009). Safety, tolerability and Pharmacokinetics of the HCV polymerase inhibitor VCH-222 following single dose administration in healthy volunteers and antiviral activity in HCV-infected individuals. 44th annual EASL meeting (Copenhagen, Denmark).

Cooper, C., Lawitz, E.J., Ghali, P., Rodriguez-Torres, M., Anderson, F.H., Lee, S.S., Bedard, J., Chauret, N., Thibert, R., Boivin, I., *et al.* (2009). Evaluation of VCH-759 monotherapy in hepatitis C infection. J. Hepatol. *51*, 39–46.

Flint, M., Mullen, S., Deatly, A.M., Chen, W., Miller, L.Z., Ralston, R., Broom, C., Emini, E.A., and Howe, A.Y. (2009). Selection and characterization of hepatitis C virus replicons dually resistant to the polymerase and protease inhibitors HCV-796 and boceprevir (SCH 503034). Antimicrob. Agents Chemother. *53*, 401–411.

Hammond, J. (2008a). Safety, tolerability and pharmacokinetics of the HCV polymerase inhibitor PF-00868554 following multiple dose administration in healthy volunteers. 59th annual AASLD meeting (San Francisco, CA).

Hammond, J. (2008b). Antiviral activity of the HCV polymerase inhibitor PF-00868554 administered as monotherapy in HCV genotype 1 infected subjects. 59th annual AASLD meeting (San Francisco, CA).

Hirashima, S., Oka, T., Ikegashira, K., Noji, S., Yamanaka, H., Hara, Y., Goto, H., Mizojiri, R., Niwa, Y., Noguchi, T., *et al.* (2007). Further studies on hepatitis C virus NS5B RNA-dependent RNA polymerase inhibitors toward improved replicon cell activities: benzimidazole and structurally related compounds bearing the 2-morpholinophenyl moiety. Bioorg. Med. Chem. Lett. *17*, 3181–3186.

Howe, A.Y., Cheng, H., Johann, S., Mullen, S., Chunduru, S.K., Young, D.C., Bard, J., Chopra, R., Krishnamurthy, G., Mansour, T., and O'Connell, J. (2008). Molecular mechanism of hepatitis C virus replicon variants with reduced susceptibility to a benzofuran inhibitor, HCV-796. Antimicrob. Agents Chemother. *52*, 3327–3338.

Jacobson, I. (2009). Antiviral activity of filibuvir in combination with pegylated interferon alfa-2a and ribavirin for 28 days in treatment-naive patients chronically infected with HCV genotype 1. 44th annual EASL meeting (Copenhagen, Denmark).

Klein, C.E., Cohen, D., Menon, R., Hazan, L., Awni, W.M., Chiu, Y.L., Dumas, E.O., Larsen, L.M., Podsadecki, T.J., Nada, A., and Berntein, B.M. (2009). Safety, tolerability and antiviral activity of the HCV polymerase inhibitor ABT-072 following single and multiple dosing in healthy adult volunteers and two days of dosing in treatment-naive HCV genotype-1-infected subjects. Hep DART, 6–10 December (Kohala Coast, Hawaii), Abstract # 56, 5, 52–53.

Klein, C.E., Menon, R., Cohen, D., Awnt, W.M., and Bernstein, B. (2009). Pharmacokinetics of a polymerase inhibitor, ABT-333, in treatment-naive genotype-1 HCV infected subjects following treatment with 2 days of ABT-333 followed by 26 days of ABT-333 plus pegylated Interferon and ribavirin. Hep DART, 6–10 December, (Kohala Coast, Hawaii), Abstract # 59, 5(Suppl. 1), 54–55.

Kneteman, N.M., Howe, A.Y., Gao, T., Lewis, J., Pevear, D., Lund, G., Douglas, D., Mercer, D.F., Tyrrell, D.L., Immermann, F., *et al.* (2009). HCV796: A selective nonstructural protein 5B polymerase inhibitor with potent anti-hepatitis C virus activity in vitro, in mice with chimeric human livers, and in humans infected with hepatitis C virus. Hepatology *49*, 745–752.

Lawitz, E., Rodriguez-Torres, M., Cohen, D., Tokimoto, D., Xiong, J., Larsen, L.M., Menon, R., Collins, C., and Bernstein, B. (2009). Dose-dependent decrease in HCV viral load following two-day monotherapy with ABT-333 in treatment-naive, HCV genotype 1-infected subjects. 4th International Workshop on Hepatitis C resistance and new compounds, 25–26 June (Boston, USA), Abstract# 18.

Lawitz, E., Rodriguez-Torres, M., DeMicco, M., Nguyen, T., Godofsky, E., Appleman, J.R., Rahimy, M., Crowley, C., and Freddo, J. (2009). Antiviral activity of ANA598, a potent non-nucleoside polymerase inhibitors, in chronic Hepatitis C patients. 44th Annual Meeting of the European Association for the Study of the Liver (EASL) (Copenhagen, Denmark), Absract# 1055. J. Hepatol. *50*, S384.

Lesburg, C.A., Cable, M.B., Ferrari, E., Hong, Z., Mannarino, A.F., and Weber, P.C. (1999). Crystal structure of the RNA-dependent RNA polymerase from hepatitis C virus reveals a fully encircled active site. Nat. Struct. Biol. *6*, 937–943.

Li, H., Tatlock, J., Linton, A., Gonzalez, J., Jewell, T., Patel, L., Ludlum, S., Drowns, M., Rahavendran, S.V., Skor, H., *et al.* (2009). Discovery of (R)-6-cyclopentyl-6-(2-(2, 6-diethylpyridin-4-yl) ethyl)-3-((5, 7-dimethyl-[1, 2, 4]triazolo[1, 5-a] pyrimidin-2-yl)methyl)-4-hydroxy-5, 6-dihydropyran-2-one (PF-00868554) as a potent and orally available hepatitis C virus polymerase inhibitor. J Med Chem. *52*, 1255–1258.

Lindenbach, B.D. and Rice, C.M. (2005). Unravelling hepatitis C virus replication from genome to function. Nature *436*, 933–938.

Nicolas, O. (2008). Genotypic analysis of HCV NS5b variants selected from patients treated with VCH-759, and viral load declines. 43rd annual EASL meeting (Milan, Italy).

Nicolas, O. (2009). Activity and genotypic and phenotypic analysis of HCV NS5b variants selected from patients treated with VCH-916. 44th annual EASL meeting (Copenhagen, Denmark).

Proulx, L. (2008). Results of a safety, tolerability and pharmacokinetics phase I study of VCH-*916*, a novel polymerase inhibitor for HC, following single ascending doses in healthy volunteers. 43rd annual EASL meeting (Milan, Italy).

Reich, S., Golbik, R.P., Geissler, R., Lilie, H., and Behrens, S.E. (2010). Mechanisms of activity and inhibition of the hepatitis C virus RNA-dependent RNA-polymerase. J. Biol. Chem. *285*, 13685–13693.

Showalter, R.E., Thompson, P.A., Steffy, K.R., and Appleman, J.R. (2009). ANA598 displays potent in vitro antiviral activity against diverse clinical isolates of genotype 1 HCV in a transient replicon shuttle vector system.

Steffy, K.R., Kirkovsky, R.L., Lanford, R.E., Showalter, R.E., Sergeeva, M., Zhao, J., Averett, D.R., and Appleman, J.R. (2008). Antiviral efficacy of the HCV RNA polymerase inhibitor ANA598 in the chimpanzee model of HCV infection. AASLD, 4–31 October (San Francisco, USA), Hepatology *48*, 1143A–1164A.

Thompson, P.A., Patel, R.A., Showalter, R.E., Li, C., Appleman, J.R., and Steffy, K.R. (2008). In vitro studies demonstrate that combinations of antiviral agents that include HCV polymerase inhibitor ANA598 have the potential to overcome viral resistance. AASLD, 31 October – 4 November (San Francisco, USA), 1908. Hepatology *48*, 1164A.

Thompson, P.A., Patel, R.A., Steffy, K.R., and Appleman, J.R. (2009). Preclinical Studies of ANA598 combined with other ant-HCV agents demonstrated potential of combination treatment, 44th Annual Meeting of the European Association for the Study of the Liver (EASL) (Copenhagen, Denmark), Abstract# 90 J. Hepatol. *44*, S37.

Troke (2009). Genotypic characterization of HCV NS5b following 8-day monotherapy with the polymerase inhibitor Filibuvir (PF-00868554) in HCV-infected subjects. 44th annual meeting EASL (Copenhagen, Denmark).

Villano, S.A., Howe, A.Y., Raible, D., Harper, D., Speth, J., and Bichier, G. (2006). Analysis of HCV NS5B genetic variants following monotherapy with HCV-*796*, a non-nucleoside polymerase inhibitor, in treatment-naive HCV-infected patients. [Abstract]. Hepatology *44*(Suppl. 1), 607A–608A.

Villano S.A., Raible, D., Harper, D., Speth, J., Chandra, P., Shaw, P., and Bichier, G. (2007). Antiviral activity of the non-nucleoside polymerase inhibitor, HCV-*796*, in combination with pegylated interferon alfa-2b in treatment-naive patients with chronic HCV. J. Hepatol. *46*, S24.

Wagner, R., Maring, C., Flentge, P., Donner, P., Hutchinson, D., Randoph, J., Krueger, A., Rockway, T., Pratt, J., Liu, D., *et al.* (2009). Preclinical characterization of ABT-333 and ABT-072: novel non-nucleoside HCV NS5B polymerase inhibitors. HepDart, 6–10 December (Kohala Coast, Hawaii), Abstract # 108, 5, 100–101.

Hepatitis C Virus Glycoprotein-dependent Entry

17

Heidi E. Drummer and Jane A. McKeating

Introduction

Hepatitis C virus (HCV) is an important human pathogen that establishes chronic infection that leads to progressive liver disease. At the present time limited therapeutics are available and new targets in the viral lifecycle are urgently required to advance the development of antiviral agents. Viruses initiate infection by attaching to molecules or receptors at the cell surface, providing an opportunity for therapeutic intervention. HCV encodes two glycoproteins, E1 and E2, and much of our current understanding of the viral entry process has been derived from studying glycoprotein interaction(s) with host cells. HCV demonstrates a clathrin-dependent internalization process that involves several host cell molecules or receptors, including CD81, scavenger receptor class B member I (SR-BI) and the tight junction proteins claudin-1 and occludin. This chapter will review the structural and functional information on the glycoproteins and their interaction(s) with receptors in defining HCV internalization.

Glycoprotein E1 and E2 are type 1 transmembrane proteins that are cleaved from the polyprotein through the action of signal peptidases in the endoplasmic reticulum (ER) where they form disulfide linked and non disulfide linked heterodimers. The native folding of E1 and E2 is highly dependent on their coexpression, although a subdomain of E2 can be expressed in isolation that retains receptor binding properties (sE2; Dubuisson and Rice, 1996; Michalak *et al.*, 1997; Cocquerel *et al.*, 2001, 2003; Patel *et al.*, 2001; Brazzoli *et al.*, 2005). The HCV glycoproteins are distantly related to the phylogenetically related *Flavivirus* glycoproteins prM and E, although sequence comparisons together with functional and structural studies suggest significant divergence.

Glycoprotein E1

Glycoprotein E1 is encoded by amino acid residues 171–383 of the HCV open reading frame (ORF; HCV strain H77 polyprotein numbering), where the last 30 amino acids form a transmembrane domain (TMD) responsible for anchoring E1 to the virion or membrane of an infected cell (Fig. 17.1A). The E1 TMD also functions to translocate the adjoining E2 protein in the ER lumen, acting in a similar manner to a signal sequence (Cocquerel *et al.*, 2002). Glycoprotein E1 encodes five conserved sites for the addition of N-linked glycans (Asn-X-Ser/Thr) although one of these sites at position 325 (NWS) is not utilized (Meunier *et al.*, 1999). In some strains, a sixth site is found at position 250 (Fig. 17.1A). The majority of E1 bears high mannose forms of N-linked carbohydrates, consistent with its localization in the ER. Heterologous expression of E1E2 results in the transport of small amounts of E1 and E2 to the cell surface. This observation allowed the development of infectious retroviral pseudotypes where plasma membrane expressed E1E2 glycoproteins are incorporated into retrovirus particles (HCVpp) that can mediate entry into hepatic cells (Bartosch *et al.*, 2003b; Drummer *et al.*, 2003; Hsu *et al.*, 2003). Analysis of N-linked glycans of HCVpp associated E1 reveal the majority to remain sensitive to deglycosylation with

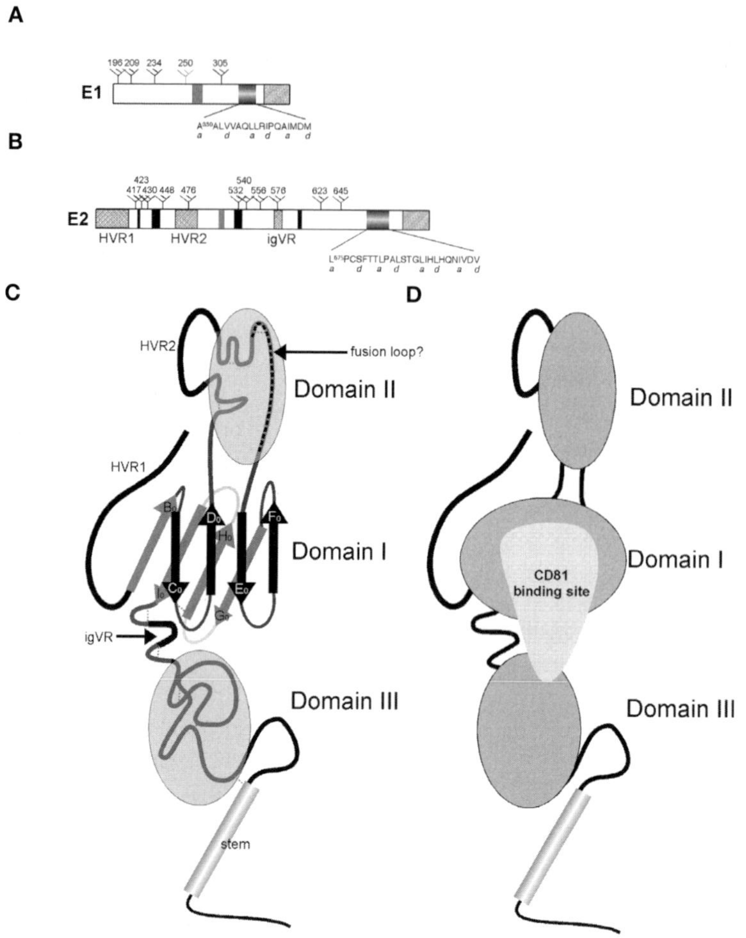

Figure 17.1 Schematic representation of the HCV glycoproteins E1 and E2. (A) Glycoprotein E1 showing the location of the 4 conserved N-linked glycosylation sites (black trees) and one non-conserved site (grey tree). The internal hydrophobic region (grey), the membrane proximal region (graded shading) and the TMD (diagonal hatch marks) are indicated. (B) Glycoprotein E2 showing the location of the 11 conserved N-linked glycosylation sites (black trees), regions involved in CD81 binding (black) the three variable regions (crossed hatch marks), the predicted fusion loop (grey) the stem region (graded shading) and the TMD (diagonal hatch marks). (C) Proposed disulfide arrangement and three dimensional structure of glycoprotein E2 (Krey *et al.*, 2010). The three domains are indicated, with domain I comprising eight antiparallel β-sheets (B0-I0). The proposed positions of HVR1, HVR2 and the igVR (thick black lines), the disulfide bonds (dotted lines) and the predicted fusion loop in domain II (dashed line) are indicated. (D) The CD81 binding site (chequered hatch marks) overlaps with residues from domains I and III.

endoglycosidase H, suggesting that only a small proportion of the high mannose carbohydrate is modified to complex or hybrid forms in the Golgi (Drummer *et al.,* 2003). N-linked glycosylation of positions 196 and 305 are essential for E1E2 dimerization, while sites 196, 209 and 305 were required to achieve wild-type levels of HCVpp entry (Goffard *et al.,* 2005). None of the glycosylation sites in E1 affect polyclonal HCV specific antibody neutralization of HCVpp (Helle *et al.,* 2007).

Expression studies in CHO-K1 cells demonstrated that the folding of an oxidized DTT-resistant form of E1 was complete by 2 hours post-synthesis whilst E2 folding required at least 6h. Native E1E2 heterodimer formation coincided with E2 folding kinetics (Brazzoli *et al.,* 2005). Other studies suggest that E1 folding is dependent on the presence of E2 (Dubuisson and Rice, 1996; Michalak *et al.*, 1997). Michalak *et al*, (1997) examined the effect(s) of C-terminal truncations on E1 localization in a mammalian expression system, where truncation at residue 311 resulted in the secretion of soluble E1. Using both mammalian and yeast expression systems, Lorent *et al.* reported on the expression of E1 ectodomain residues 192–326 of a consensus genotype 1b E1 sequence (Lorent *et al.*, 2008). Detergent was required to maintain E1 in a monomeric state and removal resulted in the formation of stable spherical particles. Preliminary structural analysis of monomeric mammalian and yeast-expressed forms of E1 reveal similar structures with approximately 20% α-helix content (Lorent *et al.*, 2008). Unlike its *Flavivirus* counterpart prM, no further post translational cleavages have been reported for E1.

E1 transmembrane domain

The E1 TMD comprises two distinct hydrophobic stretches separated by a single charged amino acid. Prior to signal peptidase cleavage, these two hydrophobic regions span the membrane twice with hairpin-like topology. Following cleavage, the C-terminal portion is reoriented towards the cytosol where the TMD assumes a single span topology (Cocquerel *et al.*, 2002). Both E1E2 glycoproteins localize to the ER via their TMDs despite the absence of classical retention signals such as a dibasic motif or KDEL sequence (Dubuisson *et al.*, 1994; Duvet *et al.*, 1998; Cocquerel *et al.*, 1999, 2000; Flint and McKeating, 1999). A single charged lysine residue (K370) in the E1 TMD is conserved in all HCV strains and is a major determinant of ER localization. Mutation of K370 to Ala results in E1 trafficking from the ER to Golgi apparatus, where high mannose carbohydrates are converted into endoglycosidase H resistant complex and hybrid forms (Cocquerel *et al.*, 2000). Additional determinants of ER localization have also been described in the juxtamembrane region of E1 (Mottola *et al.*, 2000).

Alanine insertion mutagenesis showed the central and N-terminal portion of the E1 TMD to be essential for dimerization with glycoprotein E2 (Op De Beeck *et al.*, 2000). Alanine insertions at residues 355, 358, 361 or 369 reduced E1E2 dimerization to approximately 10% of that observed for wild-type E1E2. Tryptophan substitution studies confirmed this observation and identified three amino acids as essential for achieving wild-type levels of E1E2 dimerization, G354W, G358W and K370W (Ciczora *et al.*, 2007). Although mutations G354W and K370W reduced virus entry, virions encoding G358W entered cells at near wild-type levels (Ciczora *et al.*, 2007). The glycine residues at positions 354 and 358 form a GXXXG oligomerization motif which is believed to facilitate transmembrane helix interactions primarily through van der Waals contacts. The GXXXG motif has been found in numerous homo and heterodimeric protein complexes including integrins (Schneider and Engelman, 2004), G-protein coupled receptors (Overton *et al.*, 2003) and ErbB receptor (Gerber *et al.*, 2004). However, a similar motif is not found in the E2 TMD and it remains unclear how an E1 GXXXG motif functions in dimerization with E2.

E1 membrane proximal region

Adjacent to the TMD of E1 is a weakly conserved heptad repeat sequence spanning amino acids 330–347 (Drummer *et al.*, 2007; Fig. 17.1A). Mutagenesis of the *a* and *d* positions reveal that they are not essential for dimerization with E2. However, position specific effects on virus entry were found, with V333, L337 and M347 being essential for virus entry. None of the mutations tested affected E2 folding as assessed using a conformation-dependent antibody and receptor binding assays. Recent NMR structural studies of a synthetic peptide corresponding to residues 314–342 revealed two helical regions between residues 319–323 and 329–338 (Spadaccini *et al.*, 2009). The peptide interacted with anionic phospholipids, where the presence of cholesterol affected peptide–bilayer interactions and favoured a close positioning of the peptide with the bilayer surface (Spadaccini *et al.*, 2009). These data suggest that the membrane proximal region of E1 plays a role during virus entry and may be involved in membrane destabilization and viral fusion.

E1 membrane interacting regions

Numerous studies, using synthetic peptides, site directed mutagenesis and molecular modelling have suggested a role for E1 in the membrane fusion process. Synthetic peptides encoding residues 268–298, 317–339 and 360–382 mediated hemifusion and complete fusion of model liposomes at both pH 5.4 and pH 7.4 (Perez-Berna *et al.*, 2006). The ability of the peptide-spanning residues 360–382 to interact with membranes is consistent with its role as the TMD of E1. Amino acids 268–298 comprise an internal hydrophobic region and molecular modelling studies suggest a putative internal fusion peptide (Flint *et al.*, 1999b; Garry and Dash, 2003; Fig. 17.1A). Alanine mutagenesis of key hydrophobic and aromatic amino acids in this region (Y276, V277, F285 and L286) revealed that only F285 was essential for virus entry (Drummer *et al.*, 2007). Subsequently, Lavillette showed that mutation of Y276 to phenylalanine reduced both virus entry and virus mediated liposome fusion (Lavillette *et al.*, 2007). In addition, amino acids G282 and G288 were essential for HCVpp entry and low pH mediated fusion (Lavillette *et al.*, 2007; Li *et al.*, 2009). The role of this region in membrane fusion is unlikely to be resolved in the absence of supporting three-dimensional structural information. The function of E1 as a fusion protein is not supported from the recent disulfide mapping studies of E2 glycoprotein, suggesting it represents a class II fusion protein and contains the fusion peptide or loop (Krey *et al.*, 2010).

Glycoprotein E2

The HCV glycoprotein E2 is encoded by amino acids 384–746 (HCV polyprotein numbering) and heavily modified by the addition of N-linked carbohydrates at 11 sites (Fig. 17.1B). Rodent and human anti-E2 monoclonal antibodies (mAbs) have provided important insights into antigenic differences between recombinant forms of E2, insect cell derived virus-like particles and HCVpp incorporated E2. Conformation sensitive E2 specific mAbs preferentially immunoprecipitate E1E2 complexes, suggesting they recognize native forms of E1E2 (Deleersnyder *et al.*, 1997). E2 comprises a C-terminal TMD (residues 716–746), a highly conserved membrane proximal heptad repeat region (residues 675–699), and a receptor binding domain (RBD; residues 384–661). Three variable regions with distinct properties and functions are located within the RBD, these can be deleted from the RBD without affecting CD81 binding or overall antigenic structure (McCaffrey *et al.*, 2007).

E2 transmembrane domain

The C-terminal 30 amino acids of E2 anchors the glycoprotein to the membrane of the virion or infected cell and comprises two hydrophobic stretches separated by two highly conserved charged amino acids, Asp728 and Arg730. In addition to membrane anchoring, the C-terminal TMD translocates and integrates the adjoining p7 protein into the ER lumen, similar in function to a signal peptide (Cocquerel *et al.*, 2002). The E2 TMD causes ER retention and mutation of Asp728 or Arg730 to Ala abolishes ER retention (Cocquerel *et al.*, 1998, 2000; Ciczora *et al.*, 2005). Similar to the E1 TMD, the TMD of E2 is likely to maintain a hairpin like conformation prior to signal peptide cleavage and adopts a single spanning conformation post cleavage (Cocquerel *et al.*, 2002). The E2 TMD is a major determinant of E1E2 dimerization. Although alanine insertions within E2 TMD had smaller effects on E1E2 dimerization than observed for the TMD of E1 (Op De Beeck *et al.*, 2000) tryptophan mutagenesis revealed that Asp728 is a key residue involved in E1E2 dimerization (Ciczora *et al.*, 2007).

The E2 membrane proximal region

The membrane proximal region of E2 contains a hydrophobic heptad repeat sequence (residues 675–699), where mutation of *a* and *d* positions to alanine or proline at residues Leu-675, Ser-678, Leu-689 and Leu-692 decreased E1E2 dimerization, with a concomitant reduction in virus entry (Fig. 17.1B). Mutation of two conserved proline residues at consecutive *b* positions, Pro 676 and Pro 683, to Alanine reduced virus entry with minimal effect(s) on E1E2 dimerization (Drummer and Poumbourios, 2004). The location, structure and function of the E2 heptad repeat region is similar to that found in the phylogenetically related glycoprotein E of *Flavivirus*

TBE. Glycoprotein E comprises a 496 amino acid type 1 membrane protein that has receptor binding and membrane fusion function. Glycoprotein E can be readily cleaved from the surface of virions with trypsin to yield a 395-residue fragment (Rey *et al.*, 1995). This 395-residue fragment is a dimer at neutral pH and its structure was solved to 2Å resolution (Rey *et al.*, 1995). A 50 amino acid 'stem' connects this soluble fragment to the TMD. Within the glycoprotein E stem region are two putative α-helical regions. The membrane proximal region is encoded by a heptad repeat and contributes to the stability of prM-E heterodimers at neutral pH, while the N-terminal helix promotes the conversion of glycoprotein E dimers to trimers at low pH (Allison *et al.*, 1999). Crystal structures of both the neutral (prefusion) and low pH (post-fusion) forms of glycoprotein E suggest that the stem region packs onto the outside of the E trimer driving the apposition of the lipid bound fusion peptide and TMD, facilitating membrane fusion (Bressanelli *et al.*, 2004; Modis *et al.*, 2004). It is possible that the heptad repeat region between HCV E2 RBD and TMD functions as a 'stem' region in low pH-dependent fusion.

E2 hypervariable regions

Glycoprotein E2 contains three regions whose sequences vary widely between HCV strains. These regions localize to the extreme N-terminus of E2, termed hypervariable region 1 (HVR1, residues 384–410), hypervariable region 2 (HVR2, residues 460–485) and the intergenotypic variable region (igVR, residues 570–580). Each of these regions appear to have distinct functional properties and will be discussed separately.

HVR1

Hypervariable region 1 is the most variable region within the HCV genome although there is strong selective pressure to maintain its length at 27 amino acids (Casino *et al.*, 1999; Fig. 17.1B). *In vitro* studies demonstrate that antibodies directed towards HVR1 are neutralizing but highly type-specific and diversification of the HVR1 sequence occurs rapidly through immune escape (Farci *et al.*, 1996, 2000; von Hahn *et al.*, 2007). Although HVR1 shows a high degree of variability, close examination of the sequence polymorphism reveals a conservation of basic charge, with the frequency of basic residues at each position being genotype dependent (Penin *et al.*, 2001; Callens *et al.*, 2005). Increasing the number of basic residues within HVR1 promotes HCVpp entry, whilst replacement of basic residues with uncharged amino acids reduced virus entry to levels observed for the HVR1 deletion (Callens *et al.*, 2005). Thus, the degree of positive charge in the HVR1 may modulate HCV entry. Removal of HVR1 reduces HCVpp entry and attenuates viral replication in chimpanzees (Forns *et al.*, 2000; Callens *et al.*, 2005). Deletion of HVR1 from HCV reduced the infectivity of low buoyant density particles (1.02–1.08 g/ml) and increased their ability to bind recombinant forms of CD81 (Bankwitz *et al.*, 2010). Of note, infectivity was restored by an adaptive mutation at position 347 (347L) within the E1 membrane-proximal heptad repeat. Roccasecca *et al.* demonstrated that deletion of the HVR (384–410) increased sE2 binding to Molt-4 T-cells that are known to express CD81 (Roccasecca *et al.*, 2003). In contrast, Callens *et al.* reported that deleting the HVR reduced HCVpp incorporated E1E2 glycoprotein interactions with CD81 (Callens *et al.*, 2005). Taken together these findings suggest that HVR1 does not directly participate in CD81 binding but may shield the CD81 binding region of E2.

The high degree of positive charge in the HVR1 suggests a likely role for interaction(s) with charged cell surface glycosaminoglycans (GAGs). HCV entry has been reported to be dependent on GAGs (Basu *et al.*, 2007; Morikawa *et al.*, 2007). Contradictory findings on how HCV interacts with GAGs have been published and most likely relate to the forms of E1 and E2 being used. Although intracellular forms of E2 (Callens *et al.*, 2005), E1E2 expressed in insect cells and purified as virus-like particles and recombinant E2 (Barth *et al.*, 2003) interact with heparin, HCVpp incorporated E1E2 have variable interactions with heparin (Callens *et al.*, 2005; Basu *et al.*, 2007). Furthermore, the number of basic HVR1 residues within intracellular forms of E1E2 did not alter binding to GAGs (Callens *et al.*, 2005). In summary, these data suggest that HVR1 contributes to HCV interactions with cell surface GAGs.

HVR2

Hypervariable region 2 is located between residues 460–480 and is bound by two conserved cysteine residues (Fig. 17.1B). Amino acid substitutions rapidly accumulate within HVR2, although there is limited evidence for antibody-driven selective pressures on the evolution of its sequence. Limited studies have been performed on sequence changes within HVR2 during HCV infection. A function for HVR2 in HCV entry has not been described. Replacement of HVR2 with a Gly-Ser-Ser-Gly linker did not alter the ability of E2 RBD to fold or interact with CD81 (McCaffrey *et al.*, 2007). However, when both the HVR1 and HVR2 were replaced with the corresponding regions of other HCV genotypes, these substitutions were found to cooperatively affect CD81 interactions (Roccasecca *et al.*, 2003). These studies suggests that HVR1 and HVR2 may act cooperatively to shield the CD81 binding site on E2 in a strain-dependent manner.

Intergenotypic variable region

The most recently described variable region of E2, the igVR does not display the same degree of sequence variability as seen in HVR1 or HVR2 (McCaffrey *et al.*, 2007; Fig. 17.1B). Within an infected patient, and within genotypes, its sequence is largely conserved. However, between genotypes, the igVR varies widely both in length and in sequence composition, ranging from 10 to 15 amino acids in length. Like HVR2, the igVR is flanked by conserved cysteines. Replacement of the igVR with a flexible Gly-Ser-Ser-Gly linker did not alter the ability of the E2 RBD to fold correctly or to interact with CD81. The role of the igVR in HCV biology and the mechanism of its evolution have not been described.

Receptor binding domain

The N-terminal region of glycoprotein E2, residues 384–661, comprises a discrete receptor-binding domain (RBD). Due to the ease of expression and purification of the E2 RBD, it provides an important tool for studying HCV-receptor interactions and is an ideal candidate for crystallization studies to resolve the three dimensional structure. The CD81 binding site comprises four discontinuous sequences with direct CD81 interactions mediated via Trp420, Trp437, Leu438, Leu441, Phe442, Tyr527, Trp529, Gly530, Asp535 and amino acids in the region spanning residues 613–618 (Rocasecca *et al.*, 2003; Drummer *et al.*, 2006; Owsianka *et al.*, 2006; Fig. 17.1B). Soluble forms of E2 can interact with full-length and recombinant forms of CD81 and have been invaluable to study HCV–CD81 interactions. However, mutagenesis of glycoprotein E2 reveal important differences in the exposure of the CD81 binding site between intracellular and virion incorporated forms (Drummer *et al.*, 2006). Mutation of residues G436A/S, L438V, G440A, F442W, Y443A/W induced a more pronounced effect on HCVpp interactions with CD81 than intracellular forms of the glycoproteins. These data suggest that E2 undergoes conformational maturation of the CD81 binding site during its biosynthesis in the secretory pathway.

McCaffrey *et al.* examined the role of the three variable regions on the folding and domain organization of the E2 RBD (McCaffrey *et al.*, 2007). Replacement of each variable region with a gly-ser-ser-gly linker had minimal effect on expression or secretion. None of the variable regions contributed individually to E2–CD81 interaction and all three variable regions could be simultaneously removed without affecting CD81 binding. However, the simultaneous deletion of HVR1 and HVR2 from E2 RBD resulted in a 50% reduction in binding to full-length CD81 on the surface of transfected CHO cells (McCaffrey *et al.*, 2007). In support of this, Rocasecca *et al.* demonstrated that HVR1 and HVR2 cooperatively modulate CD81 binding (Roccasecca *et al.*, 2003).

Three immunogenic domains were identified in the E2 RBD (Hadlock *et al.*, 2000; Keck *et al.*, 2004, 2005, 2007; Owsianka *et al.*, 2008; Perotti *et al.*, 2008). Antibodies directed towards domain B and C inhibit E2–CD81 interactions and HCV infectivity (Keck *et al.*, 2004, 2007). Domain B is immunodominant and epitope-mapping studies identified the region of reactivity of MAbs to reside between residues 523 and 540, involving residues Gly523, Trp529, Gly530, Asp535, Phe437, Asp431 and several residues surrounding the glycan at position 645 (Helle *et al.*, 2007;

Johansson *et al.*, 2007; Keck *et al.*, 2008a,b; Owsianka *et al.*, 2008). This region corresponds to amino acids previously identified to be a component of the CD81 binding site (Owsianka *et al.*, 2006). Antibodies that recognize domain A are non-neutralizing and have minimal effect(s) on E2 interaction with CD81 (Keck *et al.*, 2004). Domain A appears to be highly flexible and to be sensitive to changes in pH, where the B_{max} of domain A MAbs to low pH treated HCVpp increased by 46%, whereas domain C increased by only 23% and domain B decreased by 12% (Keck *et al.*, 2005). Domains A and C are likely to be in close proximity on the surface of virus particles. In addition, these data suggest that low pH increases the number of domain A epitopes available for binding without changing the overall conformation of this domain.

Soluble E2 RBD can inhibit HCV infection of hepatoma cells (Whidby *et al.*, 2009), suggesting that important epitopes required for virus infection are present within recombinant sE2. E2 RBD has the propensity to form disulfide linked dimers (Whidby *et al.*, 2009). Removal of cysteine (C565S, C652S using H77 polyprotein numbering) increased the amount of non-disulfide linked E2 but did not abrogate SDS resistant dimerization, suggesting that other cysteine residues may be involved in intermolecular disulfide linkages (Whidby *et al.*, 2009). Krey *et al.* expressed E2 RBD (residues 384–715) in the presence of glycoprotein E1 to facilitate native folding. Far-UV CD and FTIR spectra of purified monomeric E2 RBD revealed approximately 28% β-sheet and 5% helix compared with 42% β-sheet and 5% helix observed for other class II fusion proteins. Tryptic digests of deglycosylated E2 and mass spectrometry assigned the disulfide bonding arrangement. Using the observed disulfide arrangement of E2 together with known features of E2 antigenicity and CD81 binding the authors proposed a HCV E2 structure based on the known structural information for other class II fusion proteins (Fig. 17.1C). Domain I of HCV E2 is likely to comprise eight antiparallel β strands and is connected to domain III by the igVR. Domain I contains the N-terminal HVR1 and CD81 binding residues are located on β strands assigned as C_0, D_0, E_0 and F_0 (Fig. 17.1D). Domain I also contains 5 of the 11 glycosylation sites of E2 that surround the CD81 binding region. Amino acids observed to form epitopes of domain B specific MAbs map to this region suggesting that domain I is the equivalent of domain B.

Domain II comprises amino acids 444–523 and its structure is maintained through three intramolecular disulfides formed between consecutive Cys residues 452–459, 486–494 and 503–508 and incorporates HVR2 (Fig. 17.1C). The absence of long-range disulfides is likely to make this region highly flexible, consistent with the ability to delete HVR2. The proposed structure does not provide further insights into the role of HVR2 in HCV entry. Domain II is likely to be the equivalent of the highly flexible, pH sensitive domain A identified using human mAbs.

Domain III comprises amino acids 581–651 which in other flaviviruses has an immunoglobulin-like fold. The ability to replace the igVR region connecting domains I and III with a GSSG linker suggest they are in close apposition prior to membrane fusion. Domain III contains a portion of the CD81 binding region in close proximity to DI (Roccasecca *et al.*, 2003). This is in agreement with mAb studies showing that the CD81 binding site overlaps with domains B and C (Fig. 17.1D). Overall the proposed structure of E2 suggests that DI and DIII must move apart during fusogenic conformational change to appose the TMD with the fusion loop. Since both DI and DIII are bound to CD81, a requirement for membrane fusion may be CD81 dissociation at low pH. The remaining ectodomain (residues 652–715) comprises the stem region, stabilized by a disulfide bond that creates a loop between residues 652 and 677 and is consistent with the experimental data obtained for the membrane proximal region as a component of this stem region.

In the model proposed by Krey *et al.*, the fusion loop is likely to reside within DII between amino acids 502–520, a region that is strictly conserved in all HCV genotypes (Fig. 17.1C). This region is rich in hydrophobic amino acids, glycines and prolines. Lavillette *et al.* observed that mutations G504D, Y507A and F509A reduced viral entry although all of these mutations reduced E1E2 incorporation into particles (Lavillette *et al.*, 2007). It may therefore be difficult to examine the effect of these mutations on virus fusion to verify its role as a fusion peptide. Further structural studies of E2 are required to elucidate the precise location and structure of the fusion peptide in glycoprotein E2.

Role of E2 glycosylation

Glycoprotein E2 contains eleven sites for N-linked glycosylation essential for the correct folding in the ER and secretory pathway and for the formation of its antigenic structure. Ten of the eleven sites are conserved in all HCV genotypes, while the site at position 540 is absent in genotype 3 and the majority of genotype 6 sequences. Intracellular forms of E2 contain high mannose forms of N-linked carbohydrate. Mutation of the asparagine residue in the Asn-X-Ser/Thr codon to glutamine revealed that N-linked glycosylation occurs at all of the eleven potential sites in E2 (Goffard *et al.*, 2005). Addition of N-linked carbohydrate at positions 540, 556 and 623 were essential for E1E2 dimerization. Addition of N-linked carbohydrate at positions 556 and 623 were also required CD81 binding. Glycosylation at positions 423, 448, 556 and 623 were required to achieve wild-type levels of virus entry (Goffard *et al.*, 2005). A similar study performed by Falkowska *et al*, (2007) in which the asparagine residue of the Asn-X-Ser/Thr motif were mutated to alanine showed that N-linked carbohydrate at position 417, 423, 448, 532, 556 and 645 were essential for HCVpp entry. In contrast to the results obtained by Goffard *et al.*, N623A reduced HCVpp entry by only 20% relative to wild-type. It is possible that both amino acid side chains and glycosylation of a sequence contribute to glycoprotein function. Taken together the studies suggest that N-linked carbohydrates at positions 423, 448 and 556 are essential for E2 function while both the local amino acid sequence and N-linked glycosylation are important for E2 function at residues 417, 532, 623 and 645.

The role of glycosylation in shielding the E2 glycoprotein from neutralizing antibodies has been studied (Helle *et al.*, 2007). Removal of a single N-linked carbohydrate at positions 417, 532 or 645 increased the sensitivity of HCVpp to neutralization by polyclonal and anti-E2 monoclonal antibodies and increased the sensitivity of HCVpp to inhibition by recombinant forms of CD81. This result suggests that N-linked carbohydrates at positions 417, 532 and 645 shield important neutralization epitopes and the CD81 binding site on the viral glycoprotein complex. By contrast, removal of N-linked carbohydrate at positions 476 and 540 increased neutralization resistance suggesting a change in local conformation reduces the affinity of neutralizing antibodies for their epitopes (Helle *et al.*, 2007).

Systems to study HCV entry

Like many aspects of HCV research, investigations into particle attachment and entry were hampered by the lack of *in vitro* systems. For over fifteen years after the HCV genome was identified, researchers were unable to propagate the virus in cultured cells. One of the primary challenges has been the low replication efficiency of serum derived virus in primary human hepatocytes or hepatoma derived cell lines (reviewed in Farquhar and McKeating, 2008). Although hepatocytes have been shown to support low levels of HCV replication, PCR detection of negative strand viral RNA was required to verify active replication (Fournier *et al.*, 1998; Castet *et al.*, 2002; Wakita *et al.*, 2005; Molina *et al.*, 2008). The identification of molecular clones that were capable of infecting chimpanzees (Kolykhalov *et al.*, 1997) led to the development of subgenomic and full-length genomes or replicons that replicated *in vitro*. However these systems also failed to release infectious progeny virus. In spite of this limitation, replicon systems have been widely used to study the cellular requirements for HCV replication.

A well established method to study viral glycoprotein interactions with the host cell is to express the glycoproteins in isolation and to perform cell–cell fusion assays. Although this approach has been invaluable for studying the fusion requirements of many viruses, the active retention of the HCV glycoproteins in the ER limited the usefulness of this approach (reviewed in Lavie *et al.*, 2007). In contrast, soluble forms of E2 RBD bound to human cells and enabled the identification of CD81 and SR-BI as putative receptors (see below). HCV E1E2 glycoproteins exist as a heterodimer on the particle surface and sE2 association with host cell proteins is unlikely to fully recapitulate the function(s) of particle associated E1E2 glycoproteins (Deleersnyder *et al.*, 1997; Flint *et al.*, 2000; Op De Beeck *et al.*, 2004). However, recent publications demonstrating that sE2 can inhibit HCV infection of hepatoma cells, suggest that sE2 interaction(s) with the cell surface recapitulate some aspects of virus–receptor engagement (Whidby *et al.*, 2009; Krey *et al.*, 2010).

HCV pseudoparticles

In 2003 several laboratories reported the development of infectious retroviral pseudoparticles bearing E1E2 glycoproteins (HCVpp; Bartosch *et al.*, 2003a; Drummer *et al.*, 2003; Hsu *et al.*, 2003). HCVpp show a restricted infection of liver derived cells and both glycoproteins are required for the production of infectious pseudoparticles, supporting a model that E1E2 heterodimers are the functional unit defining particle entry. HCVpp have played an integral role in dissecting the mechanism(s) of viral entry, providing a single cycle model where the glycoproteins induce receptor-dependent internalization and retroviral proteins direct other stages of replication. Retrovirus particles bud from the plasma membrane and incorporate host cell or exogenous proteins into their envelope. HCVpp can be designed to express diverse glycoproteins (McKeating *et al.*, 2004; Lavillette *et al.*, 2005). HCVpp infect hepatocytes and hepatoma derived cell lines implying that virus glycoprotein–receptor interactions may in part define HCV tropism for the liver (Bartosch *et al.*, 2003a; Hsu *et al.*, 2003). This system has allowed investigators to define the role of cellular proteins in the HCV internalization process (reviewed in Budkowska, 2009).

The JFH-1 strain of HCV

Studying HCV entry in the context of an authentic particle was finally realized in 2005 when a strain of HCV was reported to replicate and release infectious particles in cell culture (HCVcc; Lindenbach *et al.*, 2005; Wakita *et al.*, 2005; Zhong *et al.*, 2005). This strain was cloned from a patient with fulminant hepatitis and is termed Japanese fulminant hepatitis-1 (JFH-1). This unique strain is able to replicate in cultured hepatoma cells without acquiring mutations and to assemble particles that are infectious *in vitro* and *in vivo* (Wakita *et al.*, 2005). Chimeric JFH-1 viruses expressing the structural proteins from diverse HCV genotypes enable experiments to study the role of glycoprotein diversity in virus entry (Lindenbach *et al.*, 2006; Gottwein *et al.*, 2007, 2009; Scheel *et al.*, 2008). The JFH-1 strain allows studies on the full life cycle from entry to egress, not only in hepatocyte derived cell lines but in cultures of primary human hepatocytes (Molina *et al.*, 2008; Ploss *et al.*, 2010). The high efficiency of JFH-1 replication makes it possible to detect viral proteins by antibody-based assays such as immunohistochemistry and flow cytometry. JFH-1 particles are infectious for chimpanzees as well as for UpA-SCID chimeric mice, which contain transplanted human hepatocytes (Wakita *et al.*, 2005; Zhong *et al.*, 2006). Virus recovered from JFH-1cc-infected animals can replicate *in vitro* (Lindenbach *et al.*, 2006). Serum derived HCV from an experimentally infected animal shows a different density profile from HCVcc propagated *in vitro* (Lindenbach *et al.*, 2006), reinforcing earlier reports that plasma derived HCV associates with lipoproteins (Thomssen *et al.*, 1993; Nielsen *et al.*, 2004). Recent studies demonstrating that infectious particle assembly and/or secretion is linked to the lipoprotein synthesis machinery of the hepatocyte suggest an important role for particle associated lipoproteins in the HCV life cycle (Popescu and Dubuisson, 2009). We recently reported that particle density associates with the sensitivity of virus to neutralization by glycoprotein specific antibodies, with more dense particles with lower lipid content being more sensitive to neutralization (Grove *et al.*, 2008).

Discovery of HCV entry factors or receptors

All viruses must transverse the barrier of the cellular membrane and this is achieved by interaction with cell surface proteins or receptors that mediate viral binding and entry. Virus–receptor interactions are specific and confer host and cellular tropism. For HCV, entry is a complex process involving a number of host factors, where initial adsorption of the virus may be facilitated by low affinity attachment factors like GAGs (Barth *et al.*, 2003), lectins (Lozach *et al.*, 2003; Pohlmann *et al.*, 2003; Cormier *et al.*, 2004a) and low density lipoprotein receptor (Agnello *et al.*, 1999; Molina *et al.*, 2007; Owen *et al.*, 2009). Four entry factors – tetraspanin CD81 (Pileri *et al.*, 1998; Bartosch *et al.*, 2003b; Hsu *et al.*, 2003; Cormier *et al.*, 2004b; Flint *et al.*, 2006), SR-BI (Catanese *et al.*, 2010; Scarselli *et al.*, 2002; Grove *et al.*, 2007) and tight junction proteins claudin-1 and occludin (Evans *et al.*, 2007; Zheng *et al.*, 2007; Meertens *et al.*, 2008; Yang *et al.*, 2008; Liu *et al.*, 2009; Ploss *et al.*, 2009) – are essential for clathrin-dependent endocytosis of virus particles (Blanchard *et al.*, 2006; Tscherne *et al.*, 2006). Ectopic expression of human CD81, SR-BI, claudin-1 and occludin render cell lines permissive for HCV entry, suggesting that they constitute the minimal receptor requirement for entry (Ploss *et al.*, 2009). Current evidence suggests that species-specific interaction of HCV with these entry factors may restrict virus entry into non-human cells and may contribute to the narrow species tropism observed for HCV. Further information on the role of the individual proteins will be discussed below:

CD81

CD81 is member of the tetraspanin superfamily of type III transmembrane proteins that contain four transmembrane domains and conserved signature amino acid residues, including a conserved CCG motif and 4 to 6 cysteine residues that form critical disulfide bonds within the second large extracellular loop (LEL or EC2; Fig. 17.2). Tetraspanins associate with a complex array of tetraspanin and non-tetraspanin proteins at cholesterol enriched microdomains and exert a diverse array of biological functions including cell–cell adhesion, cell migration and proliferation (Hemler, 2005). The first evidence for the involvement of CD81 in HCV entry was gained by screening a human cDNA library expressed in mouse fibroblasts to bind sE2 (Pileri *et al.*, 1998). More recent studies have demonstrated the importance of the CD81–E2 interaction in the

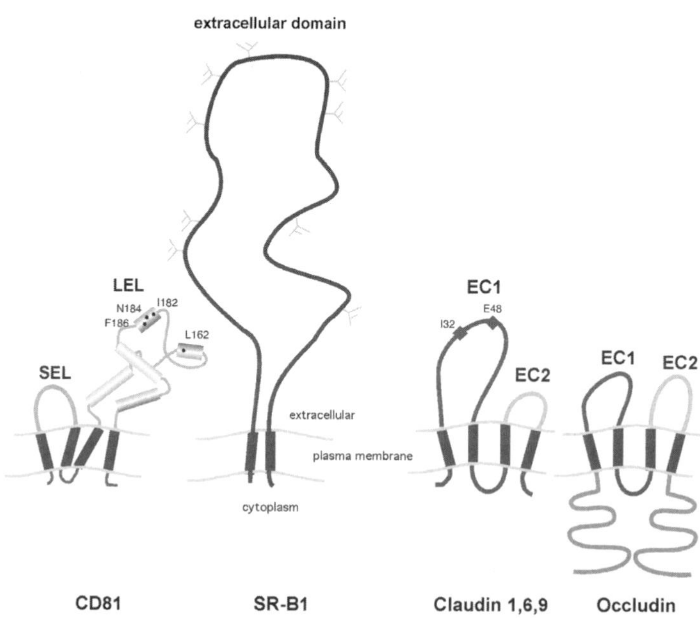

Figure 17.2 Schematic representation of the four molecules required for HCV entry. The transmembrane domains are shown as grey rods and the nomenclature of the extracellular domains is indicated. Amino acids shown to be important for binding sE2 on CD81 and amino acids in the EC1 domain of claudin-1 essential for claudin–CD81-occludin interactions and HCV entry are indicated. The glycosylation sites in the extracellular domain of SR-B1 are also shown (light grey trees).

viral lifecycle by inhibiting HCV infectivity with monoclonal antibodies specific for CD81 (Zhang *et al.*, 2004). siRNA silencing of CD81 expression in hepatoma cells prevented HCVpp and HCVcc entry whilst the introduction of CD81 into HepG2 cells, which are a CD81 negative hepatocyte cell line, rendered them permissive for HCVpp and HCVcc infection, demonstrating the essential role of CD81 in viral entry (Lavillette *et al.*, 2005; Lindenbach *et al.*, 2005; Zhang *et al.*, 2004). A soluble form of CD81 LEL binds and neutralizes HCVpp and HCVcc infectivity (McKeating *et al.*, 2004; Zhang *et al.*, 2004). The LEL structure was solved to 1.8Å resolution by Kitadokoro *et al.* and revealed a dimer with each monomer comprising five α helices labelled A to E (Kitadokoro *et al.*, 2001). Helix A and E form a stalk subdomain that projects the globular head region comprising helix B, C and D. CD81 is likely to exist as a dimer on the cell surface (Drummer *et al.*, 2005) and disruption of dimerization via a series of specific mutations within CD81-LEL, reduced sE2 binding (Drummer *et al.*, 2005). Consistent with these observations dimeric forms of CD81-LEL are 10-fold more efficient at binding sE2 compared with monomeric CD81-LEL (Nakajima *et al.*, 2005).

CD81 is universally required by HCV glycoproteins from all 6 genotypes to mediate entry using the HCVpp entry system (McKeating *et al.*, 2004) and the HCVcc system (Akazawa *et al.*, 2007; Gottwein *et al.*, 2009). A recent study validated the role of CD81 in HCV infection of SCID-uPA mouse demonstrating that prophylactic administration of anti-CD81 antibodies could protect mice against HCV challenge *in vivo* (Meuleman *et al.*, 2008). The level of expression of CD81 at the cell surface is a critical determinant for the permissiveness of Huh-7 hepatoma cells for HCVpp entry with a minimum threshold of 7×10^4 molecules/cell required for productive infection (Koutsoudakis *et al.*, 2007). Studies using serum derived HCV particles from genotype 1a and 3a confirm that infection of primary human hepatocytes is CD81 dependent (Molina *et al.*, 2008). Although binding studies between recombinant sE2 and CD81 show that the interaction is highly specific, discrepancies exist between this *in vitro* data and the requirement of full-length CD81 for entry. sE2 binds human and chimpanzee CD81 but not mouse, rat or African green monkey CD81, closely reflecting the species restriction of HCV infection (Allander *et al.*, 2000; Higginbottom *et al.*, 2000; Meola *et al.*, 2000; Drummer *et al.*, 2002). Amino acids that are directly involved in the binding of CD81 to sE2 map to the globular head domain and include Phe186, Ile182, Asn184 and Leu162 providing an 806Å surface of interaction (Flint *et al.*, 1999a; Higginbottom *et al.*, 2000; Meola *et al.*, 2000; Drummer *et al.*, 2002; Fig. 17.2). Transduction of HepG2 cells to express full-length CD81 from each of these species revealed that HCVpp and HCVcc can utilize any of these receptors to mediate entry, although a clear preference for primate and human CD81 was observed. Transduction of HepG2 cells to express CD81 proteins containing single and multiple mutation of residues within the LEL known to disrupt sE2 binding had minimal effect(s) on HCV entry (Flint *et al.*, 2006). These results suggest that there is a degree of plasticity in the interaction between E2 and CD81 and that other molecules may define the species restriction of HCV entry.

The CD81 partner protein EWI-2wint is expressed in a variety of cell types but not hepatoma cell lines. EWI-2wint is a 55kDa protein that corresponds to EWI-2 lacking the first N-terminal Ig domain. HCVpp or E1E2 glycoproteins fail to bind CD81 in the presence of EWI-2wint (Rocha-Perugini *et al.*, 2008). Furthermore, heterologous expression of EWI-2wint in hepatoma cell lines reduces their ability to support HCVpp or HCVcc infection (Rocha-Perugini *et al.*, 2008). The authors conclude that EWI-2wint controls the accessibility, conformation, localization or signalling pathways controlled by CD81, restricting the ability of HCV to access its receptor and mediate entry.

Scavenger receptor class B type I

The initial evidence for a role of SR-BI in HCV entry came from the observation that HepG2 cells, which do not express CD81, bound sE2, this was shown to be a specific interaction with SR-BI (Scarselli *et al.*, 2002), a scavenger protein involved with lipoprotein metabolism (Fig. 17.2). Lipoproteins comprise a cholesterol ester core surrounded by lipids and apoproteins. These structures transport triglycerides and cholesterol around the body and are categorized as high-density (HDL), low-density (LDL) or very low-density lipoproteins (VLDL; Krieger, 1999). The major role of SR-BI is the selective uptake of cholesterol from HDL particles bound to SR-BI via the apoprotein ApoAI (Fidge, 1999). SR-BI has been shown to interact with a variety of ligands including VLDL, LDL, HDL and chemically modified forms, such as oxidized (oxLDL) and acetylated LDL (Krieger, 1997). HCV infection can be inhibited by oxLDL and serum amyloid-(von Hahn *et al.*, 2006; Lavie *et al.*, 2006; Cai *et al.*, 2007). In contrast, HDL has been shown to enhance HCV entry and reduce the efficacy of neutralizing antibodies (Voisset *et al.*, 2005). This enhancement can be prevented by blocking the movement of cholesterol from bound HDL into the cell, suggesting that the cholesterol content of the membrane plays an important role in HCV entry. SR-BI has been reported to interact with the E2 HVR1, as deletion of this region ablates sE2–SR-BI interaction(s) (Scarselli *et al.*, 2002). Furthermore, high concentrations of anti-HVR mAb reduce sE2 interaction with SR-BI. Silencing of SR-BI expression or treatment of cells with antibodies specific to SR-BI have variable effects on HCV entry (Lavillette *et al.*, 2005; von Hahn *et al.*, 2006; Grove *et al.*, 2007; Dreux *et al.*, 2009), whereas overexpression of SR-BI enhanced virus infection, suggesting that SR-BI expression in Huh-7 cells is limiting (Grove *et al.*, 2008). Taken together these facts demonstrate an important role for SR-BI in HCV viral entry.

The role of SR-BI in the HCV entry process is not fully defined: does it serve as a structural receptor (i.e. to mediate virus attachment to the cell surface) or is the lipid transport function critical? sE2 interaction with SR-BI suggests that some degree of protein–protein interaction is essential (Scarselli *et al.*, 2002). However, several mutations in SR-BI that ablate sE2 interaction still confer HCV infection to SR-BI defective cells (Catanese *et al.*). Previous studies have demonstrated that HDL enhances HCVpp and HCVcc infection (Bartosch *et al.*, 2005; Voisset *et al.*, 2005). SR-BI-mediated cholesterol transfer into cell membranes has been shown to modulate CD81 enrichment in tetraspanin-enriched microdomains (TEMs) that are important for *Plasmodium* infection of hepatocytes (Yalaoui *et al.*). It is possible that an analogous role for SR-BI in creating a local microenvironment in the cell membrane (i.e. allowing associations between entry factors) is conducive for HCV entry.

Claudins

Since many cell lines express CD81 and SR-BI and are non-permissive for HCVpp entry (Bartosch *et al.*, 2003a; Hsu *et al.*, 2003), other factors were predicted to be required for viral entry. A cDNA library derived from a permissive hepatocellular carcinoma cell line was cloned into non-permissive 293T cells, which expresses both CD81 and SR-BI. Permissive clones were identified to express claudin-1 (Evans *et al.*, 2007). Further work by Zheng *et al.* demonstrated that HCV could infect the claudin-1 negative cell line Bel7402, due to the expression of claudin-9, where silencing claudin-9 ablated HCVpp entry. A screen of several claudin family members showed that claudin-9 and the closely related claudin-6, which are relatively poorly expressed in the liver (Mee *et al.*, 2008) could mediate HCVpp entry into 293-T cells, whereas none of the other Claudin family members showed receptor activity (Zheng *et al.*, 2007; Fig. 17.2).

The genesis of chimeric proteins between receptor active and inactive claudin-1 and -7 molecules identified claudin-1 EC1 as the minimal requirement for HCV entry into 293T cells (Evans *et al.*, 2007). Exchanging residues in claudin-1 and claudin-7 EC1 at 2 positions where they differ (codons 32 and 48) resulted in a loss and a gain of receptor activity, respectively (Evans *et al.*, 2007). These positions were also identified by Liu *et al.* who found further involvement of residues F148 and R158 (Liu *et al.*, 2009) in claudin-1 EC2 loop (Liu *et al.*, 2009). A more extensive analysis by Cukierman *et al.* demonstrated that mutation of an EC1 motif conserved among all claudins, W30-GLW51-C54-C64, as well as mutation of codon D38 ablated receptor activity (Cukierman *et al.*, 2009). Mutational analysis of claudin-9 demonstrated that residues S38 and V45 are essential for receptor activity (Zheng *et al.*, 2007). Although no clear mechanism for the receptor inactivity was shown in these publications, the authors suggest that *cis*-interaction(s) between claudin molecules on opposing cells may be important for HCV entry, with receptor inactive claudin-1 mutants showing reduced localization at cell–cell contacts (Liu *et al.*, 2009).

To date, a direct interaction between the HCV particle and claudin-1, -6 or -9 has not been demonstrated, suggesting an indirect role for Claudins in the HCV internalization process. We and others have reported that claudin-1 associates with CD81 (Kovalenko *et al.*, 2007; Harris *et al.*, 2008; Cukierman *et al.*, 2009), suggesting a role for the co-receptor complex in virus entry. Recent data from our laboratory demonstrates a relationship between receptor active Claudins and their association and organization with CD81 at the plasma membrane by fluorescent resonance energy transfer (FRET) and imaging methodologies (Harris *et al.*, 2010). In contrast, there was no relationship between receptor active Claudins and their association with occludin. Mutation of residues 32 and 48 in claudin-1 EC1 ablated CD81 and occludin association and HCV receptor activity. Molecular modelling of CD81–claudin-1 loop interactions, suggest that these mutations induce a conformational change in the claudin-1 loop consistent with its altered antigenicity. Mutation of the same residues in the receptor inactive claudin-7 molecule enable CD81 complex formation and virus entry in the absence of occludin interactions, suggesting that the claudin–CD81 association is essential for virus entry (Harris *et al.*, 2010). Modulation of Claudin–1-CD81 complexes by cholesterol depletion or ligation with receptor-specific antibodies inhibits HCV entry (Krieger *et al.*, 2009; Harris *et al.*, 2010), reinforcing an essential role for claudin–CD81 complexes in HCV infection.

Occludin

Ectopic expression of CD81, SR-BI and claudin-1 in non-human cells failed to confer HCV entry. This led Ploss *et al.* (2009) to repeat the cDNA library screening process in the non-permissive mouse cell line (NIH3T3) engineered to express human CD81, SR-BI and claudin-1, leading to the identification of occludin as a fourth entry factor (Fig. 17.2). The authors demonstrated that expression of occludin in the non-permissive renal carcinoma cell line, 786-O, or in TZM cells, a Hela cell derivative, allowed viral entry, and siRNA mediated silencing of occludin expression in Huh-7.5 hepatoma cells significantly reduced HCV infection. Finally, the authors demonstrate that rodent SR-BI and claudin-1 proteins could confer HCV entry since NIH3T3 or CHO cells expressing human CD81 and occludin were permissive for HCVpp entry (Ploss *et al.*, 2009).

Using an independent approach Liu and colleagues demonstrated a role for occludin in HCV entry (Liu *et al.*, 2009). These authors siRNA silenced proteins known to associate with claudin-1, including occludin, ZO-1, JAM-A and coxsackie and adenovirus receptor (CAR). The authors demonstrated that reduction of claudin-1, occludin or ZO-1 expression inhibited HCVpp entry. Silencing claudin-1 or occludin did not alter CD81 or SR-BI expression, however silencing ZO-1 led to significantly reduced claudin-1 expression. Chimeric proteins expressing extracellular loops of claudin-1 or occludin demonstrated that proteins expressing claudin-1 EC1 allowed HCV entry. This result is consistent with previous chimeric claudin-1/7 data from Evans and colleagues, concluding that the C-terminus of claudin-1 is dispensable for receptor activity (Evans *et al.*, 2007). Similar to claudin-1, there is minimal data for an interaction of HCV particles with occludin (Benedicto *et al.*, 2008).

Occludin was first identified by Furuse *et al.* as an integral membrane protein located at tight junctions. Structurally it is similar to the Claudins, with four transmembrane domains and two cytoplasmic regions at the N- and C- terminal (Furuse *et al.*, 1993). Studies on knockout mice demonstrate that tight junctions could still form in the absence of occludin with no gross phenotypic abnormalities (Saitou *et al.*, 2000), suggesting that occludin is not essential for tight junction integrity. The C-terminal region of occludin interacts with the guanylate kinase region of ZO-1 (Haskins *et al.*, 1998; Feldman *et al.*, 2005; Paris *et al.*, 2008). As part of the regulation of tight junctions occludin internalizes with claudin-1 in a variety of caveolae and clathrin-dependent processes (Matsuda *et al.*, 2004; Shen *et al.*, 2008; Stamatovic *et al.*, 2009).

HCV cell–cell transmission

Viruses disseminate by diffusion of cell-free particles and directed cell–cell transfer (Martin and Sattentau, 2009; Mothes *et al.*, 2010). The process of diffusion can significantly limit virus dissemination within an infected host, especially when particles are exposed to the humoral immune system. Many viruses have evolved strategies to maximize transmission within a host by the direct movement of particles from infected to naïve cells. The mechanisms fall broadly into two modes of transmission; delivery of virions across a defined cellular contact (e.g. virological synapse of HIV) or the lateral movement of cell associated particles towards adjacent cells (e.g. actin tails induced by vaccinia). Both processes can confer resistance to nAbs by shielding virus entry events from circulating Ab or by expediting virion penetration and reducing the 'window of opportunity' for antibody neutralization.

We (Timpe *et al.*, 2008) and others (Witteveldt *et al.*, 2009; Jones *et al.*, 2010) have reported that HCV can spread to neighbouring cells in the presence of glycoprotein-specific neutralizing antibodies (nAbs), suggesting cell–cell transmission. We recently documented that nAb resistant HCV transmission is dependent on contact between infected producer and naïve target cells and is at least 30-fold more efficient than cell-free particle infection (Syder *et al.*, 2010). Current experiments with Huh-7 derived hepatoma cell lines demonstrate that all four entry factors and gp expressing particles are required for cell contact-mediated antibody-resistant transmission, implying that the pathway most likely resembles conventional cell-free particle fusion. Importantly, antibodies or small molecule inhibitors targeting SR-BI were more effective at blocking HCV cell–cell transmission than cell-free particles, suggesting a potential therapeutic target (Syder *et al.*, 2010).

Tight junction proteins and virus entry

Polarized endothelia and epithelia form a barrier to pathogen access to the underlying tissue. Tight junctions regulate the movement of solutes, ions and water across such polarized cell sheets (Van Itallie *et al.*, 2008; Guttman and Finlay, 2009; Krause *et al.*, 2009). Several viruses have evolved mechanisms to utilize TJ proteins to enter and infect polarized epithelia, including adenovirus, coxsackievirus-B (CVB) and reovirus (Guttman and Finlay, 2009). CVB entry into polarized cells is dependent upon decay-accelerating factor (DAF) and tight junction proteins CAR and occludin (Coyne and Bergelson, 2006; Coyne *et al.*, 2007). CVB engagement of apically localized DAF activates tyrosine kinase AbI, which induces a Rac GTPase-dependent localization of virus–DAF complex to the TJ. Once at the TJ virus–CAR interactions prime internalization via an occludin-dependent manner. CVB does not appear to interact with occludin, suggesting that the virus internalizes by caveolae-dependent endocytic pathways normally involved in tight junction remodelling. Parallels between CVB and HCV entry have been drawn, however, several important differences exist: (1) CVB interaction with CAR is essential for virus internalization, in contrast, there is limited evidence for claudin-1 or occludin direct association with HCV; (2) CVB enters via macropinocytosis and HCV internalizes by clathrin-dependent endocytosis; and finally (3) CVB targets the apical epithelial surface(s) whereas HCV encounters the basolateral hepatocellular surface. Further work is required to ascertain whether HCV enters polarized cells in a CVB-like manner or by other, distinct mechanism(s).

Polarization of HepG2-CD81 hepatoma cells limits HCV infection, where polarizing agents such as human oncostatin M and Protein Kinase A (PKA) antagonists reduce HCVpp entry. In contrast, vascular endothelial growth factor or phorbol ester, a Protein Kinase C (PKC) activator, reduced polarity and increased HCV entry (Mee *et al.*, 2009a,b). One noticeable effect of increasing HepG2 polarity was a clear redistribution of claudin-1 to apically located tight junctions, with a concomitant reduction in basolateral pools of claudin-1 (Mee *et al.*, 2009b). Staining human liver tissue demonstrates that claudin-1 is predominantly detected at tight junctions; however, basal and lateral hepatocyte surfaces stain for claudin-1 (Reynolds *et al.*, 2008). Expression of fluorophore-labelled claudin-1 in HepG2 cells demonstrates discrete pools of claudin-1 at the tight junction and basal/lateral surfaces (Mee *et al.*, 2009b). Importantly, claudin-1 was only found to associate with basolateral pools of CD81, with minimal complexes detected at the tight junction (Harris *et al.*, 2010). Since HCV enters the liver via the sinusoidal blood the virus will encounter receptors expressed on the sinusoidal or basal surface of the hepatocyte. The location of claudin-1–CD81 complexes at the basolateral surface of polarized hepatoma cells supports a model where virus engagement of claudin-1–CD81 at the basal membrane may initiate the particle internalization process. The role of occludin in HCV entry into polarized cells is poorly defined and is hindered by a lack of antibodies to extracellular expressed epitopes. HCV internalizes via a clathrin-dependent process and fusion is believed to occur within the early endosomes (Blanchard *et al.*, 2006; Tscherne *et al.*, 2006). At present it is unknown whether any of the viral receptors, including CD81 and claudin-1, are endocytosed with HCV and further research on the trafficking and endocytic routing of receptor complexes and virus particles (Coller *et al.*, 2009) in polarized hepatocytes is required to fully appreciate the complex entry process of HCV in the liver.

HCV entry inhibitors

As we continue to learn more about HCV receptors, viral entry is becoming an increasingly attractive target for antiviral drug development due to the conserved entry mechanism observed across different genotypes. Inhibiting HCV entry can be accomplished by targeting the viral glycoproteins or the host cell receptors. Amphipathic DNA polymers were recently reported to inhibit HCV entry (Matsumura *et al.*, 2009) along with a range of other pathogens including HIV-1, human cytomegalovirus, varicella zoster virus, arenaviruses, and scrapie (Kocisko *et al.*, 2006; Vaillant *et al.*, 2006; Guzman *et al.*, 2007; Bernstein *et al.*, 2008; Lee *et al.*, 2008). Although the exact mechanism of action for these compounds is not fully elucidated, they are thought to prevent viral attachment via interaction with helices of the viral glycoproteins, such as HIV-1 gp41. Further experimentation is required to investigate their mode of action for HCV which will most likely fuse via a type II fusion process.

The crystal structure of soluble CD81 LEL (Kitadokoro *et al.*, 2001) is available to aid the design of small molecules targeting the CD81–E2 interface(Holzer *et al.*, 2008; Ziegler *et al.*, 2009). Studies from our laboratory to identify CD81 residues that interact with HCV demonstrate that mutations in CD81 LEL that ablate sE2 interaction have negligible effect(s) when engineered in the native molecule, suggesting that sCD81 may not be an accurate mimic for the native molecule and a poor model for the screening of inhibitors (Flint *et al.*, 2006). Recent studies to express full-length CD81 in yeast provide a promising tool for structure–function studies of CD81 and its interaction(s) with HCV (Jamshad *et al.*, 2008). Further studies are required to ascertain whether CD81 or CD81–CLDN1 complexes will yield as targets for therapeutic intervention without any cellular toxicity.

Recent reports demonstrate that a potent small molecule HCV entry inhibitor, ITX5061, targets SR-BI and we eagerly await the outcome of clinical trials (Syder *et al.*, 2010). Targeting CLDN1 and OCLN may be complicated given their essential role in tight junction maintenance. A recent study reported that antibodies targeting extracellular epitopes of CLDN1 inhibit HCV infection and had minimal effect(s) on HepG2 polarity or tight junction function (Krieger, 2009). Reports that HCV can infect polarized and non-polarized Caco-2 and HepG2 cells, suggest that tight junction proteins do not need to be recruited into a functional tight junction to allow viral entry (Mee, 2008). Therefore, it may be possible to target non-junctional pools of these entry factors without disrupting tight junctions, but the feasibility of this approach remains to be determined. A greater understanding of the mechanisms of HCV entry will be required before receptor-specific therapeutics can be fully realized.

Conclusions

For years, HCV research was severely restricted by the inability to propagate the virus in cultured cells. The recent development of the JFH-1 HCVcc system has enabled rapid discoveries in all areas of HCV research, including the discovery of viral entry factors. However, the detailed mechanism(s) of HCV entry into polarized cells awaits further investigation. It will be interesting to see the role of entry inhibitors in future combination therapies for the treatment of chronic HCV infection.

Acknowledgements

We thank Peter Balfe for critical reading of the manuscript. We thank PHS grant AI-400034–14, Hepacivac, MRC, Wellcome Trust and the NH and MRC for funding our research and the many helpful discussions with colleagues.

References

Agnello, V., Abel, G., Elfahal, M., Knight, B., and Zhang, Q.X. (1999). Hepatitis C virus and other flaviviridae viruses enter cells via low density lipoprotein receptor. Proc. Natl. Acad. Sci. U.S.A. *96*, 12766–12771.

Akazawa, D., Date, T., Morikawa, K., Murayama, A., Miyamoto, M., Kaga, M., Barth, H., Baumert, T.F., Dubuisson, J., and Wakita, T. (2007). CD81 expression is important for the permissiveness of Huh7 cell clones for heterogeneous hepatitis C virus infection. J. Virol. *81*, 5036–5045.

Allander, T., Forns, X., Emerson, S.U., Purcell, H., and Bukh, J. (2000). Hepatitis C Virus Envelope Protein E2 Binds to CD81 of Tamarins. Virology *277*, 358–367.

Allison, S.L., Stiasny, K., Stadler, K., Mandl, W., and Heinz, F.X. (1999). Mapping of functional elements in the stem-anchor region of tick-borne encephalitis virus envelope protein E. J. Virol. *73*, 5605–5612.

Bankwitz, D., Steinmann, E., Bitzegeio, J., Ciesek, S., Friesland, M., Hermann, E., Zeisel, M.B., Baumert, T.F., Keck, Z.Y., Foung, S.K., *et al.* (2010). Hepatitis C Virus Hypervariable Region 1 Modulates Receptor Interactions, Conceals the CD81 Binding Site and Protects Conserved Neutralizing Epitopes. J. Virol. *84*, 5751–5763.

Barth, H., Schafer, C., Adah, M.I., Zhang, F., Linhardt, R.J., Toyoda, H., Kinoshita-Toyoda, A., Toida, T., Van Kuppevelt, T.H., Depla, E., *et al.* (2003). Cellular binding of hepatitis C virus envelope glycoprotein E2 requires cell surface heparan sulfate. J. Biol. Chem. *278*, 41003–41012.

Bartosch, B., Dubuisson, J., and Cosset, F.L. (2003a). Infectious Hepatitis C Virus Pseudo-particles Containing Functional E1-E2 Envelope Protein Complexes. J. Exp. Med. *197*, 633–642.

Bartosch, B., Vitelli, A., Granier, C., Goujon, C., Dubuisson, J., Pascale, S., Scarselli, E., Cortese, R., Nicosia, A., and Cosset, F.L. (2003b). Cell entry of hepatitis C virus requires a set of co-receptors that include the CD81 tetraspanin and the SR-B1 scavenger receptor. J. Biol. Chem. *278*, 41624–41630.

Bartosch, B., Verney, G., Dreux, M., Donot, P., Morice, Y., Penin, F., Pawlotsky, J.M., Lavillette, D., and Cosset, F.L. (2005). An interplay between hypervariable region 1 of the hepatitis C virus E2 glycoprotein, the scavenger receptor BI, and high-density lipoprotein promotes both enhancement of infection and protection against neutralizing antibodies. J. Virol. *79*, 8217–8229.

Basu, A., Kanda, T., Beyene, A., Saito, K., Meyer, K., and Ray, R. (2007). Sulfated homologues of heparin inhibit hepatitis C virus entry into mammalian cells. J. Virol. *81*, 3933–3941.

Benedicto, I., Molina-Jimenez, F., Barreiro, O., Maldonado-Rodriguez, A., Prieto, J., Moreno-Otero, R., Aldabe, R., Lopez-Cabrera, M., and Majano, P.L. (2008). Hepatitis C virus envelope components alter localization of hepatocyte tight junction-associated proteins and promote occludin retention in the endoplasmic reticulum. Hepatology *48*, 1044–1053.

Bernstein, D.I., Goyette, N., Cardin, R., Kern, E.R., Boivin, G., Ireland, J., Juteau, M., and Vaillant, A. (2008). Amphipathic DNA polymers exhibit antiherpetic activity in vitro and in vivo. Antimicrob. Agents Chemother. *52*, 2727–2733.

Blanchard, E., Belouzard, S., Goueslain, L., Wakita, T., Dubuisson, J., Wychowski, C., and Rouille, Y. (2006). Hepatitis C virus entry depends on clathrin-mediated endocytosis. J. Virol. *80*, 6964–6972.

Brazzoli, M., Helenius, A., Foung, S.K., Houghton, M., Abrignani, S., and Merola, M. (2005). Folding and dimerization of hepatitis C virus E1 and E2 glycoproteins in stably transfected CHO cells. Virology *332*, 438–453.

Bressanelli, S., Stiasny, K., Allison, S.L., Stura, E.A., Duquerroy, S., Lescar, J., Heinz, X., and Rey, F.A. (2004). Structure of a flavivirus envelope glycoprotein in its low-pH-induced membrane fusion conformation. EMBO J. *23*, 728–738.

Budkowska, A. (2009). Mechanism of cell infection with hepatitis C virus (HCV)--a new paradigm in virus-cell interaction. Pol. J. Microbiol *58*, 93–98.

Cai, Z., Cai, L., Jiang, J., Chang, K.S., van der Westhuyzen, R., and Luo, G. (2007). Human serum amyloid A protein inhibits hepatitis C virus entry into cells. J. Virol. *81*, 6128–6133.

Callens, N., Ciczora, Y., Bartosch, B., Vu-Dac, N., Cosset, F.L., Pawlotsky, J.M., Penin, F., and Dubuisson, J. (2005). Basic residues in hypervariable region 1 of hepatitis C virus envelope glycoprotein e2 contribute to virus entry. J. Virol. *79*, 15331–15341.

Casino, C., McAllister, J., Davidson, F., Power, J., Lawlor, E., Yap, P.L., Simmonds, P., and Smith, D.B. (1999). Variation of hepatitis C virus following serial transmission: multiple mechanisms of diversification of the hypervariable region and evidence for convergent genome evolution. J. Gen. Virol. *80*, 717–725.

Castet, V., Fournier, C., Soulier, A., Brillet, R., Coste, J., Larrey, D., Dhumeaux, D., Maurel, P., and Pawlotsky, J.M. (2002). Alpha interferon inhibits hepatitis C virus replication in primary human hepatocytes infected in vitro. J. Virol. *76*, 8189–8199.

Catanese, M.T., Ansuini, H., Graziani, R., Huby, T., Moreau, M., Ball, J.K., Paonessa, G., Rice, C.M.,

Cortese, R., Vitelli, A., *et al.* (2010). Role of scavenger receptor class B type I in hepatitis C virus entry: kinetics and molecular determinants. J. Virol. *84*, 34–43.

Ciczora, Y., Callens, N., Montpellier, C., Bartosch, B., Cosset, F.L., Op de Beeck, A., and Dubuisson, J. (2005). Contribution of the charged residues of hepatitis C virus glycoprotein E2 transmembrane domain to the functions of the E1E2 heterodimer. J. Gen. Virol. *86*, 2793–2798.

Ciczora, Y., Callens, N., Penin, F., Pecheur, I., and Dubuisson, J. (2007). Transmembrane domains of hepatitis C virus envelope glycoproteins: residues involved in E1E2 heterodimerization and involvement of these domains in virus entry. J. Virol. *81*, 2372–2381.

Cocquerel, L., Meunier, J.C., Pillez, A., Wychowski, C., and Dubuisson, J. (1998). A retention signal necessary and sufficient for endoplasmic reticulum localization maps to the transmembrane domain of hepatitis C virus glycoprotein E2. J. Virol. *72*, 2183–2191.

Cocquerel, L., Duvet, S., Meunier, J.C., Pillez, A., Cacan, R., Wychowski, C., and Dubuisson, J. (1999). The transmembrane domain of hepatitis C virus glycoprotein E1 is a signal for static retention in the endoplasmic reticulum. J. Virol. *73*, 2641–2649.

Cocquerel, L., Wychowski, C., Minner, F., Penin, F., and Dubuisson, J. (2000). Charged residues in the transmembrane domains of hepatitis C virus glycoproteins play a major role in the processing, subcellular localization, and assembly of these envelope proteins. J. Virol. *74*, 3623–3633.

Cocquerel, L., Meunier, J.C., Op de Beeck, A., Bonte, D., Wychowski, C., and Dubuisson, J. (2001). Coexpression of hepatitis C virus envelope proteins E1 and E2 in cis improves the stability of membrane insertion of E2. J. Gen. Virol. *82*, 1629–1635.

Cocquerel, L., Op de Beeck, A., Lambot, M., Roussel, J., Delgrange, D., Pillez, A., Wychowski, C., Penin, F., and Dubuisson, J. (2002). Topological changes in the transmembrane domains of hepatitis C virus envelope glycoproteins. EMBO J. *21*, 2893–2902.

Cocquerel, L., Kuo, C.C., Dubuisson, J., and Levy, S. (2003). CD81-dependent binding of hepatitis C virus E1E2 heterodimers. J. Virol. *77*, 10677–10683.

Coller, K.E., Berger, K.L., Heaton, N.S., Cooper, J.D., Yoon, R., and Randall, G. (2009). RNA interference and single particle tracking analysis of hepatitis C virus endocytosis. PLoS Pathog. *5*, e1000702.

Cormier, E.G., Durso, R.J., Tsamis, F., Boussemart, L., Manix, C., Olson, W.C., Gardner, P., and Dragic, T. (2004a). L-SIGN (CD209L) and DC-SIGN (CD209) mediate transinfection of liver cells by hepatitis C virus. Proc. Natl. Acad. Sci. U.S.A. *101*, 14067–14072.

Cormier, E.G., Tsamis, F., Kajumo, F., Durso, R.J., Gardner, P., and Dragic, T. (2004b). CD81 is an entry coreceptor for hepatitis C virus. Proc. Natl. Acad. Sci. U.S.A. *101*, 7270–7274.

Coyne, B., and Bergelson, J.M. (2006). Virus-induced Abl and Fyn kinase signals permit coxsackievirus entry through epithelial tight junctions. Cell *124*, 119–131.

Coyne, C.B., Shen, L., Turner, R., and Bergelson, J.M. (2007). Coxsackievirus entry across epithelial tight junctions requires occludin and the small GTPases Rab34 and Rab5. Cell Host Microbe *2*, 181–192.

Cukierman, L., Meertens, L., Bertaux, C., Kajumo, F., and Dragic, T. (2009). Residues in a highly conserved claudin-1 motif are required for hepatitis C virus entry and mediate the formation of cell–cell contacts. J. Virol. *83*, 5477–5484.

Deleersnyder, V., Pillez, A., Wychowski, C., Blight, K., Xu, J., Hahn, Y.S., Rice, M., and Dubuisson, J. (1997). Formation of native hepatitis C virus glycoprotein complexes. J. Virol. *71*, 697–704.

Dreux, M., Dao Thi, V.L., Fresquet, J., Guerin, M., Julia, Z., Verney, G., Durantel, D., Zoulim, F., Lavillette, D., Cosset, F.L., *et al.* (2009). Receptor complementation and mutagenesis reveal SR-BI as an essential HCV entry factor and functionally imply its intra- and extracellular domains. PLoS Pathog. *5*, e1000310.

Drummer, H.E., and Poumbourios, P. (2004). Hepatitis C virus glycoprotein E2 contains a membrane-proximal heptad repeat sequence that is essential for E1E2 glycoprotein heterodimerization and viral entry. J. Biol. Chem. *279*, 30066–30072.

Drummer, H.E., Wilson, A., and Poumbourios, P. (2002). Identification of the hepatitis C virus e2 glycoprotein binding site on the large extracellular loop of CD81. J. Virol. *76*, 11143–11147.

Drummer, H.E., Maerz, A., and Poumbourios, P. (2003). Cell surface expression of functional hepatitis C virus E1 and E2 glycoproteins. FEBS Lett. *546*, 385–390.

Drummer, H.E., Wilson, A., and Poumbourios, P. (2005). Determinants of CD81 dimerization and interaction with hepatitis C virus glycoprotein E2. Biochem. Biophys. Res. Commun. *328*, 251–257.

Drummer, H.E., Boo, I., Maerz, L., and Poumbourios, P. (2006). A conserved Gly436-Trp-Leu-Ala-Gly-Leu-Phe-Tyr motif in hepatitis C virus glycoprotein E2 is a determinant of CD81 binding and viral entry. J. Virol. *80*, 7844–7853.

Drummer, H.E., Boo, I., and Poumbourios, P. (2007). Mutagenesis of a conserved fusion peptide-like motif and membrane-proximal heptad-repeat region of hepatitis C virus glycoprotein E1. J. Gen. Virol. *88*, 1144–1148.

Dubuisson, J., and Rice, C.M. (1996). Hepatitis C virus glycoprotein folding: disulfide bond formation and association with calnexin. J. Virol. *70*, 778–786.

Dubuisson, J., Hsu, H.H., Cheung, R.C., Greenberg, H.B., Russell, G., and Rice, C.M. (1994). Formation and intracellular localization of hepatitis C virus envelope glycoprotein complexes expressed by recombinant vaccinia and Sindbis viruses. J. Virol. *68*, 6147–6160.

Duvet, S., Cocquerel, L., Pillez, A., Cacan, R., Verbert, A., Moradpour, D., Wychowski, C., and Dubuisson, J. (1998). Hepatitis C virus glycoprotein complex localization in the endoplasmic reticulum involves a determinant for retention and not retrieval. J. Biol. Chem. *273*, 32088–32095.

Evans, M.J., von Hahn, T., Tscherne, D.M., Syder, A.J., Panis, M., Wolk, B., Hatziioannou, T., McKeating, J.A.,

Bieniasz, D., and Rice, C.M. (2007). Claudin-1 is a hepatitis C virus co-receptor required for a late step in entry. Nature *446*, 801–805.

Falkowska, E., Kajumo, F., Garcia, E., Reinus, J., and Dragic, T. (2007). Hepatitis C virus envelope glycoprotein E2 glycans modulate entry, CD81 binding, and neutralization. J. Virol. *81*, 8072–8079.

Farci, P., Shimoda, A., Wong, D., Cabezon, T., De Gioannis, D., Strazzera, A., Shimizu, Y., Shapiro, M., Alter, J., and Purcell, R.H. (1996). Prevention of hepatitis C virus infection in chimpanzees by hyperimmune serum against the hypervariable region 1 of the envelope 2 protein. Proc. Natl. Acad. Sci. U.S.A. *93*, 15394–15399.

Farci, P., Shimoda, A., Coiana, A., Diaz, G., Peddis, G., Melpolder, J.C., Strazzera, A., Chien, D.Y., Munoz, S.J., Balestrieri, A., *et al.* (2000). The outcome of acute hepatitis C predicted by the evolution of the viral quasispecies. Science *288*, 339–344.

Farquhar, J., and McKeating, J.A. (2008). Primary hepatocytes as targets for hepatitis C virus replication. J. Viral Hepat. *15*, 849–854.

Feldman, G.J., Mullin, M., and Ryan, M.P. (2005). Occludin: structure, function and regulation. Adv. Drug Deliv. Rev. *57*, 883–917.

Fidge, N.H. (1999). High density lipoprotein receptors, binding proteins, and ligands. J. Lipid Res. *40*, 187–201.

Flint, M., and McKeating, J.A. (1999). The C-terminal region of the hepatitis C virus E1 glycoprotein confers localization within the endoplasmic reticulum. J. Gen. Virol. *80*, 1943–1947.

Flint, M., Maidens, C., Loomis-Price, L.D., Shotton, C., Dubuisson, J., Monk, P., Higginbottom, A., Levy, S., and McKeating, J.A. (1999a). Characterization of hepatitis C virus E2 glycoprotein interaction with a putative cellular receptor, CD81. J. Virol. *73*, 6235–6244.

Flint, M., Thomas, J.M., Maidens, C.M., Shotton, C., Levy, S., Barclay, S., and McKeating, J.A. (1999b). Functional analysis of cell surface-expressed hepatitis C virus E2 glycoprotein. J. Virol. *73*, 6782–6790.

Flint, M., Dubuisson, J., Maidens, C., Harrop, R., Guile, G.R., Borrow, P., and McKeating, J.A. (2000). Functional characterization of intracellular and secreted forms of a truncated hepatitis C virus E2 glycoprotein. J. Virol. *74*, 702–709.

Flint, M., von Hahn, T., Zhang, J., Farquhar, M., Jones, C.T., Balfe, P., Rice, M., and McKeating, J.A. (2006). Diverse CD81 proteins support hepatitis C virus infection. J. Virol. *80*, 11331–11342.

Forns, X., Thimme, R., Govindarajan, S., Emerson, S.U., Purcell, R.H., Chisari, V., and Bukh, J. (2000). Hepatitis C virus lacking the hypervariable region 1 of the second envelope protein is infectious and causes acute resolving or persistent infection in chimpanzees. Proc. Natl. Acad. Sci. U.S.A. *97*, 13318–13323.

Fournier, C., Sureau, C., Coste, J., Ducos, J., Pageaux, G., Larrey, D., Domergue, J., and Maurel, P. (1998). In vitro infection of adult normal human hepatocytes in primary culture by hepatitis C virus. J. Gen. Virol. *79*, 2367–2374.

Furuse, M., Hirase, T., Itoh, M., Nagafuchi, A., Yonemura, S., and Tsukita, S. (1993). Occludin: a novel integral membrane protein localizing at tight junctions. J. Cell Biol. *123*, 1777–1788.

Garry, F., and Dash, S. (2003). Proteomics computational analyses suggest that hepatitis C virus E1 and pestivirus E2 envelope glycoproteins are truncated class II fusion proteins. Virology *307*, 255–265.

Gerber, D., Sal-Man, N., and Shai, Y. (2004). Two motifs within a transmembrane domain, one for homodimerization and the other for heterodimerization. J. Biol. Chem. *279*, 21177–21182.

Goffard, A., Callens, N., Bartosch, B., Wychowski, C., Cosset, F.L., Montpellier, C., and Dubuisson, J. (2005). Role of N-linked glycans in the functions of hepatitis C virus envelope glycoproteins. J. Virol. *79*, 8400–8409.

Gottwein, J.M., Scheel, T.K., Hoegh, A.M., Lademann, J.B., Eugen-Olsen, J., Lisby, G., and Bukh, J. (2007). Robust hepatitis C genotype 3a cell culture releasing adapted intergenotypic 3a/2a (S52/JFH1) viruses. Gastroenterology *133*, 1614–1626.

Gottwein, J.M., Scheel, T.K., Jensen, T.B., Lademann, J.B., Prentoe, J.C., Knudsen, M.L., Hoegh, M., and Bukh, J. (2009). Development and characterization of hepatitis C virus genotype 1–7 cell culture systems: role of CD81 and scavenger receptor class B type I and effect of antiviral drugs. Hepatology *49*, 364–377.

Grove, J., Huby, T., Stamataki, Z., Vanwolleghem, T., Meuleman, P., Farquhar, M., Schwarz, A., Moreau, M., Owen, J.S., Leroux-Roels, G., *et al.* (2007). Scavenger receptor BI and BII expression levels modulate hepatitis C virus infectivity. J. Virol. *81*, 3162–3169.

Grove, J., Nielsen, S., Zhong, J., Bassendine, M.F., Drummer, H.E., Balfe, P., and McKeating, J.A. (2008). Identification of a residue in hepatitis C virus E2 glycoprotein that determines scavenger receptor BI and CD81 receptor dependency and sensitivity to neutralizing antibodies. J. Virol. *82*, 12020–12029.

Guttman, A., and Finlay, B.B. (2009). Tight junctions as targets of infectious agents. Biochim. Biophys. Acta *1788*, 832–841.

Guzman, E.M., Cheshenko, N., Shende, V., Keller, M.J., Goyette, N., Juteau, J.M., Boivin, G., Vaillant, A., and Herold, B.C. (2007). Amphipathic DNA polymers are candidate vaginal microbicides and block herpes simplex virus binding, entry and viral gene expression. Antivir. Ther. *12*, 1147–1156.

Hadlock, K.G., Lanford, R.E., Perkins, S., Rowe, J., Yang, Q., Levy, S., Pileri, P., Abrignani, S., and Foung, S.K. (2000). Human monoclonal antibodies that inhibit binding of hepatitis C virus E2 protein to CD81 and recognize conserved conformational epitopes. J. Virol. *74*, 10407–10416.

Harris, H.J., Farquhar, M.J., Mee, C.J., Davis, C., Reynolds, G.M., Jennings, A., Hu, K., Yuan, F., Deng, H., Hubscher, S.G., *et al.* (2008). CD81 and claudin 1 coreceptor association: role in hepatitis C virus entry. J. Virol. *82*, 5007–5020.

Harris, H.J., Davis, C., Mullins, J.G.L., Hu, K., Goodall, M., Farquhar, M.J., Mee, C.J., Balfe, P., and McKeating, J.A. (2010). Claudin association with CD81 defines hepatitis C virus entry. J. Biol. Chem. *285*, 21092–21102.

Haskins, J., Gu, L., Wittchen, E.S., Hibbard, J., and Stevenson, B.R. (1998). ZO-3, a novel member of the MAGUK protein family found at the tight junction, interacts with ZO-1 and occludin. J. Cell Biol. *141*, 199–208.

Helle, F., Goffard, A., Morel, V., Duverlie, G., McKeating, J., Keck, Z.Y., Foung, S., Penin, F., Dubuisson, J., and Voisset, C. (2007). The neutralizing activity of anti-hepatitis C virus antibodies is modulated by specific glycans on the E2 envelope protein. J. Virol. *81*, 8101–8111.

Hemler, M.E. (2005). Tetraspanin functions and associated microdomains. Nat. Rev. Mol. Cell Biol. *6*, 801–811.

Higginbottom, A., Quinn, E.R., Kuo, C.C., Flint, M., Wilson, L.H., Bianchi, E., Nicosia, A., Monk, P.N., McKeating, A., and Levy, S. (2000). Identification of amino acid residues in CD81 critical for interaction with hepatitis C virus envelope glycoprotein E2. J. Virol. *74*, 3642–3649.

Holzer, M., Ziegler, S., Albrecht, B., Kronenberger, B., Kaul, A., Bartenschlager, R., Kattner, L., Klein, D., and Hartmann, R.W. (2008). Identification of terfenadine as an inhibitor of human CD81-receptor HCV-E2 interaction: synthesis and structure optimization. Molecules *13*, 1081–1110.

Hsu, M., Zhang, J., Flint, M., logvinoff, C., Cheng-Mayer, C., Rice, M., and McKeating, J.A. (2003). Hepatitis C virus glycoproteins mediate pH-dependent cell entry of pseudotyped retroviral particles. Proc. Natl. Acad. Sci. U.S.A. *100*, 7271–7276.

Jamshad, M., Rajesh, S., Stamataki, Z., McKeating, J.A., Dafforn, T., Overduin, M., and Bill, R.M. (2008). Structural characterization of recombinant human CD81 produced in Pichia pastoris. Protein Expr. Purif. *57*, 206–216.

Johansson, D.X., Voisset, C., Tarr, A.W., Aung, M., Ball, J.K., Dubuisson, J., and Persson, M.A. (2007). Human combinatorial libraries yield rare antibodies that broadly neutralize hepatitis C virus. Proc. Natl. Acad. Sci. U.S.A. *104*, 16269–16274.

Jones, C.T., Catanese, M.T., Law, L.M., Khetani, S.R., Syder, A.J., Ploss, A., Oh, T.S., Schoggins, J.W., MacDonald, M.R., Bhatia, S.N., *et al.* (2010). Real-time imaging of hepatitis C virus infection using a fluorescent cell-based reporter system. Nat. Biotechnol. *28*, 167–171.

Keck, Z.Y., Op De Beeck, A., Hadlock, K.G., Xia, J., Li, T.K., Dubuisson, J., and Foung, S.K. (2004). Hepatitis C virus e2 has three immunogenic domains containing conformational epitopes with distinct properties and biological functions. J. Virol. *78*, 9224–9232.

Keck, Z.Y., Li, T.K., Xia, J., Bartosch, B., Cosset, F.L., Dubuisson, J., and Foung, S.K. (2005). Analysis of a highly flexible conformational immunogenic domain a in hepatitis C virus E2. J. Virol. *79*, 13199–13208.

Keck, Z.Y., Xia, J., Cai, Z., Li, T.K., Owsianka, A.M., Patel, A.H., Luo, G., and Foung, S.K. (2007). Immunogenic and functional organization of hepatitis C virus (HCV) glycoprotein E2 on infectious HCV virions. J. Virol. *81*, 1043–1047.

Keck, Z.Y., Li, T.K., Xia, J., Gal-Tanamy, M., Olson, O., Li, S.H., Patel, A.H., Ball, J.K., Lemon, M., and Foung, S.K. (2008a). Definition of a conserved immunodominant domain on hepatitis C virus E2 glycoprotein by neutralizing human monoclonal antibodies. J. Virol. *82*, 6061–6066.

Keck, Z.Y., Olson, O., Gal-Tanamy, M., Xia, J., Patel, A.H., Dreux, M., Cosset, F.L., Lemon, M., and Foung, S.K. (2008b). A point mutation leading to hepatitis C virus escape from neutralization by a monoclonal antibody to a conserved conformational epitope. J. Virol. *82*, 6067–6072.

Kitadokoro, K., Bordo, D., Galli, G., Petracca, R., Falugi, F., Abrignani, S., Grandi, G., and Bolognesi, M. (2001). CD81 extracellular domain 3D structure: insight into the tetraspanin superfamily structural motifs. EMBO J. *20*, 12–18.

Kocisko, D.A., Vaillant, A., Lee, K.S., Arnold, K.M., Bertholet, N., Race, R.E., Olsen, E.A., Juteau, M., and Caughey, B. (2006). Potent antiscrapie activities of degenerate phosphorothioate oligonucleotides. Antimicrob. Agents Chemother. *50*, 1034–1044.

Kolykhalov, A.A., Agapov, E.V., Blight, K.J., Mihalik, K., Feinstone, M., and Rice, C.M. (1997). Transmission of hepatitis C by intrahepatic inoculation with transcribed RNA. Science *277*, 570–574.

Koutsoudakis, G., Herrmann, E., Kallis, S., Bartenschlager, R., and Pietschmann, T. (2007). The level of CD81 cell surface expression is a key determinant for productive entry of hepatitis C virus into host cells. J. Virol. *81*, 588–598.

Kovalenko, O.V., Yang, H., and Hemler, M.E. (2007). A novel cysteine cross-linking method reveals a direct association between claudin-1 and tetraspanin CD9. Mol. Cell Proteomics *6*, 1855–1867.

Krause, G., Winkler, L., Piehl, C., Blasig, I., Piontek, J., and Muller, S.L. (2009). Structure and function of extracellular claudin domains. Ann. N.Y. Acad. Sci. *1165*, 34–43.

Krey, T., d'Alayer, J., Kikuti, C.M., Saulnier, A., Damier-Piolle, L., Petitpas, I., Johansson, D.X., Tawar, R.G., Baron, B., Robert, B., *et al.* (2010). The disulfide bonds in glycoprotein e2 of hepatitis C virus reveal the tertiary organization of the molecule. PLoS Pathog. *6*, e1000762.

Krieger, M. (1997). The other side of scavenger receptors: pattern recognition for host defense. Curr. Opin. Lipidol. *8*, 275–280.

Krieger, M. (1999). Charting the fate of the "good cholesterol": identification and characterization of the high-density lipoprotein receptor SR-BI. Annu. Rev. Biochem. *68*, 523–558.

Krieger, S.E., Zeisel, M.B., Davis, C., Thumann, C., Harris, H.J., Schnober, E.K., Mee, C., Soulier, E., Royer, C., Lambotin, M., *et al.* (2009). Inhibition of hepatitis C virus infection by anti-claudin-1 antibodies is mediated by neutralization of E2-CD81-Claudin-1 associations. Hepatology *51*, 1144–1157.

Lavie, M., Voisset, C., Vu-Dac, N., Zurawski, V., Duverlie, G., Wychowski, C., and Dubuisson, J. (2006). Serum amyloid A has antiviral activity against hepatitis C virus by inhibiting virus entry in a cell culture system. Hepatology *44*, 1626–1634.

Lavie, M., Goffard, A., and Dubuisson, J. (2007). Assembly of a functional HCV glycoprotein heterodimer. Curr. Issues Mol. Biol. *9*, 71–86.

Lavillette, D., Tarr, A.W., Voisset, C., Donot, P., Bartosch, B., Bain, C., Patel, A.H., Dubuisson, J., Ball, K., and Cosset, F.L. (2005). Characterization of host-range and cell entry properties of the major genotypes and subtypes of hepatitis C virus. Hepatology *41*, 265–274.

Lavillette, D., Pecheur, E.I., Donot, P., Fresquet, J., Molle, J., Corbau, R., Dreux, M., Penin, F., and Cosset, F.L. (2007). Characterization of fusion determinants points to the involvement of three discrete regions of both E1 and E2 glycoproteins in the membrane fusion process of hepatitis C virus. J. Virol. *81*, 8752–8765.

Lee, A.M., Rojek, J.M., Gundersen, A., Stroher, U., Juteau, J.M., Vaillant, A., and Kunz, S. (2008). Inhibition of cellular entry of lymphocytic choriomeningitis virus by amphipathic DNA polymers. Virology *372*, 107–117.

Li, H.F., Huang, C.H., Ai, L.S., Chuang, K., and Chen, S.S. (2009). Mutagenesis of the fusion peptide-like domain of hepatitis C virus E1 glycoprotein: involvement in cell fusion and virus entry. J. Biomed. Sci. *16*, 89.

Lindenbach, B.D., Evans, M.J., Syder, A.J., Wolk, B., Tellinghuisen, T.L., Liu, C.C., Maruyama, T., Hynes, R.O., Burton, D.R., McKeating, J.A., *et al.* (2005). Complete replication of hepatitis C virus in cell culture. Science *309*, 623–626.

Lindenbach, B.D., Meuleman, P., Ploss, A., Vanwolleghem, T., Syder, A.J., McKeating, J.A., Lanford, R.E., Feinstone, S.M., Major, M.E., Leroux-Roels, G., *et al.* (2006). Cell culture-grown hepatitis C virus is infectious in vivo and can be recultured in vitro. Proc. Natl. Acad. Sci. U.S.A. *103*, 3805–3809.

Liu, S., Yang, W., Shen, L., Turner, J.R., Coyne, B., and Wang, T. (2009). Tight junction proteins claudin-1 and occludin control hepatitis C virus entry and are downregulated during infection to prevent superinfection. J. Virol. *83*, 2011–2014.

Lorent, E., Bierau, H., Engelborghs, Y., Verheyden, G., and Bosman, F. (2008). Structural characterisation of the hepatitis C envelope glycoprotein E1 ectodomain derived from a mammalian and a yeast expression system. Vaccine *26*, 399–410.

Lozach, P.Y., Lortat-Jacob, H., de Lacroix de Lavalette, A., Staropoli, I., Foung, S., Amara, A., Houles, C., Fieschi, F., Schwartz, O., Virelizier, J.L., *et al.* (2003). DC-SIGN and L-SIGN are high affinity binding receptors for hepatitis C virus glycoprotein E2. J. Biol. Chem. *278*, 20358–20366.

McCaffrey, K., Boo, I., Poumbourios, P., and Drummer, H.E. (2007). Expression and characterization of a minimal hepatitis C virus glycoprotein E2 core domain that retains CD81 binding. J. Virol. *81*, 9584–9590.

McKeating, J.A., Zhang, L.Q., logvinoff, C., Flint, M., Zhang, J., Yu, J., Butera, D., Ho, D.D., Dustin, L.B., Rice, C.M., *et al.* (2004). Diverse hepatitis C virus glycoproteins mediate viral infection in a CD81-dependent manner. J. Virol. *78*, 8496–8505.

Martin, N., and Sattentau, Q. (2009). Cell-to-cell HIV-1 spread and its implications for immune evasion. Curr Opin HIV AIDS *4*, 143–149.

Matsuda, M., Kubo, A., Furuse, M., and Tsukita, S. (2004). A peculiar internalization of claudins, tight junction-specific adhesion molecules, during the intercellular movement of epithelial cells. J. Cell Sci. *117*, 1247–1257.

Matsumura, T., Hu, Z., Kato, T., Dreux, M., Zhang, Y.Y., Imamura, M., Hiraga, N., Juteau, J.M., Cosset, F.L., Chayama, K., *et al.* (2009). Amphipathic DNA polymers inhibit hepatitis C virus infection by blocking viral entry. Gastroenterology *137*, 673–681.

Mee, C.J., Grove, J., Harris, H.J., Hu, K., Balfe, P., and McKeating, J.A. (2008). Effect of cell polarization on hepatitis C virus entry. J. Virol. *82*, 461–470.

Mee, C.J., Farquhar, M.J., Harris, H.J., Hu, K., Ramma, W., Ahmed, A., Maurel, P., Bicknell, R., Balfe, P., and McKeating, J.A. (2009a). Hepatitis C Virus Infection Reduces Hepatocellular Polarity in a Vascular Endothelial Growth Factor-Dependent Manner. Gastroenterology. *138*, 1134–1142.

Mee, C.J., Harris, H.J., Farquhar, M.J., Wilson, G., Reynolds, G., Davis, C., van, I.S.C., Balfe, P., and McKeating, J.A. (2009b). Polarization restricts hepatitis C virus entry into HepG2 hepatoma cells. J. Virol. *83*, 6211–6221.

Meertens, L., Bertaux, C., Cukierman, L., Cormier, E., Lavillette, D., Cosset, L., and Dragic, T. (2008). The tight junction proteins claudin-1, -6, and -9 are entry cofactors for hepatitis C virus. J. Virol. *82*, 3555–3560.

Meola, A., Sbardellati, A., Bruni Ercole, B., Cerretani, M., Pezzanera, M., Ceccacci, A., Vitelli, A., Levy, S., Nicosia, A., Traboni, C., *et al.* (2000). Binding of hepatitis C virus E2 glycoprotein to CD81 does not correlate with species permissiveness to infection. J. Virol. *74*, 5933–5938.

Meuleman, P., Hesselgesser, J., Paulson, M., Vanwolleghem, T., Desombere, I., Reiser, H., and Leroux-Roels, G. (2008). Anti-CD81 antibodies can prevent a hepatitis C virus infection in vivo. Hepatology *48*, 1761–1768.

Meunier, J.C., Fournillier, A., Choukhi, A., Cahour, A., Cocquerel, L., Dubuisson, J., and Wychowski, C. (1999). Analysis of the glycosylation sites of hepatitis C virus (HCV) glycoprotein E1 and the influence of E1 glycans on the formation of the HCV glycoprotein complex. J. Gen. Virol. *80*, 887–896.

Michalak, J.P., Wychowski, C., Choukhi, A., Meunier, J.C., Ung, S., Rice, M., and Dubuisson, J. (1997). Characterization of truncated forms of hepatitis C virus glycoproteins. J. Gen. Virol. *78*, 2299–2306.

Modis, Y., Ogata, S., Clements, D., and Harrison, S.C. (2004). Structure of the Dengue virus envelope protein after membrane fusion. Nature *427*, 313–319.

Molina, S., Castet, V., Fournier-Wirth, C., Pichard-Garcia, L., Avner, R., Harats, D., Roitelman, J., Barbaras, R., Graber, P., Ghersa, P., *et al.* (2007). The low-density lipoprotein receptor plays a role in the infection of primary human hepatocytes by hepatitis C virus. J. Hepatol. *46*, 411–419.

Molina, S., Castet, V., Pichard-Garcia, L., Wychowski, C., Meurs, E., Pascussi, J.M., Sureau, C., Fabre, J.M., Sacunha, A., Larrey, D., *et al.* (2008). Serum-derived hepatitis C virus infection of primary human hepatocytes is tetraspanin CD81 dependent. J. Virol. *82*, 569–574.

Morikawa, K., Zhao, Z., Date, T., Miyamoto, M., Murayama, A., Akazawa, D., Tanabe, J., Sone, S., and Wakita, T. (2007). The roles of CD81 and glycosaminoglycans in the adsorption and uptake of infectious HCV particles. J. Med. Virol. *79*, 714–723.

Mothes, W., Sherer, N.M., Jin, J., and Zhong, P. (2010). Virus cell-to-cell transmission. J. Virol. *84*, 8360–8368.

Mottola, G., Jourdan, N., Castaldo, G., Malagolini, N., Lahm, A., Serafini-Cessi, F., Migliaccio, G., and Bonatti, S. (2000). A new determinant of endoplasmic reticulum localization is contained in the juxtamembrane region of the ectodomain of hepatitis C virus glycoprotein E1. J. Biol. Chem. *275*, 24070–24079.

Nakajima, H., Cocquerel, L., Kiyokawa, N., Fujimoto, J., and Levy, S. (2005). Kinetics of HCV envelope proteins' interaction with CD81 large extracellular loop. Biochem. Biophys. Res. Commun. *328*, 1091–1100.

Nielsen, S.U., Bassendine, M.F., Burt, A.D., Bevitt, J., and Toms, G.L. (2004). Characterization of the genome and structural proteins of hepatitis C virus resolved from infected human liver. J. Gen. Virol. *85*, 1497–1507.

Op De Beeck, A., Montserret, R., Duvet, S., Cocquerel, L., Cacan, R., Barberot, B., Le Maire, M., Penin, F., and Dubuisson, J. (2000). The transmembrane domains of hepatitis C virus envelope glycoproteins E1 and E2 play a major role in heterodimerization. J. Biol. Chem. *275*, 31428–31437.

Op De Beeck, A., Voisset, C., Bartosch, B., Ciczora, Y., Cocquerel, L., Keck, Z., Foung, S., Cosset, L., and Dubuisson, J. (2004). Characterization of functional hepatitis C virus envelope glycoproteins. J. Virol. *78*, 2994–3002.

Overton, M.C., Chinault, L., and Blumer, K.J. (2003). Oligomerization, biogenesis, and signaling is promoted by a glycophorin A-like dimerization motif in transmembrane domain 1 of a yeast G protein-coupled receptor. J. Biol. Chem. *278*, 49369–49377.

Owen, D.M., Huang, H., Ye, J., Gale, M., Jr. (2009). Apolipoprotein E on hepatitis C virion facilitates infection through interaction with low-density lipoprotein receptor. Virology *394*, 99–108.

Owsianka, A.M., Timms, J.M., Tarr, A.W., Brown, R.J., Hickling, T.P., Szwejk, A., Bienkowska-Szewczyk, K., Thomson, B.J., Patel, H., and Ball, J.K. (2006). Identification of conserved residues in the E2 envelope glycoprotein of the hepatitis C virus that are critical for CD81 binding. J. Virol. *80*, 8695–8704.

Owsianka, A.M., Tarr, A.W., Keck, Z.Y., Li, T.K., Witteveldt, J., Adair, R., Foung, S.K., Ball, K., and Patel, A.H. (2008). Broadly neutralizing human monoclonal antibodies to the hepatitis C virus E2 glycoprotein. J. Gen. Virol. *89*, 653–659.

Paris, L., Tonutti, L., Vannini, C., and Bazzoni, G. (2008). Structural organization of the tight junctions. Biochim. Biophys. Acta *1778*, 646–659.

Patel, J., Patel, H., and McLauchlan, J. (2001). The transmembrane domain of the hepatitis C virus E2 glycoprotein is required for correct folding of the E1 glycoprotein and native complex formation. Virology *279*, 58–68.

Penin, F., Combet, C., Germanidis, G., Frainais, P.O., Deleage, G., and Pawlotsky, J.M. (2001). Conservation of the conformation and positive charges of hepatitis C virus E2 envelope glycoprotein hypervariable region 1 points to a role in cell attachment. J. Virol. *75*, 5703–5710.

Perez-Berna, A.J., Moreno, M.R., Guillen, J., Bernabeu, A., and Villalain, J. (2006). The membrane-active regions of the hepatitis C virus E1 and E2 envelope glycoproteins. Biochemistry *45*, 3755–3768.

Perotti, M., Mancini, N., Diotti, R.A., Tarr, A.W., Ball, J.K., Owsianka, A., Adair, R., Patel, A.H., Clementi, M., and Burioni, R. (2008). Identification of a broadly cross-reacting and neutralizing human monoclonal antibody directed against the hepatitis C virus E2 protein. J. Virol. *82*, 1047–1052.

Pileri, P., Uematsu, Y., Campagnoli, S., Galli, G., Falugi, F., Petracca, R., Weiner, A.J., Houghton, M., Rosa, D., Grandi, G., *et al.* (1998). Binding of hepatitis C virus to CD81. Science *282*, 938–941.

Ploss, A., Evans, M.J., Gaysinskaya, V.A., Panis, M., You, H., de Jong, P., and Rice, C.M. (2009). Human occludin is a hepatitis C virus entry factor required for infection of mouse cells. Nature *457*, 882–886.

Ploss, A., Khetani, S.R., Jones, C.T., Syder, A.J., Trehan, K., Gaysinskaya, V.A., Mu, K., Ritola, K., Rice, C.M., Bhatia, S.N. (2010). Persistent hepatitis C virus infection in microscale primary human hepatocyte cultures. Proc. Natl. Acad. Sci. U.S.A. *107*, 3141–3145.

Pohlmann, S., Zhang, J., Baribaud, F., Chen, Z., Leslie, G.J., Lin, G., Granelli-Piperno, A., Doms, R.W., Rice, M., and McKeating, J.A. (2003). Hepatitis C virus glycoproteins interact with DC-SIGN and DC-SIGNR. J. Virol. *77*, 4070–4080.

Popescu, C.I., Dubuisson, J. (2009). Role of lipid metabolism in hepatitis C virus assembly and entry. Biol. Cell *102*, 63–74.

Rey, F.A., Heinz, F.X., Mandl, C., Kunz, C., and Harrison, S.C. (1995). The envelope glycoprotein from tick-borne encephalitis virus at 2 A resolution. Nature *375*, 291–298.

Reynolds, G.M., Harris, H.J., Jennings, A., Hu, K., Grove, J., Lalor, P.F., Adams, D.H., Balfe, P., Hubscher, G., and McKeating, J.A. (2008). Hepatitis C virus receptor expression in normal and diseased liver tissue. Hepatology *47*, 418–427.

Roccasecca, R., Ansuini, H., Vitelli, A., Meola, A., Scarselli, E., Acali, S., Pezzanera, M., Ercole, B.B., McKeating, J., Yagnik, A., *et al.* (2003). Binding of the hepatitis C virus E2 glycoprotein to CD81 is strain specific and is modulated by a complex interplay between hypervariable regions 1 and 2. J. Virol. *77*, 1856–1867.

Rocha-Perugini, V., Montpellier, C., Delgrange, D., Wychowski, C., Helle, F., Pillez, A., Drobecq, H., Le Naour, F., Charrin, S., Levy, S., *et al.* (2008). The CD81 partner EWI-2wint inhibits hepatitis C virus entry. PLoS One *3*, e1866.

Saitou, M., Furuse, M., Sasaki, H., Schulzke, J.D., Fromm, M., Takano, H., Noda, T., and Tsukita, S. (2000). Complex phenotype of mice lacking occludin, a component of tight junction strands. Mol. Biol. Cell *11*, 4131–4142.

Scarselli, E., Ansuini, H., Cerino, R., Roccasecca, R.M., Acali, S., Filocamo, G., Traboni, C., Nicosia, A., Cortese, R., and Vitelli, A. (2002). The human scavenger receptor class B type I is a novel candidate receptor for the hepatitis C virus. EMBO J. *21*, 5017–5025.

Scheel, T.K., Gottwein, J.M., Jensen, T.B., Prentoe, J.C., Hoegh, A.M., Alter, H.J., Eugen-Olsen, J., and Bukh, J. (2008). Development of JFH1-based cell culture systems for hepatitis C virus genotype 4a and evidence for cross-genotype neutralization. Proc. Natl. Acad. Sci. U.S.A. *105*, 997–1002.

Schneider, D., and Engelman, D.M. (2004). Involvement of transmembrane domain interactions in signal transduction by alpha/beta integrins. J. Biol. Chem. *279*, 9840–9846.

Shen, L., Weber, R., and Turner, J.R. (2008). The tight junction protein complex undergoes rapid and continuous molecular remodeling at steady state. J. Cell Biol. *181*, 683–695.

Spadaccini, R., D'Errico, G., D'Alessio, V., Notomista, E., Bianchi, A., Merola, M., and Picone, D. (2009). Structural characterization of the transmembrane proximal region of the hepatitis C virus E1 glycoprotein. Biochim. Biophys. Acta. *1798*, 344–353.

Stamatovic, S.M., Keep, R.F., Wang, M.M., Jankovic, I., and Andjelkovic, A.V. (2009). Caveolae-mediated internalization of occludin and claudin-5 during CCL2-induced tight junction remodeling in brain endothelial cells. J. Biol. Chem. *284*, 19053–19066.

Syder, A.J., Lee, H., Zeisel, M.B., Grove, J., Soulier, E., Macdonald, J., Chow, S., Chang, J., Baumert, T.F., McKeating, J.A., *et al.* (2010). Small molecule scavenger receptor BI antagonists are potent HCV entry inhibitors. J. Hepatol. [Epub ahead of print].

Thomssen, R., Bonk, S., and Thiele, A. (1993). Density heterogeneities of hepatitis C virus in human sera due to the binding of beta-lipoproteins and immunoglobulins. Med. Microbiol. Immunol. *182*, 329–334.

Timpe, J.M., Stamataki, Z., Jennings, A., Hu, K., Farquhar, M.J., Harris, H.J., Schwarz, A., Desombere, I., Roels, G.L., Balfe, P., *et al.* (2008). Hepatitis C virus cell–cell transmission in hepatoma cells in the presence of neutralizing antibodies. Hepatology *47*, 17–24.

Tscherne, D.M., Jones, C.T., Evans, M.J., Lindenbach, B.D., McKeating, A., and Rice, C.M. (2006). Time- and temperature-dependent activation of hepatitis C virus for low-pH-triggered entry. J. Virol. *80*, 1734–1741.

Vaillant, A., Juteau, J.M., Lu, H., Liu, S., Lackman-Smith, C., Ptak, R., and Jiang, S. (2006). Phosphorothioate oligonucleotides inhibit human immunodeficiency virus type 1 fusion by blocking gp41 core formation. Antimicrob. Agents Chemother. *50*, 1393–1401.

Van Itallie, C.M., Betts, L., Smedley, J.G., 3rd, McClane, A., and Anderson, J.M. (2008). Structure of the claudin-binding domain of Clostridium perfringens enterotoxin. J. Biol. Chem. *283*, 268–274.

Voisset, C., Callens, N., Blanchard, E., Op De Beeck, A., Dubuisson, J., and Vu-Dac, N. (2005). High density lipoproteins facilitate hepatitis C virus entry through the scavenger receptor class B type I. J. Biol. Chem. *280*, 7793–7799.

von Hahn, T., Lindenbach, B.D., Boullier, A., Quehenberger, O., Paulson, M., Rice, M., and McKeating, J.A. (2006). Oxidized low-density lipoprotein inhibits hepatitis C virus cell entry in human hepatoma cells. Hepatology *43*, 932–942.

von Hahn, T., Yoon, J.C., Alter, H., Rice, C.M., Rehermann, B., Balfe, P., and McKeating, J.A. (2007). Hepatitis C virus continuously escapes from neutralizing antibody and T-cell responses during chronic infection in vivo. Gastroenterology *132*, 667–678.

Wakita, T., Pietschmann, T., Kato, T., Date, T., Miyamoto, M., Zhao, Z., Murthy, K., Habermann, A., Krausslich, H.G., Mizokami, M., *et al.* (2005). Production of infectious hepatitis C virus in tissue culture from a cloned viral genome. Nat. Med. *11*, 791–796.

Whidby, J., Mateu, G., Scarborough, H., Demeler, B., Grakoui, A., and Marcotrigiano, J. (2009). Blocking hepatitis C virus infection with recombinant form of envelope protein 2 ectodomain. J. Virol. *83*, 11078–11089.

Witteveldt, J., Evans, M.J., Bitzegeio, J., Koutsoudakis, G., Owsianka, A.M., Angus, A.G., Keck, Z.Y., Foung, S.K., Pietschmann, T., Rice, C.M., *et al.* (2009). CD81 is dispensable for hepatitis C virus cell-to-cell transmission in hepatoma cells. J. Gen. Virol. *90*, 48–58.

Yalaoui, S., Huby, T., Franetich, J.F., Gego, A., Rametti, A., Moreau, M., Collet, X., Siau, A., van Gemert, G.J., Sauerwein, R.W., *et al.* (2008). Scavenger receptor BI boosts hepatocyte permissiveness to Plasmodium infection. Cell Host Microbe *4*, 283–292.

Yang, W., Qiu, C., Biswas, N., Jin, J., Watkins, S.C., Montelaro, R.C., Coyne, B., and Wang, T. (2008). Correlation of the tight junction-like distribution of Claudin-1 to the cellular tropism of hepatitis C virus. J. Biol. Chem. *283*, 8643–8653.

Zhang, J., Randall, G., Higginbottom, A., Monk, P., Rice, M., and McKeating, J.A. (2004). CD81 is required for hepatitis C virus glycoprotein-mediated viral infection. J. Virol. *78*, 1448–1455.

Zheng, A., Yuan, F., Li, Y., Zhu, F., Hou, P., Li, J., Song, X., Ding, M., and Deng, H. (2007). Claudin-6 and claudin-9 function as additional coreceptors for hepatitis C virus. J. Virol. *81*, 12465–12471.

Zhong, J., Gastaminza, P., Cheng, G., Kapadia, S., Kato, T., Burton, D.R., Wieland, S.F., Uprichard, S.L., Wakita,

T., and Chisari, F.V. (2005). Robust hepatitis C virus infection in vitro. Proc. Natl. Acad. Sci. U.S.A. *102*, 9294–9299.

Zhong, J., Gastaminza, P., Chung, J., Stamataki, Z., Isogawa, M., Cheng, G., McKeating, A., and Chisari, F.V. (2006). Persistent hepatitis C virus infection in vitro: coevolution of virus and host. J. Virol. *80*, 11082–11093.

Ziegler, S., Kronenberger, B., Albrecht, B.A., Kaul, A., Gamer, A.L., Klein, D., and Hartmann, R.W. (2009). Development and evaluation of a FACS-based medium-throughput assay for HCV entry inhibitors. J. Biomol. Screen *14*, 620–626.

A Perspective on Host Cell Targets and Inhibitors

18

Kai Lin

Abstract

HCV drug discovery efforts have been mainly focusing on two viral targets, NS3 protease and NS5B polymerase, which are essential for viral replication. However, due to the high heterogeneity and high replication and mutation rates of the virus, resistance emerges quickly in patients treated with specific inhibitors of these viral enzymes. A complementary strategy is to target cellular proteins that are required for viral replication, which may have the advantage of higher genetic barrier to resistance and broader genotype coverage. Through siRNA screen and chemogenetic approach, a number of potential host targets have been identified at essentially every step of viral life cycle, including viral entry, genome replication, virion assembly and release. Some are engaged in viral replication through direct interaction with HCV proteins, such as cyclophilins. Others modulate viral life cycle via various cellular pathways, such as lipid metabolism and membrane trafficking. There exist both opportunities and challenges in developing host targeting antivirals. Currently the most advanced compounds are cyclophilin inhibitors in Ph II. The combination of host and viral targeting inhibitors presents a highly attractive strategy to suppress the emergence of resistance and maximize antiviral efficacy.

Introduction

HCV has a positive, single-stranded RNA genome of 9.6 kb that encodes only 10 individual viral proteins. So, the choice of viral proteins as direct antiviral targets is quite limited since not all of them are considered druggable. Given the success of antiviral development targeting HIV-1 protease and reverse transcriptase as well as HBV and herpesvirus polymerases, drug discovery effort on HCV has been mainly focusing on two viral targets with well-defined enzymatic functions, NS3 protease and NS5B polymerase. Indeed, potent inhibitors have been identified for these targets, some of which have shown promising results in the clinical studies. However, it has become apparent that resistance can develop quickly both *in vitro* and in patients with these specific inhibitors of viral targets. A complementary approach is to target host proteins that are also essential for viral replication, which may present a higher genetic barrier to resistance and can be used in combination with inhibitors of viral targets.

There are many potential opportunities to treat HCV infection by targeting host factors, as the virus relies heavily on host machinery and cellular proteins at essentially every step of the viral life cycle. HCV, an enveloped virus, enters the cells through interaction with a number of cellular proteins including surface receptor CD81 (Pileri *et al.*, 1998), human scavenger receptor class B type I (SR-B1; Scarselli *et al.*, 2002), tight junction protein claudin-1 (Evans *et al.*, 2007) and occludin (Ploss *et al.*, 2009). The viral nucleocapsid is released in the cytoplasm following fusion between viral and cellular membranes, which is mediated by clathrin-dependent endocytosis. After decapsidation, the positive-strand viral RNA serves as the template for the synthesis of the HCV polyprotein in the cytoplasm. The 5′-UTR of HCV contains an internal ribosomal entry site (IRES), which mediates cap-independent initiation of protein translation by recruiting a number of cellular proteins including the eukaryotic initiation factors. The HCV RNA genome encodes a single large open reading frame corresponding to a polyprotein precursor of about 3000 amino acids, which is proteolytically cleaved into ten individual proteins, in the order of C-E1-E2-p7-NS2-NS3-NS4A-NS4B-NS5A-NS5B. Host signal peptidase and signal peptide peptidase are responsible for the cleavage at the junctions of core–E1, E1–E2, E2–p7 and p7–NS2, while viral NS2–3 cysteine protease and NS3–4A serine protease are responsible for cleaving the other junctions. The viral non-structural proteins then form the replication complex with cellular components and nascent RNA strands on an endoplasmic reticulum (ER) membrane derived structure named membranous web. Following viral RNA replication, the virions are assembled on the lipid droplets (LD) associated with ER-derived membranes, where the HCV core protein is localized. HCV non-structural (NS) proteins are also recruited to the LD-associated membrane and are involved in the viral assembly. The viral particle is probably enveloped through budding into the ER lumen and then transported through Golgi to be released.

Thus, host factors greatly expand the list of potential antiviral targets. Also for many of cellular targets, inhibitors have already been discovered and developed for other indications, which provide useful pharmacological tools to confirm the findings *in vitro* and even present opportunities for a faster development path to assess efficacy in patients. Most importantly, compared with viral targets, host targets have the perceivable advantage of presenting a higher genetic barrier to resistance, as the host genes themselves are unlikely to change over the course of therapy. Mutations in viral genes can affect the interaction of viral proteins to host factors, however such a selection is likely to be much less efficient and the changes are likely to be less effective since the inhibitors do not bind to the viral proteins directly. For the same reasons, host targeting inhibitors also have the advantage of broader genotype coverage than viral targeting inhibitors, since host factors are less likely affected by the heterogeneity of the virus.

Due to the number of host factors identified and limited space for discussion here, this chapter will focus on those that either have been explored for drug discovery or may potentially be considered as novel antiviral targets. Also, receptors for viral entry and immunomodulators will be discussed separately in Chapters 17 and 19, respectively.

Discovery of host targets

Both rational and screening approaches have been taken to identify novel host targets for HCV. A better understanding of the viral life cycle and how the virus hijacks various cellular pathways has suggested some of the potential entry points for interfering viral replication, such as host factors involved in membrane trafficking (Hanoulle *et al.*, 2009). There have also been a number of efforts to screen cellular proteins that directly interact with a specific viral protein or RNA (Yen *et al.*, 1995; Wang *et al.*, 2006; Taguwa *et al.*, 2008; de Chassey *et al.*, 2008). However, methods such as yeast–two hybrid often result in hits that do not necessarily have any functional relevance. Thus, it is important to confirm the finding through loss (e.g. siRNA) or gain (e.g. ectopic expression) of function study. Whole cell-based phenotypic screens of compound libraries have also generated a lot of inhibitors with unknown targets or mechanism (Chockalingam *et al.*, 2010; Gastaminza *et al.*, 2010). The challenge is to deconvolute the pathway and to identify the specific target using chemogenetics approach. There have also been a number of siRNA screens performed, which greatly expand the list of potential host targets (Ng *et al.*, 2007; Randall *et al.*, 2007; Li *et al.*, 2009; Tai *et al.*, 2009; Vaillancourt *et al.*, 2009). However, the percentage of overlapping hits from these screens is surprisingly low, possibly due to different assay conditions or high rate of false positive. In addition, transcriptional and proteomic profiling comparing viral infected cells vs. naive cells and specific inhibitor treated cells vs. untreated cells has also generated rich information on host–virus interactions (Su *et al.*, 2002). Pathway analysis is helpful in connecting the body of knowledge generated through these efforts. Specific examples are provided below, and summarized in Table 18.1, highlighting the most promising host factors and pathways identified to date as potential novel antiviral targets.

Cyclophilins

Cyclophilins are a family of highly conserved cellular peptidyl-prolyl *cis-trans* isomerases (PPIase) that are involved in many cellular processes such as protein folding and trafficking. There are over 16 human cyclophilins with different subcellular localization and functions (Wang and Heitman, 2005). Cyclophilin A (CypA) is the cytosolic protein previously known as the binding partner of cyclosporin A (CsA; Handschumacher *et al.*, 1984), an immunosuppressive drug blocking T-cell activation and IL-2 production through the inhibition of calcineurin and NF-AT, although the PPIase function of CypA is not involved in the immunosuppressive activity of CsA (Zydowsky *et al.*, 1992). CypB and CypC both have N-terminal signal peptide sequences that anchor them to the ER and can be secreted upon binding to CsA. CypD is localized in the mitochondria and regulates mitochondria transition pore. CypE is in the nucleus and is involved in mRNA splicing.

It was known previously that CsA has inhibitory activities against a number of viruses including HIV-1, HSV, VSV and HBV (Karpas *et al.*, 1992; Vahlne *et al.*, 1992; Damaso and Keller, 1994; Rosenwirth *et al.*, 1994; Billich *et al.*, 1995; Bose *et al.*, 2003; Bouchard *et al.*, 2003). Watashi *et al.* were the first to show that CsA suppressed HCV replication directly *in vitro* (Watashi *et al.*, 2003). The anti-HCV activity of CsA was also observed in the clinical settings as the combination of CsA and IFN-2b resulted in a significantly improved antiviral response than IFN-2b alone in a controlled trial (Inoue *et al.*, 2003). These reports triggered immense interest in investigating the roles of host cyclophilins in HCV replication and exploring non-immunosuppressive cyclosporin analogues for the treatment of HCV infection. First, it was demonstrated that the anti-HCV effect of cyclosporin analogues was mediated through cyclophilins, as their HCV inhibitory activity correlated with cyclophilin-binding affinity, but not calcineurin or P-gp inhibitory activity (Ma *et al.*, 2006). Knockdown of cyclophilins in the replicon cells using specific siRNAs also led to inhibition of HCV replication (Watashi *et al.*, 2003; Nakagawa *et al.*, 2004; Nakagawa *et al.*, 2005; Gaither *et al.*, 2010). However, there have been some discrepancies on specifically which cyclophilins are required and how they are involved in HCV replication. Watashi *et al.* reported that CypB but not CypA was required for HCV replication through its interaction with NS5B (Watashi *et al.*, 2005). This finding was confirmed by Heck *et al.*, who further demonstrated that CypB stimulated RNA synthesis by NS5B *in vitro* and the PPIase activity of CypB was not required for this function (Heck *et al.*, 2009). In contrast, Yang *et al.* (2008) suggested that CypA was the main mediator of HCV replication. Moreover, the PPIase activity of CypA was essential for its function in HCV replication (Chatterji *et al.*, 2009; Liu *et al.*, 2009). NS5A domain 2 directly interacted with CypA and CypB and was a substrate of CypA and CypB PPIase *in vitro* (Hanoulle *et al.*, 2009). Sequence analysis revealed that mutations in NS5A and perhaps NS5B contributed to resistance against cyclophilin inhibitors *in vitro* (Fernandes *et al.*, 2007; Robida *et al.*, 2007; Puyang *et al.*, 2010). So, it remains to be elucidated whether the anti-HCV activity of cyclophilin inhibitors is mainly mediated by CypA or CypB or both and whether NS5A or NS5B or both are involved. Different cellular systems and assay conditions used by various groups could have contributed to some of the inconsistent findings in theses studies. Nevertheless, as illustrated in Fig. 18.1, increasing evidences suggest that cyclophilins are involved in HCV replication by (1) interacting directly with viral proteins as part of the replication complex and/or (2) mediating the correct folding and trafficking of viral proteins to the site of replication (cytosolic side of ER membrane) through their PPIase activity. Cyclophilin inhibitors block the interaction of cyclophilins with HCV proteins and thus the formation of a functional viral replication complex. In addition, there have also been data suggesting cyclophilin inhibitors may interfere with the self-cleavage of NS2–NS3 (Ciesek *et al.*, 2009) and the cleavage of NS5A–NS5B by NS3 (Kaul *et al.*, 2009), however whether such an effect is specific to JFH-1 or genotype 2a remains to be determined.

Table 18.1 Summary of cellular pathways and proteins involved in HCV replication, proposed mechanisms, and inhibitors with demonstrated antiviral effect

Pathways	Targets	Proposed mechanisms	Inhibitors
Cyclophilins	CypA, CypB	Bind to NS5A and NS5B, involved in protein folding, replication complex formation, NS2-3 and NS5A-5B cleavages	Alisporivir (Debio-025), NIM811, SCY-635
Nucleotide metabolism	IMPDH	Purine biosynthesis, required for viral RNA replication	Ribavirin, VX-497, mycophenolic acid
	CAD	Pyrimidine biosynthesis, required for viral RNA replication	
	DHODH	Pyrimidine biosynthesis, required for viral RNA replication	Leflunomide
	CKB	Bind to NS4A, regulate level of ATP and NS3 helicase activity	
Lipid metabolism	HMG-CoA	Synthesis of cholesterol and non-sterol isoprenoid products, required for the formation of viral replication complex	Statins
	MVD	Mevalonate synthesis, required for the formation of viral replication complex	
	SPT	Sphingolipid biosynthesis, required for lipid raft and replication complex	Myriocin, NA255, NA808
	ApoB	VLDL and viral particle assembly and secretion	
	ApoE	VLDL and viral particle assembly and secretion	
	MTP	VLDL and viral particle assembly and secretion	Naringenin, CP-346086, BMS-2101038
	OSBP	Bind to NS5A, Golgi trafficking and HCV particle release	
	PPAR-	Formation of membranous web	2-Chloro-5-nitro-N-(pyridyl) benzamide

Kinases and signalling pathways	PKA	Modulate core localization and viral assembly, also bind to NS3	
	PKC	Modulate core localization, also bind to NS3 and CD81	
	PRK2	Bind and phosphorylate NS5B	HA1077 (fasudil), Y27632
	Ras-ERK	Modulate IRES-dependent translation and viral RNA replication	U0126, PD98059 (MEK inhibitors)
	PI3K-Akt-mTOR	Stimulate the pathway through binding to NS5A and prevent apoptosis	LY294002 (PI3K inhibitor)
Membrane vesicle trafficking	VAP-A/B	Bind to NS5A/NS5B, required for the formation of replication complex	
	Rab/GAP/ARF	Required for the formation of membranous web and replication complex	Brefeldin A
	PI4K	Bind to NS5A, required for the formation of membranous web and replication complex	PIK93
	ESCRT	Release of viral particles	
Protein translation, modification	La/PTB/hnRNP/HuR/eIF	Bind to HCV IRES and activate IRES-dependent protein translation	Nitazoxanide (eIF2)
	miR122	Bind to 5′-UTR and stimulate translation	SPC3649
	-Glucosidase I	Glycosylation of E1 and E2	Iminosugars, Celgosivir
	HSP90/FKBP38/hB-ind1	Form complex with NS5A, required for correct protein folding in the membranous web	Geldanamycin (17-AAG), Radicicol
Autophagy	Atg5/Atg7/LC3	Maintain cellular homeostasis for sustained viral replication	Chloroquine

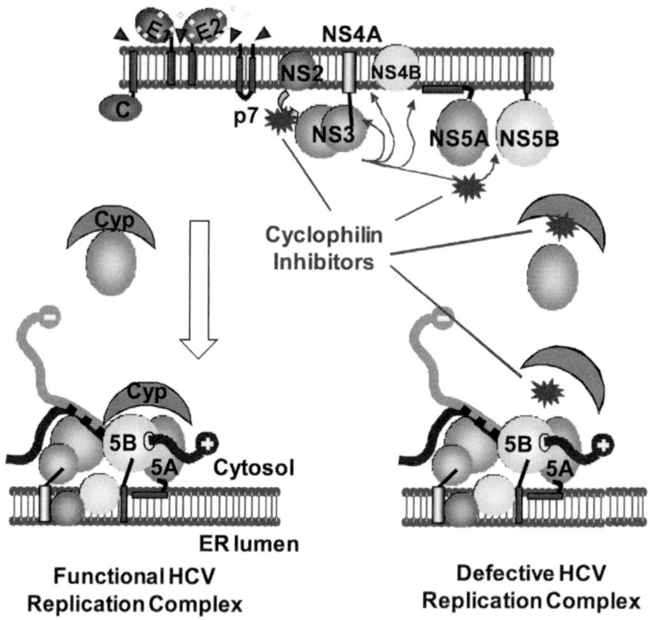

Figure 18.1 Proposed mechanisms of cyclophilin inhibitors in blocking HCV replication.

Three cyclophilin inhibitors have entered clinical development for the treatment of hepatitis C. Alisporivir (Debio-025), a non-immunosuppressive CsA analogue (Paeshuyse *et al.*, 2006), was the first to demonstrate direct antiviral effect in patients. In a phase I trial with HIV–HCV co-infected patients, 1200 mg twice daily of alisporivir monotherapy resulted in a 3.4 $\log_{10}$ reduction in HCV RNA after 14 days (Flisiak *et al.*, 2008). In a phase II combination study, 600 mg once daily of alisporivir in combination with PEG-IFN-α-2a led to a 4.6 $\log_{10}$ viral load reduction after 28 days in genotypes 1 and four patients, and the effect was even more pronounced in genotypes 2 and 3 patients with a 5.9 $\log_{10}$ viral load reduction after 28 days (Flisiak *et al.*, 2009). Not only the compound has broad genotype coverage, but also the rate of viral breakthrough due to emergence of resistance appears to be much lower compared with those with viral protease of polymerase inhibitors, which is consistent with what has been observed *in vitro*. NIM811, another non-immunosuppressive cyclosporin analogue (Ma *et al.*, 2006), also achieved clinical efficacy in combination with PEG-IFN-α-2a. In a phase Ib trial in genotype 1 HCV patients who had relapsed in prior interferon therapy, 600 mg b.i.d. NIM811 plus PEG-IFN-α-2a resulted in an HCV viral load reduction of 2.78 $\log_{10}$ compared with only 0.58 $\log_{10}$ with PEG-IFN-α-2a alone (Lawitz E. *et al.*, 2009). Recently, a third cyclosporin analogue, SCY-635, reported a 2.2 $\log_{10}$ viral load reduction in a 15-day monotherapy trial in genotype 1 HCV patients (Hopkins *et al.*, 2010).

Because of the high resistance barrier with cyclophilin inhibitors and the lack of cross-resistance with inhibitors of viral targets (Mathy *et al.*, 2008; Puyang *et al.*, 2010), these compounds present an attractive opportunity for combination therapy. Indeed, the combination of cyclophilin inhibitors with NS3 protease or polymerase inhibitors not only resulted in additive to synergistic antiviral effect *in vitro* but also significantly helped to prevent the emergence of resistance (Mathy *et al.*, 2008; Coelmont *et al.*, 2009).

Nucleotide metabolism

Since HCV has an extraordinary high replication rate, it is likely very sensitive to the supply of the building blocks of its genome, i.e. nucleotides, which may be explored as a potential antiviral target. Inosine-5′-monophosphate dehydrogenase (IMPDH), which catalyses the conversion of IMP to XMP, an essential step in the de novo biosynthesis of guanine nucleotides, represents such an opportunity. Inhibition of IMPDH leads to a depletion of intracellular GTP and dGTP pools. Ribavirin, an essential component of current standard therapy for HCV in combination with interferon alpha (IFN-α), is a competitive inhibitor of IMPDH, which has long been speculated as one of its antiviral mechanisms. Mycophenolic acid (MPA, Cellcept®) and VX-497 (merimepodib) are structurally unrelated, reversible, uncompetitive inhibitors of IMPDH. VX-497 has broad-spectrum activity against a number of viruses (Markland *et al.*, 2000). The anti-HCV effect of VX-497 *in vitro* can be reversed in the presence of guanosine *in vitro*, suggesting that the activity is mediated via IMPDH inhibition. Henry *et al.* showed that MPA also inhibited HCV replication *in vitro* although the effect seemed to be guanosine independent (Henry *et al.*, 2006). Another proposed mechanism for ribavirin is that it can be misincorporated into viral RNA which leads to an error catastrophe of genome replication. Interestingly, Zhou *et al.* showed that VX-497 or MPA increased the viral replication error rate by two-fold when used in combination with ribavirin, the effect of which can also be reversed by the addition of guanosine (Zhou *et al.*, 2003).

VX-497, a much more potent inhibitor of IMPDH than ribavirin, has been pursued in clinical development for the treatment of hepatitis C. In a double-blind, placebo-controlled trial, patients were treated with IFN-α alone of in combination with 100 mg or 300 mg VX-497 every 8 hours for 4 weeks. There appeared to be a greater HCV RNA reduction in the 100 mg VX-497 combination group compared with the IFN-α alone group (1.78 $\log_{10}$ vs. 0.86 $\log_{10}$, P=0.037) in treatment-compliant patients. However, there was no significant difference between the groups when considering all intent-to-treat population (McHutchison *et al.*, 2005). Larger and longer trials preferably including ribavirin would be necessary to determine whether this class of drugs indeed has significant antiviral activity *in vivo* and could provide additional benefit to current standard of care.

Similarly, enzymes that are involved in pyrimidine biosynthesis have also been shown to be required for viral replication. SiRNAs specifically targeting a tripartite enzyme CAD (carbamoylphosphate synthetase 2, aspartate transcarbamylase, and dihydroorotase), which catalyses the first three steps of pyrimidine biosynthesis, blocked the replication of both genotype 1b HCV replicon and genotype 2a JFH-1 virus. The antiviral effect can be reversed by the addition of 100 μM uridine and cytidine to the culture media, which essentially allowed the virus to bypass the need for de novo pyrimidine biosyntheses (Borawski *et al.*, 2009). Leflunomide, a known inhibitor of dihydroortate dehydrogenase (DHODH), another key enzyme involved in pyrimidine synthesis, also inhibited HCV replication *in vitro* (Stuyver *et al.*, 2003).

In addition, creatine kinase B (CKB), a key ATP-generating enzyme that regulates the level of ATP in subcellular compartments of non-muscle cells, was shown to interact with HCV NS4A protein and form a complex with NS3–4A (Hara *et al.*, 2009). CKB accelerated NS3–4A mediated unwinding of RNA and DNA *in vitro* and enhanced the replicase activity of HCV in permeabilized cells. Thus, recruitment of CKB to the HCV replication complex might be important for efficient replication of the viral genome, which has high and fluctuating energy demands.

Lipid metabolism

Host lipid metabolism plays an important role in multiple steps of HCV life cycle including viral replication, virion assembly and secretion. There has been a strong interest in this particular area partly because there are already drugs like statins available that can modulate host lipid pathways and potentially can be used for HCV indication in combination with other direct antiviral agents.

Mevalonate pathway and geranylgeranylation

HCV replication is associated with intracellular membranes, the composition of which is subject to modification through host lipid biosynthesis. The mevalonate pathway is responsible for the synthesis of cholesterol and non-sterol isoprenoid products including farnesyl and geranylgeranyl pyrophosphates, which are covalently attached to proteins and essential for their membrane association. Treatment of HCV replicon cells with lovastatin, an inhibitor of 3-hydroxy-3-methyglutaryl CoA reductase (HMG-CoA) and an inhibitor of protein geranylgeranyl transferase I, but not an inhibitor of farnesyl transferase induced the dissolution of the HCV replication complex (Ye *et al.*, 2003). Moreover, the addition of geranylgeraniol, but not farnesol, to the lovastatin-treated cells, restored the assembly of the replication complex and viral replication. Interestingly, the effect of geranylgeranylation is dependent on the content of fatty acid in the cell. Saturated and monounsaturated fatty acids are required for efficient HCV replication, most likely by maintaining optimal membrane structure (Kapadia and Chisari, 2005). FBL2, a geranylgeranylated protein, was shown to interact with HCV NS5A protein directly through its CAAX motif. A dominant-negative FBL2 protein inhibited replication of HCV RNA when overexpressed in the replicon cells, the effect of which can be overcome by co-expression of NS5A. Knockdown of FBL2 by siRNA led to a reduction of HCV RNA, which can also be overcome by expression of a wobble mutant of FBL2 not targeted by the siRNA (Wang *et al.*, 2005).

siRNAs targeting the mevalonate (diphospho) decarboxylase (MVD) enzyme, which catalyses the formation of mevalonate, resulted in an inhibition of HCV replication *in vitro* (Borawski *et al.*, 2009). While MVD is not a rate-limiting enzyme in cholesterol synthesis, its inhibition would likely lead to an accumulation of phosphorylated intermediates (such as mevalonic acid pyrophosphate), which may regulate the phosphorylation state of lipid biogenesis enzymes. These findings suggest that modulation of lipid biosynthesis through mevalonate pathway is a potential antiviral approach.

Statins and HMG-CoA

There has been a strong interest in exploring the use of statins for HCV indication since these drugs have been used in human. Ikeda *et al.* compared the antiviral potency of five types of statins *in vitro*. Fluvastatin exhibited the strongest anti-HCV activity with an EC_{50} of 0.9 μM. Atorvastatin and simvastatin showed moderate inhibitory effects. Iovastatin had only weak activity, whereas pravastatin showed no activity at all. The anti-HCV activities of all statins can be reversed by the addition of mevalonate or geranylgeraniol. Fluvastatin also exhibited a synergistic effect in combination with IFN-α (Ikeda *et al.*, 2006). Another recent study confirmed the anti-HCV activity of statins, although the ranking of potency was different: mevastatin and simvastatin had the strongest effect, lovastatin and fluvastatin exhibited moderate inhibitory activity, whereas pravastatin had no activity. Combination of mevastatin or simvastatin with HCV polymerase or protease inhibitors resulted in an efficient clearance of HCV RNA from the replicon cells. Furthermore, mevastatin prevented the emergence of resistance to a non-nucleoside inhibitor of HCV polymerase (Delang *et al.*, 2009).

However, the effect of statins on HCV in the clinical settings has been controversial. In a prospective study, fluvastatin significantly reduced HCV RNA in patients, however the effect was modest, variable, and often transient (Bader *et al.*, 2008). In another open-label pilot study, 10/15 (67%) patients who received 48 weeks of 20 mg daily fluvastatin along with PEG-IFN-α/ribavirin achieved a sustained virological response (SVR; Sezaki *et al.*, 2009). On the other hand, no significant effect on HCV viral load was observed in other studies using simvastatin and atorvastatin (O'Leary *et al.*, 2007; Forde *et al.*, 2009). An actual increase of HCV RNA was reported in one study in HIV-HCV co-infected patients after 4 weeks of fluvastatin treatment (Milazzo *et al.*, 2009). It was hypothesized that statins could upregulate the expression of low-density lipoprotein (LDL) receptor, which may facilitate HCV entry to the hepatocytes. Large, blinded, placebo-controlled trials would be necessary to determine whether statins indeed could provide any benefit to HCV patients, particularly in combination with standard of care.

While statins appear to inhibit HCV replication through non-sterol products, depletion of cholesterol in the cells following treatment with β-cyclodextrin directly disrupts membranous web and inhibits HCV replication (Sagan *et al.*, 2006). Moreover, depletion of cholesterol from the virus abolished HCV infectivity, which can be rescued by supplementation of exogenous cholesterol. Interestingly, cholesterol-depleted virus is normal in cell attachment but is defective in internalization (Aizaki *et al.*, 2008). Bezafibrate, an LDL-lowering drug for treating hyperlipidaemia, also demonstrated anti-HCV effects both *in vitro* (Kanda *et al.*, 2004) and in patients (Fujita *et al.*, 2006).

Sphingomyelin

HCV RNA replication depends on the association of viral proteins with lipid raft membrane. Lipid rafts are mainly composed of sphingomyelin (SM), cholesterol and glycosphingolipids (GSL). Hydrolysis of virion-associated sphingomyelin using bacterial SMase completely abolished HCV (JFH-1) infection (Aizaki *et al.*, 2008). Serine palmitoyltransferase (SPT) is the first-step enzyme in the sphingolipid biosynthetic pathway. Myriocin, an inhibitor of SPT, suppressed replication of both subgenomic HCV replicon and infectious virus (JFH-1). Combination of myriocin and IFN-α or simvastatin synergistically inhibited HCV RNA replication in the replicon cells (Amemiya *et al.*, 2008). NA255, a fungal metabolite structurally similar to myriocin, also inhibited SPT and blocked the de novo synthesis of sphingolipids, ceramide and sphingomyelin. Treatment of HCV replicon cells with NA255 resulted in a decrease in the major lipid raft components and a disruption of the association among HCV non-structural proteins on the lipid rafts. Furthermore, NS5B protein has a sphingolipid-binding motif, which was shown to interact directly with sphingomyelin (Sakamoto *et al.*, 2005). The antiviral effect of myriocin and another SPT inhibitor NA808 were demonstrated *in vivo* using chimeric mice with humanized liver infected with genotype 1a or 1b HCV. Myriocin was able reduce the levels of HCV RNA in serum and liver by 1–2 $\log_{10}$ after 8 days. Furthermore, combined of myriocin and PEG-IFN-α reduced the HCV RNA by >2 $\log_{10}$ (Umehara *et al.*, 2006).

ApoB, ApoE, MTP and VLDL

At least a portion of HCV is known to circulate in plasma in complex with the triglyceride-rich very low-density lipoproteins (VLDL), both of which are secreted uniquely by hepatocytes. The assembly and maturation of HCV occur in the ER and post-ER compartments, respectively, which depend on microsomal transfer protein (MTP) and apolipoprotein B (apoB) in a manner that parallels the formation of VLDL (Gastaminza *et al.*, 2008). The membrane vesicles in which HCV replicates are highly enriched in proteins that are also required for VLDL assembly, including apoB, apoE and MTP. An MTP inhibitor and an apoB-specific siRNA were both able to block HCV production *in vitro* (Huang *et al.*, 2007). In addition, a known inhibitor of VLDL secretion, grapefruit flavonoid naringenin, inhibited MTP activity as well as HCV secretion in infected cells (Nahmias *et al.*, 2008).

However, several other studies suggested that apoE rather than apoB is required for HCV production. The level of apoE correlated with HCV infectivity. ApoE-specific monoclonal antibodies efficiently neutralized HCV infectivity (Chang *et al.*, 2007). Knock-down of apoE by specific siRNAs reduced the levels of both intracellular and secreted HCV virions (Owen *et al.*, 2009). These findings suggested that apoE is required for both HCV production and infection. HCV virions are assembled as apoE-enriched lipoprotein particles, and apoE may facilitate HCV infection through the interaction with LDL receptor. Furthermore, Jiang *et al.* developed a single-cycle HCV growth assay and showed that siRNA knockdown of apoE suppressed the formation of HCV particles, which can be restored by expressing apoE ectopically. In contrast, apoB-specific antibodies and siRNAs had no significant effect in the same assay (Jiang and Luo, 2009). In addition, MTP inhibitors did not affect HCV production despite that the secretion of apoB-containing lipoproteins was efficiently blocked. Jiang *et al.* argued that the positive effect observed with MTP inhibitors in other studies was due to the inhibition of apoE expression and secretion at relatively high concentrations of these compounds. Interestingly, it was also shown that apoE specifically interacted with HCV NS5A protein and that co-localized in JFH-1 infected cells. The functional relevance of the apoE–NS5A interaction for production of viral particles was demonstrated by the loss of apoE-NS5A binding in assembly-defective viral mutants, suggesting that recruitment of apoE by NS5A is important for viral assembly and release of infectious viral particles (Benga *et al.*, 2010).

OSBP

Oxysterol binding protein (OSBP) binds to hydroxycholesterol in the cytosol and then translocates to the Golgi. It was shown that OSBP interacted with the N-terminal domain I of NS5A and co-localized in the Golgi. An OSBP-specific short hairpin RNA (shRNA) inhibited the release of HCV particles in culture supernatant but had little effect on viral RNA replication, suggesting that OSBP is involved in HCV maturation possibly though Golgi trafficking (Amako *et al.*, 2009). Interestingly, OSBP binds to VAP-A through its FFAT domain. It also binds to phosphatidylinositol 4-phosphate and phosphatidylinositol 4,5-biphosphate through its pleckstrin homology (PH) domain. A PH-deletion mutant OSBP failed to localize to the Golgi apparatus and inhibited the HCV particle release.

PPAR-α

An antagonist of peroxisome proliferator-activated receptor (PPAR), 2-chloro-5-nitro-N-(pyridyl)benzamide (BA), inhibited the replication of HCV replicon RNA in Huh-7 cells. Downregulation of PPAR-α but not PPAR-γ inhibited HCV replication. There was a clear build-up of lipids upon treatment with BA, which was consistent with the misregulation of lipid metabolism, phospholipid secretion, cholesterol catabolism, and triglyceride clearance (Rakic *et al.*, 2006). The inhibition of PPAR-α caused a rapid dispersion of HCV RNA from replication sites and simultaneously induced lipid storage and increased in lipid droplet size (Lyn *et al.*, 2009). These results suggest that a PPAR-α antagonist can inhibit HCV replication by interfering lipidation of host proteins associated with the HCV replication complex and disrupting the formation of membranous web where HCV replicates. Interestingly, although it appears that PPAR- but not PPAR-γ is involved in HCV replication, the PPAR-γ antagonist Pioglitazone has been investigated as an insulin-sensitizing agent to improve treatment response because the frequent association of insulin resistance in HCV patients (Overbeck *et al.*, 2008; Khattab *et al.*, 2009).

Kinases and signalling pathways

In order for the virus to hijack the cellular machinery and control various cellular processes such as cell growth, proliferation and apoptosis, one of the most effective ways is to modulate the signalling pathways that control these processes. Kinases play an essential role in almost all the signal transduction pathways and have been subject to extensive drug discovery in various therapeutic areas. So, a potential advantage of targeting kinases is that many inhibitors are already available to help probing the function of specific kinases in viral life cycle. In some cases, such inhibitors can even be tested in patients if they have already been developed for other indications. On the other hand, most of the kinase inhibitors are ATP analogues and are likely to inhibit multiple kinases with various potencies. Thus, caution should be taken regarding the specificity of the inhibitors and interpretation of the result. Below are some of the kinases and pathways that have been followed up more extensively in this area. Many more have been reported in a number of kinome or genome-wide siRNA screen but won't be discussed here due to limitation of space.

PKA and PKC

The HCV core protein was shown to be phosphorylated by protein kinase A (PKA) at Ser^{53}/Ser^{99} and Ser^{116} and protein kinase C (PKC) at Ser^{116} both *in vitro* (Shih *et al.*, 1995) and in Huh-7 cells (Lu and Ou, 2002). Mutation of Ser^{116} to aspartic acid led to an increased nuclear localization of the core protein, which suggested that PKC may play a role in regulating the nuclear localization of the core protein. Core protein has been implicated in the development of hepatocellular carcinoma as overexpression of core can transform fibroblast cells *in vitro*. The activity of phospholipase D (PLD) was elevated in NIH3T3 cells overexpressing core while the expression levels of PLD and PKC were unchanged. Phorbol 12-myristate 13-acetate (PMA), a known activator of PKC, stimulated PLD activity significantly more in core-transformed cells, which suggested that PLD might be involved in core-induced transformation (Kim *et al.*, 2004a). In addition, the NS3 protein of HCV inhibited nuclear transport and the activity of PKA, which was mediated by an arginine-rich domain of NS3 (Borowski *et al.*, 1999). The arginine-rich domain also interacted with the catalytic site of PKC and was a substrate as well as a competitive inhibitor of PKC, which may disrupt the PKC-mediated signal transduction (Borowski *et al.*, 2000).

In turn, the phosphorylation of PKA can modulate the level of HCV core in infected cells as well as nucleocapsid assembly and degradation (Majeau *et al.*, 2007). HCV infection of Huh-7.5 cells increased the levels of cAMP and phosphorylated PKA substrates, supporting a model where infection activates PKA in a cAMP-dependent manner to promote virus release and transmission. The inhibition of PKA with H89, cyclic AMP (cAMP) antagonists, or the protein kinase inhibitor peptide reduced the level of CLDN1 on the plasma membrane and thus HCV entry into Huh-7.5 cells (Farquhar *et al.*, 2008).

PKC was also shown to interact with CD81 (Zhang *et al.*, 2001). Hepatocytes are highly polarized with tight junctions (TJ) separating basolateral and apical domains. Using polarized HepG2 cells expressing CD81 and HCV pseudoparticle to study entry, Mee *et al.* showed that there was an inverse association between cell polarization and HCV entry. Treatment of the cells with PKA antagonist Rp-8-Br-cyclic AMPs (cAMPs) reduced the expression of CLDN1 at the basal membrane and inhibited HCV infection. In contrast, activation of PKC with phorbol ester reduced TJ integrity, ablated HepG2 polarity, and stimulated HCV entry (Mee *et al.*, 2009).

Interestingly, through screening of a random peptide library, it was identified that protein kinase C-related kinase 2 (PRK2) interacted with NS5B, which was confirmed by coimmunoprecipitation and colocalization of the proteins. PRK2 specifically phosphorylated NS5B by interacting with the N-terminal finger domain of NS5B. Knock-down of PRK2 using siRNA inhibited HCV RNA replication, whereas overexpression of PRK2 enhanced HCV RNA replication (Kim *et al.*, 2004b). Moreover, two selective inhibitors of PRK2, HA1077 (fasudil) and Y27632, blocked NS5B phosphorylation and inhibited HCV replication (Kim *et al.*, 2009).

RAS-ERK pathway

Ras-Erk signalling plays an important role in the viral life cycle. HCV NS5A protein attenuated the activation of the Ras-Erk signalling pathway via a conserved polyproline motif located towards the C terminus of the protein (Macdonald *et al.*, 2003). An inhibitor of the ERK signalling pathway, PD98059, specifically enhanced HCV IRES-dependent protein synthesis but not the cellular cap-dependent translation (Murata *et al.*, 2005). Furthermore, inhibition ERK signalling not only increased HCV gene expression but also stimulated viral RNA synthesis and virus production (Ndjomou *et al.*, 2009). Such an effect was independent of IFN-α pathway and did not require prior activation of MEK/ERK. Interestingly, it appeared that MEK and ERK-2 but not ERK-1 was involved.

However, more recently it was shown that the role of Ras-Erk signalling may be more complicated. Incubation in the presence of two inhibitors (U0126 and PD184352) resulted in a decrease in the levels of RNA replication, conversely incubation with inhibitor PD98059 resulted in a modest increase in replication (Gretton *et al.*, 2010). The transient replicon RNA was co-electroporated with a dominant-negative mutant of MEK1. This resulted in a reduction in replication, confirming data seen with U0126 and PD184352. These data suggested that a low level Ras-Erk signalling activity is required for RNA replication, while a complete inhibition of Ras-Erk signalling is inhibitory.

NS5A also bound to growth factor receptor-bound protein 2 (Grb2) adaptor protein, which can be induced by EGF and appeared to require cooperative interaction to both N-terminal and C-terminal SH3 domains of Grb2. Phosphorylation of ERK1/2 was inhibited in cells expressing NS5A (Tan *et al.*, 1999). While investigating the mechanism of interferon gamma, Huang *et al.* discovered that the type

II transcriptional response to IFN-gamma and its anti-HCV effect were mediated through Ras-MAPK signalling pathway. Expression of a constitutive active form of Ras inhibited HCV replication, whereas a dominant-negative mutant of Ras enhanced replication. In addition, treatment of cells with EGF suppressed HCV protein expression in the replicon cells, which can be rescued by an MEK1/2 inhibitor U0126. Inhibition of MEK1/2 kinase activity correlated with reduced phosphorylation of the HCV NS5A protein and enhanced HCV replication (Huang *et al.*, 2006).

PI3K-Akt-mTOR pathway

NS5A not only inhibits mitogenic signalling of EGF-ERK1/2 pathway through interaction with Grb2 but also activates the PI3K-AKT cell survival pathway, which may be important for viral persistence and pathogenesis. It was shown that the N-terminal region of NS5A interacted with the p85 subunit of phosphatidylinositol 3-kinase (PI3K), which led to increased tyrosine phosphorylation of PI3K, activation of the downstream serine/threonine kinase AKT, and increased phosphorylation of BAD, a proapoptotic factor and an AKT substrate (He *et al.*, 2002). Street *et al.* further demonstrated that NS5A bound to the SH3 domain of the PI3K p85 regulatory subunit and enhanced the phosphotransferase activity of the p110 catalytic subunit both *in vitro* and in cells harbouring a subgenomic HCV replicon or expressing NS5A alone. NS5A-mediated activation of PI3K resulted in increased phosphorylation and activity of Akt and concomitantly provided protection against the induction of apoptosis in these cells (Street *et al.*, 2004). The expression of Ras as well as the activities of PI3K, Akt and mTOR was found to be increased in the HCV-replicating cells. Consistent with their functions in promoting cell survival, HCV replicon cells were resistant to serum starvation-induced apoptosis. Treating cells with siRNA targeting Ras or mTOR or an inhibitor of PI3K, LY294002, resulted in increased replication of HCV. These data suggest that stimulation of the pro-survival PI3K-Akt pathway and mTOR by HCV not only protect cells against apoptosis but also contribute to the maintenance of steady-state levels of HCV replication, which may help establishing persistent infection of HCV (Mannova and Beretta, 2005).

Activation of PI3K/Akt by expression of the HCV polyprotein led to the phosphorylation and inhibition of the Akt substrates forkhead transcription factor and glycogen synthase kinase-3beta (GSK-3beta; Street *et al.*, 2005). NS5A, either alone or in complex with p85, was able to bind directly to beta-catenin. It appeared that NS5A activated beta-catenin via a novel mechanism that involved a direct interaction between the two proteins and was augmented by PI3K activity, which may contribute to the development of hepatocellular carcinoma (Milward *et al.*, 2010).

Src family kinases

NS5A contains a highly conserved polyproline motif at C-terminus, which was shown to interact with Src homology domain 3 (SH3) of several members of Src family kinases, including Fyn, Lyn, Hck and Lck but not with c-Src itself (Tan *et al.*, 1999; Macdonald *et al.*, 2004). Interestingly, NS5A differentially modulated activity of these kinases, inhibiting Hck, Lck and Lyn but activating Fyn. Shelton and Harris confirmed that NS5A bound to Fyn with high affinity, suggesting that it was able to compete with cellular Fyn ligand. Interestingly, another kinase Csk, which negatively regulates Fyn, was identified to be involved in HCV replication in a siRNA screen (Supekova *et al.*, 2008).

CKI-α and NS5A phosphorylation

NS5A is a phosphoprotein, and a switch between its basally phosphorylated form (p56) and hyperphosphorylated form (p58) seems to play an important role in regulating HCV replication. The presence of high amounts of NS5A-p58 caused an inhibition of HCV RNA replication, which can be relieved by kinase inhibitors (Neddermann *et al.*, 2004). Knock-down CKI-alpha with siRNA reduced NS5A hyperphosphorylation and HCV replication. Rescue of inhibition of NS5A-p58 was achieved by CKI-alpha overexpression (Quintavalle *et al.*, 2006). NS5A was shown to be a direct substrate of CKI-alpha. Ser-2204 NS5A was the preferred substrate residue for

CKI-alpha after prephosphorylation of Ser-2201. Moreover, *in vitro* phosphorylation of NS5A by CKI-alpha resulted in the production of basal and hyperphosphorylated forms resembling those produced in cells (Quintavalle *et al.*, 2007). There are likely other kinase involved in phosphorylating and regulating NS5A function, which remain to be further investigated.

Membrane trafficking

Like most positive-strand RNA viruses, HCV replicates its genome on the surface of rearranged membranes and induce a 'membranous web' structure which is believed to be the site of viral replication. Thus, HCV has to redirect host cell membrane trafficking to facilitate the formation of replication complex.

VAP-A/B

Vesicle-associated membrane protein (VAMP)-associated protein (VAP) subtype A (VAP-A) and subtype B (VAP-B) are localized on the cytoplasmic side of ER and Golgi and are involved in the regulation of membrane trafficking, lipid transport and metabolism, and unfolded protein response. VAP-A and VAP-B form homo- and heterodimers through their C-terminal transmembrane domains and bind to both NS5A and NS5B through a coiled-coil motif and a major sperm protein (MSP) domain, respectively (Tu *et al.*, 1999; Hamamoto *et al.*, 2005). The expression of dominant-negative mutants and small interfering RNA of VAP-A (previously known as hVAP-33) in HCV replicon cells resulted in the relocation of NS5B from detergent-resistant to detergent-sensitive membranes and the reduction of HCV RNA and proteins, indicating that VAP-A is critical for the formation of HCV replication complex and RNA replication (Gao *et al.*, 2004). The interaction of VAP-A and NS5A was also shown to be essential for viral replication as mutations in NS5A that blocked interaction with hVAP-A also strongly reduced HCV RNA replication in the replicon cells. Interestingly, there appeared an inverse correlation between NS5A phosphorylation and hVAP-A interaction, suggesting that hyperphosphorylation of NS5A disrupts interaction with hVAP-A and negatively regulates viral RNA replication (Evans *et al.*, 2004). Overexpression of VAP-B, but not of a mutant lacking its transmembrane domain, enhanced the expression of NS5A and NS5B and the replication of HCV RNA in replicon cells. In the HCV replicon cells, the knockdown of endogenous VAP-B by siRNA decreased expression of NS5B, but not of NS5A (Hamamoto *et al.*, 2005). The third subtype of VAP, VAP-C, is an alternative spliced isoform of VAP-B, which contains only the N-terminal half of MSP domain plus 29 subtype-specific amino acids generated by frameshift. VAP-C was capable of binding to NS5B but not NS5A, VAP-A and VAP-B due to the lack of the coiled-coil and transmembrane domains. Overexpression of VAP-C inhibited interaction of VAP-A and VAP-B with NS5B and reduced the replication of HCV replicon and JFH-1 infectious virus (Kukihara *et al.*, 2009). The MSP domain of VAP-A was shown to interact with TPR1, which has a TPR (tetratricopeptide repeat) domain and forms the chaperon complex with HSP90. Thus, VAP-A may also play a role in recruiting HSP90 to the replication complex (Lotz *et al.*, 2008). VAP-A is also involved in the regulation of biosynthesis or trafficking of sterols and lipids, which are also important for HCV replication.

Rabs

Another family of proteins that play important roles in regulating vesicular membrane trafficking pathways are the Rab proteins. Rabs belong to the Ras superfamily of small GTPases. Rabs cycles between a GTP-bound active form and an GDP-bound inactive form, which depends on the guanine nucleotide exchange factors (GEFs) to catalyse GTP binding and the GTPase-activating proteins (GAPs) to hydrolyse GTP. There are as many as 70 Rabs and 50 GAPs. In a two-hybrid screen, NS5A was shown to interact with a host cell protein TBC1D20, which contains a TBC (Tre-2, Bub2, and Cdc16) domain present in most known Rab-GAPs (Sklan *et al.*, 2007b). TBC1D20 was shown to be the GAP for Rab1, which is involved in regulating of anterograde traffic between ER and the Golgi. Depletion of either TBC1D20 or Rab1 led to significant reduction of HCV RNA levels (Sklan *et al.*, 2007a). In another specific siRNA screen of host membrane-trafficking genes in involved in HCV replication and infectious virus production, Rab5A and Rab7L1 were shown to be required for HCV replication. Specifically, NS5A colocalized with Rab5A and partially with Rab7L1 in the replication complex (Berger *et al.*, 2009).

HCV NS4B protein induces the 'membranous web' structure which is believed to be site of HCV replication (Egger *et al.*, 2002; Gosert *et al.*, 2003; Moradpour *et al.*, 2003). The early endosome (EE) protein Rab5 was found to be associated with NS4B and colocalized in the ER. Transfection of Rab5 siRNAs resulted in a significant decrease in HCV replication. These studies suggest that EE proteins, including Rab5, may play a role in HCV genome replication or web formation (Stone *et al.*, 2007). In another recent study, it was found that the isolated NS4B-bound subcellular fraction, which supports HCV RNA replication, was selectively associated with Rab5 and Rab7 proteins. Knock-down of Rab5 or Rab7 resulted in a significant decrease in HCV replication. Expression of dominant negative Rab5 or Rab7 led to a reticular NS4B subcellular distribution, while expression of dominant negative Rab6 impaired Rab5 recruitment into NS4B foci (Manna *et al.*, 2010).

COPI and ARF

COPI is a multisubunit protein that coats vesicles transporting proteins from the *cis* end of the Golgi complex to the rough ER, i.e. retrograde transport. SiRNA targeting COPZ1 resulted in an inhibition of HCV replicon (Tai *et al.*, 2009). The function of COPI depends on ADP ribosylation factor (ARF), a GTPase involved in membrane traffic. HCV infection was inhibited by brefeldin A (BFA), an inhibitor of several member of the ARF family that involves in various secretory and endocytic pathways. Individual knockdown of the cellular targets of BFA using siRNA identified GBF1, a guanine nucleotide exchange factor for small GTPases of the ARF family, as a host factor critically involved in HCV replication. Furthermore, overexpression of a BFA-resistant GBF1 mutant rescued HCV replication in BFA-treated cells. BFA did not block the formation of membranous web-like structures, suggesting that GBF1 is probably not involved in the formation of HCV replication complex but in its activity (Goueslain *et al.*, 2010).

PI4K

The involvement of PI3K pathway has been discussed above. Another family of lipid kinases, class III phosphatidylinositol 4-kinases (PI4K), appears to be involved in HCV replication in a completely different manner. Several groups have reported that PI4K is involved in HCV replication (Borawski *et al.*, 2009; Berger *et al.*, 2009; Tai *et al.*, 2009; Trotard *et al.*, 2009; Vaillancourt *et al.*, 2009). In a siRNA screen of kinome and druggable genome, both PI4KIII-α and PI4KIII-β were identified as novel host factors required for HCV replication. Knockdown of PI4KIII-α and PI4KIII-β led to inhibition of the replication of genotypes 1b and 1a replicons as well as genotype 2a infectious virus (JFH1). Moreover, a small molecule inhibitor of PI4K, PIK93, was also effective in blocking viral replication (Borawski *et al.*, 2009). Previously, PI4K has been reported to interact with the intracellular domain of cellular receptor CD81 (Berditchevski *et al.*, 1997) as well as HCV NS5A protein (Ahn *et al.*, 2004). Recently it was shown that PI4KIII-α co-localized with NS5A and double-stranded RNA in the detergent-resistant membrane fraction (Berger *et al.*, 2009). Treatment of replicon cells with PI4KIII-α siRNAs greatly reduced the accumulation of membranous web structures, the site of HCV replication, which suggested that PI4KIII-α plays an essential role in membrane alterations leading to the formation of HCV replication complexes. In another siRNA screen using HCV pseudoparticles (HCVpp), it was identified that both PI4KIII-α and PI4KIII-β can block infection by HCVpp. The effect was also confirmed by using cell-culture grown HCV (JFH-1; Trotard *et al.*, 2009).

ESCRT

The precise mechanism by which HCV viral particles are assembled and released has not been well characterized. Many enveloped viruses utilize the host endosomal pathway to acquire the membrane for assembly and then release by membrane scission. This is facilitated by the formation of multivesicular body (MVB), in which several multiprotein complexes termed endosomal sorting complex required for transport (ESCRT) 0, I, II and III act sequentially to traffic target proteins through the cells. It was shown that dominant-negative forms of ESCRT–III complex or Vps4 (the ATPase that recycles the ESCRT complexes) specifically blocked the release of infectious viral particle but had no effect on viral replication or the accumulation of intracellular infectious particles. In contrast, no significant effect was observed with dominant-negative ESCRT-I or Alix, an ESCRT associated protein. These data suggested that HCV requires late components of the ESCRT pathway for release of infectious virus particles (Corless *et al.*, 2010).

Protein translation, modification and maturation

La, PTB, hnRNP, HuR and eIF

HCV 5′UTR contains an internal ribosomal entry site (IRES) that is responsible for the initiation of cap-independent translation of HCV polyprotein. Many cellular proteins have been shown to interact with the 5′UTR (Yen *et al.*, 1995) such as the La antigen (Ali and Siddiqui, 1997; Spangberg *et al.*, 1999), the polypyrimidine tract binding protein (PTB; Ali and Siddiqui, 1995), hnRNP L (Hahm *et al.*, 1998; Hwang *et al.*, 2009), and eIF3 (Sizova *et al.*, 1998). Knock-down of these proteins with siRNA resulted in inhibition of HCV IRES-dependent translation. A synthetic cell-permeable peptide corresponding to the N-terminal 18 amino acids of the La protein efficiently blocked the replication of infectious JFH1 virus in cell culture. The peptide appeared to inhibit HCV IRES-mediated translation by forming complexes with PTB and PCBP2 and sequestering them from IRES, as the inhibitory effect could be fully rescued by over-expressing PCB and PCBP2 (Fontanes *et al.*, 2009).

Hu antigen R (HuR) is a member of the ELAV-like protein, which binds specifically to A/U-rich regions of mRNAs and has been implicated in oncogene and growth factor mRNA decay. HuR and hnRNP C were shown to bind to the 3′ ends of hepatitis C virus RNA of both polarities (Spangberg *et al.*, 2000). HuR functioned as an activator of HCV IRES (Rivas-Aravena *et al.*, 2009). siRNAs directed against HuR or proteasome alpha-subunit 7 (PSMA7) led to an inhibition of HCV protein translation and RNA replication in the replicon cells (Korf *et al.*, 2005). HuR is a potentially druggable target as low-molecular-weight inhibitors for HuR have been reported (Meisner *et al.*, 2007).

Nitazoxanide is a drug previously approved for parasitic infestations. Recently, it was shown to inhibit HCV through inducing phosphorylation of eukaryotic initiation factor 2, a known mediator of host antiviral defence (Korba *et al.*, 2008; Elazar *et al.*, 2009). In a phase II study in genotype 4 patients in Egypt, the combination of nitazoxanide with PEG-IFN-α-2a and ribavirin resulted in 79% SVR vs. 50% with PEG-IFN-α-2a and ribavirin alone (Rossignol *et al.*, 2008).

MiR-122

MicroRNAs (miRNAs) are small RNA molecules that regulate the intracellular abundance and translation of target mRNAs. MicroRNA 122 (miR-122) is specifically expressed and highly abundant in the human liver. Sequestering of miR-122 by modified antisense oligonucleotides resulted in a significant loss of HCV RNA in the cells. Mutational analysis suggested HCV 5′-UTR as the binding site for miR-122 (Jopling *et al.*, 2005). Moreover, exogenous expression of miR-122 could enhance HCV replication in non-hepatic cells (Chang *et al.*, 2008). MiR-122 stimulated HCV translation in liver cell lines as well as the non-liver cells and rabbit reticulocyte lysate by enhancing the association of ribosomes with the viral RNA at an early initiation stage. The effect was independent of viral RNA replication (Henke *et al.*, 2008). MiR-122 is also known to stimulate the mevalonate pathway and isoprenoid biosynthesis, but the activity of which seems to be independent of its effect on HCV (Norman and Sarnow, 2010). The liver-specific miR-122 could be a contributing factor for the liver-tropism of HCV infection. Interestingly, there is a decreased levels of microRNA miR-122 in HCV patients responding poorly to interferon therapy (Sarasin-Filipowicz *et al.*, 2009).

The positive role of miR-122 in viral replication suggests that this miRNA could be targeted for antiviral therapy. Indeed, treatment of HCV-infected chimpanzees with a locked nucleic acid (LNA)-modified oligonucleotide (SPC3649) complementary to miR-122 led to a long-lasting suppression of HCV viraemia (Lanford *et al.*, 2010). Moreover, there was no evidence of viral resistance or side-effect in the treated animals, which is consistent with the fact that the miR-122 targeted HCV 5′-UTR is highly conserved.

α-Glucosidase I

The HCV envelope glycoproteins E1 and E2 are highly N-glycosylated. Alpha-glucosidase I is involved in glycoprotein processing and is important for viral maturation and release. Inhibition of alpha glucosidase with iminosugars derivatives such as N-butyl-deoxynojirimycin (NB-DNJ) and N-nonyl-DNJ (NN-DNJ) led to misfolding of viral envelop proteins thus blocked viral assembly and release (Ouzounov *et al.*, 2002; Zitzmann *et al.*, 1999; Steinmann *et al.*, 2007; Durantel, 2009). Celgosivir (6-O-butanoyl castanospermine), a more potent inhibitor of alpha-glucosidase I (Whitby *et al.*, 2004), has been tested in clinical trials but showed only modest antiviral effect as monotherapy. In a 12-week phase II trial, only 2 out of 35 patients had greater than 1 $\log_{10}$ viral load reduction (Yoshida *et al.*, 2006). Combination of 400 mg celgosivir with PEG-IFN/ribavirin resulted in greater 2 $\log_{10}$ viral load reduction in 45% patients vs. 10% with PEG-IFN-/ribavirin only (Kaita *et al.*, 2007). Further combination trials with different dosing regime are currently on-going (Durantel, 2009).

Hsp90, FKBP38 and hB-ind1

The cellular chaperone, heat-shock protein 90 (Hsp90), was first shown to interact with the NS2/3 protease of HCV and was required for its activity. Geldanamycin and radicicol, two ATP-competitive inhibitors of Hsp90, inhibited the cleavage of NS2/3 *in vitro* and in mammalian cells expressing the protein (Waxman *et al.*, 2001). Knockdown of endogenous Hsp90 expression with siRNA led to inhibition of HCV replication. The effect of Hsp90 inhibitors on HCV replication can be prevented by over-expression of Hsp90 in the replicon cells. Combination of an Hsp90 inhibitor and PEG-IFN- was more effective than PEG-IFN- alone in reducing viral load in HCV-infected chimeric mice with humanized liver (Nakagawa *et al.*, 2007). Interestingly, in replicon cells treated with 17-allylaminogeldanamycin (17-AAG), an inhibitor of Hsp90, the only HCV protein degraded was NS3. A direct interaction of NS3 with Hsp90 was confirmed by co-immunoprecipitation, which suggested that 17-AAG suppressed HCV RNA replication by interfering Hsp90 chaperone and destabilizing NS3 (Ujino *et al.*, 2009).

FKBP38, a 38 kDa immunosuppressant FK506-binding protein, was identified to bind to NS5A in a yeast two-hybrid screen. The interaction was confirmed both *in vitro* and *in vivo* and was mapped to the amino acids 148–236 of NS5A containing a BH domain (Bcl-2 homology domain). Both proteins were found to localize in mitochondria and ER. Moreover, NS5A stably expressing Huh7 cells showed more resistance to apoptosis and such inhibition of apoptosis could specifically be abrogated by siRNA targeting FKBP38, which suggested that HCV NS5A inhibits apoptosis through interaction with FKBP38 (Wang *et al.*, 2006). Knockdown of FKBP8 using siRNA suppressed HCV RNA replication in the replicon cells, which can be reversed by the expression of an siRNA-resistant FKBP8 mutant. Furthermore, immunoprecipitation analyses revealed that FKBP8 forms a complex with Hsp90 and NS5A (Okamoto *et al.*, 2006). FKBP8 is partially colocalized with NS5A in the membranous web (Okamoto *et al.*, 2008).

In a separate yeast two-hybrid screen, another protein, human butyrate-induced transcript 1 (hB-ind1), was identified to interact with NS5A. SiRNA-mediated knockdown of hB-ind1 suppressed the replication of HCV replicons and the production of infectious virus (JFH1). Interestingly, hB-ind1 binds to both FKBP8 and Hsp90 (Taguwa *et al.*, 2008). Furthermore, hB-ind1 colocalizes with NS5A, FKBP8, and double-stranded RNA in the HCV replicon cells, which suggests that hB-ind1 recruits Hsp90 and FKBP8 to the replication complex through interaction with NS5A. The cochaperone-like activity of hB-ind1 might be important in regulating Hsp90 chaperone function. HCV replicon cells exhibited a more potent unfolded-protein response than the parental cells. An Hsp90-dependent chaperone pathway incorporating hB-ind1 may be necessary to ensure correct protein folding in the membranous web and robust HCV replication (Taguwa *et al.*, 2009).

Autophagy

Autophagy is important to maintain cellular homeostasis by removing long-lived proteins and damaged organelles. It also serves as innate immunity to remove intracellular pathogens. To counteract, viruses have evolved various mechanisms to evade, subvert, or even exploit autophagy. Sir *et al.* demonstrated that HCV induced an accumulation of autophagosomes in cells without enhancing autophagic protein degradation, which was dependent on the unfolded protein response (UPR), as the suppression of UPR signalling pathways suppressed HCV-induced lipidation of the microtubule-associated protein light chain 3 (LC3) protein, a necessary step for the formation of autophagosomes. The suppression of UPR or the suppression of expression of LC3 or Atg7, a protein that mediates LC3 lipidation, suppressed HCV replication (Sir *et al.*, 2008a). HCV can also suppress autophagic protein degradation by suppressing the fusion between autophagosomes and lysosomes. This suppression causes the accumulation of autophagosomes and enhances HCV RNA replication. Perturbation of autophagic pathway by HCV (Sir *et al.*, 2008b). Serial passage of a genotype 1a (H77) virus in immortalized human hepatocytes induced an autophagic response as demonstrated by autophagic vacuole formation and localization of autophagic markers, microtubule-associated protein-1 light chain-3 and Apg5, on the vacuoles of HCV-infected hepatocytes (Ait-Goughoulte *et al.*, 2008). The prolonged perturbation of the autophagic pathway by HCV likely plays an important role in HCV pathogenesis.

Dreux and Chisari proposed that autophagy factors might support the delivery of incoming RNA to the translation apparatus and/or the recruitment of cellular factors required to initiate HCV translation. Autophagy proteins (i.e. Beclin-1, Atg4B, Atg5, and Atg12) are proviral factors required for translation of incoming hepatitis C virus (HCV) RNA and, thereby, for initiation of HCV replication, but they are not required once infection is established (Dreux *et al.*, 2009). Tanida *et al.* observed that when Atg7 or Beclin 1 was knocked down by siRNA in HCV-infected Huh7.5.1 cells, the levels of intracellular HCV core, NS3, NS5A proteins, and HCV RNA remained unchanged, while the level of infectious HCV particles released in the medium was decreased, which suggests that autophagy contributes to the effective production of HCV particles, but little to the intracellular production of HCV proteins and RNA in a naive HCV particles-infection system (Tanida *et al.*, 2009).

Silencing of autophagy-related genes Atg5, Atg7 and light-chain (LC)-3 by siRNA blocked the replication of HCV replicon. Treatment of replicon cells with an autophagic proteolysis inhibitor chloroquine suppressed the replication of the HCV replicon in a dose-dependent manner. Furthermore, combination of chloroquine and IFN--enhanced antiviral effect (Mizui *et al.*, 2010). These results suggest that HCV utilizes machinery involving cellular autophagic proteolysis, which may provide a new therapeutic target.

Specific challenges in drug development

Host targets open the door to a lot of opportunities in drug discovery. On the other hand, there are specific challenges in developing inhibitors of host proteins, since they most likely have certain 'normal' functions in the cells, the inhibition of which might lead to target-related toxicity. There are several possibilities that such an issue could be avoided. First of all, a rapidly dividing virus likely has a very different threshold for the requirement of certain targets than normal cells. This is reflected by the fact that many cellular pathways and genes are up-regulated in viral infected cells. So, to knock these targets down to their 'normal' level may be sufficient to block viral replication while maintaining normal cellular functions. Secondly, there are many redundant pathways or genes for certain functions. Knock-down of a particular target unique for the virus could be compensated by another protein in the cells with overlapping function. Thirdly, there are certainly many non-essential genes in the cells, inhibition of which might be more tolerable depending on the dose/concentration of inhibitors. So, while there are risks of inducing target related toxicity, one should be mindful that this is not different for drug discovery in any therapeutic area. The objective remains the same, i.e. to identify the optimal doses that can achieve therapeutic effect without incurring significant toxicity. In this case, there should be sufficient therapeutic window between the dose that blocks viral replication and the dose that might affect normal host functions. This approach has been validated with the approval of the CCR5 inhibitor maraviroc for HIV-1. And cyclophilin inhibitors have so far shown great promises as potential novel therapeutic agent for HCV.

Another challenge in host target discovery is related to the limited *in vitro* tools and animal models available for HCV. The *in vitro* HCV cell culture system is mainly limited to one specific cell line, Huh-7. It is quite possible that host factors involved in viral replication in a hepatoma cell line can be significantly different from those in primary hepatocytes. Also caution should be taken when using the *in vitro* model only. For example, IMPDH inhibitors are very potent in the *in vitro* assays. In contrast, their efficacy *in vivo* (patients) has been much more variable. One might speculate that the level and supply of nucleotides and the dependency on *de novo* synthesis could be very different *in vivo* from *in vitro*, which could significantly affect the efficacy of the inhibitors. Because of these limitations, there will be surprises in some cases in translating *in vitro* activity to *in vivo* efficacy.

Lastly, the mechanism of action of host targeting inhibitors is typically a lot more challenging to determine than inhibitors of viral targets. The mechanism of resistance is also likely more complex. For example, multiple mutations need to be acquired to confer resistance to cyclophilin inhibitors, and it is likely that mutations in multiple regions of viral genome are involved. In contrast, resistance mutations to viral protease or polymerase inhibitors occur at specific binding pockets and are much more defined. The impact of host polymorphism also needs to be analysed for host targeting inhibitors. These tasks present additional challenges in drug development but also provide opportunities to help understand host–virus interactions.

Future trends

The first generation of orally active direct acting antiviral agents (DAA), specifically NS3 protease inhibitors, are expected to be approved as early as 2011 to be used in combination with current standard of care (SoC), i.e. PEG-IFN-α and ribavirin. These new triple combination regimens will significantly improve treatment response over current SoC. However, the side-effects caused by these new oral agents in addition to those associated with IFN-α and ribavirin suggest that perhaps even less patients can tolerate the new therapies. Moreover, the first generation DAA has short half-life and requires frequent dosing (q8h). Patient compliance could become a real issue which may lead to emergence of resistance and compromise the efficacy and utility of the drugs. The host target inhibitors may provide a better alternative in the near term because of their higher resistance barrier.

In the future, the ultimate objective of HCV therapy is IFN-free, oral cocktails of multiple classes of small molecule inhibitors, similar to the highly active antiretroviral therapy (HARRT) for HIV-1 infection. Several clinical trials are ongoing exploring the combinations of inhibitors of different viral targets (DAAs). It remains to be determined whether an IFN-free regimen is capable of completely clear the virus, i.e. cure HCV infection. The main issue with DAAs is the development of resistance. It has been demonstrated both *in vitro* and in patients that drug-resistant mutants can emerge quickly, even with the most potent inhibitors of viral protease or polymerase (Sarrazin and Zeuzem, 2010). This has been attributed to several factors:

1. HCV replicates at a high rate in patients, producing an estimate of 10^{10-12} virions per day (Neumann *et al.*, 1998).
2. The RNA-dependent RNA polymerase of the virus lacks proof-reading function and has an error rate of about 10^{-4} mutations per genome per replication cycle. As a result

there is an extremely high degree of heterogeneity of viral population (quasispecies) in each patient.

3 Since viral targeted inhibitors typically bind to a defined pocket of a viral protein, typically a single mutation in viral genome is sufficient to disrupt the binding of inhibitor and lead to resistance.

Theoretically all the possible single, double or even triple mutations are already pre-existing in HCV patients, therefore it may take three or more DAAs to completely suppress the emergence of resistance. Compounds with relatively high resistance barrier such as cyclophilin inhibitors could provide the key advantage in an IFN-free regimen. Thus, the combination of host and viral targeted inhibitors represents a very attractive strategy in suppressing the emergence of resistance and maximizing antiviral efficacy.

References

Ahn, J., Chung, K.S., Kim, D.U., Won, M., Kim, L., Kim, K.S., Nam, M., Choi, S.J., Kim, H.C., Yoon, M., Chae, K., and Hoe, K.L. (2004). Systematic identification of hepatocellular proteins interacting with NS5A of the hepatitis C virus. J. Biochem. Mol. Biol. *37*, 741–748.

Ait-Goughoulte, M., Kanda, T., Meyer, K., Ryerse, J.S., Ray, B., and Ray, R. (2008). Hepatitis C virus genotype 1a growth and induction of autophagy. J. Virol. *82*, 2241–2249.

Aizaki, H., Morikawa, K., Fukasawa, M., Hara, H., Inoue, Y., Tani, H., Saito, K., Nishijima, M., Hanada, K., Matsuura, Y., Lai, M.M., Miyamura, T., Wakita, T., and Suzuki, T. (2008). Critical role of virion-associated cholesterol and sphingolipid in hepatitis C virus infection. J. Virol. *82*, 5715–5724.

Ali, N., and Siddiqui, A. (1995). Interaction of polypyrimidine tract-binding protein with the 5′ noncoding region of the hepatitis C virus RNA genome and its functional requirement in internal initiation of translation. J. Virol. *69*, 6367–6375.

Ali, N., and Siddiqui, A. (1997). The La antigen binds 5′ noncoding region of the hepatitis C virus RNA in the context of the initiator AUG codon and stimulates internal ribosome entry site-mediated translation. Proc. Natl. Acad. Sci. U.S.A. *94*, 2249–2254.

Amako, Y., Sarkeshik, A., Hotta, H., Yates, J., I.I.I., and Siddiqui, A. (2009). Role of oxysterol binding protein in hepatitis C virus infection. J. Virol. *83*, 9237–9246.

Amemiya, F., Maekawa, S., Itakura, Y., Kanayama, A., Matsui, A., Takano, S., Yamaguchi, T., Itakura, J., Kitamura, T., Inoue, T., Sakamoto, M., Yamauchi, K., Okada, S., Yamashita, A., Sakamoto, N., Itoh, M., and Enomoto, N. (2008). Targeting lipid metabolism in the treatment of hepatitis C virus infection. J. Infect. Dis. *197*, 361–370.

Bader, T., Fazili, J., Madhoun, M., Aston, C., Hughes, D., Rizvi, S., Seres, K., and Hasan, M. (2008). Fluvastatin inhibits hepatitis C replication in humans. Am. J. Gastroenterol. *103*, 1383–1389.

Benga, W.J., Krieger, S.E., Dimitrova, M., Zeisel, M.B., Parnot, M., Lupberger, J., Hildt, E., Luo, G., McLauchlan, J., Baumert, F., and Schuster, C. (2010). Apolipoprotein E interacts with hepatitis C virus nonstructural protein 5A and determines assembly of infectious particles. Hepatology *51*, 43–53.

Berditchevski, F., Tolias, K.F., Wong, K., Carpenter, L., and Hemler, M.E. (1997). A novel link between integrins, transmembrane-4 superfamily proteins (CD63 and CD81), and phosphatidylinositol 4-kinase. J. Biol. Chem. *272*, 2595–2598.

Berger, K.L., Cooper, J.D., Heaton, N.S., Yoon, R., Oakland, T.E., Jordan, T.X., Mateu, G., Grakoui, A., and Randall, G. (2009). Roles for endocytic trafficking and phosphatidylinositol 4-kinase III alpha in hepatitis C virus replication. Proc. Natl. Acad. Sci. U.S.A. *106*, 7577–7582.

Billich, A., Hammerschmid, F., Peichl, P., Wenger, R., Zenke, G., Quesniaux, V., and Rosenwirth, B. (1995). Mode of action of SDZ NIM *811*, a nonimmunosuppressive cyclosporin A analog with activity against human immunodeficiency virus (HIV) type 1: interference with HIV protein-cyclophilin A interactions. J. Virol. *69*, 2451–2461.

Borawski, J., Troke, P., Puyang, X., Gibaja, V., Zhao, S., Mickanin, C., Leighton-Davies, J., Wilson, C.J., Myer, V., Cornellataracido, I., Baryza, J., Tallarico, J., Joberty, G., Bantscheff, M., Schirle, M., Bouwmeester, T., Mathy, J.E., Lin, K., Compton, T., Labow, M., Wiedmann, B., and Gaither, L.A. (2009). Class III phosphatidylinositol 4-kinase alpha and beta are novel host factor regulators of hepatitis C virus replication. J. Virol. *83*, 10058–10074.

Borowski, P., Schulze zur, W.J., Resch, K., Feucht, H., Laufs, R., and Schmitz, H. (1999). Protein kinase C recognizes the protein kinase A-binding motif of nonstructural protein 3 of hepatitis C virus. J. Biol. Chem. *274*, 30722–30728.

Borowski, P., Resch, K., Schmitz, H., and Heiland, M. (2000). A synthetic peptide derived from the nonstructural protein 3 of hepatitis C virus serves as a specific substrate for PKC. Biol. Chem. *381*, 19–27.

Bose, S., Mathur, M., Bates, P., Joshi, N., and Banerjee, A.K. (2003). Requirement for cyclophilin A for the replication of vesicular stomatitis virus New Jersey serotype. J. Gen. Virol. *84*, 1687–1699.

Bouchard, M.J., Puro, R.J., Wang, L., and Schneider, R.J. (2003). Activation and inhibition of cellular calcium and tyrosine kinase signaling pathways identify targets of the HBx protein involved in hepatitis B virus replication. J. Virol. *77*, 7713–7719.

Chang, J., Guo, J.T., Jiang, D., Guo, H., Taylor, M., and Block, T.M. (2008). Liver-specific microRNA miR-122 enhances the replication of hepatitis C virus in nonhepatic cells. J. Virol. *82*, 8215–8223.

Chang, K.S., Jiang, J., Cai, Z., and Luo, G. (2007). Human apolipoprotein e is required for infectivity and production of hepatitis C virus in cell culture. J. Virol. *81*, 13783–13793.

Chatterji, U., Bobardt, M., Selvarajah, S., Yang, F., Tang, H., Sakamoto, N., Vuagniaux, G., Parkinson, T., and Gallay, P. (2009). The isomerase active site of cyclophilin a is critical for HCV replication. J. Biol. Chem. *284*, 16998–170005

Chockalingam, K., Simeon, R.L., Rice, M., and Chen, Z. (2010). A cell protection screen reveals potent inhibitors of multiple stages of the hepatitis C virus life cycle. Proc. Natl. Acad. Sci. U.S.A. *107*, 3764–3769.

Ciesek, S., Steinmann, E., Wedemeyer, H., Manns, M.P., Neyts, J., Tautz, N., Madan, V., Bartenschlager, R., von, T., and Pietschmann, T. (2009). Cyclosporine A inhibits hepatitis C virus nonstructural protein 2 through cyclophilin A. Hepatology *50*, 1638–1645.

Coelmont, L., Kaptein, S., Paeshuyse, J., Vliegen, I., Dumont, J.M., Vuagniaux, G., and Neyts, J. (2009). Debio *025*, a cyclophilin binding molecule, is highly efficient in clearing hepatitis C virus (HCV) replicon-containing cells when used alone or in combination with specifically targeted antiviral therapy for HCV (STAT-C) inhibitors. Antimicrob. Agents Chemother. *53*, 967–976.

Corless, L., Crump, C.M., Griffin, D., and Harris, M. (2010). Vps4 and the ESCRT–III complex are required for the release of infectious hepatitis C virus particles. J. Gen. Virol. *91*, 362–372.

Damaso, C.R., and Keller, S.J. (1994). Cyclosporin A inhibits vaccinia virus replication in vitro. Arch. Virol. *134*, 303–319.

de Chassey, B., Navratil, V., Tafforeau, L., Hiet, M.S., Aublin-Gex, A., Agaugue, S., Meiffren, G., Pradezynski, F., Faria, B.F., Chantier, T., *et al.* (2008). Hepatitis C virus infection protein network. Mol. Syst. Biol. *4*, 230.

Delang, L., Paeshuyse, J., Vliegen, I., Leyssen, P., Obeid, S., Durantel, D., Zoulim, F., Op de, A., and Neyts, J. (2009). Statins potentiate the in vitro anti-hepatitis C virus activity of selective hepatitis C virus inhibitors and delay or prevent resistance development. Hepatology *50*, 6–16.

Dreux, M., Gastaminza, P., Wieland, F., and Chisari, F.V. (2009). The autophagy machinery is required to initiate hepatitis C virus replication. Proc. Natl. Acad. Sci. U.S.A. *106*, 14046–14051.

Durantel, D. (2009). Celgosivir, an alpha-glucosidase I inhibitor for the potential treatment of HCV infection. Curr. Opin. Investig. Drugs *10*, 860–870.

Egger, D., Wolk, B., Gosert, R., Bianchi, L., Blum, H.E., Moradpour, D., and Bienz, K. (2002). Expression of hepatitis C virus proteins induces distinct membrane alterations including a candidate viral replication complex. J. Virol. *76*, 5974–5984.

Elazar, M., Liu, M., McKenna, S.A., Liu, P., Gehrig, E.A., Puglisi, J.D., Rossignol, F., and Glenn, J.S. (2009). The anti-hepatitis C agent nitazoxanide induces phosphorylation of eukaryotic initiation factor 2alpha via protein kinase activated by double-stranded RNA activation. Gastroenterology *137*, 1827–1835.

Evans, M.J., Rice, M., and Goff, S.P. (2004). Phosphorylation of hepatitis C virus nonstructural protein 5A modulates its protein interactions and viral RNA replication. Proc. Natl. Acad. Sci. U.S.A. *101*, 13038–13043.

Evans, M.J., von, H.T., Tscherne, D.M., Syder, A.J., Panis, M., Wolk, B., Hatziioannou, T., McKeating, J.A., Bieniasz, D., and Rice, C.M. (2007). Claudin-1 is a hepatitis C virus co-receptor required for a late step in entry. Nature *446*, 801–805.

Farquhar, M.J., Harris, H.J., Diskar, M., Jones, S., Mee, C.J., Nielsen, S.U., Brimacombe, C.L., Molina, S., Toms, G.L., Maurel, P., Howl, J., Herberg, F.W., van I.Jzendoorn, S.C., Balfe, P., and McKeating, J.A. (2008). Protein kinase A-dependent step(s) in hepatitis C virus entry and infectivity. J. Virol. *82*, 8797–8811.

Fernandes, F., Poole, D.S., Hoover, S., Middleton, R., Andrei, A.C., Gerstner, J., and Striker, R. (2007). Sensitivity of hepatitis C virus to cyclosporine A depends on nonstructural proteins NS5A and NS5B. Hepatology *46*, 1026–1033.

Flisiak, R., Horban, A., Gallay, P., Bobardt, M., Selvarajah, S., Wiercinska-Drapalo, A., Siwak, E., Cielniak, I., Higersberger, J., Kierkus, J., *et al.* (2008). The cyclophilin inhibitor Debio-025 shows potent anti-hepatitis C effect in patients coinfected with hepatitis C and human immunodeficiency virus. Hepatology *47*, 817–826.

Flisiak, R., Feinman, S.V., Jablkowski, M., Horban, A., Kryczka, W., Pawlowska, M., Heathcote, J.E., Mazzella, G., Vandelli, C., Nicolas-Metral, V., *et al.* (2009). The cyclophilin inhibitor Debio 025 combined with PEG IFNalpha2a significantly reduces viral load in treatment-naive hepatitis C patients. Hepatology *49*, 1460–1468.

Fontanes, V., Raychaudhuri, S., and Dasgupta, A. (2009). A cell-permeable peptide inhibits hepatitis C virus replication by sequestering IRES transacting factors. Virology *394*, 82–90.

Forde, K.A., Law, C., O'Flynn, R., and Kaplan, D.E. (2009). Do statins reduce hepatitis C RNA titers during routine clinical use? World J. Gastroenterol. *15*, 5020–5027.

Fujita, N., Kaito, M., Kai, M., Sugimoto, R., Tanaka, H., Horiike, S., Konishi, M., Iwasa, M., Watanabe, S., and Adachi, Y. (2006). Effects of bezafibrate in patients with chronic hepatitis C virus infection: combination with interferon and ribavirin. J. Viral Hepat. *13*, 441–448.

Gaither, L.A., Borawski, J., Anderson, L.J., Balabanis, K.A., Devay, P., Joberty, G., Rau, C., Schirle, M., Bouwmeester, T., Mickanin, C., Zhao, S., Vickers, C., Lee, L., Deng, G., Baryza, J., Fujimoto, R.A., Lin, K., Compton, T., and Wiedmann, B. (2010). Multiple cyclophilins involved in different cellular pathways mediate HCV replication. Virology *397*, 43–55.

Gao, L., Aizaki, H., He, W., and Lai, M.M. (2004). Interactions between viral nonstructural proteins and host protein hVAP-33 mediate the formation of hepatitis C virus RNA replication complex on lipid raft. J. Virol. *78*, 3480–3488.

Gastaminza, P., Cheng, G., Wieland, S., Zhong, J., Liao, W., and Chisari, F.V. (2008). Cellular determinants of hepatitis C virus assembly, maturation, degradation, and secretion. J. Virol. *82*, 2120–2129.

Gastaminza, P., Whitten-Bauer, C., and Chisari, F.V. (2010). Unbiased probing of the entire hepatitis C virus life cycle identifies clinical compounds that target multiple aspects of the infection. Proc. Natl. Acad. Sci. U.S.A. *107*, 291–296.

Gosert, R., Egger, D., Lohmann, V., Bartenschlager, R., Blum, H.E., Bienz, K., and Moradpour, D. (2003). Identification of the hepatitis C virus RNA replication complex in Huh-7 cells harboring subgenomic replicons. J. Virol. *77*, 5487–5492.

Goueslain, L., Alsaleh, K., Horellou, P., Roingeard, P., Descamps, V., Duverlie, G., Ciczora, Y., Wychowski, C., Dubuisson, J., and Rouille, Y. (2010). Identification of GBF1 as a cellular factor required for hepatitis C virus RNA replication. J. Virol. *84*, 773–787.

Gretton, S., Hughes, M., and Harris, M. (2010). Hepatitis C virus RNA replication is regulated by Ras-Erk signalling. J. Gen. Virol. *91*, 671–680.

Hahm, B., Kim, Y.K., Kim, J.H., Kim, Y., and Jang, S.K. (1998). Heterogeneous nuclear ribonucleoprotein L interacts with the 3′ border of the internal ribosomal entry site of hepatitis C virus. J. Virol. *72*, 8782–8788.

Hamamoto, I., Nishimura, Y., Okamoto, T., Aizaki, H., Liu, M., Mori, Y., Abe, T., Suzuki, T., Lai, M.M., Miyamura, T., Moriishi, K., and Matsuura, Y. (2005). Human VAP-B is involved in hepatitis C virus replication through interaction with NS5A and NS5B. J. Virol. *79*, 13473–13482.

Handschumacher, R.E., Harding, M.W., Rice, J., Drugge, J., and Speicher, D.W. (1984). Cyclophilin: a specific cytosolic binding protein for cyclosporin A. Science *226*, 544–547.

Hanoulle, X., Badillo, A., Wieruszeski, J.M., Verdegem, D., Landrieu, I., Bartenschlager, R., Penin, F., and Lippens, G. (2009). Hepatitis C virus NS5A protein is a substrate for the peptidyl-prolyl cis/trans isomerase activity of cyclophilins A and B. J. Biol. Chem. *284*, 13589–13601.

Hara, H., Aizaki, H., Matsuda, M., Shinkai-Ouchi, F., Inoue, Y., Murakami, K., Shoji, I., Kawakami, H., Matsuura, Y., Lai, M.M., Miyamura, T., Wakita, T., and Suzuki, T. (2009). Involvement of creatine kinase B in hepatitis C virus genome replication through interaction with the viral NS4A protein. J. Virol. *83*, 5137–5147.

He, Y., Nakao, H., Tan, S.L., Polyak, S.J., Neddermann, P., Vijaysri, S., Jacobs, L., and Katze, M.G. (2002). Subversion of cell signaling pathways by hepatitis C virus nonstructural 5A protein via interaction with Grb2 and P85 phosphatidylinositol 3-kinase. J. Virol. *76*, 9207–9217.

Heck, J.A., Meng, X., and Frick, D.N. (2009). Cyclophilin B stimulates RNA synthesis by the HCV RNA dependent RNA polymerase. Biochem. Pharmacol. *77*, 1173–1180.

Henke, J.I., Goergen, D., Zheng, J., Song, Y., Schuttler, C.G., Fehr, C., Junemann, C., and Niepmann, M. (2008). microRNA-122 stimulates translation of hepatitis C virus RNA. EMBO J. *27*, 3300–3310.

Henry, S.D., Metselaar, H.J., Lonsdale, R.C., Kok, A., Haagmans, B.L., Tilanus, H.W., van der Laan, L.J. (2006). Mycophenolic acid inhibits hepatitis C virus replication and acts in synergy with cyclosporin A and interferon-alpha. Gastroenterology *131*, 1452–1462.

Hopkins, S., Scorneaux, B., Huang, Z., Murray, M.G., Wring, S., Smitley, C., Harris, R., Erdmann, F., Fischer, G., and Ribeill, Y. (2010). SCY-635, a novel nonimmunosuppressive analog of cyclosporine that exhibits potent inhibition of hepatitis C virus RNA replication in vitro. Antimicrob. Agents Chemother. *54*, 660–672.

Huang, H., Sun, F., Owen, D.M., Li, W., Chen, Y., Gale, M., Jr. and Ye, J. (2007). Hepatitis C virus production by human hepatocytes dependent on assembly and secretion of very low-density lipoproteins. Proc. Natl. Acad. Sci. U.S.A. *104*, 5848–5853.

Huang, Y., Chen, X.C., Konduri, M., Fomina, N., Lu, J., Jin, L., Kolykhalov, A., and Tan, S.L. (2006). Mechanistic link between the anti-HCV effect of interferon gamma and control of viral replication by a Ras-MAPK signaling cascade. Hepatology *43*, 81–90.

Hwang, B., Lim, J.H., Hahm, B., Jang, K., and Lee, S.W. (2009). hnRNP L is required for the translation mediated by HCV IRES. Biochem. Biophys. Res. Commun. *378*, 584–588.

Ikeda, M., Abe, K., Yamada, M., Dansako, H., Naka, K., and Kato, N. (2006). Different anti-HCV profiles of statins and their potential for combination therapy with interferon. Hepatology *44*, 117–125.

Inoue, K., Sekiyama, K., Yamada, M., Watanabe, T., Yasuda, H., and Yoshiba, M. (2003). Combined interferon alpha2b and cyclosporin A in the treatment of chronic hepatitis C: controlled trial. J. Gastroenterol. *38*, 567–572.

Jiang, J., and Luo, G. (2009). Apolipoprotein E but not B is required for the formation of infectious hepatitis C virus particles. J. Virol. *83*, 12680–12691.

Jopling, C.L., Yi, M., Lancaster, A.M., Lemon, M., and Sarnow, P. (2005). Modulation of hepatitis C virus RNA abundance by a liver-specific MicroRNA. Science *309*, 1577–1581.

Kaita, K., Yoshida, E., Kunimoto, D., Anderson, F., Morris, S., Marrotta, P., Scully, L. Pertekian, K., Enns, R., Diaz-Mitoma, F., *et al.* Ph II proof of concept study of celgosivir in combination with peginterferon alfa-2b and ribavirin in chronic hepatitis C genotype 1 non-responder patients [Abstract 127]. J. Hepatol. *46*, S56.

Kanda, T., Yokosuka, O., Imazeki, F., Tanaka, M., Shino, Y., Shimada, H., Tomonaga, T., Nomura, F., Nagao, K., Ochiai, T., and Saisho, H. (2004). Inhibition of subgenomic hepatitis C virus RNA in Huh-7 cells: ribavirin induces mutagenesis in HCV RNA. J. Viral Hepat. *11*, 479–487.

Kapadia, S.B., and Chisari, F.V. (2005). Hepatitis C virus RNA replication is regulated by host geranylgeranylation and fatty acids. Proc. Natl. Acad. Sci. U.S.A. *102*, 2561–2566.

Karpas, A., Lowdell, M., Jacobson, K., and Hill, F. (1992). Inhibition of human immunodeficiency virus and growth of infected T cells by the immunosuppressive drugs cyclosporin A and FK 506. Proc. Natl. Acad. Sci. U.S.A. *89*, 8351–8355.

Kaul, A., Stauffer, S., Berger, C., Pertel, T., Schmitt, J., Kallis, S., Zayas, M., Lohmann, V., Luban, J., and Bartenschlager, R. (2009). Essential role of cyclophilin A for hepatitis C virus replication and virus production and possible link to polyprotein cleavage kinetics. PLoS. Pathog. *5*, e1000546.

Khattab, M., Emad, M., Abdelaleem, A., Eslam, M., Atef, R., Shaker, Y., and Hamdy, L. (2009). Pioglitazone improves virological response to peginterferon alpha-2b/ribavirin combination therapy in hepatitis C genotype 4 patients with insulin resistance. Liver Int. *30*, 447–454.

Kim, J., Choi, B.H., Jang, L., and Min, D.S. (2004a). Phospholipase D activity is elevated in hepatitis C virus core protein-transformed NIH3T3 mouse fibroblast cells. Exp. Mol. Med. *36*, 454–460.

Kim, S.J., Kim, J.H., Kim, Y.G., Lim, S., and Oh, J.W. (2004b). Protein kinase C-related kinase 2 regulates hepatitis C virus RNA polymerase function by phosphorylation. J. Biol. Chem. *279*, 50031–50041.

Kim, S.J., Kim, J.H., Sun, J.M., Kim, G., and Oh, J.W. (2009). Suppression of hepatitis C virus replication by protein kinase C-related kinase 2 inhibitors that block phosphorylation of viral RNA polymerase. J. Viral Hepat. *16*, 697–704.

Korba, B.E., Montero, A.B., Farrar, K., Gaye, K., Mukerjee, S., Ayers, S., and Rossignol, J.F. (2008). Nitazoxanide, tizoxanide and other thiazolides are potent inhibitors of hepatitis B virus and hepatitis C virus replication. Antiviral Res. *77*, 56–63.

Korf, M., Jarczak, D., Beger, C., Manns, P., and Kruger, M. (2005). Inhibition of hepatitis C virus translation and subgenomic replication by siRNAs directed against highly conserved HCV sequence and cellular HCV cofactors. J. Hepatol. *43*, 225–234.

Kukihara, H., Moriishi, K., Taguwa, S., Tani, H., Abe, T., Mori, Y., Suzuki, T., Fukuhara, T., Taketomi, A., Maehara, Y., and Matsuura, Y. (2009). Human VAP-C negatively regulates hepatitis C virus propagation. J. Virol. *83*, 7959–7969.

Lanford, R.E., Hildebrandt-Eriksen, E.S., Petri, A., Persson, R., Lindow, M., Munk, M.E., Kauppinen, S., and Orum, H. (2010). Therapeutic silencing of microRNA-122 in primates with chronic hepatitis C virus infection. Science *327*, 198–201.

Lawitz E., Rouzier, R., Nguyen, T., Huang, M., Ke, J., Praestgaard, J., Serra, D., Koziel, M., Evans, T. (2009). Safety and antiviral efficacy of 14 days of the cyclophilin inhibitor NIM811 in combination with pegylated interferon α2A in relapsed genotype 1 HCV infected patients. 44th Annual Meeting of the European Association for the Study of the Liver. 22 April 2009 (Boston, MA).

Li, Q., Brass, A.L., Ng, A., Hu, Z., Xavier, R.J., Liang, J., and Elledge, S.J. (2009). A genome-wide genetic screen for host factors required for hepatitis C virus propagation. Proc. Natl. Acad. Sci. U.S.A. *106*, 16410–16415.

Liu, Z., Yang, F., Robotham, M., and Tang, H. (2009). A Critical Role of Cyclophilin A and its Prolyl-Peptidyl Isomerase Activity in the Structure and Function of the HCV Replication Complex. J. Virol. *83*, 6554–6565.

Lotz, G.P., Brychzy, A., Heinz, S., and Obermann, W.M. (2008). A novel HSP90 chaperone complex regulates intracellular vesicle transport. J. Cell Sci. *121*, 717–723.

Lu, W., and Ou, J.H. (2002). Phosphorylation of hepatitis C virus core protein by protein kinase A and protein kinase C. Virology *300*, 20–30.

Lyn, R.K., Kennedy, D.C., Sagan, S.M., Blais, D.R., Rouleau, Y., Pegoraro, A.F., Xie, X.S., Stolow, A., and Pezacki, J.P. (2009). Direct imaging of the disruption of hepatitis C virus replication complexes by inhibitors of lipid metabolism. Virology *394*, 130–142.

Ma, S., Boerner, J.E., TiongYip, C., Weidmann, B., Ryder, N.S., Cooreman, P., and Lin, K. (2006). NIM*811*, a cyclophilin inhibitor, exhibits potent in vitro activity against hepatitis C virus alone or in combination with alpha interferon. Antimicrob. Agents Chemother. *50*, 2976–2982.

Macdonald, A., Crowder, K., Street, A., McCormick, C., and Harris, M. (2004). The hepatitis C virus NS5A protein binds to members of the Src family of tyrosine kinases and regulates kinase activity. J. Gen. Virol. *85*, 721–729.

Macdonald, A., Crowder, K., Street, A., McCormick, C., Saksela, K., and Harris, M. (2003). The hepatitis C virus non-structural NS5A protein inhibits activating protein-1 function by perturbing ras-ERK pathway signaling. J. Biol. Chem. *278*, 17775–17784.

McHutchison, J.G., Shiffman, M.L., Cheung, R.C., Gordon, S.C., Wright, T.L., Pottage, J.C., Jr., Mcnair, L., Ette, E., Moseley, S., and Alam, J. (2005). A randomized, double-blind, placebo-controlled dose-escalation trial of merimepodib (VX-497) and interferon-alpha in previously untreated patients with chronic hepatitis C. Antivir. Ther. *10*, 635–643.

Majeau, N., Bolduc, M., Duvignaud, J.B., Fromentin, R., and Leclerc, D. (2007). Effect of cAMP-dependent protein kinase A (PKA) on HCV nucleocapsid assembly and degradation. Biochem. Cell Biol. *85*, 78–87.

Manna, D., Aligo, J., Xu, C., Park, W.S., Koc, H., Do, W., and Konan, K.V. (2010). Endocytic Rab proteins are required for hepatitis C virus replication complex formation. Virology *398*, 21–37.

Mannova, P., and Beretta, L. (2005). Activation of the N-Ras-PI3K-Akt-mTOR pathway by hepatitis C virus: control of cell survival and viral replication. J. Virol. *79*, 8742–8749.

Markland, W., McQuaid, T.J., Jain, J., and Kwong, A.D. (2000). Broad-spectrum antiviral activity of the IMP dehydrogenase inhibitor VX-497: a comparison with ribavirin and demonstration of antiviral additivity with alpha interferon. Antimicrob. Agents Chemother. *44*, 859–866.

Mathy, J.E., Ma, S., Compton, T., and Lin, K. (2008). Combinations of cyclophilin inhibitor NIM811 with hepatitis C Virus NS3–4A Protease or NS5B polymerase inhibitors enhance antiviral activity and suppress the emergence of resistance. Antimicrob. Agents Chemother. *52*, 3267–3275.

Mee, C.J., Harris, H.J., Farquhar, M.J., Wilson, G., Reynolds, G., Davis, C., van I. Jzendoorn, S.C., Balfe, P., and McKeating, J.A. (2009). Polarization restricts hepatitis C virus entry into HepG2 hepatoma cells. J. Virol. *83*, 6211–6221.

Meisner, N.C., Hintersteiner, M., Mueller, K., Bauer, R., Seifert, J.M., Naegeli, H.U., Ottl, J., Oberer, L., Guenat, C., Moss, S., Harrer, N., Woisetschlaeger, M., Buehler, C., Uhl, V., and Auer, M. (2007). Identification and mechanistic characterization of low-molecular-weight inhibitors for HuR. Nat. Chem. Biol. *3*, 508–515.

Milazzo, L., Meroni, L., Galazzi, M., Cesari, M., Caramma, I., Marchetti, G., Galli, M., and Antinori, S. (2009). Does fluvastatin favour HCV replication in vivo? A pilot study on HIV-HCV coinfected patients. J. Viral Hepat. *16*, 479–484.

Milward, A., Mankouri, J., and Harris, M. (2010). Hepatitis C virus NS5A protein interacts with beta-catenin and stimulates its transcriptional activity in a phosphoinositide-3 kinase-dependent fashion. J. Gen. Virol. *91*, 373–381.

Mizui, T., Yamashina, S., Tanida, I., Takei, Y., Ueno, T., Sakamoto, N., Ikejima, K., Kitamura, T., Enomoto, N., Sakai, T., Kominami, E., and Watanabe, S. (2010). Inhibition of hepatitis C virus replication by chloroquine targeting virus-associated autophagy. J. Gastroenterol. *45*, 195–203.

Moradpour, D., Gosert, R., Egger, D., Penin, F., Blum, E., and Bienz, K. (2003). Membrane association of hepatitis C virus nonstructural proteins and identification of the membrane alteration that harbors the viral replication complex. Antiviral Res. *60*, 103–109.

Murata, T., Hijikata, M., and Shimotohno, K. (2005). Enhancement of internal ribosome entry site-mediated translation and replication of hepatitis C virus by PD98059. Virology *340*, 105–115.

Nahmias, Y., Goldwasser, J., Casali, M., van, P.D., Wakita, T., Chung, T., and Yarmush, M.L. (2008). Apolipoprotein B-dependent hepatitis C virus secretion is inhibited by the grapefruit flavonoid naringenin. Hepatology *47*, 1437–1445.

Nakagawa, M., Sakamoto, N., Enomoto, N., Tanabe, Y., Kanazawa, N., Koyama, T., Kurosaki, M., Maekawa, S., Yamashiro, T., Chen, C.H., Itsui, Y., Kakinuma, S., and Watanabe, M. (2004). Specific inhibition of hepatitis C virus replication by cyclosporin A. Biochem. Biophys. Res. Commun. *313*, 42–47.

Nakagawa, M., Sakamoto, N., Tanabe, Y., Koyama, T., Itsui, Y., Takeda, Y., Chen, C.H., Kakinuma, S., Oooka, S., Maekawa, S., Enomoto, N., and Watanabe, M. (2005). Suppression of hepatitis C virus replication by cyclosporin a is mediated by blockade of cyclophilins. Gastroenterology *129*, 1031–1041.

Nakagawa, S., Umehara, T., Matsuda, C., Kuge, S., Sudoh, M., and Kohara, M. (2007). Hsp90 inhibitors suppress HCV replication in replicon cells and humanized liver mice. Biochem. Biophys. Res. Commun. *353*, 882–888.

Ndjomou, J., Park, I.W., Liu, Y., Mayo, D., and He, J.J. (2009). Up-regulation of hepatitis C virus replication and production by inhibition of MEK/ERK signaling. PLoS One. *4*, e7498.

Neddermann, P., Quintavalle, M., Di, P.C., Clementi, A., Cerretani, M., Altamura, S., Bartholomew, L., and De, F.R. (2004). Reduction of hepatitis C virus NS5A hyperphosphorylation by selective inhibition of cellular kinases activates viral RNA replication in cell culture. J. Virol. *78*, 13306–13314.

Neumann, A.U., Lam, N.P., Dahari, H., Gretch, D.R., Wiley, T.E., Layden, J., and Perelson, A.S. (1998). Hepatitis C viral dynamics in vivo and the antiviral efficacy of interferon-alpha therapy. Science *282*, 103–107.

Ng, T.I., Mo, H., Pilot-Matias, T., He, Y., Koev, G., Krishnan, P., Mondal, R., Pithawalla, R., He, W., Dekhtyar, T., Packer, J., Schurdak, M., and Molla, A. (2007). Identification of host genes involved in hepatitis C virus replication by small interfering RNA technology. Hepatology *45*, 1413–1421.

Norman, K.L., and Sarnow, P. (2010). Modulation of hepatitis C virus RNA abundance and the isoprenoid biosynthesis pathway by microRNA miR-122 involves distinct mechanisms. J. Virol. *84*, 666–670.

O'Leary, J.G., Chan, J.L., McMahon, M., and Chung, R.T. (2007). Atorvastatin does not exhibit antiviral activity against HCV at conventional doses: a pilot clinical trial. Hepatology *45*, 895–898.

Okamoto, T., Nishimura, Y., Ichimura, T., Suzuki, K., Miyamura, T., Suzuki, T., Moriishi, K., and Matsuura, Y. (2006). Hepatitis C virus RNA replication is regulated by FKBP8 and Hsp90. EMBO J. *25*, 5015–5025.

Okamoto, T., Omori, H., Kaname, Y., Abe, T., Nishimura, Y., Suzuki, T., Miyamura, T., Yoshimori, T., Moriishi, K., and Matsuura, Y. (2008). A single-amino-acid mutation in hepatitis C virus NS5A disrupting FKBP8 interaction impairs viral replication. J. Virol. *82*, 3480–3489.

Ouzounov, S., Mehta, A., Dwek, R.A., Block, M., and Jordan, R. (2002). The combination of interferon alpha-2b and n-butyl deoxynojirimycin has a greater than additive antiviral effect upon production of infectious bovine viral diarrhea virus (BVDV) in vitro: implications for hepatitis C virus (HCV) therapy. Antiviral Res. *55*, 425–435.

Overbeck, K., Genne, D., Golay, A., and Negro, F. (2008). Pioglitazone in chronic hepatitis C not responding to pegylated interferon-alpha and ribavirin. J. Hepatol. *49*, 295–298.

Owen, D.M., Huang, H., Ye, J., Gale, M., Jr. (2009). Apolipoprotein E on hepatitis C virion facilitates infection through interaction with low-density lipoprotein receptor. Virology *394*, 99–108.

Paeshuyse, J., Kaul, A., De, C.E., Rosenwirth, B., Dumont, J.M., Scalfaro, P., Bartenschlager, R., and Neyts, J.

(2006). The non-immunosuppressive cyclosporin DEBIO-025 is a potent inhibitor of hepatitis C virus replication in vitro. Hepatology *43*, 761–770.

Pileri, P., Uematsu, Y., Campagnoli, S., Galli, G., Falugi, F., Petracca, R., Weiner, A.J., Houghton, M., Rosa, D., Grandi, G., and Abrignani, S. (1998). Binding of hepatitis C virus to CD81. Science *282*, 938–941.

Ploss, A., Evans, M.J., Gaysinskaya, V.A., Panis, M., You, H., de Jong, P., and Rice, C.M. (2009). Human occludin is a hepatitis C virus entry factor required for infection of mouse cells. Nature *457*, 882–886.

Puyang, X., Poulin, D.L., Mathy, J.E., Anderson, L.J., Ma, S., Fang, Z., Zhu, S., Lin, K., Fujimoto, R., Compton, T., and Wiedmann, B. (2010). Mechanism of Resistance of HCV Replicons to Structurally Distinct Cyclophilin Inhibitors. Antimicrob. Agents Chemother. *54*, 1981–1987.

Quintavalle, M., Sambucini, S., Di, P.C., De, R., and Neddermann, P. (2006). The alpha isoform of protein kinase CKI is responsible for hepatitis C virus NS5A hyperphosphorylation. J. Virol. *80*, 11305–11312.

Quintavalle, M., Sambucini, S., Summa, V., Orsatti, L., Talamo, F., De, R., and Neddermann, P. (2007). Hepatitis C virus NS5A is a direct substrate of casein kinase I-alpha, a cellular kinase identified by inhibitor affinity chromatography using specific NS5A hyperphosphorylation inhibitors. J. Biol. Chem. *282*, 5536–5544.

Rakic, B., Sagan, S.M., Noestheden, M., Belanger, S., Nan, X., Evans, C.L., Xie, S., and Pezacki, J.P. (2006). Peroxisome proliferator-activated receptor alpha antagonism inhibits hepatitis C virus replication. Chem. Biol. *13*, 23–30.

Randall, G., Panis, M., Cooper, J.D., Tellinghuisen, T.L., Sukhodolets, K.E., Pfeffer, S., Landthaler, M., Landgraf, P., Kan, S., Lindenbach, B.D., Chien, M., Weir, D.B., Russo, J.J., Ju, J., Brownstein, M.J., Sheridan, R., Sander, C., Zavolan, M., Tuschl, T., and Rice, C.M. (2007). Cellular cofactors affecting hepatitis C virus infection and replication. Proc. Natl. Acad. Sci. U.S.A. *104*, 12884–12889.

Rivas-Aravena, A., Ramdohr, P., Vallejos, M., Valiente-Echeverria, F., Dormoy-Raclet, V., Rodriguez, F., Pino, K., Holzmann, C., Huidobro-Toro, J.P., Gallouzi, I.E., and Lopez-Lastra, M. (2009). The Elav-like protein HuR exerts translational control of viral internal ribosome entry sites. Virology *392*, 178–185.

Robida, J.M., Nelson, H.B., Liu, Z., and Tang, H. (2007). Characterization of hepatitis C virus subgenomic replicon resistance to cyclosporine in vitro. J. Virol. *81*, 5829–5840.

Rosenwirth, B., Billich, A., Datema, R., Donatsch, P., Hammerschmid, F., Harrison, R., Hiestand, P., Jaksche, H., Mayer, P., Peichl, P., and. (1994). Inhibition of human immunodeficiency virus type 1 replication by SDZ NIM *811*, a nonimmunosuppressive cyclosporine analog. Antimicrob. Agents Chemother. *38*, 1763–1772.

Rossignol, J.F., Kabil, S.M., El-Gohary, Y., Elfert, A., and Keeffe, E.B. (2008). Clinical trial: randomized, double-blind, placebo-controlled study of nitazoxanide monotherapy for the treatment of patients with chronic hepatitis C genotype 4. Aliment. Pharmacol. Ther. *28*, 574–580.

Sagan, S.M., Rouleau, Y., Leggiadro, C., Supekova, L., Schultz, P.G., Su, I., and Pezacki, J.P. (2006). The influence of cholesterol and lipid metabolism on host cell structure and hepatitis C virus replication. Biochem. Cell Biol. *84*, 67–79.

Sakamoto, H., Okamoto, K., Aoki, M., Kato, H., Katsume, A., Ohta, A., Tsukuda, T., Shimma, N., Aoki, Y., Arisawa, M., Kohara, M., and Sudoh, M. (2005). Host sphingolipid biosynthesis as a target for hepatitis C virus therapy. Nat. Chem. Biol. *1*, 333–337.

Sarasin-Filipowicz, M., Krol, J., Markiewicz, I., Heim, H., and Filipowicz, W. (2009). Decreased levels of microRNA miR-122 in individuals with hepatitis C responding poorly to interferon therapy. Nat. Med. *15*, 31–33.

Sarrazin, C., and Zeuzem, S. (2010). Resistance to direct antiviral agents in patients with hepatitis C virus infection. Gastroenterology *138*, 447–462.

Scarselli, E., Ansuini, H., Cerino, R., Roccasecca, R.M., Acali, S., Filocamo, G., Traboni, C., Nicosia, A., Cortese, R., and Vitelli, A. (2002). The human scavenger receptor class B type I is a novel candidate receptor for the hepatitis C virus. EMBO J. *21*, 5017–5025.

Sezaki, H., Suzuki, F., Akuta, N., Yatsuji, H., Hosaka, T., Kobayashi, M., Suzuki, Y., Arase, Y., Ikeda, K., Miyakawa, Y., and Kumada, H. (2009). An open pilot study exploring the efficacy of fluvastatin, pegylated interferon and ribavirin in patients with hepatitis C virus genotype 1b in high viral loads. Intervirology *52*, 43–48.

Shih, C.M., Chen, C.M., Chen, Y., and Lee, Y.H. (1995). Modulation of the trans-suppression activity of hepatitis C virus core protein by phosphorylation. J. Virol. *69*, 1160–1171.

Sir, D., Chen, W.L., Choi, J., Wakita, T., Yen, S., and Ou, J.H. (2008a). Induction of incomplete autophagic response by hepatitis C virus via the unfolded protein response. Hepatology *48*, 1054–1061.

Sir, D., Liang, C., Chen, W.L., Jung, U., and Ou, J.H. (2008b). Perturbation of autophagic pathway by hepatitis C virus. Autophagy. *4*, 830–831.

Sizova, D.V., Kolupaeva, V.G., Pestova, T.V., Shatsky, N., and Hellen, C.U. (1998). Specific interaction of eukaryotic translation initiation factor 3 with the 5′ nontranslated regions of hepatitis C virus and classical swine fever virus RNAs. J. Virol. *72*, 4775–4782.

Sklan, E.H., Serrano, R.L., Einav, S., Pfeffer, S.R., Lambright, G., and Glenn, J.S. (2007a). TBC1D20 is a Rab1 GTPase-activating protein that mediates hepatitis C virus replication. J. Biol. Chem. *282*, 36354–36361.

Sklan, E.H., Staschke, K., Oakes, T.M., Elazar, M., Winters, M., Aroeti, B., Danieli, T., and Glenn, J.S. (2007b). A Rab-GAP TBC domain protein binds hepatitis C virus NS5A and mediates viral replication. J. Virol. *81*, 11096–11105.

Spangberg, K., Goobar-Larsson, L., Wahren-Herlenius, M., and Schwartz, S. (1999). The La protein from human liver cells interacts specifically with the U-rich region in the hepatitis C virus 3′ untranslated region. J. Hum. Virol. *2*, 296–307.

Spangberg, K., Wiklund, L., and Schwartz, S. (2000). HuR, a protein implicated in oncogene and growth factor mRNA decay, binds to the 3′ ends of hepatitis C virus RNA of both polarities. Virology *274*, 378–390.

Steinmann, E., Whitfield, T., Kallis, S., Dwek, R.A., Zitzmann, N., Pietschmann, T., and Bartenschlager, R. (2007). Antiviral effects of amantadine and iminosugar derivatives against hepatitis C virus. Hepatology *46*, 330–338.

Stone, M., Jia, S., Heo, W.D., Meyer, T., and Konan, K.V. (2007). Participation of rab5, an early endosome protein, in hepatitis C virus RNA replication machinery. J. Virol. *81*, 4551–4563.

Street, A., Macdonald, A., Crowder, K., and Harris, M. (2004). The Hepatitis C virus NS5A protein activates a phosphoinositide 3-kinase-dependent survival signaling cascade. J. Biol. Chem. *279*, 12232–12241.

Street, A., Macdonald, A., McCormick, C., and Harris, M. (2005). Hepatitis C virus NS5A-mediated activation of phosphoinositide 3-kinase results in stabilization of cellular beta-catenin and stimulation of beta-catenin-responsive transcription. J. Virol. *79*, 5006–5016.

Stuyver, L.J., McBrayer, T.R., Tharnish, P.M., Hassan, A.E., Chu, C.K., Pankiewicz, K.W., Watanabe, K.A., Schinazi, F., and Otto, M.J. (2003). Dynamics of subgenomic hepatitis C virus replicon RNA levels in Huh-7 cells after exposure to nucleoside antimetabolites. J. Virol. *77*, 10689–10694.

Su, A.I., Pezacki, J.P., Wodicka, L., Brideau, A.D., Supekova, L., Thimme, R., Wieland, S., Bukh, J., Purcell, R.H., Schultz, G., and Chisari, F.V. (2002). Genomic analysis of the host response to hepatitis C virus infection. Proc. Natl. Acad. Sci. U.S.A. *99*, 15669–15674.

Supekova, L., Supek, F., Lee, J., Chen, S., Gray, N., Pezacki, J.P., Schlapbach, A., and Schultz, P.G. (2008). Identification of human kinases involved in hepatitis C virus replication by small interference RNA library screening. J. Biol. Chem. *283*, 29–36.

Taguwa, S., Okamoto, T., Abe, T., Mori, Y., Suzuki, T., Moriishi, K., and Matsuura, Y. (2008). Human butyrate-induced transcript 1 interacts with hepatitis C virus NS5A and regulates viral replication. J. Virol. *82*, 2631–2641.

Taguwa, S., Kambara, H., Omori, H., Tani, H., Abe, T., Mori, Y., Suzuki, T., Yoshimori, T., Moriishi, K., and Matsuura, Y. (2009). Cochaperone activity of human butyrate-induced transcript 1 facilitates hepatitis C virus replication through an Hsp90-dependent pathway. J. Virol. *83*, 10427–10436.

Tai, A.W., Benita, Y., Peng, L.F., Kim, S.S., Sakamoto, N., Xavier, J., and Chung, R.T. (2009). A functional genomic screen identifies cellular cofactors of hepatitis C virus replication. Cell Host Microbe *5*, 298–307.

Tan, S.L., Nakao, H., He, Y., Vijaysri, S., Neddermann, P., Jacobs, B.L., Mayer, J., and Katze, M.G. (1999). NS5A, a nonstructural protein of hepatitis C virus, binds growth factor receptor-bound protein 2 adaptor protein in a Src homology 3 domain/ligand-dependent manner and perturbs mitogenic signaling. Proc. Natl. Acad. Sci. U.S.A. *96*, 5533–5538.

Tanida, I., Fukasawa, M., Ueno, T., Kominami, E., Wakita, T., and Hanada, K. (2009). Knockdown of autophagy-related gene decreases the production of infectious hepatitis C virus particles. Autophagy. *5*, 937–945.

Trotard, M., Lepere-Douard, C., Regeard, M., Piquet-Pellorce, C., Lavillette, D., Cosset, F.L., Gripon, P., and Le, S.J. (2009). Kinases required in hepatitis C virus entry and replication highlighted by small interference RNA screening. FASEB J. *23*, 3780–3789.

Tu, H., Gao, L., Shi, S.T., Taylor, D.R., Yang, T., Mircheff, A.K., Wen, Y., Gorbalenya, A.E., Hwang, B., and Lai, M.M. (1999). Hepatitis C virus RNA polymerase and NS5A complex with a SNARE-like protein. Virology *263*, 30–41.

Ujino, S., Yamaguchi, S., Shimotohno, K., and Takaku, H. (2009). Heat-shock protein 90 is essential for stabilization of the hepatitis C virus nonstructural protein NS3. J. Biol. Chem. *284*, 6841–6846.

Umehara, T., Sudoh, M., Yasui, F., Matsuda, C., Hayashi, Y., Chayama, K., and Kohara, M. (2006). Serine palmitoyltransferase inhibitor suppresses HCV replication in a mouse model. Biochem. Biophys. Res. Commun. *346*, 67–73.

Vahlne, A., Larsson, P.A., Horal, P., Ahlmen, J., Svennerholm, B., Gronowitz, S., and Olofsson, S. (1992). Inhibition of herpes simplex virus production in vitro by cyclosporin A. Arch. Virol. *122*, 61–75.

Vaillancourt, F.H., Pilote, L., Cartier, M., Lippens, J., Liuzzi, M., Bethell, R.C., Cordingley, G., and Kukolj, G. (2009). Identification of a lipid kinase as a host factor involved in hepatitis C virus RNA replication. Virology *387*, 5–10.

Wang, C., Gale, M., Jr., Keller, B.C., Huang, H., Brown, M.S., Goldstein, L., and Ye, J. (2005). Identification of FBL2 as a geranylgeranylated cellular protein required for hepatitis C virus RNA replication. Mol. Cell *18*, 425–434.

Wang, J., Tong, W., Zhang, X., Chen, L., Yi, Z., Pan, T., Hu, Y., Xiang, L., and Yuan, Z. (2006). Hepatitis C virus non-structural protein NS5A interacts with FKBP38 and inhibits apoptosis in Huh7 hepatoma cells. FEBS Lett. *580*, 4392–4400.

Wang, P., and Heitman, J. (2005). The cyclophilins. Genome Biol. *6*, 226.

Watashi, K., Hijikata, M., Hosaka, M., Yamaji, M., and Shimotohno, K. (2003). Cyclosporin A suppresses replication of hepatitis C virus genome in cultured hepatocytes. Hepatology *38*, 1282–1288.

Watashi, K., Ishii, N., Hijikata, M., Inoue, D., Murata, T., Miyanari, Y., and Shimotohno, K. (2005). Cyclophilin B is a functional regulator of hepatitis C virus RNA polymerase. Mol. Cell *19*, 111–122.

Waxman, L., Whitney, M., Pollok, B.A., Kuo, C., and Darke, P.L. (2001). Host cell factor requirement for hepatitis C virus enzyme maturation. Proc. Natl. Acad. Sci. U.S.A. *98*, 13931–13935.

Whitby, K., Taylor, D., Patel, D., Ahmed, P., and Tyms, A.S. (2004). Action of celgosivir (6 O-butanoyl castanospermine) against the pestivirus BVDV: implications for the treatment of hepatitis C. Antivir. Chem. Chemother. *15*, 141–151.

Yang, F., Robotham, J.M., Nelson, H.B., Irsigler, A., Kenworthy, R., and Tang, H. (2008). Cyclophilin A is an essential cofactor for hepatitis C virus infection and the principal mediator of cyclosporine resistance in vitro. J. Virol. *82*, 5269–5278.

Ye, J., Wang, C., Sumpter, R., Jr., Brown, M.S., Goldstein, J.L., Gale, M., Jr. (2003). Disruption of hepatitis C virus RNA replication through inhibition of host protein geranylgeranylation. Proc. Natl. Acad. Sci. U.S.A. *100*, 15865–15870.

Yen, J.H., Chang, S.C., Hu, C.R., Chu, S.C., Lin, S.S., Hsieh, S., and Chang, M.F. (1995). Cellular proteins specifically bind to the 5′-noncoding region of hepatitis C virus RNA. Virology *208*, 723–732.

Yoshida, E., Kunimoto, D., Lee, S. S., *et al.* Results of a phase 2 dose ranging study of orally administered celgosivir as monotherapy in chronic hepatitis C genotype 1 patients. Gastroenterology 2006; 130: A784. 2006.

Zhang, X.A., Bontrager, L., and Hemler, M.E. (2001). Transmembrane-4 superfamily proteins associate with activated protein kinase C (PKC) and link PKC to specific beta integrins. J. Biol. Chem. *276*, 25005–25013.

Zhou, S., Liu, R., Baroudy, B.M., Malcolm, A., and Reyes, G.R. (2003). The effect of ribavirin and IMPDH inhibitors on hepatitis C virus subgenomic replicon RNA. Virology *310*, 333–342.

Zitzmann, N., Mehta, A.S., Carrouee, S., Butters, T.D., Platt, F.M., McCauley, J., Blumberg, B.S., Dwek, A., and Block, T.M. (1999). Imino sugars inhibit the formation and secretion of bovine viral diarrhea virus, a pestivirus model of hepatitis C virus: implications for the development of broad spectrum anti-hepatitis virus agents. Proc. Natl. Acad. Sci. U.S.A. *96*, 11878–11882.

Zydowsky, L.D., Etzkorn, F.A., Chang, H.Y., Ferguson, S.B., Stolz, L.A., Ho, I., and Walsh, C.T. (1992). Active site mutants of human cyclophilin A separate peptidyl-prolyl isomerase activity from cyclosporin A binding and calcineurin inhibition. Protein Sci. *1*, 1092–1099.

Targeting Innate Immunity for HCV Antiviral Therapy

19

Michael Gale, Jr.

Abstract

The ability of HCV to mediate persistent, lifelong infection in its human host is linked to a poor response rate to the current interferon-based therapy for treatment of infection. Future antiviral therapy against HCV protein products will most certainly be met with challenges due to virus resistance. Molecular studies of HCV–host interactions have revealed how innate immune programmes of the host can suppress HCV infection and support immune amplification. Therapeutic strategies to target these host features and factors may provide potent new approaches to HCV treatment through immune enhancement while minimizing the likelihood that HCV will resist such actions.

Exposure to hepatitis C virus (HCV) typically leads to persistent infection. Unless interrupted by the current interferon-alpha (IFN)-based therapy, infection will most often associate with a chronic disease course. The ability of HCV to mediate persistent, life-long infection in its human host is linked to the evasive nature of the virus to thwart the host immune system and to resist the antiviral actions of IFN-based therapy. Future antiviral therapy against HCV protein products will also be met with challenges due to virus resistance and evasion from therapeutic actions. Molecular studies of HCV–host interactions have demonstrated several levels of viral-mediated immune regulation and at the same time have revealed novel signalling pathways of innate and adaptive immunity whose effector actions can suppress infection. This chapter will provide an overview of potential host pathways and factors of the innate immune response that present possible therapeutic targets for viral control.

HCV infection is a problem of viral genetic diversity

As is the case with all RNA viruses, the viral polymerase of HCV lacks proofreading function. In the course of persistent infection, error-prone virus replication generates a repertoire of highly related but genetically distinct viral variants or 'quasispecies'. This is most problematic for the infected patient because quasispecies variation affords remarkable adaptive potential to HCV and has been already been implicated in evasion and control the response to IFN therapy and resistance to novel viral protease inhibitor compounds that are destined for the clinic (Pereira and Jacobson, 2009) In models of HCV infection and replication *in vitro*, infectious HCV or HCV replicon variants transduced into human hepatoma cells or even maintained in long-term culture have demonstrated a genetic adaptation or genetic drift that ultimately produced stable highly adapted viral sequence variants that exhibit robust viral replication and cell to cell virus spread (Sakamuro *et al.*, 1995; Farci *et al.*, 2000; Farci, 2001). Additional selective pressure imposed by virus-directed small molecule therapy, such as treatment with HCV protease inhibitor compounds, has been shown to render the out growth therapy-resistant HCV replicon sequences *in vitro* within a very short time frame (Bartenschlager, 2002; Sumpter *et al.*, 2004; Bartenschlager and Sparacio, 2007; Pawlotsky, 2009). Thus, with the constant potential to undergo genetic adaptation, therapeutic designs to target viral products of infection are invariably weakened by the high potential for HCV to resist therapy.

In HCV-infected patients viral genetic changes similarly occur. These changes lead to quasispecies diversification that can ultimately influence virus sensitivity to IFN therapy and to therapies that target specifically viral-encoded factors or that influence viral factor function (Pawlotsky, 2009). Clinical trials of HCV protease inhibitor compounds have been exciting and have reported a dramatic suppression of HCV RNA levels in patient blood shortly after therapy is administered and leading to an improved virological response rate when combined with IFN-based treatment (reviewed by Anderson *et al.*, 2009; Pawlotsky, 2009; Chary and Holodniy, 2010). Importantly, viral protease inhibitor therapy also supports the outgrowth of resistant viral quasispecies that are generated or selected during therapy or that are even present in the patient before therapy begins. In the latter case viral protease inhibitor therapy has already been shown to select for the outgrowth of preexisting quasispecies variants that will then predominate as the major variant in the infected patient (Anderson *et al.*, 2009; Chary and Holodniy, 2010). This scenario is dangerous because as such variants persist and gain increased fitness through further genetic adaptation they present the potential for being transmitted throughout the at-risk population, thus possibly rendering an HCV-infected patient pool who will be overall refractory to protease inhibitor antiviral therapy. Similar resistance outcome can be expected for the global HCV patient population in response to NS5A inhibitors as well as to other virus-directed inhibitor molecules. It should be noted that this dramatic example has already played out in the case of viral-directed therapy against HIV, wherein the virus rapidly acquired wide spread resistance to treatment of patients with any single therapy. This resistance has been partially alleviated through the advent of combining several virus-directed drugs together, which forms the basis of highly active retroviral therapy (HAART) for HIV infection (Hammer *et al.*, 2008). However, the contemporary reality for HCV infection is that combination virus-directed small molecule therapy is years away at best and will certainly associate with an outgrowth of resistant virus within treated patients who are refractory to the therapy.

Current IFN-based therapy requires assistance

The current approved therapy for HCV infection is parenteral administration of IFN-α in combination with oral ribavirin, a nucleoside analogue (McHutchison and Patel, 2002). IFN treatment of HCV infection was first described in 1986 by Hoofnagle *et al.* (1986). In this early study, recombinant human IFN-α2b was administered to HCV-infected patients for up to 12 months. It was found that serum liver enzyme levels [alanine amino transferase (ATL)], an indicator of liver damage, were decreased in HCV patients as a result of IFN therapy; however, in some patients, elevated serum ALT levels promptly returned after therapy was stopped. Patients who demonstrated a sustained virological response (SVR) after therapy ended (non-detectable serum HCV levels) exhibited sustained normal ALT levels, demonstrating that IFN therapy can offer both antiviral and disease-protective benefit. Problematically, today only 55% of patients achieve an SVR.

Contemporary IFN-based therapeutic regimens most often utilize IFN-α2a or 2b, although consensus IFN (CIFN), composed of the consensus sequence derived from the different IFN-α isoforms, is also approved for treatment of HCV infection (McHutchison and Patel, 2002). Therapeutic IFN is injected by the patient either subcutaneously or by intramuscular injection one to three times weekly, depending on the formulation used. The therapeutic effects are greatly enhanced with daily intake of ribavirin capsules, wherein ribavirin is incorporated into the viral RNA during processes of genome replication. The ribavirin incorporation is thought to drive viral mutation to error catastrophe that results in suppressed HCV levels (Pawlotsky, 2009). Consensus interferon and IFN-λ have been shown to have better antiviral activities *in vitro* and in brief clinical trials; however, these treatments are not widely used owing to their requirement for increased injection schedules that are unpleasant to patients (Erickson *et al.*, 2008). It has also been shown that IFN-λ could inhibit HCV replication *in vitro* (Robek *et al.*, 2005). Since IFN-λ is a type III IFN and utilizes different receptors from IFNα/β, the addition of IFN-λ to current standard interferon treatment may possibly activate another group of antiviral genes and therefore enhance or extend the effectiveness of antiviral therapy for HCV infection. Of importance is that *Ifn-λ* gene polymorphisms have been implicated in the outcome of HCV infection, suggesting an important role for this gene product in HCV immunity (Ge *et al.*, 2009; Suppiah *et al.*, 2009; Tanaka *et al.*, 2009; Thomas *et al.*, 2009; Rallon *et al.*, 2010; Rauch *et al.*, 2010; Thompson *et al.*, 2010). Overal, the current IFN-based therapy for HCV infection present side-effects to the patient of skin reactions such as redness, rash, itching, a lump, swelling, or bruising at the site of injection. Moreover, the system side-effects of therapy are at times too difficult for patients to fully comply with their treatment schedule. Indeed, a major side-effect is presented as 'flu-like symptoms' (including fever, chills, muscle aches, joint pain, headaches), along with possible upset stomach (like nausea, taste changes, diarrhoea), hair loss, and mental depression. IFN therapy is difficult to experience. The balance between the effectiveness of the treatment and the quality of life of the HCV patient remains a very challenging area of clinical translational efforts to effectively treat HCV infection. Efforts to develop new therapeutic strategies to combine with IFN and render synergistic antiviral potency are currently under way or are being considered.

Therapeutic strategies that target host factors are essential components for consideration of anti-HCV therapeutics. Host factor targeting offers the benefit of alleviating issues of viral resistance, especially by targeting host factor function essential to the HCV life cycle and should be amenable to combining with the current IFN-based therapy. Moreover, while host cells can alter their expression of specific factors and even drug uptake during the course of a therapy regimen, the chances of host-mediated resistance to the therapy are much smaller than that of viral-mediated resistance. This reduces the chances that HCV genetic adaptation will play a role in viral resistance to a host target. Various host factor targets of therapeutic consideration have been presented in earlier chapters of this book, including host factors essential for virus entry, viral replication, host cell membrane alteration that supports infection, and factors involved in virus maturation and egress. While these therapeutic strategies to target the steps of the HCV infection and replication programmes should suppress viral replication directly, they will not provide protection of bystander cells to resist infection through viral spread nor will they likely serve to enhance the immune response to infection. Thus, it is important to consider novel therapeutic strategies to target host cell and systemic immune response programmes for immune induction, immune enhancement, and antiviral actions to suppress HCV infection, cell to cell virus spread, and person to person viral transmission. Immune-targeted therapeutic approaches offer the benefit of directing pleiotropic immune actions that can suppress virus infection through multiple and varied effector mechanisms, thus making it less likely that HCV would resist such multi-effector therapeutic actions. In particular, the innate immune response directs antiviral programming of hepatocytes, the target cell of HCV infection, and serves to promote antigen presentation by dendritic cells, effector actions of natural killer (NK) cells, and the influx of proinflammatory cells that are essential for clearance of virus infection (Loo and Gale, Jr., 2007). The innate immune response serves an important role therefore to directly suppress viral infection and to drive the maturation and evolution of a successful adaptive immune response to render viral clearance.

The need for immune-targeted therapy for HCV infection

HCV infection and disease are intimately linked to immune regulation. Acute HCV infection can trigger innate immune programmes of the infected cell but these are often overcome and suppressed through viral-directed processes (described below). In this sense HCV is actually detected by the host cell as 'non-self', resulting in immune triggering and a response aimed at suppressing viral replication within the infected hepatocyte. Non-self signalling then result in the production of products that impart adaptive immune response modulation. However, during the course of chronic HCV infection the T cell arm of the adaptive immune response becomes 'exhausted' or anergic and thereby largely attenuated though processes that are still not well defined. Moreover, the humeral immune response to HCV infection is left without a productive innate immune response, which is likely problematic for the proper programming of antibody development and diversification. In addition, the constant genetic adaptation by HCV often renders the outgrowth of viral variants that can escape cytotoxic T cell responses and/or antibody responses through processes of constant epitope drift and selection (Bigger *et al.*, 2004). The end result of these virus–host immune interaction and evasion programmes is that HCV can replicate and spread with little interruption from the host immune response. Thus, there is a major need for immune-stimulatory or enhancing therapies to control HCV infection. Such therapeutic strategies that trigger and enhance innate antiviral immunity would serve to suppress HCV replication and stimulate effective and possibly long-lasting adaptive immunity that could operate in tandem with direct antiviral drugs or even contemporary IFN-based therapy.

Innate intracellular immune defences

In response to virus infection, signalling pathways within mammalian cells direct a variety of intracellular events that generate an antiviral state directly within the infected cell. This antiviral response, termed the 'host response' to virus infection, represents our first line of immune defence against virus infection. If this response is successful, it will serve to suppress viral replication and spread, and will drive the proper maturation of the adaptive immune response to neutralize infection and develop long-lasting immune memory to combat future infections. It is the hepatic innate immune response that imposes initial immune defences against HCV infection (Gale, Jr., 2003). This response is triggered when the infected cell recognizes a molecular signature within the invading virus. This signature, known as a pathogen-associated molecular pattern (PAMP), is typically a physical characteristic of the virus that is recognized and engaged by specific PAMP receptor proteins within the host cell (O'Neill, 2004). The PAMP/PAMP receptor interaction initiates signalling cascades that induce the expression of antiviral effector genes (Sen, 2001). For RNA viruses, protein and nucleic acid products of infection comprise an array of PAMP signatures that can engage specific PAMP receptors, including Toll-like receptors (TLRs) and nucleic acid binding proteins (Cook *et al.*, 2004; Iwasaki and Medzhitov, 2004). The HCV RNA contains specific PAMP signatures, including poly uridine motifs and stem-loop double stranded RNA (dsRNA) structures within its single-stranded RNA genome (Tuplin *et al.*, 2002; Simmonds *et al.*, 2004). HCV RNA is sufficient to induce the innate immune response in cultured hepatocyte-derived cell lines and *in vivo* in mouse liver (McCormick *et al.*, 2004; Sumpter *et al.*, 2005). The product of retinoic acid inducible gene I (RIG-I), which has been defined as a viral RNA PAMP receptor, is critical for host response signalling induced by HCV RNA (Yoneyama *et al.*, 2004; Sumpter *et al.*, 2005). In hepatocytes RIG-I has been shown to direct the innate immune response to virus infection (Li *et al.*, 2005a; Sumpter *et al.*, 2005; Loo *et al.*, 2006, 2008), while TLR3 has been suggested to drive a response to dsRNA PAMPs (Li *et al.*, 2005a,b).

PAMP/PAMP receptor engagement signals the activation of a variety of transcription factors in the infected hepatocyte. PAMP-driven transcription factor activation results in the immediate expression of innate and adaptive immune response genes (Malmgaard, 2004). Interferon regulatory factor (IRF)-3 and nuclear factor-kappa B (NF-κB) are triggered in response to virus infection, and each are activated upon cellular recognition of the HCV PAMP (Au *et al.*, 1995; Lin *et al.*, 1999; Fredericksen *et al.*, 2001; Richmond, 2002; Prabhu *et al.*, 2004). Their activation proceeds through PAMP-responsive signalling pathways of the cell that promote their nuclear translocation and transactivation functions. Other IRF family members, including IRF-5 and IRF-7 are essential components of the innate and adaptive immune responses to virus infection (Barnes *et al.*, 2001; Kawai *et al.*, 2004), though the specific role of each in HCV infection has not been characterized. Virus-induced signalling events that activate ATF-2 and further direct chromatin remodelling contribute to the building of a virus-triggered enhancesome on the IFN-β promoter. The IFN-β enhancesome includes IRF-3 and NF-κB, which produce a transcriptional response resulting in IFN-β expression and secretion from the infected cell (Sen, 2001). NF-κB activation and function is central to the chemokine and proinflammatory cytokine response to virus infection, which functions side by side with IFN to modulate inflammation and the ensuing adaptive immune response (Tai *et al.*, 2000). Secreted levels of IFN-α drive autocrine and paracrine signalling processes by binding the IFN-α/β receptors of the infected cell and local surrounding tissue, respectively. This results in activation of the Jak-STAT pathway. Here, receptor-associated Jak and Tyk1 protein kinases phosphorylate signal transducer and activator of transcription (STAT) proteins on critical serine and tyrosine residues. This phosphorylation confers STAT activation, STAT association with IRF-9, and nuclear localization of the resulting ISGF3 transcription factor complex. ISGF3 is the central transcription factor that promotes high level expression of the interferon stimulated genes (ISGs) by binding to the IFN-stimulated response element (ISRE) within the ISG promoter/enhancer region. IFN binding to its receptor and signalling of the Jak-STAT pathway drives a second wave of transcriptional activity initiated by virus infection and denoted by the expression of ISGs. Therapeutic strategies designed to enhance innate immune programming will therefore impart innate and adaptive immune potency that could limit HCV infection.

ISG products have antiviral functions and represent potential therapeutic targets

The human genome encodes hundreds of ISGs (Der *et al.*, 1998). The ISG products direct regulatory functions that control virus infection, but the function of most ISGs is not yet characterized. However, based on their overall induction by innate immune programmes and their known potency as antivirals or immunomodulators, ISGs targeting, induction, and mimicry through small molecules will offer a unique and specific therapeutic potential to suppress HCV infection and overall enhance antiviral immunity. ISGs have been shown to interrupt HCV replication through processes that include suppression of viral protein synthesis and synthesis inhibition of the viral negative strand replicative intermediate (Guo *et al.*, 2001; Shimazaki *et al.*, 2002; Wang *et al.*, 2002; Prabhu *et al.*, 2004). The main feature resulting from the secretion of IFN from infected cells into the local tissue is the paracrine induction of a tissue-wide host response that blocks cell to cell spread of the virus. Many components of the host response pathways are themselves ISGs, and though expressed basally at a low level that facilitates surveillance and response to virus infection, their abundance will increase in response to IFN signalling. In the liver, paracrine signalling of IFN serves to enhance overall responsiveness of cellular signalling pathways to potentiate the host response to infection in a tissue-wide manner. IFN signalling thus provides an amplification loop to further promote the host response against HCV (Foy *et al.*, 2005). Recent studies have shown that this amplification loop is dependent upon the transcriptional activity of IRF-7 (Honda *et al.*, 2005). IRF-7 is an ISG and its expression in the liver is IFN dependent (Smith *et al.*, 2003; Honda *et al.*, 2005). IRF-7 promotes the expression of the various IFN-α subtypes, wherein IFN-α production and secretion mediate further amplification of the host response and prolonged IFN production (Honda *et al.*, 2005). IFN-based therapy for HCV infection exploits these actions of IFN-α to limit HCV replication and spread (McHutchison and Patel, 2002). Importantly, since IFN also signals the maturation of immune effector cells, antigen presenting cells and dendritic cells, and it potentiates the production of other proinflammatory cytokines by resident hepatic cells to indirectly modulate cell-mediated defences and adaptive immunity (Biron, 1999). Therapeutic strategies aimed at triggering the innate immune response may therefore offer the potential to serve as immune adjuvants for control of virus infection and robust stimulation of immune response function and long-term immune memory through enhancement of the cross-talk between innate and adaptive immune programmes.

Innate and adaptive immune cross-talk

IFN is a potent immunomodulator that influences the onset of the cellular immune response and adaptive immunity to HCV infection. IFN modulates NK cell activity towards lysis of infected target cells by promoting NK cell activation and proliferation, and supporting their survival through the induction of IL-15 production (Loza and Perussia, 2004; Biron, 1999). Activated NK cells produce IFN (Shoukry *et al.*, 2004). In the HCV-infected liver, NK cell homing and local secretion of IFN-β may serve to limit HCV replication directly. Indeed, IFN has been shown to mediate direct antiviral effects against HCV RNA replication *in vitro* (Frese *et al.*, 2002), and its production by immune effector cells in the liver associates with viral clearance in the chimpanzee model of HCV infection (Thimme *et al.*, 2001; Su *et al.*, 2002). IFN also influences the maturation of dendritic cells and modulates their presentation of viral antigens (Colonna *et al.*, 2004; Barth *et al.*, 2005). Antigen presentation by dendritic cells under the influence of IFN contributes to the differentiation of CD4 T cells towards the Th1 phenotype and importantly, a Th1-predominate response is associated with clearance of HCV infection (Shoukry *et al.*, 2004). IFN production during HCV infection may therefore indirectly contribute towards directing the maturation of CD4 T cells to the Th1 phenotype. Dendritic cells are also involved in the cross-priming and IFN production of CD8 T cells. In this context the co-stimulatory signals presented at the time of antigen cross-presentation to CD8 T cells can determine whether the cells are cross-primed for a cytotoxic response or cross-tolerized for anergy (Cooper *et al.*, 1999; Shoukry *et al.*, 2004). TLR3 plays an essential role in cross-priming by dendritic cells, in which it signals the production of IFN triggered by viral PAMPs within products of phagocytosis (Schulz *et al.*, 2005). IFN produced by the dendritic cell in this manner induces the expression of co-stimulatory molecules and cytokines to promote the cross-priming and activation of CD8 T cells, thus linking host response signalling processes to induction of the adaptive immune response. Enhancing the link between innate, IFN-induced antiviral pathways and the adaptive immune response therefore presents an attractive therapeutic strategy for the control of HCV infection and immunity.

Hepatic defences are triggered by HCV: a lesson for targeted therapeutics

Functional genomic analyses from cohorts of patients infected with HCV have shown that infection associates with a hepatic gene expression profile marked by ISGs whose expression levels vary widely among patients (Smith *et al.*, 2003). These observations indicate that HCV can induce and regulate the hepatic host response to infection and provide evidence that therapeutic strategies to mimic this triggering and immune induction should be possible. Importantly, gene profiling studies of HCV-infected chimpanzees have demonstrated that acute resolving HCV infection is associated with a host response characterized by high level hepatic ISG expression (Bigger *et al.*, 2001), suggesting that therapy to induce innate immune defences could serve to render HCV into a self-limiting infection controlled by the host innate defences. In similar studies, 'outcome predictor' gene sets were identified by the overall hepatic expression level of certain ISGs and virus-responsive genes. This gene set was accurately defined as those virus-responsive genes whose high expression associated with low viral load and effective viral clearance but whose low expression correlated with progression to chronic HCV infection (Su *et al.*, 2002). Like many virus-responsive genes and ISGs, the various products of the 'outcome predictor' gene set are known to interact with components of T cell immunity, again demonstrating the complex cross-talk that goes on during the host response to infection and between parameters of innate and adaptive immunity. Thus, strategies to target the induced expression of such genes should prove useful for the control of HCV infection. These observations also demonstrate that the hepatic host response is triggered during HCV infection but can be differentially regulated in association with disease course, and that HCV might mediate persistence through its own strategies to regulate and/or evade the antiviral actions of this response.

Triggering the innate immune response to HCV infection

The processes by which HCV initiates and controls the innate immune response have been addressed in cell culture models of HCV RNA replication and viral protein expression. Genome-length or specific subgenomic fragments of HCV RNA are sufficient to trigger IFN-β expression and secretion when introduced into cultured human hepatoma cells (Fredericksen *et al.*, 2001; McCormick *et al.*, 2004), and this effect has been attributed in part to the 5′ and 3′ non-translated region (NTR) of the HCV genome. These regions encode a series of highly conserved stem-loop/dsRNA structures, and the 3′ NTR includes a variable length poly u region (Tuplin *et al.*, 2002). In cultured cells the HCV NTRs present PAMP structures that serve as potent agonists of the innate immune response. This suggests that during infection these RNA motifs are recognized and engaged by PAMP receptor(s) that trigger the host response (Sumpter *et al.*, 2005), thus providing insights into strategies to trigger innate immunity through targeted therapeutic approaches aimed at engaging viral PAMP receptors and specifically HCV PAMP receptors.

The nature of at least one HCV RNA PAMP receptor was revealed through studies of the Huh7-derived (human hepatoma) cell line, termed Huh7.5. This cell line does not mount a host response to virus infection or transfected HCV RNA and it is highly permissive for HCV RNA replication (Blight *et al.*, 2002; Sumpter *et al.*, 2005). Complementation studies identified RIG-I as an HCV RNA PAMP receptor that binds HCV dsRNA motifs and signals the activation of IRF-3 and NF-κB. This was shown to induce IFN-α expression and onset of the innate immune response (Sumpter *et al.*, 2005). RIG-I is an RNA helicase and a member of the DEx/D box RNA helicase family. It contains amino-terminal regions of homology to the caspase activation and recruitment domain (CARD; Yoneyama *et al.*, 2004). RIG-I signalling is mediated by the CARD homology motifs while the helicase domain imparts PAMP recognition, RNA-binding and regulation of signalling (Yoneyama *et al.*, 2004). Signalling by RIG-I directs a response that suppresses HCV RNA replication (Foy *et al.*, 2005; Sumpter *et al.*, 2005). The permissiveness of Huh7.5 cells for HCV RNA replication has been attributed to a point mutation in the RIG-I CARD motifs that ablated downstream IRF-3 phosphorylation and NF-κB activation (Sumpter *et al.*, 2005). Virus-induced signalling by RIG-I and the resulting immune response may therefore influence the outcome of HCV infection and therapies aimed at triggering this response will present a viable approach for immune enhancement. TLR3 is a dsRNA PAMP receptor that signals a host response after its engages the PAMP ligand (Alexopoulou *et al.*, 2001). TLR3 signals the activation of IRF-3 and NF-κB through MyD88-independent processes that require the Toll-IL-1 receptor domain-containing adaptor inducing IFN (TRIF) protein (Yamamoto *et al.*, 2003). TLR3 is essential for cross-presentation of antigens by certain classes of dendritic cells. Strategies to target and engage TLR3 actions therapeutically could prove beneficial to drive both antiviral and adaptive immune programming.

Protein products of infection may also stimulate the innate immune response to HCV such that therapeutic mimicry of these interactions by small molecules may also serve to drive immune response programmes. Expression of the HCV NS5A protein induces cellular stress signalling pathways to activate STAT3, which can serve to drive proinflammatory response that can suppress infection (Gong *et al.*, 2001). Similar to the IFN receptor signalling, STAT3 promotes gene expression through processes that involve the Jak-STAT pathway (Sarcar *et al.*, 2004). This results in a gene expression profile that includes ISGs and proinflammatory cytokines that may influence the overall level of HCV RNA replication (Zhu *et al.*, 2003). Moreover, the HCV core protein has been shown to activate the catalytic activity of protein kinase R (PKR), an ISG effector component of the host response to virus infection (Delhem *et al.*, 2001). PKR is a dsRNA binding protein and innate immune effector. Its activation is induced by binding to dsRNA PAMPs, resulting in eukaryotic initiation factor 2 phosphorylation and inhibition of local protein synthesis. Active PKR also signals the DNA binding activity of NF-κB and induces IRF-1 transcription-effector action (Williams, 1999). Activation of PKR by the HCV core protein is most likely attributed to the 5′ NTR viral RNA binding property of the core protein (Tanaka *et al.*, 2000), which would provide the PKR activator dsRNA PAMP required for kinase activation. Cell interaction with virus particles may also trigger signalling events that induce IFN production. HCV pseudo particle binding to dendritic cells has been shown to mediate particle uptake and dendritic cell activation (Barth *et al.*, 2005). Since dendritic cell subsets constitute a major source of IFN production during viral infection, small molecule therapeutic modulation of dendritic cell function may influence HCV infection by modulation local or systemic IFN levels (Colonna *et al.*, 2004).

Innate immune-targeted therapeutics against HCV

Modified IFNs

IFN-based therapy has a level of proven success as a treatment for HCV infection and will therefore likely remain as the mainstay for patient treatment. Thus, most new small molecule compounds of novel HCV therapy will likely be combined with IFN as their treatment regimen. In addition, there is a substantial effort underway to constantly improve IFN therapy itself. This has evolved through the use of IFN monotherapy to combination with ribavirin, to combination therapy using pegylated IFNs that have longer lasting effects (Pereira and Jacobson, 2009). Current efforts include a major focus to further modify the IFN molecule to enhance potency, duration of response, and to reduce side-effects of treatment. These efforts include developing oral administered forms of IFN and molecules with enhanced potency. In terms of the latter, DNA shuffling has been used to produce MAXY-a or R7025, which is a human IFN-α derivative that has been enhanced through a complex genetic engineering process. However, while this molecule exhibited enhanced potency of the IFN response it has not performed well in clinical trials (reviewed by Pereira and Jacobson, 2009). Other recombinant forms of IFN, including CIFN have been similarly shown to have enhanced potency of signalling, and in some cases enhanced therapeutic benefit in HCV patients (Erickson *et al.*, 2008; Witthoft, 2008). Moreover, albumin IFN or Albuferon, a novel recombinant protein fused to human albumin, has a long *in vivo* life span that abrogates the need for multiple injections each week, and has been demonstrated in clinical trials to mediate an increased response rate to therapy in HCV patients who previously failed the standard of care IFN therapy regimen (previously reviewed by Pereira and Jacobson, 2009). IFN-λ is also the focus of new therapeutic development, as it is effective in driving an antiviral response against HCV to suppress virus replication, and may offer fewer side-effects to treated patients (Miller *et al.*, 2009; O'Brien, 2009). Though *Ifn-λ* polymorphisms have been linked to the outcome of HCV infection it is not at all clear how IFN-λ functions in the overall immune response to infection, and further studies are needed to assess the immunomodulatory actions of therapeutic IFN-λ (Ge *et al.*, 2009; Rauch *et al.*, 2010; Suppiah *et al.*, 2009; Tanaka *et al.*, 2009; Thomas *et al.*, 2009; Thompson *et al.*, 2010). Taken together, these aforementioned studies demonstrate that approaches to enhance IFN antiviral actions could lead to redesigning the mainstay therapy. Finally, it should be noted that improvements and refinements to ribavirin, the current therapy partner, are constantly underway in which the focus is to reduce drug toxicity while improving antiviral function and bioavailability. In this sense, a ribavirin analogue, Taribavirin, a prodrug of ribavirin that accumulates specifically in the liver, has been shown to have reduced side-effects while exhibiting therapy enhancement in weight-based dosing trials (Shields and Pockros, 2009). Improved IFN and ribavirin therapy will serve as the foundation for future anti-HCV therapy combinations as drug development proceeds.

TLR agonists

TLRs are expressed variably in human tissues and have been shown to be expressed in hepatocytes and immune cells, including T cells, B cells and dendritic cells (Takeda and Akira, 2005). Treatment of HCV-infected hepatocytes with TLR agonist compounds has been shown to drive an innate immune response that has the ability to suppress viral replication, suggesting that such approaches may serve to limit HCV infection *in vivo* (Pereira and Jacobson, 2009). Indeed, clinical studies have shown the potential for TLR agonists as immunotherapeutics for HCV treatment. Treatment of chronic HCV patients with the TLR7 agonist, isatoribine, induced an immune response, including innate immune activation, that mediated a moderate suppression of viral load in patient blood (Horsmans *et al.*, 2005). TLR3 agonists, including Hiltonol and Aplifen have been shown to drive immune adjuvant effects that enhance anti-tumor immunity, and it is likely that their application towards HCV infection will result in similar immune enhancement for a suppression of viral load (Pereira and Jacobson, 2009). A major advantage of TLR3 agonists for antiviral therapy is that they should be expected to induce or enhance cross-priming of antigen presentation by dendritic cells (Akira and Takeda, 2004).

RLR agonists

The RLRs, RIG-I and MDA5 represent novel therapeutic targets that have the potential to induce a potent innate immune response. The potential of RLR targeting and activation to suppress virus infection is underscored by the several studies of transgenic knockout mice or immune evasion that define these factors as essential determinants of immune function that can suppress virus infection and spread. The RLRs have recently been implicated as important components that modulate the adaptive immune response, including effecting humeral immunity and driving the expansion and regulation of regulatory T cells (Suthar *et al.*, 2010). These observations indicate the potential power of targeting and activating the RLRs for immunotherapy against HCV. Whereas HCV PAMP RNA serves as a potent trigger for RIG-I signalling *in vivo*, other RIG-I ligand RNAs have also been identified. In recent studies it was shown that triggering RIG-I signalling activation in a therapeutic manner using either HCV PAMP RNA or synthetic 5′ triphosphate RNA as a ligand lead to suppression of HCV infection while the latter could suppress pandemic H1N1 influenza virus infection *in vitro* (Saito *et al.*, 2008; Chakravarthy *et al.*, 2010). Moreover, others have shown that stimulation of MDA5 with long dsRNA or activation of RIG-I with short dsRNA or short 3′ overhang RNA could serve to stimulate innate immune programmes that could suppress infection by a variety of RNA viruses (Hornung *et al.*, 2006; Kato *et al.*, 2006, 2008; Cui *et al.*, 2008; Pichlmair *et al.*, 2009). Now that the structure of the ligand binding region of each RLR has been defined it should be possible to develop structure-based small molecule ligands that target specific RLRs for immune activation (Cui *et al.*, 2008; Takahasi *et al.*, 2009). Since RLR signalling can induce IFN production as well as elicit proinflammatory cytokines the ensuing pleiotropic response would be less likely to drive the outgrowth of viral resistant variants but instead would impose a variety of blockades to HCV replication and spread and serve to enhance overall immunity to the infection.

ISGs

ISGs represent new potential effectors for therapeutic targeting and are revealed by functional genomics analyses. It is the biochemical actions of the ISG products of an innate immune response that would confer the actions of any innate immune therapeutic that engages the IFN arm of this response. Of principle importance to this response for controlling HCV replication are those ISGs expressed in hepatocytes, the principle HCV target cells. Studies that measured the kinetics and magnitude of decay in serum HCV RNA levels within IFN-treated patients indicate that IFN, and hence ISGs, impose an acute blockade upon de novo virus production and may potentiate the death of HCV-infected cells. (Neumann *et al.*, 1999) Thus, targeted therapies to induce or enhance the actions of specific ISGs may have potent antiviral potential against HCV. Results from studies that have utilized microarray approaches to categorize ISGs in cultured cells and in human liver undergoing HCV infection indicate that the human genome encodes hundreds, if not thousands of functionally diverse ISGs, some of which are known to direct antiviral actions.(de Veer *et al.*, 2001) However, the complete spectrum of ISGs has not been defined nor have the ISGs that impart control of HCV replication been fully identified.

The refinement and further improvement of IFN therapy or immune-based therapy development for chronic HCV infection will no doubt come from a better understanding of how IFN affects the target cell. This requires defining the spectrum and kinetics of ISG expression under *in vivo* conditions of IFN therapy (preferably within liver tissue), and categorizing ISG expression patterns within patients who respond to therapy and in those patients who fail IFN therapy. *In vivo* studies that couple analyses of viral decay kinetics with microarray profiling of ISG expression during specific phases of IFN therapy will provide the opportunity to correlate ISG expression patterns with antiviral effect. Such an approach should facilitate the identification of an ISG subset or 'short list' whose expression patterns positively correlate with viral decay and whose actions are likely to influence HCV replication. Further mechanistic studies to evaluate protein levels, biochemical actions, and anti-HCV potential of ISG products encoded in this 'short list' could then define those ISGs that are the important players of IFN action against HCV. Results from such studies could lend to further refinement of IFN dosing regimens that are biased to maximize the expression and extent of action of these specific anti-HCV effector ISGs. Several studies have been conducted using genomics approaches to define the spectrum of ISG that are important for HCV control. Overall, the gene expression profiles that were reported and obtained from *in vivo* and *ex vivo* exposure to IFN therapy or HCV infection are remarkably similar. On the average, more than 500 ISGs have been identified after *in vivo* or *ex vivo* IFN treatment, while only a small number of genes were identified whose expression was suppressed in response to IFN therapy. Thus, RLR, TLR or other innate immune therapies have the potential to trigger the onset of a wide variety of expressed ISGs. This could be a potent therapy approach or could also be dangerous in terms of immune toxicity induction through the many ISGs expressed.

Consistent with most genomics studies of ISG analysis, the identified ISGs represent diverse functional groups, ranging from genes encoding immunomodulatory proteins to those encoding metabolic regulators. Using stringent statistical analyses, a 'short list' of ISGs were identified whose expression increased *in vivo* and *in vitro* within hours of HCV infection or IFN treatment, thereby preceding the onset of viral decline that first follows innate immune induction.(Neumann *et al.*, 1999; Thimme *et al.*, 2001; Su *et al.*, 2002; Lau *et al.*, 2008). Assuming a temporal relationship in which ISG expression and action precede the decline of HCV load in the se cases, it is likely that ISGs within this 'short list' contribute to the therapeutic effects of the innate immune response against HCV. The results presented in these studies mark a positive step forward towards identifying the spectrum of ISGs in human cells that respond to innate immune induction and could thereby serve as antiviral effectors for the control of HCV infection. ISGs in the category include IFIT1, Viperin, PKR, ISG15, USP18, OAS1, ISG20, and other (reviewed by Gale, Jr. and Foy, 2005 and by Loo *et al.*, 2006). Structure–function studies of these gene products and assessment of their regulatory programmes of function will allow for the consideration of targeted therapeutics where a specific ISG could be triggered or mimicked by small molecules to suppress HCV infection and/or enhance anti-HCV immunity.

By the classical definition ISGs are the genetic component of the cellular response to IFN. However, recent work in the IFN field has demonstrated that during virus infection the overall complexity of the ISG repertoire expressed by the host cell is heavily influenced by the activation state of IRFs that are regulated independently of IFN(Levy, 2002). In particular, and as described above, RNA PAMPs and other products of virus infection and replication are potent inducers of IRFs which cooperate with other transcription coactivators to stimulate the expression of specific IRF target genes(Sen, 2001). The nature of these target genes has been revealed, in part, by microarray studies to identify dsRNA or IRF-3 targets within cultured human cells.(Geiss *et al.*, 2001; Grandvaux *et al.*, 2002) Importantly, many known ISGs were among the list of identified PAMP RNA or IRF-3 target genes, demonstrating that expression of the ISG repertoire is influenced through cellular pathways that can be signalled independently of the type I IFN-receptor. Thus, as we learn more of how these pathways and genes cooperatively function to mediate antiviral immunity it should become possible to target various effector pathways through small molecule therapeutics with the goal of stimulating specific effector actions of the innate immune response to suppress HCV infection.

MicroRNA

MicroRNAs (miRNAs) represent an untapped therapeutic potential for HCV infection and innate immunity. miRNAs are small RNA molecules that target specific cellular or even viral mRNA resulting in suppression of the target RNA expression (Gantier *et al.*, 2007; Gantier, 2010; Skalsky and Cullen, 2010). In this sense the miRNAs of the cell represent potent and novel antiviral sources. Indeed, HCV replication has been shown to be dependent on a host miRNA termed miR122, wherein suppression of miR122 expression resulted in loss of permissiveness for HCV RNA replication to hepatoma cells (Jopling *et al.*, 2005). Conversely, more recent studies have identified a subset of miRNAs that are induced upon IFN treatment of cells and that have immunomodulatory activities to suppress HCV infection either directly through viral RNA targeting or by altering the host cell profile to suppress viral permissiveness (Pedersen and David, 2008; Pedersen *et al.*, 2007). In either case the targeting of miRNAs through specific induction or reduced expression showed that such approaches could prove useful for anti-HCV therapy either as antiviral approaches or as innate immune enhancing approaches. Therapeutic targeting of miRNA metabolism pathway of the host cell has also been shown to be useful for modulation of cellular functions (Jackson and Linsley, 2010). Thus, miRNAs represent a potential target for development of novel antiviral and immune-modifying therapies.

Conclusion

The future for antiviral therapy development to treat HCV infection and other viral infections presents many exciting opportunities as new immune response pathways are revealed and as canonical pathways and their cross-talk programmes are characterized alone and in the context of HCV infection. Future HCV therapeutics will most likely been practiced in combination with IFN and ribavirin, thus forming a therapy cocktail analogous to HAART for HIV treatment. Such a cocktail approach increases the potency of antiviral treatment while reducing the chances that HCV will mutate and adapt to the therapy application to become refractory to treatement. Thus, a multi-targeted therapeutic regimen will one day serve as the mainstay for the treatment of HCV, a curable infection.

References

Akira, S., and Takeda, K. (2004). Functions of toll-like receptors: lessons from KO mice. C. R. Biol. *327*, 581–589.

Alexopoulou, L., Holt, A.C., Medzhitov, R., and Flavell, R.A. (2001). Recognition of double-stranded RNA and activation of NF-kappaB by Toll-like receptor 3. Nature *413*, 732–738.

Anderson, J., Schiffer, C., Lee, K., and Swanstrom, R. (2009). Viral protease inhibitors. Handb. Exp. Pharmacol. 85–110.

Au, W.C., Moore, P.A., Lowther, W., Juang, T., and Pitha, P.M. (1995). Identification of a member of the interferon regulatory factor family that binds to the interferon-stimulated response element and activates expression of interferon-induced genes. Proc. Natl. Acad. Sci. U.S.A. *92*, 11657–11661.

Barnes, B.J., Moore, A., and Pitha, P.M. (2001). Virus-specific activation of a novel interferon regulatory factor, IRF-5, results in the induction of distinct interferon alpha genes. J. Biol. Chem. *276*, 23382–23390.

Bartenschlager, R. (2002). Hepatitis C virus replicons: potential role for drug development. Nat. Rev. Drug Discov. *1*, 911–916.

Bartenschlager, R., and Sparacio, S. (2007). Hepatitis C virus molecular clones and their replication capacity in vivo and in cell culture. Virus Res. *127*, 195–207.

Barth, H., Ulsenheimer, A., Pape, G.R., Diepolder, H.M., Hoffmann, M., Neumann-Haefelin, C., Thimme, R., Henneke, P., Klein, R., Paranhos-Baccala, G., *et al.* (2005). Uptake and presentation of hepatitis C virus-like particles by human dendritic cells. Blood. *105*, 3605–3614.

Bigger, C.B., Brasky, M., and Lanford, R.E. (2001). DNA microarray analysis of chimpanzee liver during acute resolving hepatitis C virus infection. J. Virol. *75*, 7059–7066.

Bigger, C.B., Guerra, B., Brasky, K., Hubbard, G., Beard, M.R., Luxon, B.A., Lemon, M., and Lanford, R. (2004). Intrahepatic gene expression during chronic hepatits C vrius infection in chimpanzees. J. Virol. *78*, 13779–13792.

Biron, C.A. (1999). Initial and innate responses to viral infections – pattern setting in immunity or disease. Curr. Opin. Microbiol. *2*, 374–381.

Blight, K.J., McKeating, A., and Rice, C.M. (2002). Highly permissive cell lines for subgenomic and genomic hepatitis C virus RNA replication. J. Virol. *76*, 13001–13014.

Chakravarthy, K.V., Bonoiu, A.C., Davis, W.G., Ranjan, P., Ding, H., Hu, R., Bowzard, J.B., Bergey, E.J., Katz, J.M., Knight, P.R., *et al.* (2010). Gold nanorod delivery of an ssRNA immune activator inhibits pandemic H1N1 influenza viral replication. Proc. Natl. Acad. Sci. U.S.A. *107*, 10172–10177.

Chary, A., and Holodniy, M. (2010). Recent Advances in Hepatitis C Virus Treatment: Review of HCV Protease Inhibitor Clinical Trials. Rev. Recent Clin. Trials. *5*, 158–173.

Colonna, M., Trinchieri, G., and Liu, Y.J. (2004). Plasmacytoid dendritic cells in immunity. Nat. Immunol. *5*, 1219–1226.

Cook, D.N., Pisetsky, S., and Schwartz, D.A. (2004). Toll-like receptors in the pathogenesis of human disease. Nat. Immunol. *5*, 975–979.

Cooper, S., Erickson, A.L., Adams, E.J., Kansopon, J., Weiner, A.J., Chien, D.Y., Houghton, M., Parham, P., and Walker, C.M. (1999). Analysis of a successful immune response against hepatitis C virus. Immunity. *10*, 439–449.

Cui, S., Eisenacher, K., Kirchhofer, A., Brzozka, K., Lammens, A., Lammens, K., Fujita, T., Conzelmann, K.K., Krug, A., and Hopfner, K.P. (2008). The C-terminal regulatory domain is the RNA 5′-triphosphate sensor of RIG-I. Mol. Cell *29*, 169–179.

Delhem, N., Sabile, A., Gajardo, R., Podevin, P., Abadie, A., Blaton, M.A., Kremsdorf, D., Beretta, L., and Brechot, C. (2001). Activation of the interferon-inducible protein kinase PKR by hepatocellular carcinoma derived-hepatitis C virus core protein. Oncogene *20*, 5836–5845.

Der, S.D., Zhou, A., Williams, G., and Silverman, R.H. (1998). Identification of genes differentially regulated by interferon alpha, beta, or gamma using oligonucleotide arrays. Proc. Natl. Acad. Sci. U.S.A. *95*, 15623–15628.

Erickson, A.K., Seiwert, S., Gale, M., Jr. (2008). Antiviral potency analysis and functional comparison of consensus interferon, interferon-alpha2a and pegylated interferon-alpha2b against hepatitis C virus infection. Antivir. Ther. *13*, 851–862.

Farci, P. (2001). Hepatitis C virus. The importance of viral heterogeneity. Clin. Liver Dis. *5*, 895–916.

Farci, P., Shimoda, A., Coiana, A., Diaz, G., Peddis, G., Melpolder, J.C., Strazzera, A., Chien, D.Y., Munoz, S.J., Balestrieri, A., *et al.* (2000). The outcome of acute hepatitis C predicted by the evolution of the viral quasispecies. Science *288*, 339–344.

Foy, E., li, K., Sumpter, R., Jr., Loo, Y.M., Johnson, C.L., Wang, C., Fish, P.M., Yoneyama, M., Fujita, T., Lemon, S.M., *et al.* (2005). Control of antiviral defenses through hepatitis C virus disruption of retinoic acid-inducible gene-I signaling. Proc. Natl. Acad. Sci. U.S.A. *102*, 2986–2991.

Fredericksen, B., Akkaraju, G., Foy, E., Wang, C., Pflugheber, J., Chen, Z.J., Gale, M., Jr. (2001). Activation of the inteferon-β promoter during hepatitis C virus RNA replication. Viral Immunol. *15*, 29–40.

Frese, M., Schwarzle, V., Barth, K., Krieger, N., Lohmann, V., Mihm, S., Haller, O., and Bartenschlager, R. (2002). Interferon-gamma inhibits replication of subgenomic and genomic hepatitis C virus RNAs. Hepatology *35*, 694–703.

Gale, M., Jr. (2003). Effector genes of interferon action against hepatitis C virus. Hepatology *37*, 975–978.

Gale, M., Jr. and Foy, E.M. (2005). Evasion of intracellular host defense by hepatitis C virus. Nature *436*, 939–945.

Gantier, M.P. (2010). New perspectives in MicroRNA regulation of innate immunity. J. Interferon Cytokine Res. *30*, 283–289.

Gantier, M.P., Sadler, J., and Williams, B.R. (2007). Fine-tuning of the innate immune response by microRNAs. Immunol. Cell Biol. *85*, 458–462.

Ge, D., Fellay, J., Thompson, A.J., Simon, J.S., Shianna, K.V., Urban, T.J., Heinzen, E.L., Qiu, P., Bertelsen, A.H., Muir, A.J., *et al.* (2009). Genetic variation in IL28B predicts hepatitis C treatment-induced viral clearance. Nature *461*, 399–401.

Geiss, G.K., Jin, G., Guo, J., Bumgarner, R.E., Katze, G., and Sen, G.C. (2001). A comprehensive view of regulation of gene expression by double-stranded RNA-mediated cell signaling. J. Biol. Chem. *276*, 30178–30182.

Gong, G., Waris, G., Tanveer, R., and Siddiqui, A. (2001). Human hepatitis C virus NS5A protein alters intracellular calcium levels, induces oxidative stress, and activates STAT-3 and NF-kappa B. Proc. Natl. Acad. Sci. U.S.A. *98*, 9599–9604.

Grandvaux, N., Servant, M.J., tenOever, B., Sen, G.C., Balachandran, S., Barber, G.N., Lin, R., and Hiscott, J. (2002). Transcriptional profiling of interferon regulatory factor 3 target genes: direct involvement in the regulation of interferon-stimulated genes. J. Virol. *76*, 5532–5539.

Guo, J., Bichko, V., and Seeger, C. (2001). Effect of alpha interferon on the hepatitis C virus replication. J. Virol. *75*, 8516–8523.

Hammer, S.M., Eron, J.J., Jr., Reiss, P., Schooley, R.T., Thompson, M.A., Walmsley, S., Cahn, P., Fischl, M.A., Gatell, J.M., Hirsch, M.S., *et al.* (2008). Antiretroviral treatment of adult HIV infection: 2008 recommendations of the International AIDS Society-USA panel. JAMA *300*, 555–570.

Honda, K., Yanai, H., Negishi, H., Asagiri, M., Sato, M., Mizutani, T., Shimada, N., Ohba, Y., Takaoka, A., Yoshida, N., *et al.*. (2005). IRF-7 is the master regulator of type-I interferon-dependent immune responses. Nature. *434*, 772–777.

Hornung, V., Ellegast, J., Kim, S., Brzozka, K., Jung, A., Kato, H., Poeck, H., Akira, S., Conzelmann, K.K., Schlee, M., Endres, S., and Hartmann, G. (2006). 5′-Triphosphate RNA Is the Ligand for RIG-I. Science. *314*, 994–997.

Horsmans, Y., Berg, T., Desager, J.P., Mueller, T., Schott, E., Fletcher, S.P., Steffy, K.R., Bauman, L.A., Kerr, M., and Averett, D.R. (2005). Isatoribine, an agonist of

TLR7, reduces plasma virus concentration in chronic hepatitis C infection. Hepatology *42*, 724–731.

Iwasaki, A., and Medzhitov, R. (2004). Toll-like receptor control of the adaptive immune responses. Nat. Immunol. *5*, 987–995.

Jackson, A., and Linsley, P.S. (2010). The therapeutic potential of microRNA modulation. Discov. Med. *9*, 311–318.

Jopling, C.L., Yi, M., Lancaster, A.M., Lemon, M., and Sarnow, P. (2005). Modulation of hepatitis C virus RNA abundance by a liver-specific MicroRNA. Science *309*, 1577–1581.

Kato, H., Takeuchi, O., Sato, S., Yoneyama, M., Yamamoto, M., Matsui, K., Uematsu, S., Jung, A., Kawai, T., Ishii, K.J., *et al.* (2006). Differential roles of MDA5 and RIG-I helicases in the recognition of RNA viruses. Nature *441*, 101–105.

Kato, H., Takeuchi, O., Mikamo-Satoh, E., Hirai, R., Kawai, T., Matsushita, K., Hiiragi, A., Dermody, T.S., Fujita, T., and Akira, S. (2008). Length-dependent recognition of double-stranded ribonucleic acids by retinoic acid-inducible gene-I and melanoma differentiation-associated gene 5. J. Exp. Med. *205*, 1601–1610.

Kawai, T., Sato, S., Ishii, K.J., Coban, C., Hemmi, H., Yamamoto, M., Terai, K., Matsuda, M., Inoue, J., Uematsu, S., *et al.* (2004). Interferon-alpha induction through Toll-like receptors involves a direct interaction of IRF7 with MyD88 and TRAF6. Nat. Immunol. *5*, 1061–1068.

Lau, D.T., Fish, P.M., Sinha, M., Owen, D.M., Lemon, S.M., Gale, M., Jr. (2008). Interferon regulatory factor-3 activation, hepatic interferon-stimulated gene expression, and immune cell infiltration in hepatitis C virus patients. Hepatology *47*, 799–809.

Levy, D.E. (2002). Whence interferon? Variety in the production of interferon in response to viral infection. J. Exp. Med. *195*, F15–F18.

Li, K., Chen, Z., Kato, N., Gale, M., Jr. and Lemon, S.M. (2005a). Distinct poly-I: C and virus-activated signaling pathways leading to interferon-beta production in hepatocytes. J. Biol. Chem. *280*, 16739–16747.

Li, K., Foy, E., Ferreon, J.C., Nakamura, M., Ferreon, A.C., Ikeda, M., Ray, S.C., Gale, M., Jr. and Lemon, S.M. (2005b). Immune evasion by hepatitis C virus NS3/4A protease-mediated cleavage of the Toll-like receptor 3 adaptor protein TRIF. Proc. Natl. Acad. Sci. U.S.A. *102*, 2992–2997.

Lin, R., Heylbroeck, C., Genin, P., Pitha, M., and Hiscott, J. (1999). Essential role of interferon regulatory factor 3 in direct activation of RANTES chemokine transcription. Mol. Cell Biol. *19*, 959–966.

Loo, Y.M., and Gale, M., Jr. (2007). Viral regulation and evasion of the host response. Curr. Top. Microbiol. Immunol. *316*, 295–313.

Loo, Y.M., Owen, D.M., li, K., Erickson, A.K., Johnson, C.L., Fish, P.M., Carney, D.S., Wang, T., Ishida, H., Yoneyama, M., *et al.* (2006). Viral and therapeutic control of IFN-beta promoter stimulator 1 during hepatitis C virus infection. Proc. Natl. Acad. Sci. U.S.A. *103*, 6001–6006.

Loo, Y.M., Fornek, J., Crochet, N., Bajwa, G., Perwitasari, O., Martinez-Sobrido, L., Akira, S., Gill, M.A., Garcia-Sastre, A., Katze, M.G., *et al.* (2008). Distinct RIG-I and MDA5 signaling by RNA viruses in innate immunity. J. Virol. *82*, 335–345.

Loza, M.J., and Perussia, B. (2004). Differential regulation of NK cell proliferation by type I and type II IFN. Int. Immunol. *16*, 23–32.

McCormick, C.J., Challinor, L., Macdonald, A., Rowlands, J., and Harris, M. (2004). Introduction of replication-competent hepatitis C virus transcripts using a tetracycline-regulable baculovirus delivery system. J. Gen. Virol *85*, 429–439.

McHutchison, J.G., and Patel, K. (2002). Future therapy of hepatitis C. Hepatology *36*, S245-S252.

Malmgaard, L. (2004). Induction and regulation of IFNs during viral infections. J. Interferon Cytokine Res. *24*, 439–454.

Miller, D.M., Klucher, K.M., Freeman, J.A., Hausman, D.F., Fontana, D., and Williams, D.E. (2009). Interferon lambda as a potential new therapeutic for hepatitis C. Ann. N. Y. Acad. Sci. *1182*, 80–87.

Neumann, A.U., Lam, N.P., Dahari, H., Gretch, D.R., Wiley, T.E., Layden, J., and Perelson, A.S. (1999). Hepatitis C virus dynamic in vivo and the antiviral efficacy of interferon-a therapy. Science *282*, 103–107.

O'Brien, T.R. (2009). Interferon-alfa, interferon-lambda and hepatitis C. Nat. Genet. *41*, 1048–1050.

O'Neill, L.A. (2004). Immunology. After the toll rush. Science *303*, 1481–1482.

Pawlotsky, J.M. (2009). Hepatitis: HCV variability, the immune system and resistance to antiviral drugs. Nat. Rev. Gastroenterol. Hepatol. *6*, 383–385.

Pedersen, I. and David, M. (2008). MicroRNAs in the immune response. Cytokine *43*, 391–394.

Pedersen, I.M., Cheng, G., Wieland, S., Volinia, S., Croce, C.M., Chisari, V., and David, M. (2007). Interferon modulation of cellular microRNAs as an antiviral mechanism. Nature *449*, 919–922.

Pereira, A.A., and Jacobson, I.M. (2009). New and experimental therapies for HCV. Nat. Rev. Gastroenterol. Hepatol. *6*, 403–411.

Pichlmair, A., Schulz, O., Tan, C.P., Rehwinkel, J., Kato, H., Takeuchi, O., Akira, S., Way, M., Schiavo, G., Reis e Sousa (2009). Activation of MDA5 requires higher-order RNA structures generated during virus infection. J. Virol. *83*, 10761–10769.

Prabhu, R., Joshi, V., Garry, R.F., Bastian, F., Haque, S., Regenstein, F., Thung, S., and Dash, S. (2004). Interferon alpha-2b inhibits negative-strand RNA and protein expression from full-length HCV1a infectious clone. Exp. Mol. Pathol. *76*, 242–252.

Rallon, N.I., Naggie, S., Benito, J.M., Medrano, J., Restrepo, C., Goldstein, D., Shianna, K.V., Vispo, E., Thompson, A., McHutchison, J., and Soriano, V. (2010). Association of a single nucleotide polymorphism near the interleukin-28B gene with response to hepatitis C therapy in HIV/hepatitis C virus-coinfected patients. AIDS *24*, F23-F29.

Rauch, A., Kutalik, Z., Descombes, P., Cai, T., Di Iulio, J., Mueller, T., Bochud, M., Battegay, M., Bernasconi,

E., Borovicka, J., *et al.* (2010). Genetic variation in IL28B is associated with chronic hepatitis C and treatment failure: a genome-wide association study. Gastroenterology 138, 1338–45, 1345.

Richmond, A. (2002). Nf-kappa B, chemokine gene transcription and tumour growth. Nat. Rev. Immunol. *2*, 664–674.

Robek, M.D., Boyd, S., and Chisari, F.V. (2005). Lambda interferon inhibits hepatitis B and C virus replication. J. Virol. *79*, 3851–3854.

Saito, T., Owen, D.M., Jiang, F., Marcotrigiano, J., Gale, M., Jr. (2008). Innate immunity induced by composition-dependent RIG-I recognition of hepatitis C virus RNA. Nature *454*, 523–527.

Sakamuro, D., Furukawa, T., and Takegami, T. (1995). Hepatitis C virus nonstructural protein NS3 transforms NIH 3T3 cells. J. Virol. *69*, 3893–3896.

Sarcar, B., Ghosh, A.K., Steele, R., Ray, R., and Ray, R.B. (2004). Hepatitis C virus NS5A mediated STAT3 activation requires co-operation of Jak1 kinase. Virology *322*, 51–60.

Schulz, O., Diebold, S.S., Chen, M., Naslund, T.I., Nolte, M.A., Alexopoulou, L., Azuma, Y.T., Flavell, R.A., Liljestrom, P., Reis e Sousa (2005). Toll-like receptor 3 promotes cross-priming to virus-infected cells. Nature *433*, 887–892.

Sen, G.C. (2001). Viruses and interferons. Annu. Rev. Microbiol. *55*, 255–281.

Shields, W.W., and Pockros, P.J. (2009). Ribavirin analogs. Clin. Liver Dis. *13*, 419–427.

Shimazaki, T., Honda, M., Kaneko, S., and Kobayashi, K. (2002). Inhibition of internal ribosomal entry site-directed translation of HCV by recombinant IFN-alpha correlates with a reduced La protein. Hepatology *35*, 199–208.

Shoukry, N.H., Cawthon, G., and Walker, C.M. (2004). Cell-mediated immunity and the outcome of hepatitis C virus infection. Annu. Rev. Microbiol. *58*, 391–424.

Simmonds, P., Tuplin, A., and Evans, D.J. (2004). Detection of genome-scale ordered RNA structure (GORS) in genomes of positive-stranded RNA viruses: Implications for virus evolution and host persistence. RNA. *10*, 1337–1351.

Skalsky, R.L., and Cullen, B.R. (2010). Viruses, microRNAs, and Host Interactions. Annu. Rev. Microbiol. *64*, 123–141.

Smith, M.W., Yue, Z.N., Korth, M.J., Do, H.A., Boix, L., Fausto, N., Bruix, J., Carithers, R.L., Jr. and Katze, M.G. (2003). Hepatitis C virus and liver disease: global transcriptional profiling and identification of potential markers. Hepatology *38*, 1458–1467.

Su, A.I., Pezacki, J.P., Wodicka, L., Brideau, A.D., Supekova, L., Thimme, R., Wieland, S., Bukh, J., Purcell, R.H., Schultz, G., *et al.* (2002). Genomic analysis of the host response to hepatitis C virus infection. Proc. Natl. Acad. Sci. U.S.A. *99*, 15669–15674.

Sumpter, R., Wang, C., Foy, E., Loo, Y.-M., and Gale, M.J. (2004). Viral evolution and interferon resistance of hepatitis C virus RNA replication in a cell culture model. J. Virol. *78*, 11591–11604.

Sumpter, R., Loo, Y.-M., Foy, E., li, K., Yoneyama, M., Fujita, T., Lemon, M., and Gale, M.J. (2005). Regulating intracellular anti-viral defense and permissiveness to hepatitis C virus RNA replication through a cellular RNA helicase, RIG-I. J. Virol. *79*, 2689–2699.

Suppiah, V., Moldovan, M., Ahlenstiel, G., Berg, T., Weltman, M., Abate, M.L., Bassendine, M., Spengler, U., Dore, G.J., Powell, E., *et al.* (2009). IL28B is associated with response to chronic hepatitis C interferon-alpha and ribavirin therapy. Nat. Genet. *41*, 1100–1104.

Suthar, M.S., Ma, D.Y., Thomas, S., Lund, J.M., Zhang, N., Daffis, S., Rudensky, A.Y., Bevan, M.J., Clark, E.A., Kaja, M.K., *et al.* (2010). IPS-1 is essential for the control of West Nile virus infection and immunity. PLoS. Pathog. *6*, e1000757.

Tai, D.I., Tsai, S.L., Chen, Y.M., Chuang, Y.L., Peng, C.Y., Sheen, I.S., Yeh, C.T., Chang, K.S., Huang, S.N., Kuo, C., *et al.* (2000). Activation of nuclear factor kappaB in hepatitis C virus infection: implications for pathogenesis and hepatocarcinogenesis. Hepatology *31*, 656–664.

Takahasi, K., Kumeta, H., Tsuduki, N., Narita, R., Shigemoto, T., Hirai, R., Yoneyama, M., Horiuchi, M., Ogura, K., Fujita, T., *et al.* (2009). Solution structures of cytosolic RNA sensor MDA5 and LGP2 C-terminal domains: identification of the RNA recognition loop in RIG-I-like receptors. J. Biol. Chem. *284*, 17465–17474.

Takeda, K., and Akira, S. (2005). Toll-like receptors in innate immunity. Int. Immunol. *17*, 1–14.

Tanaka, Y., Shimoike, T., Ishii, K., Suzuki, R., Suzuki, T., Ushijima, H., Matsuura, Y., and Miyamura, T. (2000). Selective binding of hepatitis C virus core protein to synthetic oligonucleotides corresponding to the 5′ untranslated region of the viral genome. Virology *270*, 229–236.

Tanaka, Y., Nishida, N., Sugiyama, M., Kurosaki, M., Matsuura, K., Sakamoto, N., Nakagawa, M., Korenaga, M., Hino, K., Hige, S., *et al.* (2009). Genome-wide association of IL28B with response to pegylated interferon-alpha and ribavirin therapy for chronic hepatitis C. Nat. Genet. *41*, 1105–1109.

Thimme, R., Oldach, D., Chang, K.M., Steiger, C., Ray, C., and Chisari, F.V. (2001). Determinants of viral clearance and persistence during acute hepatitis C virus infection. J. Exp. Med. *194*, 1395–1406.

Thomas, D.L., Thio, C.L., Martin, M.P., Qi, Y., Ge, D., O'Huigin, C., Kidd, J., Kidd, K., Khakoo, S.I., Alexander, G., *et al.* (2009). Genetic variation in IL28B and spontaneous clearance of hepatitis C virus. Nature *461*, 798–801.

Thompson, A.J., Muir, A.J., Sulkowski, M.S., Ge, D., Fellay, J., Shianna, K.V., Urban, T., Afdhal, N.H., Jacobson, I.M., Esteban, R., *et al.* (2010). Interleukin-28B polymorphism improves viral kinetics and is the strongest pre-treatment predictor of sustained virlogic response in genotype 1 hepatitis C virus. Gastroenterology *139*, 120–129.

Tuplin, A., Wood, J., Evans, D.J., Patel, H., and Simmonds, P. (2002). Thermodynamic and phylogenetic prediction of RNA secondary structures in the coding region of hepatitis C virus. RNA. *8*, 824–841.

de Veer, M.J., Holko, M., Frevel, M., Walker, E., Der, S., Paranjape, J.M., Silverman, H., and Williams, B.R. (2001). Functional classification of interferon-stimulated genes identified using microarrays. J. Leukoc. Biol. *69*, 912–920.

Wang, C., Pflugheber, J., Sumpter, R., Sodora, D., Hui, D., Sen, G.C., Gale, M., Jr. (2002). Alpha interferon induces distinct translational control programs to suppress hepatitis C virus RNA replication. J. Virol. *77*, 3898–3912.

Williams, B.R. (1999). PKR; a sentinel kinase for cellular stress. Oncogene *18*, 6112–6120.

Witthoft, T. (2008). Review of consensus interferon in the treatment of chronic hepatitis C. Biologics *2*, 635–643.

Yamamoto, M., Sato, S., Hemmi, H., Hoshino, K., Kaisho, T., Sanjo, H., Takeuchi, O., Sugiyama, M., Okabe, M., Takeda, K., *et al.* (2003). Role of adaptor TRIF in the MyD88-independent toll-like receptor signaling pathway. Science *301*, 640–643.

Yoneyama, M., Kikuchi, M., Natsukawa, T., Shinobu, N., Imaizumi, T., Miyagishi, M., Taira, K., Akira, S., and Fujita, T. (2004). The RNA helicase RIG-I has an essential function in double-stranded RNA-induced innate antiviral responses. Nat. Immunol. *5*, 730–737.

Zhu, H., Zhao, H., Collins, C.D., Eckenrode, S.E., Run, Q., McIndoe, R.A., Crawford, J.M., Nelson, D.R., She, X., and Liu, C. (2003). Gene expression associated with interferon alfa antiviral activity in an HCV replicon cell line. Hepatology *37*, 1180–1188.

Index

A

B

C

D

E

F

G

H

V

W

Y

Current Books of Interest

Emerging Trends in Antibacterial Discovery: Answering the Call to Arms	2011
Epigenetics: A Reference Manual	2011
Metagenomics: Current Innovations and Future Trends	2011
Nitrogen Cycling in Bacteria: Molecular Analysis	2011
The Biology of Paramyxoviruses	2011
Helicobacter pylori	2011
Microbial Bioremediation of Non-metals: Current Research	2011
Lactic Acid Bacteria and Bifidobacteria: Current Progress in Advanced Research	2011
Viruses and Interferon: Current Research	2011
Essentials of Veterinary Parasitology	2011
Hepatitis C: Antiviral Drug Discovery and Development	2011
Streptomyces: Molecular Biology and Biotechnology	2011
Alphaherpesviruses: Molecular Virology	2011
Recent Advances in Plant Virology	2011
Vaccine Design: Innovative Approaches and Novel Strategies	2011
Salmonella: From Genome to Function	2011
PCR Troubleshooting and Optimization: The Essential Guide	2011
Insect Virology	2010
Environmental Microbiology: Current Technology and Water Applications	2011
Sensory Mechanisms in Bacteria: Molecular Aspects of Signal Recognition	2010
Bifidobacteria: Genomics and Molecular Aspects	2010
Molecular Phylogeny of Microorganisms	2010
Nanotechnology in Water Treatment Applications	2010
Iron Uptake and Homeostasis in Microorganisms	2010
Caliciviruses: Molecular and Cellular Virology	2010
Epstein-Barr Virus: Latency and Transformation	2010
Anaerobic Parasitic Protozoa: Genomics and Molecular Biology	2010
Lentiviruses and Macrophages: Molecular and Cellular Interactions	2010
Microbial Population Genetics	2010
Borrelia: Molecular and Cellular Biology	2010
Influenza: Molecular Virology	2010
RNA Interference and Viruses: Current Innovations and Future Trends	2010
Retroviruses: Molecular Biology, Genomics and Pathogenesis	2010
Metagenomics: Theory, Methods and Applications	2010
Aspergillus: Molecular Biology and Genomics	2010
Environmental Molecular Microbiology	2010
Neisseria: Molecular Mechanisms of Pathogenesis	2010
Frontiers in Dengue Virus Research	2010
ABC Transporters in Microorganisms	2009
Pili and Flagella: Current Research and Future Trends	2009
Lab-on-a-Chip Technology: Biomolecular Separation and Analysis	2009

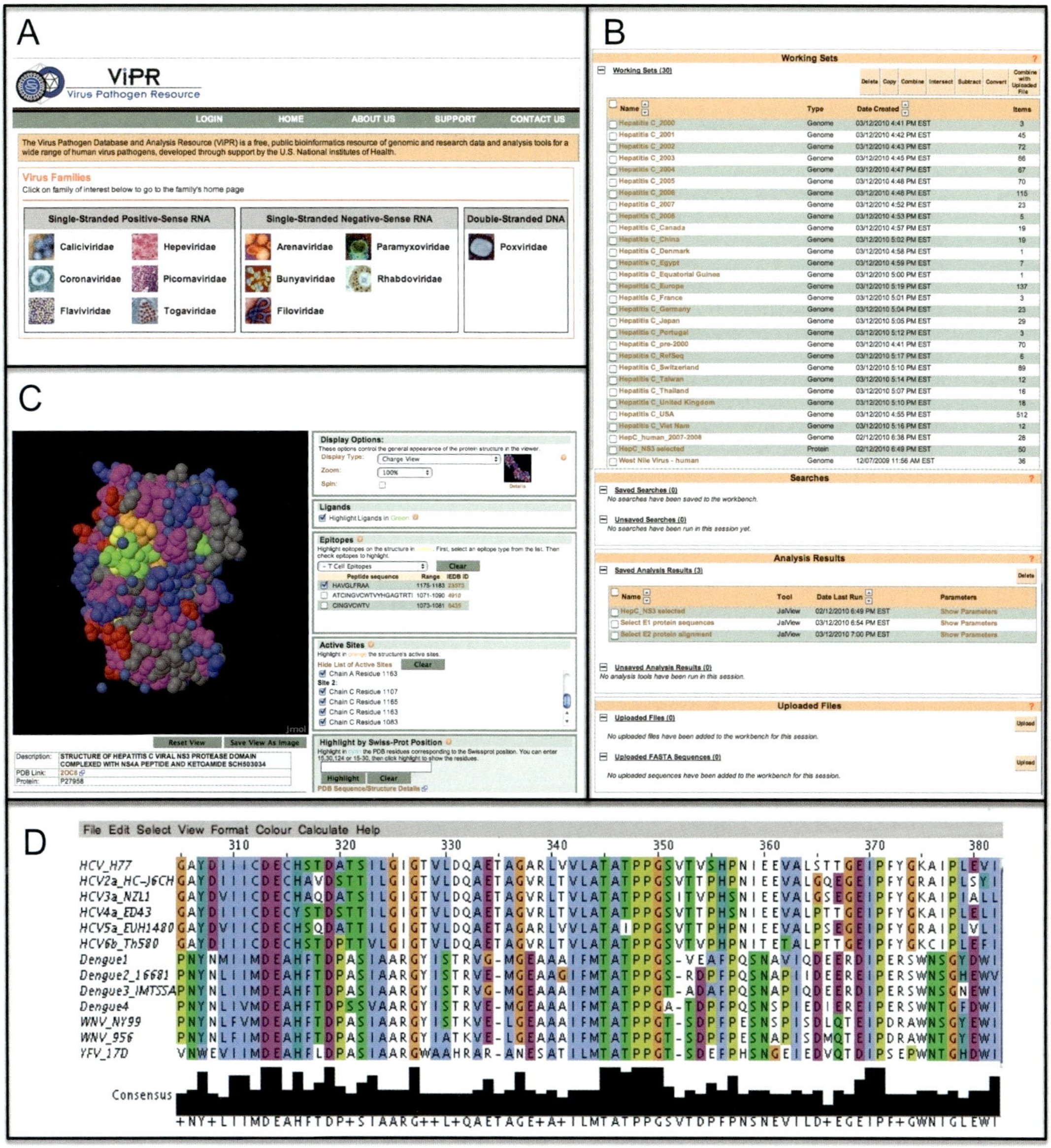

Plate 4.1 Virus Pathogen Database and Analysis Resource (ViPR). (A) ViPR provide a comprehensive resource of research data about a wide variety of different human pathogenic viruses, including HCV and related members of the Flaviviridae family. (B) An example of a private workbench containing saved working sets of sequence records and analysis results is shown. HCV complete genome sequence sets for viruses isolated in recent years or in defined country locations can be maintained. Working sets can be combined to generate unions (Combine), intersections (Intersect) and differences (Subtract) sets (e.g. Hepatitis C_Europe). (C) An example of data integration using the ViPR 3D protein structure viewer is shown. The viewer allows users to customize the display to highlight different protein features, including the location of enzyme active sites (orange), bound ligands (green) and immune epitopes (yellow). (D) Access to data about all flaviviruses allows for the comparative analysis within a virus family. The alignment of NS3 protein sequences of NCBI Reference Sequences for HCV, Dengue, West Nile virus and Yellow Fever virus is shown together with the level of sequence conservation (bar graph) and the resulting consensus sequence.

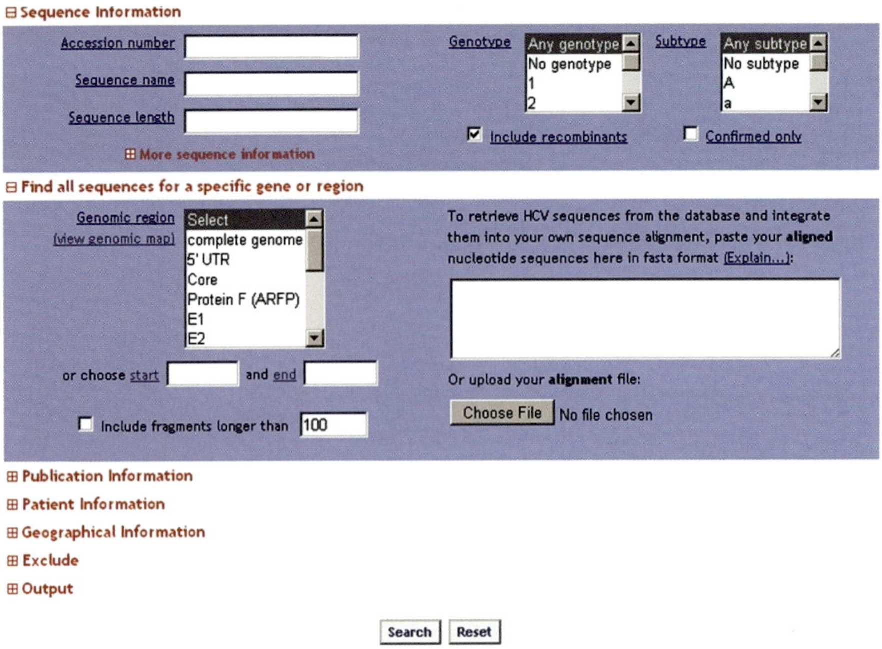

Plate 4.3 HCV database search interface, default view.

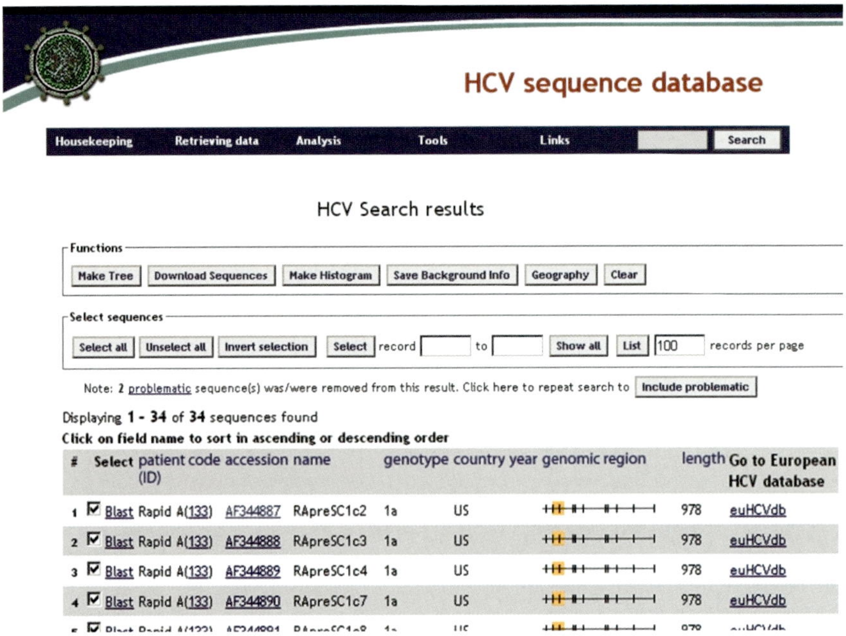

Plate 4.4 LANL HCV database search output.

FindModel Results

Initial tree constructed using Weighbor

MODELS CONSIDERED:

Model name	AIC	LnL
JC: Jukes-Cantor (model 1)	16733.206244	-8366.603122
JC: Jukes-Cantor plus Gamma (model 3)	16203.129808	-8100.564904
F81: Felsenstein 1981 (model 5)	16688.910728	-8341.455364
F81: Felsenstein 1981 plus Gamma (model 7)	16129.619562	-8060.809781
K80: Kimura 2-parameter (model 9)	16154.530594	-8076.265297
K80: Kimura 2-parameter plus Gamma (model 11)	15514.6553	-7755.327650
HKY: Hasegawa-Kishino-Yano (model 13)	16083.32436	-8037.662180
HKY: Hasegawa-Kishino-Yano plus Gamma (model 15)	15404.524192	-7697.262096
TrN: Tamura-Nei (model 21)	16083.255278	-8036.627639
TrN: Tamura-Nei plus Gamma (model 23)	15405.534532	-7696.767266
GTR: General Time Reversible (model 53)	16057.575514	-8020.787757
GTR: General Time Reversible plus Gamma (model 55)	15393.817872	-7687.908936

Click one of the model names in the table to display the parameter matrix for that model.

AIC-SELECTED MODEL: **GTR: General Time Reversible plus Gamma** (model 55)
LnL = -7687.908936
AIC = 15393.817872
Parameter details

The matrix on the right illustrates the different models considered by Findmodel.
Rate Parameters with the same color are assumed to have the same value.
Base Frequencies are the same when they are shown as f_N, and different otherwise.

General Time Reversible + γ

	T	C	A	G
T	f_T	a	b	c
C	a	f_C	d	e
A	b	d	f_A	f
G	c	e	f	f_G

Plate 4.6 FindModel output showing the tested models, their scores, and a dynamic matrix showing the features of each selected model.

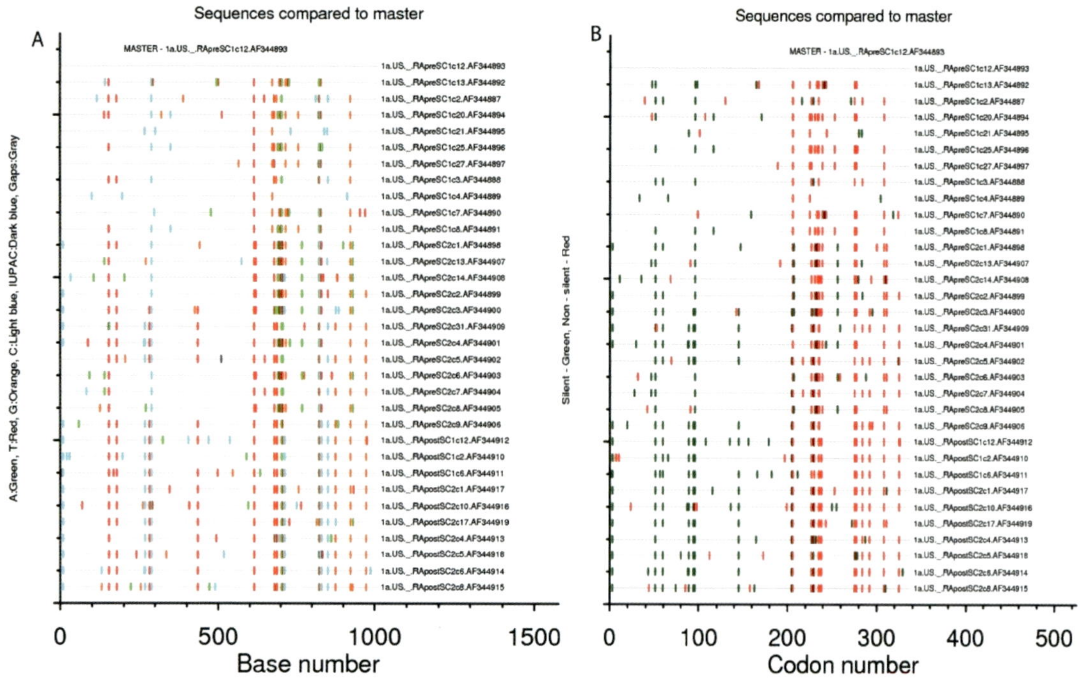

Plate 4.7 Highlighter output; left: standard settings; right: silent (green) and non-silent (red) changes.

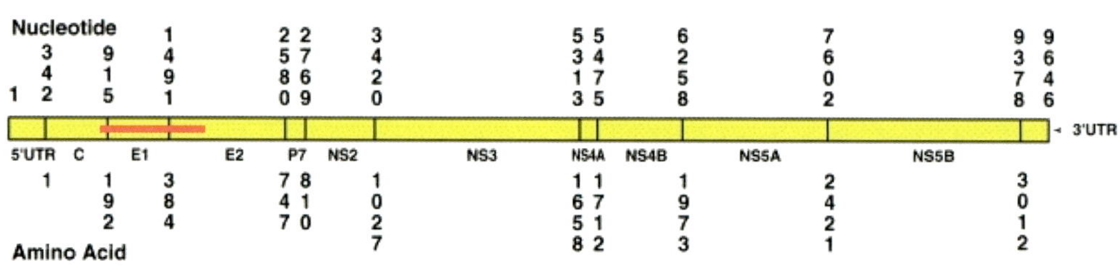

Plate 4.8 Sequence Locator output showing the location of the query sequence in the alignment.

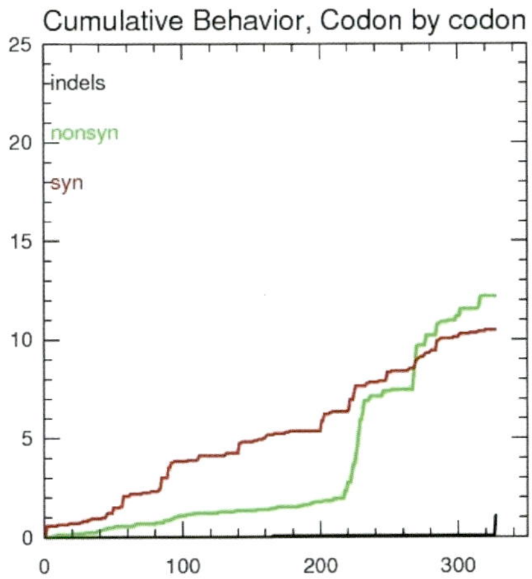

Plate 4.9 Graphical output of SNAP, the Silent/Nonsilent Analysis Program.

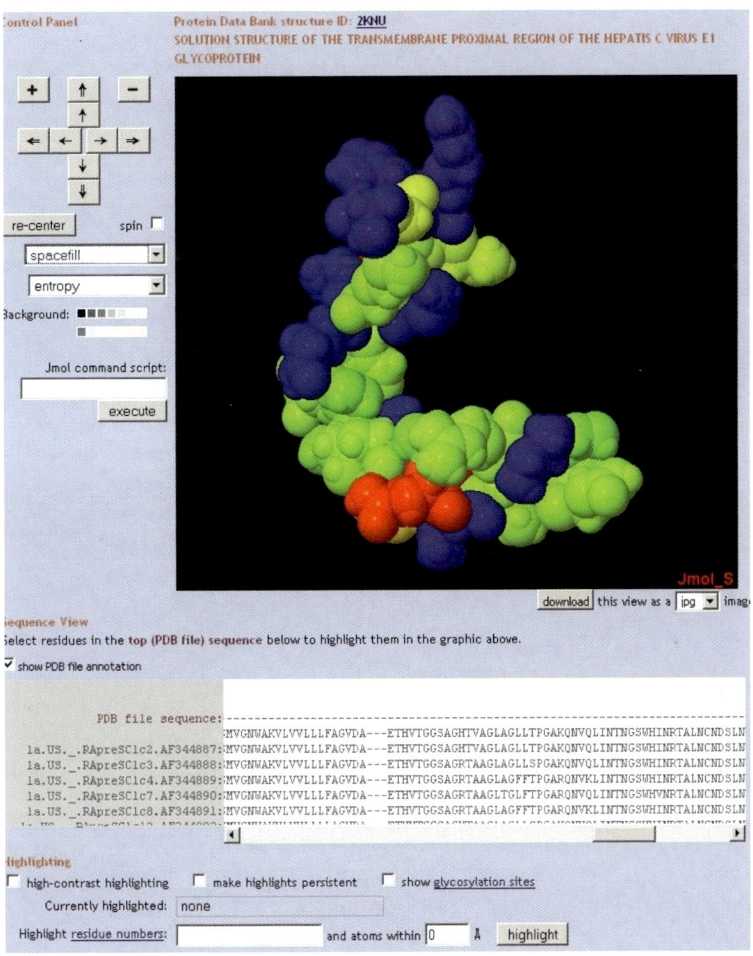

Plate 4.11 'Protein Feature Accent', the protein visualization and annotation tool, showing the entropy source alignment and the calculated entropy, mapped onto an E2 fragment.

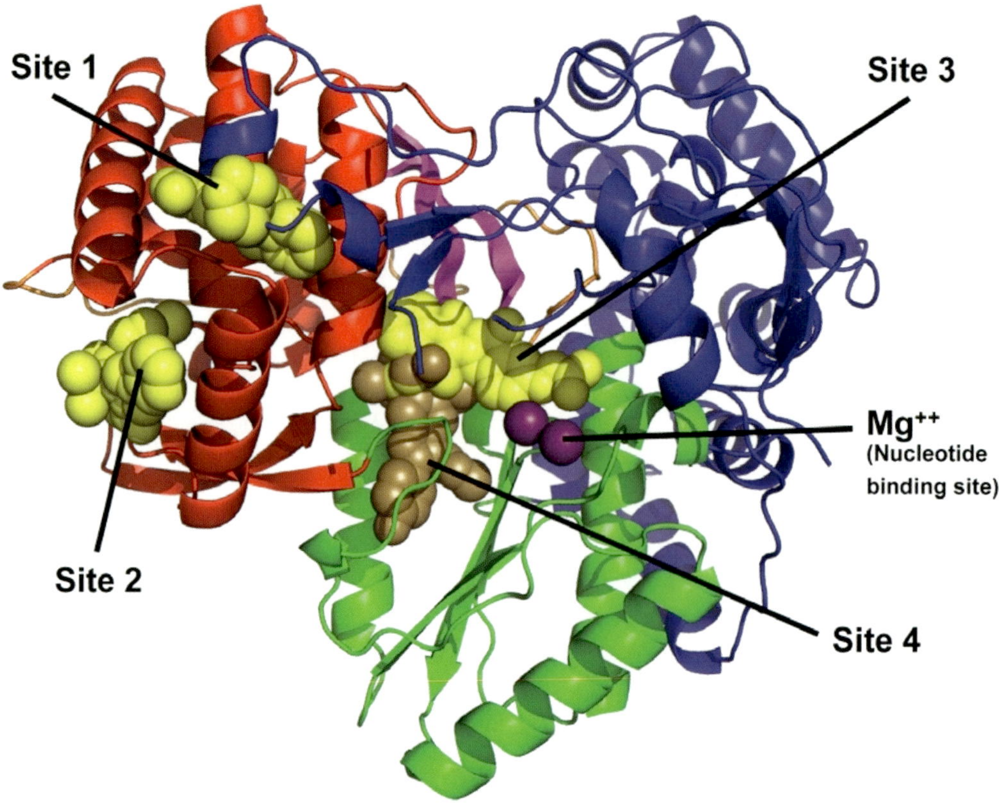

Plate 16.1 Crystal structure of NS5B and NNI binding sites (Cooper *et al.*, 2009).